AF903447

Data Handling in Science and Technology
Volume 30

Resolving Spectral Mixtures

With Applications from Ultrafast Time-Resolved Spectroscopy to Super-Resolution Imaging

Data Handling in Science and Technology

Volume 30

Resolving Spectral Mixtures

With Applications from Ultrafast Time-Resolved Spectroscopy to Super-Resolution Imaging

Edited by

Cyril Ruckebusch
Université de Lille, Sciences et Technologies, LASIR CNRS, LILLE, France

Series Editors

Beata Walczak
Lutgarde Buydens

ELSEVIER

AMSTERDAM • BOSTON • HEIDELBERG • LONDON • NEW YORK • OXFORD
PARIS • SAN DIEGO • SAN FRANCISCO • SINGAPORE • SYDNEY • TOKYO

Elsevier
Radarweg 29, PO Box 211, 1000 AE Amsterdam, Netherlands
The Boulevard, Langford Lane, Kidlington, Oxford OX5 1GB, United Kingdom
50 Hampshire Street, 5th Floor, Cambridge, MA 02139, United States

First edition 2016

Copyright © 2016 Elsevier B.V. All rights reserved.

No part of this publication may be reproduced or transmitted in any form or by any means, electronic or mechanical, including photocopying, recording, or any information storage and retrieval system, without permission in writing from the publisher. Details on how to seek permission, further information about the Publisher's permissions policies and our arrangements with organizations such as the Copyright Clearance Center and the Copyright Licensing Agency, can be found at our website: www.elsevier.com/permissions.

This book and the individual contributions contained in it are protected under copyright by the Publisher (other than as may be noted herein).

Notices
Knowledge and best practice in this field are constantly changing. As new research and experience broaden our understanding, changes in research methods, professional practices, or medical treatment may become necessary.

Practitioners and researchers must always rely on their own experience and knowledge in evaluating and using any information, methods, compounds, or experiments described herein. In using such information or methods they should be mindful of their own safety and the safety of others, including parties for whom they have a professional responsibility.

To the fullest extent of the law, neither the Publisher nor the authors, contributors, or editors, assume any liability for any injury and/or damage to persons or property as a matter of products liability, negligence or otherwise, or from any use or operation of any methods, products, instructions, or ideas contained in the material herein.

ISBN: 978-0-444-63638-6
ISSN: 0922-3487

For information on all Elsevier publications visit our website at https://www.elsevier.com/

ELSEVIER Book Aid International Working together to grow libraries in developing countries
www.elsevier.com • www.bookaid.org

Publisher: John Fedor
Acquisition Editor: Kathryn Morrissey
Editorial Project Manager: Jill Cetel
Production Project Manager: Vijayaraj Purushothaman
Cover Designer: Victoria Pearson

Typeset by SPi Global, India

Contents

Contributors

Numbers in Parentheses indicate the pages on which the author's contributions begin.

E. Aguado Sarrió (519), Multivariate Statistical Engineering Group, Universitat Politècnica de València, Valencia, Spain

S. Aloïse (353), Université de Lille, Sciences et Technologies, LASIR CNRS, LILLE, France

Y. Altmann (185), School of Engineering and Physical Sciences, Heriot-Watt University, Riccarton, Edinburgh, EH14 4AS, United Kingdom

J.M. Amigo (613), Faculty of Sciences, University of Copenhagen, Frederiksberg C, Denmark

A. Bogomolov (53), Art Photonics GmbH, Berlin, Germany

D. Brie (279), Centre de Recherche en Automatique de Nancy, Université de Lorraine, CNRS, Vandœuvre-lès-Nancy, France

N. Brun (185), Laboratoire de Physique des Solides, CNRS, Univ. Paris-Sud, Univ. Paris-Saclay, 91405 Orsay Cedex, France

R. Calvini (613), University of Modena and Reggio Emilia, Reggio Emilia, Italy

C. Carteret (279), Laboratoire de Chimie Physique et Microbiologie pour l'Environnement, Université de Lorraine, CNRS, Villers-lès-Nancy, France

M. Cocchi (409), University of Modena and Reggio Emilia, Modena, Italy

A. de Juan (5), Universitat de Barcelona, Barcelona, Spain

O. Devos (353, 453), Université de Lille, Sciences et Technologies, LASIR CNRS, LILLE, France

N. Dobigeon (185), University of Toulouse, IRIT/INP-ENSEEIHT, 31071 Toulouse Cedex 7, France

M. Dossot (279), Laboratoire de Chimie Physique et Microbiologie pour l'Environnement, Université de Lorraine, CNRS, Villers-lès-Nancy, France

L. Duponchel (477), LASIR CNRS UMR 8516, Université Lille 1, Sciences et Technologies, Villeneuve d'Ascq Cedex, France

A. Ferrer (519), Multivariate Statistical Engineering Group, Universitat Politècnica de València, Valencia, Spain

C. Gomez (579), IRD, UMR LISAH (INRA-IRD-SupAgro), Montpellier, France

N. Gorretta (579), IRSTEA UMR ITAP, Montpellier Cedex 5, France

D.M. Haaland (381), Spectral Resolutions, Albuquerque, NM, United States

P. de B. Harrington (311), Clippinger Laboratories, Center for Intelligent Chemical Instrumentation, Ohio University, Athens, OH, United States

S. Hugelier (453), Université de Lille, Sciences et Technologies, LASIR CNRS, LILLE, France

H.D.T. Jones (381), HyperImage Solutions, Rio Rancho, NM, United States

D. Jouan-Rimbaud Bouveresse (225), UMR Ingénierie Procédés Aliments, AgroParisTech, INRA, Université Paris-Saclay, Massy, France

A. Jürß (135), Universität Rostock, Institut für Mathematik, Rostock, Germany

S. Kucheryavskiy (53), Aalborg University, Aalborg, Denmark

M. Li Vigni (409), University of Modena and Reggio Emilia, Modena, Italy

A. Malik (101), Institute of Environmental Assessment and Water Research (IDÆA) SPANISH COUNCIL OF SCIENTIFIC RESEARCH (CSIC), Barcelona, Spain

S. Miron (279), Centre de Recherche en Automatique de Nancy, Université de Lorraine, CNRS, Vandœuvre-lès-Nancy, France

S. Moussaoui (185, 279), Ecole Centrale de Nantes, IRCCyN, UMR CNRS 6597, Nantes Cedex 3, France

R. Métivier (353), PPSM, ENS Cachan, CNRS, Cachan Cedex, France

K. Neymeyr (135), Universität Rostock, Institut für Mathematik; Leibniz-Institut für Katalyse, Rostock, Germany

M. Offroy (477), LASIR CNRS UMR 8516, Université Lille 1, Sciences et Technologies, Villeneuve d'Ascq Cedex, France

K.M. Pierce (333), Seattle Pacific University, Seattle, WA, United States

D.K. Pinkerton (333), University of Washington, Seattle, WA, United States

J.-P. Placial (353), PPSM, ENS Cachan, CNRS, Cachan Cedex, France

J.M. Prats-Montalbán (519), Multivariate Statistical Engineering Group, Universitat Politècnica de València, Valencia, Spain

D. Roberts (551), University of California, Santa Barbara, CA, United States

C. Ruckebusch (1, 353, 453), Université de Lille, Sciences et Technologies, LASIR CNRS, LILLE, France

D.N. Rutledge (225), UMR Ingénierie Procédés Aliments, AgroParisTech, INRA, Université Paris-Saclay, Massy, France

M. Sawall (135), Universität Rostock, Institut für Mathematik, Rostock, Germany

H. Schröder (135), Universität Rostock, Institut für Mathematik, Rostock, Germany

M. Sliwa (353), Université de Lille, Sciences et Technologies, LASIR CNRS, LILLE, France

B. Somers (551), Division of Forest, Nature, and Landscape, KU Leuven, Leuven, Belgium

R.E. Synovec (333), University of Washington, Seattle, WA, United States

R. Tauler (5, 101), Institute of Environmental Assessment and Water Research (IDÆA) SPANISH COUNCIL OF SCIENTIFIC RESEARCH (CSIC), Barcelona, Spain

J.A. Timlin (381), Sandia National Laboratories, Albuquerque, NM, United States

L. Tits (551), Flemish Institute for Technological Research (VITO), Remote Sensing Unit, Boeretang 200, BE-2400 Mol, Belgium

A. Ulrici (613), University of Modena and Reggio Emilia, Reggio Emilia, Italy

E. Wetherley (551), University of California, Santa Barbara, CA, United States

W. Windig (53), Eigenvector Research Inc., Manson, WA, United States

Preface

This book deals with ways and means to resolve spectral mixtures, a longstanding but topical issue that is tackled in many research fields, with an unlimited number of possible applications. In chemistry, spectral mixture resolution investigates situations where the spectral signatures of a sample result from the superposition of the various chemicals or materials composing it, as observed in process spectroscopy or hyperspectral imaging. More broadly, resolving a spectral mixture is an inverse problem that is addressed in different ways in chemistry and in the different fields of science and technology. This has led to the development of a wealth of methods.

Spectral mixture resolution remains a hot issue in chemometrics statistics or signal and image processing. However, the diversity of methods (multivariate curve resolution, blind source separation, linear unmixing, etc.) can be a serious impediment to wider understanding and dissemination of the spectral mixture problem. By taking a multi-angle and cross-disciplinary approach, this book has the ambition to turn this limitation into an asset and to provide a comprehensive and comprehensible description of the current state of the art.

The book is multi-authored, written as a collection of independent chapters that provide different perspectives and applications. Expressed altogether, these perspectives translate into a real interplay between the chapters. Basic concepts and main methods are presented in the first part of the book, whereas the second part is more oriented toward applications in chemistry and remote sensing. Some chapters are written as tutorials, others are reviews. Overall, the book presents an extensive bibliography which is incomplete—considering the amount of scientific literature on the subject—but good list to start with.

Written by invited authors that are recognized experts in their field, the book is addressed to graduate students, researchers, and practitioners in analytical science, emphasizing applications dealing with any type of spectral data. Some background in data analysis (chemometrics, statistics, signal or image processing, etc.) and knowledge of the basics of linear algebra will help readability. However, each chapter was meant to contain enough information to be, by itself, sufficient and to be read independently of the other chapters.

Lastly, I would like to express my gratitude to all the invited authors that have accepted to cooperate and contributed to this book, and to the numerous people who, one way or another, were involved in this project.

Cyril Ruckebusch
Lille, January 2016

Foreword

The topic of this volume is, of course, very interesting and it is the right time for its publication as a rapidly growing number of researchers rely on the methods described here for the analysis of their data. Very likely this number will increase significantly in the near future.

All work presented here deals with a series of measured spectra, and the goal of the analyses is the resolution of this collection of spectra into the product of contributions of the components in terms of their concentrations and their spectral responses. The task is, in some ways, both surprisingly simple and surprisingly difficult.

All analyses are based on the model that the overall signal at a particular wavelength or channel is the sum over the contributions of all components where each component contribution is the product of its concentration times its signal for unit concentration. This is most clearly realized in solution absorption spectroscopy where Beer–Lambert's law reigns; at each wavelength the absorption is the sum over the products of the component concentrations times their molar absorptivity at the particular wavelength. It is, therefore, not surprising that most methods presented in this volume have been developed for the resolution of series of absorption spectra, typically an HPLC chromatogram equipped with a diode array detector. This volume demonstrates that the methodologies have found much wider application. However, the following explanations will be based on the language of absorption spectroscopy.

Eq. (1) describes the relationship for the absorption reading of spectrum i measured at wavelength j, $\mathbf{D}_{i,j}$ is the sum over the products of concentration of species k, $\mathbf{C}_{i,k}$, times its molar absorptivity at wavelength j, $\mathbf{A}_{k,j}$:

$$\mathbf{D}_{i,j} = \sum_k \mathbf{C}_{i,k}\mathbf{A}_{k,j} \tag{1}$$

The Equation (1) can most conveniently be written as a matrix equation (Eq. 2)

$$\mathbf{D} = \mathbf{CA} \tag{2}$$

where **D** contains row-wise the spectra taken typically during a process, it is known; the matrix **C** contains the concentrations of the components and **A** contains their spectra or responses. These matrices are unknown and the goal is to determine them. It is useful to represent the situation graphically:

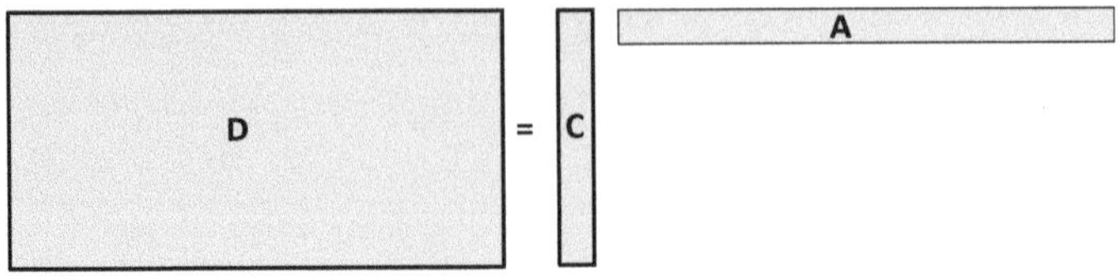

Each element of **D** represents an equation of type (1), and the elements of **C** and **A** are the unknowns. A prerequisite for all analyses presented here is that there must be more elements in **D** (equations) than unknowns in **C** and **A**; in fact, in most instances the area of **D** is much larger than the areas of **C** and **A** combined.

The first question one has to ask is: as there are more equations than unknowns, is there a solution at all? The answer is yes, and there must be a solution, the true concentrations, and molar absorptivities. Second, is the solution unique? As all elements of **C** and **A** must be positive, there is hope that there is a unique solution. However, in most case the solution is not unique, and there are a range of feasible solutions.

In the following chapters, several issues related to the task of decomposing a measured data set into two meaningful matrices are discussed:

How can the solution be computed?
(Chapters 2, 3, 7, 8, and 10)

This is clearly a central issue as the computation of the unique solution or of one of a range of feasible solutions is not a trivial task. In the chemometrics literature the alternating least squares, ALS, algorithm is most commonly used; it is based on an elegant and simple algorithm.

What if the structure of the data is more complex than the matrix equation (2)*?*
(Chapters 2, 10, and all image chapters)

Eq. (2) is very simple, if the structure of the data is more complex; e.g., in all hyperspectral images, the data can be rearranged in such a way that they fit into the structure of Eq. (2), and this is also the case for higher order data sets where individual matrices are concatenated into the basic structure of Eq. (2).

Is there a unique solution?
(Chapters 4, 5, and 8)

Even if there are many more equations than unknowns, for most data sets the solution is not unique; there are a range of feasible solutions for the matrices **C** and **A** which resolve the data matrix **D** with exactly the same accuracy. We call this rotational ambiguity. There are readily available methods that either approximate this range (Chapter 4) or perform a comprehensive analysis (Chapter 5).

Can additional constraints be applied?
(Chapters 3, 14, 17, and 18)

Nonnegativity of the elements of the matrices **C** and **A** is the universally applied constraint. Rotational ambiguity can be reduced or even eliminated altogether by application of other constraints or assumptions, such as maximum smoothness of or dissimilarity between the results.

What if the signal is not strictly following Eqs. (1) and (2)?
(Chapter 6, all image chapters)

Eqs. (1) and (2) describe the ideal case of linear signal contribution; quite often, this is not strictly the case, and it certainly is only an approximation in hyperspectral images. This does not render the analysis irrelevant; the result is an approximation and often a very useful one. Chapter 6 specifically explores the possibility of a nonlinear relationship between component concentration and signal.

What if the data set is huge?
(Chapters 9, 12, and 13)

Wavelet transformation is an option for the reduction of the size of the original data matrix **D**. Other corrections like baseline subtraction, etc., can also be useful.

Can the spatial resolution be improved?
(Chapter 15)

Hyperspectral images have a certain given spacial resolution; surprisingly, it can be improved by clever algorithms.

How widely can the methods be applied?
(Chapters 11 and 16)

There are two chapters that report on the application of the methods presented here in very different fields of chemical research.

Are the results qualitative or quantitative?
(Chapter 2)

The typical result of most analyses presented here is a set of concentration profiles/distributions and response vectors/spectra. They can be represented in graphs. Primarily, the results do not allow quantitative interpretation. The extension of the methods toward quantitative analyses is most important.

To conclude this foreword I would like to add an important comment: Every scientist knows that a reported value for a parameter should be further characterized by an error estimate, e.g., $K = 350 \pm 25$. Similarly, a spectral mixture analysis without analysis of rotational ambiguity should be rejected. In my experience, too many users take the analysis result as a fact, rather than one of a range of possible solution. Implementing further constraints may result in unique solutions, but if the additional constraint is not based on actual facts but rather on wishful thinking, the unique outcome may well not be the true one. The authors in this volume are the leaders in the field and should educate the novice or casual user about this problem and lead with the proverbial good example. (I will not list the offenders.).

As we have seen, there are many difficulties, should the present methods not be used? The answer is a resounding no! The methods are very powerful and while they might not give ultimate and detailed answers they nonetheless deliver a lot of very useful information which would be difficult if not impossible to gather otherwise. More work will be done in the field, there is no doubt.

Marcel Maeder
University of Newcastle, Australia

Chapter 1

Introduction

C. Ruckebusch[1]
Université de Lille, Sciences et Technologies, LASIR CNRS, Lille, France
[1]Corresponding author: e-mail: cyril.ruckebusch@univ-lille1.fr

Chapter Outline

1 INTRODUCTION

This chapter introduces first very basic information about the topic of the book and sets the overall context. It provides broad definitions and clarifies some points regarding the terminology. The second part provides information about the organization of the book. A first insight into the content of the 19 chapters composing the book, and their interplay, is given. The intention of these few words of introduction is mainly the presentation of the issues that will be tackled more comprehensively along the chapters of the book. These questions can be roughly put as follows:

- What is a spectral mixture?
- What does resolving a spectral mixture mean?
- What are the different ways to tackle the spectral mixture issues?
- What difficulties remain?
- And what are the perspectives?

2 THE SPECTRAL MIXTURE PROBLEM

A spectral mixture is a data that results from the observation of a chemical system composed of (mixed) individual components and submitted to some variation. This variation is related to the change of an external factor, which is usually a physical or chemical variable. It can be for example sampling time, position, or pH. The spectral data thus consist of a superposition, or mixture, of the pure spectra of the individual components and their associated proportions. When dealing with evolving systems such as chemical reactions or processes, these proportions correspond to concentration profiles.

Data Handling in Science and Technology, Vol. 30. http://dx.doi.org/10.1016/B978-0-444-63638-6.00001-2
Copyright © 2016 Elsevier B.V. All rights reserved.

Spectral mixture data are usually arranged in a matrix with columns as spectral variables (wavelength, wavenumbers, etc.) and objects (time, position, etc.) as rows. Objects can be of different nature, but should always be clearly related to the state of the before mentioned physical or chemical variable. Ideally, the variations contained in the spectral data translate what is supposed to be relevant information for the problem at hand. Spectral mixture resolution aims to decompose the variations of the spectral data into a model of the contributions from the individual unknown components. These components are composed of source proportions and spectral signatures. It is important to realize that, more than often, this decomposition is aimed at situations for which little a priori information is available. It should also be noted that, in practice, some physical perturbations or chemical interferences may complicate the ideal situation.

In chemistry, spectral mixture resolution corresponds to the resolution of complex mixture spectra into pure contributions, consisting of concentration distributions and spectra of the different chemical components. The basic model underlying this decomposition, usually termed multivariate curve resolution (MCR) in chemometrics, corresponds to the Lambert-Beer law written in a matrix form. This factorial model states a bilinear relation between the matrix of observations and the two matrices of contributions containing concentration profiles and spectra, respectively. It should be noted that this extends to the analysis of spectral and hyperspectral images when investigating a specimen (in microscopy) or a scene (in remote sensing). Also, the bilinear model can be extended for the analysis of multiple data sets that are meant to connect different experiments together. Overall, MCR can be applied in situations where a reasonable approximation of the bilinear model, or any other fundamental basic equation that has the same mathematical structure, holds.

Application of MCR methods is broad, quite straightforward, and provides results which are readily chemically/physically interpretable. These assets explain why MCR has spread in the chemical literature and in many other scientific fields. However, considering the mathematical conditions for exact resolution of the MCR problem, some theoretical issues remain and are currently the subject of intensive research. The most puzzling of these issues is the so-called rotational ambiguity of the resolution. In more common words, this translates into the fact that a unique solution cannot be obtained in general. Then, particular attention should be paid to the initial condition, or to the constraints applied during resolution, and it is important to assess the extent of rotational ambiguity before any definitive conclusion to be drawn. Considering these aspects, one may notice a certain antagonism in MCR between wide applicability and high interpretability on the one hand and mathematical complexity of the resolution on the other hand. This explains to a large extend the continuous development of this topic into a proper research field, still very much in progress.

Taking a broader perspective, spectral unmixing enters the more general category of inverse problems, important, and ubiquitous problems in analytical science and data analysis. From a set of (spectral) observations, one aims to extract the unknown sources that produced the data but could not be observed directly. Mixture analysis, MCR, blind source separation, linear unmixing, etc. are methods that share this objective but were developed in different scientific fields, chemistry, statistics, or signal and image processing.

3 BOOK CONTENT AND ORGANIZATION

The book starts with Chapter 2 that introduces the key concepts and provides an overview of the progress in MCR with an emphasis on applications to spectroscopic data. Focus is on constraints, multiset analysis, and quantitative aspects in multivariate curve resolution alternating least squares (MCR-ALS). Next, Chapter 3 revisits the concept of variable purity, with purity defined as the observation of a nonzero contribution from one and only one of the mixture components. Issues and solutions relative to rotational ambiguity of the MCR solutions, currently a very active research topic, are then discussed in Chapters 4 and 5. Chapter 4 sets the basis of the problem and focuses on a nonlinear constrained optimization approach for the direct calculation of maximum and minimum band boundaries of feasible solutions. In contrast, Chapter 5 aims to provide a systematic introduction to the concept of area of feasible solutions, from which feasible solutions can be derived. With Chapter 6, spectral unmixing and spectral mixture analysis are introduced. These methods aim at extracting the spectral characteristics and quantifying the spatial distribution over a spectral image. This chapter goes beyond the state of art by introducing nonlinear approaches to SU which allows to take into consideration more complex mixing process or spectral variability of the sources. Chapter 7 covers the basic of independent component analysis, a source separation method initially developed in the field of telecommunications and now applied in different domains including chemometrics and spectroscopy. Chapter 8 deals with a Bayesian positive source separation approach of the MCR problem which is motivated by the search of unique solutions. The second part of the book, oriented more towards applications, starts with Chapter 9. It introduces a wavelet compression strategy that facilitates the application of MCR to large data sets. Chapter 10 deals with chromatography coupled with spectral detection, the type of data which originally motivated development of MCR, and extends to the application of trilinear approaches. With Chapter 11, the focus is on the application of MCR-ALS for ultrafast time-resolved absorption spectroscopy data. Chapter 12 tackles the analysis of hyperspectral images of biological samples with the use of automated data preprocessing and improved MCR methods, increasing the sensitivity and accuracy of the chemical images obtained. In Chapter 13, the integration of wavelet transform with multivariate image analysis in a multiresolution analysis approach opens the possibility of simultaneously accomplishing denoising and feature selection. With Chapter 14, a new constraint that allows forcing some information related to the low-frequency character of the components profiles and distribution maps in MCR-ALS is introduced. Chapter 15 discusses the potential of super-resolution in vibrational spectroscopy imaging, merging instrumental and algorithmic developments. Chapter 16 deals with the current topic of biomarker imaging for early cancer detection applying MCR to magnetic resonance images. Chapters 17 and 18 provide ways and means to deal with remotely sensed data. Chapter 17 focuses on the use of spectral libraries for spectral mixtures analysis. How to compose, handle, and optimize endmember libraries are the issues discussed in detail. Chapter 18 reviews the recent developments of spectral unmixing algorithms that incorporate spatial information, termed spatial-spectral unmixing. To close, Chapter 19 presents a sparse approach for spectral unmixing of hyperspectral images, which provides better interpretability of the results obtained.

Chapter 2

Multivariate Curve Resolution-Alternating Least Squares for Spectroscopic Data

A. de Juan*,[1] and R. Tauler†

*Universitat de Barcelona, Barcelona, Spain

†Institute of Environmental Assessment and Water Research (IDÆA) SPANISH COUNCIL OF SCIENTIFIC RESEARCH (CSIC), Barcelona, Spain

[1]Corresponding author: e-mail: anna.dejuan@ub.edu

Chapter Outline

1 MCR: THE CONCEPT AND THE LINK WITH SPECTROSCOPIC DATA

Multivariate curve resolution (MCR) is the generic denomination of a family of methods meant to solve the mixture analysis problem, i.e., able to provide a chemically (scientifically) meaningful additive bilinear model of pure contributions from the sole information of an original data matrix including a mixed measurement [1–7]. Thus, MCR analysis describes the total signal of a multicomponent data set (**D**) as the sum of the signal contributions coming from

Data Handling in Science and Technology, Vol. 30. http://dx.doi.org/10.1016/B978-0-444-63638-6.00002-4
Copyright © 2016 Elsevier B.V. All rights reserved.

each of the pure components, $\mathbf{D}_i$ (Eq. 1). Each term of this additive model can be expressed by the product of a dyad of profiles, $\mathbf{c}_i\mathbf{s}_i^{\mathrm{T}}$, where $\mathbf{s}_i^{\mathrm{T}}$ is the pure response of the component weighted by the related concentration profile, $\mathbf{c}_i$ (Eq. 2). Finally, the classical compact expression of the bilinear additive model is shown in Eq. (3):

$$\mathbf{D} = \sum_i \mathbf{D}_i + \mathbf{E} \tag{1}$$

$$\mathbf{D} = \sum_i \mathbf{c}_i\mathbf{s}_i^{\mathrm{T}} + \mathbf{E} \tag{2}$$

$$\mathbf{D} = \mathbf{C}\mathbf{S}^{\mathrm{T}} + \mathbf{E} \tag{3}$$

where $\mathbf{C}$ contains the concentration profiles of all components and $\mathbf{S}^{\mathrm{T}}$ the related pure responses. $\mathbf{E}$ is the matrix expressing the error or variance unexplained by the bilinear model in all earlier equations.

The expressions shown in Eqs. (1)–(3) remind clearly the natural bilinear model of spectroscopic measurements, the Beer–Lambert law, where the variation of a data table containing the spectra of a multicomponent system (**D**) can be expressed through the product of the pure spectra (absorptivities) ($\mathbf{S}^{\mathrm{T}}$) of the components by their related concentration profiles (**C**). Then, it is not surprising that the first applications of MCR were devoted to analyze spectroscopic data because of the identical underlying models of the measurement and the method [1,5]. In difference to other bilinear decomposition methods, MCR provides meaningful profiles because chemical properties related to the concentration profiles and spectra are actively incorporated in the optimization of the bilinear model.

There are many multicomponent systems described by spectroscopic data. MCR adapts to this diversity by tuning the way of application of the method according to the characteristics of the spectroscopic technique and the concentration profiles. In a very general way, we can distinguish between process and mixture data sets. Process data sets show very structured concentration profiles, displaying a smooth variation as a function of a process variable. Typical examples may be an HPLC–DAD chromatographic data set, where the concentration (elution) profiles have a peak-shaped signal or reactions monitored by spectroscopic techniques (see Fig. 1A). Instead, the concentration profiles of a mixture data set can vary in a nonpatterned way. This less structured variation can come from the nature of the data, e.g., a set of independent samples, or from the data set configuration needed for MCR application, e.g., hyperspectral images (HSIs) are treated by using a data table of pixel spectra, which does not preserve the spatial organization of the original measurement (see Section 4 and chapters 6,12,13,15,17 and 19 in this book for more detail on this kind of measurements). MCR applies in all the scenarios described, but specificities of the different data set typologies should be taken into account in the data set configuration and in the different steps of application of the algorithm.

Before describing the details and application of MCR algorithms, a necessary comment has to be made regarding one of the fundamental assumptions

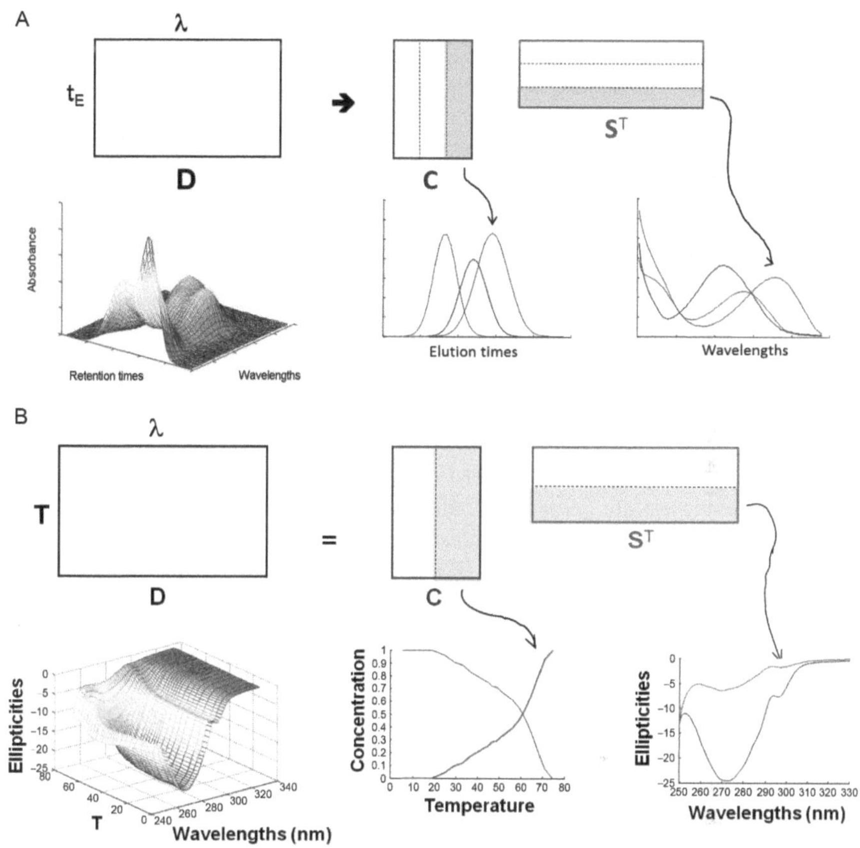

FIG. 1 (A) Bilinear model of an HPLC–DAD chromatogram and (B) bilinear model of a thermal-dependent process monitored by circular dichroism.

of MCR, i.e., the bilinearity of spectroscopic data. Indeed, the natural fulfillment of the Beer–Lambert law may be affected by signal artifacts, such as scattering in near infrared spectroscopy [8,9], or fluorescence contributions in Raman spectra [10]. Most of the times, the bilinear behavior is easily recovered by suitable preprocessing, e.g., scatter or baseline correction, adapted to the spectroscopic measurement of interest [11–14]. Because of the nature of the spectroscopy used, some techniques need more dedicated and intensive preprocessing, like ultrafast spectroscopy measurements (see Chapter 11) [15,16] and, in extreme cases, nonlinear unmixing methods can be applied (see Chapter 6).

2 MCR-ALS: ALGORITHM AND DATA SET CONFIGURATION

Multivariate Curve Resolution-Alternating Least Squares (MCR-ALS) is an iterative algorithm that solves the bilinear MCR model by optimizing concentration profiles and pure spectra in an alternating least squares way under constraints [5,6,17–20]. To trigger the optimization, a set of initial estimates of spectral or concentration profiles are needed. The general minimization step in the iterative optimization can be expressed as follows:

$$\min \left\| \mathbf{D} - \mathbf{C}\mathbf{S}^{\mathrm{T}} \right\| \tag{4}$$

taking into account that $\mathbf{C}$ and $\mathbf{S}^{\mathrm{T}}$ are calculated in each iterative cycle.

The main assets of this algorithm are the variety of information that can be included in the optimization process and the capability to work with single data tables and complete and incomplete multisets [3,6,20–22].

Multisets are defined as data structures formed by more than one data table. Multiset structures can be extremely flexible and can be formed by data matrices of different size and meaning. All multiset structures present a bilinear model and the obvious requirement is that the different subsets share components in common. Different multiset typologies can be obtained depending on how the data subsets are connected. Multisets can be structured as: (a) row-wise augmented matrices, when data tables are appended besides each other, e.g., multitechnique process monitoring (Eq. 5), (b) column-wise augmented matrices, when data tables are one on top of each other, e.g., different batches monitored in a production process (Eq. 6), and (c) row- and column-wise augmented matrices, when the multiset extends in the row and column direction (Eq. 7).

$$[\mathbf{D}_1\, \mathbf{D}_2\, \mathbf{D}_3 \ldots \mathbf{D}_{\mathrm{L}}] = \mathbf{C}[\mathbf{S}_1^{\mathrm{T}}\, \mathbf{S}_2^{\mathrm{T}}\, \mathbf{S}_3^{\mathrm{T}} \ldots \mathbf{S}_{\mathrm{L}}^{\mathrm{T}}] + [\mathbf{E}_1\, \mathbf{E}_2\, \mathbf{E}_3 \ldots \mathbf{E}_{\mathrm{L}}] = \mathbf{C}\mathbf{S}_{\mathrm{aug}}^{\mathrm{T}} \tag{5}$$

$$\begin{pmatrix} \mathbf{D}_1 \\ \mathbf{D}_2 \\ \mathbf{D}_3 \\ \ldots \\ \mathbf{D}_K \end{pmatrix} = \begin{pmatrix} \mathbf{C}_1 \\ \mathbf{C}_2 \\ \mathbf{C}_3 \\ \ldots \\ \mathbf{C}_K \end{pmatrix} \mathbf{S}^{\mathrm{T}} + \begin{pmatrix} \mathbf{E}_1 \\ \mathbf{E}_2 \\ \mathbf{E}_3 \\ \ldots \\ \mathbf{E}_K \end{pmatrix} = \mathbf{C}_{\mathrm{aug}}\mathbf{S}^{\mathrm{T}} + \mathbf{E}_{\mathrm{aug}} \tag{6}$$

$$\begin{bmatrix} \mathbf{D}_{11} & \mathbf{D}_{12} & \mathbf{D}_{13} & \ldots & \mathbf{D}_{1\mathrm{L}} \\ \mathbf{D}_{21} & \mathbf{D}_{22} & \mathbf{D}_{23} & \ldots & \mathbf{D}_{2\mathrm{L}} \\ \mathbf{D}_{31} & \mathbf{D}_{32} & \mathbf{D}_{33} & \ldots & \mathbf{D}_{3\mathrm{L}} \\ \ldots & \ldots & \ldots & \ldots & \ldots \\ \mathbf{D}_{\mathrm{K}1} & \mathbf{D}_{\mathrm{K}2} & \mathbf{D}_{\mathrm{K}3} & \ldots & \mathbf{D}_{\mathrm{KL}} \end{bmatrix} = \begin{pmatrix} \mathbf{C}_1 \\ \mathbf{C}_2 \\ \mathbf{C}_3 \\ \ldots \\ \mathbf{C}_{\mathrm{K}} \end{pmatrix} [\mathbf{S}_1^{\mathrm{T}}\ \mathbf{S}_2^{\mathrm{T}}\ \mathbf{S}_3^{\mathrm{T}}\ \ldots\ \mathbf{S}_{\mathrm{L}}^{\mathrm{T}}]$$
$$+ \begin{pmatrix} \mathbf{E}_{11} & \mathbf{E}_{12} & \mathbf{E}_{13} & \ldots & \mathbf{E}_{1\mathrm{L}} \\ \mathbf{E}_{21} & \mathbf{E}_{22} & \mathbf{E}_{23} & \ldots & \mathbf{E}_{2\mathrm{L}} \\ \mathbf{E}_{31} & \mathbf{E}_{32} & \mathbf{E}_{33} & \ldots & \mathbf{E}_{3\mathrm{L}} \\ \ldots & \ldots & \ldots & \ldots & \ldots \\ \mathbf{E}_{\mathrm{K}1} & \mathbf{E}_{\mathrm{K}2} & \mathbf{E}_{\mathrm{K}3} & \ldots & \mathbf{E}_{\mathrm{KL}} \end{pmatrix} = \mathbf{C}_{\mathrm{aug}}\mathbf{S}_{\mathrm{aug}}^{\mathrm{T}} + \mathbf{E}_{\mathrm{aug}} \tag{7}$$

The minimization function expressed in Eq. (4) also holds, just replacing the $\mathbf{C}$ and/or $\mathbf{S}^{\mathrm{T}}$ matrices by their augmented versions when needed.

Looking at the different expressions in Eqs. (5)–(7), it can be understood that the data tables appended in a multiset should have, at least, a common mode (the concentration or the spectral mode). This means that the size (dimensionality) and profile behavior must be the same. For instance, when a process is monitored simultaneously with different analytical techniques, i.e., giving a row-wise augmented matrix, the measurements (rows) collected should refer to the same process stages in all data matrices because the single concentration matrix will define the process behavior for all data tables treated together. Likewise, when different experiments are carried out using the same spectroscopic technique, all of them should be performed using the same spectral range and spectral resolution, because a single matrix of pure spectral signatures has to be valid for all experiments.

A recent particular situation is the one presented by incomplete multisets [22]. In this case, a regular multiset structure cannot be built and missing blocks are present (see Fig. 2B).

To cope with this kind of data structure, the minimization problem is slightly modified. The incomplete structure can be generally defined by several complete multisets with their related bilinear models ($\mathbf{D}_A$ and $\mathbf{D}_B$ in

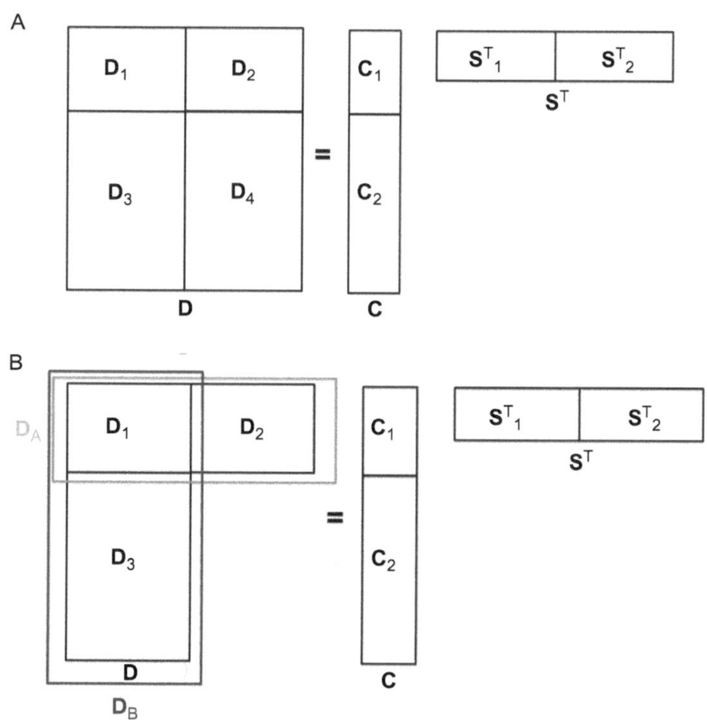

FIG. 2 Multiset examples and related bilinear models. (A) Complete row- and column-wise augmented matrix and (B) incomplete row- and column-wise augmented matrix.

Fig. 2B). The regular multisets are optimized in each iterative cycle and the minimization function optimizes all residuals from all calculated bilinear models in a simultaneous way (see Eq. 8).

$$\min\left(\left\|\mathbf{D}_A - \mathbf{C}_A\mathbf{S}_A^T\right\| + \left\|\mathbf{D}_B - \mathbf{C}_B\mathbf{S}_B^T\right\|\right) \tag{8}$$

Averaged profiles or selected profiles from the different regular multisets optimized are provided as the resolved profiles in each iterative cycle or as the final MCR solution [22,23].

Although the equations and minimizations presented take as a basis an ordinary least squares optimization, the same algorithm can be used when information about the noise level and structure is available. In this case, weighted MCR-ALS variants are also available [24–28].

The flexibility of data configurations and the possibility to introduce general information in MCR-ALS models is outstanding, but all MCR algorithms are affected by the so-called ambiguity phenomenon. This means that different combinations of concentration profiles and spectra can describe equally well, in terms of model fit, the original data set [19,29]. Indeed, three main modalities of this phenomenon can be encountered

(a) *Permutation ambiguity*. There is no sorting order on the MCR components. Therefore, they can be shuffled in the concentration and spectra matrix (always keeping the right correspondence of the dyads) and provide identical result.

(b) *Intensity ambiguity*. Dyads of profiles having the same shape but different relative scales between concentration profile and spectrum reproduce equally well the original data set. Indeed, concentration values and pure spectra intensities in $\mathbf{c}_i$ and $\mathbf{s}_i^T$ profiles are always in arbitrary units unless reference information on real intensities is available and actively used in the resolution process. This can be seen taking Eq. (2) as initial expression of the bilinear model skipping the error term. Eq. (2) can be easily transformed into Eq. (9) by introducing an unknown and different scaling factor for each i-th component, k_i, to define the related dyad $\mathbf{c}_i\mathbf{s}_i^T$.

$$\mathbf{D} = \sum_i \mathbf{c}_i\mathbf{s}_i^T \tag{2}$$

$$\mathbf{D} = \sum_i (\mathbf{c}_i k_i)\left(\mathbf{s}_i^T \frac{1}{k_i}\right) \tag{9}$$

(c) *Rotational ambiguity*. Sets of concentration profiles and spectra with different shapes can reproduce the original data set with the same fit quality. This is the most relevant kind of ambiguity and can be expressed taking as a basis the bilinear model in Eq. (3) and modifying it into Eq. (10), by introducing a **T** transformation matrix, as follows:

$$\mathbf{D} = \mathbf{C}\mathbf{S}^{\mathrm{T}} + \mathbf{E} \quad (3)$$

$$\mathbf{D} = (\mathbf{C}\mathbf{T})\left(\mathbf{T}^{-1}\mathbf{S}^{\mathrm{T}}\right) + \mathbf{E} \quad (10)$$

Different ways to assess the extent and location of ambiguity in an MCR solution will be explained in Chapters 4 and 5 [30–33]. The way to decrease or suppress this phenomenon is by introducing chemical or mathematical constraints to model the resolved profiles or by working with more information related to the same chemical system, i.e., with multiset structures.

2.1 MCR-ALS Algorithm: Steps

The earlier section described the general concept behind the MCR-ALS algorithm, i.e., the iterative alternating least squares optimization of concentration profiles and pure spectra under constraints. Below, the general steps in the MCR-ALS algorithm are detailed

1. Determination of the number of components.
2. Generation of initial estimates of **C** or $\mathbf{S}^{\mathrm{T}}$.
3. Iterative alternating least squares optimization of **C** and $\mathbf{S}^{\mathrm{T}}$ under constraints until convergence is achieved.

The number of components in a data set can be known beforehand or determined by auxiliary means, such as principal component analysis (PCA). Both methods aim at describing the maximum variance of the data set, and the assumption is that a data set can be described with a number of MCR contributions equal to the number of significant principal components used for the same purpose. In this context, examination of the magnitude of singular values is typically used, i.e., singular values related to chemical contributions are large, whereas singular values related to noise are small and similar among them. This diagnostic can be complemented by looking at the emergence of noisy patterns in scores and loadings profiles, typical in noise-related components. When in doubt, several MCR models with different number of components can be calculated. The final model will be selected as the smallest one providing an optimal model fit and chemically meaningful resolved profiles.

Initial estimates in MCR-ALS can be concentration profiles or spectra. A good hint in the generation of estimates is starting with sensible initial guesses that present the same general properties sought in the profiles to be recovered. This means that initial random values are generally of no help because the profiles expected do not have this numerical structure [4–7,34]. Initial estimates can be based on previous knowledge, e.g., spectra of components known to be present in the data set, spectra at maximum of chromatographic peaks, or can be obtained by auxiliary methods. The selection of methods to obtain initial estimates will depend essentially on the structure of the data set, particularly on the presence or absence of systematic behavior in the concentration direction (process data vs. mixture or image data).

A general family of methods apt to generate initial estimates for any kind of data sets are the so-called purest variable selection methods, such as SIMPLISMA [35], orthogonal projection approach [36], and key set factor analysis [37,38]. All these methods have the common goal of selecting the most dissimilar rows or columns in the original data set **D** and provide, as a consequence, initial estimates of spectra or concentration profiles, respectively. They will be explained in more detail in Chapter 3. A good property of these methods is that, when selective rows or columns for a particular component exist, the related pure spectrum or concentration profile is readily recovered. However, it is generally incorrect associating the profiles extracted by these methods with pure component profiles because selectivity cannot be always ensured. Methods related to the generation of initial estimates for process data will be explained in Section 3.

Once initial estimates are obtained, optimization takes place until convergence is achieved. To control convergence, figures of merit related to the model fit are used, such as the lack of fit (% LOF) or variance explained (in Eqs. 11 and 12, respectively).

$$\%\ \mathrm{LOF} = 100\sqrt{\frac{\sum_{i,j} e_{ij}^2}{\sum_{i,j} d_{ij}^2}} \tag{11}$$

$$\%\ \mathrm{var.expl.} = 100\left(1 - \frac{\sum_{i,j} e_{ij}^2}{\sum_{i,j} d_{ij}^2}\right) \tag{12}$$

where e_{ij} is the residual associated with the MCR reproduction of the related element in the original data set, $d_{ij.}$

The optimization is usually finished when the difference of model fit between consecutive iterations does not improve significantly (e.g., a difference of less than 0.1% among the lack of fit between consecutive iterations). Other possibilities are using a maximum number of iterations or criteria based on the comparison of shapes of the optimized profiles among iterations.

2.2 Constraints

Constraints are the essential part of the MCR-ALS algorithm and they have two main objectives: (a) introducing chemical and mathematical knowledge to provide chemical meaning to the concentration profiles and pure spectra and (b) reducing or suppressing the ambiguity related to the MCR solutions [4–7,19].

A constraint can be defined as any chemical or mathematical property systematically present in the component profiles of the data set. Within each iterative cycle, the computed profiles by least squares are modified so that they obey the conditions defined by the preselected constraints (see Fig. 3).

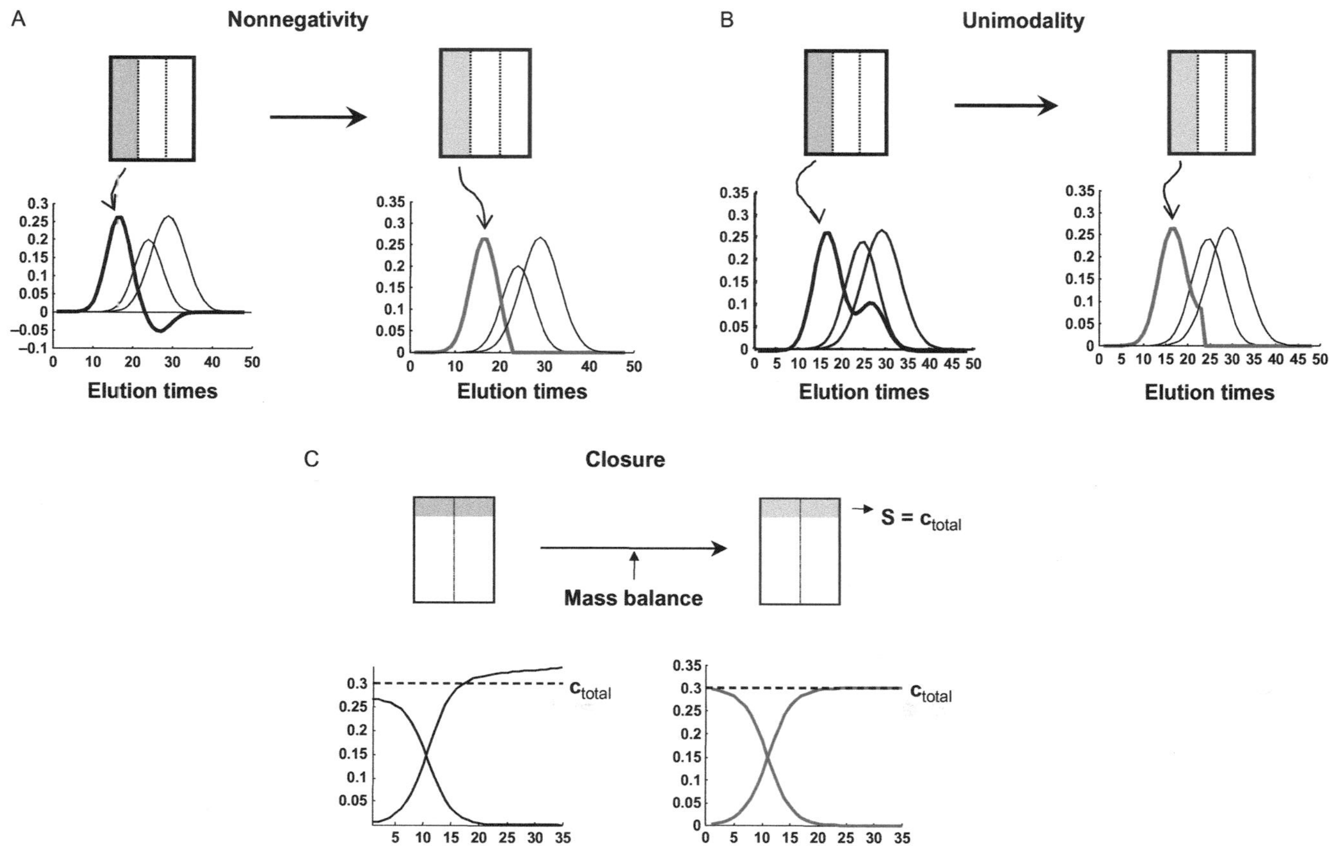

FIG. 3 Examples of constraints. (A) Nonnegativity, (B) unimodality, and (C) closure. Unconstrained profile is *black bold* and constrained profile is *red bold* (*light gray bold* in print version).

There are many approaches to implement constraints: smooth least squares [39,40] or penalty-based [41,42] procedures and other methodologies that work directly replacing erroneous profiles or elements in a profile by different values. The degree of tolerance in the fulfillment of constraints can also be tuned according to the properties of the data set to be treated, e.g., depending on the noise level of the data.

In the MCR-ALS framework, flexibility in the application of constraints is a key element. There is no default list of constraints, because default chemical data sets do not exist. Selection of constraints is an active step that MCR users have to be responsible for since they know the properties of their data sets and of the profiles to be resolved. Hence, application of all constraints is optional and they can be used differently (a) mode-wise, i.e., the concentration and the spectral profiles may have different properties, (b) compound-wise, i.e., the different compounds (profiles) within the $\mathbf{C}$ or $\mathbf{S}^T$ matrix may behave in a different way, and (c) subset-wise, i.e., the properties of different submatrices in $\mathbf{C}$ or $\mathbf{S}^T$ in a multiset structure may be very diverse.

The most common constraints, applicable to any kind of data set structure are

(a) *Nonnegativity*. Forces the profiles to be formed by positive values. Can be implemented replacing negative values by zeros or with softer algorithms, such as nonnegative least squares or fast nonnegative least squares [39]. It applies to all concentration profiles and to many spectroscopic responses. It should be avoided in certain kinds of spectroscopic profiles, such as those coming from circular dichroism or electronic paramagnetic resonance, which provide natural negative measurements, or when working with derivative spectra.

(b) *Unimodality*. Allows the presence of a single maximum per profile. It refers to peak-shaped profiles, such as elution profiles in chromatography, and to monotonic profiles, i.e., steadily increasing or decreasing, like some reaction profiles. Least squares implementations of this constraint exist, usually linked simultaneously to nonnegativity [40]. In constraints working by replacement of nonunimodal values, variable degrees of tolerance on the fulfillment of the constraint can also be set [43]. When working with multisets, unimodality is applied separately to the profiles of each of the subsets of the augmented $\mathbf{C}$ or $\mathbf{S}^T$ matrix.

(c) *Closure*. It is the expression of the mass balance condition. Applies to some concentration profiles of reaction systems and rescales the concentration profiles so that the total sum of the concentration of compounds in the closure amounts to a constant value during all the reaction. It can be applied as a strict equality or using a least squares implementation [44]. When working with multisets, closure is applied separately to the profiles of each of the subsets of the augmented $\mathbf{C}$ matrix.

(d) *Selectivity and local rank*. This constraint reflects that the local rank in some concentration windows or spectral ranges can be lower than the total rank of the data set, i.e., not all compounds are present everywhere in the data set. When selective concentration or spectral regions for a particular compound exist, the rest of compounds are set to zero and the pure spectrum or concentration profile for that compound is readily recovered. Local rank constraints apply when only some compounds are absent in a certain concentration or spectral range; however, these constraints also play a very relevant role in the data set to decrease and, in appropriate conditions, suppress ambiguity. In next sections, it will be described how local rank and selective regions can be gathered from a process or an image data set.

(e) *Equality constraints.* They apply when some elements in a profile or a full profile are known, e.g., a pure spectrum of a compound is known. They set the profile or the known elements in it to be equal to the known information. Local rank and selectivity can be considered equality constraints when absent compounds are set equal to zero. Likewise, nonnegativity can be applied as an equality constraint when negative values are set to be equal to zero.

Some additional constraints are associated only with multiset structures:

(a) *Correspondence of species.* This constraint extends the idea of local rank and selectivity to a multiset context and has the same beneficial effect. In this case, for column-wise augmented multisets, this constraint expresses the presence/absence of compounds in the original experiments. Therefore, full concentration profiles in certain **C** submatrices are forced to have null profiles and a correct correspondence between the compounds present and absent in the different experiments is explicitly defined [5,6,20,21].

(b) *Model constraints.* Classical MCR works under the assumption of a bilinear model to describe the data. However, there may be multisets that behave according to multiway models, such as trilinear or multilinear models [20,21,45,46], or factor interaction (Tucker) models [21,47]. These more complex models can also be implemented under the form of constraints in profiles from augmented **C** or $\mathbf{S}^T$ matrices, e.g., trilinearity can be imposed in profiles of augmented matrices by forcing them to have the same shape and preserving their natural scale. To do so, as can be seen in Fig. 4, the profiles of a certain compound in the different **C** submatrices, $\mathbf{c}_i^n$, are arranged one besides the other in a single data matrix. Singular value decomposition is then applied to this matrix. The first score profile is adopted as the basic shape of this compound for all **C** submatrices. To recover the properly scaled constrained profile for each **C** submatrix, the common score profile is multiplied by the related loading, which includes the scale information.

The main difference with this constraint and the approach used by natural multiway multilinear methods, such as parallel factor analysis (PARAFAC) [48] or direct trilinear decomposition (DTD) [49], is the

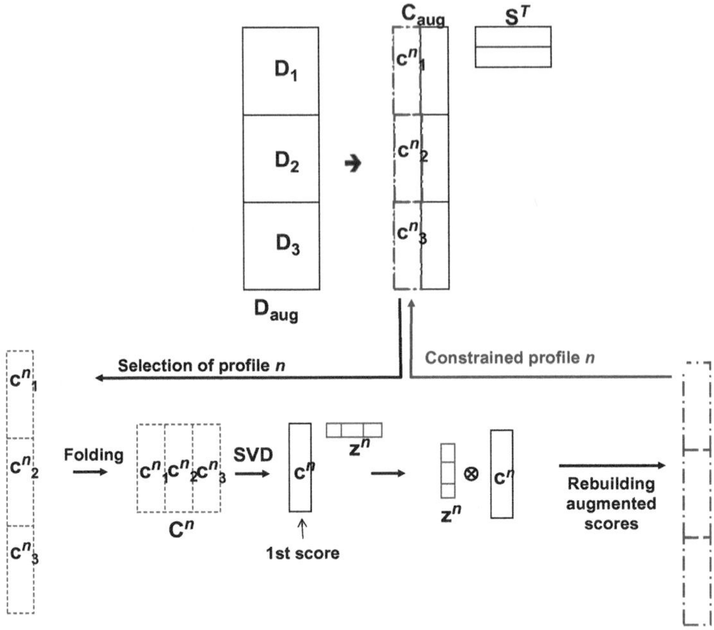

FIG. 4 Implementation of trilinearity constraint.

compound-wise application of the model constraint. Therefore, hybrid models, which have some compounds behaving in a bilinear way and others in a trilinear or multilinear way, can coexist [50,51]. When trilinearity or multilinearity is applied as constraints in MCR, the related compounds are resolved in a unique way, without ambiguities [19,29].

Finally, there are some constraints linked to particular tasks performed often by other data analysis methods, such as hard-modeling methods or multivariate calibration approaches. These are

(a) *Hard-modeling constraint.* The mechanism or the physicochemical law driving a reaction can be included as an additional constraint. Process profiles can be fitted with a hard model and the related model parameters, i.e., reaction constants, can also be retrieved. More detail on this constraint is given in Section 3, devoted to process analysis [21,52].

(b) *Correlation constraint.* This constraint includes an inner calibration model within MCR to provide quantitative information, in real concentration units, to the resolved concentration profiles of the components of interest. More detail on this constraint will be given in Section 5, dedicated to MCR for quantitative analysis [34,53].

In principle, any systematic property of a concentration or spectral profile can be potentially encoded as a constraint. The possibilities grow to give response to new kinds of problems and measurements. Clear examples of new constraints are related to image analysis, where the spatial pattern of the image starts to be taken into account (see Chapters 14 and 18), or to particular kinds of response or concentration profiles that present properties closely defined by mathematical conditions, such as sparseness (see Chapter 19).

3 MCR-ALS APPLIED TO PROCESS ANALYSIS

MCR was born to provide a solution to describe the evolution of multicomponent processes. The term process was applied to define a wide family of problems, all of them sharing the fact of having a continuous nonrandom evolution in the concentration direction. Therefore, a process could be from a classical reaction-driven system to an evolving system with no chemical transformation involved, like a chromatographic elution, a blending process, or a conformational or polymorphic transition. In all these systems, the concentration profiles are never a set of random numbers; instead, they reflect the smooth evolution of a particular compound as a function of a certain process variable (time, temperature, etc.). The structured concentration direction in processes enabled the use of particular kinds of information under the form of constraints to improve the description of these systems by MCR.

3.1 Encoding Process Information: Sequentiality and Physicochemical Models

3.1.1 Sequentiality

A very common fact in processes is the sequentiality in the evolution of components, i.e., the fact that a system can be defined by a set of components that emerge and decay in an ordered way. This property explains the frequent use of certain constraints, such as unimodality, in a process context. Indeed, a single emergence-decay curve for a component is linked to peak-shaped signals, such as elution profiles in chromatography, or to monotonic and peak-shaped concentration profiles in processes, i.e., the initial compound in a process can only decay, the intermediate ones emerge and decay and the last product keeps increasing until stabilizing at the end of the process.

The sequential structure of many processes is also the core assumption of auxiliary local rank analysis methods, such as evolving factor analysis (EFA), that provide information on the sequential emergence-decay concentration windows of the different components in a process [54,55]. Taking advantage of this knowledge, initial estimates of concentration profiles can be generated and selectivity and local rank constraints, i.e., concentration windows of presence and absence of certain compounds along the process direction, can be set [4,19].

Indeed, at the very beginning of MCR methods, many algorithms were based on the proper setting of local rank subspaces to solve in a correct and unique way the concentration profiles and related pure spectra of the compounds involved in a process [3,56–59]. In 1995, Manne proposed two theorems that could define when a data set could be uniquely and correctly resolved based only on the patterned local rank structure in the concentration direction [44]. These theorems stated

(a) The concentration profile of a compound can be correctly resolved if all components inside its concentration window are also present outside,

(b) The pure spectrum of a compound can be correctly recovered if its concentration window is not fully embedded into the window of another component.

These pioneering theorems refute the false and too common statement saying that selectivity for all compounds is needed for unique resolution in MCR. They also warn about the importance of the active use of local rank information for a proper recovery of concentration profiles and pure spectra in process analysis.

The retrieval of local rank information from a data set can be done with local rank analysis methods, such as EFA [55,59]. Modifications of this parent algorithm are fixed size moving window-EFA (FSMW-EFA) [60], more powerful to detect the presence of minor compounds, or exhaustive EFA [61], devoted to check the validity of the assumption of sequentiality in a process.

EFA can be easily defined as a chemometric process monitoring. Thus, EFA performs many local rank analyses including every time a new process observation (spectrum) in the data subset analyzed. This is done from the beginning to the end of the process (forward EFA) and in the opposite direction (backward EFA; see Fig. 5A). The information of all these analyses is displayed in two plots, related to forward and backward EFA. In these plots, the eigenvalues of each local rank analysis are plotted in a vertical line as a function of the process variable designing the last row included in the subset analyzed (being an elution time, temperature, etc.). All 1st, 2nd, …, n-th eigenvalues are connected with a line. When the complete data set is considered, at one extreme of the plot, the number of lines above the noise level indicates the total rank of the system. Inside the plot, the point (time, temperature, etc.) where an eigenvalue rises from the noise level indicates the emergence (forward EFA plot) or the decay (backward EFA plot) of a new process contribution.

Joining the forward and backward EFA plot in a single picture and connecting the first eigenvalue emerging (in the forward EFA plot) with the first decaying (in the backward EFA plot), the second emerging with the second decaying, and so forth, initial estimates of concentration profiles for a sequential process can be obtained. Besides, the definition of the concentration window for each compound allows for setting related local rank constraints, i.e., zones of absence for each particular compound in the concentration direction (see Fig. 5B).

3.1.2 *Physicochemical Models*

Chemical process modeling was classically done fitting physicochemical models to experimental data [62,63]. Actually, physicochemical models based

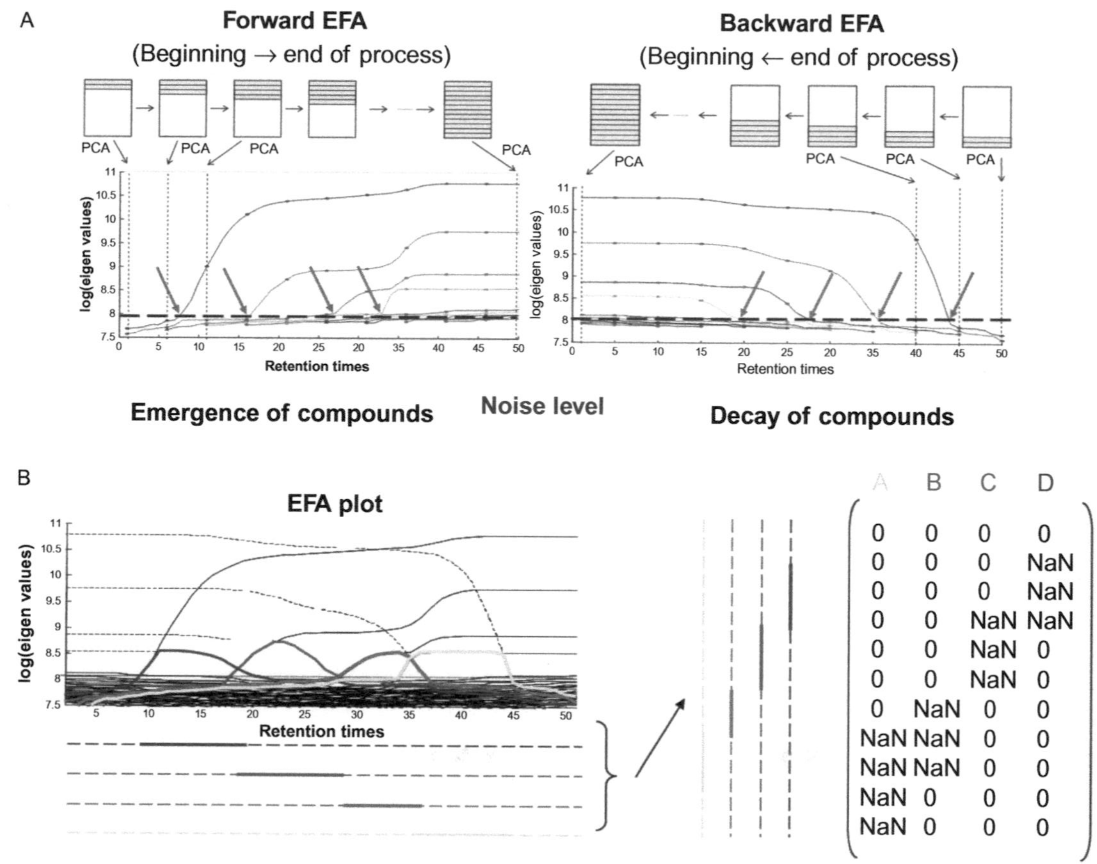

FIG. 5 *Modus cperandi* of evolving factor analysis (EFA). (A) Forward and backward EFA plots, (B) joint forward (*solid black lines*) and backward EFA plot (*dashed black lines*), and derived concentration profiles. At the bottom, concentration windows of components (in *bold colored* (*different bold gray shades* in the print version) *lines*) and (C) translation of concentration window information into a mask matrix of local rank information.

on first principles are the best way to describe a process evolution in an accurate way as long as a reliable model can be found to describe the phenomenon under study.

Classical hard-modeling works considering that concentration profiles can be described by mathematical functions based on the use of suitable process parameters, e.g., rate constants [63]. Tuning these parameters appropriately with iterative optimization algorithms, the process evolution can be properly modeled. Classical hard-modeling approaches also take into account implicitly the bilinear model of spectroscopic data. Thus, the function to be minimized is the same as in Eq. (4), conveniently transformed to the form presented in Eq. (14):

$$\min \left\| \mathbf{D} - \mathbf{C}\mathbf{S}^{\mathrm{T}} \right\| \tag{4}$$

$$\mathbf{S}^{\mathrm{T}} = \left(\mathbf{C}^{\mathrm{T}}\mathbf{C}\right)^{-1}\mathbf{C}^{\mathrm{T}}\mathbf{D} = \mathbf{C}^{+}\mathbf{D} \tag{13}$$

$$\min \left\| \mathbf{D} - \mathbf{C}\mathbf{C}^{+}\mathbf{D} \right\| = \min \left\| \mathbf{D}(\mathbf{I} - \mathbf{C}\mathbf{C}^{+}) \right\|, \text{where } \mathbf{C} = \mathrm{f}(k_i) \tag{14}$$

The form of Eq. (14) implies that only the parameters linked to the function describing the concentration profiles need to be fitted. Pure spectra of the process compounds can be afterward calculated using Eq. (13). In classical hard-modeling approaches, there is a single set of parameters, k_i, providing concentration profiles that optimally fit the variation of the experimental data and, therefore, a unique answer can be generally obtained. However, the necessary condition to ensure the successful application of hard-modeling approaches is that all contributions to the signal measured must be described by the physicochemical model. This condition is mandatory because the model fits the raw spectroscopic data, **D**.

Since physicochemical models are very accurate expressions to describe the shapes of process concentration profiles, they can also be used in MCR under the form of a hard-modeling constraint [52,64,65]. In this case, as any other constraint, the physicochemical model will act on the concentration profiles, **C**, and the function to be minimized will be

$$\min \left\| \mathbf{C}_{\mathrm{ALS}} - \mathbf{C}_{\mathrm{HM}} \right\| \tag{15}$$

$\mathbf{C}_{\mathrm{ALS}}$ meaning the concentration profiles calculated by least squares within the MCR optimization and $\mathbf{C}_{\mathrm{HM}}$ the profiles shaped according to a physicochemical model. In this constraint, the related parameters of the model will be iteratively optimized until $\mathbf{C}_{\mathrm{ALS}}$ is optimally fitted by $\mathbf{C}_{\mathrm{HM}}$. The constrained profiles will be equal to $\mathbf{C}_{\mathrm{HM}}$ and, as with any other hard-modeling fitting, the optimized parameters will be obtained as an additional result.

The main advantage of using hard-modeling as a constraint within MCR can be understood by comparing the two minimization functions in Eqs. (14) and (15). In classical hard-modeling, the raw measurement **D** is fitted using the model, whereas the model fits the concentration profiles, **C**, in MCR.

Since MCR allows selecting the compounds to be constrained, only those related to the process to be modeled will be hard-modeled [66–69]. This allows letting outside the hard-modeling constraint interference contributions that affect the signal measured, but not the process behavior (see Fig. 6).

This separation of the information is not possible when the raw signal as such is fitted, as it occurs in classical hard-modeling. An additional value of the hard-modeling constraint is that the fitted concentration profiles lack ambiguity and, as a consequence, the whole description of the system improves [29,70]. Advantages of the hard-modeling constraint linked to multiset analysis will be mentioned in Section 3.2.

3.2 Multiset Analysis: Multiexperiment Analysis and Data Fusion

Multiset analysis is a very useful strategy in process analysis. Indeed, most chemical systems need several experiments done in different initial conditions or varying as a function of different process variables to depict the global scenario studied. Likewise, there is seldom a single instrumental technique that allows "seeing" all the events that take place during a process. Data fusion, i.e., monitoring the same process using different sensors, is required. Techniques are complementary and data fusion plays the same role as having a powerful multiresponse instrument.

Multiset analysis always outperforms single data analysis. The starting information used is more diverse and, as a consequence, a more integral description of the system is obtained and more reliable (less ambiguous) concentration profiles and pure spectra are retrieved [19,21,29].

Multiset analysis is always recommended, but it is mandatory when the information of a single data set is insufficient to describe the chemical system. This problem can be mainly linked to two different situations: (a) the information coming from different techniques/experiments is complementary and using analyses of single data matrices will provide partial descriptions of the system or (b) when correlation among concentration profiles or pure spectra in a single experiment prevents modeling all contributions linked to the process, causing the so-called rank-deficiency phenomenon [71].

Rank-deficiency occurs when the number of contributions to be modeled by MCR is smaller than the total number of compounds present in a process. Actually, a necessary condition to differentiate all the compounds in a system is that they should not show profiles linearly related to each other. This condition needs to be fulfilled by both the concentration and the spectral profiles. Usually, rank-deficiency is associated with either the concentration or the spectral direction and the rank-deficient direction determines the number of contributions that will be finally modeled [71]. An example of rank-deficiency in the concentration direction would be associated with a kinetic process driven by the mechanism $A+B \rightarrow C$. In this case, three compounds are involved in the process, but the concentration profiles of A and B have

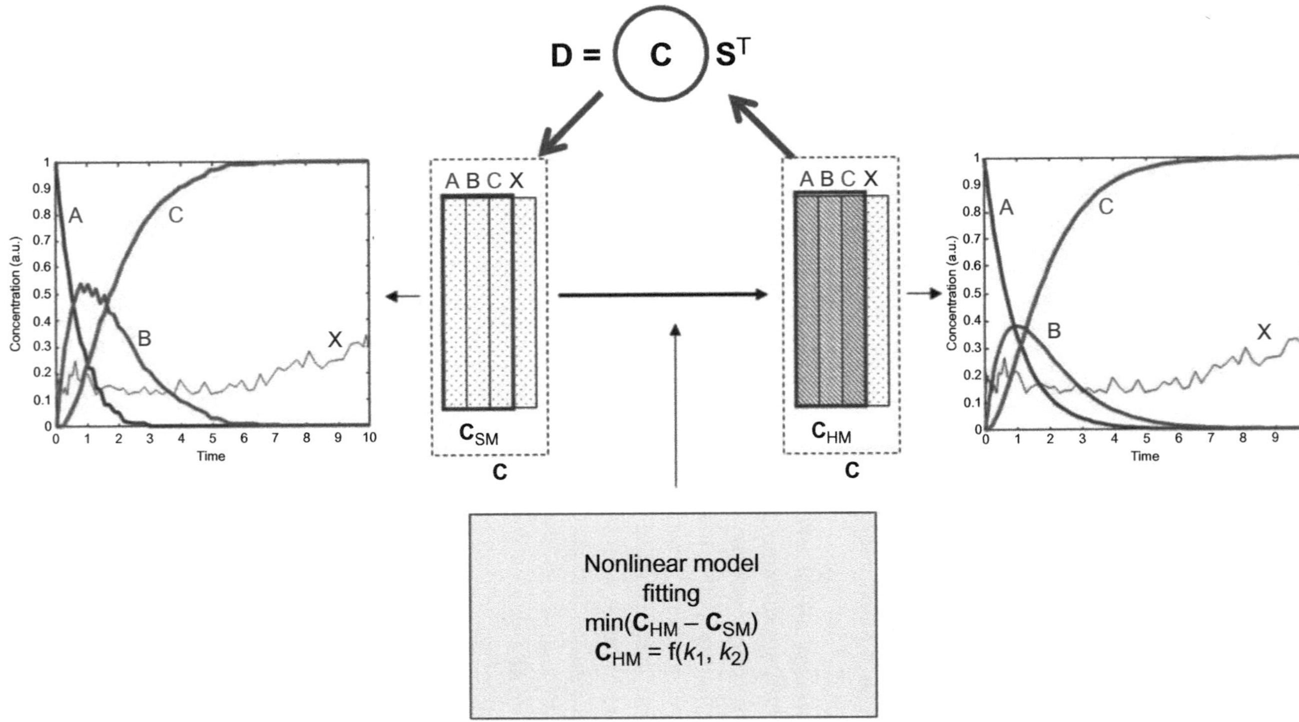

FIG. 6 Hard-modeling constraint. *Left plot*: unconstrained profiles (A, B, and C, process profiles; X, interferences). *Right plot*: constrained profiles (*bold*) according to the first-order consecutive reaction model $A \rightarrow B \rightarrow C$.

the same shape, i.e., [A] = k[B], where k is a scaling factor. Even if the pure spectra of the three compounds are clearly different, only two contributions will be detected and modeled. Rank-deficiency in the spectral direction could be found if two compounds present the same spectral signature or if their signatures differ only in intensity, but not in shape. In this situation, only one contribution would be detected and modeled.

The rank-deficiency problem cannot be solved if a single data set is analyzed. To break the rank-deficiency, multiset structures are needed that append information in the rank-deficient direction in such a way that the linear relationship between profiles disappears (see Fig. 7) [71–73].

There are many possible ways to do that. Following with the previous examples, rank-deficiency in the concentration direction for reaction A+B → C can be suppressed by adding an experiment in which A and B are in a concentration ratio different from the first experiment. In such a case, the linear relationship [A] = k[B] will no longer hold for the augmented concentration profiles including both experiments. A simpler way to break this rank-deficiency would be by appending a pure spectrum of A (or B) to the original data set. As can be seen in Fig. 7A, the augmented direction (**C**) is the rank-deficient direction. Likewise, the example of spectral rank-deficiency would be solved by adding data coming from the same experiment monitored with a different spectroscopic technique, for which the spectral shapes of compounds were different. In this case, the solution comes from augmenting the spectral direction ($\mathbf{S}^T$), as shown in Fig. 7B.

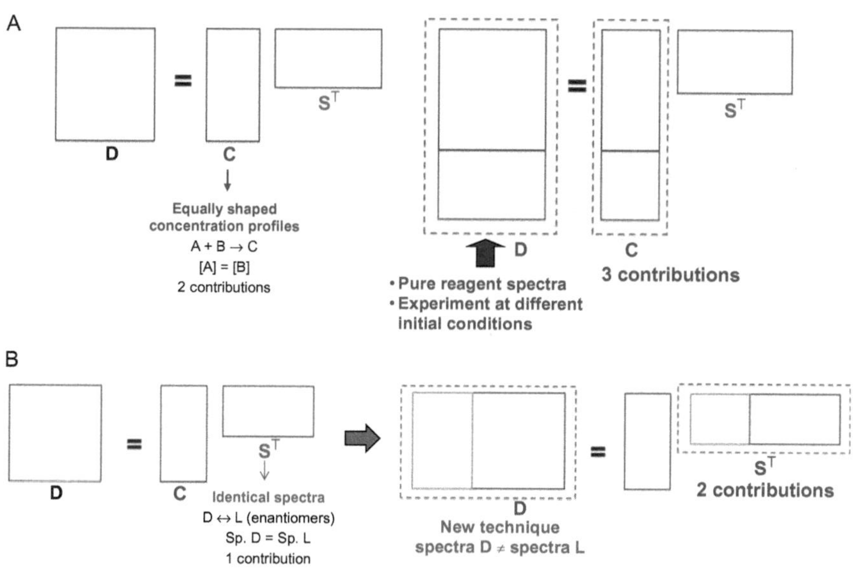

FIG. 7 Rank-deficiency problem (A) in concentration direction and (B) in spectral direction. Appending information in the rank-deficient direction suppresses the rank-deficiency problem.

Additional aspects related to the complementariness and value of using multiset analysis in a process analysis context will be commented below. For the sake of simplicity, specificities linked to multiexperiment and multitechnique analysis will be described separately. Using multiexperiment/multitechnique multisets will have all benefits mentioned for the simpler parent multiset structures.

3.2.1 Multiexperiment Analysis

Going back to the bilinear model associated with column-wise augmented matrices (Eq. 6), this kind of multiset only requires a common spectral mode among experiments. The concentration mode is free and allows for introducing completely different kinds of experiments and information. The bilinear model presents a single spectral matrix, valid for all the experiments analyzed simultaneously and an augmented concentration matrix that contains different subsets of concentration profiles related to the experiment analyzed. This bilinear model ensures consistency in the spectral definition of the compounds in the chemical system under study and, at the same time, respects the natural variability in the evolution of concentration profiles in all the experiments analyzed together [5–7,20,21].

This multiset structure allows covering different scenarios, such as the description of chemical or biological systems, where the same compound is submitted to modifications of different variables, such as pH, temperature, addition of other reagents, or to sets of experiments performed in different initial conditions, which may enhance the detection of minor compounds or intermediate species [5,7,15,74–77].

In an industrial context, multibatch analysis is an example of multiexperiment analysis. Batches performed in different conditions, e.g., following an experimental design to see the influence of different factors in a process, can be analyzed together to understand how the trajectories of evolution of compounds get modified. In this case, the column-wise augmented matrix configuration is crucial to let the batch mode free and preserve the natural variability of the different experiments [78–80].

The possibility to constrain differently the profiles of the **C** submatrices related to each coupled experiment is particularly interesting when using the hard-modeling constraint. This flexibility allows fitting all the experiments with a single global physicochemical model, fitting subsets of experiments with different models or even every experiment with a particular model [15,66,69,75,81]. The most interesting option is, however, combining hard-modeled experiments and model-free experiments [82]. Fig. 8 shows a multiset combining spectroscopic and chromatographic data to study a complex kinetic degradation process [82].

In this example, the information provided by the chromatograms coming from solutions collected during the process helped to differentiate similarly evolving compounds in the kinetic process. In this case, the elution profiles

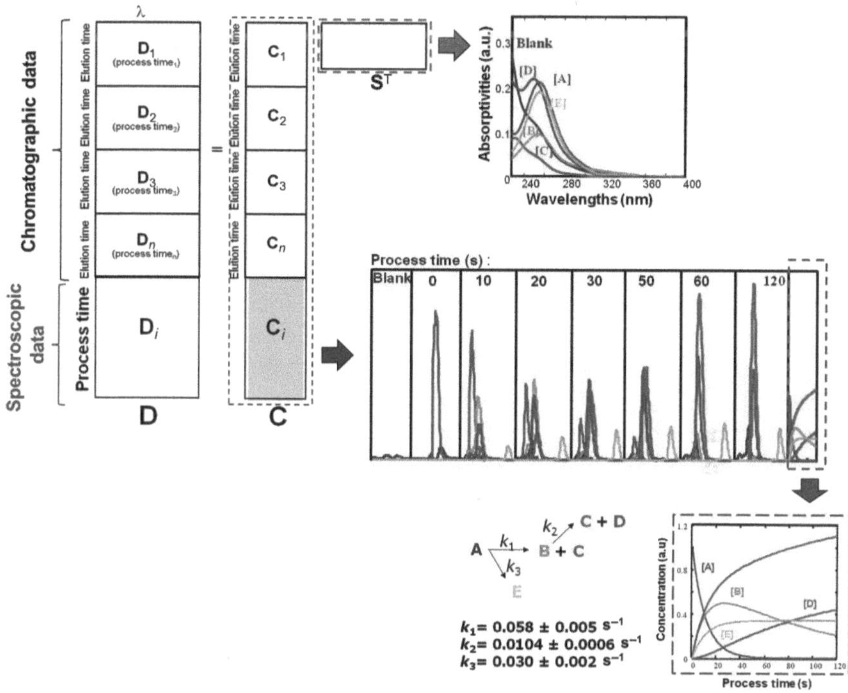

FIG. 8 Column-wise augmented data set coupling chromatographic and spectroscopic experiments and related bilinear model. *Top right plot*, pure spectra; *top center plot*, multiset concentration profiles; *bottom plot*, zoomed hard-modeled process profiles of spectroscopic experiment.

of the $\mathbf{C}_i$ chromatogram submatrices were modeled using only nonnegativity and unimodality constraints, whereas the $\mathbf{C}_i$ submatrix from the spectroscopic degradation experiment was hard-modeled.

To finish, one of the most used multiexperiment multisets relates to quantitative analysis. In this case, experiments related to calibration and unknown samples are treated together. The correspondence of species constraint plays a fundamental role, since the composition of the calibration samples, which can be formed by pure standards, or controlled mixtures of compounds, is known [5,6,21,83]. More on this topic will be detailed in Section 5, devoted to quantitative analysis by MCR.

3.2.2 *Multitechnique Analysis (Data Fusion)*

Multitechnique analysis is performed in multisets formed by a single experiment monitored with several techniques. In this situation, the bilinear model in Eq. (5) holds and the original multiset is described by a single concentration matrix, with profiles describing the evolution of the process, and an augmented matrix $\mathbf{S}^T$, which contains the extended spectral signatures for each one of the process compounds, formed by the concatenated pure responses of all coupled spectroscopic techniques. To perform data fusion, the process mode has to be common and, therefore, the measurements obtained by the different techniques should refer to the same process stages. When this is not the case, synchronization is required to provide a reliable common process mode. Since the appended data sets in matrix **D** may come from very different spectroscopic techniques, scaling may be needed to make the intensity of the different signals comparable. It is also advisable that the number of spectral channels be similar among techniques to keep a similar weight among the different kinds of measurements [5,6,21,73,76,82,84].

Data fusion has been proven to be a very good strategy to improve process description. As clear advantages: (a) it allows taking advantage of the complementary information provided by the different techniques, (b) can compensate for the lack of signal of certain techniques for certain compounds, and (c) can solve situations of spectral rank-deficiency.

Data fusion is particularly useful in very complex processes that incorporate many events of different nature, such as those linked to biological molecules like proteins or nucleic acid structures [73,76,85]. To illustrate the gain of using multiset analysis toward individual single data analysis, a brief description of a process linked to pH-induced hemoglobin transitions is described Ref. [136]. This process was monitored using UV, fluorescence, and circular dichroism in the far UV, near UV, and Soret region. These techniques were sensitive to changes in one or several protein structural levels of the protein and to changes in the binding of the heme group to the protein structure. Table 1 shows the number of components detected by each single technique analysis. As can be seen, the number of resolved components, in this case protein species or conformations, ranges from two to four and the pH transitions detected are not always the same.

When multiset analysis is performed on the row-wise augmented data set formed by all techniques, four process profiles are clearly defined with their

TABLE 1 Results Obtained from the Single Analysis of the Techniques Used to Monitor pH-Dependent Hemoglobin Conformational Transitions

Spectrometric technique	Nr. of modeled species	pH crossing point transitions	Variance explained (%)
Far UV CD	2	4.0	99.7
Near UV CD	2	4.5	93.9
CD Soret region	2	7.8	99.8
Fluorescence	3	4.2/8.7	99.9
UV	4	2.8/3.9/8.5	99.7

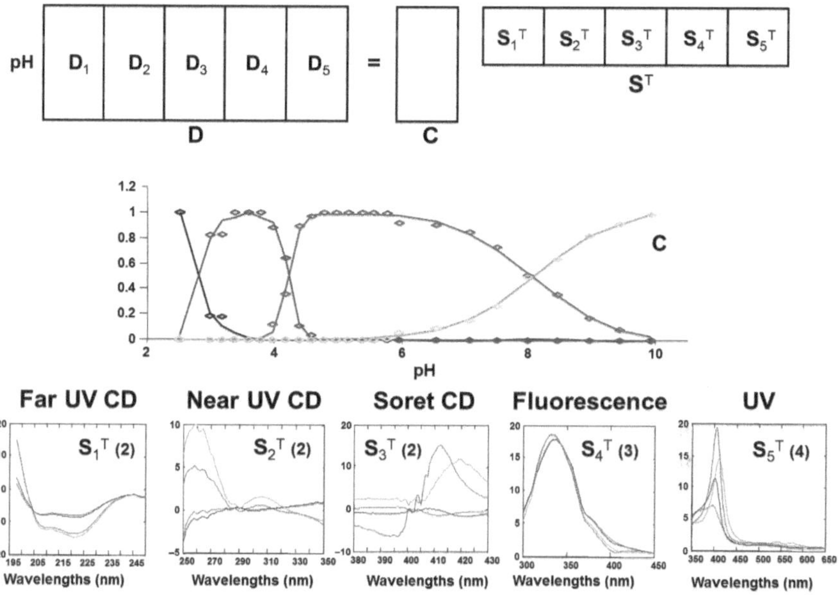

FIG. 9 Data fusion related to the study of pH-dependent transitions of hemoglobin. (A) Bilinear model related to the use of five spectroscopic techniques for process monitoring and (B) single set of concentration profiles (*top plot*) and pure spectral signatures for protein conformations related to the different techniques used in the multiset (*bottom plot*). The figure in *parenthesis* in each plot indicates the number of contributions detected when single technique analysis is carried out.

related spectral signatures (see Fig. 9). These compounds could be described as one conformation found only at a very acidic pH value, only detected by UV spectroscopy, with no heme group bound and denatured protein structure; a second one found at moderate acidic pH values, with no heme group and denatured structure; a third structure around neutral pH value with native secondary and tertiary structure and heme group bound; and a fourth structure at moderate basic pH values, with native secondary and tertiary structure and oxyheme group bound.

Several problems were solved with the multitechnique analysis and explain why single techniques could not model properly the process:

(a) *Lack of signal.* The Soret region is only sensitive to protein conformations with the heme group bound to the protein. This is why only the third and fourth conformations were detected when using CD in this spectral range.

(b) *Rank-deficiency in the spectral direction.* CD at far and near UV only detected changes between native and denatured protein forms (at a secondary structural level in far UV CD and at the tertiary structural level in near UV CD). This is why only two species were detected by these techniques (the two denatured forms at acidic pH could not be distinguished and the two native forms, one related to the heme group bound and the other to the oxyheme group bound, did not induce changes in the native protein structure and could not be differentiated either). Likewise, fluorescence spectroscopy could distinguish only three protein species because the two denatured forms at acidic pH values presented the same fluorescence spectrum.

(c) *Improvement of the global signal differences among compounds.* Although the UV technique is able to detect the four protein conformations, the differences in spectral signatures among some of them are very small. The definition of all transitions among protein species benefit from the fact that CD spectroscopy shows much larger spectral differences among the species that this technique is able to distinguish.

This example shows clearly the advantages of data fusion, a still underexploited strategy that should be taken into account in areas like on-line multisensor process monitoring in industrial contexts [86].

4 MCR-ALS APPLIED TO HSI ANALYSIS

HSIs are spectroscopic measurements that contain spatial and spectral information [87,88]. Despite of the fact that an image is often displayed as a cube of data (with spatial coordinates and spectral dimension), this measurement is nothing else than hundreds or thousands of spectra and, as such, the mathematical description is also the bilinear Beer–Lambert law. To understand the bilinear model related to HSI, the original image cube must be unfolded into a data table containing all pixel spectra. The bilinear model is formed by a matrix $\mathbf{S}^{\mathrm{T}}$, related to the spectral signatures of the pure constituents in the image, and a matrix **C** formed by stretched profiles of pixel concentration values. The concentration profiles, conveniently refolded to recover the spatial structure of the image, display the distribution maps of the image constituents (see Fig. 10) [89].

As in the previous spectroscopic examples, MCR-ALS is a suitable algorithm to deal with this kind of data, irrespectively from the spectroscopic technique used and the origin of the image analyzed (biological, pharmaceutical, remote sensing, etc.) [90–95]. The basic steps of application of MCR-ALS are identical to those described above. Some specificities linked to constraints and multiset analysis for image data sets are explained in following sections.

4.1 Encoding Image Information: The Spatial Dimension

When MCR-ALS is applied to an HSI, the straightforward result is the pure spectra of image constituents and the related concentration profiles, which are arrays of numbers that contain pixel concentration values. The stretched concentration profiles related to the image constituents cannot accommodate many constraints because the natural two-dimensional structure of the image

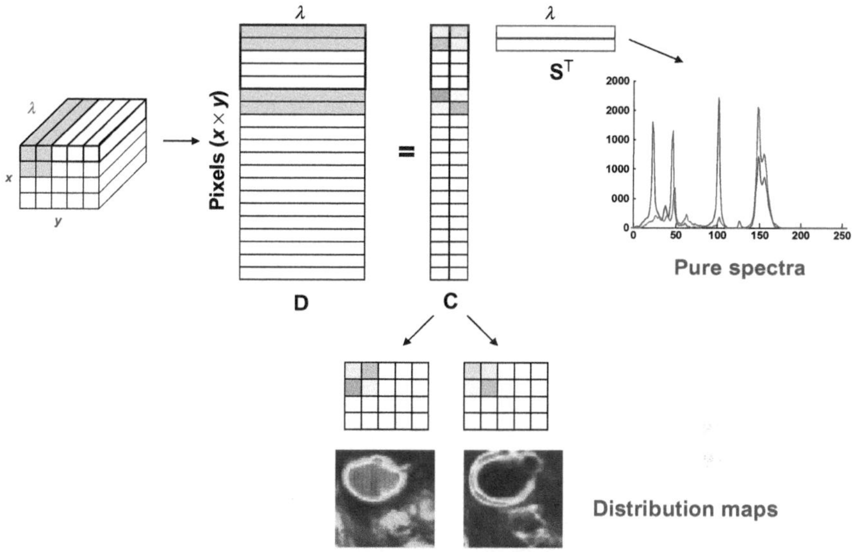

FIG. 10 Bilinear model related to hyperspectral image analysis.

is lost, i.e., a pixel is no longer surrounded by all its neighbors in the original spectroscopic image data table analyzed. In practice, nonnegativity seems to be the only obvious constraint to be applied in this context.

However, the spatial information is the specific characteristic of images and, at short or long term, there exists a relationship between the composition of a pixel and its neighbors. This local relationship may be more obvious in images coming from structured samples, such as biological samples, where pixels belonging to the same tissue are spatially ordered and have a very similar composition, or remote sensing images, where the landscape scanned is formed by large regions linked to clear environmental compartments or objects, e.g., lakes, fields, and houses. However, even when images come from samples lacking apparently spatial structure, for instance, pharmaceutical powder samples, at a very local level, neighboring pixels tend to have similar composition.

The connection between pixel neighborhood and composition has promoted the birth of a new generation of constraints that take into account the spatial information of the image [89]. Some of these constraints act directly on the distribution map of each constituent, i.e., the refolded concentration profile, and they can take into account features related to the two-dimensional description of the variation of the concentration. Some of these constraints work enforcing smoothness or roughness in the concentration values of the distribution map, as it will be seen in Chapter 14 [96,97]. Other possibilities, such as including topological information or object-shape information, can be envisioned, always taking into account the pure two-dimensional distribution map as the starting information to be constrained. As it happens with processes, the applicability of these constraints will depend on the natural spatial structure of the image.

The concept of local rank can also be adapted to images and takes into account their spatial structure and the pixel neighborhood concept. To do so, an adaptation of the FSMW-EFA algorithm can be used [60]. Fixed size image window-EFA (FSIW-EFA) is a local rank analysis method that can be defined as a chemometric image point scanning [98,99]. Thus, many local rank analyses are performed moving a small two-dimensional (for pixel images) or three-dimensional (for voxel images) window including a pixel and its immediate spatial neighbors across the full image (see Fig. 11A). Singular value (or eigenvalue) maps are displayed in which the local composition complexity of the image can be seen (see Fig. 11B). By setting a threshold that separates significant from noise-related eigenvalues, a local rank map can be derived, which shows the spatial compound overlap of the image at a pixel level (see Fig. 11C). More detail can be found in [100].

For EFA results linked to sequential processes, it was straightforward associating local rank information with presence or absence of particular compounds in the concentration profiles. In the case of images, an additional step is needed because there is no general ordered pattern in the presence/absence of compounds in the different spatial regions of the image. To do so, reference spectral information of the image constituents and local rank information must be combined. The reference spectral information can be obtained by using pure variable selection methods or adopting as reference information resolved spectra from previous MCR analyses applying softer constraints, such as nonnegativity. Thus, for a particular pixel area, the number of absent compounds is obtained as the total rank in the image minus the local rank at that pixel area. Once this is known, the absent compounds will be identified as those presenting the lowest correlation coefficient between the pixel spectrum studied and the reference spectra related to the different image constituents. Please note that not all the pixels need to be constrained, only those with a clearly defined rank, i.e., clear number of absent compounds, and a clear correlation coefficient diagnostic. For more detail, see [100]. Recently, other ways to set local rank constraints have been proposed, either based also on local rank analyses (for systems with unknown composition) [101] or using orthogonal projection methodologies (when the identity of all compounds in the image can be known) [102].

As mentioned in Section 2, a suitable local rank structure ensures uniqueness in the MCR solutions [44]. Manne's theorems translated to the image analysis context would state that if distribution maps and pure spectra of a particular constituent have to be recovered, this constituent should be absent in some pixels and should never be overlapping with the same compound [89].

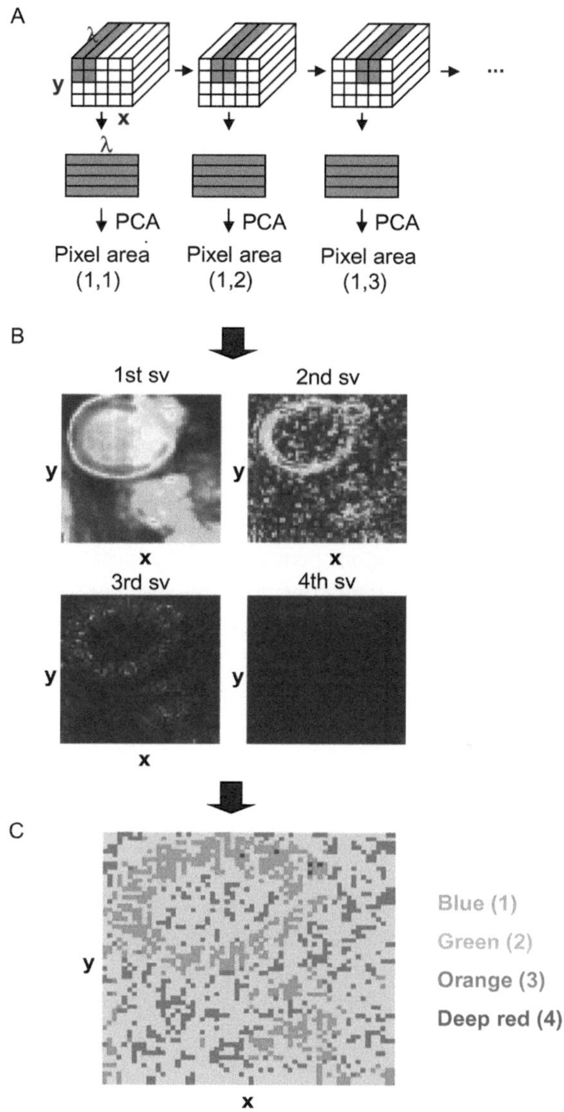

FIG. 11 Scheme of application of fixed size image window-evolving factor analysis (FSIW-EFA). (A) Window definition and local rank analysis screening along the image surface, (B) singular value maps, and (C) local rank map.

4.2 Image Multiset Analysis

Multiset analysis has also a very clear value in image analysis [89,99]. Thus, it makes sense to analyze several images together because they belong to the same sample or because they have a clear relationship among them, e.g., a process monitored by images. Image multiset analysis covers two different scenarios: the work with several images acquired with the same technique, which provide a column-wise augmented multiset, and the work with images collected on the same sample with different spectroscopic platforms, which give a row-wise augmented multiset. These two kinds of multisets provide the structures and bilinear models shown in Fig. 12.

In the case of column-wise augmented multisets, the spectral mode is common and the resolved spectra are valid for all the images treated together. This configuration ensures that the distribution maps related to a particular component in the different images share the same spectral signature, which gives consistency to the description of the full system. Such a property is an obvious need when images combined together refer to image layers collected at different depths or in different spatial areas of the same sample [89], or when images come from process monitoring.

An additional advantage coming from this kind of multiset structure is that the pixel mode is completely free. This implies that images treated together can have different geometries, total number of pixels or different spatial resolution (pixel size). Such a freedom provides a very large number of scenarios that can be tackled using this data configuration. To mention a few, we could

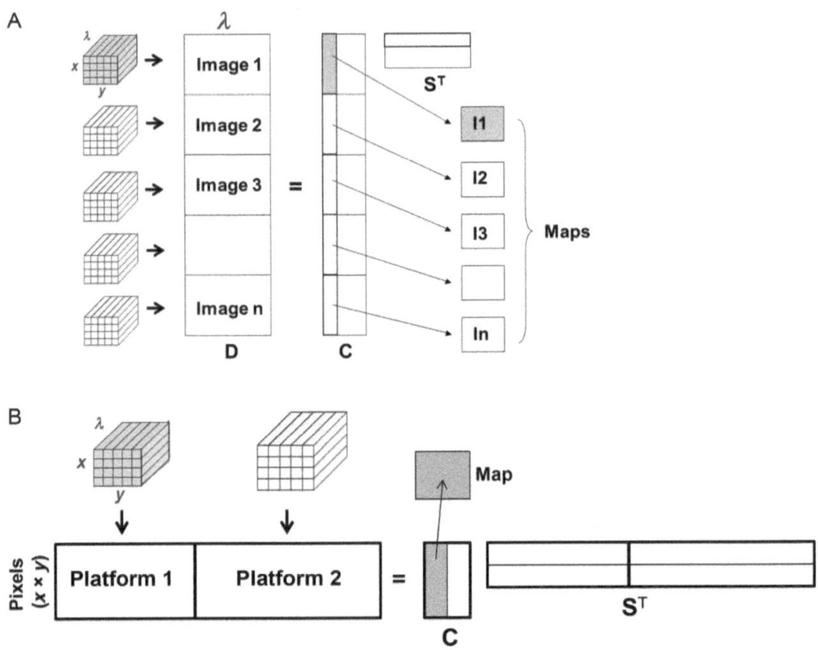

FIG. 12 Bilinear models linked to multiset analysis of images. (A) Multiset formed by images coming from the same technique. (B) Multiset formed by images coming from different techniques on the same sample (data fusion).

think of images coming from similar samples, e.g., a population of images collected from tissues of different patients in the same health condition, which have different sizes, or the joint analysis of images coming from the same sample with different spatial resolution, combining a global image of the sample with lower spatial resolution with images on zoomed details, i.e., scanned with a much higher spatial resolution.

Multiset analysis on images collected with the same technique does not have more complexity than single image analysis and has been applied in many instances. The most common fields are analysis of multilayer images [89], analysis of images belonging to the same sample population [103], quantitative image analysis [89], where standard images and images of unknown samples are combined (see Section 5 for more detail) and process monitoring, where a spatial and global description of the process is obtained [104,105]. As an example of the last application, Fig. 13 shows the resolution of a thermal polymorphic transformation of carbamazepine, monitored by Raman images collected at different temperatures [104].

Pure spectra describe the polymorphic forms of the compound in the process. Distribution maps show the spatial changes induced by the process at a local level. The global process evolution is obtained by displaying the average values of the distribution maps of the image constituents as a function of the temperature. In this example, the inclusion of spatial information for process monitoring is essential, since the two initial polymorphs, thermally evolving in the same way, would have never been distinguished by classical spectroscopy (rank-deficiency phenomenon). However, they could be differentiated because of the different spatial pattern of the two compounds in the sample analyzed.

Image multiset analysis for images obtained on the same sample with different spectroscopic platforms is a challenging field [88]. This implies image data fusion and has the same benefits as when it is carried out in process analysis. As seen in Fig. 12B, the free mode in this multiset structure is the spectral one, which means that different techniques covering different number of spectral channels can be combined, but the concentration direction (the pixel mode) has to be common. Pixel congruency among images is a challenging issue because most of the times images from different platforms can show shifts and rotations among them and may also have different spatial resolution. Therefore, preprocessing is an issue to take into account these problems. There are different references in the literature oriented to propose procedures of image matching that enclose spatial transformations (shifts/rotations) when images have the same pixel size [106,107].

To perform image data fusion, a strategy has been recently proposed, based on the use of MCR-ALS for incomplete multisets, which aims at treating images from different platforms with different spatial resolution [108]. The starting point is treating two images from different techniques with different spatial resolution. The goal of this approach is taking advantage of the

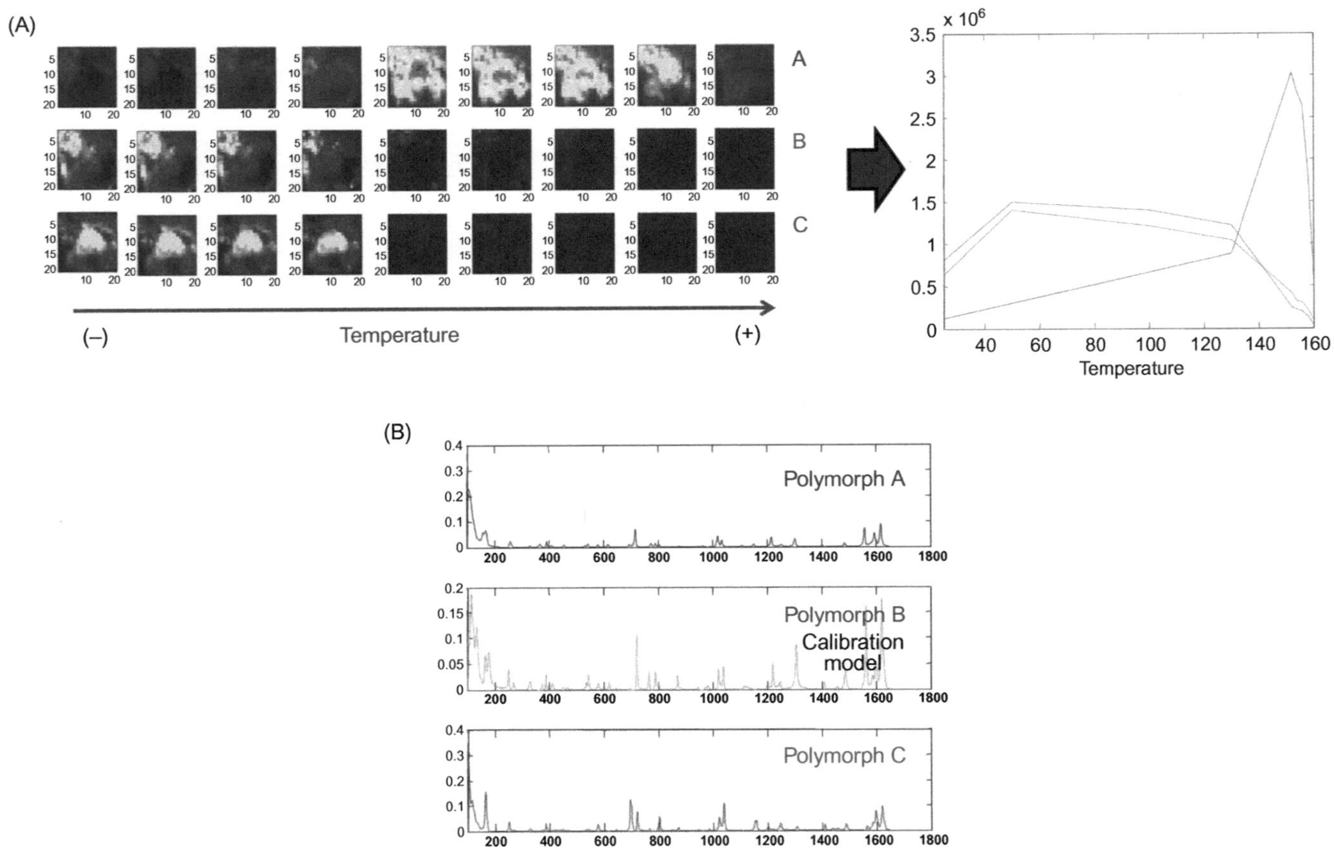

FIG. 13 Resolution results from the multiset linked to the thermal-induced polymorphic transformation of carbamazepine. A set of images is collected at the different temperatures. (A) Distribution maps related to the polymorphic forms and derived thermal global process profiles and (B) spectral signatures of polymorphic forms.

different properties of the spectroscopies used, keeping the highest possible spatial resolution. Simple downsampling of the image with highest spatial resolution is not an option and oversampling the lowest spatial resolution image could induce spurious variations in the image, linked to artifacts related to an unreliable interpolation. To solve this problem, working with an incomplete image multiset, as shown in Fig. 2B, is proposed. In this structure, two regular multisets can be drawn: $\mathbf{D}_A$, which contains the images from the two platforms, with the lowest spatial resolution and spatially matched (we avoid explaining the procedure of spatial matching for brevity [108]) and $\mathbf{D}_B$, which contains the image with highest spatial resolution, column-wised appended to its downsampled version. In each MCR-ALS iteration, both multisets $\mathbf{D}_A$ and $\mathbf{D}_B$ are solved as described in Section 2. The final outcome is spectral signatures for each image constituent, which benefit from the information of the coupled spectroscopic techniques, and a set of distribution maps with the highest possible spatial resolution (see Fig. 14). This strategy and similar ones under development are needed to fully exploit data fusion in image analysis.

4.3 MCR Postprocessing

The most common final goal of MCR applied to image analysis is obtaining the distribution maps and pure spectra of the image constituents. However, it should not be disregarded that the information contained in distribution maps (or the **C** matrix) and pure spectra can be an excellent starting point for further data analysis.

Indeed, one of the common problems in image data analysis is the size and, often, the noise linked to these data sets. To alleviate this problem, compression methods are used many times before the analysis. Those methods may perform a selection of subsets of spectral variables or pixels, representative of the variation in the full image, or may use compressed representations of the pixel space, such as those coming from PCA scores.

Linking the last idea to the MCR bilinear model, we need to consider that the **C** and $\mathbf{S}^T$ matrices are excellent compressed representations of the variation of the pixel space and the spectral space of an HSI, respectively, and, as such, they can be used for further data analysis tasks. As positive features, **C** and $\mathbf{S}^T$ are noise-filtered representations of the information in the raw data set and they contain separate compound-specific information in their profiles [89].

As examples of the use of MCR information, we can mention the use of the space defined by the $\mathbf{S}^T$ matrix for spectral identification of image constituents by projecting library spectra on to it [109]. Once the closest candidates to this space are found, definitive identification of compounds is straightforward, since the library spectra selected correlate well with the meaningful MCR resolved spectra.

MCR scores (or **C** profiles) have even wider applicability. A very good option is using these scores as starting information for segmentation

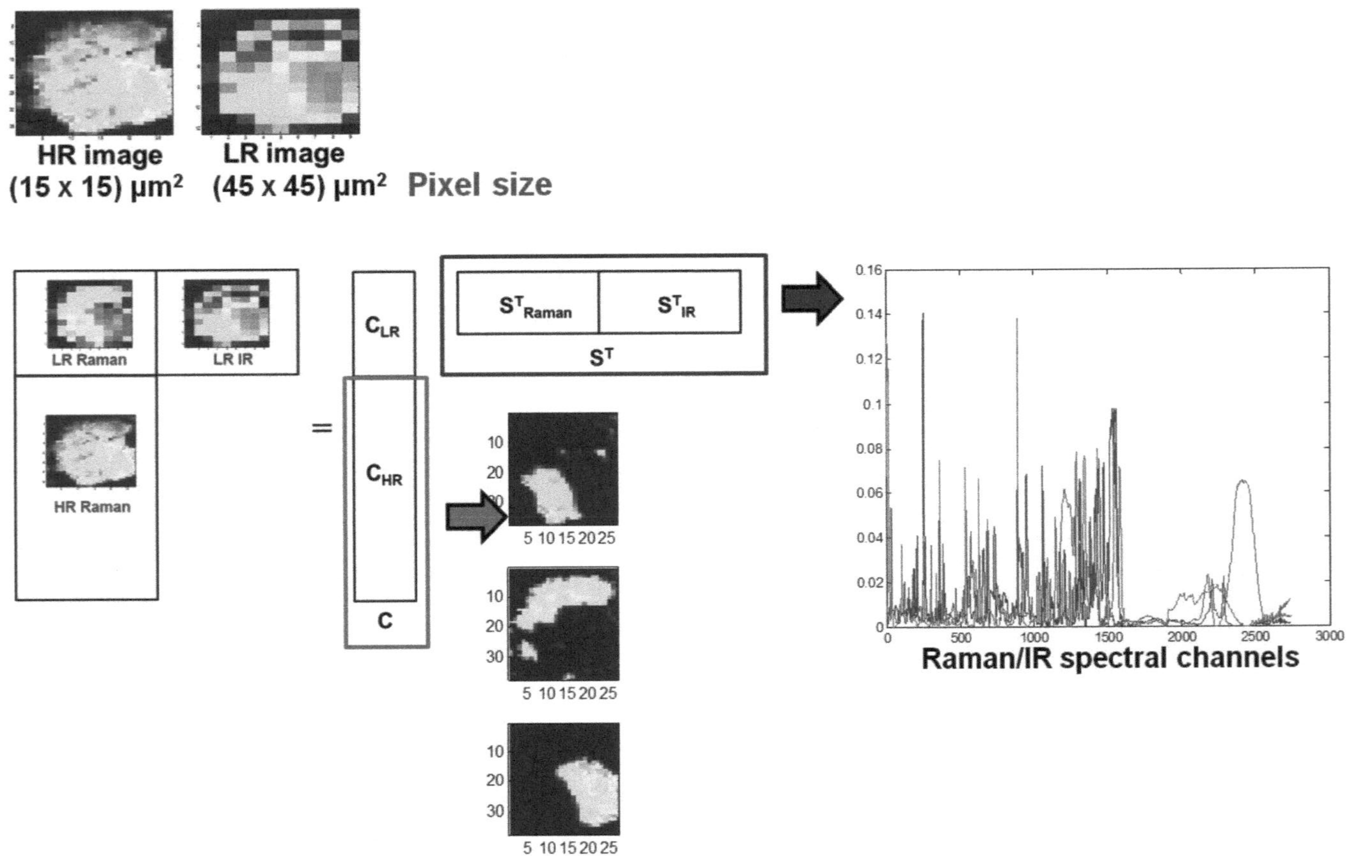

FIG. 14 Data fusion of hyperspectral images from different techniques and different spatial resolution, solved via MCR applied to an incomplete multiset. Bilinear model and final distribution maps and pure spectra.

approaches, i.e., to find classes of pixels with similar properties. In this context, the use of MCR scores offers many advantages, namely: (a) low computation time as compared with using full pixel spectra, (b) possibility to select only some compounds (concentration profiles) to perform segmentation, e.g., skipping a background contribution or uninteresting chemical compounds, (c) possibility to balance the importance of major and major compounds by scaling appropriately their concentration profiles, and (d) obtaining centroids (cluster averages) easily interpretable, because they are related to the relative abundance of the different image constituents (see Fig. 15). It is also important to note that the use of MCR scores for segmentation can be used in segmentation of single images [95,109] or in a multiset segmentation mode [103].

MCR scores and related distribution maps can also be used as seeding information for super-resolution approaches, i.e., distribution maps coming from a resolved multiset formed by related low spatial resolution images can be combined to reconstruct maps with a much higher spatial resolution [110] (see Chapter 15 for more detail). Besides, distribution maps can also be used for quantitative analysis and heterogeneity studies, as it will be detailed in the next section.

5 MCR-ALS AND QUANTITATIVE ANALYSIS

The main purpose of MCR-ALS applied to spectroscopic data sets is usually recovering the correct shape of the concentration profiles and pure spectra of the components present in a mixture data set. In absence of reference quantitative information, these resolved profiles are in arbitrary units, due to the intensity ambiguity linked to the MCR bilinear models.

When quantitative analysis is also required, the use of reference information in real concentration units is mandatory. This use can be performed introducing this information within the MCR optimization or using it, in an external way, once the MCR concentration profiles are obtained [83].

5.1 Second-order Calibration

Quantitative analysis by MCR was traditionally linked to second-order calibration problems, i.e., to situations where every sample provided a full data matrix of measurements, e.g., a chromatogram with DAD detection or an HSI. In these particular situations, calibration problems were addressed by using a column-wise augmented multiset, where the data sets related to calibration samples (standards) and unknown samples were put one under the other one in the same data structure. The MCR results provided a single matrix of pure spectra and an augmented **C** matrix, with as many submatrices as samples in the data set [5,6,21,111–113].

In second-order data, the quantitative information of a sample is related to the global shape of the concentration profile and can be well expressed by

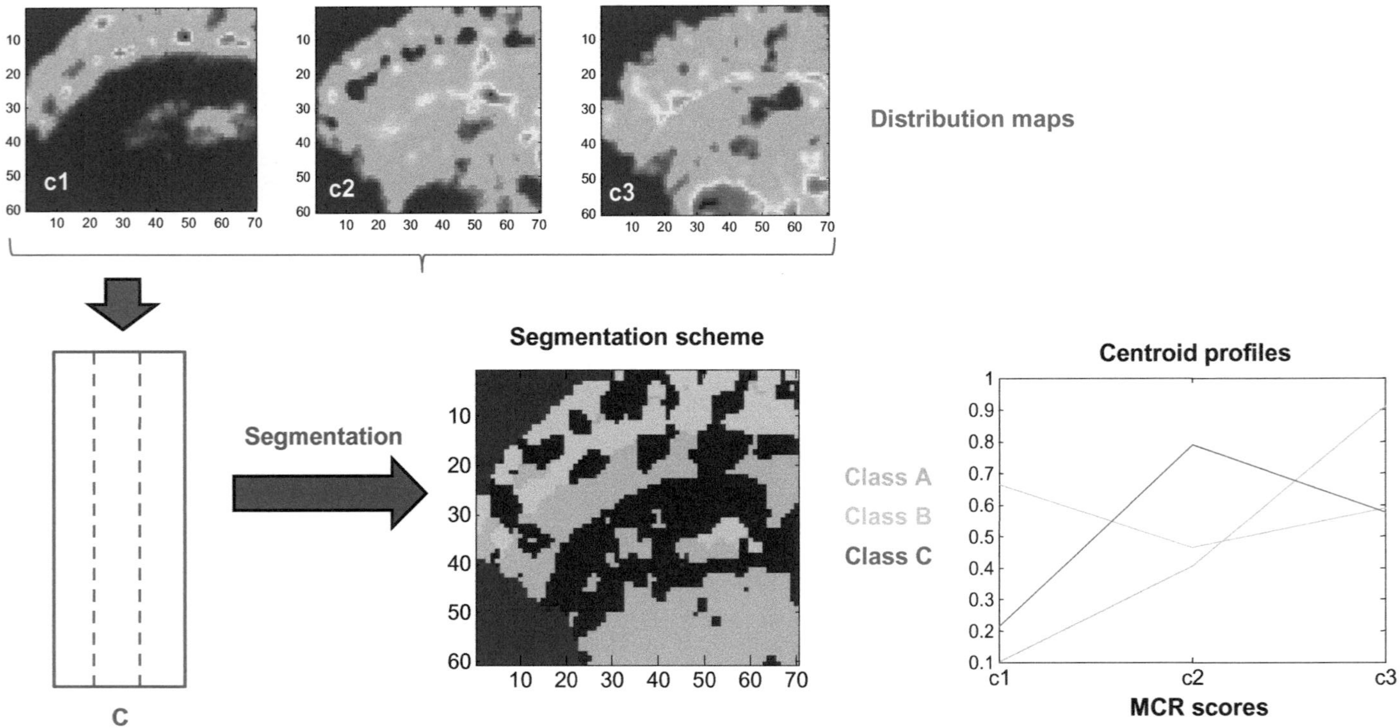

FIG. 15 Use of concentration profiles (MCR scores) for segmentation purposes in the analysis of a kidney stone image.

profile-related parameters, such as the peak area (in chromatography) or the mean value of a distribution map (in HSIs). Since MCR expresses the information related to the different compounds in separate profiles, pseudounivariate calibration models per each compound of interest can be obtained when reference information about the concentration of calibration samples exists. These models are built by regressing the areas (or mean concentration values) of the resolved concentration profiles of the calibration samples toward their related reference concentration values. The real concentrations of unknown samples are retrieved by using the area (or mean concentration value) of their concentration profiles and the calibration line (see Fig. 16). In all these cases, the so-called second-order advantage is fulfilled, i.e., the concentration of a compound in a sample can be known in the presence of interfering compounds, even if those are absent in the calibration samples [114,115].

MCR quantitative analysis based on chromatographic data or on excitation–emission spectroscopy can benefit from typical analytical calibration strategies linked to univariate calibration, such as the use of internal standards or standard addition methods [116]. When working with HSIs, quantitative information can be obtained at a global and local level. Bulk image concentration is obtained when mean values of resolved MCR distribution maps of unknown samples are used for prediction in the calibration model. Distribution maps in real

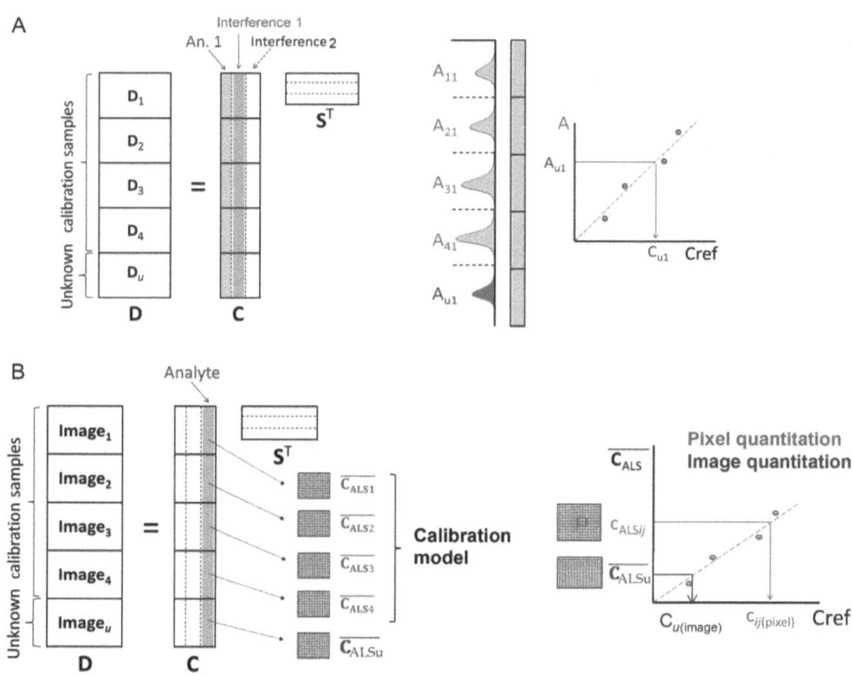

FIG. 16 MCR for second-order quantitative analysis. (A) Example of chromatographic data and (B) example of hyperspectral image data.

concentration units are obtained when the individual concentration values obtained by MCR for each particular pixel are used in the prediction step [83,89,111].

It is important to note that second-order calibration problems yield often unique concentration profiles and spectra of analytes as long as a proper design of the calibration set and a selection of suitable constraints are carried out. Thus, pure spectra of analytes are easily recovered when using pure standards of these compounds. And, according to Manne's resolution theorems [44], concentration profiles of analytes in unknown samples can be correctly recovered as long as all compounds inside its concentration window are also present outside. In this sense, it should be reminded that we refer to the augmented concentration profiles in the multiset structure, formed by the submatrices from the calibration samples and the test samples. This means that, by doing a proper design of the calibration samples to be included in the multiset structure and using explicitly the information of presence/absence of compounds through the constraints of correspondence of species (when referred to full concentration profiles of particular samples) or local rank (when referred to a particular region of the concentration direction within a sample), correct concentration profiles for the analytes can be obtained. This includes difficult situations, such as those where the concentration profile of analytes in test samples is completely embedded under interference profiles. In some other instances, like calibration problems based on excitation–emission fluorescence measurements, the properties of this spectroscopy allow for the application of trilinearity constraint to the analyte profiles and, as a consequence, for the retrieval of correct unique solutions [19,29,51,117].

In absence of ambiguity in the resolved concentration profiles of the analytes, the same figures of merit used in univariate calibration were traditionally employed to assess the predictive ability of MCR pseudounivariate calibration models [118–120]. Nowadays, there are more elaborated approaches that take also into account the presence of interferences in the unknown samples to assess the quality of the predicted concentration. Some methodologies use methods of resampling to empirically assess the error in prediction linked to calibration [121,122]. Other approaches derive analytical figures of merit, such as the sensitivity of a calibration model, taking into account the spectral overlap of the analyte with the other compounds in the data set. The sensitivity parameter is included afterward in the calculations of other relevant calibration parameters, such as the limits of detection and quantification [123].

5.2 First-order Calibration: Correlation Constraint

Although quantitative analysis in MCR is traditionally linked to second-order calibration problems, MCR has been recently adapted to cope with first-order calibration problems. In this context, a sample is represented by a single spectrum and a data matrix is formed by a set of calibration and test samples. The quantitative information is now linked to the individual elements of the concentration profiles because each element is related to a particular sample (row in the data matrix).

The calibration model between the elements of the concentration profiles and the related reference quantitative information is now built and incorporated in the MCR optimization through the correlation constraint [53,124,125]. Within each iterative cycle of MCR, one or more univariate calibration models can be obtained. Each calibration model is built by regressing the MCR concentration values of a particular compound in the calibration samples toward the related real reference concentrations, as shown in Fig. 17. The updated constrained profile contains the reference concentrations

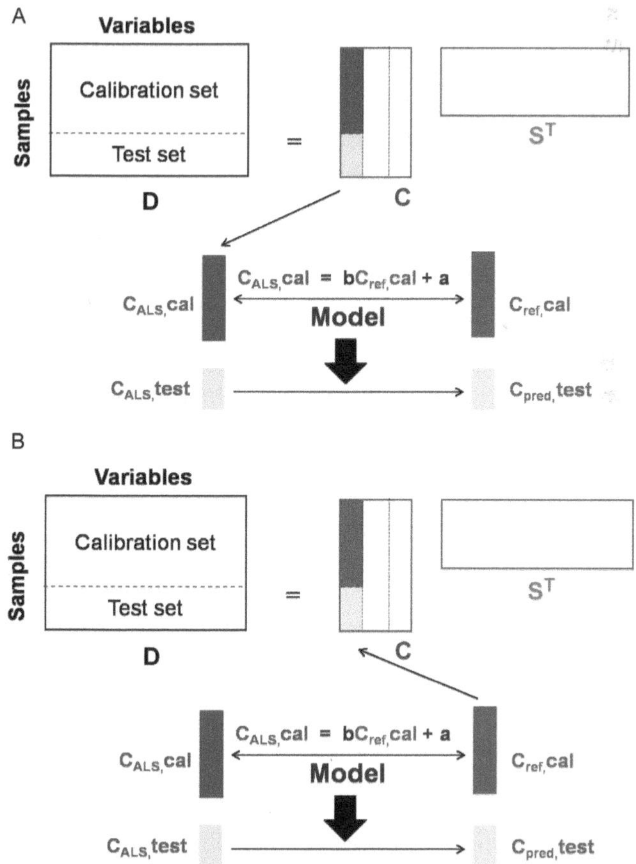

FIG. 17 Application of the correlation constraint. (A) Setting the inner calibration model between ALS concentration values and reference concentration values for calibration samples and (B) updating constrained profiles, by using reference concentration values in calibration samples and predicting real concentrations in test samples with the calibration model.

for the calibration samples and the predicted values using the internal calibration line for the unknown test samples.

As a result, the concentration profiles of analytes contain concentration values in real units and the definition of the concentration profiles and pure spectra of analytes becomes less ambiguous. The correlation constraint has been applied to different kinds of data sets, coming from different analytical responses and samples. The classical implementation of the correlation constraint can be applied to a single data matrix [53,124,125] or to a multiset problem [79], as long as the information related to concentration is contained in the individual elements of a concentration profile and not in parameters linked to the global profile. The problems studied until now have already shown the flexibility of application of the correlation constraint, which allows for the determination of one or several analytes in the same MCR analysis or for the use of a single calibration model for all samples or the use of different calibration models for sample subsets. This last possibility has enabled correcting sample matrix effects in quantitative analysis [126,127]. In general, the application of the correlation constraint to first-order calibration problems enables obtaining real concentration values for the analytes of interest in the samples together with a meaningful description of the rest of the compounds in the samples analyzed, through the recovery of their concentration profiles and spectra.

Recently, developments are also done in the direction of applying the correlation constraint to second-order problems. In this case, the inner calibration model built during the MCR optimization would be done regressing peak areas (or profile-related parameters) vs. reference concentrations. As a consequence of these calibration models, full concentration profiles would be conveniently rescaled to match real concentration units [83]. This new kind of approach will help in cases where a certain degree of ambiguity is left in the resolution of the concentration profiles.

6 MCR-ALS AND OTHER BILINEAR DECOMPOSITION METHODS

Spectroscopic data behave according to a bilinear model based on the Beer–Lambert law. Therefore, it is not surprising that many algorithms based on bilinear models be used to deal with this kind of measurement. The suitability of the different algorithms depends basically on the goal of the study and on the nature of the data set. The diversity of algorithms used to tackle the mixture analysis problem in spectroscopy is also linked to the different background knowledge existing in the applied domains where they were born.

Exploratory analysis can easily be carried out by classical bilinear decomposition methods, such as PCA [128] or independent component analysis (ICA) [129]. PCA provides a bilinear description of relevant spectral or concentration features, through the profiles of a set of principal components, which are orthogonal to each other. ICA, coming from a signal processing domain, expresses these variations through components, which are statistically independent. Both methods provide useful exploratory information, but may not be adequate when the goal is retrieving real spectral and concentration profiles of chemical compounds, which are seldom orthogonal or independent from each other [130]. This being said, these methods may provide quite acceptable results for unmixing problems when the spectroscopies used show fine structure, there are large spectral differences among compounds and the samples do not present strongly patterned concentration profiles, such as some Raman images of powdered samples. In these instances, concentration profiles and pure spectra may be close to the orthogonal or independent character of the profiles provided by PCA or ICA.

Other families of methods devoted to solve the mixture analysis problem in spectroscopic data are the so-called unmixing methods, coming from the remote sensing community [131]. The unmixing problem is looked upon using other kinds of algorithms, often more based on geometrical interpretations of the measurement, such as seeing all the pixel spectra of an image as enclosed in a simplex where the vertices (so-called endmembers) represent the pure spectra of the image constituents. This basic idea would correspond to methods, such as vertex component analysis (VCA) [132] or iterative constrained endmembers (ICE) [133]. Optimizations work then based on the minimization of the volume of this simplex or on similar approaches. It is interesting to note that some relevant ideas linked to the proper resolution of images, such as the local rank pattern needed in an image for a correct resolution, are formulated in their own terms in the remote sensing community, saying that there should be pixels in all simplex sides (note that pixel sides have always some image constituents missing) to allow for a correct recovery of endmembers. When mixing problems present good local rank conditions for resolution, MCR-ALS and unmixing algorithms provide very similar final results. In more difficult instances, MCR-ALS allows for an easier implementation and application of additional constraints.

Within chemometrics, there are also multiway resolution methods, with a close analogy with multiset analysis in MCR-ALS, since they work with several data sets structured in cubes or hypercubes. These methods are based on multilinear models, such as DTD [49] or PARAFAC [48], or on Tucker models, which allow for factor interaction among the modes of the multiway structures [134] Again, the suitability of these methods depends on the natural structure of the data set, which may accommodate a trilinear or multilinear model for description or may need a more relaxed augmented bilinear or hybrid model to cope with sample-to-sample variation of concentration profiles [134,135].

Some of the methods mentioned will be described in more detail in next chapters and analogies and differences with MCR-ALS will be more clearly understood.

REFERENCES

[1] Lawton WH, Sylvestre EA. Self modeling curve resolution. Technometrics 1971;13:617–33.

[2] Rutan SC, de Juan A, Tauler R. Introduction to multivariate curve resolution. In: Brown S, Tauler R, Walczak R, editors. Comprehensive chemometrics, vol. 2. Oxford: Elsevier; 2009. p. 249–59.

[3] Liang Y. Two-way data analysis: multivariate curve resolution—non-iterative methods. In: Brown S, Tauler R, Walczak R, editors. Comprehensive chemometrics, vol. 2. Oxford: Elsevier; 2009. p. 309–24.

[4] de Juan A, Rutan SC, Tauler R. Two-way data analysis: multivariate curve resolution–iterative resolution methods. In: Brown S, Tauler R, Walczak R, editors. Comprehensive chemometrics, vol. 2. Oxford: Elsevier; 2009. p. 325–44.

[5] de Juan A, Tauler R. Chemometrics applied to unravel multicomponent processes and mixtures. Revisiting latest trends in multivariate resolution. Anal Chim Acta 2003;500:195–210.

[6] de Juan A, Tauler R. MCR from 2000: progress in concepts and applications. Crit Rev Anal Chem 2006;36:163–76.

[7] Ruckebusch C, Blanchet L. Multivariate curve resolution: a review of advanced and tailored applications and challenges. Anal Chim Acta 2013;765:28–36.

[8] Geladi P, MacDougall D, Martens H. Linearization and scatter-correction for near-infrared reflectance spectra of meat. Appl Spectrosc 1985;39:491–500.

[9] Afseth NK, Kohler A. Extended multiplicative signal correction in vibrational spectroscopy, a tutorial. Chemometr Intell Lab Sys 2012;117:92–9.

[10] Eilers P. Parametric time warping. Anal Chem 2004;76:404–11.

[11] Engel J, Gerretzen J, Szymańska E, Jansen JJ, Downey G, Blanchet L, et al. Breaking with trends in pre-processing? TrAC Trends Anal Chem 2013;50:96–106.

[12] Gerretzen J, Szymańska E, Jansen JJ, Bart J, van Manen HJ, van den Heuvel ER, et al. Simple and effective way for data preprocessing selection based on design of experiments. Anal Chem 2015;87(24):12096–103.

[13] Brown S, Tauler R, Walczak R, editors. Section data preprocessing. In: Comprehensive chemometrics, Oxford: Elsevier; 2009.

[14] Jones HD, Haaland DM, Sinclair MB, Melgaard DK, Collins AM, Timlin JA. Preprocessing strategies to improve MCR analyses of hyperspectral images. Chemom Intell Lab Syst 2012;117:149–58.

[15] Ruckebusch C, Sliwa M, Pernot P, De Juan A, Tauler R. Comprehensive data analysis of femtosecond transient absorption spectra: a review. J Photochem Photobiol C: Photochem Rev 2012;13(1):1–27.

[16] de Rooi JJ, Devos O, Sliwa M, Ruckebusch C, Eilers PH. Mixture models for two-dimensional baseline correction, applied to artifact elimination in time-resolved spectroscopy. Anal Chim Acta 2013;771:7–13.

[17] Gampp H, Maeder M, Meyer CJ, Zuberbühler AD. Calculation of equilibrium constants from multiwavelength spectroscopic data—IV: model-free least-squares refinement by use of evolving factor analysis. Talanta 1986;33(12):943–51.

[18] Tauler R, Casassas E. Application of principal component analysis to the study of multiple equilibria systems: study of copper (II)/salicylate/mono-, di- and triethanolamine systems. Anal Chim Acta 1989;223:257–68.

[19] Tauler R, Smilde AK, Kowalski BR. Selectivity, local rank, three-way data analysis and ambiguity in multivariate curve resolution. J Chemometr 1995;9:31–58.

[20] Tauler R. Multivariate curve resolution applied to second order data. Chemom Intell Lab Syst 1995;30:133–46.

[21] Tauler R, Maeder M, de Juan A. Two-way extended curve resolution. In: Brown S, Tauler R, Walczak R, editors. Comprehensive chemometrics, vol. 2. Oxford: Elsevier; 2009. p. 473–505.

[22] Alier M, Tauler R. Multivariate curve resolution of incomplete data multisets. Chemometr Intell Lab Sys 2013;127:17–28.

[23] De Luca M, Ragno G, Ioele G, Tauler R. Multivariate curve resolution of incomplete fused multiset data from chromatographic and spectrophotometric analyses for drug photostability studies. Anal Chim Acta 2014;837:31–7.

[24] Wentzell PD, Karakach TK, Roy S, Martinez MJ, Allen CP, Werner-Washburne M. Multivariate curve resolution of time course microarray data. BMC Bioinformatics 2006;7(1):343.

[25] Wentzell PD. Other topics in soft-modeling: maximum likelihood based soft-modeling methods. In: Brown SD, Tauler R, Walczak B, editors. Comprehensive chemometrics, vol. 2. Amsterdam: Elsevier; 2009. p. 507–58.

[26] Jaumot J, Piña B, Tauler R. Application of multivariate curve resolution to the analysis of yeast genome-wide screens. Chemom Intell Lab Syst 2010;104:53–64.

[27] Blanchet L, Réhault J, Ruckebusch C, Huvenne JP, Tauler R, de Juan A. Chemometrics description of measurement error structure: study of an ultrafast absorption spectroscopy experiment. Anal Chim Acta 2009;642:19–26.

[28] Dadashi M, Abdollahi H, Tauler R. Maximum likelihood principal component analysis as initial projection step in multivariate curve resolution analysis of noisy data. Chemom Intell Lab Syst 2012;118:33–40.

[29] Tauler R, Maeder M. Two-way data analysis: multivariate curve resolution—error in curve resolution. In: Brown S, Tauler R, Walczak R, editors. Comprehensive chemometrics, vol. 2. Oxford: Elsevier; 2009. p. 345–63.

[30] Tauler R. Calculation of maximum and minimum band boundaries of feasible solutions for species profiles obtained by multivariate curve resolution. J Chem Theory Comput 2001;15:627–46.

[31] Abdollahi H, Maeder M, Tauler R. Calculation and meaning of feasible band boundaries in multivariate curve resolution of a two-component system. Anal Chem 2009;81:2115–22.

[32] Sawall M, Kubis C, Selent D, Börner A, Neymeyr K. A fast polygon inflation algorithm to compute the area of feasible solutions for three-component systems. I: concepts and applications. J Chemometr 2013;27:106–16.

[33] Rajkó R, István K. Analytical solution for determining feasible regions of self-modeling curve resolution (SMCR) method based on computational geometry. J Chem Theory Comput 2005;19:448–63.

[34] de Juan A, Jaumot J, Tauler R. Multivariate curve resolution (MCR). Solving the mixture analysis problem. Anal Methods 2014;6(14):4964–76.

[35] Windig W, Guilment J. Interactive self-modeling mixture analysis. Anal Chem 1991;63:1425–32.

[36] Cuesta Sánchez F, van den Bogaert B, Rutan SC, Massart DL. Multivariate peak purity approaches. Chemom Intell Lab Syst 1996;34:139–71.

[37] Malinowski ER. Obtaining the key set of typical vectors by factor analysis and subsequent isolation of component spectra. Anal Chim Acta 1982;134:129–37.

[38] Windig W. Two-way data analysis: detection of purest variables. In: Brown S, Tauler R, Walczak R, editors. Comprehensive chemometrics, vol. 2. Oxford: Elsevier; 2009. p. 275–307.

[39] Bro R, De Jong S. A fast non-negativity-constrained least squares algorithm. J Chemom 1997;11:393–401.
[40] Bro R, Sidiropoulos ND. Least squares algorithms under unimodality and non-negativity constraints. J Chem Theory Comput 1998;12:223–47.
[41] Gemperline PJ, Cash E. Advantages of soft versus hard constraints in self-modeling curve resolution problems. Alternating least squares with penalty functions. Anal Chem 2003;75:4236–43.
[42] Gallagher NB, Shaver JM, Martin EB, Morris J, Wise BM, Windig W. Curve resolution for multivariate images with applications to TOF-SIMS and Raman. Chemom Intell Lab Syst 2004;73(1):105–17.
[43] Jaumot J, de Juan A, Tauler R. MCR-ALS GUI 2.0: new features and applications. Chemom Intell Lab Syst 2015;140:1–12.
[44] Manne R. On the resolution problem in hyphenated chromatography. Chemom Intell Lab Syst 1995;27:89–94.
[45] Tauler R, Marqués I, Casassas E. Multivariate curve resolution applied to three-way trilinear data: study of a spectrofluorimetric acid-base titration of salicylic acid at three excitation wavelengths. J Chem Theory Comput 1998;12:55–75.
[46] Malik A, Tauler R. Extension and application of multivariate curve resolution-alternating least squares to four-way quadrilinear data-obtained in the investigation of pollution patterns on Yamuna River, India-A case study. Anal Chim Acta 2013;794:20–8.
[47] Alier M, Felipe M, Hernández I, Tauler R. Trilinearity and component interaction constraints in the multivariate curve resolution investigation of NO and O-3 pollution in Barcelona. Anal Bioanal Chem 2011;399:2015–29.
[48] Bro R. PARAFAC. Tutorial and applications. Chemom Intell Lab Syst 1997;38:149–71.
[49] Sanchez E, Kowalski BR. Tensorial resolution: a direct trilinear decomposition. J Chemom 1990;4:29–45.
[50] Terrado M, Barceló D, Tauler R. Quality assessment of the multivariate curve resolution alternating least squares method for the investigation of environmental pollution patterns in surface water. Environ Sci Technol 2009;43:5321–6.
[51] Elcoroaristizábal S, de Juan A, García JA, Durana N, Alonso L. Comparison of second-order multivariate methods for screening and determination of PAHs by total fluorescence spectroscopy. Chemom Intell Lab Syst 2014;132:63–74.
[52] de Juan A, Maeder M, Martínez M, Tauler R. Combining hard- and soft-modelling to solve kinetic problems. Chemom Intell Lab Syst 2000;54:123–41.
[53] Azzouz T, Tauler R. Application of multivariate curve resolution alternating least squares (MCR-ALS) to the quantitative analysis of pharmaceutical and agricultural samples. Talanta 2008;74:1201–10.
[54] Maeder M, Zilian A. Evolving factor analysis, a new multivariate technique in chromatography. Chemom Intell Lab Syst 1988;3:205–13.
[55] Maeder M, de Juan A. Two-way data analysis: evolving factor analysis. In: Brown S, Tauler R, Walczak R, editors. Comprehensive chemometrics, vol. 2. Oxford: Elsevier; 2009. p. 261–74.
[56] Malinowski ER. Window factor analysis: theoretical derivation and application to flow injection analysis data. J Chemom 1992;6:29–40.
[57] Kvalheim OM, Liang YZ. Heuristic Evolving Latent projections-resolving 2-way multicomponent data. 1. Selectivity, latent projective graph, datascope, local rank and unique resolution. Anal Chem 1992;64:936–46.
[58] Manne R, Shen H, Liang Y. Subwindow factor analysis. Chemom Intell Lab Syst 1999;45:171–6.

[59] Maeder M, Zuberbühler AD. The resolution of overlapping chromatographic peaks by evolving factor-analysis. Anal Chim Acta 1986;181:287–91.
[60] Keller HR, Massart DL. Peak purity control in liquid chromatography with photodiode-array detection by a fixed size moving window evolving factor analysis. Anal Chim Acta 1991;246:379–90.
[61] Whitson AC, Maeder M. Exhaustive evolving factor analysis (E-EFA). J Chemom 2001;15:475–84.
[62] Maeder M, Neuhold YM. Practical data analysis in chemistry, vol. 26. Amsterdam: Elsevier; 2007.
[63] Maeder M, Zuberbuehler AD. Nonlinear least-squares fitting of multivariate absorption data. Anal Chem 1990;62:2220–4.
[64] Bezemer E, Rutan SC. Multivariate curve resolution with non-linear fitting of kinetic profiles. Chemom Intell Lab Syst 2001;59:19–31.
[65] Sawall M, Börner A, Kubis C, Selent D, Ludwig R, Neymeyr K. Model-free multivariate curve resolution combined with model-based kinetics: algorithm and applications. J Chemom 2012;26(10):538–48.
[66] Mas S, de Juan A, Lacorte S, Tauler R. Photodegradation study of decabromodiphenyl ether by UV spectrophotometry and a hybrid hard-and soft-modelling approach. Anal Chim Acta 2008;618(1):18–28.
[67] Diewok J, de Juan A, Maeder M, Tauler R, Lendl B. Application of a combination of hard and soft modeling for equilibrium systems to the quantitative analysis of pH-modulated mixture samples. Anal Chem 2003;75:641–7.
[68] Amigo JM, de Juan A, Coello J, Maspoch S. A mixed hard- and soft-modelling approach to study and monitor enzymatic systems in biological fluids. Anal Chim Acta 2006; 567:236–54.
[69] Blanchet L, Ruckebusch C, Mezzetti A, Huvenne JP, de Juan A. Monitoring and interpretation of photoinduced biochemical processes by rapid-scan FTIR difference spectroscopy and hybrid hard and soft modeling. J Phys Chem B 2009;113(17):6031–40.
[70] Golshan A, Abdollahi H, Maeder M. The reduction of rotational ambiguity in soft-modeling by introducing hard models. Anal Chim Acta 2012;709:32–40.
[71] Amrhein M, Srinivasan B, Bonvin D, Schumacher MM. On the rank deficiency and rank augmentation of the spectral measurement matrix. Chemom Intell Lab Syst 1996;33(1): 17–33.
[72] Saurina J, Hernández-Cassou S, Tauler R, Izquierdo-Ridorsa A. Multivariate resolution of rank-deficient spectrophotometric data from first-order kinetic decomposition reactions. J Chemom 1998;12(3):183–203.
[73] Navea S, de Juan A, Tauler R. Modeling temperature-dependent protein structural transitions by combined near-IR and mid-IR spectroscopies and multivariate curve resolution. Anal Chem 2003;75:5592–601.
[74] Cutler P, Gemperline PJ, de Juan A. Experimental monitoring and data analysis tools for protein folding: study of steady-state evolution and modeling of kinetic transients by multitechnique and multiexperiment data fusion. Anal Chim Acta 2009;632:52–62.
[75] Mouton N, Devos O, Sliwa M, de Juan A, Ruckebusch C. Multivariate curve resolution-alternating least squares applied to the investigation of ultrafast competitive photoreactions. Anal Chim Acta 2013;788:8–16.
[76] Jaumot J, Escaja N, Gargallo R, González C, Pedroso E, Tauler R. Multivariate curve resolution: a powerful tool for the analysis of conformational transitions in nucleic acids. Nucleic Acids Res 2002;30(17):e92.

[77] Garrido M, Rius FX, Larrechi MS. Multivariate curve resolution-alternating least squares (MCR-ALS) applied to spectroscopic data from monitoring chemical reactions processes. Anal Bioanal Chem 2008;390:2059–66.
[78] Tauler R, Kowalski B, Fleming S. Multivariate curve resolution applied to spectral data from multiple runs of an industrial process. Anal Chem 1993;65(15):2040–7.
[79] Jaumot J, Igne B, Anderson CA, Drennen JK, de Juan A. Blending process modelling and control by multivariate curve resolution. Talanta 2013;117:492–504.
[80] Ma B, Gemperline PJ, Cash E, Bosserman M, Comas E. Characterizing batch reactions with in situ spectroscopic measurements, calorimetry and dynamic modeling. J Chemom 2003;17(8–9):470–9.
[81] Amigo JM, de Juan A, Coello J, Maspoch S. A mixed hard-and soft-modelling approach for the quantitative determination of oxipurines and uric acid in human urine. Anal Chim Acta 2006;567(2):236–44.
[82] Mas S, Tauler R, de Juan A. Chromatographic and spectroscopic data fusion analysis for interpretation of photodegradation processes. J Chromatogr A 2011;1218:9260–8.
[83] de Juan A, Tauler R. Multivariate curve resolution for quantitative analysis. In: de la Peña AM, Goicoechea HC, Escandar GM, Olivieri AC, editors. Fundamentals and analytical applications of multiway calibration. Elsevier Ltd; 2015. p. 247–92.
[84] Fernández C, Callao MP, Larrechi MS. UV–visible-DAD and 1H-NMR spectroscopy data fusion for studying the photodegradation process of azo-dyes using MCR-ALS. Talanta 2013;117:75–80.
[85] Navea S, Tauler R, de Juan A. Monitoring and modeling of protein processes using mass spectrometry, circular dichroism, and multivariate curve resolution methods. Anal Chem 2006;78:4768–78.
[86] Peré-Trepat E, Tauler R. Analysis of environmental samples by application of multivariate curve resolution on fused high-performance liquid chromatography–diode array detection mass spectrometry data. J Chromatogr A 2006;1131:85–96.
[87] Grahn HF, Geladi P. Techniques and applications of hyperspectral image analysis. Chichester: John Wiley; 2007.
[88] Salzer R, Siesler HW, editors. Infrared and Raman spectroscopic imaging: second, completely revised and updated edition. Weinheim: Wiley-VCH Verlag GmbH Co. KGaA; 2014.
[89] de Juan A, Piqueras S, Maeder M, Hancewicz T, Duponchel L, Tauler R. Chemometric tools for image analysis. In: Salzer R, Siesler HW, editors. Infrared and Raman spectroscopic imaging: second, completely revised and updated edition. Weinheim: Wiley-VCH Verlag GmbH Co. KGaA; 2014. p. 57–110.
[90] Amigo JM, Cruz J, Bautista M, Maspoch S, Coello J, Blanco M. Study of pharmaceutical samples by NIR chemical-image and multivariate analysis. TrAC Trends Anal Chem 2008;27(8):696–713.
[91] Gendrin C, Roggo Y, Collet C. Pharmaceutical applications of vibrational chemical imaging and chemometrics: a review. J Pharm Biomed Anal 2008;48:533–53.
[92] Felten J, Hall H, Jaumot J, Tauler R, de Juan A, Gorzsás A. Vibrational spectroscopic image analysis of biological material using multivariate curve resolution-alternating least squares (MCR-ALS). Nat Protoc 2015;10(2):217–40.
[93] Vermaas WF, Timlin JA, Jones HD, Sinclair MB, Nieman LT, Hamad SW, et al. In vivo hyperspectral confocal fluorescence imaging to determine pigment localization and distribution in cyanobacterial cells. Proc Natl Acad Sci 2008;105(10):4050–5.

[94] Zhang X, Tauler R. Application of multivariate curve resolution alternating least squares (MCR-ALS) to remote sensing hyperspectral imaging. Anal Chim Acta 2013;762:25–38.
[95] Tits L, Somers B, Saeys W, Coppin P. Site-specific plant condition monitoring through hyperspectral alternating least squares unmixing. IEEE J Sel Topics Appl Earth Observ Remote Sens 2014;7(8):3606–18.
[96] Siewert H, Devos O, Ruckebusch C. On the implementation of spatial constraints in multivariate curve resolution alternating least squares for hyperspectral image analysis. J Chemom 2015;29:557–61.
[97] Hugelier S, Devos O, Ruckebusch C. Constraining shape smoothness in multivariate curve resolution-alternating least squares. J Chemom 2015;29:448–56.
[98] de Juan A, Maeder M, Hancewicz T, Tauler R. Local rank analysis for exploratory spectroscopic image analysis. Fixed size image window-evolving factor analysis. Chemom Intell Lab Syst 2005;77(1):64–74.
[99] de Juan A, Tauler R, Dyson R, Marcolli C, Rault M, Maeder M. Spectroscopic imaging and chemometrics: a powerful combination for global and local sample analysis. TrAC Trends Anal Chem 2004;23(1):70–9.
[100] de Juan A, Maeder M, Hancewicz T, Tauler R. Use of local rank-based spatial information for resolution of spectroscopic images. J Chemom 2008;22(5):291–8.
[101] Zhang X, de Juan A, Tauler R. Local rank-based spatial information for improvement of remote sensing hyperspectral imaging resolution. Talanta 2016;146:1–9.
[102] Boiret M, de Juan A, Gorretta N, Ginot YM, Roger JM. Setting local rank constraints by orthogonal projections for image resolution analysis: application to the determination of a low dose pharmaceutical compound. Anal Chim Acta 2015;892:49–58.
[103] Piqueras S, Krafft C, Beleites C, Egodage K, von Eggeling F, Guntinas-Lichius O, et al. Combining multiset resolution and segmentation for hyperspectral image analysis of biological tissues. Anal Chim Acta 2015;881:24–36.
[104] Piqueras S, Duponchel L, Tauler R, de Juan A. Monitoring polymorphic transformations by using in situ Raman hyperspectral imaging and image multiset analysis. Anal Chim Acta 2014;819:15–25.
[105] Alexandrino GL, Khorasani MR, Amigo JM, Rantanen J, Poppi RJ. Monitoring of multiple solid-state transformations at tablet surfaces using multi-series near-infrared hyperspectral imaging and multivariate curve resolution. Eur J Pharm Biopharm 2015;93:224–30.
[106] Bocklitz T, Bräutigam K, Urbanek A, Hoffmann F, von Eggeling F, Ernst G, et al. Novel workflow for combining Raman spectroscopy and MALDI-MSI for tissue based studies. Anal Bioanal Chem 2015;407(26):7865–73.
[107] Schaaff TG, McMahon JM, Todd PJ. Semiautomated analytical image correlation. Anal Chem 2002;74(17):4361–9.
[108] Piqueras, S, Maeder, M, Tauler, R, de Juan, A. (submitted).
[109] Piqueras S, Duponchel L, Tauler R, de Juan A. Resolution and segmentation of hyperspectral biomedical images by multivariate curve resolution-alternating least squares. Anal Chim Acta 2011;705(1):182–92.
[110] Piqueras S, Duponchel L, Offroy M, Jamme F, Tauler R, de Juan A. Chemometric strategies to unmix information and increase the spatial description of hyperspectral images: a single-cell case study. Anal Chem 2013;85(13):6303–11.
[111] Piqueras S, Burger J, Tauler R, de Juan A. Relevant aspects of quantification and sample heterogeneity in hyperspectral image resolution. Chemom Intell Lab Syst 2012;117:169–82.

[112] Cruz J, Bautista M, Amigo JM, Blanco M. Nir-chemical imaging study of acetylsalicylic acid in commercial tablets. Talanta 2009;80(2):473–8.

[113] de Juan A, Tauler R. Factor analysis of hyphenated chromatographic data. Exploration, resolution and quantification of multicomponents systems. J Chromatogr A 2007;1158: 184–95.

[114] Booksh K, Kowalski BK. Theory of analytical chemistry. Anal Chem 1994;66:782A–791A.

[115] Escandar GM, Faber NM, Goicoechea HC, Muñoz de la Peña A, Olivieri AC. Second- and third-order multivariate calibration: data, algorithms and applications. TrAC Trends Anal Chem 2007;2007(26):752–65.

[116] Peré-Trepat E, Lacorte S, Tauler R. Alternative calibration approaches for LC-MS quantitative determination of coeluted compounds in complex environmental mixtures using multivariate curve resolution. Anal Chim Acta 2007;595:228–37.

[117] Tauler R, Barceló D. Multivariate curve resolution applied to liquid-chromatography diode-array detection. TrAC Trends Anal Chem 1993;12:319–27.

[118] Ortiz MC, Sarabia L. Quantitative determination in chromatographic analysis based on n-way calibration strategies. J Chromatogr A 2007;1158:94–110.

[119] Rodríguez-Cuesta MJ, Boqué R, Rius FX, Vidal JLM, Frenich AG. Development and validation of a method for determining pesticides in groundwater from complex overlapped HPLC signals and multivariate curve resolution. Chemom Intell Lab Syst 2005;77:251–60.

[120] Danzer K, Currie LA. Guidelines for calibration in analytical chemistry—Part 1. Fundamentals and single component calibration (IUPAC recommendations. Pure Appl Chem 1998;70:993–1014.

[121] Olivieri AC, Faber NM. A closed-form expression for computing the sensitivity in second-order bilinear calibration. J Chemom 2005;19:583–92.

[122] Olivieri AC. Computing sensitivity and selectivity in parallel factor analysis and related multiway techniques: the need for further developments in net analyte signal theory. Anal Chem 2005;77:4936–46.

[123] Saurina J, Leal C, Compañó R, Granados M, Prat MD, Tauler R. Estimation of figures of merit using univariate statistics for quantitative second-order multivariate curve resolution. Anal Chim Acta 2001;432:241–51.

[124] Antunes MC, Simão JEJ, Duarte AC, Tauler R. Multivariate curve resolution of overlapping voltammetric peaks: quantitative analysis of binary and quaternary metal mixtures. Analyst 2002;127:809–17.

[125] Mas S, Miguel C, Melo MJ, Lopes JA, de Juan A. Screening and quantification of proteinaceous binders in medieval paints based on μ-Fourier transform infrared spectroscopy and multivariate curve resolution alternating least squares. Chemom Intell Lab Syst 2014;134:148–57.

[126] Lyndgaard LB, van den Berg F, de Juan A. Quantification of paracetamol through tablet blister packages by Raman spectroscopy and multivariate curve resolution-alternating least squares. Chemom Intell Lab Syst 2013;125:58–66.

[127] Rocha de Oliveira R, de Lima KMG, Tauler R, de Juan A. Application of correlation constrained multivariate curve resolution alternating least-squares methods for determination of compounds of interest in biodiesel blends using NIR and UV-visible spectroscopic data. Talanta 2014;125:233–41.

[128] Jolliffe I. Principal component analysis. New York: John Wiley Sons Ltd; 2002.

[129] Rutledge DN, Jouan-Rimbaud Bouveresse D. Independent components analysis with the JADE algorithm. TrAC Trends Anal Chem 2013;50:22–32.

[130] Parastar H, Halali-Heravi M, Tauler R. Is independent component analysis appropriate for multivariate resolution in analytical chemistry? TrAC Trends Anal Chem 2012;31:134–43.

[131] Bioucas-Dias JM, Plaza A, Dobigeon N, Parente M, Du Q, Gader P, et al. Hyperspectral unmixing overview: geometrical, statistical, and sparse regression-based approaches. IEEE J Sel Topics Appl Earth Observ Remote Sens 2012;5(2):354–79.

[132] Nascimento JM, Bioucas-Dias JM. Vertex component analysis: A fast algorithm to unmix hyperspectral data. IEEE Trans Geosci Remote Sens 2005;2005(43):898–910.

[133] Berman M, Phatak A, Lagerstrom R, Wood BR. ICE: a new method for the multivariate curve resolution of hyperspectral images. J Chemomet 2009;23:101–16.

[134] Smilde A, Bro R, Geladi P. Multi-way analysis. Applications in the chemical sciences. West Sussex: John Wiley Sons Ltd; 2004.

[135] de Juan A, Tauler R. Comparison of 3-way resolution methods for non-trilinear data sets. J Chemom 2001;15:749.

[136] Muñoz G, de Juan A. pH-and time-dependent hemoglobin transitions: a case study for process modelling. Anal Chim Acta 2007;595:198–208.

Chapter 3

Spectral Unmixing Using the Concept of Pure Variables

S. Kucheryavskiy*[,1], W. Windig[†] and A. Bogomolov[‡]
*Aalborg University, Aalborg, Denmark
†Eigenvector Research Inc., Manson, WA, United States
‡Art Photonics GmbH, Berlin, Germany
1Corresponding author: e-mail: svk@bio.aau.dk

Chapter Outline

1 INTRODUCTION

The subject we are going to introduce falls under self-modeling mixture analysis [1]. Self-modeling mixture analysis is a series of techniques that calculate estimates of the pure chemical components and their contributions for mixture datasets without any prior knowledge. The term "contributions" is often used for concentrations since there are unknown factors between the original concentrations of mixtures and the results of self-modeling mixture analysis.

The methods to be considered in this chapter are all based on the same approach—a pure variable. The term "pure" is used for a spectral variable (i.e., data values at a specific spectral band), which experiences a nonzero contribution from one and only one mixture component. Provided that the data obeys the Beer–Lambert law of absorbance or a similar linear behavior, the values of this pure variable are directly proportional to the concentrations.

Data Handling in Science and Technology, Vol. 30. http://dx.doi.org/10.1016/B978-0-444-63638-6.00003-6
Copyright © 2016 Elsevier B.V. All rights reserved.

Knowing the mixture data and the concentrations enable us to calculate the spectra of the individual components.

We will introduce one of the methods, SIMPLISMA (SIMple-to-use Interactive Self-modeling Mixture Analysis), in full details with many illustrations, but will also discuss similar methods and papers that led to the pure variable approach toward self-modeling mixture analysis.

As the practical cases considered in the chapter are limited to the optical spectroscopic data, we will mostly use "spectral unmixing" instead of a more general "curve resolution" term. This terminological simplification, however, does not restrict the application of the pure variable approach. The methods can be equally used for resolving mixture data of any other, nonspectroscopic nature.

The chapter is organized as follows. Section 2 describes datasets used to illustrate different aspects of the methods and their performances. Section 3 introduces the concept of pure variables explaining how their presence (and having them known) allows carrying out the spectral unmixing. Sections 4 and 5 discuss the SIMPLISMA algorithm with an emphasis on the approach it uses to look for the most optimal pure variables. Section 6 provides a review of other methods utilizing the same principle and compares them with SIMPLISMA. Section 7 considers pure variable application in multivariate curve resolution using alternating least squares (MCR-ALS) that is probably the most widely used algorithm for spectral unmixing. Section 8 is mainly devoted to a discussion and general conclusions.

The datasets and R code with implementation of the algorithms as well as with datasets and the examples shown in this chapter can be freely downloaded as a package from a GitHub repository (https://github.com/svkucheryavski/supure). Use command "?supure" when the package is installed to get started.

The expressions and equations in the chapter use the following notation. Matrices are shown using capital letters and a boldface, **D**, while vectors are represented with small caps and boldface, **d**. A letter with regular face and subscript indices denotes an element of a matrix (or a vector), where the first index is for a row and second for a column (e.g., D_{ij} is an element of matrix **D** at the *i*th row and *j*th column).

2 CASE STUDIES

We will use four different datasets to show how the methods and their different modifications perform. The first two datasets were simulated by applying a bilinear model to pure spectra of three chemical components with different concentration levels. The other two contain real experimental spectra.

The dataset *A* represents Raman spectra of fructose, lactose, and ribose and their mixtures truncated to the range from 200 to 1600 cm^{-1}. The individual spectra were downloaded from publicly available SPECARB library [2] created by S.B. Engelsen. The concentrations of the components followed a simplex lattice design with four levels. Some noise calculated as a random number uniformly distributed between 0% and 3% of maximum initial intensity was added to each spectrum of the dataset individually.

Fig. 1 presents individual spectra and resulting simulated mixture dataset. Despite the high degree of peak overlap, which is typical for a mixture of homologically close compounds, some signals in the mixture spectra look more or less selective, i.e., unique for the individual components. In order to facilitate the explanation, we have selected two groups of peaks. One of them, denoted as P1, includes the following characteristic bands: 818 cm^{-1} for lactose, 357 cm^{-1} for fructose, and 542 cm^{-1} for ribose. Another group, P2, consists of the signals at 626, 477, and 884 cm^{-1} characteristic of fructose, lactose, and ribose, respectively. The peaks are indicated with colored dashed lines in the upper plot.

The dataset *B* consists of UV/vis spectra of three polyaromatic carbohydrates (C1, 2-aceto-phenanthrene; C2, 2-acetylaminophenanthrene; and C3, 3-acetylaminophenanthrene) and their mixtures. Component concentration values (in arbitrary units) in the mixtures were generated randomly, using a

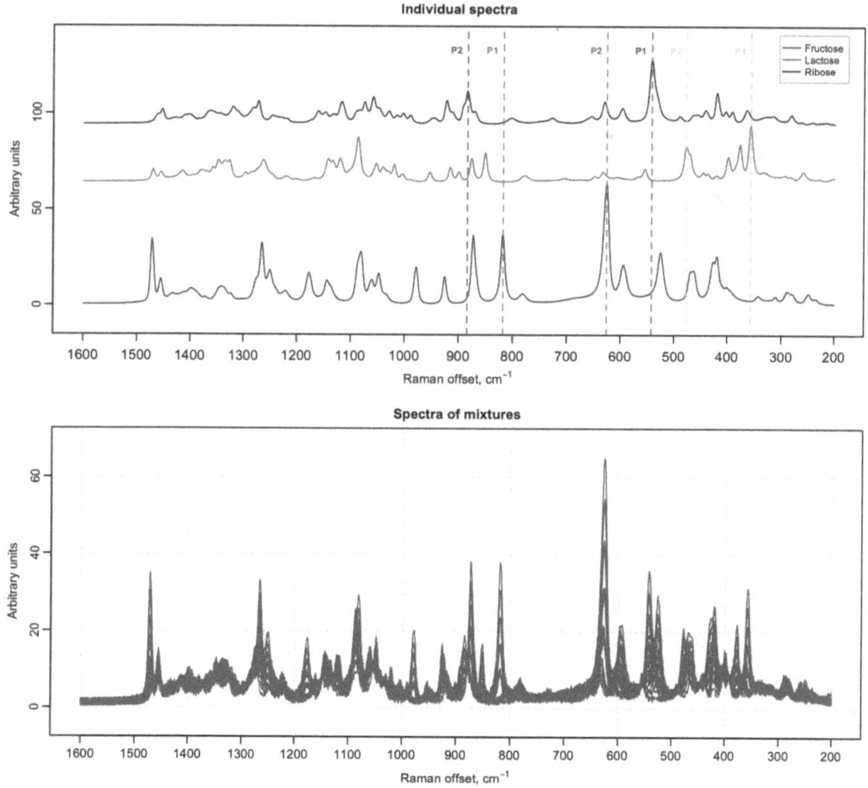

FIG. 1 Individual spectra (top) and spectra of mixtures (bottom) for dataset *A*.

uniform distribution from 0 to 1. Fig. 2 shows the individual spectra and the spectra of the mixtures. As it could be expected for the UV/vis range, the spectra of individual components are pretty similar (especially for C1 and C2) and therefore extremely overlapped in the mixture data. No unique (i.e., selective) peaks are observed for the individual components. For the analysis, we have selected a peak set at 267 nm for C1, 303 nm for C2, and 311 nm for C3 (dashed lines in the top plot of Fig. 2).

The dataset *C* contains mid-infrared (MIR) spectra of binary aqueous solution of ethanol and glucose prepared at laboratory conditions. The component concentrations were chosen in accordance to the "diagonal design" scheme and varied from 0 to 144 g/L for ethanol and from 0 to 288 g/L for glucose.

The MIR spectra in the range from 680 to 2900 cm^{-1} with spectral resolution of about 2 cm^{-1} (Fig. 3, upper plot) were acquired with a matrix-MF FT-IR spectrophotometer (Bruker AG, Ettlingen, Germany) using a diamond-crystal attenuated total reflection probe by Art Photonics GmbH (Berlin, Germany). More details about the data and spectra acquisition procedure can be found in Bogomolov *et al.* [3]. The data rows (samples) were

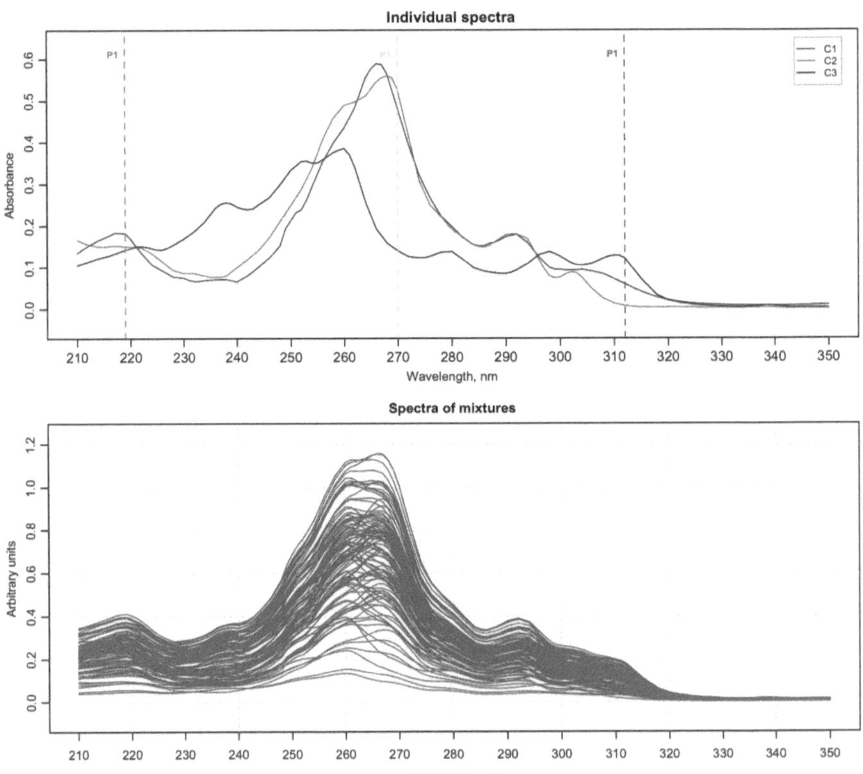

FIG. 2 Individual spectra (top) and spectra of mixtures (bottom) for dataset *B*.

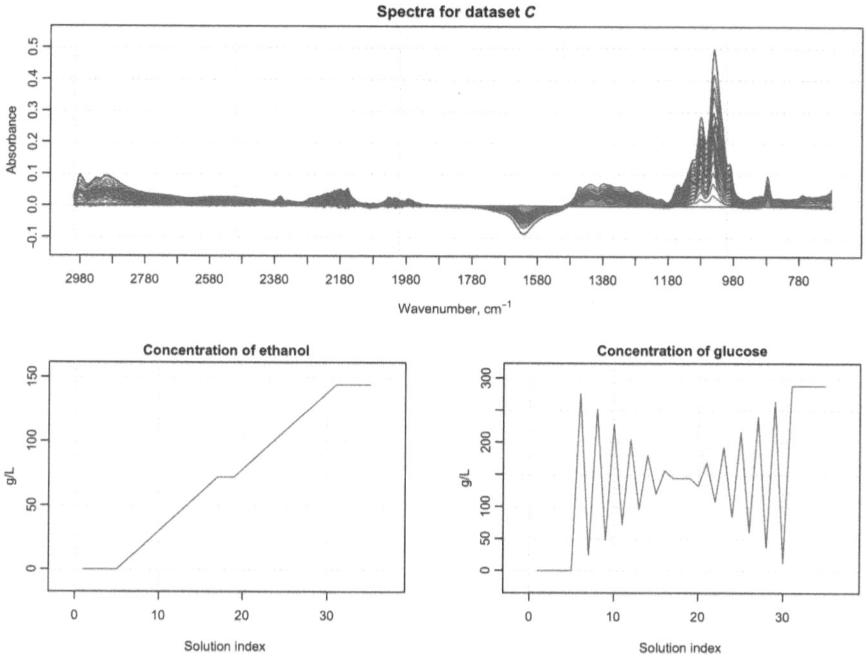

FIG. 3 Spectral plot for dataset *C*.

resorted in order to get easily to recognize concentration profiles, i.e., gradually increasing ethanol concentration and a butterfly like glucose profile (Fig. 3, lower plots).

Ethanol has a pair of two distinct characteristic peaks in the fingerprint region of the spectra, at about 1090 and 1050 cm^{-1} plus a smaller peak at 880 cm^{-1}. Glucose in this region is recognized by a complex combination of poorly resolved bands. Two most selective bands observed at about 1060 and 1030 cm^{-1}. Since spectrum of water was used as a reference, the spectra shown in Fig. 3 have negative signals appearing in a range that correspond to the water absorption bands.

The fourth dataset, *D*, is a hyperspectral image of an oil-in-water emulsion. The spectra for the image were obtained using Instruments SA Explorer with some modifications of optical system, coupled with motorized stage and a Raman spectrometer with excitation source worked at 633 nm. More details about the system and acquisition details can be found in Andrew *et al.* [4].

As it was described in Andrew *et al.* [4] the depicted area consists of an aqueous phase (with mostly glycerol and in a smaller concentration polawax), two droplet phases (one is polawax and another with parabens, polawax and some panthenol) and a structural phase with a crystalline polawax. The image was used in several papers, including ones devoted to applications of MCR [5,6].

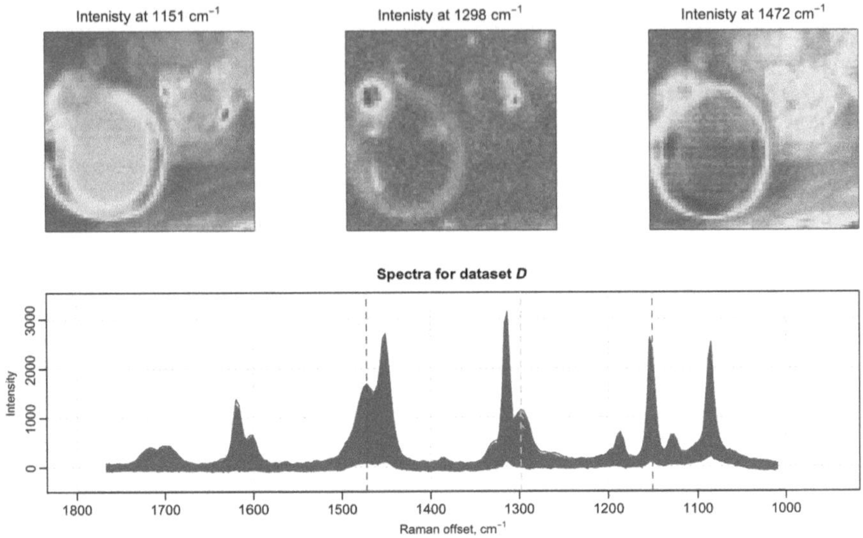

FIG. 4 Spectral plot and pseudocolor images for the selected wavebands for dataset *D*.

Fig. 4 shows the Raman spectra (bottom) as well as an intensity of three selected Raman peaks (1151, 1298, and 1472 cm^{-1}) refolded to a matrix and shown as pseudocolor images. The color gradient in the images reflects the spectral intensity at the selected band for each pixel (from blue for small values to the red for large intensities).

3 SPECTRAL UNMIXING WITH PURE VARIABLES

Mathematically speaking, spectral unmixing implies decomposition of original mixture spectra, to a product of individual spectra of the components and contributions (concentrations), which reflects a relative weight of each component in the mixtures. Assuming that the original spectra can be well described by the bilinear additive model (Lambert–Beer law) the decomposition can be written in a matrix form as following:

$$\mathbf{D} = \mathbf{C}\mathbf{S}^{\mathrm{T}} + \mathbf{E} \tag{1}$$

Here **D** is a matrix of original spectra, where the rows are individual spectral measurements and columns correspond to individual spectral variables, e.g., wavelengths; **S** is a matrix of pure component spectra in columns, **C** is a matrix of concentrations, and **E** is a matrix of residuals (or error matrix).

Finding matrices **C** and **S** from original data **D** is an inverse problem and, in general case, it has no unique solution. To find a proper one, several assumptions should be made on the spectral data. However, if one of the matrices is known, the second one can be easily found by using simple linear algebra operations. Let us assume that we know concentrations of the components, **C**. Then the pure spectra, **S**, can be estimated by classical least squares method:

$$\hat{\mathbf{S}} = \mathbf{D}^{\mathrm{T}}\mathbf{C}\left(\mathbf{C}^{\mathrm{T}}\mathbf{C}\right)^{-1} \tag{2}$$

The procedure is very similar to multiple linear regression (MLR) [7], where **D** is a response matrix, **C** is a matrix of predictors, and **S** represents the regression coefficients. Similar to MLR, we can evaluate the quality of the solution by making "predictions" of the matrix **D**:

$$\hat{\mathbf{D}} = \mathbf{C}\hat{\mathbf{S}}^{\mathrm{T}} \tag{3}$$

and analyzing residuals, i.e., differences between the initial values and their estimates:

$$\mathbf{E} = \mathbf{D} - \hat{\mathbf{D}} \tag{4}$$

We will talk about the residuals later, now the main question is how to find the matrix with concentrations, **C**? The idea was already mentioned in Section 1: we have to find a pure variable (wavenumber, wavelength, frequency, etc.)—the one, which mainly experiences a contribution of a single chemical component. Let us illustrate this approach using one of the simulated datasets.

As we discussed in Section 2 and one also can see from Fig. 1, some peaks in individual spectra of the components for dataset *A* are very well resolved, especially for the set P1. Let us try to use these peaks as pure variables. Since we assume that height of each selected peak depends on concentration of only one chemical component, the data values at the chosen wavelengths can be taken as a first estimate of **C**. In fact, pure concentration profiles at that stage are presented by a reduced data matrix $\mathbf{D}_R$ that is composed of three columns of the initial dataset **D** corresponding to our pure variables at 818, 357, and 542 cm^{-1}. Then Eq. (2) can be rewritten as:

$$\hat{\mathbf{S}} = \mathbf{D}^{\mathrm{T}}\mathbf{D}_R\left(\mathbf{D}_R^{\mathrm{T}}\mathbf{D}\right)^{-1} \tag{5}$$

The estimated spectra $\hat{\mathbf{S}}$ are then used to get the matrix with estimated concentrations $\hat{\mathbf{C}}$ by:

$$\hat{\mathbf{C}} = \mathbf{D}\hat{\mathbf{S}}\left(\hat{\mathbf{S}}^{\mathrm{T}}\hat{\mathbf{S}}\right)^{-1} \tag{6}$$

On the final stage, the estimated spectra and concentrations have to be weighted in order to keep the correct relative difference among concentration of the pure components. The weighting coefficients can be found using an assumption that concentration of every component in a mixture influences shape of a whole spectrum. So if we calculate a column-vector **f**, which represents a sum of all spectral values for each mixture:

$$\mathbf{f} = \sum_{j=1}^{n} D_{ij} \tag{7}$$

where n is a number of variables in **D**, then the relationship can be written as:

$$\hat{\mathbf{C}}\mathbf{a} = \mathbf{f} \tag{8}$$

Here **a** is a vector with the required scaling coefficients, which can be found by:

$$\mathbf{a} = \left(\hat{\mathbf{C}}^{\mathrm{T}}\hat{\mathbf{C}}\right)^{-1}\hat{\mathbf{C}}^{\mathrm{T}}\mathbf{f} \tag{9}$$

The concentrations now can be scaled by multiplying each column to the corresponding coefficient from **a** and the spectra by multiplying them to the inverse values of the coefficients:

$$\begin{aligned} \hat{\mathbf{C}}_{\mathrm{s}} &= \hat{\mathbf{C}}\mathbf{A} \\ \hat{\mathbf{S}}_{\mathrm{s}} &= \hat{\mathbf{S}}\mathbf{A}\left(\mathbf{A}^{\mathrm{T}}\mathbf{A}\right)^{-1} \end{aligned} \tag{10}$$

Here **A** is a diagonal matrix with values from **a** in the main diagonal.

The right part of Fig. 5 shows a chunk of R code implementing the whole procedure (we assume that the original spectral data and vector with pure variables, denoted in code as `D` and `purevars`, are defined) in just few lines.

The only problem here is that the values in $\mathbf{D}_R$ are not in the same units as the concentration values in **C**, and, therefore, even after weighting, the estimated concentrations and spectra will not precisely correspond to the real

```
# initial estimated for concentrations
Dr = D[, purevars, drop = F]

# resolve spectra and concentrations
St = t(D) %*% Dr %*% solve(t(Dr) %*% Dr)
Ct = D %*% St %*% solve(t(St) %*% St)

# weight the solution
f = as.matrix(colSums(D))
a = solve(t(Ct) %*% Ct) %*% t(Ct) %*% f
A = diag(as.vector(a))

Ct = Ct %*% A
St = St %*% A %*% solve(t(A) %*% A)
```

```
f = 0.05

nvar = ncol(D)
r = runif(nvar * ncomp)
S = matrix(r, ncol = ncomp, nrow = nvar)

for (i in 1:500)
{
  # normalize S
  len = sqrt(colSums(S  2))
  S = sweep(S, 2, len, '/')

  # add small portion of mean to S
  m = apply(S, 1, mean)
  S = sweep((1 - f) * S, 1, f * m, '+')

  # calculate C and apply non-negativity
  C = D %*% S %*% solve(t(S) %*% S)
  C[C < 0] = 0

  # calculate S and apply non-negativity
  S = t(D) %*% C %*% solve(t(C) %*% C)
  S[S < 0] = 0
}
```

FIG. 5 Examples of R code implementing least sqaures unmixing (left) and MCR-ALS with angle constraint (right).

values. So the solution is rather qualitative than quantitative. In order to obtain the real quantitative component spectra and profiles, the component concentrations should be directly determined in at least one mixture. This approach results in a new one-point calibration method (MCR regression) that can be used in practice instead of traditional (e.g., partial least squares regression) calibration methods to avoid tedious calibration experiments [8].

Alternatively, the spectra (or concentration profiles) can be scaled to the unit vector lengths or unit maximum intensity or area. As one part of the **C-S** pair is scaled, another one should be "symmetrically" recalculated as in Eq. (10). Simply speaking, if any pure spectrum is rescaled with the factor of 10, its corresponding concentration profile should be multiplied by 0.1.

The comparison between the original spectra of individual components in the dataset *A* and the spectra resolved by manually selected pure variables (set P1 as shown on the top plot of Fig. 1) for the dataset *A* is given in Fig. 6.

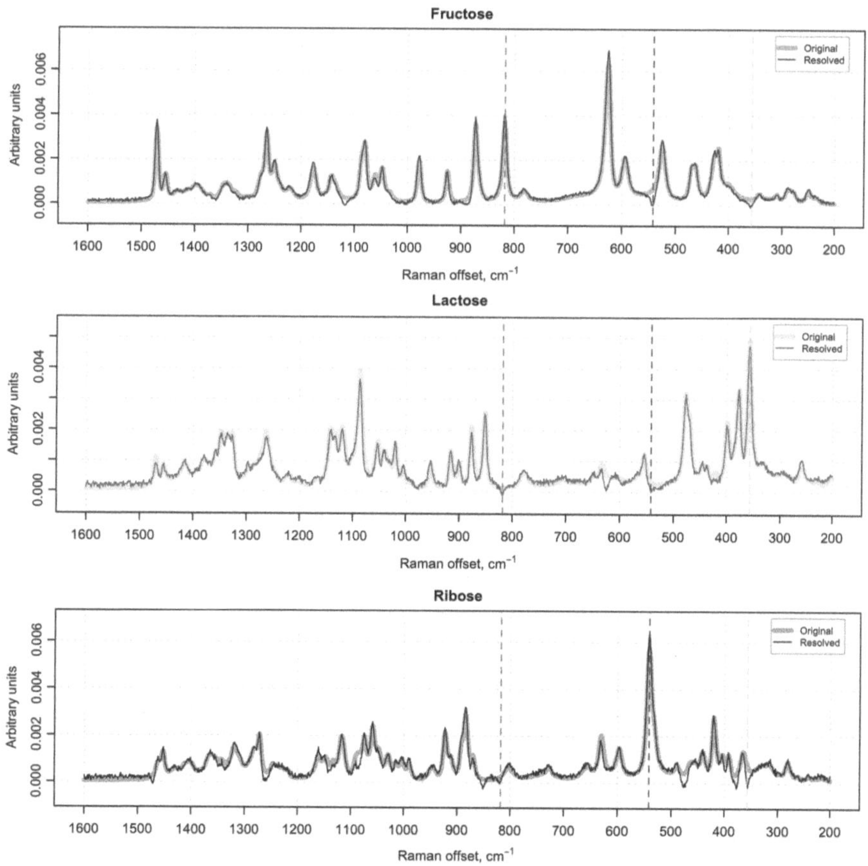

FIG. 6 Results of spectral unmixing of dataset *A* using the first set of pure variables.

The original and the resolved spectra were normalized to a unit area prior to plotting for better comparison. As one can see, the resolved spectra fit the originals very well. Several disturbances appear mostly at the wavebands that correspond to the selected pure variables. The main assumption of the algorithm is that only one component has nonzero contribution at a chosen pure variable is rarely fulfilled completely. In reality, the pure variables often experience the effect of "tails" of the closest neighbor spectral signals of another component or even multiple components and this leads to the observed disturbances.

Fig. 7 illustrates the result of unmixing using the second set of the pure variables, P2. As expected, the results are slightly worse than for the first set, since the selected peaks have a larger overlap degree that can be seen when comparing the spectra of the individual components. Nevertheless, overall resolution quality is still quite good, especially for the lactose.

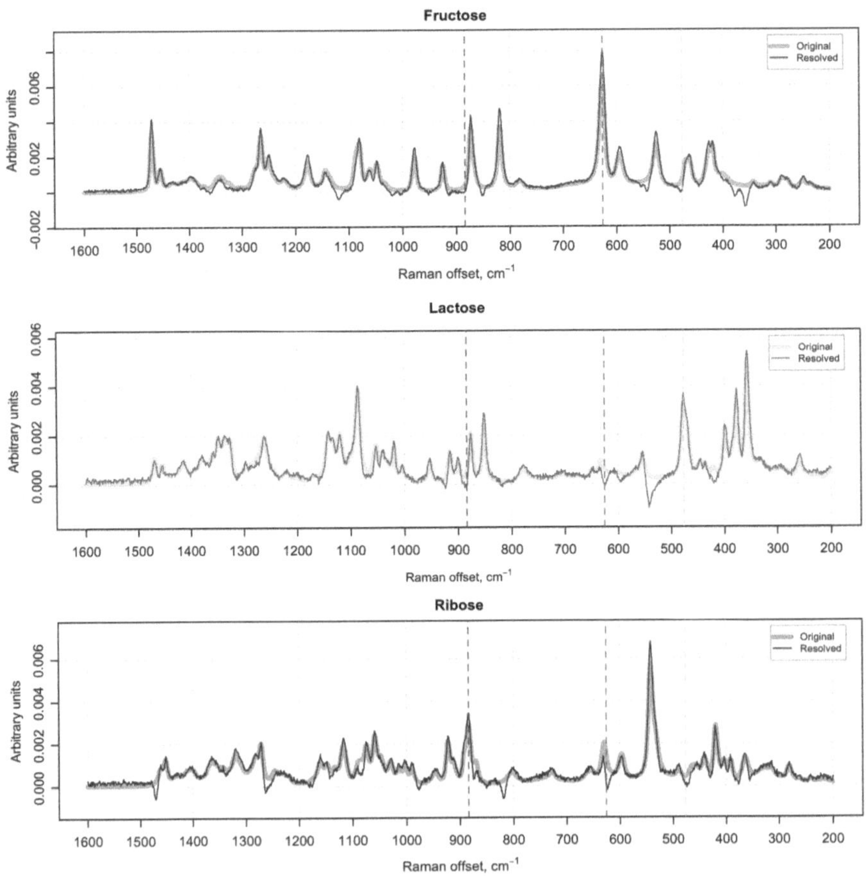

FIG. 7 Results of spectral unmixing of dataset *A* using the second set of pure variables.

In all cases, the characteristic peaks of the components have been correctly resolved.

Table 1 shows the original concentrations **C**, used for simulating the spectra in dataset *A*, in comparison with two estimated concentration matrices, $\hat{\mathbf{C}}$, obtained for the both sets of pure variables (P1 and P2). As it was mentioned before, the values for estimated concentrations have different magnitude if compare with original ones. However, the relationship among the values is kept very well. Thus for C1, estimated value around 8500 corresponds to 1.0 from the original matrix, value around 6800 corresponds to 0.8, and so on. Table 2 contains correlation coefficients calculated for the estimated values vs the original concentrations and rounded up to two decimals. The values clearly show a very strong relationship between the originals and the estimates.

What happens if we choose a wrong number of components, while the location of pure variables is appropriate? For example, let us resolve two instead of three components from the dataset *A*. In this case the solution can still be found, however the resolved spectra may show some artifacts to cover total data variation as full as possible. As it can be seen in Fig. 8, which shows the results of unmixing of the dataset *A* by using only pure variables for fructose and lactose, the resolved spectrum for lactose has an unexpected peak at 542 cm^{-1} (the strongest peak for ribose spectrum) and some disturbance at 884 cm^{-1} (the second strongest peak for ribose). The resolved concentrations are also quite different from the reality in this case. Also, if one incorrectly

TABLE 1 Original and Estimated Concentrations for First Eight Records of Dataset *A*

Original concentrations			Estimated for set P1			Estimated for set P2		
C1	*C2*	*C3*	*C1*	*C2*	*C3*	*C1*	*C2*	*C3*
1.0	0.0	0.0	8500.5	539.0	984.6	7377.7	1673.3	973.1
0.8	0.2	0.0	6838.4	1551.8	840.7	5990.8	2310.5	929.6
0.6	0.4	0.0	5176.3	2564.5	696.8	4603.8	2947.7	886.2
0.4	0.6	0.0	3514.2	3577.3	552.9	3216.8	3584.9	842.7
0.2	0.8	0.0	1852.1	4590.1	409.0	1829.9	4222.1	799.2
0.0	1.0	0.0	190.0	5602.9	265.1	442.9	4859.3	755.8
0.8	0.0	0.2	6871.5	612.4	1728.8	6102.7	1466.2	1643.7
0.6	0.2	0.2	5209.4	1625.1	1584.9	4715.8	2103.4	1600.2

TABLE 2 Correlation Coefficients for the Original and Estimated Concentrations (Rounded to Two Decimals)

	C1	C2	C3
Estimated for set P1	1.00	1.00	0.99
Estimated for set P2	1.00	0.98	0.99

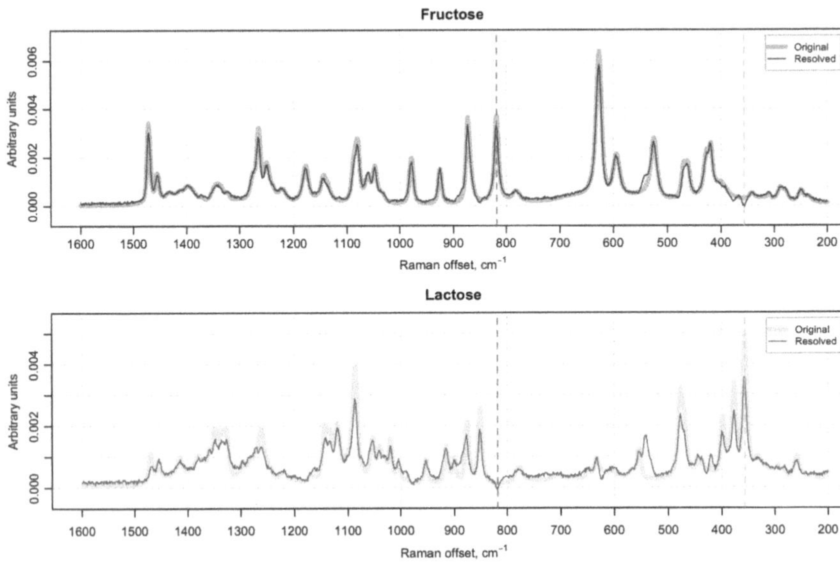

FIG. 8 Results of spectral unmixing of dataset *A* using two pure variables from the first set.

chooses more components than exist, a solution can be found only if noise is present in the data.

Dataset *B* presents a more challenging case, where the spectra have low selectivity and thus the signals of individual components are heavily overlapped. In this case, purity-based resolution of the component contributions is problematic and the solution may be essentially distorted or be entirely wrong. Thus if we apply the ordinary least squares to resolve the individual spectra for dataset *B* using given set of manually selected pure variables the solution will be pretty much unacceptable as shown in Fig. 9. This problem can be tackled by using a second derivative of spectra when one gets the initial estimates for concentrations, $\mathbf{D}_R$, instead of the original spectral data [9].

If peaks are heavily but not completely overlapped, the local maximum (where curve changes its slope from positive to negative) on one spectral curve will correspond to a relatively smooth slope on the other curve. It means

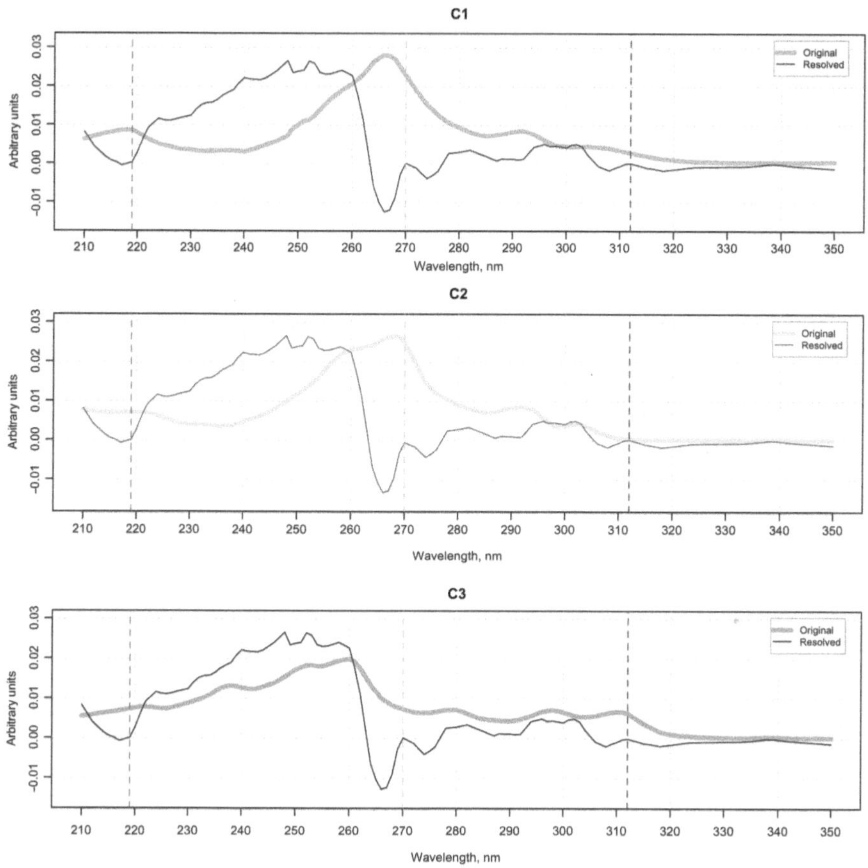

FIG. 9 Results of spectral unmixing of dataset *B* using the original data.

if we take a second derivative at this point it will have a negative value for the first curve (since peak is a local maximum) and, at the same time, have value around zero for the second. By multiplying the second derivative values to -1 and plotting the values as a line plot, we can reveal the partially overlapped peaks as extrema in positive part of the plot. It can be also useful to set negative values in the inverted derivative to zero prior to plotting as we are not interested in local minima of our spectra.

Fig. 10 illustrates this approach on the dataset *B*. The second derivative was calculated using Savitzky–Golay filter [10] with preliminary smoothing of the spectra by quadratic polynomial with filter width equal to five points. As we can see, the pure variables in the inverted second derivative plot indeed correspond to the parts of spectra where only one component influences the signal. Using the preprocessed data for estimated pure contributions, $\mathbf{D}_R$, and the original spectra as matrix $\mathbf{D}$, gives much better results as shown in Fig. 11.

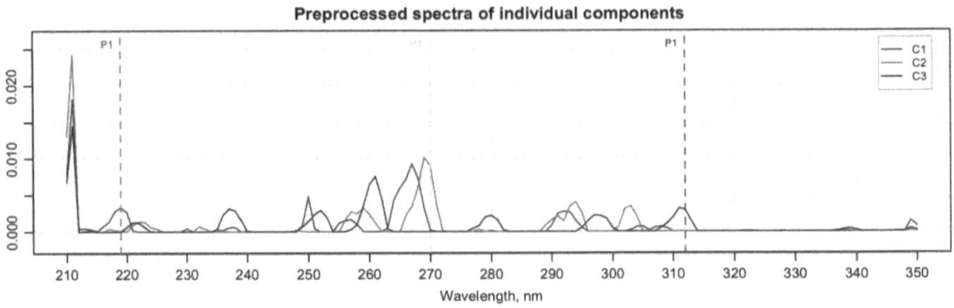

FIG. 10 Preprocessed spectra of individual components from dataset *B*.

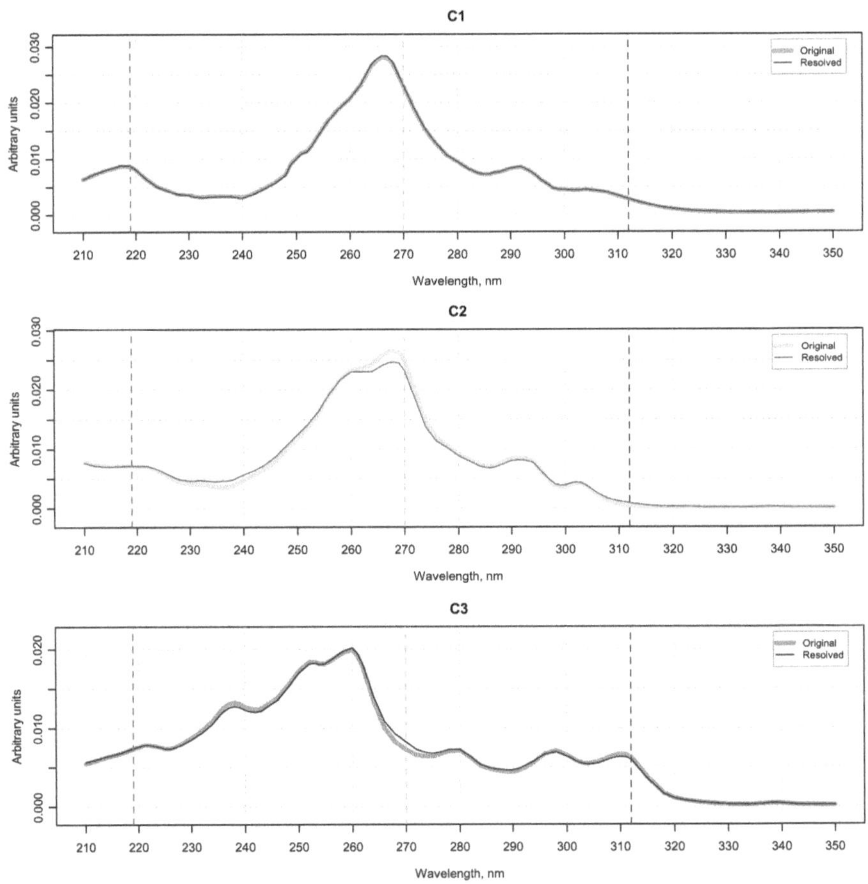

FIG. 11 Results of spectral unmixing of dataset *B* using the preprocessed data.

Summing up we can conclude that by knowing number and location of pure variables—spectra variables, which mostly experience influence from individual components, it is possible to obtain the pure spectra using simple ordinary least squares approach. If spectra of individual components are severely overlapped, this may cause significant challenges in finding proper solution, which, in some cases, can be overcome by using a second derivative of the spectra [9].

3.1 Nonnegativity Constraint

Unmixing data using the variable purity approach may sometimes result in negative intensities showing up in the resolved spectra. This may seem confusing, since the original data are all positive. The question is often asked why the algorithms like SIMPLISMA do not offer any nonnegativity constraint in order to improve the results of the unmixing.

The answer is directly related to the main reason why negative intensities may show up. This effect is quite common for the case when peak shifts occur in the mixture dataset. Another possibility is the widening/narrowing of peaks, which also causes a similar artifact in the results [9].

In order to demonstrate the effect of peak shifts, we have prepared a simulated data with two components as shown in the top plot of Fig. 12. The left peak in the data is a Gaussian curve with the standard deviation of five and mean equal to 25. The height of the peak is proportional to 100 random numbers uniformly distributed between zero and two. The second peak is another Gaussian curve with a standard deviation of five but with mean varied from 74.5 to 75.5. The variation of mean is directly proportional to the intensity of the first peak. In other words, the shift of the second peak depends on the intensity of the first peak. The height of the second peak corresponds to 100 random numbers distributed from 0 to 10. Besides that, a noise was added

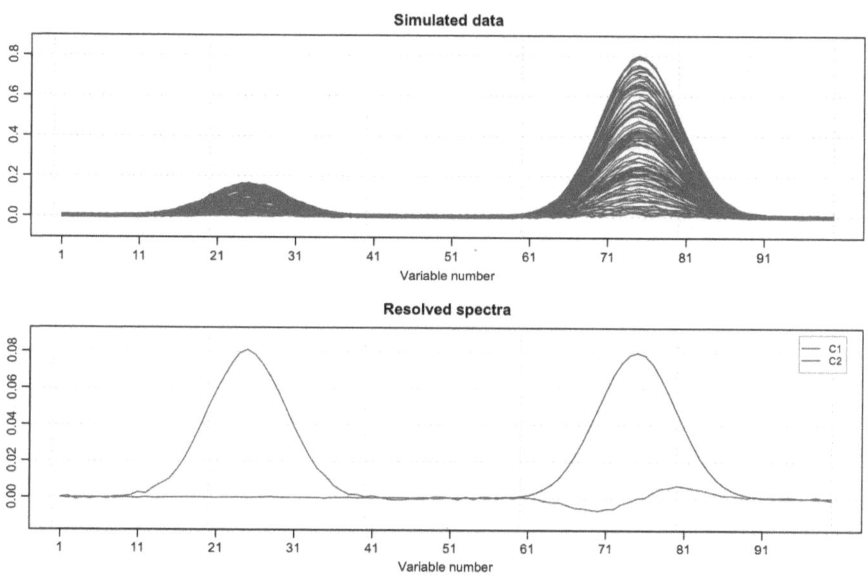

FIG. 12 Influence of a peak shift on the resolved spectra.

to each generated spectrum as uniformly distributed random values with a range of $\pm 1\%$ of the maximum intensity in the whole dataset.

Resolving this dataset with pure variables corresponding to 25 and 75 gives two pure components, as shown in a bottom plot of Fig. 12. A clear derivative type of pattern is present in the resolved component related to the first peak, due to the relation between the concentrations of the first peak and the peak shift of the second peak. This pattern is very distinctive, although the shift is not really noticeable in the original data due to the presence of noise.

When ignoring the negative part, it is easy to think that an extra peak is present. However, showing the negative values reveal a clear pattern that an experienced spectroscopist will quickly recognize as an artifact typical to a peak shift. Therefore, careful interpretation of the resolved spectra and residuals, including their negative parts, helps to avoid errors caused by incorrect determination of the number of components. Negative peaks in resolved spectra of a component can also be a compensation of an "overresolution" of another component that can be caused by a nonlinearity or either component response or by an overlap of their signals at the chosen pure variable.

Patterns of this type are also observed in actual data. For example, see the unmixed Raman spectra of a time resolved reaction [11] in Fig. 13. The pattern is clearly present in the third component (the range where the pattern appears is indicated on all plot by a gray semitransparent rectangle) and probably also in the second component. The width and location of the derivative pattern indicate that the shift is present in the fourth resolved component. As noted earlier, it is not easy to visually observe the shift due to noise in the original data.

4 CHASING THE PURE VARIABLES

4.1 Identifying the First Pure Variable

In this section, we will discuss how to find pure variables and correctly identify the number of individual components to be resolved, based on the method named as SIMPLISMA and first presented in Windig and Guilment [11]. As it was mentioned in Section 1, there are several other methods to define the pure variables. They will be briefly described and compared in the next section.

Let us start with some intuition using a very simple example. Spectral intensity of a mixture of two components, A and B, depends on two things—a concentration of the components in the mixture and an intensity of the individual spectra of the components at a certain wavelength. For a particular combination of concentration values i and for variable number j, this be formally expressed as follows:

$$D_{ij} = C_{iA}S_{Aj} + C_{iB}S_{Bj} \tag{11}$$

Which is a particular case of more general Eq. (1) in the case when noise is absent. Apparently, if a spectral variable does not contain any signal of

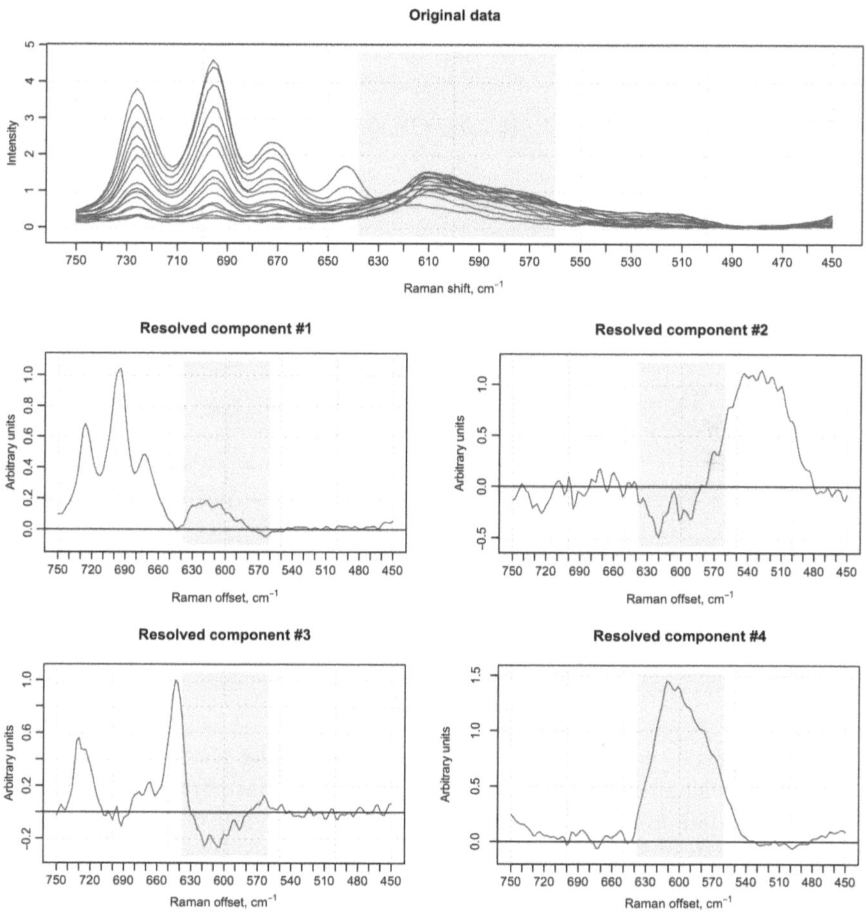

FIG. 13 Possible shift of a peak in a real Raman spectra and its influence on the resolved spectra.

either component, the intensity values will be quite small and vary very little regardless the variation of the concentrations of the components. But what if we consider two variables (i.e., wavelength), n and m, so n experiences a contribution from both components and m contains only response of component A (so spectrum of B at the waveband m is almost 0)? Let us try to calculate how the spectral intensities will vary in this case. We start with the mean, assuming we have f mixtures with different concentrations of the components:

$$\mu_j = \frac{1}{f}\sum_{i=1}^{f} D_{ij} = S_{Aj}\frac{1}{f}\sum_{i=1}^{f} C_{iA} + S_{Bj}\frac{1}{f}\sum_{i=1}^{f} C_{iB} = S_{Aj}\mu(C_A) + S_{Bj}\mu(C_B) \qquad (12)$$

Here $\mu(C_A)$ and $\mu(C_B)$ are average concentration values of the components in the mixtures. Now we can also calculate a variance for the values:

$$\begin{aligned}\sigma_j^2 &= \frac{1}{f}\sum_{i=1}^{f}\left(D_{ij}-\mu_j\right)^2 = \frac{1}{f}\sum_{i=1}^{f}\left(C_{iA}S_{Aj}+C_{iB}S_{Bj}-S_{Aj}\mu(C_A)-S_{Bj}\mu(C_B)\right)^2 \\ &= \frac{1}{f}\sum_{i=1}^{f}\left(S_{Aj}(C_{iA}-\mu(C_A))+S_{Bj}(C_{iB}-\mu(C_B))\right)^2 \end{aligned} \tag{13}$$

For the sake of simplicity let us assume that the concentration values are relative and sum up to one (so mixture is closed), which means the average value lies between 0 and 1. Then the variance can be calculated as:

$$\begin{aligned}\sigma_j^2 &= \frac{1}{f}\sum_{i=1}^{f}\left(S_{Aj}(C_{iA}-\mu(C_A))-S_{Bj}(C_{iA}-\mu(C_B))\right)^2 \\ &= \left(S_{Aj}-S_{Bj}\right)^2\frac{1}{f}\sum_{i=1}^{f}(C_{iA}-\mu(C_A))^2 = \left(S_{Aj}-S_{Bj}\right)^2\sigma^2(C_A)\end{aligned} \tag{14}$$

which lead us to the following expression for mean and standard deviation of variable j:

$$\mu_j = S_{Aj}\mu(C_A)+S_{Bj}(1-\mu(C_A)) \tag{15}$$

$$\sigma_j = \left|S_{Aj}-S_{Bj}\right|\sigma(C_A) \tag{16}$$

Both statistics depend on the difference between the spectral intensities at a given wavelength, j, as well as on distribution of concentration values. The standard deviation is generally higher when pure spectral intensity of one of the components at a specific wavelength is significantly larger than the intensity of the another component. However, the standard deviation does not reflect the individual contribution of each spectrum, e.g., intensities $S_{Aj}=0.5$ $S_{Bj}=0$ and $S_{Aj}=1.0$ $S_{Bj}=0.5$ will give the same standard deviation of the variable j, even though in the first case we clearly have a pure variable.

The mean value is in fact a weighted mean for the two intensities, which takes into account a relative concentration of the components. Thus if the component A has lower concentration values in the mixture (<0.5) the weight for S_{Aj} will be smaller and vice versa.

Since the mean reflects the individual pure spectral intensities, we can use it as a scale factor for the standard deviation, trying to reveal the cases, where relatively low intensity values of pure component spectra are giving large variation in the spectra of mixtures.

Let us define a statistic p_j, which we will call a *purity* of variable j, as follows:

$$p_j = \frac{\sigma_j}{\mu_j} = \frac{\left|S_{Aj}-S_{Bj}\right|\sigma(C_A)}{S_{Aj}\mu(C_A)+S_{Bj}(1-\mu(C_A))} \tag{17}$$

and investigate its behavior for a different combinations of pure intensity values and average concentration of the components.

Fig. 14 shows three colorized contour plots, where color gradient represents p values depending on combinations of pure spectral intensities (each varies from 0 to 5) at three average concentration levels: $C_A=0.2$ ($C_B=0.8$), $C_A=0.8$ ($C_B=0.2$), and $C_A=0.5$ ($C_B=0.5$) at the unit standard deviation. As one can clearly see, in all plots the values for p are increasing when difference between two intensities grows and the purity approaches its maximum when one of the intensities gets close to zero.

Fig. 15 further illustrates the idea behind the purity using simulated spectra. The plots on the left side show the spectra of pure components (solid gray curves) as well as spectra of 100 mixtures (light gray shaded areas). The concentration of C_A in the mixtures varied randomly: for the top plot from 0 to 0.3, for the middle plot from 0.7 to 1, and for the bottom plot from 0 to 1. The concentration of C_B for each case was selected to have a sum of the concentrations equal to one. The dashed lines show four selected wavenumbers, denoted as p_1–p_4. The variable p_1 corresponds to a pure variable for the component A. Variable p_2 and p_3 correspond to variables, where one of the components dominates over another. The variable p_4 corresponds to the part of the spectra where both components have a very low intensity. The right part of Fig. 15 shows box and whiskers plots for the selected variables, where the axis labels contain the variable name as well as the calculated purity value for each of them. In all cases the p_1 is an obvious winner.

The purity concept can also be represented from the geometrical point of view, as shown in Windig and Guilment [11]. If we consider variables from the matrix **D** as points in the objects space (objects here are rows of the matrix **D** corresponding to different concentration levels in the mixtures), then we can define a squared length of each variable by:

$$\lambda_j^2 = \frac{1}{f}\sum_{i=1}^{f} D_{ij}^2 \tag{18}$$

On the other hand, the variable variance can be calculated as:

$$\sigma_j^2 = \frac{1}{f}\left(\sum_{i=1}^{f} D_{ij}^2 - \frac{1}{f}\left(\sum_{i=1}^{f} D_{ij}\right)^2\right) = \frac{1}{f}\sum_{i=1}^{f} D_{ij}^2 - \left(\frac{\sum_{i=1}^{f} D_{ij}}{f}\right)^2 = \lambda_j^2 - \mu_j^2 \tag{19}$$

Now we can rewrite expression (19) as:

$$\lambda_j^2 = \sigma_j^2 + \mu_j^2 \tag{20}$$

It seems like the λ, μ, and σ are accordingly a hypotenuse and two catheti of a right triangle. And in this case the purity value p, defined in (17) is a tangent of angle between λ and σ. Why do the pure variables have the largest

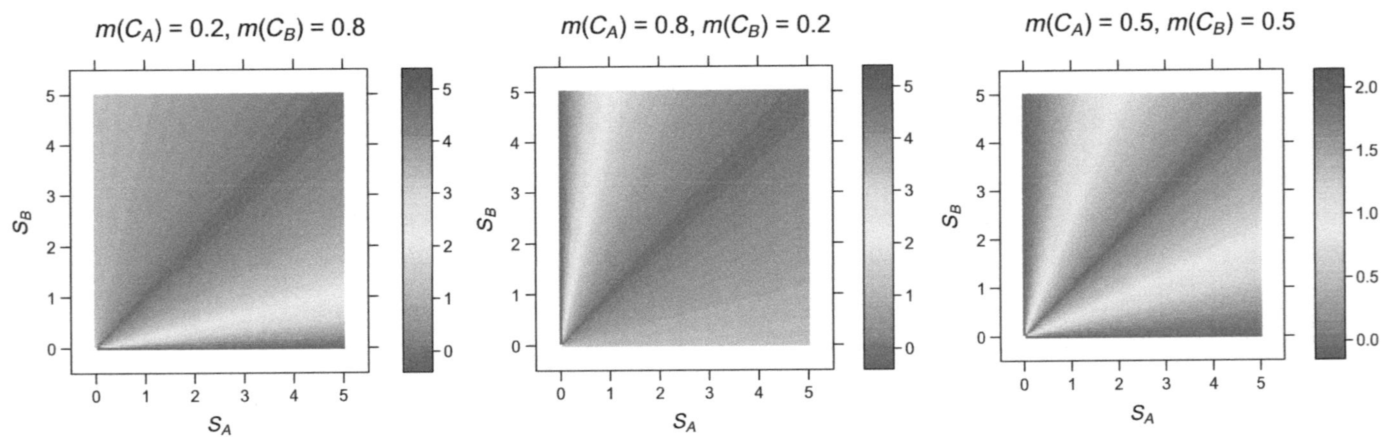

FIG. 14 Illustration of the purity values depending on relationship between spectral intensities for a given waveband at different average concentration of pure components.

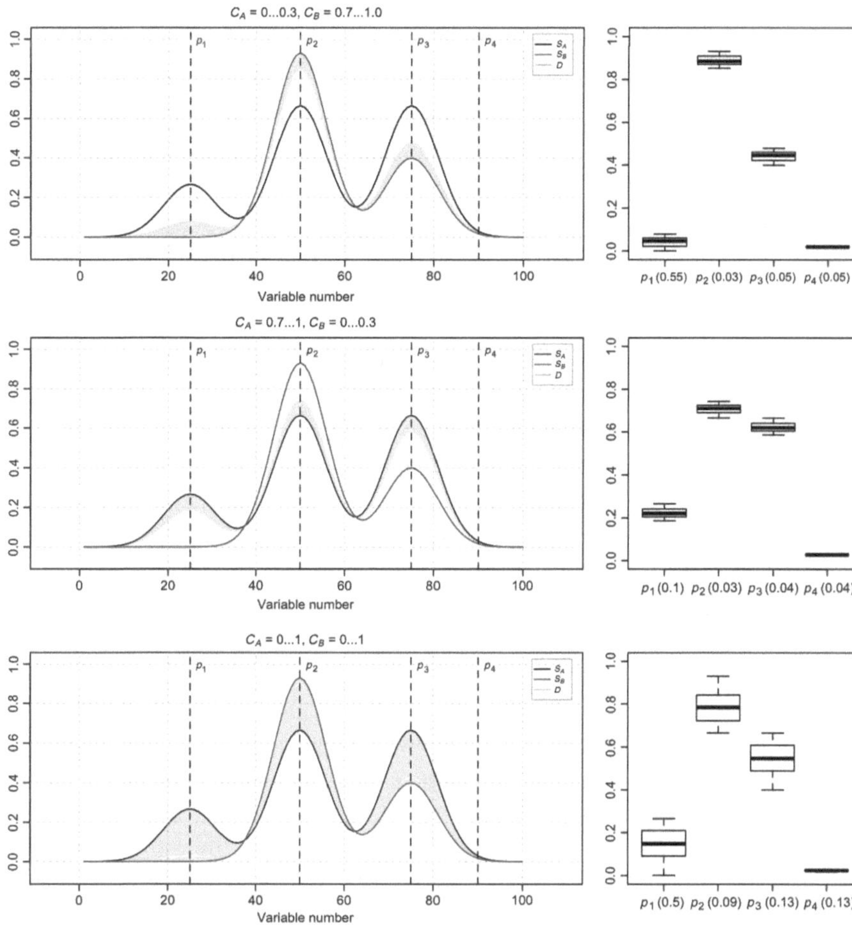

FIG. 15 Illustration of the purity approach—mean and variation of spectral intensity values.

angle? This can be illustrated by using a simple case with two concentration levels. Let us take the same pure spectra of components *A* and *B* that are shown on plots in Fig. 15 and generate a dataset with concentrations [1, 0] and [0, 1] for each of the components, so that we have two mixtures only. Since the matrix **D** in this case has only two rows, we can illustrate the variables from **D** as points in two-dimensional space as illustrated on the left plot in Fig. 16. The gray points are the variables, whereas the colored points are the selected variable numbers, denoted in Fig. 14 as p_1–p_4.

The dashed line with the unit slope corresponds to the case when concentrations of both components are equal, so if we project a variable to this line we will get a mean value, μ as illustrated for the variable p_2 (green color). The distance between this line and the variable is the standard deviation, σ.

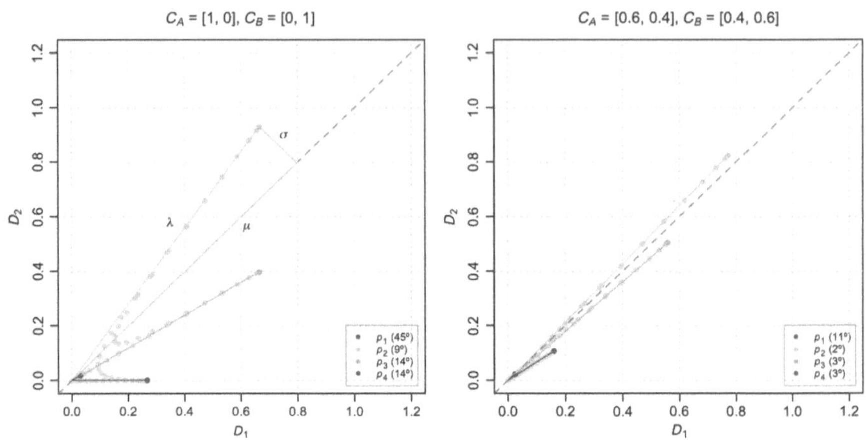

FIG. 16 Geometrical interpretation of the purity concept.

Because the coordinate axes in this case correspond to the concentrations of pure components, a pure variable should be located very close to the only one of the axes, and therefore having large angle between the direction of the variable and the dashed line, exactly as the variable p_1 looks like. In the plot legend the calculated angle (in degrees) is shown for each of the selected variables, and the p_1 has the expected value of 45 degree.

In the right par of Fig. 16 a similar plot was made for the data, where each component varied only from 0.4 to 0.6. Apparently in this case, the variables are oriented much closer to the dashed line, due to a lack of variance, however the p_1 has the largest angle anyway.

What happens if the mean value for a specific spectral variable is very close to zero and standard deviation is not? This may occur, e.g., if we observe a part of a spectrum without peaks but with presence of a random noise. Apparently in this case the purity value will become infinitely huge. To avoid this problem, Eq. (17) should be modified by adding a fixed value, α, to the denominator as suggested by Windig and Guilment [11]:

$$p_j = \frac{\sigma_j}{\mu_j + \alpha} \tag{21}$$

The value should be small enough to avoid any competition with the standard deviation for determining the purity of informative data regions, but, at the same time, not too small to avoid any overestimation of the purity value. To achieve this on practice, the value of α can be taken as 1–5% of the maximum mean value among the variables.

The vector p_j, calculated for all spectral variables is called a *purity spectrum* and can be plotted in the same way as a conventional spectral plot. But before we investigate this option, let us consider how to find the further pure variables starting from the second one.

4.2 Identifying Second and Further Pure Variables

How to find the second and all subsequent pure variables? Since we have defined the variable purity as experiencing a contribution of a single chemical component and thus it is not influenced by other components, then the pure variables should be mutually orthogonal (or close to that) in order to be resolved. In other words, the correlation between them is close to zero.

We can use the knowledge to decrease the purity of the variables, which correlate with the ones already found as pure. This can be done by weighting the purity spectrum according to the correlation values. Thus, for calculating the purity spectrum for the second pure variable, we should use weights, which will reflect correlation between each spectral variable and the first pure variable. If the correlation is high, the weight should be close to zero, and it should approach one if the correlation is small. For the third purity spectrum, the weight should take into account the correlation among a candidate variable and two variables previously identified as pure, and so on. If spectral peaks are seriously overlapped an inversion of second derivative with all negative values set to zero can be used instead of original spectral data as it was illustrated in the previous section.

In order to implement this approach, we can utilize the variance/covariance matrix. But in this case the covariance will also depends on the variable vector length. To tackle this, every variable in the data matrix should be first scaled to the unit length:

$$D'_{ij} = D_{ij}/\lambda_j \tag{22}$$

or, taking Eq. (20) into account:

$$D'_{ij} = \frac{D_{ij}}{\sqrt{\mu_j^2 + \sigma_j^2}} \tag{23}$$

However, if length of a variable is close to zero the purity function can experience the same issue as before—the silent variables will get enormously large values after scaling. In order to avoid this, the same approach as for the purity value calculation (21) is used:

$$D'_{ij} = \frac{D_{ij}}{\sqrt{\mu_j^2 + (\sigma_j + \alpha)^2}} \tag{24}$$

The variance–covariance matrix can now be calculated as:

$$\mathbf{R} = \frac{1}{f}\mathbf{D}'(\mathbf{D}')^T \tag{25}$$

Since the data values are scaled to the unit length, at $\alpha = 0$ all diagonal elements of **R** will be equal to one and all other values will reflect the correlation between each pair of the variables from **D**. The matrix **R** is apparently symmetric. In order to calculate the weight for the second purity spectrum, $p_{j,2}$, the following determinant should be taken:

$$\omega_{j,2} = \begin{vmatrix} R_{jj} & R_{jp_1} \\ R_{p_1j} & R_{p_1p_1} \end{vmatrix} \tag{26}$$

Here p_1 is an index of the first pure variable, identified as a maximum of the original purity spectrum, p_j. In case if parameter α is zero this can be rewritten as:

$$\omega_{j,2} = \begin{vmatrix} 1 & R_{jp_1} \\ R_{jp_1} & 1 \end{vmatrix} = 1 - R_{jp_1}^2 \tag{27}$$

The $R_{jp_1}^2$ is a squared correlation between jth and p_1st columns of the matrix **D**. So if correlation is small the weight will be close to one and if it is large, it will approach zero, exactly as needed. Now the second purity spectrum can be found as:

$$p_{j,2} = \frac{\sigma_j^2}{\mu_j^2 + \alpha}\omega_{j,2} \tag{28}$$

By utilizing the same idea, weights for the third purity spectrum can be found as:

$$\omega_{j,3} = \begin{vmatrix} R_{jj} & R_{jp_1} & R_{jp_2} \\ R_{p_1j} & R_{p_1p_1} & R_{p_1p_2} \\ R_{p_2j} & R_{p_2p_1} & R_{p_2p_2} \end{vmatrix} \tag{29}$$

Again, in case of zero alpha and assuming a zero correlation between the first two pure variables gives:

$$\omega_{j,3} = \begin{vmatrix} 1 & R_{jp_1} & R_{jp_2} \\ R_{jp_1} & 1 & 0 \\ R_{jp_2} & 0 & 1 \end{vmatrix} = 1 - R_{jp_1}^2 - R_{jp_2}^2 \tag{30}$$

So it leads us to a general expression for a mth pure variable.

$$\omega_{j,m} = \begin{vmatrix} R_{jj} & R_{jp_1} & \cdots & R_{jp_{m-1}} \\ R_{p_1j} & R_{p_1p_1} & \cdots & R_{p_1p_{m-1}} \\ \cdots & \cdots & \cdots & \cdots \\ R_{p_{m-1}j} & R_{m-1p_1} & \cdots & R_{p_{m-1}p_{m-1}} \end{vmatrix} \tag{31}$$

Finally, we also have to define the $\omega_{j,m}$ value for the first pure spectrum, when $m=1$. Obviously when no pure variables are selected the weight will be simply equal to R_{jj}, which, taking into account the expressions (24) and (25), can be written as:

$$\omega_{j,1} = R_{jj} = \frac{1}{f}\sum_{i=1}^{f}\left(D'_{ij}\right)^2 = \frac{1}{f}\sum_{i=1}^{f}\frac{D_{ij}^2}{\mu_j^2 + \left(\sigma_j + \alpha_j\right)^2} = \frac{\lambda_j^2}{\mu_j^2 + \left(\sigma_j + \alpha_j\right)^2} \tag{32}$$

which leads us to the general expression for a purity:

$$p_{j,m} = \frac{\sigma_j}{\mu_j + \alpha}\omega_{j,m} = \frac{\sigma_j \omega_{j,m}}{\mu_j + \alpha} = \frac{s_{j,m}}{\mu_j + \alpha} \tag{33}$$

where $s_{j,m}$ is a weighted standard deviation (also named as *standard deviation spectrum* in Windig and Guilment [11]) for a purity variable m. The purity variables now can be chosen simply by finding a maximum of each purity spectrum $p_{j,m}$. However, the best option in this case would be to carry out an investigation of both the purity values and other statistics followed by tuning the parameters of algorithm prior to the final decision.

5 INVESTIGATION OF PURITY CHARACTERISTICS

In this section, we will discuss how to use the approach described earlier and how the unmixing results it gives can be additionally improved. We will start with the simulated datasets, *A* and *B*, but at the end of this section we will also show a real data example.

Fig. 17 illustrates the purity spectra as well as the weighted standard deviations calculated for the mixtures from dataset *A* using 5% offset. The values are plotted in a spectral form for the first four pure variables in a form of a spectrum. For ideal spectra of mixtures, it would not be possible to extract more pure components than their actual number due to a computational problem of taking an inverse matrix in the least squares algorithm. This can be done, however, for spectral data containing noise, which is ubiquitous in the real measurements.

It can be noticed that for each next pure component being extracted the magnitude of both purity and weighted standard deviation is decreasing. So does the number of local maxima in the purity spectrum, which is quite reasonable—the less components are left the easier to find the next one. However, is there any guidance, when to stop with the component resolution?

There are several ways to select a proper number of pure components. First of all, a simple visual investigation of the standard deviation and purity spectra plots for different components can give a suggestion, as they look very noisy when the unmixing model becomes overfitted. This can be clearly seen on the plot for component C4 in Fig. 17.

Another way to check the number of components is calculation of correlation between the original pure variables and the calculated contributions. The correlation values often have clear maximum for the proper number of components.

The third way to tackle this issue is to investigate the residuals—difference between the real data values, $\mathbf{D}$, and the reconstructed values, $\hat{\mathbf{D}}$,

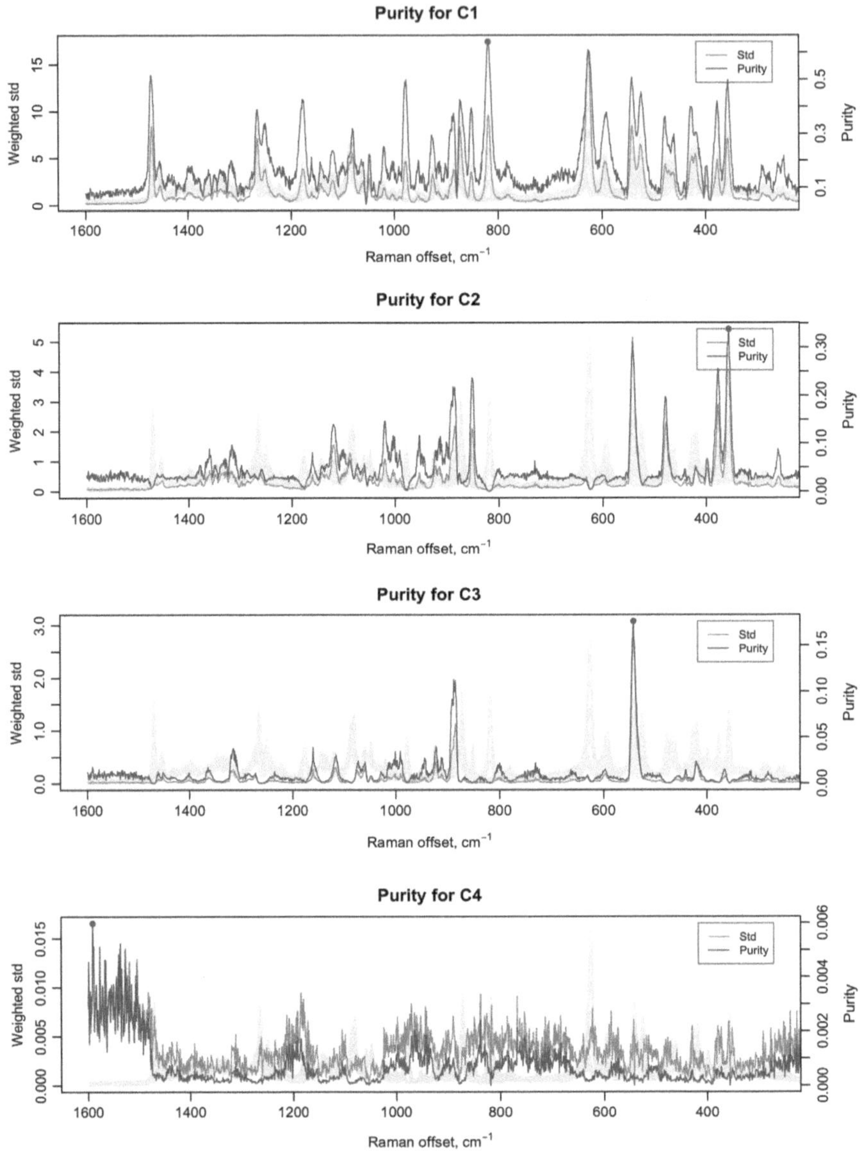

FIG. 17 Plot with purity statistics for the dataset *A* (calculated for four components).

calculated by multiplication of the matrix with pure contributions to the matrix with pure spectra, both evaluated for a certain amount of pure variables, as defined in (3) and (4). There are several ways to do that. One of the simplest way is to calculate the ratio between the residual variance and the total variance of the spectral data, similar to residual variance calculation in principal component analysis (PCA):

$$s_{res}^2 = \frac{\sum_i \sum_j \left(D_{ij} - \hat{D}_{ij}\right)^2}{\sum_i \sum_j D_{ij}^2} \tag{34}$$

The closer are the reconstructed and original data values, the closer will the residual variance be to zero. Another alternative is to use a relative root of sum of square differences [11], also known as a "lack of fit," which is a square root of the residual variance, s_{res}^2.

To identify the proper number of components one has to look at the residual variance values depending on the number of selected components as shown in Fig. 18. As one can see, if we use only one component (the first pure variable), almost 40% of the total variance in spectral data remain uncaptured. With two components this ratio reduced to 14% and with three components only 0.5% stay unexplained, which assumes that the fourth component is obsolete.

The next point is to investigate how the purity spectrum depends on the offset parameter α. Fig. 19 shows the purity and weighted standard deviation spectra for the first two components calculated without offset (top), with an offset equal to 5% of maximum average value (middle) and with an offset equal to 10% of the maximum average value (bottom). As one can see, when purity is calculated without any offset, the silent but noisy parts of the spectra have relatively large purity values (e.g., the right tail of the spectra with wavenumbers from 1500 cm^{-1} and above). Introducing the offset allows to decrease the influence of noise significantly. However, this may also change the relationship among purity values for the spectral peaks. Thus having 5%

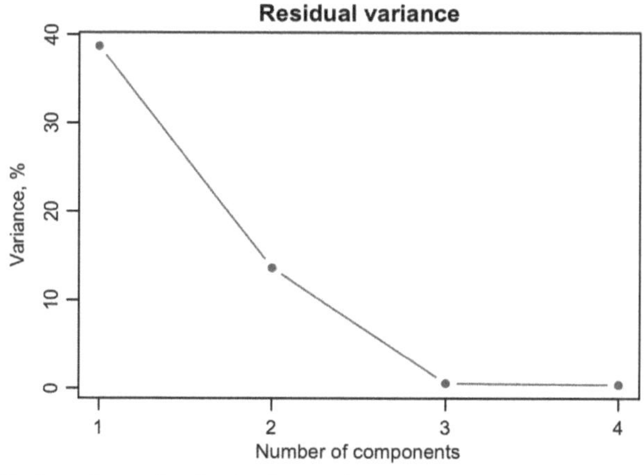

FIG. 18 Residuals variance plot for the dataset *A*.

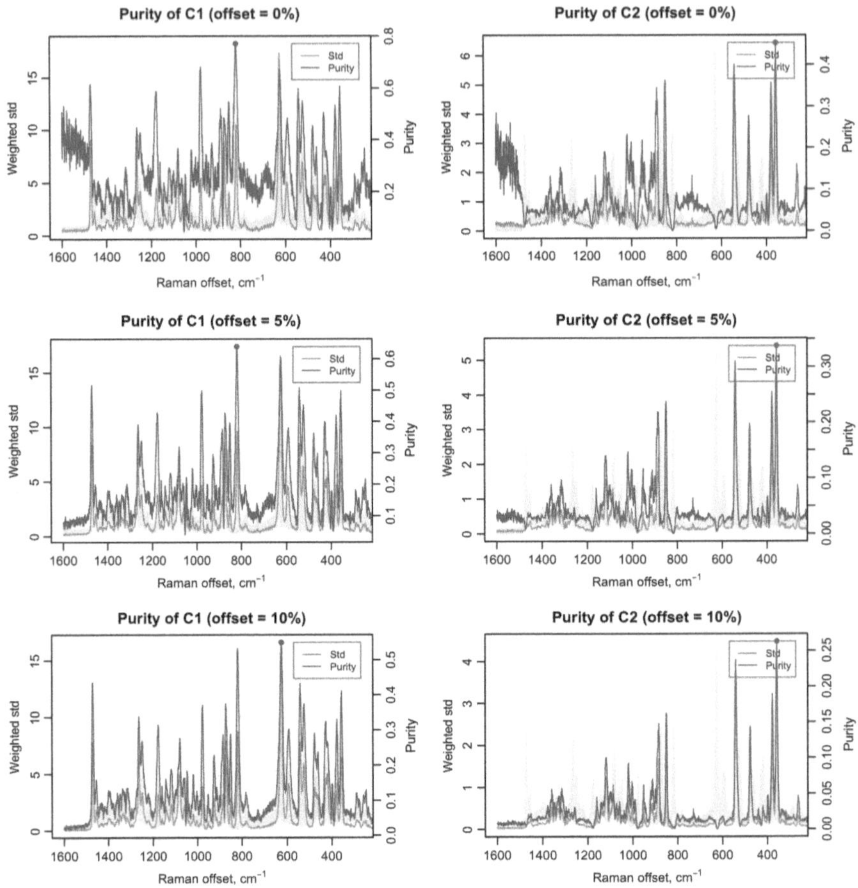

FIG. 19 Purity spectra for dataset *A* depending on an offset value.

offset keeps the ratio among the main candidates for the first pure variable, while increasing it to 10% leads to a new winner for C1 (the selected variable is shown with a red dot on the purity spectrum plot). It must be noted that when the offset is too high the detected pure variables may become less pure.

In most cases, having offset around 5% of the largest mean value allows to achieve satisfactory results. However, do not trust blindly the recommended settings, it is always a good idea to investigate the plots at varying the parameters before the final decision is made. As the method is computationally very fast, it does not take a lot of time and actually can be even done interactively, using an appropriate graphical user interface tool. Such tools exist in most of the software packages where the purity-based curve resolution is implemented. The software usually allows changing various parameters and immediately seen the results almost immediately.

Now let us compare the SIMPLISMA results for original spectra with those obtained by manual selection of pure variables. Fig. 20 shows the resolved spectra for the dataset *A*. Plots on the left and right sides of the figure contain the resolved spectra obtained using 5% and 10% offset, respectively. The location of pure variables found with 5% offset was almost identical to the ones, selected manually as a set P1 (818, 358, and 542 cm^{-1} vs 818, 357, and 542 cm^{-1}), while using 10% offset led to a different choice of the first pure variable (625 cm^{-1}), which slightly changed the unmixing results. The value in parenthesis in the title of each plot is a relative residual variance calculated by taking the ratio of sum of squared differences between the resolved and the original spectrum to the sum of squared values for the original spectrum. Thus, for the first two components the results are pretty much

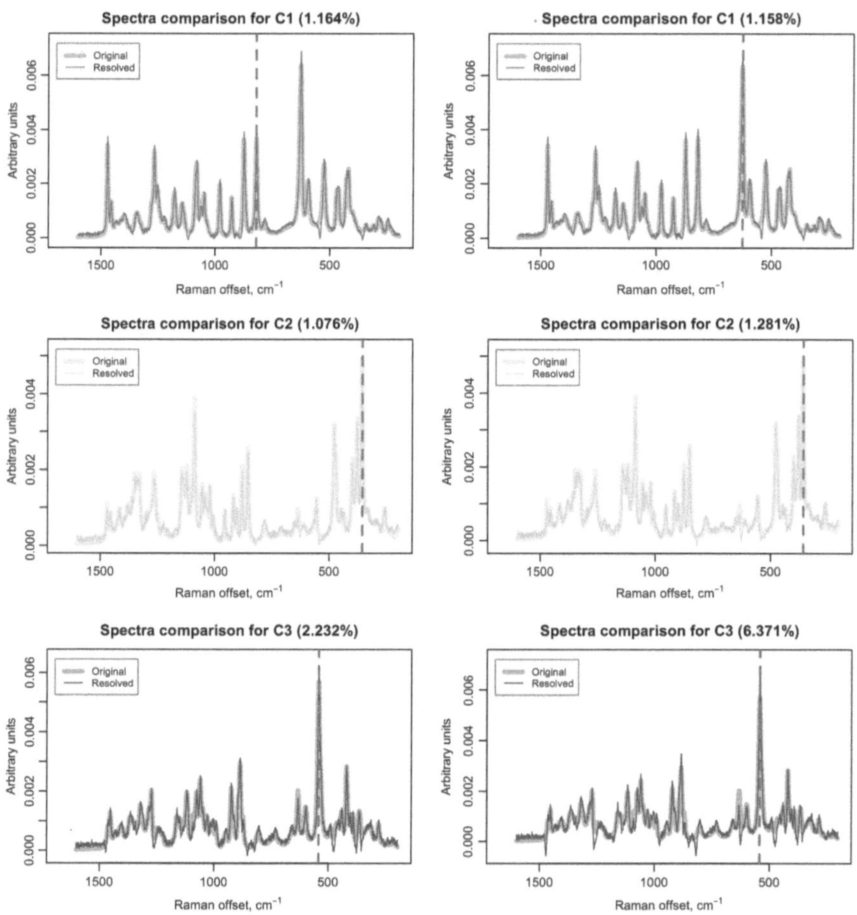

FIG. 20 Location of pure variables and results of unmixing of spectra from the dataset *A* (left for offset = 5%, right for offset = 10%).

similar, while for the third component the result obtained using pure variables estimated with 10% offset is obviously worse.

The unmixing results for the dataset *B* are shown in Fig. 21. As expected the results are not as good as for the dataset *A* and need a better tuning of parameters. The left plots show the original and resolved spectra obtained with pure variables detected with 5% offset and second derivative for the unmixing step. The detected pure variables (238, 268, and 310 nm) for the last two components were quite close to the ones selected manually (219, 270, and 312 nm). However, the difference in position of the first pure variable led to deterioration in the unmixing quality. The solution was improved by using second derivative also for the detection of pure variables with a smaller offset (3%) as shown on the right part of Fig. 21.

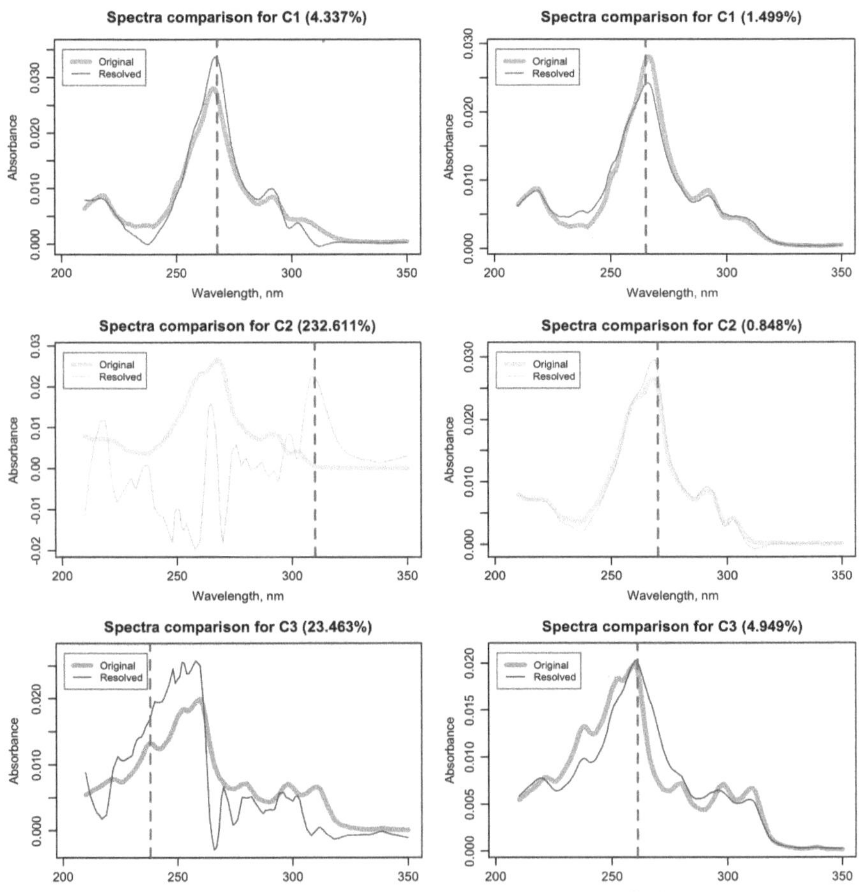

FIG. 21 Location of pure variables and results of unmixing of spectra from the dataset *B* (left for offset=5%, right for offset=3%, and second derivative for identification of the pure variables).

Fig. 22 shows the results of unmixing for the dataset *C* in a form of purity plots and the resolved concentration profiles. In this case, we have an advantage of knowing the concentration profiles of pure components, since spectra were acquired for designed concentration levels. In reality this is rarely known however in many cases, such as reaction kinetics or process monitoring, there are some qualitative expectations about the concentrations' behavior and this knowledge can be used for an assessment of unmixing quality or to evaluate the proper number of pure components.

The left plots in Fig. 22 show the results obtained using default settings (offset of 5% and no derivative). The selection of the first (the leftmost) spectral variable as a second pure variable leads to switching of components 1 and 2 as well as to a very poor resolution of one of the components. The use of the

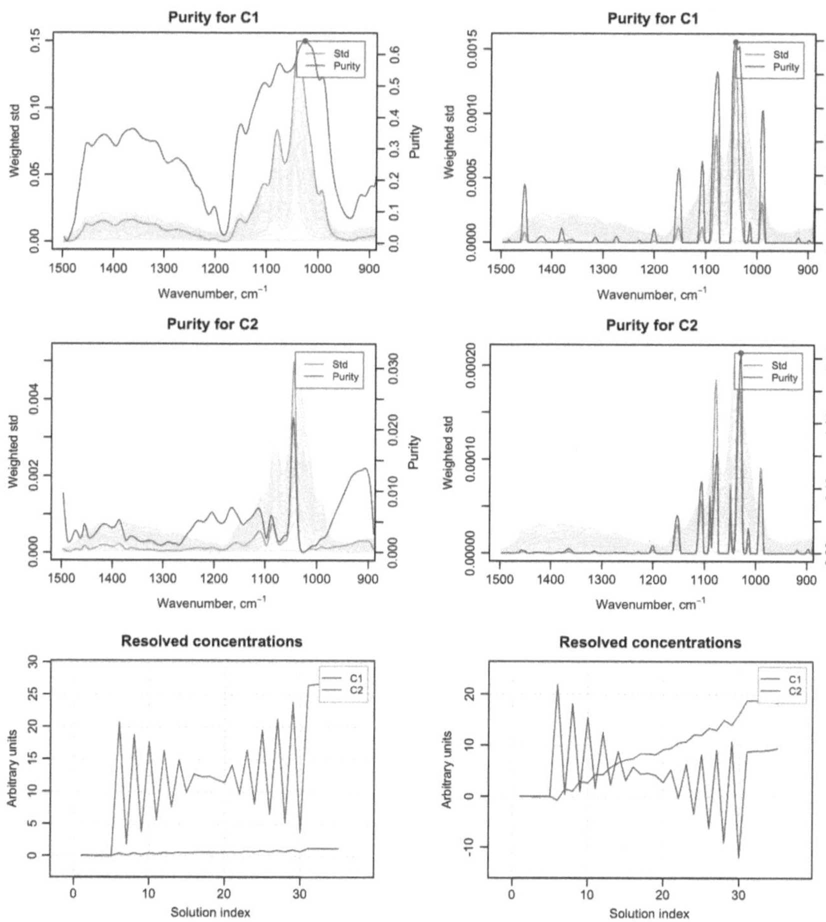

FIG. 22 Purity spectra and concentration profiles for dataset *C* calculated using default settings (left) and second derivative (right).

second derivative for locating pure variables improves the results as shown on the right part of the figure, however, the second component concentration profile still has a distinct artificial trend.

For this dataset, we can also use a trick with transposing data matrix and carry out the unmixing for concentration profiles instead of the spectra. In this case, we will select pure spectra instead of pure variables and the solution will give resolved contributions, which are as linearly independent as possible (in contrast with conventional approach when solution gives linearly independent spectra). Mathematically, it changes nothing one just has to remember that the obtained matrices with resolved spectra and resolved contributions have to be swapped. And since the concentration profiles in this data have clear regions where the component concentration is about zero it may improve the results significantly. Fig. 23 shows the purity spectra, resolved concentration profiles, and resolved spectra for this case obtained using standard settings (offset of 5% and no derivative). As one can see the, pure spectra were identified correctly and it leads to a very decent unmixing quality, all characteristic peaks are well observed in the resolved spectra.

Resolution of hyperspectral images provides another possibility for checking the results of unmixing—visualization of the resolved contribution values for pure components via concentration maps. Fig. 24 shows the residual variance for unmixing of the spectral data from dataset *D*, clearly indicating that three components should be used. Fig. 25 represents the unmixed spectra and the concentration maps for the components. The spectra of pure components correspond very closely to the spectra of structural phase (C2) and two droplet phases (C1 and C3) [4]. The concentration maps show clearly the patterns that correspond to each of the phases.

6 OTHER WAYS TO FIND THE PURE VARIABLES

The SIMPLSMA is neither the only nor the first method utilizing the pure variable approach. One of the first paper devoted to this problem was presented by Lawton and Sylvestre [12] and was based on PCA of the spectral data.

For demonstration purposes we will use four simple simulated datasets—mixtures of two components, *A* and *B*, where a pure spectrum of each component is a Gaussian curve with different variance and mean values. The concentration of component *A* (C_A) varied from 0.1 to 0.9 with a step of 0.1 resulting in nine concentration levels. The concentration of component *B*, C_B, was calculated as $C_B = 1 - C_A$.

The pure component spectra and the simulated data for all four cases are shown in Fig. 26. Five equidistant variables, p_1–p_5, were selected along the spectra for assessing their purity (not as defined in the previous section, but in a wider sense, as a measure of how small is the contribution from more than one individual component for each of the selected variables). Case 1 corresponds to a partially overlapped peaks, so the purity of each variable

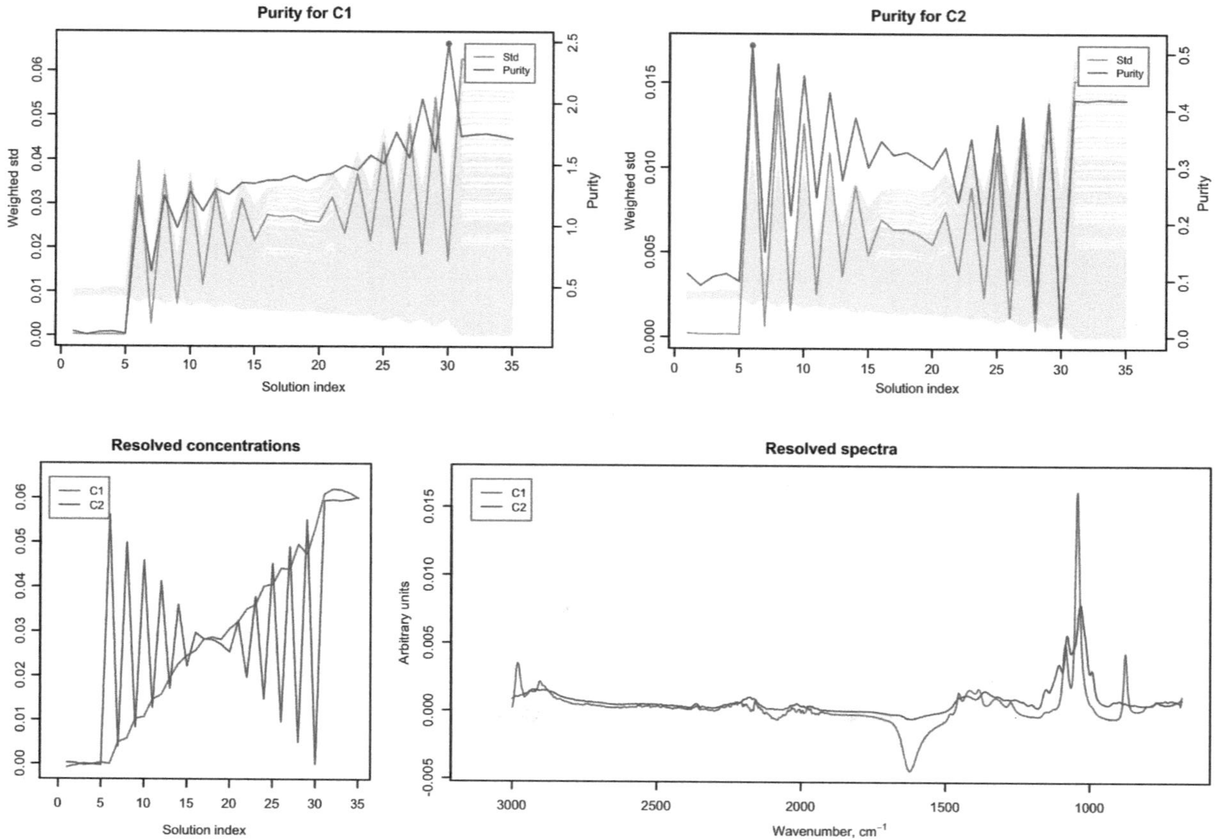

FIG. 23 Purity spectra, resolved concentration profiles, and resolved spectra for dataset *C* obtained for unmixing concentrations instead of spectra.

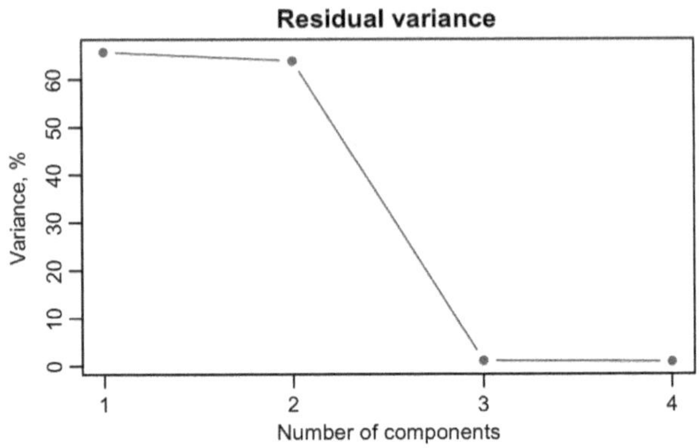

FIG. 24 The residual variance plot for dataset D.

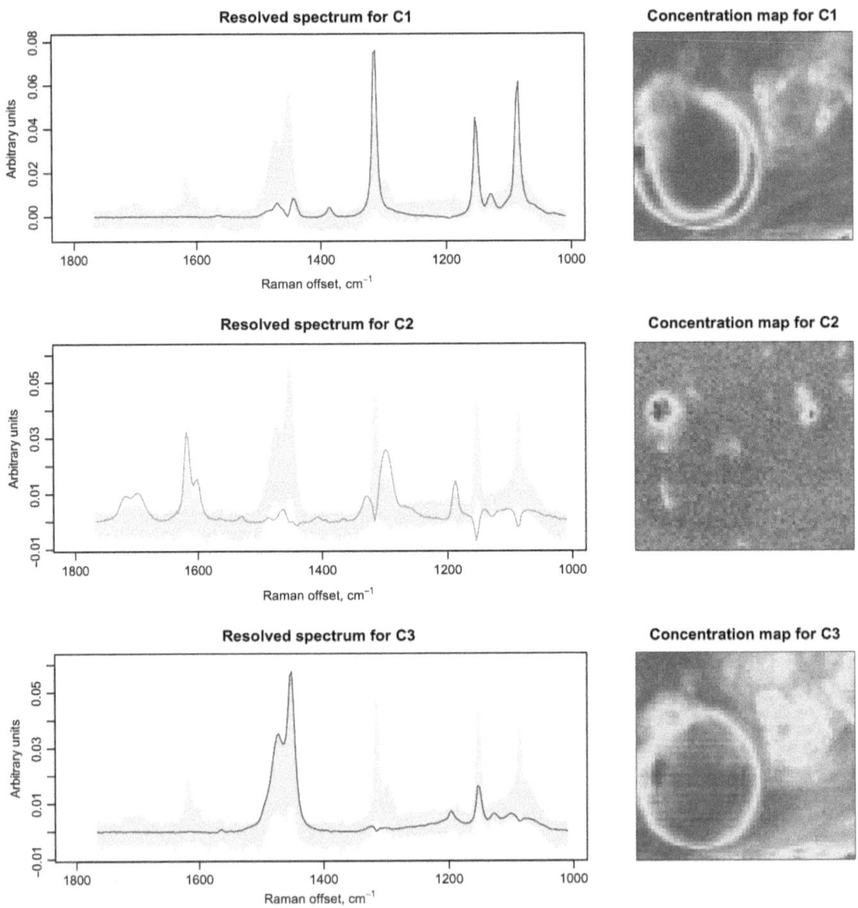

FIG. 25 Resolved spectra and concentration maps for dataset D.

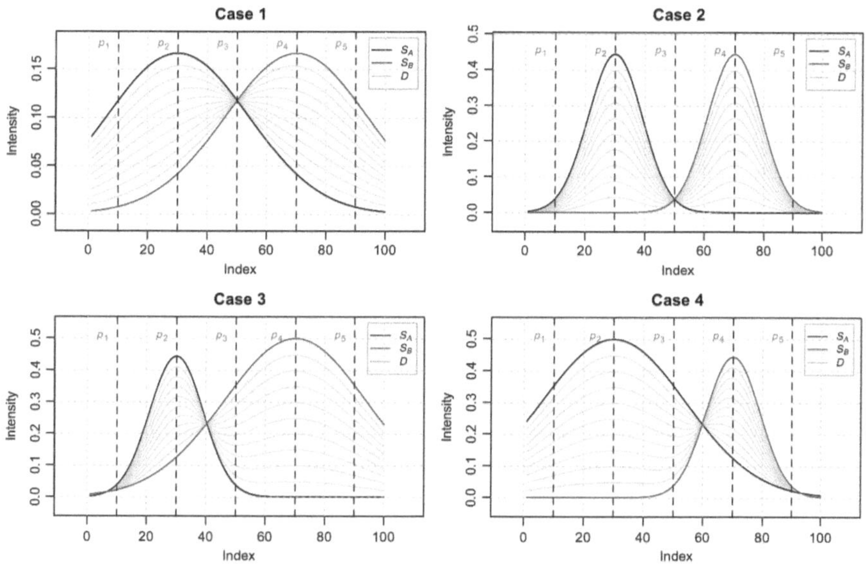

FIG. 26 Four simulated datasets for demonstration of Lawton and Sylvestre approach.

depends on the distance from the variable to the center, where spectra of pure components intersect. In case 2, peaks are clearly separated and most of the variables in the left (including p_1 and p_2) and right (including p_4 and p_5) parts of the spectra are pure. Cases 3 and 4 represent situations with overlapping peaks for one component and separated for another, so the pure variables are located only in one part of the spectra (right for case 3 and left for case 4), while the purity of the other variables vary depending on how far they are from the point of intersection of the individual components spectra.

Fig. 27 shows four PCA biplots (where scores and loadings are plotted together) for the first two components. Each column of the data matrix was scaled to the unit length and all values were multiplied to the square root of the number of rows prior to the PCA decomposition, in order to bring the scores and loadings values into the same scale.

The scores are shown in the plots as red points and the loadings are solid blue lines. The Lawton and Sylvestre discussed [12] where the pure components of the mixtures would be in this PCA space. To figure this out, the knowledge that the original concentrations and spectra are positive limits the possibilities.

When we look at the range of the scores obtained for the data from case 1 (left plot in Fig. 27), it is clear that the scores of the pure components cannot lie between the two outermost scores 1 and 9. The spectra associated with scores of 1 and 9 could be pure components of the mixtures in the dataset, where the spectrum associated with score 5 is a 50/50 mixture of scores 1 and 9. As in the earlier example, no pure component spectra are possible

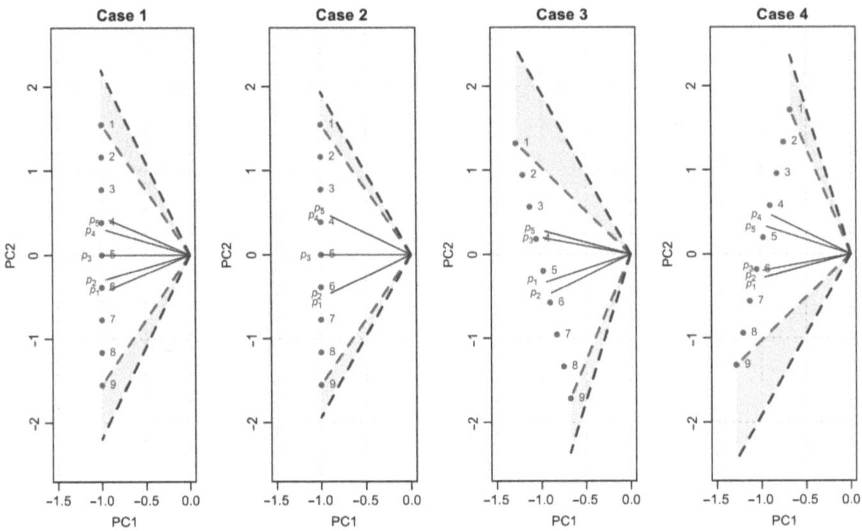

FIG. 27 Biplots with solution areas for pure spectra obtained for the simulated datasets.

between the outermost spectra (1 and 9) in the dataset. There are red dashed lines in the plots to indicate the two scores that bracket all the others. The lines form a pie shape, hence the conclusion is that positive spectra will lie outside this pie [12,13].

There is also another restriction, which we can discover by a closer study of the loadings (solid blue lines on the plots). We see that loadings of variables p_1 and p_5 on the first plot bracket the other loadings (in this case variable p_2–p_4). The reason is that variable p_1 is clearly associated with the contributions of the component *A*. Similarly, variable p_5 is clearly associated with the contributions of the component *B*. Variables p_1 and p_5 are therefore correspond to the pure(st) variables of the yet unknown pure components *A* and *B*. This concept was recognized by Lawton and Sylvestre discussing pure wavelengths in their data.

When we draw a line perpendicular to the pure(st) variable of the component *A*, we obtain one of the thick dashed blue lines. A similar line can be drawn for the component *B*. Again we have a pie shape formed by the two lines. Since these lines are perpendicular to the pure(st) variables in the data, no spectra for pure components are possible outside this blue pie shape, because the variables p_1 or p_5 will be negative then. So, the possible nonnegative spectra for the pure components will lie inside the blue pie and, as we have shown earlier, outside the red pie. In summary, the possible spectra are only nonnegative in the shaded areas shown on the plots [12].

A similar trend can be observed if we look at the biplot made for the data from case 2. However, in this case, the variables p_1 and p_2 as well as the variables p_3 and p_4 have the same position as they both correspond to the pure variables. We can also see that the shaded area becomes a bit smaller giving more certainty for the position of the pure components spectra.

The similar plots for the cases 3 and 4 show a different trend. First of all, the scores are not orthogonal to the first principal component anymore as the pure spectra do not contribute to the data symmetrically. Because of this the two pies, red and blue, are rotated relating to each other. Moreover, the purest variables that form the blue pie shape are different from the first two cases—(p_2 and p_5) for the case 3 and (p_1 and p_4) for the case 4. Which actually correspond to conclusions we can draw from exploring the spectral plots for the cases, shown in Fig. 26.

One can also evaluate the purity of each variable simply by calculating an angle between the corresponding loading and the closest side of the red pie. Table 3 shows the angle values for the selected variables. As one can see, the smallest value, 29.8 degree corresponds to the loading values of variables which do not experience any contribution from the second component, namely (p_1, p_2, p_4, p_5) for the case 2, (p_4 and p_5) for the case 3 and (p_1 and p_2) for the case 4. Also the purity of the variables in the first two cases is symmetric while in the last two is not. Having small angles leads to narrower areas between the inner (red) and outer (blue) pies and therefore giving more certainty to the location of the pure spectra.

Lawton and Sylvestre have given the pure variable solution and have shown the two resolved pure components. Papers by Ohta [14], Sylvestre *et al.* [13] extended self-modeling to three components, although no clear or easy algorithm was presented to achieve this goal. Knorr and Futrell [15] presented a technique to resolve two components with clear directions on how to do this programmatically.

Knorr and Futrell [15] gave simple directions for resolving three components using the term *pure mass*. The first pure mass variable is the one with the lowest loading on PC1, and the second pure variable is the one furthest away from the first pure variable. In this way, we find the loadings of the variables that bracket all the other ones. For the third pure variable, the loading

TABLE 3 Angles (in Degrees) Calculated for the Selected Variables

	p_1	p_2	p_4	p_5
Case 1	32.6	39.9	39.9	32.6
Case 2	29.8	29.8	29.8	29.8
Case 3	49.4	41.0	29.8	29.8
Case 4	29.8	29.8	41.0	49.4

furthest away from the average of the first two variables is taken, etc. However, a third pure variable furthest from the mean of the first two variables does not guarantee that it is furthest away from the original pure variables.

A more sophisticated method came from Malinowski [16]. The first pure variable is chosen as the one with the lowest loading on PC1. For the next pure variable, the determinants of the first pure variable with all the other variables are calculated. The (absolute value of) determinant is proportional to the sine of the angle between vectors, so the variable with the largest determinant is the next pure variable.

In practice, normalization procedures are often involved, such as normalizing the spectra to a sum of 1, and the inverse of that scaling applied to the contributions. Furthermore, the actual data with noise the contributions will be estimated by another least squares operation, similar to the one described in Section 3, so that it is based on the whole dataset and not just a limited number of variables. The method of key set was later refined by Schostack and Malinowski [17].

The problem with PCA-based approaches for identification of pure variables is that the procedure is not really interactive. For example, it assumes that when a proper pure variable is selected, e.g., a saturated measurement, one has to edit the dataset and rerun PCA. This lack of interactivity is a serious drawback for the practice, e.g., process analysis in the industry. Projects involving advanced analysis of industrial data are often based on manufacturing issues. The industrial samples are, by definition, not designed and often no replicates are available. It was one of the reasons for designing interactive methods, namely ISMA [18] and then SIMPLISMA [11] as well as further developments based on SIMPLISMA such as OPA [19] and a two-dimensional purity approach for correlation spectroscopy [20].

An alternative to SIMPLISMA approach has been developed for the commercialization of the purity approach by Eigenvector Research [21]. The main difference from the SIMPLISMA is that the purity spectra are based on the length-scaled intensities of the data rather than the standard deviation. The standard deviation is calculated based on the mean centered data. In terms used by Malinowski [22], the purity in SIMPLISMA is calculated *about the mean* and the purity in the new method is calculated *about the origin.*

This modification has some consequences. For example, when we have a series of solution mixtures, the solvent spectrum experiences no or little variations. As a consequence, the profile of this solvent spectrum will not be reflected by the purity spectrum of SIMPLISMA since the standard deviation of that component is zero. Therefore, the solvent cannot be extracted as a separate component. The spectrum of the solvent can be then present in each of the resolved component spectra. With the new purity approach, the solvent would be visible in the purity spectra and can be resolved as a separate component. Also, for the combined conventional and derivative approach of SIMPLISMA [23] components with broad spectral features are distorted in the successive series of purity spectra in SIMPLISMA, while these features stay preserved in the new approach.

7 PURE VARIABLES AND MCR-ALS

The methods discussed earlier are based on the assumption that pure variables or pure spectra do exist in the data. This assumption is often reasonable. However, most of the methods have been developed in a time, when spectroscopic datasets were generally represented by high-quality spectra with a good signal-to-noise ratio. With modern development of the instrumental image analysis techniques, there are enormous amounts of data that have a low signal-to-noise ratio due to a weak signal.

In such cases, the pure variable or pure spectrum required for SIMPLISMA or similar methods may be ill defined. First of all, because a pure variable/spectrum does not really represent a pure component. Besides that, the high rank of such a dataset may make an interactive approach very difficult to implement. Although reconstructing a dataset from a limited number of components may help, a technique like SIMPLISMA may fall short and iterative methods, such as e.g., constrained alternative least squares (MCR-ALS) [24], can be more efficient.

Assuming that a dataset can be expressed in its pure components as shown in (4), where **C** contains the resolved contributions and **S** contains the pure spectra, MCR-ALS in general works as follows:

1. Make initial estimate of **S** (e.g., with random numbers)
2. Normalize **S** (e.g., to a unit length)
3. Calculate **C**
4. Apply nonnegativity constraint
5. Calculate **S**
6. Apply nonnegativity constraint
7. Repeat steps 2–6 until a convergence criterion is reached

A similar algorithm can be started with the initial estimates of **C** where the second step in this scheme would be Calculate **S**, etc. Also additional constraints can be applied to this process [24].

As it has been explained elsewhere (see e.g., [25]) there is generally a range of solutions possible, the so-called feasible range. The assumption of a pure variable of a pure spectrum results in a solution on the extremes of the feasible bands (see also Chapter 4 in this book) and confusion about feasible ranges is avoided. As the pure variable solution or pure spectrum solution is often desirable, MCR-ALS was adapted to deliver these solutions [26].

For example, when the initial estimate is based on *a*(*n*) (imperfect) pure variable solution the solution will be biased toward a pure variable solution. Similarly, it can be biased toward a pure spectrum solution by starting with

$a(n)$ (imperfect) pure spectrum estimate. The biased solution can also be far from the true pure variable or pure spectrum, e.g., between them. For additional explanation on the trade-off between the pure spectrum and pure variable solutions, see de Juan *et al.* [27].

Now let us answer the question how to adapt this process so that it gives a pure variable solution. When discussing the pure variable approach, we have seen that pure spectra are the spectra which are most dissimilar, since they are at the outer boundaries of the feasible range with the highest contrast. If we go beyond that range, we get spectra with negative intensities thus violating the nonnegativity constraint. In other terms, the pure spectra are the spectra with the highest contrast.

There is an inverse relation between the contrast in the resolved spectra and the resolved contributions. If the contrast in the spectra is high, the contrast between the contributions is low. This relation is easy to understand realizing that the contrast in the whole dataset is of a given constant value. Which means increasing the contrast in **C** has to be compensated for by a decrease in the contrast in **S** and vice versa. Searching for a solution that gives, for example, maximum contrast in the contributions, **C**, can be done by reducing the contrast in the pure spectra as follows:

1. Make initial estimate of **S** (e.g., with random numbers)
2. Normalize **S** (e.g., to a unit length)
3. Add 5% of mean of **S** (across the rows)
4. Calculate **C**
5. Apply nonnegativity constraint
6. Calculate **S**
7. Apply nonnegativity constraint
8. Repeat steps 2–7 until a convergence criterion is reached

Thinking of the columns of **S** as vectors, adding a small part of the mean to these vectors decreases the angle between the vector representation of **S**. The angle is directly related to the contrast, therefore we achieved our goal of increasing the contrast in the contributions by decreasing the contrast in **S**. A similar procedure can be defined for increasing the contrast in pure spectra by decreasing the contrast in pure contributions (e.g., adding 5% of mean to **C** in the algorithm).

A simple implementation of the algorithm in R is shown on right part of Fig. 5 (we assume that the original spectral data and the number of components, denoted in code as **D** and **ncomp** in the code, are already defined).

Fig. 28 demonstrates the result of unmixing for the dataset *A*, where left plots were obtained using conventional MCR-ALS algorithm with nonnegativity constraint and the right plots with the adding 5% of mean to the spectra for increasing contrast between the resolved contributions. As one can notice the use of contrast correction allowed to get rid of several artifacts in the resolved spectra mostly located around overlapping peaks corresponding to

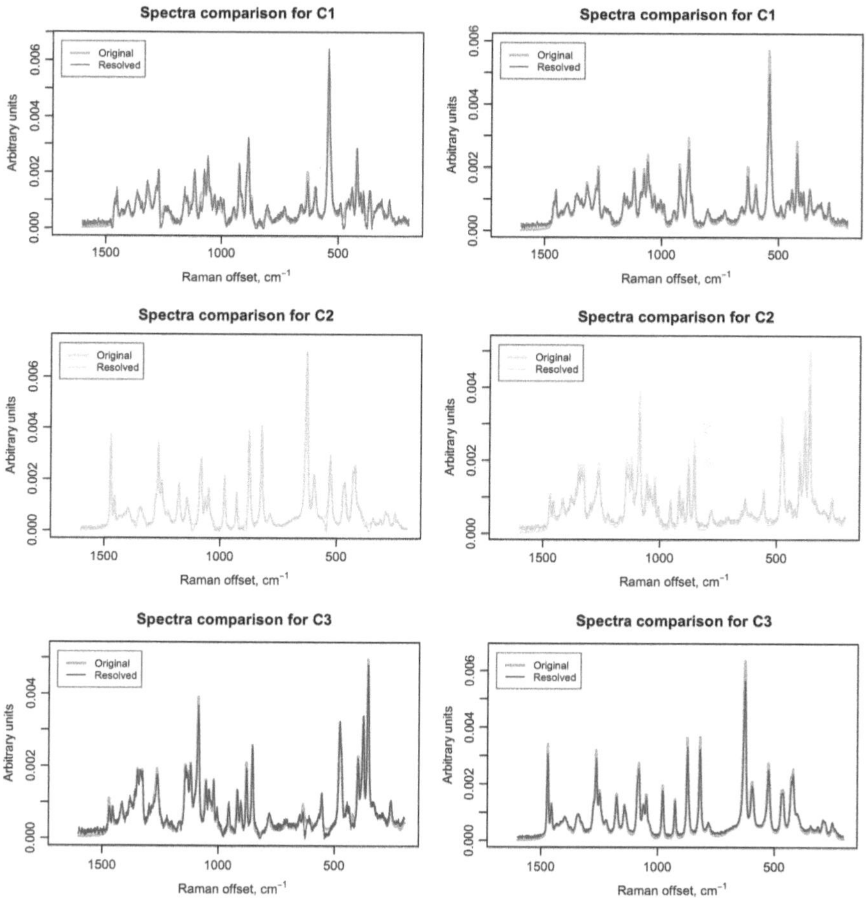

FIG. 28 Pure spectra from dataset *A* resolved using conventional MCR-ALS algorithm (left) and with angle constraint version (right).

the pure variables (look for example at the part of the resolved spectra for C1 just above 500 cm^{-1} or par of the resolved spectra for C3 below 500 cm^{-1}).

In contrast, if we add 5% mean the contribution values, this will lead to a pure variable solution (pure spectra with maximum contrast) close to what is shown in Fig. 6 with similar artifacts.

Increasing contrast among the pure contributions, **C**, has another interesting effect when one deals with hyperspectral images. Fig. 29 shows the concentration maps for the dataset *D* obtained by conventional MCR-ALS (top) and the algorithm with contrast correction (bottom). Apparently, the contrast between images has been increased. Also in this case we were able to resolve the concentration of the aqueous phase as well. Fig. 30 demonstrates the resolved spectra for this case.

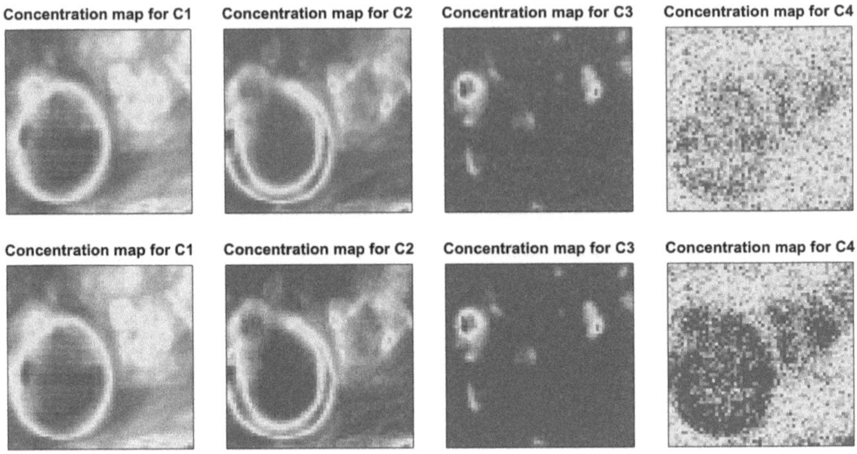

FIG. 29 Resolved concentration maps for dataset *D* obtained using conventional MCR-ALS algorithm (top) and with angle constraint version (bottom).

More details about the original angle-constrained MCR-ALS algorithm and its modifications, including numerous examples and a MATLAB code can be found in Windig and Keenan [26] and Windig *et al.* [28].

8 DISCUSSION AND CONCLUSIONS

The variable purity is a very straightforward approach to the challenging problem of curve resolution. It is particularly useful in optical spectral analysis of mixtures, where the purity is expected. Specifically, in selective methods, such as infrared (IR) or Raman spectroscopy, there is a fair probability that some wavelengths (or wavenumbers) contain prevailing variance of individual components only. The possibility to use the inversed second derivative to improve the resolution makes the purity requirement quite liberal and the algorithm—highly practicable, expanding its applicability to poorly selective data like NIR or UV/vis.

SIMPLISMA and similar algorithms perform the resolution in a gentle and graceful manner: no clumsy initial estimates, no coarse penalties to return a stray solution under constraints, no hundreds of iterations to reach a tolerable convergence. The algorithms are mainly based on very simple calculations. The only computationally intensive operation is a regression step resolving the curves for each subsequent pure variable. It is probably the fastest MCR technique that also copes well with large numbers of components, which is very important with modern large and complex data.

Interactivity is another important advantage of purity-based techniques. Solution rotational ambiguity is a known issue of the curve resolution in general, and the resolved set of components is only one from plenty of solutions, not necessarily optimal. Therefore, any solution should be critically assessed.

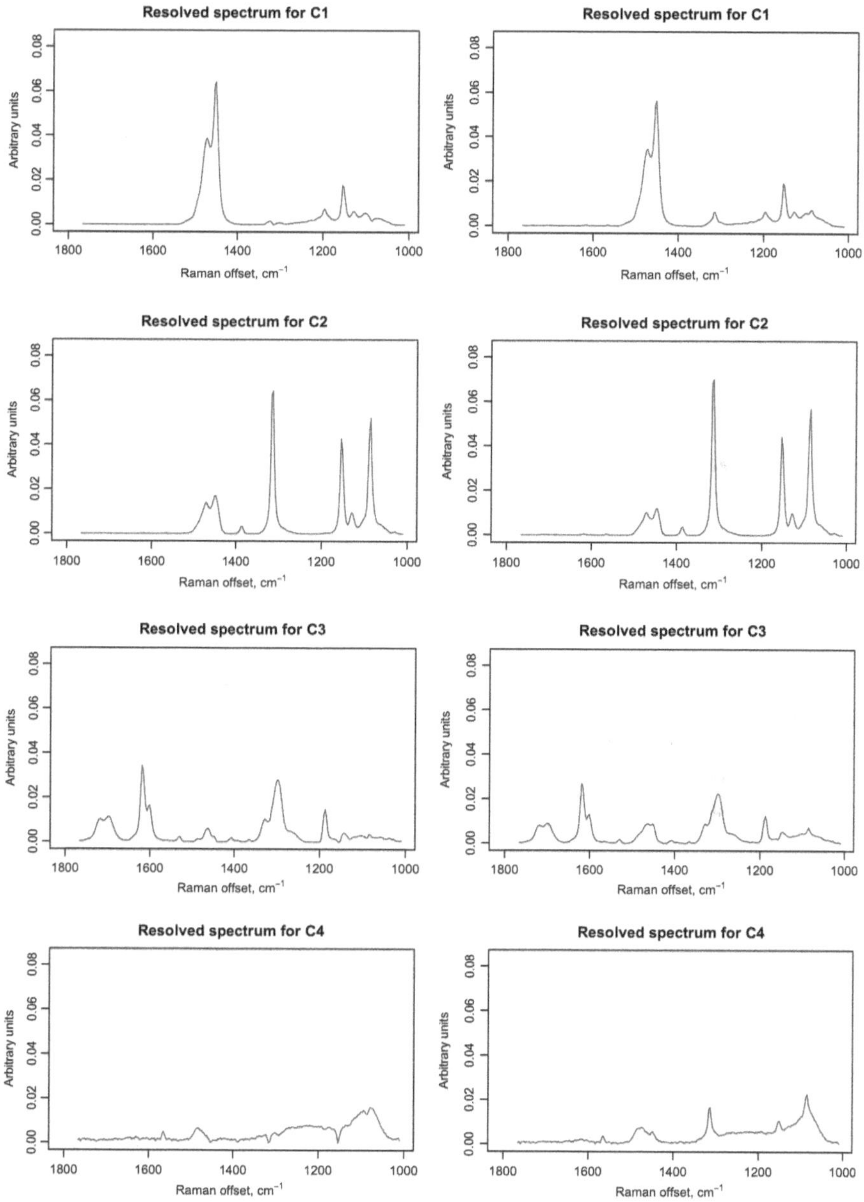

FIG. 30 Pure spectra for dataset *D* resolved using conventional MCR-ALS algorithm (left) and with angle constraint version (right).

Black-box methods do not let the user to estimate the quality of suggested solution and its correction capabilities are usually very limited. In contrast to this, algorithms of SIMPLISMA family turns the user from a witness into a participant of resolution, offering option adjustment or even manual correction of the chosen pure variable at each component resolution step. The second derivative approach can also be switched on and off for each component individually, which can be very useful to resolve narrow and broad signal, as in the case of Raman peaks "growing" on a wide and intensive fluorescence background. Some software implementations allow a dynamic resolution, when all the curves are instantly recalculated, while the pure variable cursor is drawn by a user. This option brings another quality of the resolution process making it possible to perform a fine adjustment of the whole solution that can be necessary in particularly complex cases. At the same time, interactivity is not binding; the options can be preset thus turning the algorithm into a black magic box giving out the whole solution in one click.

Another significant advantage of the SIMPLISMA like methods is that they do not require the number of components to be preliminarily determined. The resolution can be stopped at any time in a given step; there is a number of residual-based statistics to help with the decision. In e.g., MCR-ALS, the solution strictly depends on the chosen number of factors. Consequently, there is no option to start with calculating a deliberately exceeding number of components and then keep a few first ones as principal, as in PCA or in PLS regression methods. The capability to deduce the optimal number of components in the course of resolution is, therefore, another work acceleration factor of the interactive purity algorithms.

The application-driven contrast adjustment of the solution toward the spectra or concentration profile purity, considered in Section 7, is an interesting idea that turns the rotational ambiguity issue into an algorithm feature. The solution rotational adjustment itself does not depend on the resolution algorithm directly, but it was definitely inspired by the concept of purity.

Another interesting possibility illustrated in this chapter that has not been fully supported in any chemometric software is to perform the resolution using any of the two matrix dimensions providing maximum selectivity and hence potential purity of variables, e.g., chromatograms instead of spectra in HPLC-DAD data. Indeed, the matrix nature is twofold and this fact is often missed by a user concentrated on only one side of the data.

To complete the comparative analysis of purity-based algorithms let us also consider its disadvantages. SIMPLISMA and similar methods are constraint-free, and additional information, such as nonnegativity, cannot be easily integrated. At least, this has not been done yet in either software product. The purity algorithms are particularly sensitive to the background and noise present in the data. Therefore, preliminary baseline correction and sometimes smoothing are important prerequisites of successful resolution there. Purity methods and, in particular, their interactive implementations are more complex for the home programming than, e.g., MCR-ALS. For this reason, we have provided some useful R code excerpts.

SIMPLISMA-resolved components can also be used as initial estimates in MCR-ALS. The last component resolution step is then iteratively continued under chosen constraints (but with the purity conditions removed!). In some cases, this approach may improve the solution, but it can also be destructive. In an ideal case of highly selective and linear data in the absence of intensive background (i.e., high-quality IR spectra) SIMPLISMA tends to produce an exact solution. Therefore, ALS post-processing is in no way compulsory and should be applied with necessary caution.

Considering all its pluses, lesser dissemination of the purity-based curve resolution compared to other methods, in particular, MCR-ALS seems unexpected. This fact can be partially related to a higher software implementation complexity of the algorithm, to reach a high level of interactivity. This bias can also be accounted for by purely historical reasons. It seems that the approach only needs a little popularization effort to take its deserved place in the collection of favorable chemometrics tools. To help with this, this chapter was written in a tutorial style and provided with a specially developed set of curve resolution examples with simulated and real datasets in order to facilitate understanding and shorten the algorithm learning curve.

The reader could notice some terminological difficulties related to novelty and originality of the purity approach, and consequently, the necessity to create new entities, such as *purity function* or *contribution* to describe the algorithm. The latter term, widely used in the literature, is a general term to designate the MCR scores. Having the scope of practical applications limited to the spectroscopic analysis of chemical mixtures allows us replacing it with a more convenient *concentration profiles* (leaving the term *pure component spectra* for MCR loadings). Furthermore, the resolution problem can be called "spectral unmixing" in our case. Terminology simplification is very useful for the material digestion, but one should not forget, that the method itself can be used on the data of rather different nature, e.g., from the field of psychology, sociology, or economy.

The earlier mentioned advantages turn the methods presented in this chapter into a powerful tool for the analysis of process spectroscopic data, where both pure spectra and concentration profiles of the resolved components have a physical sense and an informative shape. The data from hyphenated analytical techniques, such as HPLC-DAD or thermogravimetric analysis with IR spectroscopic detection (TGA-IR) can be listed under the same category. Interesting, the original SIMPLISMA patent claims the invention "Apparatus for controlling a process wherein starting material undergoes a reaction over a period of time to produce reaction products." In many years, process analytical chemistry and technology are only now starting to exploit its potential. Unlike MCR-ALS, SIMPLISMA (and its derivatives) do not apply any crude constraining influences over the process trajectory. It is a mixture analysis algorithm, and the process is a very important example of mixture.

An option that can be specifically useful for the analysis of process data is the possibility to fix a priori known pure variables. For instance, excitation light maximum in fluorescence spectra of scattering solutions is supposed to be pure and can thus be readily used for resolution of the scattering component, e.g., biomass [29]. Fixing the variables reduces the number of meta parameters, thus, improving the solution stability, which is advantageous, especially for real-time monitoring.

Therefore, variable purity methods represent a viable alternative to other curve resolution techniques, such as widely known MCR-ALS. Both having their own advantages and drawbacks these two main algorithmic strategies are in some sense complementary. Commonly, it is recommended to try both in order to decide, which algorithm better suits to a specific unmixing problem. It can therefore be recommended for chemometric software producers to have both implementations at hand, and for the users—to try them in their practical data analysis work.

REFERENCES

[1] Windig W. 2.17—Two-way data analysis: detection of purest variables. Compr Chemometr 2009;2:275–307. Available at: http://www.sciencedirect.com/science/article/pii/B978044452701100048X.

[2] Engelsen SB, Database on Raman spectra of carbohydrates. Available at: http://www.models.life.ku.dk/~specarb/specarb.html.

[3] Bogomolov A, et al. Development and testing of mid-infrared sensors for in-line process monitoring in biotechnology. Sens Actuators B Chem 2015;221:1601–10. Available at: http://linkinghub.elsevier.com/retrieve/pii/S0925400515301751.

[4] Andrew JJ, et al. Raman imaging of emulsion systems. Appl Spectrosc 1998;52(6):790–6.

[5] Andrew JJ, Hancewicz TM. Rapid analysis of Raman image data using two-way multivariate curve resolution. Appl Spectrosc 1998;52(6):797–807. Available at: <Go to ISI>://WOS:000074610300004..

[6] de Juan A, et al. Use of local rank-based spatial information for resolution of spectroscopic images. J Chemometr 2008;22(5):291–8. Available at: http://doi.wiley.com/10.1002/cem.1099 [accessed December 2, 2010].

[7] Kalivas JH. 3.01—Calibration methodologies. Compr Chemometr 2009;3:1–32. Available at:http://www.sciencedirect.com/science/article/pii/B9780444527011000727.

[8] Tauler R, Kowalski B. Multivariate curve resolution applied to spectral data from multiple runs of an industrial process. Anal Chem 1993;65:2040–7.

[9] Windig W. The use of second-derivative spectra for pure-variable based self-modeling mixture analysis techniques. Chemometr Intell Lab Syst 1994;23(1):71–86.

[10] Rinnan A, Berg F, van den Engelsen SB. Review of the most common pre-processing techniques for near-infrared spectra. Trends Anal Chem 2009;28(10):1201–22. Available at: http://linkinghub.elsevier.com/retrieve/pii/S0165993609001629 [accessed July 23, 2010].

[11] Windig W, Guilment J. Interactive self-modeling mixture analysis. Anal Chem 1991;63(14):1425–32.

[12] Lawton WH, Sylvestre EA. Self modeling curve resolution, Technometrics 1971;13(3):617–33. Available at: http://www.jstor.org/stable/1267173 [accessed December 2, 2010].

[13] Sylvestre EA, Lawton WH, Maggio MS. Curve resolution using a postulated chemical reaction. Technometrics 1974;16(3):353. Available at: http://www.jstor.org/stable/1267665?origin=crossref.

[14] Ohta N. Estimating absorption bands of component dyes by means of principal component analysis. Anal Chem 1973;45(3):553–7. Available at: http://pubs.acs.org/doi/abs/10.1021/ac60325a010.

[15] Knorr FJ, Futrell JH. Separation of mass spectra of mixtures by factor analysis. Anal Chem 1979;51(8):1236–41. Available at: http://pubs.acs.org/doi/abs/10.1021/ac50044a030.

[16] Malinowski ER. Obtaining the key set of typical vectors by factor analysis and subsequent isolation of component spectra. Anal Chim Acta 1982;134:129–37. Available at: http://linkinghub.elsevier.com/retrieve/pii/S0003267001841842.

[17] Schostack KJ, Malinowski ER. Preferred set selection by iterative key set factor analysis. Chemometr Intell Lab Syst 1989;6(1):21–9.

[18] Windig W, et al. Interactive self-modeling multivariate analysis. Chemometr Intell Lab Syst 1990;9(1):7–30.

[19] Sánchez FC, et al. Orthogonal projection approach applied to peak purity assessment. Anal Chem 1996;68(1):79–85.

[20] Windig W, Margevich DE, McKenna WP. A novel tool for two-dimensional correlation spectroscopy. Chemometr Intell Lab Syst 1995;28(1):109–28.

[21] Windig W, et al. A new approach for interactive self-modeling mixture analysis. Chemometr Intell Lab Syst 2005;77(1–2):85–96 (Spec. Iss).

[22] Malinowski ER. Factor analysis in chemistry. 3rd ed. New York: Wiley; 2002.

[23] Windig W, et al. Combined use of conventional and second-derivative data in the SIMPLISM a self-modeling mixture analysis approach. Anal Chem 2002;74(6):1371–9.

[24] Rutan SC, de Juan A, Tauler R. Introduction to multivariate curve resolution. In: Brown SD, Tauler R, Walczak B, editors. Comprehensive chemometrics: chemical and biochemical data analysis. Elsevier; 2009. Vol 2. p. 249–59. http://www.sciencedirect.com/science/article/pii/B9780444527011000466.

[25] Tauler R, Maeder M. Two-way data analysis: multivariate curve resolution—error in curve resolution. In: Brown S, Tauler R, Walczak B, editors. Comprehensive chemometrics. Oxford: Elsevier; 2009. p. 345–63.

[26] Windig W, Keenan MR. Angle-constrained alternating least squares. Appl Spectrosc 2011;65(3):349–57.

[27] de Juan A, Rutan SC, Tauler R. Two-way data analysis: multivariate curve resolution—iterative resolution methods. In: Brown S, Tauler R, Walczak B, editors. Comprehensive chemometrics. Oxford: Elsevier; 2002. p. 325–44.

[28] Windig W, et al. Simplification of alternating least squares solutions with contrast enhancement. Chemometr Intell Lab Syst 2012;117:159–68. Available at: http://dx.doi.org/10.1016/j.chemolab.2012.01.013.

[29] Bogomolov A, Grasser T, Hessling M. In-line monitoring of Saccharomyces cerevisiae fermentation with a fluorescence probe: new approaches to data collection and analysis. J Chemometr 2011;25(7):389–99.

Chapter 4

Ambiguities in Multivariate Curve Resolution

A. Malik and R. Tauler[1]
Institute of Environmental Assessment and Water Research (IDÆA) SPANISH COUNCIL OF SCIENTIFIC RESEARCH (CSIC), Barcelona, Spain
[1]*Corresponding author: e-mail: roma.tauler@idaea.csic.es*

Chapter Outline

1 MULTIVARIATE CURVE RESOLUTION AND AMBIGUITIES

Multivariate curve resolution (MCR) techniques have been proposed for the recovery of profiles (spectra, pH profiles, time profiles, elution profiles, etc.) of more than one component in an unresolved and unknown mixture when no prior information is available about the nature and composition of

Data Handling in Science and Technology, Vol. 30. http://dx.doi.org/10.1016/B978-0-444-63638-6.00004-8
Copyright © 2016 Elsevier B.V. All rights reserved.

this mixture [1–4]. In MCR a bilinear decomposition of the experimental data matrix is performed using the model equation:

$$\mathbf{D} = \mathbf{C}\mathbf{S}^{\mathrm{T}} + \mathbf{E} \tag{1}$$

where, $\mathbf{C}$ is the matrix containing concentration profiles, $\mathbf{S}^{\mathrm{T}}$ is the matrix describing the response (or pure spectral profiles), and $\mathbf{E}$ is the residual matrix with the data variance unexplained by $\mathbf{C}\mathbf{S}^{\mathrm{T}}$. Whether the pure contributions of a data set follow a natural bilinear model or not, MCR can always be used to furnish a simple and meaningful bilinear description of the variation in the data that may include some natural properties related to the problem of interest. However, the solution of Eq. (1) for $\mathbf{C}$ and $\mathbf{S}^{\mathrm{T}}$ is ambiguous if no additional information is available; or, in other words, there is rotational and scale freedom in the solutions of Eq. (1). This problem is often called the factor analysis ambiguity problem [1–6]. Complete resolution of a two-way data set without ambiguities is only possible in some favorable cases where selectivity [6] or local rank conditions [5] are present. When these conditions are not present in the system, resolution without ambiguities is not possible even if constraints such as nonnegativity, unimodality, or closure are applied [7,8].

The problem to solve in MCR may be stated in the following way: given the data matrix $\mathbf{D}$: (1) find N, the number of chemical components or species causing the observed data variance $\mathbf{D}$; (2) find the concentration profiles of these species, i.e., matrix $\mathbf{C}$; and (3) find the pure response or spectral profiles of these species, i.e., matrix $\mathbf{S}^{\mathrm{T}}$. However, the solution of Eq. (1) for $\mathbf{C}$ and $\mathbf{S}^{\mathrm{T}}$ is ambiguous if no additional information is available. Whereas the main asset of MCR models is the simplicity and ease of interpretation because of the chemically meaningfulness of the resolved profiles, the main limitation is the uncertainty present in this model, linked specially to the phenomena of ambiguity. The concept of ambiguity means that different combinations of sets of concentration profiles in $\mathbf{C}$ and response profiles in $\mathbf{S}^{\mathrm{T}}$ can reproduce the original data set with the same fit quality [9]. Three different kinds of ambiguity can be observed in MCR solutions:

1.1 Permutation Ambiguity

There is no sorting order on the MCR components. Therefore, they can be shuffled in the $\mathbf{C}$ and $\mathbf{S}^{\mathrm{T}}$ matrix (always keeping the right correspondence of the dyads) and obtain identical results (see Fig. 1A).

1.2 Intensity or Scalar Ambiguity

Under scalar ambiguity it is understood that only the shapes of concentration profiles and spectra can be determined. Dyads of profiles having the same shape but different relative scales between concentration profile and related pure response profile reproduce equally well the original data set. Normalization

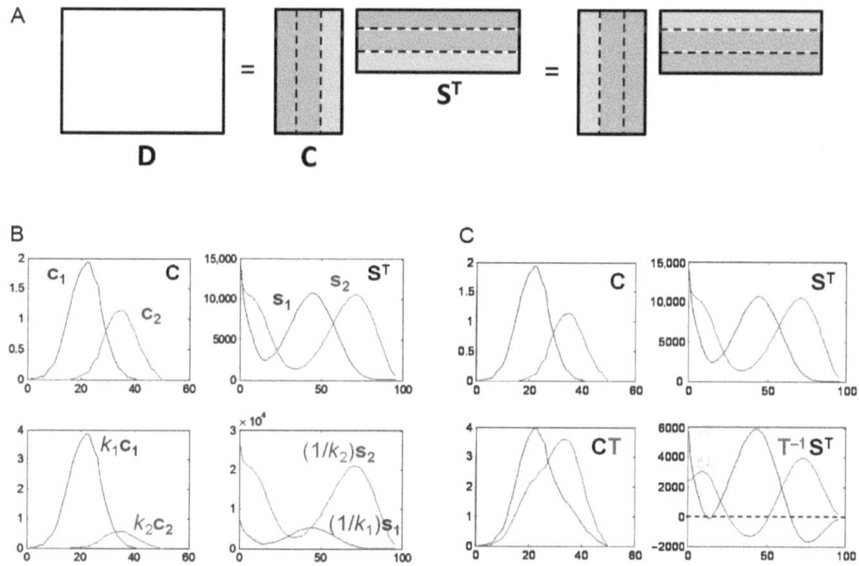

FIG. 1 Graphical representations of the ambiguity phenomenon in MCR: (A) Permutation ambiguity (dyads of profiles can be shuffled, and the data set reproduction remains invariant), (B) intensity ambiguity (*top plots*: real solution; *bottom plots*: feasible solution with scale modifications), (C) rotational ambiguity (*top plots*: real solution; *bottom plots*: feasible rotated solution). Source: *Tauler R, de Juan A. Multivariate Curve resolution for quantitative analysis. Fundamentals and Analytical Applications of Multiway Calibration. In: Muñoz de la peña A, Goicoichea H, Escandar G, Olivieri A, editors. Data Handling in Science and Technology. Elsevier, 2015.*

of concentration profiles or resolved spectra with a scaling factor, or use of reference concentration values help to suppress the intensity ambiguity (see Fig. 1B). Multiplication of any concentration profile with a scalar and the corresponding spectrum with the inverse of the scalar has no net effect on data fitting. For any component or species, $n=1\ldots N$, this can be illustrated using the following equation:

$$\mathbf{c}_n\mathbf{s}_n^{\mathrm{T}} = (\mathbf{c}_n k)\left(k^{-1}\mathbf{s}_n^{\mathrm{T}}\right) \tag{2}$$

which states that there is the scale indeterminacy in the product of concentration by spectra profiles, where k is any scalar value and $\mathbf{c}_n$ and $\mathbf{s}_n^{\mathrm{T}}$ are column and row vector profiles. If the scale of one of both, $\mathbf{c}_n$ or $\mathbf{s}_n^{\mathrm{T}}$, is known for one or several components, the scalar ambiguity (not the rotational ambiguity) problem is solved for this component. Or if, as it is often done, the scale is fixed arbitrarily, then the problem is also solved for this arbitrary selected scale.

Due to this unavoidable scale ambiguity, only the shapes of the concentration and spectra profiles can be determined and no absolute scale (quantitative) information can be obtained, in general, directly from MCR methods. However, relative quantitative information can be obtained directly by curve resolution methods, when they are applied simultaneously to multiple data sets or data matrices [10–13], for second-order calibration data (measurement of one sample gives one data matrix), like in other multiway data analysis methods [14]. In fact, the approach used in these cases is similar to the approach generally used in Analytical Chemistry, where in order to recover the scale information (or the quantitative information), the intensity or area of the resolved analyte response in an unknown sample is compared to the intensity or area of the resolved response of one or several standard samples of the same analyte at known concentration.

1.3 Rotation Ambiguities

Sets of concentration profiles and spectra with different shapes can reproduce the original data set with the same fit quality. In fact there an infinite number of possible and mathematically equivalent solutions of **C** and $\mathbf{S}^{\mathrm{T}}$, which multiplied with each other, give the same result **D***. This is the most relevant kind of ambiguity and can be expressed as follows:

$$\mathbf{D}^{*} = \mathbf{C}_{\mathrm{old}}\mathbf{S}_{\mathrm{old}}^{\mathrm{T}} = \left(\mathbf{C}_{\mathrm{old}}\mathbf{T}\right)\left(\mathbf{T}^{1}\mathbf{S}_{\mathrm{old}}^{\mathrm{T}}\right) = \mathbf{C}_{\mathrm{new}}\mathbf{S}_{\mathrm{new}}^{\mathrm{T}} \quad (3)$$

where, **T** is any transformation matrix which keeps providing profiles that still obey the constraints set in the MCR optimization (see Fig. 1C). It is important to state that rotation ambiguities are not the effect of the iterative method used to solve Eq. (1); it is a purely theoretical limitation.

It is clear that if no constraints are considered, there is an infinite number of possible solutions of Eq. (3) for any nonsingular matrix **T**, i.e., **T** should be invertible, but this is accomplished by an infinite number of matrices **T**. However, **T** is of reduced dimensions **T**(N,N), much lower than the dimensions of **C** (I,N), $\mathbf{S}^{\mathrm{T}}$ (N,J), or **D** (I,J), since $N \ll I$ and $N \ll J$. It is usually possible to reduce considerably the number of possible matrices **T** (and consequently the range of solutions for the matrices **C** and $\mathbf{S}^{\mathrm{T}}$) by means of constraints derived from the physical nature of the system and from prior knowledge of the problem under study. For instance, only positive values for the concentrations of the chemical components in the mixture have physical meaning and in many spectroscopic methods, only positive values are allowed in the spectra. In addition to the nonnegativity constraints, there are other physical constraints like peak unimodality in chromatographic elution profiles. The most useful constraints to limit drastically the number of possible solutions are the selectivity and local rank constraints. Under these constraints [5,6,15], unique solutions can be obtained. In the case that selectivity [6] or resolution local rank conditions [5,15] are present in one of the two measurement modes for every species of the unresolved mixture, the correct recovery of the concentration and/or spectra profiles for all the different species is more easily achieved. Another way to reduce ambiguities and break rank deficiency problems is to extend the bilinear model described by Eq. (1), to the simultaneous analysis of multiple runs of the process or experiment under study [6,10–13].

2 EVALUATION OF MCR AMBIGUITIES

There are several approaches to detect and assess the ambiguity in MCR solutions including manual exploration techniques [16], optimization of signal [17,18], and weighted deviations between data and fit [19]. To find the full range of feasible solutions, only few approaches exist and they cannot go beyond systems with four components [20–22]. To better illustrate the topic related to rotational ambiguity, we will focus first on a simple two-component system [23]. Thus, $N=2$ and all possible decomposition matrices are formed by two columns $\mathbf{C}$ matrices and two rows $\mathbf{S}^{\mathrm{T}}$ matrices. In this particular case the matrix $\mathbf{T}$ in Eq. (3) will be a 2×2 matrix. Only the shapes of the concentration and spectra profiles will be estimated. For this reason it is legitimate to "normalize" the matrix $\mathbf{T}$. There are different possible options, the one chosen here is to normalize the diagonal elements to one, thus

$$\mathbf{T}=\begin{bmatrix} 1 & T_{12} \\ T_{21} & 1 \end{bmatrix} \tag{4}$$

$$\begin{aligned}\mathbf{C}^{\mathrm{old}}\mathbf{T} &= \left(\mathbf{C}_1^{\mathrm{old}}\,\mathbf{C}_2^{\mathrm{old}}\right)\begin{pmatrix} 1 & T_{12} \\ T_{21} & 1 \end{pmatrix} = \left(\mathbf{C}_1^{\mathrm{old}}+T_{21}\mathbf{C}_2^{\mathrm{old}} T_{12}\mathbf{C}_1^{\mathrm{old}}+\mathbf{C}_2^{\mathrm{old}}\right) \\ &= \left(\mathbf{C}_1^{\mathrm{new}}\,\mathbf{C}_2^{\mathrm{new}}\right)=\mathbf{C}^{\mathrm{new}}\end{aligned} \tag{5}$$

The matrix $\mathbf{T}$ is defined by the two off-diagonal elements T_{12} and T_{21} and following Eq. (3), the matrix $\mathbf{C}$ of concentration profiles is multiplied by $\mathbf{T}$. Therefore, T_{21} in the first column of $\mathbf{T}$ essentially defines the first concentration profile $\mathbf{C}_1^{\mathrm{new}}$ of the product matrix. Similarly, T_{12} in the second column of $\mathbf{T}$ will define the second concentration profile $\mathbf{C}_2^{\mathrm{new}}$. Note that this is only one of many possible normalizations; it is particularly simple in the present application. Alternative options such as normalization of the concentration profiles or the spectra to a maximum of one are also common. For the calculation of $\mathbf{S}^{\mathrm{T}}$ we need to invert $\mathbf{T}$.

$$\mathbf{T}^{-1}=\frac{1}{1-T_{12}T_{21}}\times\begin{bmatrix} 1 & -T_{12} \\ -T_{21} & 1 \end{bmatrix} \tag{6}$$

$$\begin{aligned}\mathbf{T}^{-1}\mathbf{S}^{\mathrm{T,old}} &= \frac{1}{1-T_{12}T_{21}}\begin{pmatrix} 1 & -T_{12} \\ -T_{21} & 1 \end{pmatrix}\begin{pmatrix} \mathbf{s}_1^{\mathrm{T,old}} \\ \mathbf{s}_2^{\mathrm{T,old}} \end{pmatrix} \\ &= \frac{1}{1-T_{12}T_{21}}\begin{pmatrix} \mathbf{s}_1^{\mathrm{T,old}}-T_{12}\mathbf{s}_2^{\mathrm{T,old}} \\ \mathbf{s}_2^{\mathrm{T,old}}-T_{12}\mathbf{s}_1^{\mathrm{T,old}} \end{pmatrix} = \begin{pmatrix} \mathbf{s}_1^{\mathrm{T,new}} \\ \mathbf{s}_2^{\mathrm{T,new}} \end{pmatrix} = \mathbf{S}^{\mathrm{T,new}}\end{aligned} \tag{7}$$

Thus, apart from the determinant which is just a scalar, T_{12} defines the new second concentration profile and the new first spectrum, while T_{21} defines the new first concentration profile and the new second spectrum. There is an apparently surprising cross-relationship between concentration and spectra profiles of the different components, which it has been already documented in the literature [5,6].

Feasible solutions are those matrices **T** for which all constraints are fulfilled and which result in matrices **C** and $\mathbf{S}^T$ which represent the original data matrix **D** adequately, i.e., within the error limits of the data. To test a data pair (T_{12}, T_{21}) we can compute **C** and $\mathbf{S}^T$. Negative values are set to zero and any other constraints are imposed, this results in new, corrected matrices **C** and $\mathbf{S}^T$. A matrix $\mathbf{R} = \mathbf{D} - \mathbf{C} \cdot \mathbf{S}^T$ of residuals is obtained.

The sum of squares of the residuals in matrix **R** (ssq) defines the adequacy of the two parameters T_{12} and T_{21}. As **T** is defined by only two parameters we can represent the situation graphically. For the simple case of a two-component system, the sum of squares distribution as a function of the two relevant elements of **T** may be visualized as a three-dimensional contour or surface plot (Fig. 2). The range of values between maximum and minimum ssq can be very large and so it is useful to plot its logarithm.

These error maps in Fig. 2 are "landscapes" with several prominent features: (a) there is a pair of hyperbolic ridges with very high sums of squares; (b) there are two minima; (c) the minima are often not sharp or unique, and there can be two rectangular regions with identical minimal ssq; (d) there is a general "symmetry" in the figure which is due to the fact that there is no inherent order in the rows of $\mathbf{S}^T$ and the corresponding columns

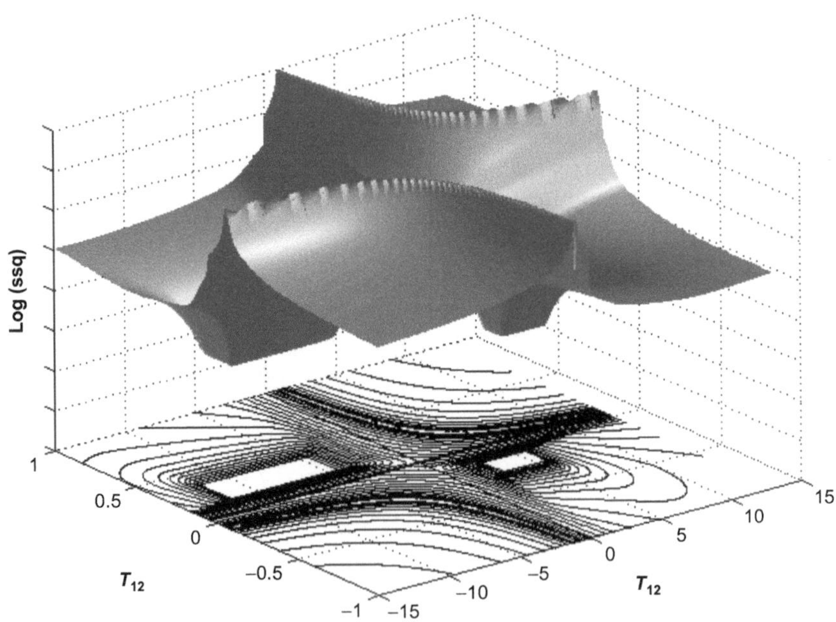

FIG. 2 Error maps. Source: *Tauler R, Maeder M. 2.20—Two-way data analysis: multivariate curve resolution—error in curve resolution. Compr Chemom 2009;345–363.*

of **C** (see the permutation ambiguity previously mentioned). Interchanging the two rows of $\mathbf{S}^T$ and columns of **C** is achieved by swapping the two rows of **T** followed by the normalization step to reset the diagonal elements to one:

$$\begin{bmatrix} 1 & T_{12} \\ T_{21} & 1 \end{bmatrix} \xrightarrow{\text{swapping}} \begin{bmatrix} T_{21} & 1 \\ 1 & T_{12} \end{bmatrix} \xrightarrow{\text{normalisation}} \begin{bmatrix} 1 & \dfrac{1}{T_{21}} \\ \dfrac{1}{T_{12}} & 1 \end{bmatrix} \tag{8}$$

Thus, each point $[T_{12}, T_{21}]$ on the error map has its equivalent point $[1/T_{21}, 1/T_{12}]$ with identical ssq. This explains the twofoldedness of the map. The hyperbolic ridges are the coordinates which produce singular matrices **T**. It occurs if $T_{12} = 1/T_{21}$ or equivalently $T_{21} = 1/T_{12}$. Note, that these singularity ridges are identical for all error maps irrespective of the data analyzed. From the present point of view the most important features are the two flat minimal regions. Any set of coordinates in these regions results in different matrices **C** and $\mathbf{S}^T$ but the same product $\mathbf{CS}^T$ and thus the solutions are not distinguishable; they cover the range of feasible solutions within the area of rotational ambiguity. If there is a unique solution these two minimal areas are reduced to two sharp minima. In some instances one of the elements, T_{12} or T_{21}, is uniquely defined while the other is not; then the rectangular area is reduced to a linear range.

This is best represented graphically, in Fig. 3(A) represents the rotational ambiguity for both elements T_{12} and T_{21} resulting in a rectangular minimal area; in (B) one of the two elements of **T** (T_{12}) is uniquely defined, while T_{21} is not; and in (C) both parameters are uniquely defined. Any point within the region of minimal ssq defines a feasible solution; the abscissa, T_{12}, defines the first spectrum and the second concentration profile; the ordinate, T_{21}, the other spectrum and concentration profile. The borders of the minimal region define the limits of the range of feasible solutions. These regions of feasible solutions may be also represented as bands (see later), limited by two lines. It is, however, very important to realize that these bands are not structure less regions within which the true solution is located. For each concentration profile or spectrum, the range of possible solutions is limited to a linear manifold of traces. These manifolds are completely defined by the range of values the appropriate matrix elements T_{12} or T_{21} can have.

Fig. 4A displays the many plausible concentration profiles which are totally within the range but are not all feasible solutions. Fig. 4B illustrates this manifold for one concentration profile, the markers represents the true solution. Note, that all profiles are normalized to a maximum of one. Fig. 4C shows two examples of profiles that are contained within the feasible range; they appear to be plausible, but in fact, they are not linear combinations of the eigenvectors and thus are not feasible solutions.

However, systematic generalization and visualization of Error Maps in a 2D plot for a larger number of components is not possible and some simpler

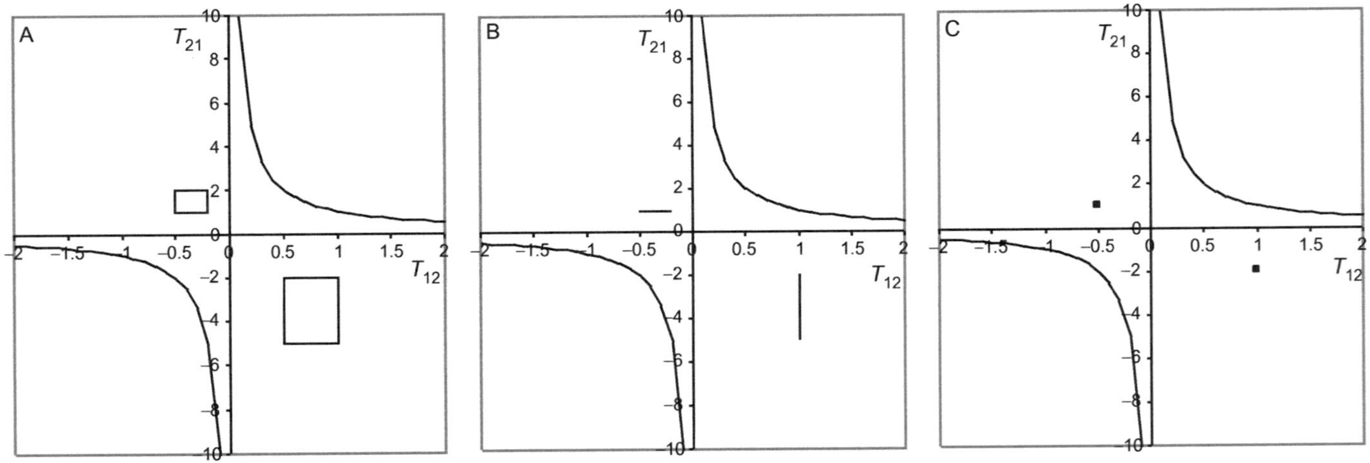

FIG. 3 Different degrees of rotational ambiguity: (A) both values for T_{12} and T_{21} not uniquely defined; (B) one value uniquely defined; (C) both values unique, sharp minimum. Source: *Tauler R, Maeder M. 2.20—Two-way data analysis: multivariate curve resolution—error in curve resolution. Compr Chemom 2009;345–363.*

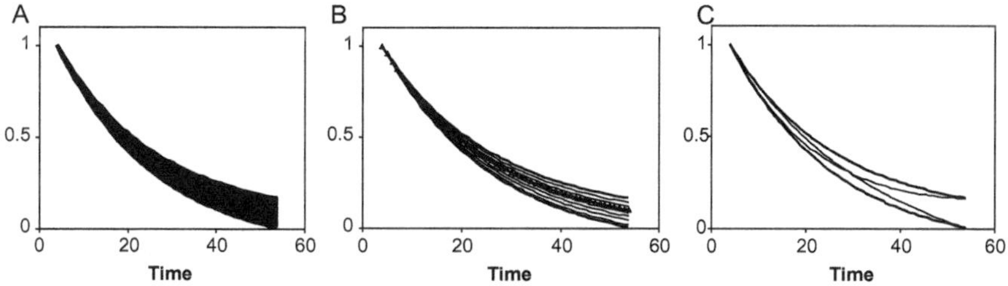

FIG. 4 Feasible regions: (A) "traditional view"; (B) series of possible solutions, the markers indicate the true solution; (C) the light lines are within the feasible region as indicated in (A), but they are not solutions. Source*: Tauler R, Maeder M. 2.20—Two-way data analysis: multivariate curve resolution—error in curve resolution. Compr Chemom 2009;345–363.*

methods for the evaluation of the extension of rotation ambiguities have been proposed. Gemperline [24] proposed a strategy based on the optimization of an objective function defined in terms of the ratio between the signal contribution of a particular species respect the whole signal contribution for the mixture of all components has been proposed. Based on this proposal, the estimation of the extension of rotation ambiguities, i.e., of the band of feasible solutions, obtained by MCR under a particular set of constraints [25] has been developed and applied to different simulated and experimental data systems [26,27].

3 ESTIMATION OF THE EXTENSION OF ROTATION AMBIGUITIES AND OF THEIR MCR FEASIBLE SOLUTIONS

After obtaining a particular MCR solution, it is of great interest to check whether the estimated solution is unique, or if not, to explore for the amount of rotational ambiguity associated with such a solution. If the MCR model is adequate (i.e., correct selection of the number of components and correct selection of constraints) and the MCR solution is unique (no rotation ambiguity), then the MCR solution can be considered the correct one (the "true" one) apart from experimental error uncertainties. In the more general case of presence of rotation ambiguities, a range of feasible MCR solutions is obtained, and there is no way to be sure that one solution is better than another one, i.e., that the obtained solution is the true one, unless more information is provided to the system. It is for this reason that the estimation of the extension of rotation ambiguities associated to a MCR solution has two goals in mind. First, it allows for the detection of nonunique MCR solutions and second, when rotation ambiguities are present, it allows for an estimation of their extension.

For a particular component or species profile the set of feasible solutions under constraints defines a range or band of feasible solutions, and this band might be considered to be delimited by extreme band profiles. These extreme profiles could be defined in different ways. One of the simplest ways for their definition is that proposed by Gemperline [24] in terms of maximum and minimum signal contributions of each species to the whole measured signal. These species contribution maximum and minimum contributions will be related to specific rotation matrices **T** for each species, which will be called $\mathbf{T}_{max}$ and $\mathbf{T}_{min}$. Considering a particular set of solutions of Eq. (3) fulfilling the constraints defined by the problem, $\mathbf{C}_{inic}$ and $\mathbf{S}^{T}_{inic}$, the maximum band profiles $\mathbf{C}_{max}$ (for the concentration profiles) and $\mathbf{S}^{T}_{max}$ (for the spectral profiles) and the minimum band profiles $\mathbf{C}_{min}$ (for the concentration profiles) and $\mathbf{S}^{T}_{min}$ (for the spectral profiles) may be defined by the equation:

$$\begin{aligned}\mathbf{D}^* &= \mathbf{C}_{inic}\mathbf{S}^{T}_{inic} = \mathbf{C}_{inic}\mathbf{T}_{min}\mathbf{T}^{-1}_{min}\mathbf{S}^{T}_{inic} = \mathbf{C}_{min}\mathbf{S}^{T}_{min} \\ &= \mathbf{C}_{inic}\mathbf{T}_{max}\mathbf{T}^{-1}_{max}\mathbf{S}^{T}_{inic} = \mathbf{C}_{max}\mathbf{S}^{T}_{max}\end{aligned} \tag{9}$$

The aim here is to find a way to calculate the values of $\mathbf{T}_{max}$ and $\mathbf{T}_{min}$ which define the species profiles in terms of maximum and minimum signal contribution under a set of constraints defined for a particular data set. A possible algorithm to determine these extreme (maximum and minimum) band profiles is based on the definition of an objective function which should be optimized, i.e., maximized and minimized as a function of **T** in order to find $\mathbf{T}_{max}$ and $\mathbf{T}_{min}$.

3.1 Optimization Problem and Method

The calculation of the extension of feasible solutions defined by the maximum and minimum values of Eq. (9) may be considered in the general frame of a nonlinear constrained optimization problem (NCP) [28]. This optimization problem is mathematically described using the following equation:

$$\underset{\mathbf{T}}{\text{minimize}}\, f_n(\mathbf{T})\, \text{subject to}\, g_{eq}(\mathbf{T}) = 0 \;\; \text{and} \;\; g_{ineq}(\mathbf{T}) \leq 0 \tag{10}$$

where, **T** is a matrix of variables, $f_n(\mathbf{T})$ is the nonlinear scalar function of **T** (given by Eq. 10), $g_{eq}(\mathbf{T})$ is the vector of equality constraints and $g_{ineq}(\mathbf{T})$ is the vector of inequality constraints. In the general framework, both $g_{eq}(\mathbf{T})$ and $g_{ineq}(\mathbf{T})$ will be considered nonlinear functions of the matrix of variables **T**.

3.2 Objective Function to Minimize

The definition of this objective function is the critical step in the implementation of the optimization algorithm. It should be a scalar function of the variables and have well-defined boundaries (maximum and minimum). For a good performance of the optimization algorithm, this optimization function should be scaled, for instance between 0 and 1. The proposed optimization function is defined as follows [24,25]:

$$f_n(\mathbf{T}) = \frac{\sum_{i=1}^{I}\sum_{j=1}^{J}\left(\mathbf{c}_{i,n}(\mathbf{T})\mathbf{s}_{j,n}(\mathbf{T})\right)^2}{\sum_{n=1}^{N}\sum_{i=1}^{I}\sum_{j=1}^{J}\left(\mathbf{c}_{i,n}(\mathbf{T})\mathbf{s}_{j,n}(\mathbf{T})\right)^2} = \frac{\left\|\mathbf{c}_n(\mathbf{T})\mathbf{s}_n^{\mathrm{T}}(\mathbf{T})\right\|}{\left\|\mathbf{CS}^{\mathrm{T}}\right\|} \quad (11)$$

The function defined by Eq. (11) gives the ratio between the signal contribution of a particular species n (numerator of Eq. 11) and the total signal contribution for all the components of the mixture (denominator of Eq. 11). The optimization of this objective function, either maximized or minimized, will give respectively an estimate of the maximum and minimum values of the species contribution defined by Eq. (11). For a particular species or component, these maximum and minimum values will define those profiles (concentration and spectra) fulfilling the constraints of the problem and giving a relative maximum and minimum signal contribution of this component ($\mathbf{c}_n\mathbf{s}_n^{\mathrm{T}}$) over the whole signal contribution from all components, $\mathbf{CS}^{\mathrm{T}}$. When this optimization is implemented as a minimization of the objective function like in Eq. (11), it is appropriate to find out the minimum of this function, whereas if the optimization is implemented as a maximization of the objective function, the function is changed of sign.

3.3 Variables to Optimize

The variables to optimize are the elements of the rotation matrix $\mathbf{T}$, which give the band of feasible solutions of Eq. (11). For each component, two matrices will be obtained, $\mathbf{T}_{max}$ which defines the maximum of the function defined by Eq. (11) and $\mathbf{T}_{min}$ which defines its minimum. Therefore, two optimizations are performed for each component. The number of variables in matrices $\mathbf{T}$, and therefore the complexity of the optimization, increases with the number of components of the system. For N components, the number of variables in the $\mathbf{T}$ matrix is equal to N^2 if no normalization is included in matrix $\mathbf{T}$.

Since this is a nonlinear optimization, initial values of the variables ($\mathbf{T}$) and of the component profiles ($\mathbf{C}_{inic}$ and $\mathbf{S}_{inic}^{\mathrm{T}}$) are required. An obvious way to facilitate the optimization is that $\mathbf{T}_{inic}$ is chosen to give a feasible solution, i.e., that $\mathbf{C}_{inic}\mathbf{T}_{inic}$ and $\mathbf{T}_{inic}^{-1}\mathbf{S}_{inic}^{\mathrm{T}}$ fulfill the constraints of the system. For instance, if $\mathbf{C}_{inic}\mathbf{S}_{inic}^{\mathrm{T}}$ are MCR solutions, $\mathbf{T}_{inic}$ can be the identity matrix, or any perturbation of it producing $\mathbf{C}_{inic}\mathbf{T}_{inic}$ and $\mathbf{T}_{inic}^{-1}\mathbf{S}_{inic}^{\mathrm{T}}$ solutions fulfilling the constraints of the system. The constrained nonlinear optimization problem described by Eq. (10) may be solved using Sequential Quadratic Programming algorithms mixed with quadratic or cubic line search methods [28] such as those implemented in MATLAB [29] and Optimization Toolbox [30] *fmincon* function.

To address the ambiguities that are inherently present in MCR analysis, the concept of chemical constraints is employed. Several constraints that are commonly employed include nonnegativity, unimodality, closure or mass balance, selectivity, and various hard modeling constraints such as those based on physicochemical models (i.e., equilibrium or kinetic) of the experimental system [6,7].

4 MCR CONSTRAINTS AND THEIR IMPLEMENTATION

Constraints implementation is a critical step in the optimization of the function in Eqs. (10) and (11). The solutions of Eqs. (10) and (11) are unbounded if no constraints are considered; i.e., there is an infinite number of possible solutions, and it is not possible to find any boundaries for them. However, in most of MCR problems, it is still possible to apply adequate constrains to decrease the extension of rotation ambiguities [6,31,32]. Commonly used constraints are schematically presented in Fig. 5. Constraints are implemented as mathematical conditions and drive the MCR optimization process to the final solutions. There are constraints linked to natural properties of the measurement of the data set and others that express mathematical conditions or models [6,9,33].

4.1 Normalization and/or Closure Constraints

It is possible to avoid the intensity or scale ambiguity in the $\mathbf{c}_n$ or $\mathbf{s}_n^{\mathrm{T}}$ profiles by using appropriate normalization and/or closure constraints. Intensity ambiguities arise because for any scalar k and species or component n profiles $\mathbf{c}_n$ and $\mathbf{s}_n^{\mathrm{T}}$,

$$\mathbf{c}_n k k^{-1} \mathbf{s}_n^{\mathrm{T}} = \mathbf{c}_n \mathbf{s}_n^{\mathrm{T}} \tag{12}$$

For a particular component n the concentration profile $\mathbf{c}_n$ can be arbitrarily increased in an unlimited way by multiplying it by an arbitrary scalar number k if at the same time its spectrum $\mathbf{s}_n^{\mathrm{T}}$ is decreased by the same amount by dividing it by the same number k. It is possible to limit the size of $\mathbf{c}_n$ or $\mathbf{s}_n^{\mathrm{T}}$ profiles using appropriate normalization and closure constraints. A spectral normalization constraint may be implemented using the norm of the species spectra, for instance forcing it to be equal to one, i.e., $||(\mathbf{s}_n(\mathbf{T}))|| = 1$. This is an equality constraint and is implemented in Eq. (10) by the equality constraint function,

$$g_{\mathrm{norm},n}(\mathbf{T}) = 1 - ||\mathbf{s}_n(\mathbf{T})|| = 0, \; n = 1, \ldots, N \tag{13}$$

If N species are present, this will give N equality constraints. Alternatively, other possible normalization constraints can be applied by constraining the signal height or maximum intensity of a profile to be equal to a constant value. Obviously this will also give N equality constraints implemented in a very similar way to the one given previously in Eq. (13).

A completely different constraint is the closure constraint, which is usually implemented on the rows of matrix $\mathbf{C}$. By this constraint the sum of the

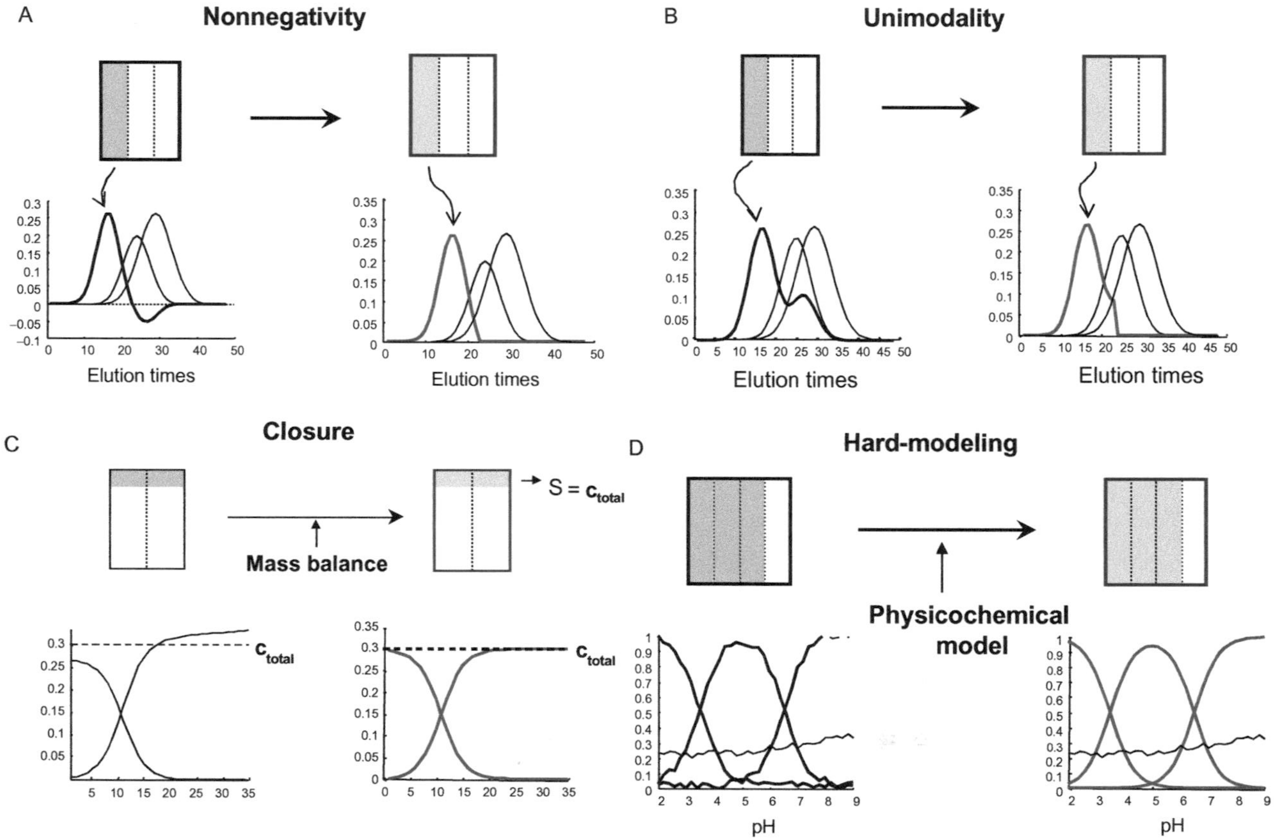

FIG. 5 Examples of common constraints used in MCR. (A) nonnegativity, (B) unimodality, (C) closure, (D) hard modeling. *Bold black* profiles on the left plots are unconstrained. *Red* (*gray* in the print version) profiles on the right are profiles that have been constrained. Source*: Tauler R, de Juan A. Multivariate Curve resolution for quantitative analysis. Fundamentals and Analytical Applications of Multiway Calibration. In: Muñoz de la peña A, Goicoichea H, Escandar G, Olivieri A, editors. Data Handling in Science and Technology. Elsevier, 2015.*

elements of each row of matrix **C** is equal to a known constant. This is the case for instance in reaction-based systems, where a mass balance equation is obeyed by the concentration profiles of the species present in the system, i.e., $\sum_{n=1}^{N} c_{i,n}(\mathbf{T}) = \text{TOT}_i$. This closure constraint is implemented by the equation:

$$g_{\text{clos},i}(\mathbf{T}) = \text{TOT}_i - \sum_{n=1}^{N} c_{i,n}(\mathbf{T}) = 0, \; i = 1, \ldots, NR \tag{14}$$

where TOT_i is the known total concentration of the species in the considered spectra (row of matrix **D**). This will give *NR* equality constraints, one for each row of matrix **C**. The number of equality constraints increases with this constraint, decreasing the degrees of freedom of the optimization.

Either normalization or a closure constraint or other normalization constraint applied during the optimization of the function defined by Eq. (11) avoid scaling problems in the variables during the optimization. If neither or both constraints are applied, the optimization will not work because of the absence of scale boundaries. On the other hand, only one of the two constraints, spectral normalization or concentration closure, is usually applied, otherwise the system becomes overly constrained. Closure constraint can only be applied if this is the case for the data, otherwise the solutions of Eq. (11) will not be correct. Conversely, when there is no closure in the concentrations, normalization of the species spectra $\mathbf{S}^{\text{T}}$ is highly recommended to achieve a successful optimization and recover the shape of the species profiles, even if the true solutions are obviously not fulfilling this normalization constraint.

4.2 Nonnegativity Constraints

Nonnegativity constraints are probably the more commonly used constraints in curve resolution from the initial works of Lawton and Sylvestre [3] and afterward [34,35]. Physical concentrations can be only positive or zero ($\mathbf{C} \geq 0$) and in many spectroscopies, spectra values can be also only positive or zero ($\mathbf{S}^{\text{T}} \geq 0$). Both constraints can be better implemented as inequality constraints:

$$g_{\text{cneg},i,n}(\mathbf{T}) = -c_{i,n}(\mathbf{T}) \leq 0 \tag{15}$$

$$g_{\text{sneg},j,n}(\mathbf{T}) = -s_{j,n}(\mathbf{T}) \leq 0 \tag{16}$$

$$i = 1, \ldots, NR, \quad j = 1, \ldots, NC, \quad n = 1, \ldots, N$$

According to Eq. (15) for nonnegative concentrations, for each component or species there will be *NR* inequality constraints, and for *N* species this will give a total of $N \times NR$ inequality constraints. According to Eq. (16) for nonnegative spectra, for each species there will be *NC* inequality constraints, and for *N* species this will give a total of $N \times NC$ inequality constraints.

Nonnegativity constraints together with the normalization/closure constraints are the more frequently applied constraints in MCR problems. However, they do not constrain the problem sufficiently to give unique solutions in general [25,32].

4.3 Selectivity and Local Rank Constraints

Selectivity and local rank constraints are often used constraints in curve resolution because they allow in many cases the partial or even the total resolution of the system and elimination of rotational ambiguities [5,6,15]. Selectivity and local rank constraints have a tremendous effect narrowing considerably the bands of feasible solutions, collapsing them into a unique solution. Selectivity and local rank constraints refer to the fact that in certain windows or regions of the data matrix, a particular species is known to exist while others are known not to exist. For instance, in some chemical evolving processes (like spectroscopic titrations or chromatography), it is a common situation that some of the species are not present at the beginning or at the end of the evolving process. Also, for spectroscopic signals, it can happen that a particular species does not have any signal in a particular spectral range. In these cases it is extremely helpful to use this information as a constraint defining for each species the concentration window where it exists and the concentration window where it does not (or where it does exist only at very negligible concentrations), the spectral window where it contributes to the measured signal and the spectral window where it does not. As it has been demonstrated [5], when the resolution conditions are appropriate, the use of the local rank information as a constraint allows the recovery of the true species profiles without ambiguities. In practice, however, these resolution conditions are often not completely achieved, as for instance in the case of embedded peaks in chromatography, or when the species profiles are extremely overlapped like in kinetic reaction-based systems.

Equality constraints are sometimes too strong during the optimization and, in fact, in many circumstances we cannot be completely sure that the values are exactly equal to zero or to another value in a certain data window. Usually it is known that a particular species does not exist at appreciable concentrations or that it does not contribute to the signal in an appreciable way. However, this does not mean that the concentration or the signal should be exactly zero in that window, but, below a threshold value. For this reason it is usually better just to select a threshold value and to force the values of the profile in the window to be lower than this threshold value. If that threshold value is equal to ε then the selectivity/local rank constraints can be described as,

$$g_{\mathrm{csel},i,n}(\mathbf{T}) = \mathrm{c}_{\mathrm{sel},i,n}(\mathbf{T}) - \varepsilon \leq 0 \tag{17}$$

$$g_{\mathrm{ssel},j,n}(\mathbf{T}) = \mathrm{s}_{\mathrm{sel},j,n}(\mathbf{T}) - \varepsilon \leq 0 \tag{18}$$

$$i = 1,\ldots,\mathrm{ncsel}_n, \quad j = 1,\ldots,\mathrm{nssel}_n, \quad n = 1,\ldots,N$$

The number of inequality constraints will depend on the number of values included in the selected windows, ncsel_n and nssel_n, for each species n considered.

Another type of possible constraints would be to fix some or all values in **C** or $\mathbf{S}^{\mathrm{T}}$ to a value exactly known. It may happen for instance, that the spectrum of one of the species be known in advance. Known spectra essentially define certain elements of the transformation matrix [36]. In that case such information can be used to simplify the optimization problem through inequality constraints like

$$g_{\text{cknown},i,n}(\mathbf{T}) = c_{i,n}(\mathbf{T}) - c_{\text{known},i,n} \leq \text{thresh} \tag{19}$$

$$g_{\text{sknown},j,n}(\mathbf{T}) = s_{j,n}(\mathbf{T}) - s_{\text{known},j,n} \leq \text{thresh} \tag{20}$$

$$i = 1,\ldots,NR, \quad j = 1,\ldots,NC, \quad n = 1,\ldots,N$$

for a preselected low threshold value, thresh. If the whole concentration profile of one species is known, this gives NR equality constraints. If the whole spectrum of one species is known, this gives NC equality constraints.

4.4 Unimodality

In a wide variety of situations, concentration profiles have unimodal peak shapes. This is the case for instance for chromatographic elution profiles or also for concentration profiles in reaction-based systems (excluding oscillating reactions). On the contrary, spectra profiles are usually not unimodal in the general case. Some other types of instrumental signals may also be unimodal, as in electrochemistry. Whenever the shape of the profiles is unimodal, an additional constraint which may be useful is unimodality. In that case, what is intended is to avoid the formation of secondary maxima. The way used to implement unimodality constraints has been algorithmic. First the highest maximum is detected and then all the departures from the unimodal condition are constrained, discarding left and right maxima: at the left of the peak maximum,

$$g_{\text{unimod},i,n} = c_{i-1,n}(\mathbf{T}) - c_{i,n}(\mathbf{T}) \leq \text{thresh} \tag{21}$$

and at the right of the peak maximum,

$$g_{\text{unimod},i,n} = c_{i+1,n}(\mathbf{T}) - c_{i,n}(\mathbf{T}) \leq \text{thresh} \tag{22}$$

for a preselected low threshold value, thresh. This constraint can also include a small local departure from unimodality tolerance because of noise, which gives the amount of local increasing concentration tolerated at the right or at the left of the maximum concentration. See Refs. [7,37] for more detailed descriptions of the unimodality constraint. The number of inequality constraints from Eqs. (21) and (22) will change during the optimization depending on the shape adopted by the concentration profile **c**(**T**).

4.5 Model or Multilinearity Constraints

These are constraints applicable to multiset data using the extension of MCR bilinear model to the analysis of augmented data matrices. Although MCR gives by default a bilinear model, the conditions of trilinearity [38,39], multilinearity [40], or factor interaction (like in Tucker models) [41] can be implemented as a constraint in a component-wise way. Therefore, completely bilinear, completely trilinear, or hybrid models can be set to be obtained in the final MCR results. Implementation of multilinear model constraint enforces decompositions to give unique solutions for the multifactor matrices (apart from scale and trivial permutation rotation ambiguities) and avoids the presence of rotational ambiguities associated with lower structured bilinear models [25]. The fulfillment of this type of constraints implies for instance that a particular component (species) has the same concentration profiles in the different data sets (matrices) simultaneously analyzed, apart than the spectra profile already enforced to be the same by the default bilinear mode. An easy way to implement this in practice is checking for the common shape of the concentration profiles of this component in the different data sets by the singular value decomposition method [50]. If the shape of the considered component concentration profile is the same in the different data sets, (fulfillment of the trilinear model), only one singular value should be larger than the noise level. A constraint can be implemented using this idea using the following equation:

$$g_{\text{tril},n} = \text{svd}(2)/\text{svd}(1) \leq \text{thresh} \tag{23}$$

where svd(2) and svd(1) refer respectively to the second and first singular value of the matrix having the k concentration profiles of the same component (n) in the k simultaneously analyzed data sets (k data matrices). thresh is a threshold value sufficiently small to constrain the considered profile of this component n to be practically invariant among different k data sets.

As shown in previous equations, final implementation of all these constraints may be performed using vectors of equality or inequality constraints. For instance spectra normalization constraints are implemented as equality constraints and nonnegativity constraints are instead implemented as inequality constraints (concentration and spectra should not be negative, but they can be zero or positive). All equality constraints are grouped together in a vector of equality constraints, g_{eq}, and all inequality constraints are grouped together in a vector of inequality constraints, g_{ineq}. See Ref. [25] for a detailed description of the implementation of these and other type of constraints.

4.6 Hard Modeling

This constraint implies performing a model curve fitting task during the iterative optimization process [38,42]. When this constraint is applied for instance as a mass action law or as a kinetic rate law, the related concentration profiles do not present ambiguity. It enforces the concentration profiles to be fitted by a parametric physicochemical model, and the parameters of the model are obtained as an additional output. It can also be applied to pure analytical responses (in $\mathbf{S}^T$ in Eq. 1), when the shape of the pure signal can be defined by a parametric equation. It is relevant to note that the hard modeling constraint and, particularly, some of the fitted parameters, e.g., total concentration in equilibrium or enzymatic models have been used for quantitative analysis purposes [43].

5 IMPLEMENTATION OF THE MCR-BANDS METHOD

The MCR-BANDS method has been implemented in a user-friendly graphical interface and a command line MATLAB computer program [51] for the evaluation of the extent of rotation ambiguities associated to MCR solutions are presented. The use of the program is shown for different examples of application including the simultaneous analysis of multiple data sets and the implementation of local rank and trilinearity constraints, basic tools to reduce and eliminate rotation ambiguities. The program allows for an easy check of the extent of rotation ambiguity remaining in MCR solutions in the investigation of a particular system, and it also allows for the checking of the effect of applied constraints. In this way, conditions and limitations to achieve optimal solutions in MCR are more easily assessed.

6 EXAMPLE OF CALCULATION OF MCR FEASIBLE SOLUTIONS USING THE MCR-BANDS METHOD

The chromatographic separation of three coeluting components using multiwavelength UV diode array detection is examined. Fig. 6 shows the plot of the data matrix **D** generated using Eq. (1) and the concentration/elution, **C**, and spectra profiles, $\mathbf{S}^T$ given also in the right of Fig. 6. Gaussian noise with a standard deviation of 0.001 units was added (matrix **E** in Eq. 1). As it is shown in Fig. 6, chromatographic elution profiles are strongly overlapped and only for the third component there was a small region at the end of the peak cluster where it was the only component present at low concentrations. This third component had also the least overlapped spectrum, with a selectivity region at one of the ends of its absorption band. Therefore, it is expected that component 3 would be the component resolved with least ambiguities and that resolution of components 1 and 2 would be more difficult.

Fig. 7 gives the results obtained in the estimation of rotation ambiguities (feasible bands) for both the elution and spectra profiles of Fig. 6 under the different scenarios. Solid lines describe the profiles at maximum and minimum values of the optimization function defined by Eq. (11) when only

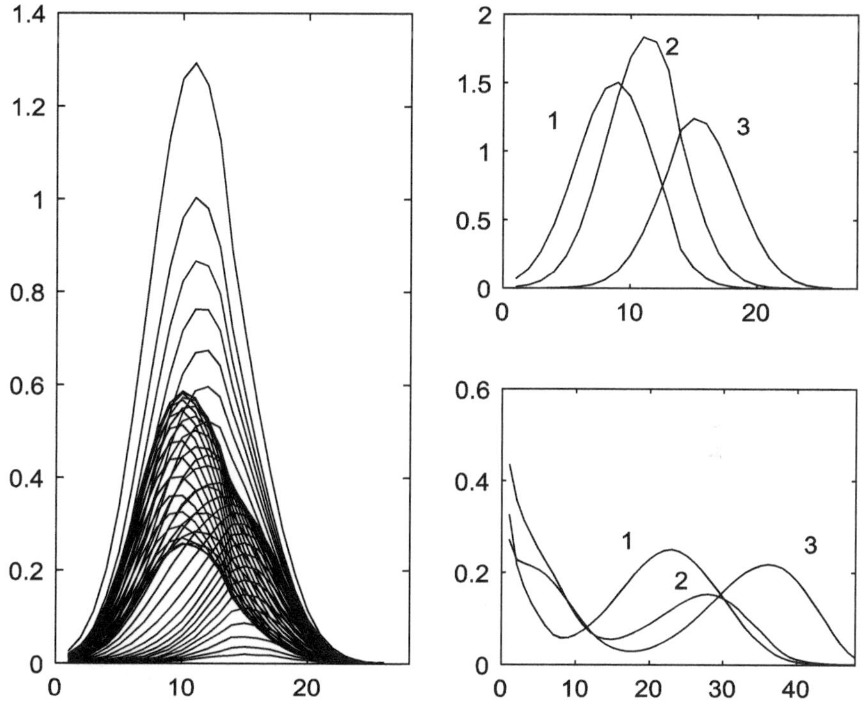

FIG. 6 Simulation of a three-component chromatographic coelution system (data matrix **D**). On the *left plot* of the 26 simulated chromatographic profiles at 48 different wavelengths. On the *right top*, plot of elution profiles (C). On the *right down*, plot of spectra of each component (S^T). Source*: Tauler R. Calculation of maximum and minimum band boundaries of feasible solutions for species profiles obtained by multivariate curve resolution. J Chemom 2001;15:627–646.*

nonnegativity (to both elution and spectra profiles) and spectra normalization constraints were applied. The broad separation obtained between these two profiles indicates that using only these two constraints, the results obtained by curve resolution are still rather ambiguous. The minimum feasible solution for the elution profile of component 1 showed a double peak, which is not acceptable for chromatographic peaks. When unimodality constraint was also implemented, in the calculation of the extreme feasible solutions, they were improved to fulfill this constraint (not shown in the figure). Finally, extreme profiles obtained using local rank constraints in addition to nonnegativity and normalization constraints are plotted using dotted lines in Fig. 7. In this particular case, the tails of the elution profiles for the first and second species were constrained to be below 0.01 units for the last four points of the elution profile of the first species and for the last three points of the elution profile of second species (see Fig. 6). Also the first three points of the front of the

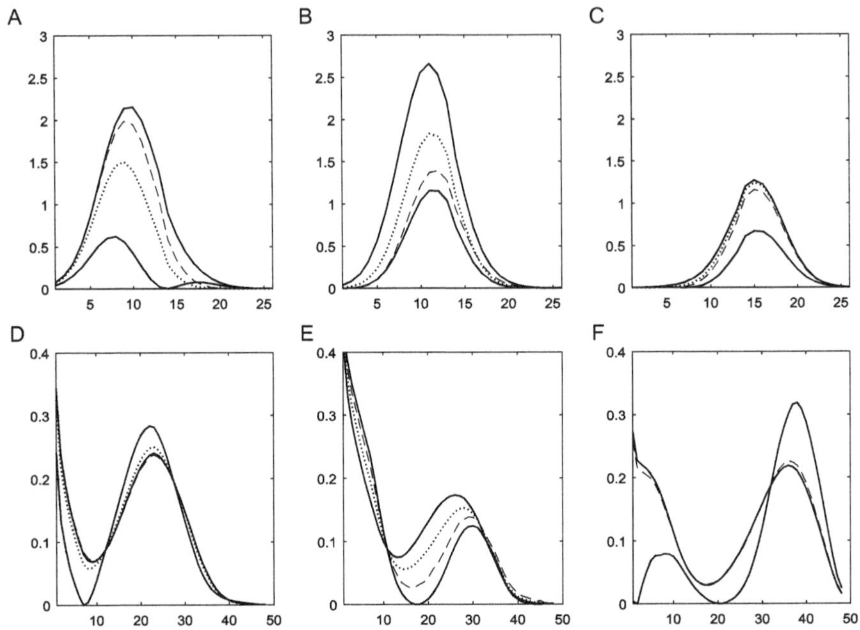

FIG. 7 Band boundaries calculated for the component profiles of data matrix **D** in Fig. 6. Solid lines give the band boundaries using the nonnegativity and spectra normalization constraints. *Dotted lines* are the band boundaries using the nonnegativity, spectra normalization, and selectivity/local rank constraints. *Dash-dotted lines* are the "true" profiles. *Dashed lines* are the profiles obtained using MCR-ALS. Panels (A)–(C) gives the elution profiles of species 1, 2, and 3, respectively, and Panels (D)–(F) the corresponding spectra profiles. Source: *Tauler R. Calculation of maximum and minimum band boundaries of feasible solutions for species profiles obtained by multivariate curve resolution. J Chemom 2001;15:627–646.*

elution profile of the third species were constrained to be below 0.01 units. In this way, the band of feasible solutions corresponding to the spectra (Fig. 7F) and elution profiles (Fig. 7C) of the third species collapsed to a unique solution, and the difference between maximum and minimum spectra profiles of the first (Fig. 7D) and second species (Fig. 7E) was now also narrower. The feasible regions for the elution profiles of first and second species, although narrower than when no selectivity/local rank constraints were applied (solid lines compared to dotted lines in Fig. 7A and B), they were still wide and showed that rotation ambiguities were still present. In Fig. 7, the $\mathbf{C}$ and $\mathbf{S}^T$ profiles used for the data generation (dot-dashed lines), and the $\mathbf{C}$ and $\mathbf{S}^T$ profiles obtained using MCR-ALS (dashed lines) are also plotted. These lines appear in most of the cases to be located well within the calculated band extreme (maximum and minimum) profiles. It may happen, however, as in Fig. 7E, that some solutions may be visually outside the band extreme plots.

These solutions have the same data fit and fulfill the constraints of the system, but they could appear visually plotted outside of these extreme profiles. Although these feasible band extreme profiles do define the maximum and minimum relative signal contributions of every component (function defined in Eq. 11), and all feasible solutions have profiles with signal contributions within these extreme profiles, their graphical representation in a two-dimensional plot could apparently appear to be outside of these extreme profiles. This is a graphical representation problem, and it does not invalidate the intrinsic meaning and interpretation of the profiles defined by the maximum and minimum extreme values of the function defined by Eq. (11), and their possible use as a means to detect the presence and extension of unresolved rotation ambiguities.

Finally the band extreme profiles obtained for the second chromatographic elution profile using different initial profile estimations ($\mathbf{C}_{\text{inic}}$ and $\mathbf{S}_{\text{inic}}^{\text{T}}$) are compared in Fig. 8. The concentration profile of this second component was recovered with larger rotational ambiguities than the others giving wider bands (in Fig. 7B and E). Solid lines in Fig. 8 give the band extreme profiles obtained under nonnegativity and normalization constraints using the true solution (dash-dotted line) as initial estimate. Solid lines with a x symbol give

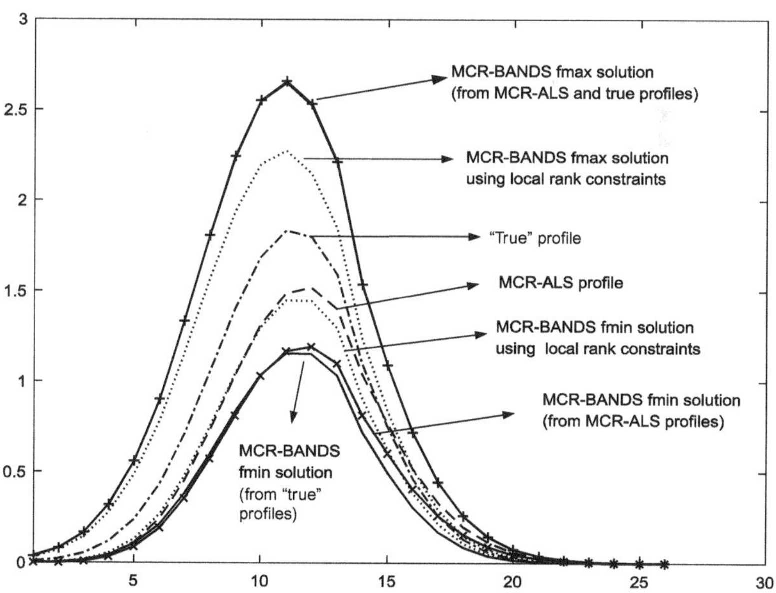

FIG. 8 Detail of band boundaries obtained for the elution profile of component 2 obtained using different initial estimates and different constraints. Source: *Tauler R. Calculation of maximum and minimum band boundaries of feasible solutions for species profiles obtained by multivariate curve resolution. J Chemom 2001;15:627–646.*

the band extreme profiles obtained using the MCR-ALS solutions (dashed line) as initial estimates. As it is shown in Fig. 8, the band extreme profiles estimated using these two initial solutions were very similar. This reinforces the reliability of the proposed method. Finally, dotted lines in Fig. 8 give the band extreme profiles obtained using the local rank constraints previously described for this system.

7 COMPARISON OF SOLUTIONS OBTAINED BY DIFFERENT MCR METHODS

The values of the function $f_n(\mathbf{T})$ of Eq. (11) for the solutions obtained by different MCR methods can be compared when applied to the same data set. In Table 1, these values are given for the MCR-ALS [44], MVSA [45], ICA [46], and MCR-FMIN [27] methods when they are applied to the data set described in Fig. 6. $f_n(\mathbf{T})$ values obtained by PCA are also given in the Table 1. In this way, results obtained by the different methods could be compared more easily.

As described previously, the MCR-BANDS method is basically based on the maximization and minimization of the $f_n(\mathbf{T})$ defined by the Eq. (11). This optimization is performed for every species separately, i.e., for the dyad of profiles (concentration and spectra) defining every species. The difference between the maximum and minimum value of this function $f_n(\mathbf{T})$ gives a measure of the extent of the rotation ambiguity associated to a particular MCR solution. When this difference is close to zero, then it means that the MCR solution is practically unique for the considered constraints. If, on the other hand, this difference approaches the unity (note that $f_n(\mathbf{T})$ values are normalized between zero and the unity), then the ambiguity of the MCR solution is very high. Plotting the profiles of the feasible solutions at the maximum and minimum values of $f_n(\mathbf{T})$ gives also an approximate graphical representation of the extent of rotation ambiguities associated to a particular MCR solution. It does not, however, describe totally the whole range of feasible solutions, simply because for a system with more than two components, this description cannot be bonded by only two limiting (f_{min} and f_{max}) profiles. In the case of a system with three or more components, the plot in two dimensions of all the feasible solutions and of their boundaries is a much more complex graphical representation in multidimensional space, with crossing lines and voids within the boundaries.

Fig. 9 gives the graphical representation of the concentration and spectral profiles resolved by the different bilinear methods are given. For comparison purposes, the pure spectra of the components were normalized to unit area, and the concentration profiles were rescaled between 0 and 1. Profiles for MCR-ALS, ICA, and true values were all within MCR-BANDS f_{min} and f_{max} profile ranges, and all of them provided nonnegative profiles and fitted the data equally well. In contrast, PCA profiles, and for some of the profiles for

TABLE 1 f_n(T) values (Eq. 11) for the three components data set of Fig. 6 (f_{true}) and using MCR-ALS ($f_{MCR\text{-}ALS}$), PCA (f_{PCA}), MVSA (f_{MVSA}), MCR-FMIN ($f_{MCR\text{-}FMIN}$), ICA (f_{ICA}), and f_{min} and f_{max} solutions calculated by MCR-BANDS, under nonnegativity constraints, for this three-component system

Component	f_{TRUE}	$f_{MCR\text{-}ALS}$	f_{PCA}	f_{MVSA}	$f_{MCR\text{-}FMIN}$	f_{ICA}	MCR-BANDS		
							f_{max}	f_{min}	$f_{max}-f_{min}$
1	0.409	0.596	**0.982**	0.536	0.433	0.511	0.728	0.298	0.430
2	0.489	0.371	**0.186**	0.358	**0.120**	0.367	0.371	0.189	0.182
3	0.328	0.267	**0.041**	**0.803**	0.666	0.356	0.620	0.185	0.435

Data values in bold font in this table correspond to values that are out of range of f_{max} and f_{min} values calculated by MCR-BANDS.

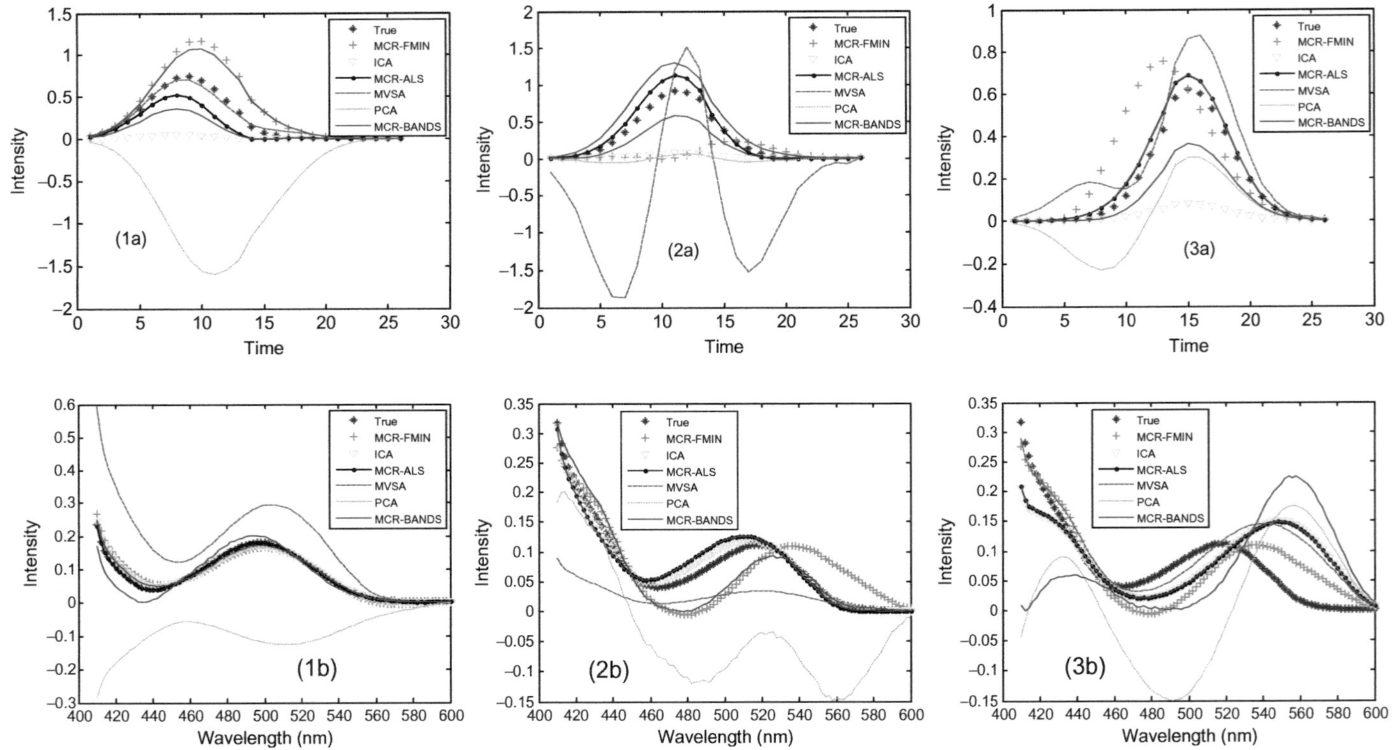

FIG. 9 Concentration and spectral profiles resolved by the different bilinear methods applied in this work (MCR-ALS, ICA, MCR-FMIN, PCA, MVSA, and profiles corresponding to $f_{\min}$ and $f_{\max}$ of MCR-BANDS) in the resolution of data set given in Figure 6. True ("*," *blue* (*dark gray* in the print version)), MCR-FMIN ("+," *green* (*gray* in the print version)), ICA ("Δ," *yellow* (*light gray* in the print version)), MCR-ALS (".-," *black*), MVSA ("— —," *purple* (*gray* in the print version)), PCA ("- -," *cyan* (*light gray* in the print version)), MCR-BANDS ("—," *red* (*dark gray* in the print version)). Source*: Zhang X, Tauler R. Measuring and comparing the resolution performance and the extent of rotation ambiguities of some bilinear modeling methods. Chemom Intell Lab Syst 2015;147:47–57.*

MVSA and MCR-MIN values, resulted to be well outside of the range between f_{min} and f_{max} MCR-BANDS values. PCA profiles were as expected to be orthogonal, and they had consequently negative values, and therefore, they were out of the MCR-BANDS (nonnegative) feasible range of solutions. MVSA method provided only solutions for the pure spectra but not for concentration profiles, which had to be estimated by an additional least squares step, and they were then always giving negative values, meaning that they did not fulfill the nonnegativity requirement for the concentration profiles, and as a consequence, they were also out of the MCR-BANDS range. MCR-FMIN had the problem of embedded noise from initial profiles estimated by PCA, which prevented having feasible nonnegative solutions. Although these MCR-FMIN results were near to the true profiles, their $f_n(\mathbf{T})$ MCR-MIN values were out of range of f_{min} and f_{max} MCR-BAND values and profiles.

8 COMPARISON OF THE RANGES OF MCR FEASIBLE SOLUTIONS OBTAINED BY DIFFERENT METHODS

In addition to the MCR-BANDS method described in previous sections, other approaches have been proposed for the calculation of the area of feasible solutions, AFS, describing the full range of feasible solutions associated to rotation ambiguities in MCR methods, and they have been reviewed recently [47].

Especially interesting is the FAC-PACK method implemented in a freely available MATLAB toolbox which is explained separately in Chapter 11. In this approach, a polygon inflation method is used to approximate the border of each AFS segment. Starting with an initial triangle whose vertices are located on the boundary of an AFS segment, all the edges are subdivided and the new vertices are moved to the boundary of the AFS. Up to now, FAC-PACK has only been shown for systems that have equal to or less than three components, although a future version for four components has been announced [48]. FAC-PACK method provides similar results to the Borgen plot method [21] and to the gridding search method [49], but its application is simpler and it is currently available as open source application for two- and three-component systems, under nonnegative constraints [22].

The AFS obtained by FAC-PACK when applied to the three-component system shown previously in Fig. 6 are shown in Fig. 10. Solutions obtained by the different MCR methods discussed in previous section 7 are also given in this figure.

FAC-PACK results give three isolated topologically disconnected subsets corresponding to the three components of the analyzed system. Marked points in the figure were projected for the profiles previously obtained by the different bilinear methods tested in this work: MCRALS, ICA, MCR-FMIN, PCA, MVSA, and f_{min} and f_{max} from MCR-BANDS. As it is shown in Fig. 10, MCR-BANDS f_{min} and f_{max} profiles appeared located in the border of

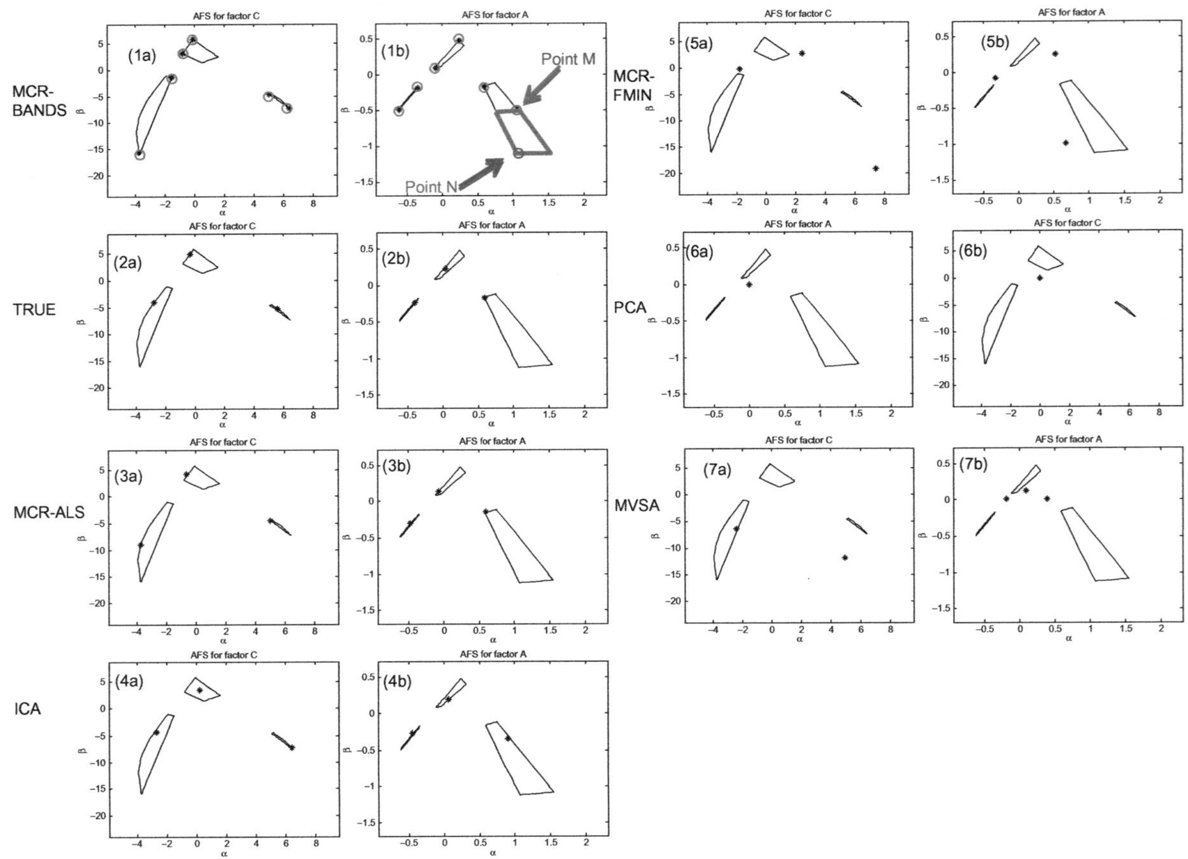

FIG. 10 See legend on opposite page.

FAC-PACK polygons, and the location of MCR-ALS, ICA, and MCR-BANDS f_{min} and f_{max} solutions were located inside the AFS calculated by FAC-PACK, whereas the location of some of the solutions provided by FMIN, PCA, MVSA solutions were located outside the AFS. Interestingly, f_{min} and f_{max} MCR-BANDS solutions projected on this figure were usually located on one of the corners/edges of the borders (marked in blue (gray in the print version) circles) of the AFS except for one of the profiles (point M in Fig. 10) marked by a red (gray in the print version) circle and a red (gray in the print version) arrow in this figure. To further study the relationship existing between MCR-BANDS solutions and FAC-PACK feasible solutions AFS, $f_n(\mathbf{T})$ values were evaluated in the region located away from this point M in Fig. 10 (shown as a blue (gray in the print version) rectangle). For instance, the feasible solution located at point N (away from M and in the lower corner of the polygon marked by the purple (dark gray in the print version) circle in Fig. 10) gave spectra and concentration profiles shown as red (gray in the print version) lines in Fig. 11.

In the upper part of Fig. 11, the locations of the projected profiles of the three components defined by point N using the FAC-PACK method are indicated. Locations corresponding to concentration profiles are marked as "o" and locations corresponding to spectra profiles are marked as "×." The corresponding concentration and spectral profiles of solutions at point N together with concentration and spectra profiles calculated for f_{min} and f_{max} of MCR-BANDS are also given in the two lower parts of Fig. 11.

Table 2 gives the $f_n(\mathbf{T})$ values (Eq. 11) for the different profiles resolved from data set of Fig. 6, MCR-BANDS f_{min} and f_{max} solutions, as well as for the "TRUE" ones, and MCR-ALS solutions with nonnegativity constraints, and, more interestingly, for the feasible solutions located at point N previously described in Fig. 11. MCR-BANDS f_{min} and f_{max} values calculated either using TRUE or MCR-ALS initials with nonnegativity constraints

FIG. 10 AFS obtained by the FAC-PACK method in the analysis of Figure 6. Polygons in each plot are the projected areas of the feasible solutions for each component calculated by FAC-PACK. Marked points (using "*") in the figure were obtained from the projection of the profiles resolved by the different bilinear methods applied in this work (MCR-ALS, ICA, MCR-FMIN, PCA, MVSA, and profiles corresponding to f_{min} and f_{max} of MCR-BANDS). The area in *blue* (*gray* in the print version) tetragon means the area of polygon from FAC-PACK which is not between the points projected by profiles corresponding to f_{min} and f_{max} of MCR-BANDS. Point M (marked by a *red* (*gray* in the print version) circle and a *red* (*gray* in the print version) arrow) is the projection of a MCR-BANDS solution which is not located on a corner of the borders of the AFS in this figure. The other related points projected by f_{min} and f_{max} of MCR-BANDS solutions are located on the corners of the polygons and they are marked in *blue* (*gray* in the print version) circles. Point N is a representative point in the lower corner of the polygon (marked by a *purple* (*dark gray* in the print version) circle), in the region located away from the point M. Source: *Zhang X, Tauler R. Measuring and comparing the resolution performance and the extent of rotation ambiguities of some bilinear modeling methods. Chemom Intell Lab Syst 2015;147:47–57.*

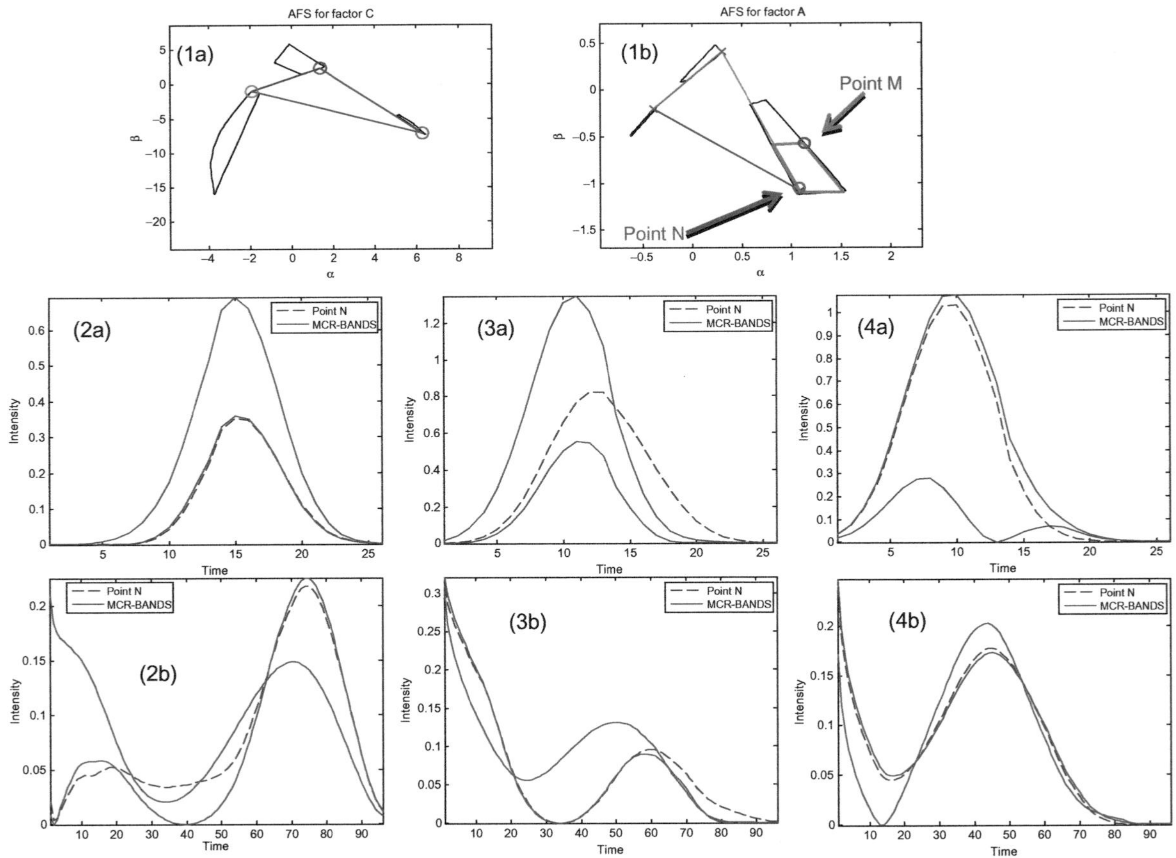

FIG. 11 Concentration and spectra profiles of the three components defined by point N in Figure 10. Locations of the concentration profiles of point N are marked as "○" in (1a) and of spectra profiles are marked as "×" in (1b)). Panels 2a,b, 3a,b, and 4a,b show the comparison between profiles obtained for f_{min} and f_{max} solutions of MCR-BANDS and concentration and spectra profiles of the three components defined by point N. Source: *Zhang X, Tauler R. Measuring and comparing the resolution performance and the extent of rotation ambiguities of some bilinear modeling methods. Chemom Intell Lab Syst 2015;147:47–57.*

TABLE 2 f_n(T) values (Eq. 11) for the different profiles resolved from data set of Fig. 6

Component[a]	f_{Point_N}[b]	$f_{MCR\text{-}ALS}$[c]	f_{TRUE}[d]	f_{max_MCR}[e]	f_{min_MCR}[f]	$f_{max\_point\ N}$[g]	$f_{min\_point\ N}$[h]	f_{max_TRUE}[i]	f_{min_TRUE}[j]
1	0.1835	0.2675	0.3283	0.3701	0.1814	0.3401	0.1780	0.3402	0.1780
2	0.4853	0.3701	0.4894	0.7359	0.2855	0.6821	0.3058	0.7399	0.3057
3	0.5831	0.5955	0.4088	0.6197	0.1423	0.6201	0.1603	0.6201	0.1586

[a] *Component number.*
[b] *$f_n(\mathbf{T})$ values for the profiles corresponding to point N (example feasible solution in Fig. 6).*
[c] *$f_n(\mathbf{T})$ values of MCR-ALS solutions with nonnegativity constraint.*
[d] *$f_n(\mathbf{T})$ values of pure component spectra used for simulation of the data set.*
[e] *Maximum $f_n(\mathbf{T})$ values obtained using MCR-BANDS from MCR resolved profiles.*
[f] *Minimum $f_n(\mathbf{T})$ values obtained using MCR-BANDS from MCR resolved profiles.*
[g] *Maximum $f_n(\mathbf{T})$ values obtained using MCR-BANDS from point N corresponding profiles.*
[h] *Minimum $f_n(\mathbf{T})$ values obtained using MCR-BANDS from point N corresponding profiles.*
[i] *Maximum $f_n(\mathbf{T})$ values obtained using MCR-BANDS from pure component spectra used for simulation of the data set.*
[j] *Minimum $f_n(\mathbf{T})$ values obtained using MCR-BANDS from pure component spectra used for simulation of the data set.*

were very similar to those obtained using initials from the point located at N in Fig. 11. All $f_n(\mathbf{T})$ values at point N for the three components were located between the range of f_{min} and f_{max} obtained by MCR-BANDS. This is an important result which confirms that $f_n(\mathbf{T})$ values of the profiles in the quadrilateral blue (gray in the print version) area of Fig. 11, away from point M, were also between the limiting f_{min} and f_{max} MCR-BANDS values. Therefore, results from MCR-BANDS and from FAC-PACK are confirmed to be in agreement, in spite of being based on completely different approaches.

As previously explained for a two-way data system [20], it is possible to define AFS graphical boundaries in two dimensions by f_{min} and f_{max} MCR-BANDS solutions to illustrate the whole set of feasible solutions. However, for higher dimensional systems (three and more components), it is not possible to illustrate graphically the whole set of feasible solutions using only the plot of two profiles, and a more complex graphical display is needed to delimit the limits of AFS plots. Therefore, MCR-BANDS cannot be used to describe precisely the boundaries of the AFS, but to give an approximate estimation of the extent of rotation ambiguity associated to a particular MCR solution.

A more detailed discussion about the different methods for the determination of ranges of feasible solutions resulting from different MCR methods applied to multivariate data can be found in a recent multiauthor review [47], where the available methods for the calculation of the AFS are compared in the analysis of simulated and experimental data sets.

9 CONCLUSIONS

Several methods have been proposed for the estimation of the range or AFS associated to ambiguous MCR solutions. Among them, the MCR-BANDS method provides an easy and flexible estimation of the extension of rotation ambiguities for any number of components and constraints. On the other hand, other more involved methods have been proposed to provide a more accurate display of the geometrical space spanned by all feasible solutions. Interestingly, however, the extension of rotation ambiguities estimated by the MCR-BANDS method and the AFS calculated by these methods are in good agreement.

REFERENCES

[1] Borgen OS, Davidsen N, Mingyang Z, Oyen O. The multivariate N-Component resolution problem with minimum assumptions. Mikrochim Acta 1986;2:63–73.

[2] Borgen OS, Kowalski BR. An extension of the multivariate component-resolution method to three components. Anal Chim Acta 1985;174:1–26.

[3] Lawton WH, Sylvestre EA. Self modeling curve resolution. Technometrics 1971;13:617–33.

[4] Sylvestre EA, Lawton WH, Maggio MS. Curve resolution using a postulated chemical reaction. Technometrics 1974;16:353–68.

[5] Manne R. On the resolution problem in hyphenated chromatography. Chemom Intell Lab Syst 1995;27:89–94.
[6] Tauler R, Smilde A, Kowalski B. Selectivity, local rank, three-way data analysis and ambiguity in multivariate curve resolution. J Chemom 1995;9:31–58.
[7] de Juan A, Vander Heyden Y, Tauler R, Massart D. Assessment of new constraints applied to the alternating least squares method. Anal Chim Acta 1997;346:307–18.
[8] Gargallo R, Tauler R, Cuesta-Sánchez F, Massart DL. Validation of alternating least-squares multivariate curve resolution for chromatographic resolution and quantitation. Trends Anal Chem 1996;15:279–86.
[9] Tauler R, Maeder M. 2.20—Two-way data analysis: multivariate curve resolution—error in curve resolution. Compr Chemom 2009;2:345–63.
[10] Peré-Trepat E, Lacorte S, Tauler R. Alternative calibration approaches for LC-MS quantitative determination of coeluted compounds in complex environmental mixtures using multivariate curve resolution. Anal Chim Acta 2007;595:228–37.
[11] Tauler R, Barcelo D. Multivariate curve resolution applied to liquid chromatography-diode array detection. Trends Anal Chem 1993;12:319–27.
[12] Tauler R, Izquierdo-Ridorsa A, Casassas E. Simultaneous analysis of several spectroscopic titrations with self-modelling curve resolution. Chemom Intell Lab Syst 1993;18:293–300.
[13] Tauler R, Kowalski B, Fleming S. Multivariate curve resolution applied to spectral data from multiple runs of an industrial process. Anal Chem 1993;65:2040–7.
[14] Smilde A, Bro R, Geladi P. Multi-way analysis: applications in the chemical sciences. West Sussex PO19 8SQ England: John Wiley & Sons; 2004.
[15] Kvalheim OM, Liang YZ. Heuristic evolving latent projections: resolving two-way multicomponent data. 1. Selectivity, latent-projective graph, datascope, local rank, and unique resolution. Anal Chem 1992;64:936–46.
[16] PAATERO P, HOPKE P, BEGUM B, BISWAS S. A graphical diagnostic method for assessing the rotation in factor analytical models of atmospheric pollution. Atmos Environ 2005;39:193–201.
[17] Jaumot J. Multivariate curve resolution: a powerful tool for the analysis of conformational transitions in nucleic acids. Nucleic Acids Res 2002;30:e92.
[18] Vandeginste BGM, Derks W, Kateman G. Multicomponent self-modelling curve resolution in high-performance liquid chromatography by iterative target transformation analysis. Anal Chim Acta 1985;173:253–64.
[19] Navea S, de Juan A, Tauler R. Detection and resolution of intermediate species in protein folding processes using fluorescence and circular dichroism spectroscopies and multivariate curve resolution. Anal Chem 2002;74:6031–9.
[20] Abdollahi H, Maeder M, Tauler R. Calculation and meaning of feasible band boundaries in multivariate curve resolution of a two-component system. Anal Chem 2009;81:2115–22.
[21] Rajkó R, István K. Analytical solution for determining feasible regions of self-modeling curve resolution (SMCR) method based on computational geometry. J Chemom 2005;19:448–63.
[22] Sawall M, Kubis C, Selent D, Börner A, Neymeyr K. A fast polygon inflation algorithm to compute the area of feasible solutions for three-component systems. I: concepts and applications. J Chemom 2013;27:106–16.
[23] Vosough M, Mason C, Tauler R, Jalali-Heravi M, Maeder M. On rotational ambiguity in model-free analyses of multivariate data. J Chemom 2006;20:302–10.
[24] Gemperline PJ. Computation of the range of feasible solutions in self-modeling curve resolution algorithms. Anal Chem 1999;71:5398–404.

[25] Tauler R. Calculation of maximum and minimum band boundaries of feasible solutions for species profiles obtained by multivariate curve resolution. J Chemom 2001;15:627–46.
[26] Garrido M, Larrechi MS, Rius FX, Tauler R. Calculation of band boundaries of feasible solutions obtained by multivariate curve resolution—alternating least squares of multiple runs of a reaction monitored by NIR spectroscopy. Chemom Intell Lab Syst 2005;76:111–20.
[27] Tauler R. Application of non-linear optimization methods to the estimation of multivariate curve resolution solutions and of their feasible band boundaries in the investigation of two chemical and environmental simulated data sets. Anal Chim Acta 2007;595:289–98.
[28] Gill PE, Murray W, Wright MH. Practical optimization. London, England: Academic Press; 1981.
[29] MATLAB computer and visualization environment. The MathWorks Inc, Natick, MA, US; 2004.
[30] MATLAB optimization Toolbox. The MathWorks Inc, Natick, MA, US; 2004.
[31] Jaumot J, Gargallo R, De Juan A, Tauler R. A graphical user-friendly interface for MCR-ALS: a new tool for multivariate curve resolution in MATLAB. Chemom Intell Lab Syst 2005;76:101–10.
[32] Tauler R. Multivariate curve resolution applied to second order data. Chemom Intell Lab Syst 1995;30:133–46.
[33] De Juan A, Rutan SC, Tauler R. Two-way data analysis: multivariate curve resolution—iterative resolution methods. Compr Chemom 2010;2:325–44.
[34] Bro R, De Jong S. A fast non-negativity-constrained least squares algorithm. J Chemom 1997;11:393–401.
[35] Van Benthem MH, Keenan MR, Haaland DM. Application of equality constraints on variables during alternating least squares procedures. J Chemom 2002;16:613–22.
[36] Mason C, Maeder M, Whitson A. Resolving factor analysis. Anal Chem 2001;73:1587–94.
[37] Bro R, Sidiropoulos ND. Least squares algorithms under unimodality and non-negativity constraints. J Chemom 1998;12:223–47.
[38] Tauler R, Maeder M, de Juan A. 2.24—Multiset data analysis: extended multivariate curve resolution. Compr Chemom 2009;2:473–505.
[39] Tauler R, Marqués I, Casassas E. Multivariate curve resolution applied to three-way trilinear data: study of a spectrofluorimetric acid-base titration of salicylic acid at three excitation wavelengths. J Chemom 1998;12:55–75.
[40] Malik A, Tauler R. Extension and application of multivariate curve resolution-alternating least squares to four-way quadrilinear data-obtained in the investigation of pollution patterns on Yamuna River, India—a case study. Anal Chim Acta 2013;794:20–8.
[41] Alier M, Felipe M, Hernández I, Tauler R. Trilinearity and component interaction constraints in the multivariate curve resolution investigation of NO and O3 pollution in Barcelona. Anal Bioanal Chem 2011;399:2015–29.
[42] De Juan A, Maeder M, Martínez M, Tauler R. Combining hard- and soft-modelling to solve kinetic problems. Chemom Intell Lab Syst 2000;54:123–41.
[43] Diewok J, De Juan A, Maeder M, Tauler R, Lendl B. Application of a combination of hard and soft modeling for equilibrium systems to the quantitative analysis of pH-modulated mixture samples. Anal Chem 2003;75:641–7.
[44] Zhang X, Tauler R. Measuring and comparing the resolution performance and the extent of rotation ambiguities of some bilinear modeling methods. Chemom Intell Lab Syst 2015;147:47–57.
[45] Lopes M, Wolff J-C, Bioucas-Dias J, Figueiredo M. Investigating counterfeit medicines—the near infrared chemical imaging picture. NIR News 2011;22:10.

[46] Parastar H, Jalali-Heravi M, Tauler R. Is independent component analysis appropriate for multivariate resolution in analytical chemistry? Trends Anal Chem 2012;31:134–43.
[47] Golshan A, Abdollahi H, Beyramysoltan S, Maeder M, Neymeyr K, Rajkó R, et al. A review of recent methods for the determination of ranges of feasible solutions resulting from soft modeling analyses of multivariate data. Anal Chim Acta 2016;911:1–13.
[48] Sawall M, Neymeyr K. A fast polygon inflation algorithm to compute the area of feasible solutions for three-component systems. II: theoretical foundation, inverse polygon inflation, and FAC-PACK implementation. J Chemom 2014;28:633–44.
[49] Beyramysoltan S, Rajkó R, Abdollahi H. Investigation of the equality constraint effect on the reduction of the rotational ambiguity in three-component system using a novel grid search method. Anal Chim Acta 2013;791:25–35.
[50] Golub GH, Van Loan CF. Matrix Computations, 2nd edition. The John Hopkins University Press; 1989.
[51] Jaumot J, Tauler R. MCR-BANDS: A user friendly MATLAB program for the evaluation of rotation ambiguities in multivariate curve resolution. Chemom Intell Lab Syst 2010;103(2):96–107.

Chapter 5

On the Analysis and Computation of the Area of Feasible Solutions for Two-, Three-, and Four-Component Systems

M. Sawall*, A. Jürß*, H. Schröder* and K. Neymeyr*,†
*Universität Rostock, Institut für Mathematik, Rostock, Germany
†Leibniz-Institut für Katalyse, Rostock, Germany

Chapter Outline

Data Handling in Science and Technology, Vol. 30. http://dx.doi.org/10.1016/B978-0-444-63638-6.00005-X
Copyright © 2016 Elsevier B.V. All rights reserved.

1 INTRODUCTION

Multivariate curve resolution (MCR) techniques serve to extract the pure component information from multivariate (spectroscopic) data. Typically, the data are taken by spectral observation of a chemical reaction system on a time $\times$ frequency grid. If k spectra are measured, each at n frequencies, then the resulting matrix D is a $k \times n$ matrix. The measured data result from a superposition of the pure component spectra. MCR methods can be applied in order to extract the pure component spectra and the concentration profiles. The basic bilinear model underlying these methods is the Lambert–Beer law. In matrix notation, the Lambert–Beer law is a relation between D and the matrix factors $C \in \mathbb{R}^{k\times s}$ and $A \in \mathbb{R}^{s\times n}$ in the form

$$D = CA. \tag{1}$$

Therein s is the number of the pure components which are assumed to be independent so that s equals the rank of the matrix D. An error matrix $E \in \mathbb{R}^{k\times n}$ with entries close to zero can be added on the right-hand side of (1) in order to allow approximate factorizations in case of perturbed or noisy data D. In general, the matrices C and A are called abstract factors. One is interested in finding a factorization $D = CA$ with chemically interpretable C and A. Then the columns of $C \in \mathbb{R}^{k\times s}$ are the concentration profiles along the time axis of the pure components. And the rows of $A \in \mathbb{R}^{s\times n}$ are the associated pure component spectra.

The aim of an MCR method is to determine the number s together with the pure component factors C and A. Sometimes no additional information on the pure components is available. Then the MCR method only uses D for the pure component decomposition within a model-free approach. The main hurdle for any MCR technique is the so-called rotational ambiguity of the solution. See, e.g., Refs. [1–3] for an introduction to the ambiguity problem. By applying additional hard or soft constraints to the pure component factorization problem, one can often determine a single solution by means of a regularized optimization problem. In case of proper constraints this solution can be the chemically correct one. A large number of successful MCR methods have been developed. Some examples are methods as MCR-ALS [4], RFA [5], SIMPLISMA [6], BTEM [7], and PCD [8]. Alternatively, one might be interested in computing not only a single factorization of D by solving a regularized optimization problem. Instead, one can follow the global approach of determining the full range of all nonnegative factorizations $D = CA$ with nonnegative rank-s matrices C and A. Such continua of possible nonnegative matrix factors can graphically be presented either by drawing the bands of possible concentration profiles together with the bands of possible spectra [9] or by plotting these sets of feasible factors by a certain low-dimensional representation, the so-called *area of feasible solutions* (AFS). The aim of this chapter is to provide a systematic introduction to the AFS concept together with many references to the chemometric literature. Important properties of the AFS are presented together with computational techniques for its numerical approximation.

1.1 Organization of the Chapter

In the remaining part of this section, we introduce the four data sets which are used within this work. Section 2 explains in a compact form the principles of MCR methods and their relation to the singular value decomposition (SVD) of the data matrix D. The techniques to compute the factors C and A are explained. In Section 3 the AFS and some of its important properties are described in detail. The rules for a classification of feasible or nonfeasible points are discussed in Section 5. In Section 6 methods for the computation of the AFS are explained with a focus on the polygon inflation method. Methods for the reduction of the AFS by using additional information or soft constraints are demonstrated in Section 8. In Section 9 we study dynamic changes of the shape of the AFS under changes in the data (e.g., variation of a shift parameter). Finally in Section 10 the software package *FACPACK* for the computation of the AFS is pointed out.

1.2 Model Data Sets and Experimental Spectral Data

In this work we use four different data sets (two model data sets and two IR-spectroscopic data sets) for all demonstrations. The data sets are as follows:

Data set 1. (FT-IR experimental data on a rhodium catalyst formation) This data set describes the two-component subsystem ($s = 2$) of the formation of a rhodium catalyst from a certain precursor; for the underlying chemical problem, see Ref. [10]. The data set includes a number of $k = 977$ spectra, each with $n = 481$ wavenumbers. Within this spectral window the dominant absorbing components are a catalyst precursor and the catalyst. The usage of a first-order reaction scheme allows to find unique factors C and A. These factors are shown in Fig. 1. For the associated areas of feasible solutions, see Fig. 9.

Data set 2. (A three-component model problem) The consecutive reaction

$$X \xrightarrow{k_1} Y \xrightarrow{k_2} Z$$

is considered with the rate constants $k_1 = 0.3$ and $k_2 = 0.1$. The initial concentrations are $c_X(0) = 1$ and $c_Y(0) = c_Z(0) = 0$, and the time interval is $t \in [0,10]$. The solution of the associated rate equation and its discretization for a number of $k = 201$ time steps results in the concentration matrix factor $C \in \mathbb{R}^{201\times 3}$. The matrix factor A is derived from the three Gaussian functions

$$a_X(x) = \exp\left(-\frac{(x-30)^2}{1250+w}\right), \quad a_Y(x) = \exp\left(-\frac{(x-50)^2}{1000+w}\right),$$

$$a_Z(x) = \exp\left(-\frac{(x-70)^2}{1000+w}\right)$$

for $x \in [0,100]$ and equidistant discretization along the frequency axis. A number of $n = 401$ spectral channels are used. The spectra depend on the real parameter w which controls the signal width. The rows of the data matrix D as well as the original factors C and A for $w = 0$ are plotted in Fig. 2. The

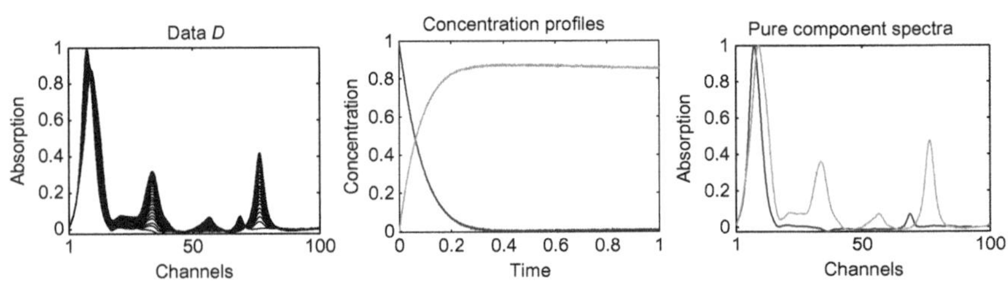

FIG. 1 Data set 1: The series of mixture spectra is shown *left* (only every tenth spectrum of the data is plotted). All ordinate axes are scaled to a maximum of 1 and the channel windows are set to [1,...,100]. By enclosing a kinetic hard model for the reaction scheme ($X \rightarrow Y$) into the pure component decomposition unique matrix factors (aside from the multiplicative ambiguity and the permutation ambiguity) have been determined. These are shown in the *centered and right subplot*.

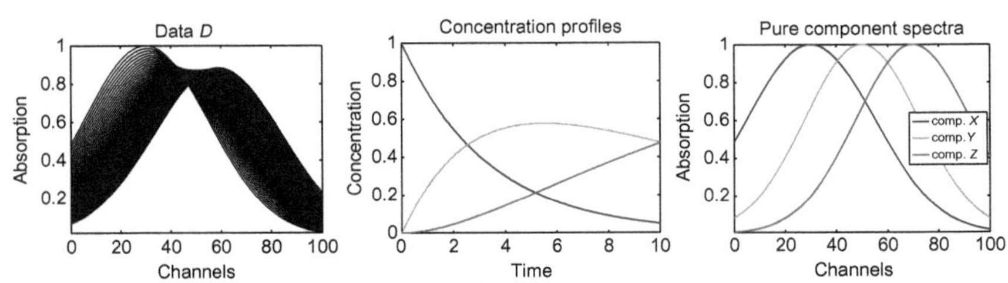

FIG. 2 Data set 2 for $w = 0$: The *leftmost plot* shows the series of the mixture spectra, i.e., the rows of the matrix D. Only every third spectrum of the data is plotted. The remaining two plots show the concentration profiles and the spectra of the three pure components.

associated areas of feasible solutions, namely the AFS for the concentration factors C and the AFS for the spectral factors A, are presented in the first column of Figs. 20 and 21. See additionally the Fig. 6 (with the data representation in the leftmost plot) and Fig. 15 for a successive approximation of the AFS by means of the polygon inflation technique.

Data set 3. (Operando FT-IR spectroscopic data from the rhodium-catalyzed hydroformylation process). This data set is described in detail in Ref. [11]. The data consist of $k = 850$ spectra, each with $n = 642$ wavenumbers. The three major absorbing components form a reaction subsystem and are an olefin component, a hydrido complex, and an acyl complex. All these chemical components are explained in Ref. [11]. The series of mixture spectra and the factors C and A are shown in Fig. 3. The Michaelis–Menten kinetic has been used as a kinetic hard model in order to find unique concentration profiles. The complete set of all nonnegative factorizations of the data matrix D is represented by the AFS for the concentration factor and by the AFS for the spectral factor. These AFS sets are plotted together with the associated bands of feasible solutions in Fig. 7.

Data set 4. (A four-component model problem) The concentration factor C of this model problem results from solving the rate equations for the reaction scheme

$$W \xrightarrow{k_1} X \xrightarrow{k_2} Y \xrightarrow{k_2} Z$$

with the kinetic constants $k_1 = 1$, $k_2 = 0.25$, and $k_3 = 0.1$. The initial concentrations are $c_W(0) = 1$ and $c_X(0) = c_Y(0) = C_Z(0) = 0$, and the time interval is $t \in [0,10]$. The time-continuous concentration functions are discretized in $k = 26$ time steps to form C.

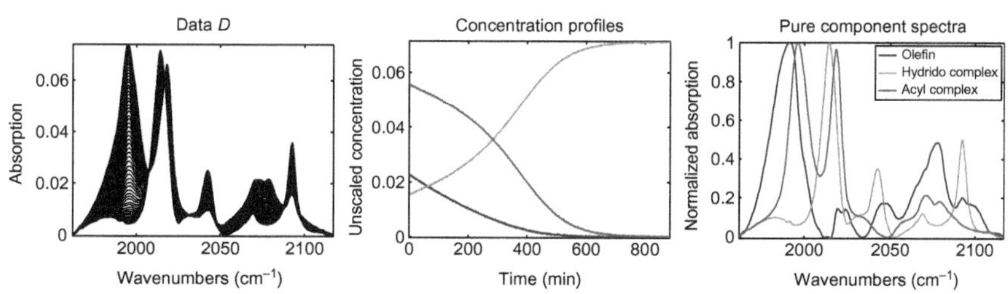

FIG. 3 Data set 3 on catalyst formation within the hydroformylation process: The *leftmost subplot* shows the spectral data, i.e., the rows of D. Only every tenth spectrum of the data is actually plotted. The remaining two subplots show the concentration profiles and the spectra of the pure components. For a successful pure component decomposition of these spectral data a Michaelis–Menten kinetic hard model has been integrated into the pure component factorization.

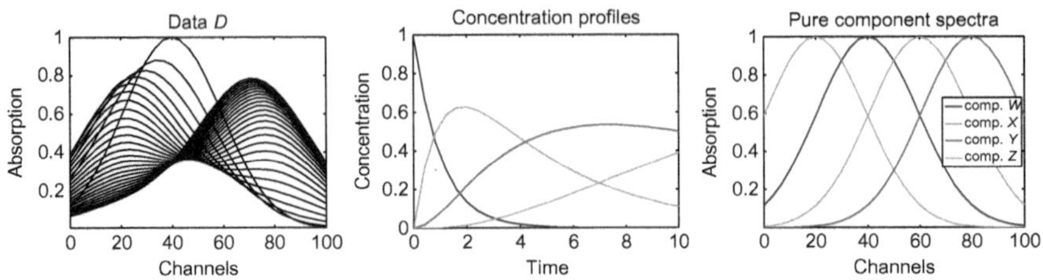

FIG. 4 Data set 4 for $\sigma = 750$: The spectral data (*left*) for the four-component model problem together with the concentration profiles of the pure components (*center*) and the pure component spectra (*right*).

The factor A derives from the Gaussian functions

$$a_W(x) = \exp\left(-\frac{(x-40)^2}{\sigma}\right), \quad a_X(x) = \exp\left(-\frac{(x-20)^2}{\sigma}\right),$$

$$a_Y(x) = \exp\left(-\frac{(x-80)^2}{\sigma}\right), \quad a_Z(x) = \exp\left(-\frac{(x-60)^2}{\sigma}\right),$$

depending on the parameter σ on the frequency interval $x \in [0,100]$. Numerical evaluation of these functions in $n = 31$ equidistant channels results in the matrix A. Thus the matrix A depends on σ. The mixture spectra, namely the rows of $D = CA$, are plotted together with the original factors C and A for $\sigma = 750$ in Fig. 4. The associated areas of feasible solutions (concentrational and spectral AFS) are shown in Fig. 19.

2 MCR METHODS

MCR methods are key tools in order to extract the pure component information from the chemical mixture data in D. The problem is to compute

1. the number of independent components s and
2. the pure component factors C and A.

Any available information on the factors can and should be integrated into the MCR computations.

2.1 The Singular Value Decomposition

The SVD, see Ref. [12], is a very powerful tool of numerical linear algebra to compute the left and right orthogonal bases for the expansion of the pure component factors $C \in \mathbb{R}^{k \times s}$ and $A \in \mathbb{R}^{s \times n}$; see, for example, Refs. [1, 2, 13, 14]. The SVD of D reads

$$D = U\Sigma V^T.$$

Therein $U \in \mathbb{R}^{k \times k}$ and $V \in \mathbb{R}^{n \times n}$ are orthogonal matrices whose columns are the left and right singular vectors. The diagonal matrix Σ contains on its diagonal the singular values σ_i in decreasing order. The singular values are real and nonnegative. For an s-component system the first s singular vectors and the associated singular values contain all information on the system. For data not including perturbations only the first s singular values are nonzero if C and A are rank-s matrices. For data including noise some additional singular values are nonzero. In such cases the SVD allows to compute optimal (with respect to least-squares) rank-s approximations of D. If in the case of noisy data the noise-to-signal ratio is not too large, then the number of independent chemical components s can often be determined from the SVD. Then the relevant and meaningful singular values are clearly larger compared to the remaining nonzero singular values which evince the influence of noise, cf. Refs. [1, 8].

2.2 Reconstruction of the Pure Component Factors

The first s singular vectors, namely the first s columns of U and the first s columns of V, are used as bases to expand the desired pure component factors C and A. For ease of notation we denote these submatrices of the SVD factors again by U and V. Then $U \in \mathbb{R}^{k \times s}$ and $V \in \mathbb{R}^{n \times s}$. The matrices C and A are formed according to

$$C = U\Sigma T^{-1}, \quad A = TV^T. \tag{2}$$

Therein $T \in \mathbb{R}^{s \times s}$ is a regular matrix which remains to be determined. MCR methods typically provide a single pure component factorization $D = CA$ and thus they explicitly or implicitly determine the matrix T. From C and A the matrix T of expansion coefficients is accessible from Eq. (2). For SVD-based MCR methods, see Refs. [1, 2, 8, 13] and references therein.

The basis expansion approach (2) drastically reduces the number of free variables of the pure component factorization problem. The crucial point is that the number of matrix elements of C and A is $(k + n)s$, whereas the representation by Eq. (2) reduces the degrees of freedom to the s^2 matrix elements of T. Hence the representation (2) is a basic ingredient for the construction of computationally effective MCR methods.

2.3 Application of Hard and Soft Constraints

Hard and soft constraints have a crucial role in the construction of MCR methods [8, 15–18]. A very restrictive and often successful hard constraint is a kinetic model of the underlying chemical reaction system. Only those concentration factors C are acceptable which are consistent with the kinetic model [1, 19–21]. Typically, the rate constants are implicitly computed as a by-product of the model fitting process.

If no kinetic model is available for C, then soft constraints can be used in order to extract (from the set of all nonnegative factorizations) solutions with special properties (see, e.g., Refs. [8, 15–18]). Typical examples of such soft constraints are those on the smoothness of the concentration profiles in C or A, constraints on a small or large integral of the spectra in A (in order to favor a solution with few and sharp peaks or alternatively those with a large number of wide peaks), criteria on the closure of the concentration data, and so on. Such soft constraints are usually added to the reconstruction functional $\| D - CA \|_F^2$ in terms of a cost function

$$g(T) = \sum_{i=1}^{p} \gamma_i \| g_i(C,A) \|_2^2 .$$

According to the representations of $C = C(T)$ and $A = A(T)$ as functions of T, the cost function $g(T)$ includes a number of p constraint functions g_i. The $\gamma_i \geq 0$ are the associated weight factors which give the user the possibility to determine a certain balance between the different constraints. A small reconstruction error $\| D - CA \|_F^2$ together with the nonnegativity of the factors C and A is of highest importance; sometimes small negative entries in C and A can be acceptable. Other constraint functions are of lower importance, e.g., on the smoothness. For these constraint functions smaller weight factors γ_i are used.

3 THE AREA OF FEASIBLE SOLUTIONS

Even with proper constraint functions and proper weight factors, MCR methods cannot always find the chemically correct or "true" solution. Thus one might follow the alternative idea to determine the set of *all* nonnegative factorizations $D = CA$. Such solutions which only fulfill the nonnegativity constraint are called *feasible* or *abstract* factors. The global approach of computing all feasible factors provides an elegant way in order to survey the complete rotational ambiguity of the pure component factorization problem. However, the sets of feasible matrices $C \in \mathbb{R}^{k \times s}$ or $A \in \mathbb{R}^{s \times n}$ are difficult to handle. The key idea to make these sets of feasible factors accessible is their low-dimensional representation in terms of the so-called *area of feasible solutions* (AFS). The AFS refers to the representation of C and A as functions of T by Eq. (2). Only a single row of T is sufficient to represent a feasible spectrum (see Section 3.2 for the details). In the following, our analysis aims at determining the AFS for the spectral factor A starting from a spectral data matrix D. This analysis can immediately be used to determine the AFS for the concentrational AFS containing the feasible factors C. Therefore we apply the procedure to the transposed data matrix D^T since in $D = CA$ the factors change their places by the transposition $D^T = A^T C^T$.

The aim of this section is to introduce the AFS and to discuss some of its important properties. We consider especially those properties which are decisive for an effective numerical computation of the AFS.

3.1 Development of the AFS Concept and Discussion of Methods for Its Numerical Computation

In this section a short overview is given on the development of the AFS concepts. These developments are closely related with the growth of effective numerical methods for its computation.

The AFS for two-component systems was first analyzed by Lawton and Sylvestre in 1971 (see Ref. [13]). The Lawton–Sylvestre plot is a 2D plot of the set of the two expansion coefficients (with respect to the basis of singular vectors) which result in nonnegative matrix factors C and A. The Lawton–Sylvestre plot for a two-component system consists of two cones whose boundaries can be computed analytically (see Refs. [22–24] and Section 4).

For three-component systems a direct analogue of the Lawton–Sylvestre plot would be a three-dimensional plot of feasible expansion coefficients. Such three-dimensional objects are somewhat more complicated to draw, to handle, and to understand. However, there is a tricky dimension reduction (by a certain scaling) which allows to represent these AFS sets for three-component systems only by two expansion coefficients (and thus by plots in 2D). This was suggested by Borgen and Kowalski, who published in 1985 [25] a geometric construction of these 2D AFS plots for three-component systems. These plots are called Borgen plots. The mentioned dimension reduction can be explained by a typical Lawton–Sylvestre plot which is shown in Fig. 5. For a two-component system the Lawton–Sylvestre plot consists of two cones. If the first expansion coefficient t_1 is fixed to 1, then the intersection of the dash-dotted line at $t_1 = 1$ with the two cones are two separated intervals. These two intervals are a 1D analogue of the Borgen plots. For a three-component system the 2D Borgen plot is the two-dimensional intersection of a plane at t_1 with a 3D generalization of a Lawton–Sylvestre plot.

The *geometric construction* of the Borgen plots is deepened by further concepts in Refs. [26, 27]. In addition to the geometric constructions of the Borgen plots, various techniques for a numerical approximation of the AFS for ($s = 3$)-component systems have been devised. These are the grid search method (for two-component systems in Refs. [22, 23]), the triangle enclosure method [28], and the polygon inflation method [29, 30]. One benefit of the numerical methods compared to the geometric methods is that the numerical methods are able to compute the AFS in the presence of noise. Recently, the classical geometric construction has been generalized in a way which allows geometric AFS constructions for noisy data [27].

For systems with more than three components the problem of AFS computations is more complex and requires large computation times. For four-component systems a generalized triangle enclosure method which uses a slicing of the 3D AFS has already been published [31]. The basic grid search algorithm and the polygon inflation method can also be extended to such higher dimensional problems.

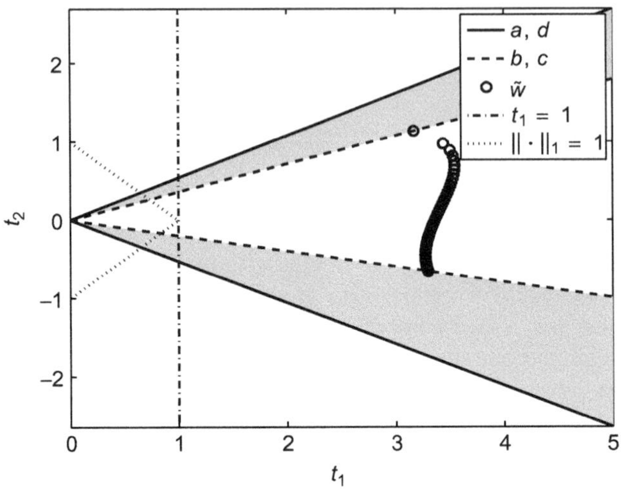

FIG. 5 The Lawton–Sylvestre plot (*gray triangle-shaped areas*) for a two-component model problem. The *dash-dotted* and the *dotted lines* define two intersections with the Lawton–Sylvestre plot. These intersections are the 1D AFS representations with respect to the FSV scaling and with respect to the $\|\cdot\|_1$ normalization. The points $\tilde{w}(i,:) = D(i,:)V(:,1:2)$, $i = 1,\ldots,n$, are drawn by *small circles*. These points represent the rows of D. Mathematically the coordinates of these points are the expansion coefficients of the rows of D with respect to the two dominant right singular vectors. For a more detailed explanation of the AFS and for the meaning of a, b, c, and d, see Section 4.

AFS computations can be used to determine the so-called *feasible bands*. To this end, one can draw the continua of concentration profiles or the continua of possible spectra which are represented by the AFS. The resulting feasible bands visualize the rotational ambiguity. The MCR-Bands toolbox, see Chapter 3 or Refs. [9, 32], pursues the same objective, namely to construct the feasible bounds. However, the MCR-Bands approach does not require a previously constructed AFS. Instead, the band boundaries are constructed by a minimization respectively maximization of a properly constructed target function.

3.2 The Set of Feasible Pure Component Spectra

Our focus is on the construction of the set of all possible pure component spectra. The set of feasible concentration profiles can be computed similarly; therefore, the computational procedure is to be applied to D^T. The permutation of the columns of C and the same permutation applied to the rows of A do not provide any new information. This fact is known as the (trivial) permutation ambiguity. A consequence of this property is that the problem to find all feasible factors A is equal to the problem to determine the set of all first rows of the feasible factors A. Hence the set of feasible pure component spectra (also called feasible bands) for an s-component system reads

$$\mathcal{A} = \{a \in \mathbb{R}^n : \text{ exist } C, A \geq 0 \text{ with } A(1,:) = a \text{ and } D = CA\}. \tag{3}$$

For the computation of $\mathcal{A}$ we prefer the SVD-based approach (2). A further reduction of the degrees of freedom is possible. This is explained next.

3.3 Reduction of the Degrees of Freedom

Eq. (2) is a representation of the $s \times n$ matrix A by the matrix T with only s^2 matrix elements. These s^2 matrix elements are the expansion coefficients with respect to the basis of right singular vectors. As shown for the derivation of (3) only the first row of $A = TV^T$ is required in order to form the set $\mathcal{A}$ of feasible spectra. The first row of A equals the first row of TV^T. Hence only the s matrix elements of the first row of T are decisive. This reduces the degrees of freedom from s^2 down to s.

These remaining s degrees of freedom for an s-component reaction system can further be reduced to $s - 1$ by a certain scaling of the rows of A. In Refs. [25, 26] the $\|\cdot\|_1$ vector norm (i.e., the sum of the absolute values of the components) is used for the normalization of the spectra. Alternatively, the maximum norm $\|\cdot\|_{\max}$ can be used. Here we follow the approach in Refs. [28–31] and use a scaling which sets the first column of T equal to ones, i.e.,

$$T(i,1) = 1 \quad \text{for all} \quad i = 1, \ldots, s. \tag{4}$$

This can be called the first-singular-vector scaling (FSV scaling) since it uses for the first right singular vector the fixed expansion coefficient 1. A precise justification for this choice is based on the Perron–Frobenius theory [33] on spectral properties of nonnegative matrices; for the details, see Theorem 2.2 in Ref. [30]. Fig. 5 shows a typical Lawton–Sylvestre plot for a two-component model problem. The dash-dotted and the dotted lines define two intersections with the Lawton–Sylvestre plot. These intersections are the 1D AFS representations with respect to the FSV scaling and with respect to the $\|\cdot\|_1$ normalization.

3.4 Definition of the AFS

With respect to the FSV scaling (4) the matrix T in (2) has the form

$$T = \begin{pmatrix} 1 & x_1 & \cdots & x_{s-1} \\ 1 & & & \\ \vdots & & S & \\ 1 & & & \end{pmatrix}. \tag{5}$$

Therein S is an $(s-1) \times (s-1)$ matrix. Only the $s - 1$ elements of the row vector $x = (x_1, \ldots, x_{s-1})$ are decisive for the representation of the set of all feasible solutions.

With these definitions the set $\mathcal{A} \subset \mathbb{R}^n$ by Eq. (3) can be represented by the associated set of expansion vectors $x \in \mathbb{R}^{s-1}$. Such a set of $(s-1)$-dimensional vectors for a chemical reaction system with s species is much easier to handle compared to the subset $\mathcal{A}$ of the higher dimensional space $\mathbb{R}^n$. The set

$$\begin{aligned}\mathcal{M} = \{x \in \mathbb{R}^{s-1} : \text{ exists } S \text{ so that } T \text{ in } (5) \text{ fulfills} \\ \operatorname{rank}(T) = s \text{ and } C, A \geq 0\}\end{aligned} \tag{6}$$

is called the AFS for the factor A or the spectral AFS.

Fig. 6 shows a typical AFS for the three-component system underlying the data set 2. This AFS consists of three isolated subsets. In Refs. [29, 30] these subsets are called segments of an AFS. Further, two series of points are marked within two segments of the AFS. The associated series of spectra, which are represented by these points, are also shown in Fig. 6. Fig. 7 displays the AFS sets of the factors C and A for the data set 3. Additionally, this figure shows the associated bands of feasible pure component spectra, i.e., the set $\mathcal{A}$. The feasible bands of concentration profiles are also plotted; this set results from computing the set $\mathcal{A}$ for the transposed data matrix D^T.

3.5 Properties of the AFS

The AFS has several interesting properties. Many of these properties derive from the Perron–Frobenius spectral theory of nonnegative matrices [33]. This theory provides (see Section 3.3) the justification for the scaling condition (4). An important property of the AFS is its boundedness (see Section 3.5.2). This property makes possible a numerical approximation of the boundary of the AFS. The AFS sets may have several shapes. For three-component systems the most important cases are AFS sets which consist of three separated segments and AFS sets which have the form of a topologically connected set with a single hole. Such a hole always contains the origin (null vector) (see, e.g., Fig. 24 and Section 3.5.3). Further explanations on the geometric construction of the

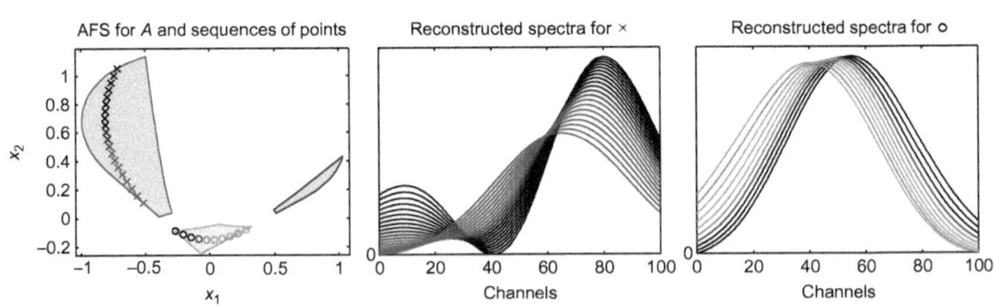

FIG. 6 Data set 2. *Left*: The spectral AFS for this three-component system consists of three isolated subsets. In two of these subsets sequences of points are marked. *Center*: The points marked by × have a color shading from *red* (*gray* in the print version) to *black*. *Right*: The points marked by ∘ have a color shading from *green* (*light gray* in the print version) to *black*. The associated series of spectra are shown in the same color shading in the remaining two subplots.

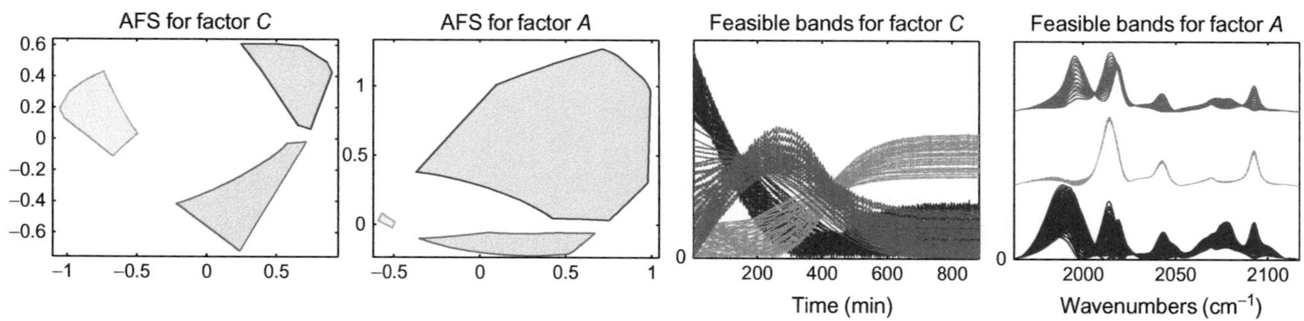

FIG. 7 Data set 3: The areas of feasible solutions, namely the AFS for the concentration factor C and the AFS for the spectral factor A, are plotted (*left*) together with the associated bands of feasible solutions (*right*).

AFS and its relationship to the theory of simplices and convex combinations are contained in Refs. [26, 27].

3.5.1 Definition of FIRPOL and INNPOL

For the further analysis the two polygons FIRPOL and INNPOL are to be introduced. First, the set

$$\mathcal{M}^{+}=\{x\in\mathbb{R}^{s-1}:(1,x)V^{T}\geq 0\} \tag{7}$$

is called FIRPOL [25–27]. All points x in $\mathcal{M}^{+}$ result in nonnegative linear combinations of the right singular vectors, i.e., $(1,x)V^{T}$. Thus FIRPOL is a superset of the set of feasible spectra. The membership of a certain x to $\mathcal{M}^{+}$ does not guarantee that the nonnegative spectrum $(1,x)V^{T}$ is part of a feasible pure component decomposition $D = CA$. The crucial point is that nonnegativity of $(1,x)V^{T}$ does not necessarily imply the nonnegativity of an associated concentration profile.

Further, the set $\mathcal{M}^{*}$ is to be introduced

$$\begin{aligned}\mathcal{M}^{*}=\{x\in\mathbb{R}^{s-1}:\ &\text{exists } S \text{ so that } T \text{ in } (5) \text{ fulfills}\\ &\operatorname{rank}(T)=s \text{ and } C\geq 0,\ A(2:s,:)\geq 0\}.\end{aligned} \tag{8}$$

The two sets $\mathcal{M}^{+}$ and $\mathcal{M}^{*}$ are supersets of the AFS $\mathcal{M}$. The definition of $\mathcal{M}^{+}$ includes only the nonnegativity constraint $A(1,:)\geq 0$. The definition of $\mathcal{M}^{*}$ includes the remaining constraints on nonnegativity and the rank condition. Thus $\mathcal{M}=\mathcal{M}^{+}\cap\mathcal{M}^{*}$. Finally, the vectors

$$w(i,:)=\frac{D(i,:)V(:,2:s)}{D(i,:)V(:,1)}\in\mathbb{R}^{s-1}\ \text{for } i=1,\ldots,n \tag{9}$$

are introduced. In Refs. [25–27] the convex hull of these points $w(i,:)$, $i=1,\ldots,n$, is called INNPOL.

3.5.2 Boundedness of the AFS

Three *numerical* approximation methods for the computation of the AFS for two- and three-component system have been described in literature. These are the grid search algorithm [22], the triangle enclosure method [28], and the polygon inflation method [29]. In order to guarantee that these algorithms terminate within finite times, the boundedness of the AFS is required. Theorem 2.4 in Ref. [30] proves that $\mathcal{M}^{+}$ is a bounded set if and only if $D^{T}D$ is an irreducible matrix. Thus the AFS $\mathcal{M}$ is also a bounded set since $\mathcal{M}\subset\mathcal{M}^{+}$. See Ref. [33] or [30] for the definition of an irreducible quadratic matrix. Practically, irreducibility is always guaranteed if the whole system does not allow a complete separation into two independent subsystems. The latter case is already a trivial case as completely isolated subsystems can be analyzed separately.

3.5.3 *The Origin Is Never Included in the AFS*

As already mentioned in Section 3.5 the origin (or null vector) is never contained in the AFS. The proof for this fact is given in Theorem 2.2 in Ref. [30]. It is based on the Perron–Frobenius theory and uses the irreducibility of the matrix D^TD. For the cases that the AFS consists of several isolated subsets, called segments, these subsets do not contain any holes. In mathematics such sets are called *simply connected.* The approach of inflating polygons can be used in order to approximate such AFS segments. The remaining case that the AFS consists of only a single set is more complicated. Such a single-segment AFS always contains a hole and this hole encloses the origin. The inverse polygon inflation algorithm is a modification of the polygon inflation algorithm which allows a fast and accurate numerical approximation of such one-segment AFS sets with a hole. See Section 4 in Ref. [30] for the details.

3.5.4 *Geometric AFS Construction*

The geometric construction of the AFS for three-component systems is introduced in Refs. [25, 34] (see also Refs. [26, 27]). The construction principles of Borgen plots are not limited to $s = 3$ but can be applied to general s-component systems. In the general case a point x is feasible if and only if there exist further $s - 1$ points y_j, $j = 1\ldots, s-1$, so that the simplex spanned up by x and the vectors y_j is enclosed in $\mathcal{M}^+$ and includes all points $w(i,:)$ given by Eq. (9). These points are the vertices of INNPOL. Consequently, the classical theory by Borgen and Kowalski works with convex equations. A weakening from *convex combinations* toward *affine combinations* allows to generalize the geometric AFS construction [27]. The resulting generalized Borgen plots can be constructed even for noisy or perturbed spectral data.

3.6 Segment Structure of the AFS

For two-component systems and with respect to the FSV scaling, the AFS (6) always consists of two separated 1D intervals. The intervals may degenerate to single points. One of these intervals is completely negative and the other one is completely positive (see Fig. 8 and in Fig. 5 the cross-section at $t_1 = 1$).

For three-component systems the AFS can consist of a single segment with a hole around the origin or of a number of $3m$ segments for $m = 1,2,\ldots$. A formal proof is planned for a forthcoming paper. For experimental data sets only the cases of a one-segment AFS and of three-segment AFS sets have been observed. Only these cases appear to be practically relevant. However, nonnegative matrices D can be constructed whose associated AFS sets consist of 6,9,... separated segments.

For s-component systems with $s \geq 4$ little information exists on the possible numbers of segments.

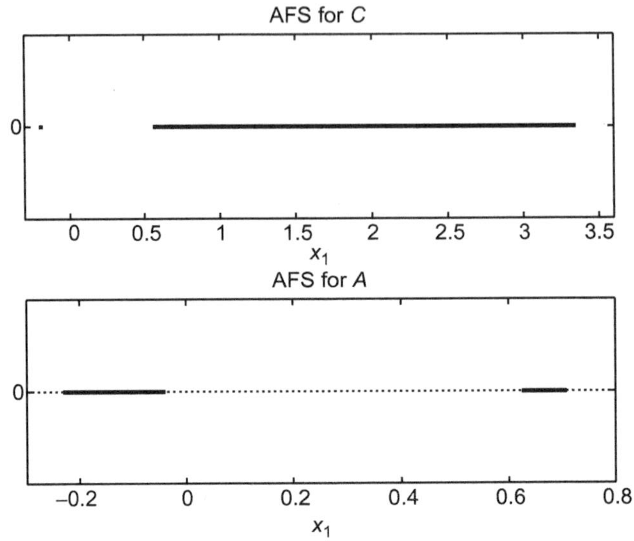

FIG. 8 Data set 1: The AFS $\mathcal{M}$ for this two-component system for the concentrational factor (*upper plot*) and for the spectral factor (*lower plot*). The control parameter ε on acceptable negative entries reads $\varepsilon = 0.035$.

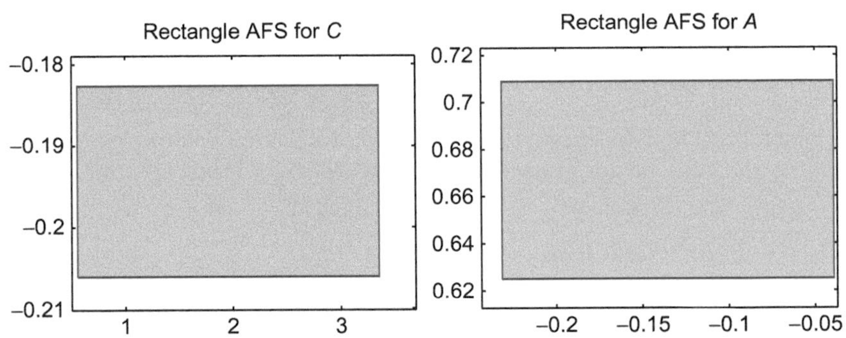

FIG. 9 Data set 1: Two-dimensional rectangle representation of the 1D AFS in which the two intervals of the one-dimensional AFS $\mathcal{M}$ are the edges of a rectangular. A certain point (with the two coordinates x and S_{11}) completely determines either C and A. The control parameter is again $\varepsilon = 0.035$.

By changing the data D continuously one can explore the resulting changes of the associated AFS sets. However, such parameter-dependent data sets could hardly be found as experimental data sets. The data sets 2 and 4 in Section 1.2 include the parameters w and σ. For each of these data sets a variation of these parameters allows to start from AFS sets with either three or four separated segments and to end in one-segment AFS sets each with a hole. See Fig. 24 for the AFS-dynamics in case of the data set 2 and Fig. 25 for the data set 4.

4 THE AFS FOR TWO-COMPONENT SYSTEMS

The AFS for systems with a number of $s = 2$ components can explicitly be described analytically. The numerical evaluation of the analytic formula results in the 1D AFS plots. Next these formulas are compiled. The starting point for the case $s = 2$ is the matrix T by (5) which together with its inverse read

$$T = \begin{pmatrix} 1 & x \\ 1 & S_{11} \end{pmatrix}, \quad T^{-1} = \frac{1}{S_{11} - x} \begin{pmatrix} S_{11} & -x \\ -1 & 1 \end{pmatrix}.$$

The nonnegativity for the factors C and A results in feasible intervals for x and S_{11} (see also Section 3.6 in Refs. [23, 30]). With

$$\begin{aligned} a &= -\min_{i \text{ with } V(i,2)>0} \frac{V_{i1}}{V_{i2}}, & d &= -\max_{i \text{ with } V(i,2)<0} \frac{V_{i1}}{V_{i2}}, \\ b &= \min_i \frac{U_{i2}\sigma_2}{U_{i1}\sigma_1}, & c &= \max_i \frac{U_{i2}\sigma_2}{U_{i1}\sigma_1} \end{aligned} \tag{10}$$

the AFS for the two-component system has the form of two separated intervals

$$\mathcal{M} = [a,b] \cup [c,d]. \tag{11}$$

This result can be interpreted in a way that both

$$x \in [a,b] \quad \text{and} \quad S_{11} \in [c,d]$$

and

$$S_{11} \in [a,b] \quad \text{and} \quad x \in [c,d]$$

result in nonnegative factors.

A certain choice

$$(x, S_{11}) \in [a,b] \times [c,d]$$

completely determines a nonnegative factorization $D = CA$. The second choice $(S_{11},x) \in [a,b] \times [c,d]$ does not provide any new information. Instead, the second solution is equal to the first solution after a row permutation in A and a column permutation in C. This fact justifies that the AFS for two-component systems is often represented by the rectangular $[a,b] \times [c,d]$ (see Refs. [22, 23, 30]).

4.1 Numerical AFS Computation for the Data Set 1

For noisy or perturbed spectral data it can be advantageous to accept small negative entries in the pure component factors. With a positive control parameter ε on the tolerated size of negative entries of C and A (see Section 5.1) one can generalize the AFS bounds as discussed above. Next numerical results generated by the *FACPACK* software [35] are presented for the two-component experimental FT-IR spectroscopic data set 1. Fig. 8 shows the one-dimensional AFS sets with $\varepsilon = 0.035$ for the concentrational factor C and for the spectral factor A. Additionally, Fig. 9 shows for the same problem the two-dimensional rectangle representation of the same AFS. The two intervals of the one-dimensional AFS form the edges of a rectangular. The advantage of such a rectangle representation is that a certain point (with the two coordinates x and S_{11}) completely determines C or A. Moreover, a known factor C allows to compute A from $D = CA$ and vice versa.

5 FEASIBILITY OF POINTS IN THE AFS

For the case of two-component systems the question of the feasibility of a certain point is solved by the analysis in Section 4 (at least for noise-free data). In the following we consider systems with three or more components. For these systems the feasibility question arises in two major ways. There is first the feasibility analysis for noise-free (model) data by considering a certain geometric construction. Alternatively, and with a focus on experimental spectral data, there is a numerical feasibility analysis which is based on the numerical solution of an optimization problem. Unfortunately, the numerical feasibility test can yield false results, if the numerical optimization procedure (e.g., due to convergence problems or poor initial estimates) is not successful.

This section explains the feasibility checks of the polygon inflation algorithm by soft constraints (in the Section 5.1), of the triangle enclosure technique as well as the grid search method (see Section 5.2), and of the geometric-constructive Borgen plot approach (see Section 5.3).

5.1 Soft Constraint-Based Feasibility Check

Soft constraints can be added to the feasibility check on nonnegativity. The aim of this approach is to compute the matrix elements of the submatrix S of (5) by solving a minimization problem for a certain target function which guarantees that CA is a good approximation of the initial matrix D. Simultaneously various constraints on C and A are to be satisfied. This feasibility test which also underlies the polygon inflation algorithm [30] is explained in the following.

First we introduce a small control parameter $\varepsilon \geq 0$ so that $-\varepsilon$ is a lower bound for the acceptable negative elements of C and A (in a relative sense related to the maximal value of a concentration profile or spectrum). Mathematically, these conditions read

$$\frac{\min_j C(j,i)}{\max_j C(j,i)} \geq -\varepsilon, \qquad \frac{\min_j A(i,j)}{\max_j A(i,j)} \geq -\varepsilon$$

for all $i = 1,\ldots,s$. The acceptance of small negative entries can often significantly stabilize the computation in case of noisy or perturbed (e.g., by a background subtraction) data.

The feasibility test for a point x is done in two steps. First, a rapid and computationally very cheap test is used in order to check whether x is contained in the set FIRPOL $\mathcal{M}^+$ (see Eq. (7)). If x is not in FIRPOL, then x cannot be an element of the AFS $\mathcal{M}$. Once again, we accept small negative entries. To this end we use an approximate FIRPOL test in order to check whether or not

$$f_0(x) := \min\left(\frac{(1,x)V^T}{\| (1,x)V^T \|_\infty} + \varepsilon, 0\right) \tag{12}$$

satisfies that $f_0(x) \geq 0$ (in a component-wise sense). Therein, $\| \cdot \|_\infty$ is the maximum vector norm, which is the largest absolute value of all components of its argument.

If this test is passed successfully, then a second much more expensive test follows. Therefore the soft constraint function

$$\begin{aligned} f(x,S) = &\sum_{i=1}^{s} \left\| \min\left(\frac{C(:,i)}{\| C(:,i) \|_\infty} + \varepsilon, 0\right) \right\|_F^2 \\ &+ \sum_{i=2}^{s} \left\| \min\left(\frac{A(i,:)}{\| A(i,:) \|_\infty} + \varepsilon, 0\right) \right\|_F^2 + \| I_s - TT^+ \|_F^2 \end{aligned} \tag{13}$$

is considered with $T = T(x,S)$ by (5). Further, C, A are computed according to (2). If

$$\min_{S \in \mathbb{R}^{(s-1)\times(s-1)}} f(x,S) \leq \varepsilon_{\text{tol}}, \tag{14}$$

then x has passed the feasibility test successfully. Therein ε_{tol} is a small positive control parameter, e.g., $\varepsilon_{\text{tol}} = 10^{-10}$.

To summarize, the approximate feasibility test with the control parameters ε and ε_{tol} results in the (approximate) AFS

$$\mathcal{M} = \{x \in \mathbb{R}^{s-1} : x \text{ fulfills } f_0(x) \geq 0 \text{ and } \min_S f(x,S) \leq \varepsilon_{\text{tol}}\}. \tag{15}$$

Fig. 10 shows the function $\|f_0(x)\| + \min_S f(x,S)$ on the grid $(x_1,x_2) \in [-1.2, 1.2] \times [-0.4, 1.4]$ for the data set 2. For these computations the control parameters are $\varepsilon = 10^{-12}$ and $\varepsilon_{\text{tol}} = 10^{-6}$. All points with $\|f_0(x)\| + \min_S f(x,S) \leq 10^{-6}$ belong to the AFS; these are the points at the valley bottom in Fig. 10.

For the data set 1, a two-component system, Fig. 11 displays the function $\|f_0(x)\| + f(x,S)$ on the grid $(x,S) \in [-0.29, -0.01] \times [0.55, 0.8]$ for $\varepsilon = 0.035$. For this two-component system no minimization is required; the arguments S and x of f are real numbers. The valley bottom is equal to the right subplot of Fig. 9. The endpoints of the intervals are approximately equal to a and b in (10) and are located on the x_1-axis. Further the endpoints c and d are located on the S-axis.

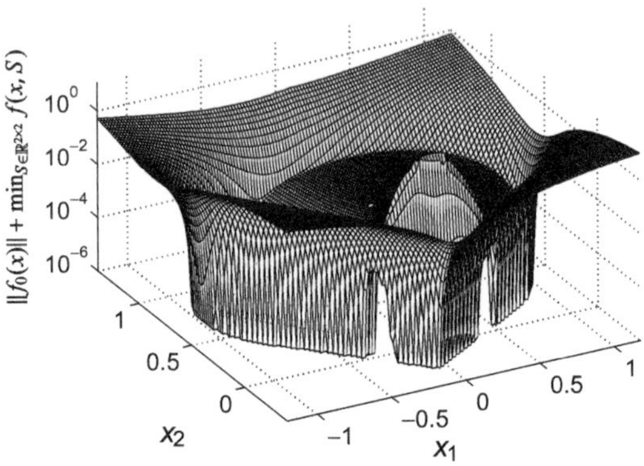

FIG. 10 Data set 2: The function $\|f_0(x)\| + \min_S f(x,S)$ is plotted. The valley bottom is the AFS.

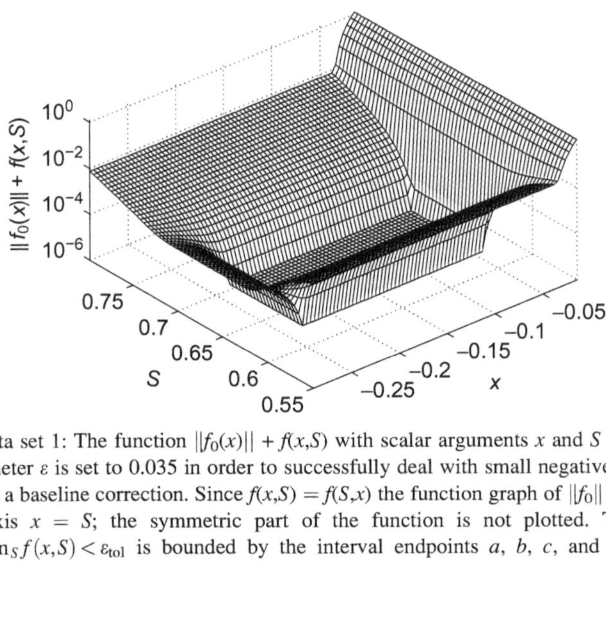

FIG. 11 Data set 1: The function $\|f_0(x)\| + f(x,S)$ with scalar arguments x and S is plotted. The control parameter ε is set to 0.035 in order to successfully deal with small negative entries which are caused by a baseline correction. Since $f(x,S) = f(S,x)$ the function graph of $\|f_0\| + f$ is symmetric to the axis $x = S$; the symmetric part of the function is not plotted. The area with $\|f_0(x)\| + \min_S f(x,S) < \varepsilon_{tol}$ is bounded by the interval endpoints a, b, c, and d as given in Eq. (10).

5.2 The *ssq*-Function-Based Feasibility Check

The *ssq*-function (ssq for sum-of-squares) approach evaluates the reconstruction functional

$$ssq(x,S) = \| D - \max(C,0) \cdot \max(A,0) \|_F^2 .$$

Therein $\max(C,0)$ and $\max(A,0)$ are the matrices whose negative entries are zeroed. The matrices C and A depend on $T = T(x,S)$ according to (5). The triangle enclosure algorithm [28] and the grid search method [22, 23] are based on the evaluation of the *ssq*-function. Computationally, the *ssq*-evaluation is relatively expensive as the computation of $\mathcal{O}(k \cdot n)$ squares is required whereas the evaluation of (13) needs only $\mathcal{O}(k+n)$ squares. For large numbers k and/or n this results in significantly different computation times; see, for example, Tables 1 and 2 in Ref. [30] for a direct comparison of the soft-constrained approach (13) compared to the *ssq*-based AFS computation.

Finally, the AFS can be written as

$$\mathcal{M} = \left\{ x \in \mathbb{R}^{s-1} : \min_{S} ssq(x,S) \leq \varepsilon_{\text{tol}} \right\}$$

for a fixed small parameter $\varepsilon_{\text{tol}} > 0$.

5.3 Geometric Constructive Feasibility Test

The fundamentals of the geometric AFS construction are shortly outlined in Section 3.5.4. Principally, these constructions are not limited to $(s = 3)$-component systems. However, the current literature does not contain any investigations for $s \geq 4$.

The geometric feasibility test of a certain point $x \in \mathcal{M}^+$ for the case $s = 3$ amounts to the following steps: First, two tangents of INNPOL are constructed which run through the given point x and which (tightly) enclose INNPOL. Next, the intersection of the first tangent with the boundary of $\mathcal{M}^+$ (the line segment between x and this point must touch at least one point w_i) is defined as P_1. The same is done for the second tangent. This results in the point P_2. Then, x is a feasible point of the AFS if and only if the triangle with the vertices x, P_1, and P_2 includes the polygon INNPOL (see Section 3.5.1). An extension of this geometric construction which is applicable to noisy or perturbed data is developed in Ref. [27].

6 AFS COMPUTATIONS FOR THREE-COMPONENT SYSTEMS

The definition of the AFS and the discussion of some of its numerous properties are now followed by a numerical algorithm for its computation. The focus is on the polygon inflation algorithm [29, 30] and its implementation in the *FACPACK* software. We also briefly discuss the geometric AFS construction, the triangle enclosure algorithm, the grid search approach, and the MCR-Bands method for the computation of upper and lower band boundaries. We do not claim to present a complete discussion of all methods for AFS computations. For example, we do not consider the particle swarm algorithm for the detection of feasible regions [36].

6.1 Borgen Plots and Computational Geometry

The geometric construction of Borgen plots has already been introduced in Section 3.5.4. This construction is purely geometric [25, 26]. The practical implementation of the construction on a computer requires methods of computational geometry. Hence a floating-point arithmetic is used so that certain approximation problems can occur. Perturbed spectral data, which result from an SVD low rank approximation or spectral data containing small negative entries (e.g., from background subtraction), cannot successfully be treated with the classical Borgen plots. In Ref. [27] a generalized Borgen plot construction has been suggested which has extended the construction principles to perturbed spectral data. Generalized Borgen plots can be constructed with the *FACPACK* software.

6.2 Grid Search

The grid search approach [22, 23] is a brute-force method to compute the AFS. It can be used for any $s \geq 2$. For two-component systems the function $f(x,S)$ is plotted on a proper grid in $\mathbb{R}^2$ (see Section 4.1 or Fig. 8). For the cases $s \geq 3$ one has to evaluate $\|f_0(x)\| + \min_S f(x,S)$ by Eq. (13) or $\min_S ssq(x,S)$ on a suitable grid. Figs. 10 and 11 show the function graphs for the data set 2 and for the data set 1. The grid search method can simply be implemented on a computer. However, the computational costs increase exponentially in the number of components s. Moreover, the grid search approach is a non-adaptive method. Any increase of the resolution by a factor κ (e.g., in each coordinate direction the number of grid points is multiplied by κ) results in an increase of the computational costs by the factor of κ^{s-1}.

6.3 Triangle Enclosure

The triangle enclosure method was introduced in 2011 for three-component systems in Ref. [28]. The idea is to approximate the boundary of the two-dimensional AFS by series of equilateral triangles which cover the boundary. Therefore an initial triangle is computed for which at least one vertex x fulfills that $x \in \mathcal{M}$ and at least one vertex y is not located in $\mathcal{M}$. The third point z can be a feasible or a nonfeasible point. Then the generation of further triangles starts. The idea is to mirror the most recently generated triangle along one of its edges which intersects the boundary of $\mathcal{M}$. This is done in a way that the triangle chain grows until the initial triangle is reached. Then the boundary of a segment of the AFS has been successfully approximated.

Fig. 12 illustrates the idea of the triangle enclosure method. The shape of an Erlenmeyer flask is approximated by a chain of equilateral triangles. This is done for different edge lengths a. The precision of the boundary approximation is equal to the edge length a. Halving the edge length in order to increase the approximation quality doubles the number of required triangles.

For each segment (or isolated subset) of the AFS a separate chain of triangles is to be computed (see Section 3.5 for the possible cases). However, if the AFS consists of only a single topologically connected set, then this set contains a hole which encloses the origin. In this case an additional run of

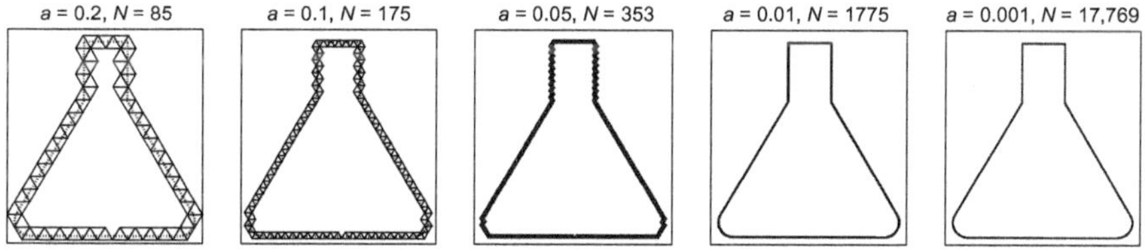

FIG. 12 Approximation of the shape of an Erlenmeyer flask by the triangle enclosure method by chains of equilateral triangles with edge lengths a. The required number of triangles is N. The start triangle is chosen in the way that its centroid is located in the middle of the bottomline. The edge length a limits the precision of the boundary approximation.

the triangle enclosure algorithm is required in order to approximate the inner boundary by a second chain of triangles. The computational costs of the triangle enclosure algorithm are significantly smaller compared to the grid search method. The reason is that feasibility tests are only required for points close to the boundary.

6.4 MCR-Bands

The MCR-Bands method, see Chapter 3 and Refs. [9, 32, 37], is not generically an AFS computation method. Instead, MCR-Bands aims at the computation of lower and upper boundaries for the feasible bands of each component. The method works for any $s \geq 1$. The band boundaries are computed by minimizing (for the lower boundaries) or maximizing (for the upper boundaries) a certain cost function. As this method does not aim at a direct computation of the AFS we refer for further explanations to Ref. [9].

The minimal and maximal band boundaries show an interesting property. If a band boundary function for the possible spectra is expanded with respect to the first s right singular vectors, then the associated expansion coefficients (after FSV-scaling) are located on the boundary of the AFS (see, e.g., Refs. [3, 38] for systems with $s = 2$ or $s = 3$ components). So far, a systematic explanation for this has not been given and is a possible topic for future work. For noisy or perturbed spectral data, the localization of the band boundaries on the boundary of the AFS does not hold strictly. The reason is that the MCR-Bands toolbox and the feasibility check by Eqs. (12)–(15) deal in different ways with perturbed or noisy data.

6.5 Polygon Inflation

The polygon inflation algorithm [29] and its algorithmic variation of inverse polygon inflation [30] are adaptive methods for the computation of the AFS for three-component systems. The idea is to approximate the boundary of each segment of the AFS by sequences of step-by-step refined polygons. A combination of a local error estimation with a local strategy for the refinement of the polygon results in a very effective, adaptive approximation scheme. Various control parameters allow to steer the approximation process and its quality. The polygon inflation method can be generalized to a polyhedron inflation procedure in order to compute the AFS for four-component systems (see Section 7). We show some first examples in this work.

The geometric idea of the boundary approximation by inflating polygons is demonstrated in Fig. 13. The shape of an Erlenmeyer flask is approximated in 2D by the polygon inflation method. Additionally the surface of a 3D Erlenmeyer flask is approximated by the polyhedron inflation algorithm. The surface of the polyhedron is a triangle mesh.

6.5.1 Steps of the Polygon Inflation Algorithm

In this section the steps of polygon inflation method are explained. The inverse polygon inflation algorithm derives from the standard polygon inflation in a way that the sets $\mathcal{M}^+$ and $\mathcal{M}^*$ by Eqs. (7) and (8) are computed separately by inflating polygons. Then the AFS $\mathcal{M}$ is computed by forming the intersection $\mathcal{M}^+ \cap \mathcal{M}^*$. Thus the costs for the inverse polygon inflation is less than twice that of the polygon inflation. For further details, see Refs. [29, 30]. Next the single steps of polygon inflation are explained for the case of an AFS consisting of three separate segments.

Step 1: Computation of an initial factorization of D

The first step of the polygon inflation method is to compute a nonnegative matrix factorization of D. According to Eqs. (2) and (5) this allows to find three points in the (spectral) AFS; the planar coordinates of these points are the matrix elements of the second and third column of T. If the AFS consists of three separated segments, then each segment contains exactly one of these points. For each of these three points an associated polygon is inflated.

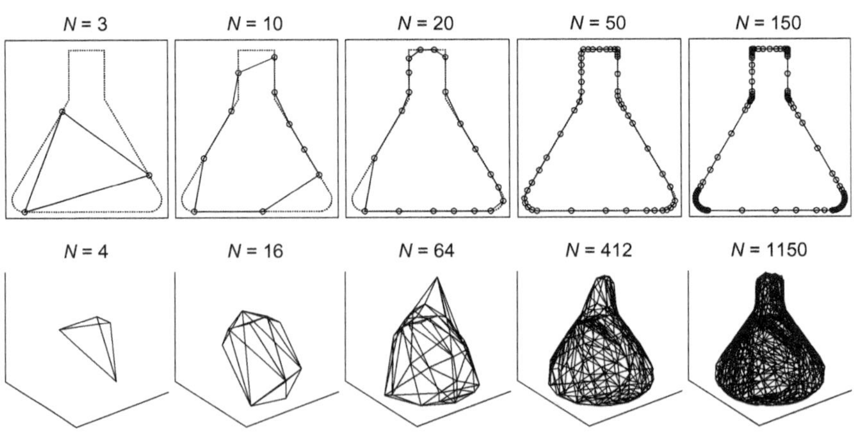

FIG. 13 *Top*: Approximation of the shape of an Erlenmeyer flask in 2D by the polygon inflation algorithm with N vertices. *Bottom*: Approximation of a 3D Erlenmeyer flask by the polyhedron inflation algorithm. The surface of the polyhedron consists of N triangles.

The following Steps 2 and 3 are executed for each of the three points from Step 1 which are located in the three AFS segments.

Step 2: Computation of an initial polygon

The starting point is used in order to construct an initial triangle. The three vertices of this triangle are located on the boundary of the AFS segment. The resulting initial triangle includes the starting point.

Step 3: Step-by-step polygon inflation by edge subdivision

Given a current polygon with m vertices, a certain edge of this polygon is selected for a refinement. The selection rules are explained later in this section. If P_i and P_{i+1} are the endpoints of the selected edge, then the mid-perpendicular of the edge is determined. The closest point of intersection[a] of this mid-perpendicular with the boundary of the AFS segment defines a new point P'_{i+1}. The vertices of the new and refined polygon read

$$[P_1, \ldots, P_i, P'_{i+1}, P_{i+1}, \ldots, P_m].$$

The introduction of the new vertex results in a gain or loss Δ_i of the area of the polygon. In most cases the area of the polygon increases, i.e., the polygon is inflated. If in rare cases the points P_i and P_{i+1} limit a nonconvex region of the boundary of the AFS segments, then the refinement process can decrease the area of the polygon.

Fig. 14 shows the three steps of the polygon refinement process for the data set 3. For the data set 2 with $w = 0$ the polygon inflation algorithm and the inverse polygon inflation algorithm are illustrated in Fig. 15 by sequences of step-by-step refined polygons.

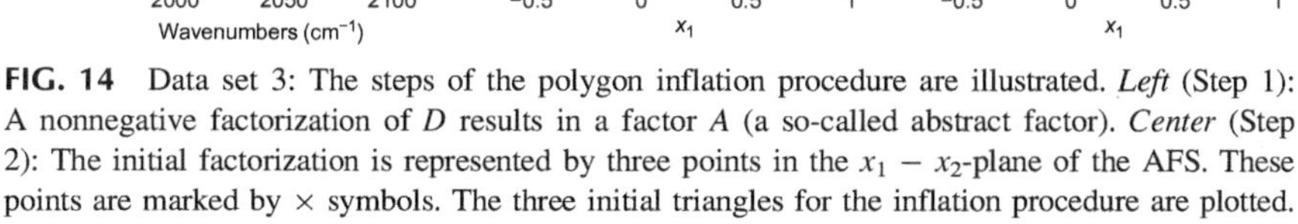

FIG. 14 Data set 3: The steps of the polygon inflation procedure are illustrated. *Left* (Step 1): A nonnegative factorization of D results in a factor A (a so-called abstract factor). *Center* (Step 2): The initial factorization is represented by three points in the $x_1 - x_2$-plane of the AFS. These points are marked by × symbols. The three initial triangles for the inflation procedure are plotted. *Right* (Step 3): The inflation procedure (applied separately to the three initial triangles) results in a final AFS approximation. This reproduces the AFS shown in the second subplot of Fig. 7.

[a]There is always a second point of intersection of the mid-perpendicular with the boundary of the AFS. However, this more distant point on the opposite side of the AFS segment would not result in a successful approximation of the entire boundary by a polygon.

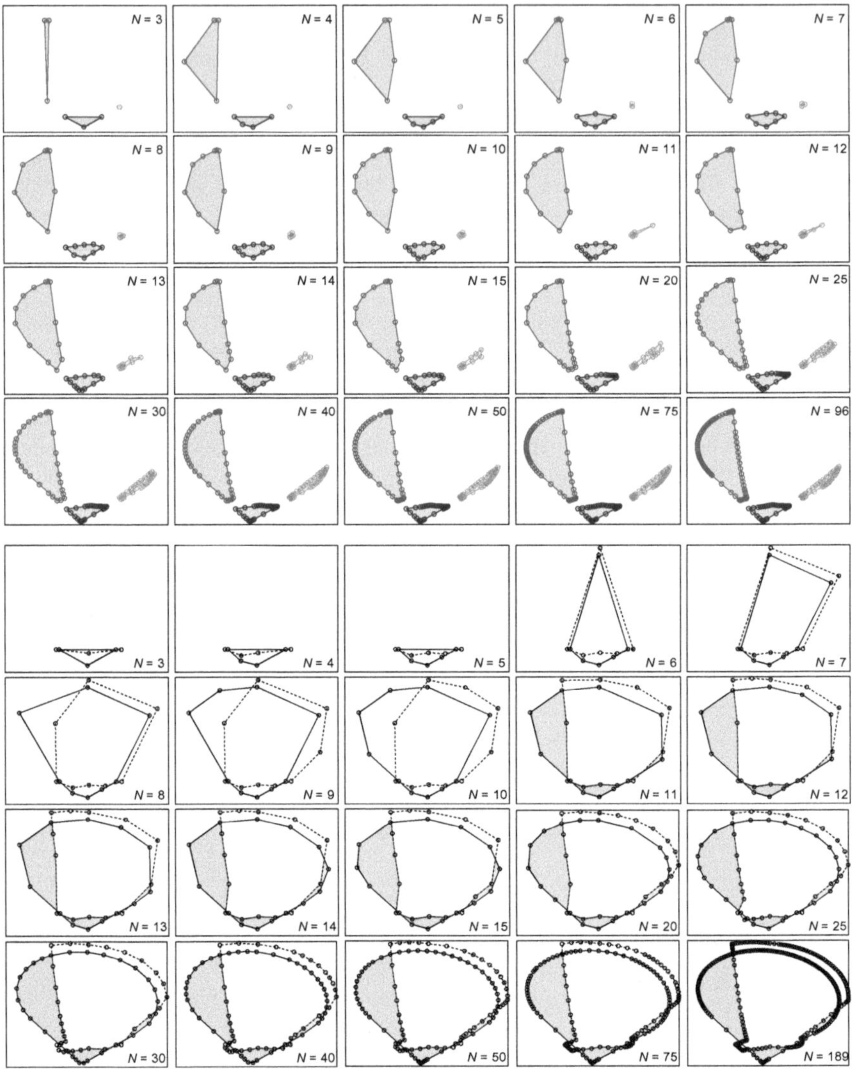

FIG. 15 Data set 2 with $w = 0$: Polygon inflation (rows 1–4) and inverse polygon inflation (rows 5–8) are illustrated for the same data matrix. In case of inverse polygon inflation the *solid lines* represent the boundary of $\mathcal{M}^+$ and the *broken lines* stand for the boundary of $\mathcal{M}^*$.

Polygon refinement

If a certain edge between the adjacent vertices P_i and P_{i+1} is marked for a refinement, then first the feasibility check, see Section 5.1, is applied to the mid-point $M = \frac{1}{2}(P_i + P_{i+1})$ of this edge. If M is feasible, then a second point $M' \notin \mathcal{M}$ is determined along the mid-perpendicular of the current edge. The new point M' is to be found in the outward direction with regard to the current polygon. If M is not a feasible point, then a second point $M' \in \mathcal{M}$ is determined along the mid-perpendicular in the inward direction with regard to the current polygon. Hence the two points M and M' on the mid-perpendicular are determined. One of these points is a feasible point and the other point is not a feasible point. In other words, the boundary of the AFS segment intersects the mid-perpendicular between M and M'. The point of intersection is to be determined. This can be done in a slow but very stable way by means of the iterative bisection method. The iteration is stopped if a final precision ε_b has been reached.

Edge selection and stopping criterion for the polygon refinement

The change-of-area variable $\Delta_i \geq 0$ as introduced in Step 3 of the polygon inflation procedure can be stored as an attribute for each edge. If each of the three edges of the initial triangle has been subdivided at least once, then Δ_i has been assigned for each edge of the polygon. The stopping control parameter $\delta > 0$ is used to stop the polygon refinement process if $\Delta_j < \delta$ for all edges of the polygon. Then the gain or loss of area in the last subdivision of every edge is bounded by δ. If this stopping condition is not fulfilled, then an edge with the index ℓ is selected for a subdivision if $\Delta_\ell = \max_j \Delta_j$.

Fig. 16 shows for the data set 2 with $w = 0$, a typical Δ_i convergence history in the *FACPACK* software. For these computations the stopping control parameter $\delta = 10^{-3}$ has been used.

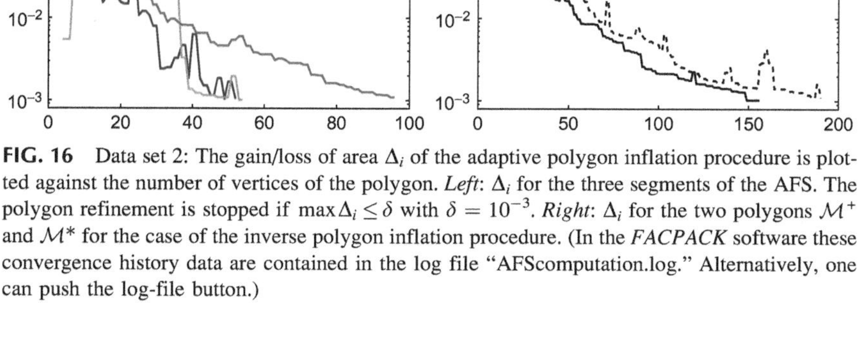

FIG. 16 Data set 2: The gain/loss of area Δ_i of the adaptive polygon inflation procedure is plotted against the number of vertices of the polygon. *Left*: Δ_i for the three segments of the AFS. The polygon refinement is stopped if $\max \Delta_i \leq \delta$ with $\delta = 10^{-3}$. *Right*: Δ_i for the two polygons $\mathcal{M}^+$ and $\mathcal{M}^*$ for the case of the inverse polygon inflation procedure. (In the *FACPACK* software these convergence history data are contained in the log file "AFScomputation.log." Alternatively, one can push the log-file button.)

6.5.2 The Control Parameters

The polygon inflation algorithm in the *FACPACK* software includes the following control parameters:

- $\varepsilon \geq 0$ is an upper bound for the acceptable relative size of negative entries in C and A (see Eqs. (12) and (13)).
- ε_{tol} is the tolerance parameter in the feasibility test (see Eqs. (14) and (15)).
- δ is the stopping parameter for the polygon refinement. The iteration is stopped if the nonnegative change of area attribute Δ_i of each edge of the polygon is bounded by δ. In the *FACPACK* software $\delta = 10^{-3}$ is the default value.
- ε_b is the control parameter on the guaranteed precision of the vertex localization close to the boundary of the AFS. This parameter should satisfy $\varepsilon_b \leq \delta$.

Suitable parameter selection for the parameter ε

The parameter ε on the acceptable relative size of negative entries in C and A considerably influences the size of the AFS. Thus ε should be selected carefully. A reasonable choice is as follows: If a computational procedure for nonnegative matrix factorizations (an NNMF routine) is only capable of producing a factorization $D = CA$ with matrices C and A with small negative entries, then ε can be selected according to

$$\varepsilon \geq \max\left(0, -\min_{i=1,\ldots,s} \frac{C(:,i)}{\|C(:,i)\|_\infty}, -\min_{i=1,\ldots,s} \frac{A(i,:)}{\|A(i,:)\|_\infty}\right). \tag{16}$$

As a rule of thumb, an increase of the value ε results in a growth of the area of the AFS. The crucial point is that increasing ε means that the factors C and A can contain (in absolute values) larger negative entries. Hence a larger number of approximate nonnegative matrix factorizations can be accepted as feasible. Thus the AFS grows. Fig. 17 illustrates this relationship for the data set 3 for four different values of ε.

6.5.3 Benefit of the Adaptive AFS Approximation

A strength of the polygon inflation method is its adaptive strategy for the boundary approximation. If the boundary of the AFS is locally more or less a straight line, then only few vertices of the polygon are necessary for a high-quality boundary approximation. For other boundary regions which are strongly curved or nonsmooth, a higher resolution of the polygon inflation method is a result of the local error estimation together with the local polygon refinement. Figs. 12 and 13 demonstrate that only few vertices are sufficient for a good approximation of rectilinear regions of the boundary.

Table 1 provides a comparison of the numbers of vertices which are necessary for an approximation with guaranteed boundary precision of the shape of an Erlenmeyer flask by the grid search approach, by the triangle enclosure method, and by the polygon inflation method. The boundary precision values

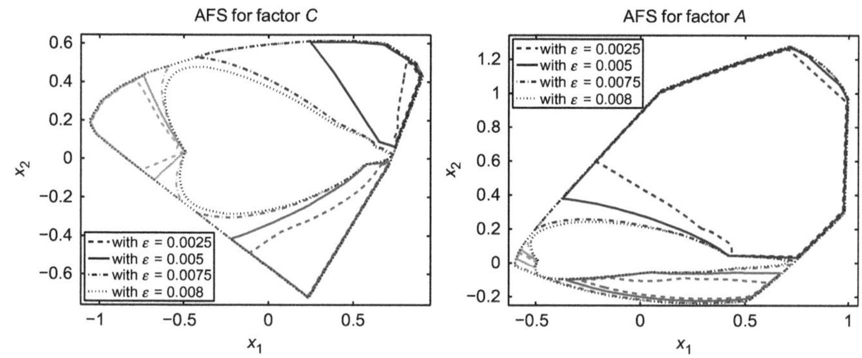

FIG. 17 Data set 3: Variation of the control parameter ε on the acceptable relative size of negative entries in *C* and *A* (see Section 6.5.2). The area of the AFS segments increases with rising ε. For ε ≥ 0.008 the initially isolated three segments are grown together into a single AFS with a hole (*black dotted lines*). The continua of feasible bands for the parameter ε = 0.005 are shown in the two rightmost plots of Fig. 7.

TABLE 1 Approximation of the Shape of an Erlenmeyer Flask by the Grid Search Approach, the Triangle Enclosure Method, and the Polygon Inflation Method

Boundary precision	Grid search	Triangle enclosure	Polygon inflation
10^{-1}	632	175	29
10^{-2}	6.32×10^4	1775	85
10^{-3}	6.32×10^6	17769	127
10^{-4}	6.32×10^8	177735	299
10^{-5}	6.32×10^{10}	1777349	901

Tabulated are the numbers of grid points (grid search approach), the total numbers of vertices of the approximating triangle chain, and the numbers of vertices of the approximating polygons. The boundary precision values are $\varepsilon = 10^{-i}$, $i = 1,\ldots,5$. These data affirm that the adaptive strategy has a clear advantage. See also Fig. 12 and the first row in Fig. 13.

are $\varepsilon = 10^{-i}$, $i = 1,\ldots,5$. Further, Table 2 in Ref. [29] contains a similar comparison of the triangle enclosure method and the polygon inflation method for a three-component model problem.

7 AFS COMPUTATIONS FOR FOUR-COMPONENT SYSTEMS

All the methods for AFS computations for three-component systems as explained in Section 6 can be extended to four-component systems. The AFS for a four-component system is a bounded subset of the three-dimensional space. In this section a short overview is given on already published methods and on an on-going research project.

7.1 The Slicing Method

In Ref. [31] the idea has been presented to compute the three-dimensional AFS by two-dimensional slices. In this chapter the triangle enclosure method has been used to approximate the boundary of the two-dimensional slices. For example, a slicing in z-direction works with a matrix T in (5) whose fourth column is fixed to a certain value ζ, i.e.,

$$T(i,4) = \zeta, \quad i = 1,\ldots,4.$$

(Additionally, the FSV scaling requires that $T(i,1) = 1$ for all i.) All other matrix elements of T are free variables and are determined within the optimization procedure. With such a choice of T, one can compute the intersection of the 3D AFS with the 2D plane with $z = \zeta$. The entire AFS is computed by setting z to $m + 1$ equidistant values in the interval $[z_b, z_u]$ with

$$\zeta_j = z_b + \frac{j}{m}(z_u - z_b), \quad j = 0,\ldots,m.$$

In each of the slices any of the AFS approximation methods from Section 6 can be applied.

Fig. 18 illustrates the slicing method for the four-component data set 4. In each of the slices the polygon inflation method has been applied in order to compute the 2D intersections (the slices) with the 3D AFS. A number of 30 slices has been used for the slicing in the coordinate directions. The achievable resolution in the slicing direction depends on the number of slices.

7.2 Polyhedron Inflation Method

The direct generalization of the adaptive polygon inflation method is the polyhedron inflation method. To this end the 3D AFS is approximated by a series of polyhedrons. Each face of the polyhedron is a triangle. Thus the AFS is approximated by a 3D triangle mesh. The principles of the triangle subdivision and of the local error estimation are similar to those of the 2D AFS, but there are various technical challenges. Once again, adaptivity is an advantage of the inflation algorithm. We will explain the method in detail in a future work.

The method is demonstrated by the approximation of the shape of a 3D Erlenmeyer flask in Fig. 13. Further, Fig. 19 shows the concentrational AFS and the spectral AFS for the four-component model system of the data set 4 with $\sigma = 750$.

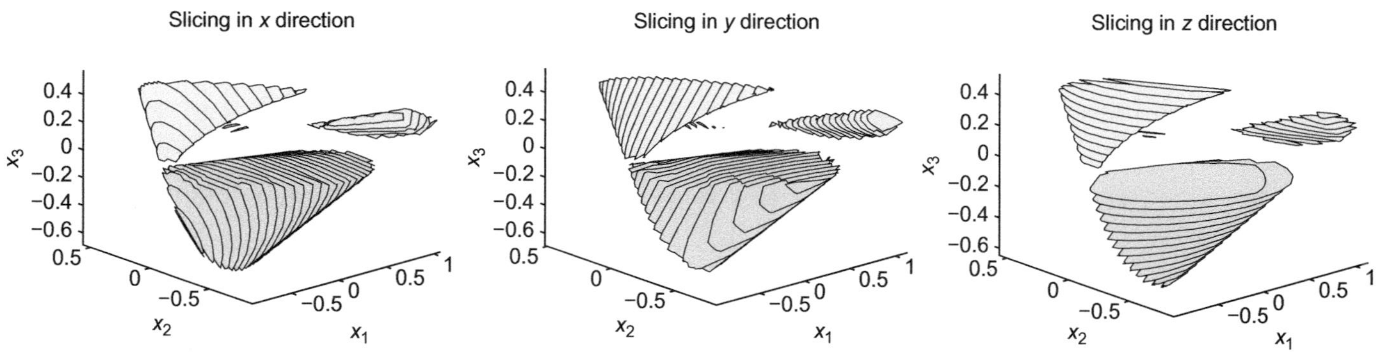

FIG. 18 Data set 4 with $\sigma = 750$: Computation of the AFS for the spectral factor A by slicing along the three coordinate axes. A number of 30 slices is used for each plot. The same AFS computed by polyhedron inflation is shown in the right subplot of Fig. 19.

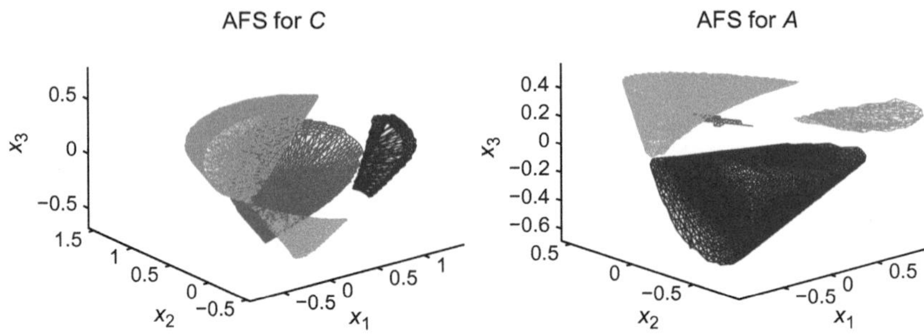

FIG. 19 Data set 4 for $\sigma = 750$: For this four-component model problem the *left plot* shows the concentrational AFS and the *right plot* is the spectral AFS. The concentrational AFS is also shown in Fig. 18.

8 REDUCTION OF THE ROTATIONAL AMBIGUITY BY SOFT CONSTRAINTS REPRESENTED IN THE AFS

The AFS construction includes the nonnegativity of the factors C and A as the one and only constraint. Sometimes additional information is available on the chemical reaction system. Such information is welcome in order to reduce the rotational ambiguity. For instance known concentration profiles or pure component spectra lead to a significant reduction of the rotational ambiguity by means of the duality- and complementarity theory [39–44]. Alternatively, soft constraints can be very useful for extracting chemically meaningful solutions from the AFS [45–47].

8.1 Equality Constraints: Locked Points in the AFS

The knowledge of a certain pure component spectrum is often called an equality constraint. It means that a certain point of the AFS is fixed (or locked). The effect on the remaining components is a reduction of the rotational ambiguity or equivalently a reduction of the area of the AFS segments. The reduced AFS can either be constructed geometrically (then one vertex in the simplex rotation algorithm [25] is fixed) or numerically (then a certain row of T is fixed). One observes that the reduction effect on the remaining segments of the AFS is relatively large if a point close to the origin is fixed.

All these concepts also apply to known concentration profiles. Then the whole procedure works with $D^T = A^T C^T$ where C and A have changed their places.

8.1.1 Complementarity Theory and the AFS

The restrictions on the factor C which results from a given pure component spectrum are theoretically well understood by the complementarity theory or duality theory [40–43, 48].

In general the AFS for an s-component system is a bounded subset of the $\mathbb{R}^{s-1}$. The complementarity theory provides a reduction of the rotational ambiguity to $(s-2)$-dimensional affine spaces. Theorem 3.1 in Ref. [43] proves that a given point x in the AFS for A forces the complementary concentration profiles represented by y to fulfill

$$\sum_{i=1}^{s-1} y_i x_i = -1.$$

Additionally locked points have an area-reduction impact on the AFS segment of the concentration factor of the same component for which the spectrum is given. All these restrictions can be combined if more than one point is locked. If s_0 points in the AFS for A are locked, then the admissible points y in the AFS for C for the complementary concentration profiles are in $(s-1-s_0)$-dimensional subspaces. This means for the important case of a three-component system that one known spectrum restricts the complementary two concentration profiles to a straight line in the AFS for C. Further the concentration profile of the chemical component whose spectrum is given is located in an AFS segment with a reduced area.

8.1.2 Equality Constraints Applied to a Three-component Model Problem

Fig. 20 illustrates for the data set 2 with $w = 0$ the reduction of the AFS for the cases of one or two given spectra. The left column shows the AFS sets for C and A only under the nonnegativity constraint. In the centered column the pure component spectrum of the component X is locked as a constraint (that is an equality constraint). The position of this spectrum is marked by a blue cross (dark gray in the print version) in the spectral AFS. This known spectrum results in a reduction of the areas of the AFS segments for the components Y and Z in the spectral AFS. Further the AFS segment for X in the concentrational AFS is reduced (black dashed lines). Due to the complementarity theory the AFS segments for the components Y and Z in the concentrational AFS are reduced to linear AFS segments (black dashed lines). In the right column of Fig. 20 a second spectrum is locked. The remaining segment in the AFS for A is reduced a second time. In the AFS for C a second line restriction has been added. The point of intersection of these two lines uniquely determines the concentration profile of the component Y (green color (light gray in the print version)).

8.2 Further Soft Constraints

The target function f by Eq. (13) or the *ssq*-function in Section 5.2 can easily be extended by additional soft constraints in order to strengthen certain desired characteristics of the solution [42, 45–47]. In other words, one hopes to extract only those points from the AFS which additionally satisfy certain

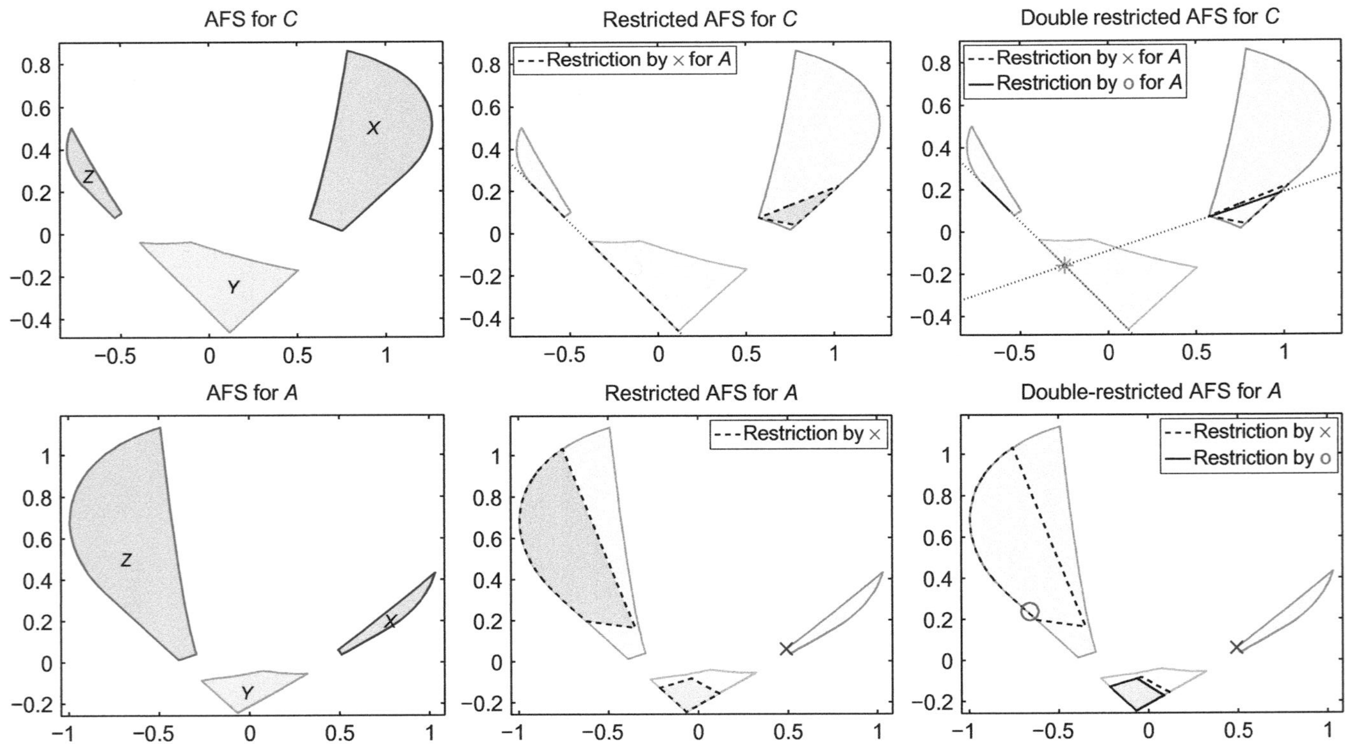

FIG. 20 Data set 2: Application of equality constraints (this is the lock mode in *FACPACK*) to the AFS for C and the AFS for A. The three segments of the AFS sets and their associated chemical components X, Y, and Z are shown in the *left column*. *Left column*: Areas of feasible solutions without restrictions. *Center column*: The pure component spectrum of the component X is locked as a constraint. The position of the spectrum in the spectral AFS is marked by a *blue cross* (*dark gray* in the print version). This known spectrum results in a reduction of the areas of the AFS segments for the components Y and Z in the spectral A. Further the AFS segment for X in the concentrational AFS is reduced (*black dashed lines*). Due to complementarity [41, 43] and duality theory [40], the AFS segments for the components Y and Z in the AFS for C are reduced to linear AFS segments (*black dashed lines*). *Right column*: If additionally the pure component spectrum for component Z is locked (*red* $\circ$ (*gray* in the print version)), then the segment for the component Y in the AFS for A is reduced a second time (*solid line*) and the concentration profile of Y is unique (*green* $*$ *symbol* (*light gray* in the print version)). Finally, the reduced AFS segments for X and Z in the AFS for C are line segments (*solid lines*).

(soft) constraints. Finally and by using various constraints, one hopes to extract only one and chemically meaningful solution. Tauler [9] uses a similar approach within the MCR-Bands method.

Possible soft constraints are the unimodality constraint for the concentration profiles or even stronger a monotonicity constraint. Alternatively, one can use assumptions on closure or windowing (see Section 4.3 in Ref. [46]). Our experiences concerning the closure constraint for AFS computations is that it induces only minor restrictions to the AFS. Closure can be expressed by the left singular vectors. Any linear combinations of the left singular vectors which contain negative components can be excluded from the AFS. In many cases the closure constraint results only in a minor reduction of the 2D geometric area of the AFS. For this reason, we have never applied the closure soft constraint for any AFS computations throughout this chapter. Fig. 21 for the data set 2 with $w = 0$ illustrates how the unimodality soft constraints can be used for a reduction of the AFS. Additionally, Fig. 21 shows the associated reductions of the feasible bands.

9 GEOMETRIC SHAPES OF THE AFS AND AFS DYNAMICS

The AFS can have various geometric shapes depending on the number s of independent components. For two-component systems the AFS always consists of two separated intervals. These intervals can degenerate to single-point intervals. One of these intervals contains only negative numbers and the other interval only positive numbers. For $s \geq 3$ the diversity of possible geometric shapes of AFS segments is much larger. In this section the focus is on the different (and partially degenerated) geometric shapes of the segments of an AFS. Furthermore the transition from an AFS with isolated segments to a one-segment AFS is demonstrated for model data sets with either three or four independent components.

9.1 Shapes of AFS Segments and Degenerated Segments

The segments of an AFS for an s-component system are subsets of the $\mathbb{R}^{s-1}$. Very often the $(s - 1)$-dimensional volume of these segments is nonzero. However, certain problems can contain one or more unique spectra or concentration profiles. A further possible case is that the nonnegative matrix factorization problem for a given D has a unique solution [49, 50]. Such unique solutions are associated with single-point AFS segments. Further *degenerated cases* are line-shaped AFS segments in the case of 2D AFS sets for three-component systems. In the case of four-component systems the possible degenerated cases are single-point, line-shaped, or planar AFS segments. These degenerated AFS segments have only been observed for model data. For perturbed or noisy experimental data the control parameter ε on tolerated negative entries inflates degenerated AFS segments (see Fig. 17).

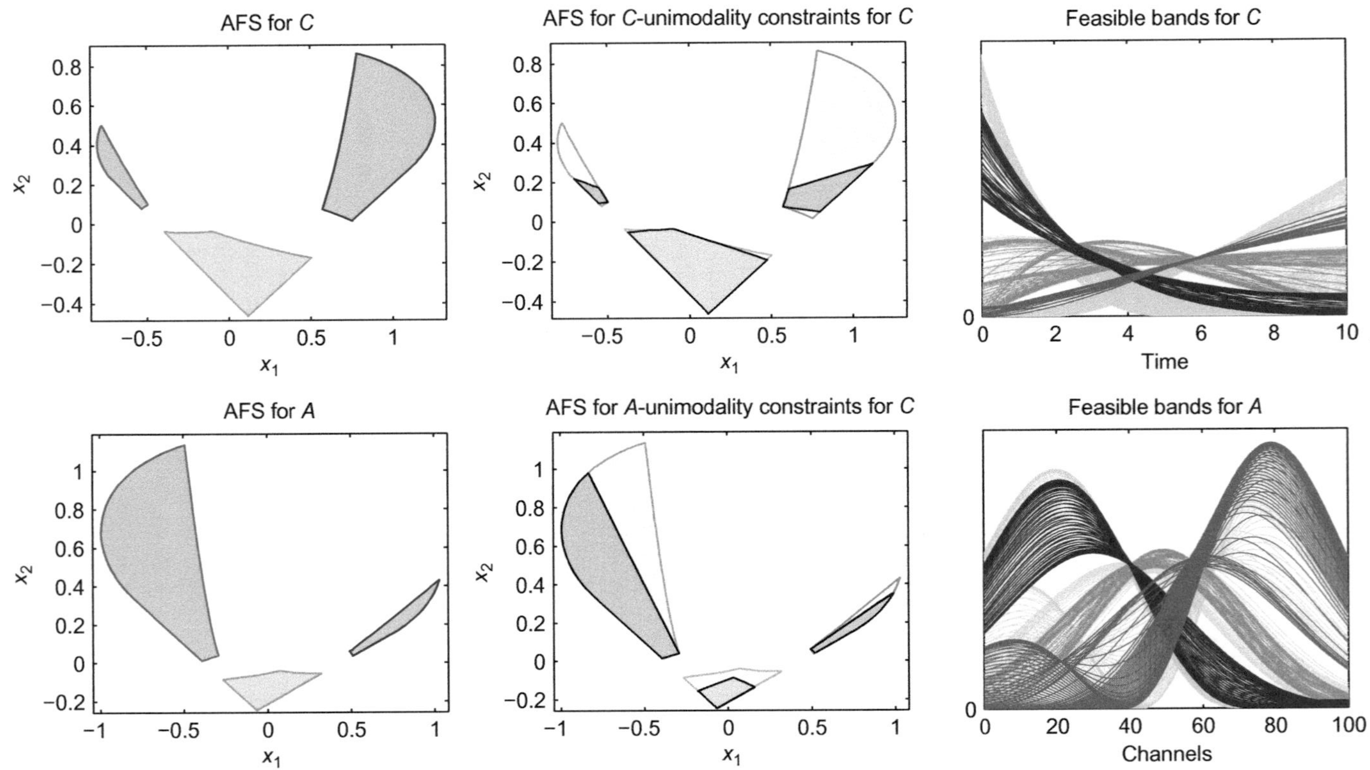

FIG. 21 Data set 2 for $w = 0$: Application of the unimodality soft constraint. *Left column*: AFS sets for C and A without restrictions. *Center column*: Application of monotonicity soft constraints only to the concentration profiles reduces the concentrational AFS and the spectral AFS. *Right column*: The associated feasible bands for C and A without unimodality soft constraint (*pale colors* (*light shades of gray* in the print version)) and with the unimodality soft constraint for the factor C (*bright colors*) (*dark shades of gray* in the print version).

Computationally, single-point AFS segments can easily be computed. The initial nonnegative matrix factorization of D provides their coordinates. Somewhat more complicated is the computation of i_0-dimensional segments with $1 < i_0 < s - 1$. (For example, for $s = 3$ the case $i_0 = 1$ refers to a line-shaped AFS segment.) For these segments the polygon inflation algorithm (and also the triangle enclosure method) (see Sections 6) as well as the polyhedron inflation method (see Section 7) is to be adapted properly. The start-up phase of the polygon inflation algorithm tries to build an initial triangle (with a nonzero area) and the polyhedron inflation algorithm tries to generate an initial nondegenerated tetrahedron (see left column of Fig. 13). In other words, the algorithms aim at the construction of an s-simplex with a nonzero volume. If this is not successful, the algorithms try to construct an $(s - 1)$-simplex with a nonzero ($(s - 1)$-dimensional) volume, and so on. The extreme and final case is that the algorithm ends in a 1-simplex, which is a unique point.

For three-component systems there is only the case of a line-shaped AFS segment between the extreme cases of a planar AFS segment with a nonzero area and a unique point. The numerical strategy for the computation of line-shaped AFS segments by polygon inflation is explained in Section 4.6 of Ref. [30]. The idea can easily be explained by means of a lighthouse. Starting from an initial feasible point x, which is considered as the position of the lighthouse, one determines a feasible beam direction so that the line

$$g_{r,x}(\varphi) = x + r\begin{pmatrix} \sin(\varphi) \\ \cos(\varphi) \end{pmatrix}$$

covers the desired line-shaped AFS segment. In a second step one determines the beam ranges in the form of bounds r_l and r_r so that $\mathcal{M} = \mathcal{L}_l \cup \mathcal{L}_r$ with

$$\mathcal{L}_l = \left\{ x + r\begin{pmatrix} \sin(\varphi - \pi) \\ \cos(\varphi - \pi) \end{pmatrix} \text{ with } r \in [0, r_l] \right\},$$

$$\mathcal{L}_r = \left\{ x + r\begin{pmatrix} \sin(\varphi) \\ \cos(\varphi) \end{pmatrix} \text{ with } r \in [0, r_r] \right\}.$$

This idea can easily be extended to the computation of planar AFS segments for the case of four-component systems; then two angles are to be determined.

Fig. 22 shows for a three-component model system that the three shapes of AFS segments can occur simultaneously. Both the AFS for C and the AFS for A contain a single-point segment, a line-shaped segment, and a planar segment with a nonzero area. Fig. 23 illustrates the analogue of this for a four-component model problem.

9.2 AFS Dynamics

Section 3.6 contains a discussion on the possible numbers of segments of an AFS. The obvious question on the inherent connection of this segment structure is discussed next. We demonstrate for the parameter-dependent data

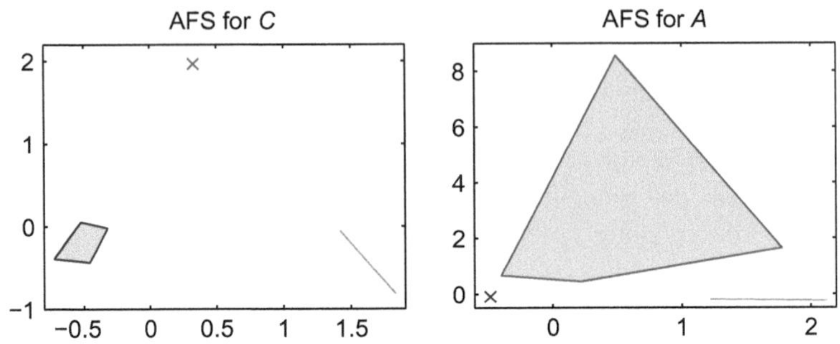

FIG. 22 The AFS sets of the factors C and A for a three-component model problem. Each AFS consists of three isolated segments. One segment is a unique point (i.e., one spectrum and one concentration profile are uniquely determined), a further segment is line shaped, and the remaining third AFS segment has a geometric area larger than zero.

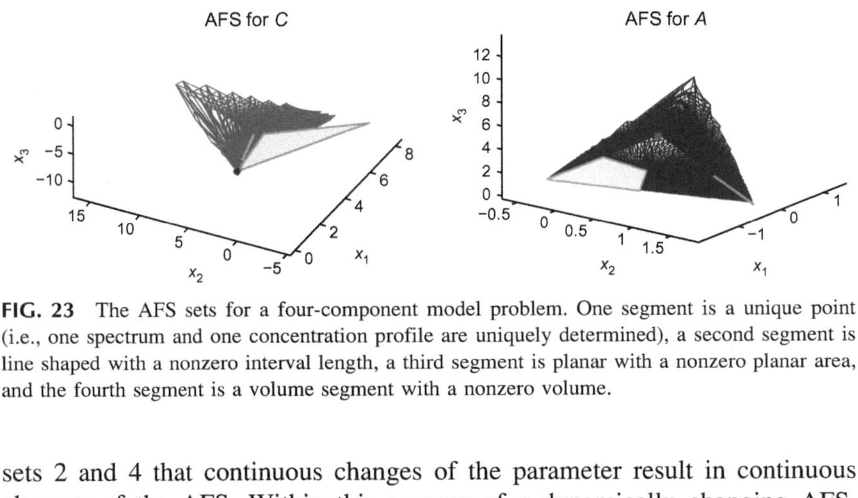

FIG. 23 The AFS sets for a four-component model problem. One segment is a unique point (i.e., one spectrum and one concentration profile are uniquely determined), a second segment is line shaped with a nonzero interval length, a third segment is planar with a nonzero planar area, and the fourth segment is a volume segment with a nonzero volume.

sets 2 and 4 that continuous changes of the parameter result in continuous changes of the AFS. Within this process of a dynamically changing AFS, the number of isolated segments of the AFS varies in a discontinuous way.

9.2.1 The Three-Component Model Data Set 2

We compute for the data set 2 the AFS sets for a sequence of parameter values $w \in [100, 240]$. For $w = 100$ the signal width and thus the overlap of the pure component spectra are relatively small. The AFS for the spectral factor contains three separated segments (see Fig. 24). For increasing w the segments show an increasing geometric area; the distances between the segments decrease. For $w \geq 170$ the AFS computation by the *FACPACK* software should better be done by the inverse polygon inflation procedure. For $w \geq 192$ the AFS consists of only one segment, which contains a hole. The point of discontinuity w for which the three-segment AFS turns into a one-segment AFS is the same for the spectral AFS and the concentrational AFS.

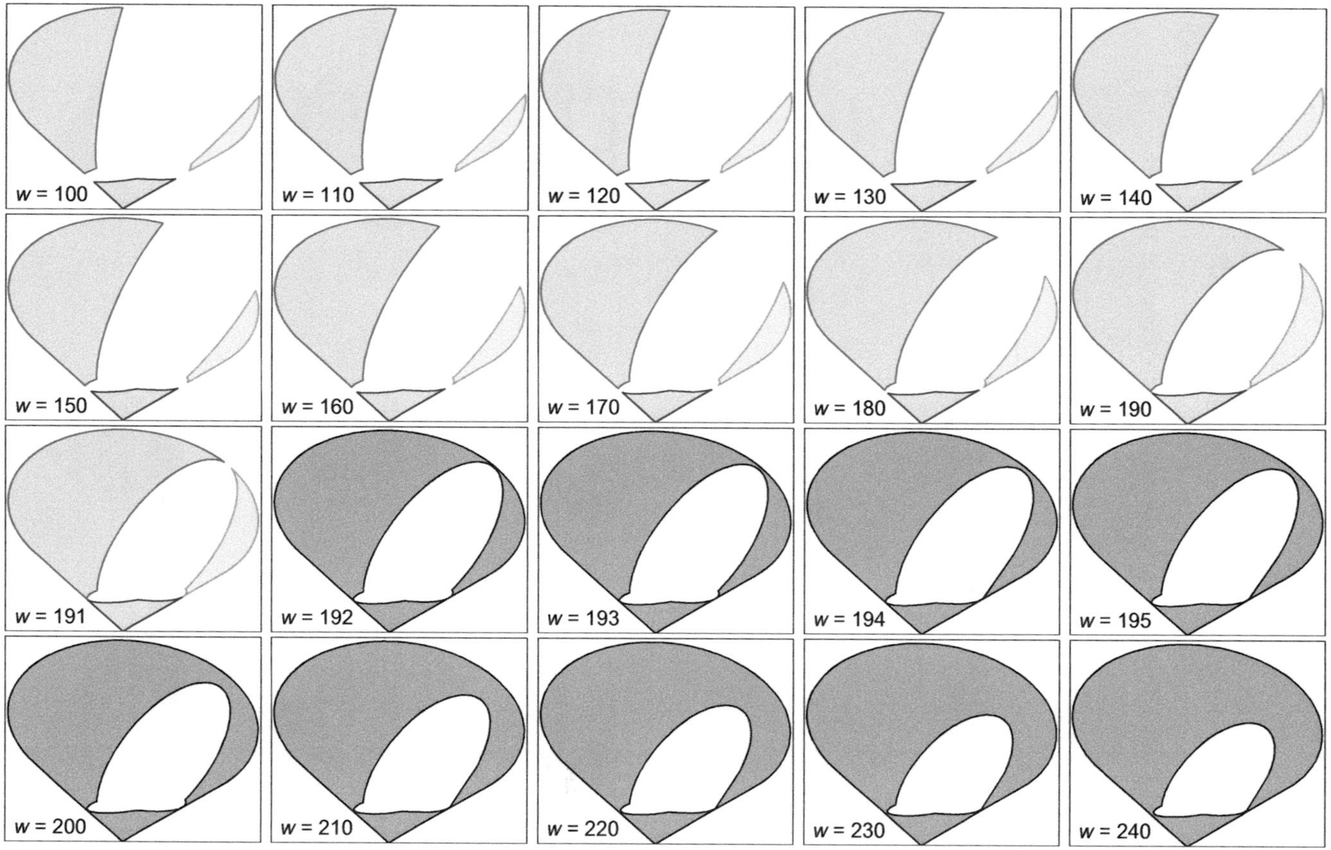

FIG. 24 Data set 2: AFS dynamics with respect to a variation of the model parameter w. The shape of the spectral AFS is changing from three isolated segments to a one-segment AFS with a hole. An AFS with exactly two isolated segments is not possible.

9.2.2 The Four-Component Model Data Set 4

Next the spectral AFS sets are computed for the model problem 4 for a sequence of parameter values σ with $\sigma \in [550, 850]$. The associated AFS sets are shown in Fig. 25. For $\sigma = 550$ the AFS consists of four isolated segments. These segments grow together for increasing values of σ. Finally, for $\sigma = 810$ all four segments are connected and form a Swiss-cheese-like 3D object with a complicated hole structure; the origin is contained in the hole.

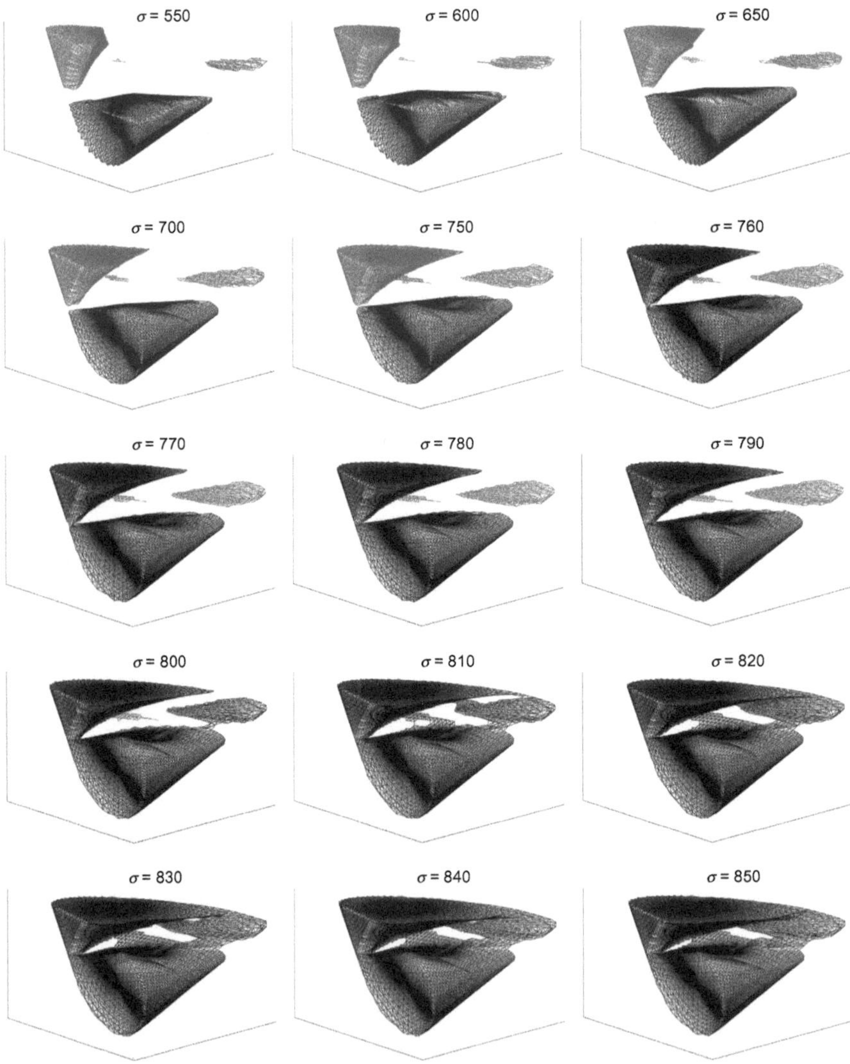

FIG. 25 Data set 4: AFS dynamics with respect to a variation of the model parameter σ. For $\sigma = 550$ the spectral AFS contains four separated segments. Finally, for $\sigma = 850$ these segments are grown together to a one-segment AFS with a hole.

Remark 1. It is not clear by the results shown in Fig. 25 if an AFS for a four-component system can consist of exactly two or exactly three segments. The resolution of the computations is not high enough. This question has to be investigated analytically.

10 THE FACPACK TOOLBOX FOR AFS COMPUTATIONS

The *FACPACK* toolbox [35] is a software package for AFS computations and can be applied to experimental spectroscopic data, to model data, or even to any nonnegative matrix D. The software contains implementations of the polygon inflation method [30], the inverse polygon inflation method, and the geometric constructive approach of generalized Borgen plots [27]. Moreover, it allows to apply the complementarity theory [43] within the AFS representation. The main functionalities are implemented in separate software modules. The current software revision is the *FACPACK* revision 1.2. Additionally, modules are contained for baseline corrections of spectroscopic data and for a library management of computed pure component spectra. The software can be downloaded from the *FACPACK* homepage: http://www.math.uni-rostock.de/facpack/.

For more details on *FACPACK*, see the manual [51]. This includes a guide on how to install and to start the software.

The *FACPACK* software comes with a graphical user interface in MATLAB. The time-consuming computational procedures are externalized in precompiled C programs. Up to now the AFS computation and the visualization are available for $s = 2$ and $s = 3$ component systems. For the future we plan to present an extension to four-component systems.

10.1 Data Import

The spectral data matrix D is to be provided in a MATLAB-compatible $*$ .mat file and is assumed to contain row-wise the k spectra. Additionally, a vector of spectral wavenumbers/frequencies can be provided by an n-component vector variable x together with a vector t of time coordinate values.

10.2 The AFS Computation Module

The core of the *FACPACK* software is the polygon inflation algorithm for three-component systems. This software is contained in the *AFS computation* module.

See Fig. 26 for a screenshot of the *AFS computation* module window with a short explanation of the possible users' operations. The first step in order to run the software is to push the button "AFS computation" in the start window. The second step is to load the data. The third step is to select whether a two- or three-component system is considered. The singular values $\sigma_1, \ldots, \sigma_4$ of D are displayed and serve indicators for the selection of the number of independent

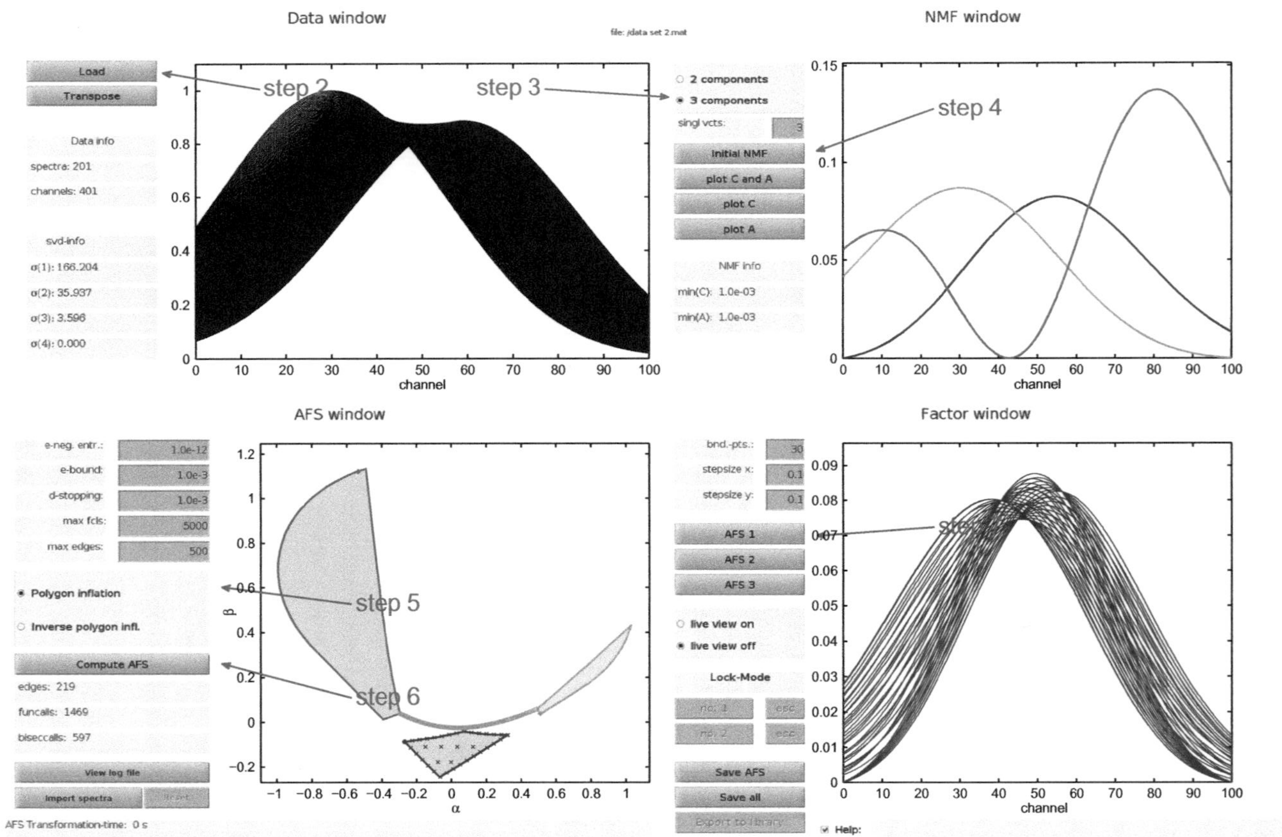

FIG. 26 See legend on opposite page.

components. In Step 4 an initial nonnegative matrix factorization is computed, which supplies the initial point for the construction of the initial triangles. Then, Step 5 allows to select either the standard polygon inflation algorithm or the inverse polygon inflation algorithm. The AFS is computed by pressing the button "Compute AFS." A special feature of the software is its "live-view mode," see Ref. [35], which allows to move the mouse pointer through the AFS and to see simultaneously the associated spectrum (or the concentration profile in case of the concentrational AFS).

The AFS computation module allows to apply equality constraints. Either one-, two-, or three pure component spectra can be locked. Then the reduced AFS can be computed (see Section 8.1). External spectra (e.g., from a spectra library) can be imported; their coordinates are marked within the AFS.

10.3 The Complementarity & AFS Module

Section 8.1.1 and in more detailed form Ref. [43] explain the (possibly drastic) reduction of the rotational ambiguity by the complementarity theory. The module "Complementarity & AFS" illustrates this within the AFS.

The module is activated by pressing the button "Complementarity & AFS" in the start window of the *FACPACK* software. After data loading the third step is to compute the AFS sets for C and A. Alternatively, it is possible to compute only the set FIRPOL $\mathcal{M}^+$. The live-view mode allows to move the mouse pointer through the AFS and to see simultaneously the associated spectra or concentration profile. By clicking the mouse button the current pointer position is locked and serves as a restriction for the complementarity theory. If three points either in AFS for C or A are locked, then a unique factorization $D = CA$ is defined. If some of these points are to be modified, then the user can repeat the Steps 4–6. All steps are illustrated in Fig. 27.

10.4 The Generalized Borgen Plot Module

The *Generalized Borgen Plot* module (see Section 6.1 for more details) allows the construction of the AFS for noisy of perturbed data by means of computational geometry methods.

FIG. 26 Quick-start in seven steps for the data set 2.

Step 1: Press the button **AFS computation** in the start window to activate this GUI.
Step 2: Load the data. Here data set 2 is used.
Step 3: Select 3 as the number of components.
Step 4: Compute an initial nonnegative matrix factorization (NNMF).
Step 5: Choose Polygon inflation.
Step 6: Compute the AFS which consists of three isolated segments.
Step 7: Plot the range of spectral factors (spectral bands), e.g., for the lowermost AFS segment.

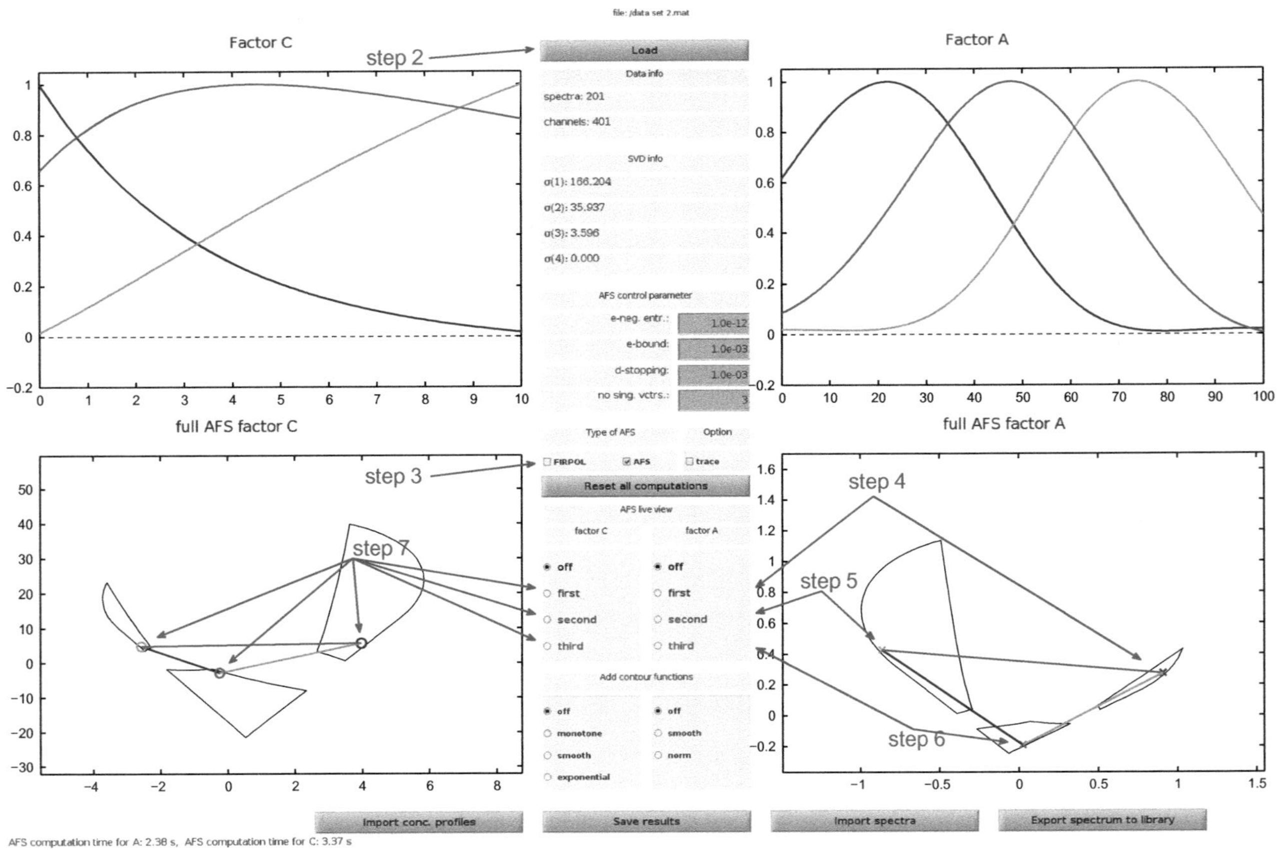

FIG. 27 See legend on opposite page.

First, the module is activated by the button “Generalized Borgen Plot” in the start window. Then the data are loaded and the desired scaling is selected. The further steps (AFS computation and live-view mode) are briefly explained in Fig. 28.

11 OUTLOOK AND OPEN PROBLEMS

The AFS analysis which was launched by Lawton and Sylvestre [13] in 1971 and Borgen *et al.* [25, 34] in 1985 has produced a continuous stream of results for a deeper understanding of the rotational ambiguity of MCR methods. Especially, in the last decade a large number of papers have been published (see e.g., Refs. [3, 27–31, 40, 42, 43, 45, 52, 53]).

Many challenges and open questions remain for the future. Some of them are:

1. How to integrate AFS methods to the toolboxes of practically working chemometricians? Up to now, AFS methods are more or less in the niche of research tools. AFS methods should be popularized as useful instruments for the investigation of the possible outcome of MCR methods.
2. How to compute and to visualize the AFS for systems with five or more independent chemical components? Then the AFS is a four- or even higher-dimensional bounded set whose graphical representation is very difficult. The derived representation by the dimension-independent feasible bands is a possible solution.
3. The nonnegative matrix factorization problem appears in various research fields as bioinformatics, text mining, neural sciences, and others. It remains to be analyzed to which extent the global AFS approach could be helpful in order to investigate the inherent ambiguity for these factorization problems.

FIG. 27 Complementarity theory and AFS implementation demonstrated for the data set 2.

Step 1: Press the button **Complementarity & AFS** to activate this GUI.

Step 2: Load the data. Here data set 2 is used.

Step 3: Either FIRPOL or the AFS sets for C and A can be computed. The computation of the AFS may be time consuming.

Step 4: After clicking the button “first”, the mouse pointer can be moved through the spectral AFS. The associated spectra are plotted in the live-view mode. A certain spectrum $A(1,:)$ can be locked by clicking in the AFS.

Step 5: Click the button “second” to repeat the spectrum selection for a second spectrum $A(2,:)$. For the remaining third component the concentration profile is shown which is uniquely determined by the complementarity theory.

Step 6: A last third spectrum $A(3,:)$ can be selected by moving the mouse pointer through the spectral AFS. The resulting predictions on the spectral factor are shown interactively.

Step 7: These three buttons and also the buttons “first”, “second”, “third” can be clicked and then a spectrum or concentration profile can be modified by moving the mouse pointer through the AFS. This allows to modify the two triangles which represent a pure component factorization of the given spectral data matrix.

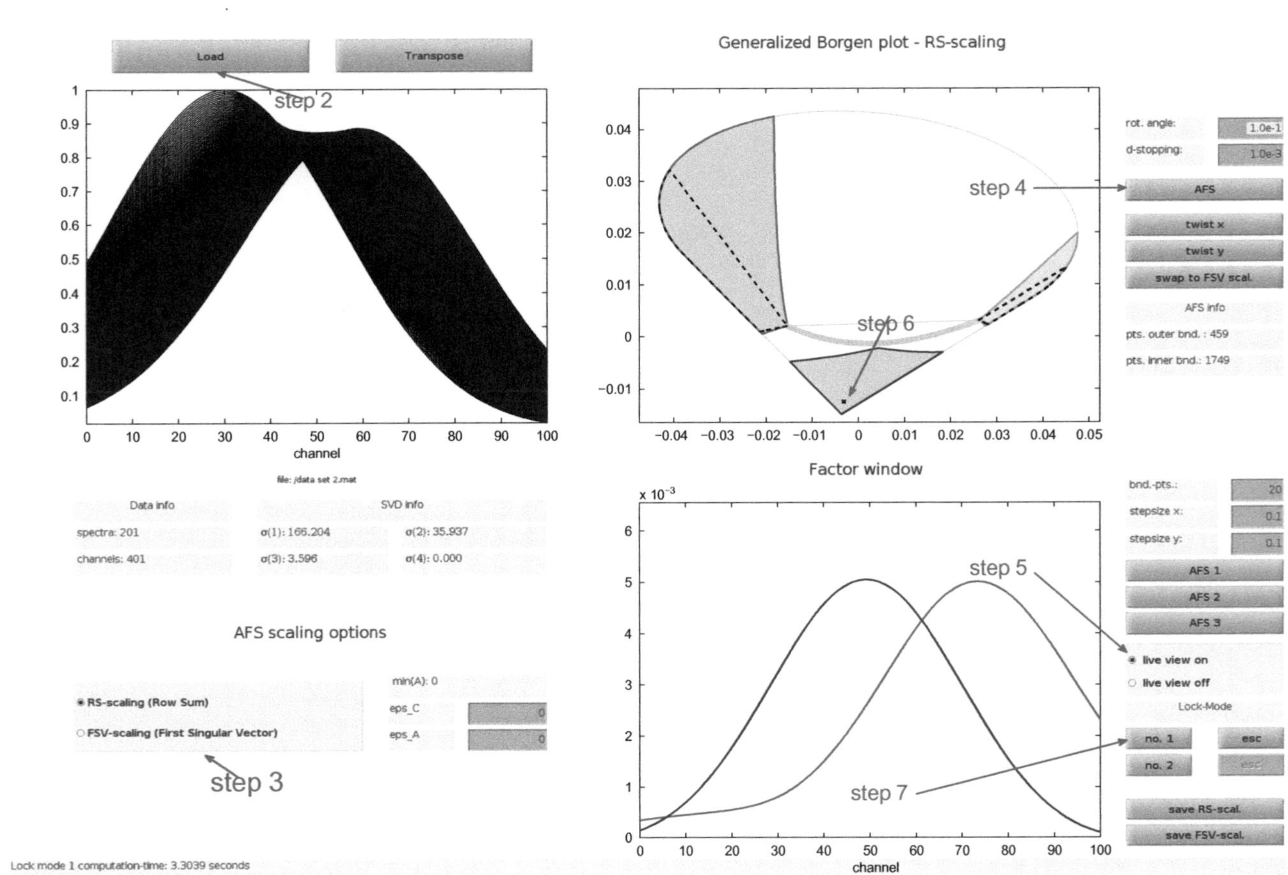

FIG. 28 See legend on opposite page.

REFERENCES

[1] Maeder M, Neuhold YM. Practical data analysis in chemistry. Amsterdam: Elsevier; 2007.
[2] Malinowski E. Factor analysis toolbox for Matlab. Sharon, MA: Applied Chemometrics; 1992.
[3] Abdollahi H, Tauler R. Uniqueness and rotation ambiguities in multivariate curve resolution methods. Chemom Intell Lab Syst 2011;108(2):100–11.
[4] Jaumot J, Gargallo R, de Juan A, Tauler R. A graphical user-friendly interface for MCR-ALS: a new tool for multivariate curve resolution in MATLAB. Chemom Intell Lab Syst 2005;76(1):101–10.
[5] Mason C, Maeder M, Whitson A. Resolving factor analysis, Anal Chem 2001;73(7):1587–94. http://dx.doi.org/10.1021/ac991141q, http://pubs.acs.org/doi/abs/10.1021/ac991141q.
[6] Windig W, Guilment J. Interactive self-modeling mixture analysis. Anal Chem 1991;63(14):1425–32. http://dx.doi.org/10.1021/ac00014a016.
[7] Chew W, Widjaja E, Garland M. Band-target entropy minimization (BTEM): an advanced method for recovering unknown pure component spectra. Application to the FT-IR spectra of unstable organometallic mixtures. Organometallics 2002;21(9):1982–90.
[8] Neymeyr K, Sawall M, Hess D. Pure component spectral recovery and constrained matrix factorizations: concepts and applications. J Chemom 2010;24:67–74.
[9] Tauler R. Calculation of maximum and minimum band boundaries of feasible solutions for species profiles obtained by multivariate curve resolution. J Chemom 2001;15(8):627–46. http://dx.doi.org/10.1002/cem.654, http://dx.doi.org/10.1002/cem.654.
[10] Kubis C, Selent D, Sawall M, Ludwig R, Neymeyr K, Baumann W, et al. Exploring between the extremes: conversion dependent kinetics of phosphite-modified hydroformylation catalysis. Chem Eur J 2012;18(28):8780–94.
[11] Kubis C, Sawall M, Block A, Neymeyr K, Ludwig R, Börner A, et al. An operando FTIR spectroscopic and kinetic study of carbon monoxide pressure influence on rhodium-catalyzed olefin hydroformylation. Chem Eur J 2014;20(37):11921–31. http://dx.doi.org/10.1002/chem.201402515, http://dx.doi.org/10.1002/chem.201402515.
[12] Golub GH, Van Loan CF. Matrix computations. Baltimore, MD: Johns Hopkins University Press; 2012. http://books.google.de/books?id=X5YfsuCWpxMC.
[13] Lawton WH, Sylvestre EA. Self modelling curve resolution. Technometrics 1971;13:617–33.
[14] Ruckebusch C, Blanchet L. Multivariate curve resolution: a review of advanced and tailored applications and challenges. Anal Chim Acta 2013;765:28–36. http://dx.doi.org/10.1016/j.aca.2012.12.028.

FIG. 28 The Borgen plot module demonstrated for the data set 2.

Step 1: Press the button **Generalized Borgen Plot** to activate this GUI.
Step 2: Load the data. Here data set 2 is used.
Step 3: The scaling options are the Row Sum scaling and the First Singular Vector scaling.
Step 4: The AFS is computed and consists of three isolated segments.
Step 5: Activate the live-view on mode.
Step 6: Move the mouse pointer through the AFS and watch the interactively computed solutions. Points in the AFS can be fixed by clicking the left mouse button.
Step 7: If a point is fixed, then the reduced AFS segments for the remaining components are computed.

[15] Haario H, Taavitsainen VM. Combining soft and hard modelling in chemical kinetics. Chemom Intell Lab Syst 1998;44:77–98.
[16] Gemperline PJ, Cash E. Advantages of soft versus hard constraints in self-modeling curve resolution problems. Alternating least squares with penalty functions. Anal Chem 2003;75:4236–43.
[17] Widjaja E, Li C, Chew W, Garland M. Band target entropy minimization. A robust algorithm for pure component spectral recovery. Application to complex randomized mixtures of six components. Anal Chem 2003;75:4499–507.
[18] Sawall M, Kubis C, Barsch E, Selent D, Börner A, Neymeyr K. Peak group analysis for the extraction of pure component spectra. J Iran Chem Soc 2016;13(2):191–205. http://dx.doi.org/10.1007/s13738-015-0727-4.
[19] de Juan A, Maeder M, Martínez M, Tauler R. Combining hard and soft-modelling to solve kinetic problems. Chemom Intell Lab Syst 2000;54:123–41.
[20] Jaumot J, Gemperline PJ, Stang A. Non-negativity constraints for elimination of multiple solutions in fitting of multivariate kinetic models to spectroscopic data. J Chemom 2005;19(2):97–106. http://dx.doi.org/10.1002/cem.914, http://dx.doi.org/10.1002/cem.914.
[21] Sawall M, Börner A, Kubis C, Selent D, Ludwig R, Neymeyr K. Model-free multivariate curve resolution combined with model-based kinetics: algorithm and applications. J Chemom 2012;26:538–48.
[22] Vosough M, Mason C, Tauler R, Jalali-Heravi M, Maeder M. On rotational ambiguity in model-free analyses of multivariate data. J Chemom 2006;20(6–7):302–10. http://dx.doi.org/10.1002/cem.1022, http://dx.doi.org/10.1002/cem.1022.
[23] Abdollahi H, Maeder M, Tauler R. Calculation and meaning of feasible band boundaries in multivariate curve resolution of a two-component system. Anal Chem 2009;81(6):2115–22.
[24] Rajkó R. Additional knowledge for determining and interpreting feasible band boundaries in self-modeling/multivariate curve resolution of two-component systems. Anal Chim Acta 2010;661(2):129–32. http://dx.doi.org/10.1016/j.aca.2009.12.025, http://www.sciencedirect.com/science/article/pii/S0003267009017358.
[25] Borgen OS, Kowalski BR. An extension of the multivariate component-resolution method to three components. Anal Chim Acta 1985;174:1–26. http://dx.doi.org/10.1016/S0003-2670(00)84361-5, http://www.sciencedirect.com/science/article/pii/S0003267000843615.
[26] Rajkó R, István K. Analytical solution for determining feasible regions of self-modeling curve resolution (SMCR) method based on computational geometry. J Chemom 2005;19(8):448–63. http://dx.doi.org/10.1002/cem.947, http://dx.doi.org/10.1002/cem.947.
[27] Jürß A, Sawall M, Neymeyr K. On generalized Borgen plots. I: from convex to affine combinations and applications to spectral data. J Chemom 2015;29(7):420–33. http://dx.doi.org/10.1002/cem.2721, http://dx.doi.org/10.1002/cem.2721.
[28] Golshan A, Abdollahi H, Maeder M. Resolution of rotational ambiguity for three-component systems. Anal Chem 2011;83(3):836–41.
[29] Sawall M, Kubis C, Selent D, Börner A, Neymeyr K. A fast polygon inflation algorithm to compute the area of feasible solutions for three-component systems. I: concepts and applications. J Chemom 2013;27:106–16.
[30] Sawall M, Neymeyr K. A fast polygon inflation algorithm to compute the area of feasible solutions for three-component systems. II: theoretical foundation, inverse polygon inflation, and FAC-PACK implementation. J Chemom 2014;28:633–44.
[31] Golshan A, Maeder M, Abdollahi H. Determination and visualization of rotational ambiguity in four-component systems. Anal Chim Acta 2013;796:20–6.

[32] Jaumot J, Tauler R. MCR-BANDS: a user friendly MATLAB program for the evaluation of rotation ambiguities in multivariate curve resolution. Chemom Intell Lab Syst 2010;103(2):96–107.

[33] Minc H. Nonnegative matrices. New York: John Wiley & Sons; 1988.

[34] Borgen OS, Davidsen N, Mingyang Z, Øyen O. The multivariate N-component resolution problem with minimum assumptions. Microchim Acta 1986;89:63–73.

[35] Sawall M, Neymeyr K. How to compute the area of feasible solutions, a practical study and users' guide to FAC-PACK. In: Khanmohammadi M, editor. Current applications of chemometrics. New York: Nova Science Publishers; 2014. p. 97–134 [chapter 6].

[36] Skvortsov AN. Estimation of rotation ambiguity in multivariate curve resolution with charged particle swarm optimization (cPSO-MCR). J Chemom 2014;28(10):727–39. http://dx.doi.org/10.1002/cem.2663, http://dx.doi.org/10.1002/cem.2663.

[37] Gemperline PJ. Computation of the range of feasible solutions in self-modeling curve resolution algorithms. Anal Chem 1999;71(23):5398–404.

[38] Zhang X, Tauler R. Measuring and comparing the resolution performance and the extent of rotation ambiguities of some bilinear modeling methods. Chemom Intell Lab Syst 2015;147:47–57. http://dx.doi.org/10.1016/j.chemolab.2015.08.005, http://www.sciencedirect.com/science/article/pii/S0169743915001963.

[39] Henry RC. Duality in multivariate receptor models. Chemom Intell Lab Syst 2005;77(1–2):59–63. http://dx.doi.org/10.1016/j.chemolab.2005.04.001, http://www.sciencedirect.com/science/article/pii/S0169743905000584.

[40] Rajkó R. Natural duality in minimal constrained self modeling curve resolution. J Chemom 2006;20(3–4):164–9. http://dx.doi.org/10.1002/cem.999, http://dx.doi.org/10.1002/cem.999.

[41] Sawall M, Fischer C, Heller D, Neymeyr K. Reduction of the rotational ambiguity of curve resolution techniques under partial knowledge of the factors. Complementarity and coupling theorems. J Chemom 2012;26:526–37.

[42] Beyramysoltan S, Rajkó R, Abdollahi H. Investigation of the equality constraint effect on the reduction of the rotational ambiguity in three-component system using a novel grid search method. Anal Chim Acta 2013;791:25–35.

[43] Sawall M, Neymeyr K. On the area of feasible solutions and its reduction by the complementarity theorem. Anal Chim Acta 2014;828:17–26.

[44] Hemmateenejad B, Shojaeifard Z, Shamsipur M, Neymeyr K, Sawall M, Mohajeri A. Solute-induced perturbation of methanol-water association. RSC Adv 2015;5:71102–8. http://dx.doi.org/10.1039/ C5RA13514A, http://dx.doi.org/10.1039/C5RA13514A.

[45] Beyramysoltan S, Abdollahi H, Rajkó R. Newer developments on self-modeling curve resolution implementing equality and unimodality constraints. Anal Chim Acta 2014;827:1–14. http://dx.doi.org/10.1016/j.aca.2014.03.019, http://www.sciencedirect.com/science/article/pii/S0003267014003225.

[46] Sawall M, Rahimdoust N, Kubis C, Schröder H, Selent D, Hess D, et al. Soft constraints for reducing the intrinsic rotational ambiguity of the area of feasible solutions. Chemom Intell Lab Syst 2015;149(Pt. A):140–50. http://dx.doi.org/10.1016/j.chemolab.2015.10.010, http://www.sciencedirect.com/science/article/pii/S0169743915002610.

[47] Rahimdoust N, Sawall M, Neymeyr K, Abdollahi H. Investigating the effect of flexible constraints on the accuracy of self-modeling curve resolution methods in the presence of perturbations. J Chemom 2016. http://dx.doi.org/10.1002/cem.2787.

[48] Neymeyr K, Sawall M. On an SVD-free approach to the complementarity and coupling theory: a note on the elimination of unknowns in sums of dyadic products. J Chemom 2016;30:30–6.

[49] Donoho D, Stodden V. When does non-negative matrix factorization give a correct decomposition into parts? 2003. Advances in Neural Information processes, NIPS 2003, Cambridge.

[50] Laurberg H, Christensen MG, Plumbley MD, Hansen LK, Jensen SH. Theorems on positive data: on the uniqueness of NMF. Computational Intelligence and Neuroscience 2008;9.

[51] Sawall M, Jürß A, Neymeyr K. FAC-PACK: a software for the computation of multi-component factorizations and the area of feasible solutions, revision 1.2; 2014. FAC-PACK homepage, http://www.math.uni-rostock.de/facpack/.

[52] Golshan A, Abdollahi H, Maeder M. The reduction of rotational ambiguity in soft-modeling by introducing hard models. Anal Chim Acta 2012;709:32–40. http://dx.doi.org/10.1016/j.aca.2011.10.057.

[53] Rajkó R, Abdollahi H, Beyramysoltan S, Omidikia N. Definition and detection of data-based uniqueness in evaluating bilinear (two-way) chemical measurements. Anal Chim Acta 2015;855:21–33. http://dx.doi.org/10.1016/j.aca.2014.12.017, http://www.sciencedirect.com/science/article/pii/S0003267014014184.

Chapter 6

Linear and Nonlinear Unmixing in Hyperspectral Imaging

N. Dobigeon[1,*], Y. Altmann[†], N. Brun[‡] and S. Moussaoui[§]
[*]University of Toulouse, IRIT/INP-ENSEEIHT, 31071 Toulouse Cedex 7, France
[†]School of Engineering and Physical Sciences, Heriot-Watt University, Riccarton, Edinburgh, EH14 4AS, United Kingdom
[‡]Laboratoire de Physique des Solides, CNRS, Univ. Paris-Sud, Univ. Paris-Saclay, 91405 Orsay Cedex, France
[§]Ecole Centrale de Nantes, IRCCyN, UMR CNRS 6597, Nantes Cedex 3, France
[1]Corresponding author: e-mail: Nicolas.Dobigeon@enseeiht.fr

Chapter Outline

1 SPECTRAL MIXTURE ANALYSIS

For several decades, development of spectral imaging has significantly contributed to numerous advances in various applicative fields, not only remote sensing and earth observation [1] but also planetology [2–4], food monitoring [5–7] and biomedical engineering [8, 9]. It consists of observing a scene of interest at various wavelengths and, contrary to conventional red-green-blue imaging, it generally offers several hundreds of spectral bands for each spatial measurement. This high spectral resolution confers hyperspectral imaging a powerful ability to discriminate distinct materials that are present in the observed area. The general concept of hyperspectral imaging as a remote sensing tool is illustrated in Fig. 1. In this specific application context, such

Data Handling in Science and Technology, Vol. 30. http://dx.doi.org/10.1016/B978-0-444-63638-6.00006-1
Copyright © 2016 Elsevier B.V. All rights reserved.

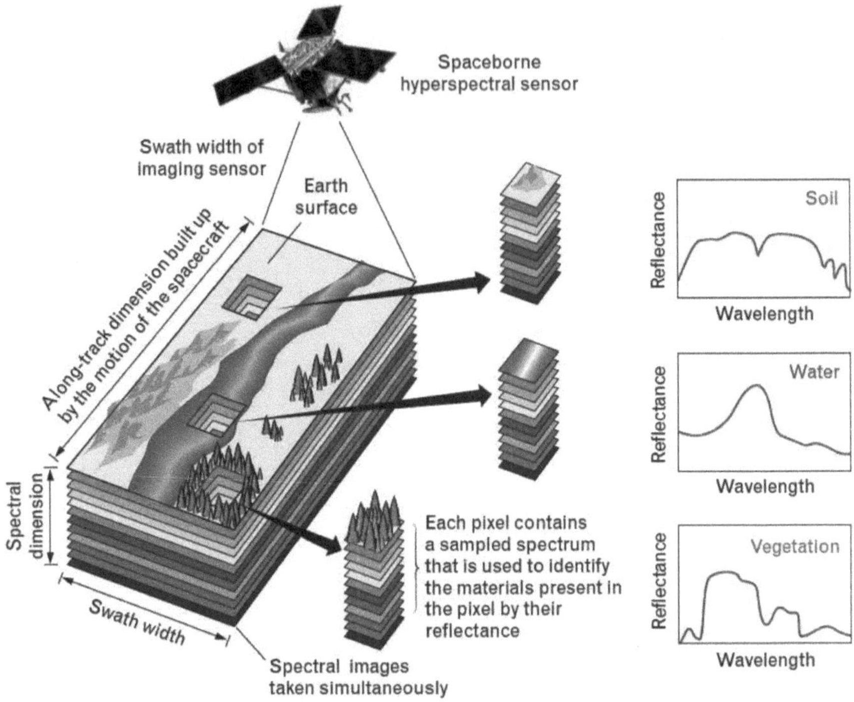

FIG. 1 Hyperspectral imaging concept in the remote sensing applicative context (from Shaw GA, HK Burke. Spectral imaging for remote sensing. Lincoln Labor J 2003;14(1):3-28).

measurement technique has shown to be a valuable modality to solve challenging issues, such as material quantification [10], anomaly detection [11, 12] and pixel classification [13–15], avoiding tedious on-ground sampling campaigns during mineralogic surveys.

At a completely different scale, hyperspectral images can be obtained by microscopy imaging techniques. Two examples are Raman microspectroscopy with a pixel size down to about 1μm and electron energy loss spectroscopy (EELS) performed in a scanning transmission electron microscope (STEM) with a sub-nanometer resolution. Measurements collected by these techniques may also be in the form of 3d-datacubes, characterized by a 2d-spatial sampling and a 1d-spectral sampling. As illustrated in Fig. 2, *per se*, spectroscopic imaging also provides hyperspectral data that share numerous similarities with remotely sensed images, e.g., resulting from a push-broom scanning system. Notably, it has demonstrated its value to address various issues, e.g., for material science and chemical analysis [16–18].

Unfortunately, due to an intrinsic trade-off between spatial and spectral sensitivities of the sensors, conventional hyperspectral images generally suffer from limited spatial resolution when compared to multispectral or

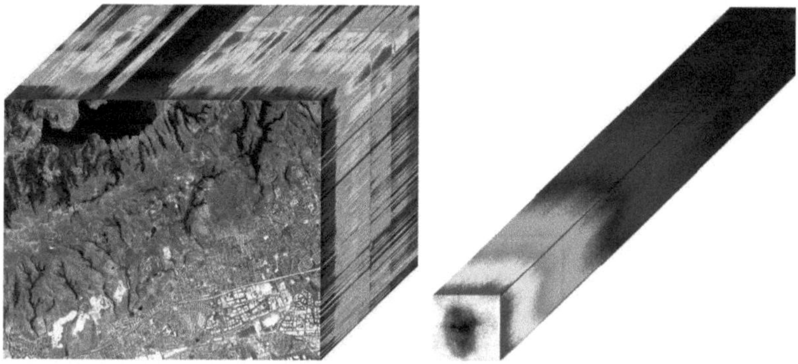

FIG. 2 *Left*: hyperspectral image acquired over Moffett Field, CA, by the Airborne Visible/Infrared Imaging Spectrometer (AVIRIS) in $L = 224$ spectral bands. *Right*: spectrum-image acquired by STEM-EELS. The top image corresponds to a synthetic RGB composition of the scene.

panchromatic images [19]. In particular, when they are acquired at a spatial resolution that is higher than the resolution of the sensor (i.e., scanner, microscope), these images are corrupted by blurring effects related to the optical response of the instrument (point spread function, PSF). To deal with this super-resolution problem, deconvolution techniques can be advocated to restore the observed data [20]. Such methods require a previous processing step for PSF estimation. When a complementary image of higher spatial resolution but lower spectral resolution is acquired simultaneously, reconstructing a high spatial and high spectral resolution image can also overcome these limitations. Various multi-band reconstruction techniques, known as hyperspectral pansharpening [21] or multi-band image fusion [22, 23], have shown promising results. However, when such complementary data of high spatial resolution is not available, alternative processing approaches should be considered. Besides, even if these blurring effects are neglected, several macroscopic materials (e.g., vegetation, minerals, and human made construction) usually contribute to the spectrum measured at each given single pixel of the hyperspectral image. This more general restriction may question the operational use of these images when an accurate mapping of the scene components is required.

Specifically, spectral unmixing (SU), or spectral mixture analysis, aims at providing comprehensive descriptions of the hyperspectral measurements. It consists of extracting the spectral signatures that are characteristics of the main materials present in the scene and quantifying their respective spatial distribution (abundance) over the image [24, 25]. More precisely, SU decomposes the P spectra $\mathbf{y}_1,\ldots,\mathbf{y}_P$ related to the hyperspectral image pixels with $\mathbf{y}_p = \left[y_{1,p},\ldots,y_{L,p}\right]^T \in \mathbb{R}^L$ (where L stands for the number of spectral bands) into a collection of R elementary spectra $\mathbf{m}_1,\ldots,\mathbf{m}_R$, and a set of P abundance vectors $\mathbf{a}_1,\ldots,\mathbf{a}_P$, with $\mathbf{a}_p = \left[a_{1,p},\ldots,a_{R,p}\right]^T$ [24, 25]. In the most common acceptation of SU, the spectral signatures $\mathbf{m}_r$ ($r = 1,\ldots,R$), usually referred to as *endmembers*, are assumed to represent the *pure* materials present in the observed scene and the abundance coefficient $a_{r,p}$ quantifies the fractional proportion of the rth material in the pth pixel.

At this stage of the SU description, one should highlight the fact that the concept of endmember and its definition highly depend on the target application or, at least, the considered problem (see the Chapter by Somers [26] of this book). For instance, in a remote sensing scenario, when observing a vegetated area composed of two distinct tree species, characterizing these species with two endmember signatures may be sufficient to assess the presence of these plants over the observed image. However, when the foliage composition is of prime interest, e.g., for ecological purpose, endmember definition needs to be refined, i.e., by considering a set of two spectral signatures associated with leaves and branches, respectively, for each tree species.

Another difficulty appears when properly defining the abundance coefficients. Generally, they are expected to quantify the presence of the endmembers in the observed pixels. In the most favorable case, i.e., when the mixing model is assumed to be linear (see Section 2), the abundances represent the relative areas occupied by the endmembers in the pixels [27]. Governed by this finding that leads to easier physical interpretations of SU, most unmixing methods incorporate non-negativity and additivity (i.e., sum-to-one) constraints imposed on the abundance coefficients

$$a_{r,p} \geq 0 \quad \forall r, \forall p \tag{1}$$

$$\sum_{r=1}^{R} a_{r,p} = 1 \quad \forall p. \tag{2}$$

Under such modeling, abundance coefficients are associated with concentration, proportion or even probability of occurrence. In particular these constraints aim at providing a full (i.e., complete) and additive part-based representation of the measurements [24, 25]. However, in more complex acquisition scenario, i.e., when the mixing model departs from the linear one, other interpretations of these abundance coefficients can be advocated, such as fractional masses of the material in the considered mixture. Depending on the observation conditions (e.g., illumination) and the nature of the underlying scene, the sum-to-one constraint might not be appropriate and may be relaxed.

After properly defining the concepts of endmembers and abundances for the considered study framework, and neglecting any side-effects, e.g., induced by the spatial convolution due to the instrument PSF, relating the observed measurement $\mathbf{y}_p$ to the explanatory variables, namely the endmember matrix $\mathbf{M} \triangleq [\mathbf{m}_1, \ldots, \mathbf{m}_R]$ and the abundance vector $\mathbf{a}_p$, can be pixelwise stated as

$$\mathbf{y}_p \approx f(\mathbf{M}, \mathbf{a}_p). \tag{3}$$

In Eq. (3), the approximation symbol is due to measurement noises and to errors resulting from the model underlying the function $f(\cdot, \cdot)$. Thus, setting SU in equation straightforwardly consists of solving the following minimization problem

$$\min_{\mathbf{M},\mathbf{A}} \sum_{p=1}^{P} D_p\left[\mathbf{y}_p | f(\mathbf{M}, \mathbf{a}_p)\right] \text{ s.t. } (1) \tag{4}$$

where $\mathbf{A} \triangleq [\mathbf{a}_p, \ldots, \mathbf{a}_P]$ and $D_p(\cdot|\cdot)$ measures the discrepancy between the observed measurement $\mathbf{y}_p$ and the predicted spectrum model $f(\mathbf{M}, \mathbf{a}_p)$. In most research works dealing with SU, the approximation in Eq. (3) takes the form of an additive noise, generally assumed to be pixel independent, centered and normally distributed with covariance matrix $\boldsymbol{\Sigma}$. Under this widely admitted hypotheses, adopting an optimization point of view leads to the solving of a weighted least squares problem. The discrepancy measure becomes quadratic

$$D_p[\mathbf{u}, \mathbf{v}] = \|\mathbf{u} - \mathbf{v}\|_{\boldsymbol{\Sigma}}^2, \quad \forall p \tag{5}$$

where $\|\mathbf{u}\|_{\boldsymbol{\Sigma}}^2 = \mathbf{u}^T \boldsymbol{\Sigma}^{-1} \mathbf{u}$ is the Mahalanobis norm induced by $\boldsymbol{\Sigma}$, which reduces to the Euclidean norm

$$D_p[\mathbf{u}, \mathbf{v}] = \|\mathbf{u} - \mathbf{v}\|_2^2, \quad \forall p$$

when the noise sequence is additionally assumed to be band-independent. As stated by Févotte and Dobigeon [28], the Euclidean distance underlying the presence of additive Gaussian noise is the more commonly used discrepancy measure advocated in hyperspectral unmixing. However, it is worthy to note that other measures of divergence can be considered [29]. More generally, within a statistical estimation framework, the discrepancy measure $D_p(\cdot|\cdot)$ could derive from the definition of a statistical risk and, in this case, it would be defined as an expectation over random processes [30].

For a long time, SU has been decomposed into two successive processing steps. In the first step, the endmember spectra are identified. This identification can result from *a priori* knowledge, by choosing these signatures from a spectral library or by picking up the spectral signatures associated with specific image pixels which are assumed to be *pure*, i.e., composed of only one endmember. When this prior knowledge is not available, one should resort to a so-called *endmember extraction algorithm* (EEA) able to identify the endmember signatures from the measurements in an unsupervised manner [31]. Subsequently, once these endmembers have been identified, the problem in Eq. (4) reduces to the so-called *inversion* step which consists of estimating the abundance vectors in each pixel by solving for $p = 1, \ldots, P$

$$\min_{\mathbf{a}_p} D_p\left[\mathbf{y}_p | f(\mathbf{M}, \mathbf{a}_p)\right] \text{ s.t. } (1). \tag{6}$$

Whatever the deterministic or statistical framework adopted to tackle the unmixing problem, a crucial issue results from the choice of the modeling function $f(\cdot,\cdot)$ that relates the observed measurement $\mathbf{y}_p$ to the parameters of interest, namely $\mathbf{M}$ and $\mathbf{a}_p$. The choice for this function generally results from approximate descriptions of the physical processes underlying the acquisition process. The linear mixing model (LMM) has been certainly the most widely used when unmixing hyperspectral data, not only because of its simplicity but also since it provides an acceptable first-order approximation of the mixing process. To address scenarios where the LMM shows to be not sufficient to properly describe the measurements, researchers have developed other nonlinear models able to take various nuisance effects into account. Besides, when accurate approximate physics-based models are unmanageable, i.e., when they lead to a problem formulated by Eq. (4) too difficult to solve, more versatile analytical models, yet not motivated by physical considerations, can be derived. This chapter draws a wide albeit partial panorama of these models, as well as the main corresponding unmixing algorithms available in the literature. It is worth noting that most of the MATLAB codes associated with these algorithms are available online on the first author's webpage[1] .

2 LINEAR UNMIXING

2.1 The Linear Mixing Model (LMM)

As noticed in Bioucas-Dias *et al.* [25] and Keshava and Mustard [24], most unmixing algorithms rely on the assumption of a linear mixing process to explain the observed spectra. More precisely, under this linear mixing model (LMM), each observed spectrum in each pixel of a given image is assumed to result from the linear combination of the R endmember spectra

$$\mathbf{y}_p^{(\mathrm{LMM})} \approx \sum_{r=1}^{R} a_{r,p}\mathbf{m}_r \approx \mathbf{M}\mathbf{a}_p \tag{7}$$

It is worthy to note that, in some sense, following the interpretation proposed by Kent and Mardia [32], SU can be cast as a soft multi-class classification since each measurement $\mathbf{y}_p$ is quantitatively and linearly associated with the endmembers through the pairs $\{\mathbf{m}_1,\mathbf{a}_{1,p}\},\ldots,\{\mathbf{m}_R,\mathbf{a}_{R,p}\}$.

This LMM has received considerable attention since it generally consists of an acceptable first-order approximation of the physical processes involved in most of the scenes of interest [24]. Consequently, it has motivated a lot of research works, mainly in the remote sensing and image processing literature, that aim at developing efficient EEAs, able to recover pure component

1. http://dobigeon.perso.enseeiht.fr/

signatures in the image, and inversion techniques to estimate the abundance coefficients for a given (estimated or *a priori* known) set of endmembers. Comprehensive overviews of these EEAs and inversion methods can be found in [25], [24], [33], and in the Chapter by Somers [26] of this book.

Interestingly, the earlier studies dedicated to linear SU were founded on geometrical interpretation of the problem. Indeed, under the widely admitted non-negativity and sum-to-one constraints in Eq. (1) imposed on the abundance vectors $\mathbf{a}_1,\ldots,\mathbf{a}_P$, in the free-noise case, i.e., when perfect equality holds in Eq. (7), the observed measurements lie in the R-polytope whose convex hull is formed by the endmember spectral signatures. More precisely, all the pixels $\mathbf{y}_p \in \mathbb{R}^L$ $(p = 1,\ldots,P)$ belong to the $(R - 1)$-simplex whose vertices are the endmembers. Fig. 3 illustrates this geometrical representation of the LMM when considering mixtures of $R = 3$ endmembers. When the sum-to-one constraint does not hold, the observed measurements belong to the positive cone defined by the endmember spectral signatures and pre-processing can be used to project the data onto a particular simplex within this cone [25].

This finding clearly shows that the number of endmembers is intrinsically related to the dimension of the subspace actually occupied by the observed measurements. Thus, identifying this signal subspace of interest appears as a relevant pre-processing step. It allows not only the number of endmembers actually involved in the mixtures to be estimated, but also opens the door to straightforward denoising techniques by projecting these measurements onto this subspace of interest. This signal subspace identification problem can be tackled by extracting a subset of the bands chosen for their a priori known high signal-to-noise ratios, as in [34]. Conversely, more sophisticated methods attempt to identify the most appropriate subspace to represent the measurements and mainly differ by the criterion designed to measure this appropriateness. Principal component analysis (PCA) [35], minimum noise

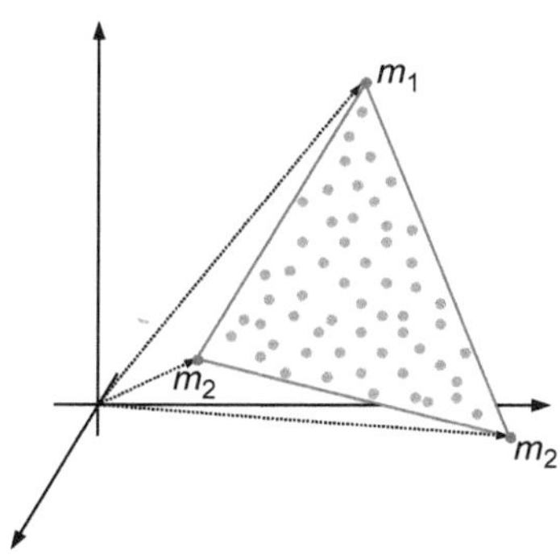

FIG. 3 Schematic representation (in $\mathbb{R}^3$) of the LMM for $R = 3$ endmembers. All the observed pixels (gray points) belong to a 2-simplex (red (gray in the print version) lines) whose vertices are the endmembers (blue (gray in the print version) points).

fraction (MNF) [36] and hyperspectral signal identification by minimum error (HySime) [37] are the most widely used subspace identification techniques.

2.2 Endmember Extraction Algorithms

In a first step of SU, the macroscopic materials that are present in the image are estimated using an EEA. The very first automated EEAs have exploited the geometrical interpretation of SU introduced in the previous section. Since the measurements are known to belong to the simplex formed by the endmembers, these pure spectral signatures can be equivalently recovered by identifying this simplex and, more precisely, its vertices. The most popular EEAs include pixel purity index (PPI) [38], N-FINDR [39], and more recently the vertex component analysis [40]. They propose to recover the vertices of the simplex of maximum volume where image pixels are potential candidate for these vertices. An assumption common to these EEAs is that they require the presence of pure pixels in the observed image, since the closer to the simplex the candidates, the better endmember estimates. As noticed by Ma *et al.* [41], a similar technique of *pure pixel pursuit* has been independently proposed by Araújo *et al.* [42] to analyze spectroscopic multicomponents in chemometrics. Other simplex volume maximization algorithms include the schemes detailed by Chan *et al.* [43] and Wu *et al.* [44]. Conversely, a reciprocal class of EEAs consists of recovering the smallest simplex that contains all the dataset [45–47]. The major advantage of these strategies is that they do not require the prior assumption of pure pixels in the measurements, in contrast to most of the simplex volume maximization techniques. Robust counterparts of these simplex volume minimization or maximization techniques include [48–50], e.g., to face with noise and/or anomaly corruption.

2.3 Inversion Algorithms

2.3.1 Supervised SU

Once the endmember signatures gathered in the matrix **M** have been identified, the second step of SU, referred to as *supervised* SU or *inversion*, consists of estimating the set of abundances (or abundance vectors) associated with each measurement. Basically, the initial formulation of SU in Eq. (4) turns out to the following P minimization problems

$$\min_{\mathbf{a}_p} D_p\left(\mathbf{y}_p|\mathbf{M}\mathbf{a}_p\right) \;\; \text{s.t.}\,(1). \tag{8}$$

When the discrepancy measure $D(\cdot|\cdot)$ is chosen as the standard Euclidean distance, the inversion reduces to the constrained least-square (LS) regression

$$\min_{\mathbf{a}_p} \left\|\mathbf{y}_p - \mathbf{M}\mathbf{a}_p\right\|_2^2 \;\; \text{s.t.}\,(1). \tag{9}$$

One difficulty arising from these optimization problems results from the constraints that the abundances should satisfy, which makes impossible any derivation of closed-form solutions. As a consequence, various algorithmic strategies have been proposed to solve the constrained optimization problems in Eq. (9). One may cite the fully constrained least-square (FCLS) algorithm proposed by Heinz and Chang [51] or the detection-oriented technique by Tu *et al.* [52]. This problem still motivated recent research works, e.g., to ensure the abundance constraints are properly satisfied [53], to reach better convergence guarantees of the algorithm [54, 55] or to cope with the possible high dimension of the problem when unmixing large hyperspectral images [56]. Statistical counterparts of these inversion methods consider the regression task as an estimation problem [57]. Such approach has been for instance considered to unmix hyperspectral data acquired by the MEx/OMEGA instrument over the Mars surface [58]. One major advantage of Bayesian techniques lies in the possibility of including additional nuisance parameters (e.g., noise variance) into the statistical model and estimating them jointly with the abundance vectors [59]. Moreover, in such framework, inference can generally be solved using Markov chain Monte Carlo (MCMC) techniques [60, 61], which allows various Bayesian estimators of the abundance vectors to be computed, e.g., the maximum a posteriori (MAP) and minimum mean square error (MMSE) estimators.

2.3.2 Semi-supervised SU

The inversion step has been also considered to deal with the specific scenario where the endmember spectra $\mathbf{m}_1,\ldots,\mathbf{m}_R$ are not perfectly identified but known to belong to an a priori available spectral library. This spectral library, also referred to as a dictionary and denoted as $S=[s_1,\ldots,s_K]$, is composed of K spectral signatures assumed to be acquired in the same conditions as the measurements to be unmixed. Note that the number K of spectral signatures in the library is much greater than the number R of endmembers actually involved in the observed mixtures. Thus, in this semi-supervised framework, SU consists of extracting from $\mathbf{S}$ a subset of signatures that can explain the measurements according to the model in Eq. (7). To cope with the large dimension of the dictionary $\mathbf{S}$ ($K \gg R$), the problem in Eq. (8) is slightly modified to include a sparsity constraint or penalization, while at the same time the sum-to-one constraint is relaxed. Under the hypothesis of quadratic cost function, semi-supervised SU consists of solving the following sparse LS regression problems

$$\min_{\mathbf{a}_p} \left\|\mathbf{y}_p - \mathbf{S}\mathbf{a}_p\right\|_2^2 \text{ s.t. } \left\|\mathbf{a}_p\right\|_0 \leq d \text{ and } a_{r,p} \geq 0 \ \forall r \tag{10}$$

or alternatively

$$\min_{\mathbf{a}_p} \left\|\mathbf{a}_p\right\|_0 \text{ s.t. } \left\|\mathbf{y}_p - \mathbf{S}\mathbf{a}_p\right\|_2^2 \leq \epsilon \text{ and } a_{r,p} \geq 0 \ \forall r. \tag{11}$$

In Eq. (10), d represents the maximum number of non-zero coefficients in $\mathbf{a}_p$ quantified by the ℓ_0-semi norm (or counting function) $\|\mathbf{a}_p\|_0$ and thus the maximum number of endmembers allowed to be involved in the considered mixture. Conversely, in Eq. (11), ϵ quantifies the tolerated reconstruction error resulting from any measurement noise or other modeling approximation. Solving such NP-hard problems requires the use of greedy algorithms, such as matching pursuit [62] and orthogonal matching pursuit [63]. Several variational formulations based on a convex relaxation of the counting function $\|\cdot\|_0$, which is merely replaced by the sparsity-inducing ℓ_1-norm, have been proposed in the literature, such as the famous SUNSAL algorithm proposed by Bioucas-Dias and Figueiredo [64]. Comprehensive analysis of ability of such algorithms to solve semi-supervised SU has been conducted by Iordache *et al.* [65]. Again, Bayesian counterparts of these semi-supervised unmixing techniques have been proposed by Themelis *et al.* [66] and Dobigeon *et al.* [61].

2.4 Unsupervised SU: Joint Extraction and Inversion

Addressing the SU problem by considering two successive steps may appear as suboptimal, e.g., as the accuracy of the inversion step highly depends on the quality of the endmembers identified by the EEA. To overcome this limitation, some approaches address the SU problem by estimating the endmember signatures $\mathbf{m}_1,\ldots,\mathbf{m}_R$ and all the abundance vectors $\mathbf{a}_1,\ldots,\mathbf{a}_P$ jointly, i.e., in a single step. In this so-called *unsupervised* framework, under the LMM assumption, SU in Eq. (3) formulated for all the pixels consists of estimating the endmember $\mathbf{M}$ and abundance $\mathbf{A} \triangleq [\mathbf{a}_1,\ldots,\mathbf{a}_P]$ matrices given the observation matrix $\mathbf{Y} = [\mathbf{y}_1,\ldots,\mathbf{y}_P]$ such that

$$\mathbf{Y} \approx \mathbf{M}\mathbf{A} \tag{12}$$

under the set of abundance constraints in Eq. (1). Since the endmember spectra generally correspond to, e.g., radiance, reflectance, absorbance or other energy measures, they can be assumed to be non-negative and thus an additional set of non-negativity constraints is included into the model, namely

$$m_{l,r} \geq 0, \quad \forall l, \quad \forall r. \tag{13}$$

As a consequence, SU can be cast as a constrained blind source separation (BSS). BSS is a well-known problem, mainly addressed in the signal processing literature for various applications, for instance audio (i.e., music and speech) analysis [67, 68]. The most popular latent variable model-based separation techniques to perform BSS include principal component analysis (PCA) and independent component analysis (ICA). For instance, PCA has been conducted to extract material signatures from EELS spectrum-image by Bonnet *et al.* [69] or Bosman *et al.* [70]. However, as noticed by Arenal *et al.* [71], PCA extracts component signatures that are difficult to be interpreted physically, mainly due to the inherent orthogonality constraint imposed by PCA to the components. Naturally, ICA has been considered as an alternative to analysis hyperspectral images acquired from, e.g., remote sensors [72, 73] or STEM-EELS [74, 75]. However, these ICA-based approaches can fail to provide meaningful interpretation of the identified signatures, specially in applications where both these signatures, as well as the abundance maps, are highly correlated, breaking the required independence assumption [76]. Besides, in their standard formulations, both PCA and ICA do not allow the constraints (1) to be satisfied. As a consequence, the end-users, from geoscience or experimental physics communities, should seize other advanced methodological tools able to offer a comprehensive and interpretable description of the hyperspectral data.

Other promising pathways can be followed once the problem of Eq. (12) has been rather interpreted as a non-negative matrix factorization (NMF). Such latent variable models only impose non-negativity constraints on the elements of the two model factors **A** and **M**, an assumption that is widely satisfied by hyperspectral data for most modalities. Since the pioneering works of Lee and Seung [77] and Paatero and Tapper [78], NMF has demonstrated its flexibility and its efficiency to infer relevant factors in various applications. Its potential was materialized by Miao and Qi [79] who combined such approach with geometrical concepts to develop, up to authors' knowledge, the first fully unsupervised NMF-based unmixing algorithm. Capitalizing on this advance, numerous works have been achieved, mainly by the remote sensing community, to design more efficient NMF-based unmixing algorithms [80–83]. As for the supervised and semi-supervised frameworks, even more flexibility can be reached by tackling the NMF task as a statistical inference problem. Efficient Bayesian source separation algorithms accounting for non-negativity constraints of the source signals and the mixing coefficients have been proposed to analyze spectral mixtures [84] and, in particular, hyperspectral images in various applicative fields, for instance remote sensing [85], food monitoring [86], planetology [87, 88], Raman [89], and near-infrared (NIR) spectroscopy [90], STEM-EELS [18, 91], and even genetics [92]. To cope with the somehow high computational complexity of the MCMC-based inference algorithms, specific algorithmic improvements may be required [93]. Interested readers are invited to refer to as the Chapter by Brie [94] of this book for a comprehensive description of Bayesian NMF.

3 NONLINEAR UNMIXING

According to Bioucas-Dias *et al.* [25], LMM consists of a reasonable model once two assumptions are satisfied. Firstly, the mixing process should occur at a macroscopic scale [95]. Secondly, the photons that reach the hyperspectral sensor must interact with only one material, as is the case in checkerboard type scenes [96]. A schematic illustration of this model is depicted in Fig. 4 (left) for a scene composed of two materials. When one of these two

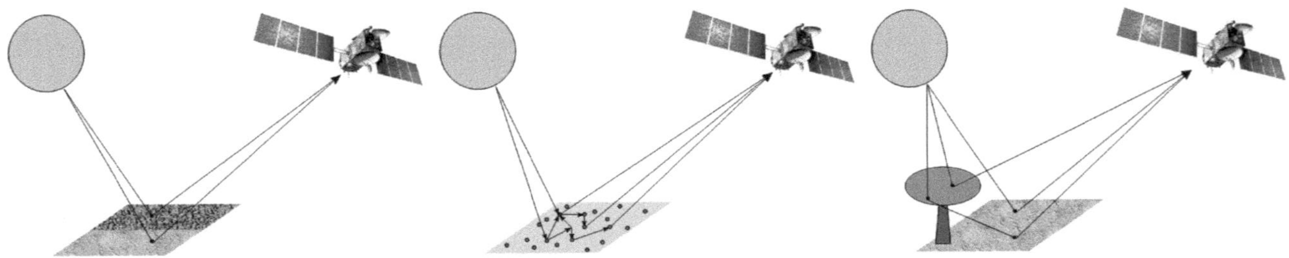

FIG. 4 *Left*, linear mixing model (LMM): the imaged pixel is composed of two materials. *Middle*, intimate mixture model (IMM): the imaged pixel is composed of a microscopic mixture of several constituents. *Right*, bilinear mixing model (BMM): the imaged pixel is composed of two endmembers, namely tree and soil. In addition to the individual contribution of each material, nonlinear (e.g., bilinear) interactions between the tree and the soil reach the sensor (from Dobigeon N, Tourneret JY, Richard C, Bermudez JCM, McLaughlin S, Hero AO, Nonlinear unmixing of hyperspectral images: Models and algorithms. IEEE Signal Process Mag 2014;31(1):89-94).

assumptions is not satisfied, the LMM may be not valid since various nonlinear effects may occur. To cope with these nonlinear interactions, more complex mixing models have been proposed, mainly in the spectroscopy and remote sensing literature. These nonlinear models have shown interesting properties to unveil meaningful information that would have been unattainable for the standard LMM [97–104].

To describe the light scattering by the materials in an observed scene, an accurate model based on radiative transfer (RT) theory has been advocated by Hapke [105]. Unfortunately, deriving SU algorithms exploiting this model would require to solve too complex problems. In particular, this model assumes linear spectral mixtures (subject to non-negativity and sum-to-one constraints) in the radiance domain instead of in the reflectance domain. To overcome these difficulties, several approximations have been proposed, mainly in the spectroscopy literature, such as the famous Hapke's bidirectional model [27]. However, these models still require highly non-linear and integral formulations that preclude any real-world implementations of SU techniques. Thus, no unsupervised nonlinear unmixing algorithm able to jointly extract the endmembers from the data and estimate their relative proportions in the pixels have been proposed to handle these models.

More recently, alternate approximate but workable nonlinear mixing models have been recently proposed in the remote sensing and image processing literature. Some of them are motivated by physical findings while others exploit a more flexible nonlinear mathematical formulation to improve unmixing performance when analyzing various types of datasets. Once a particular nonlinear model has been identified to analyze a specific dataset, preprocessing could be conducted before unmixing to mitigate the nonlinearities and to restore the linear mixing assumption. Although possibly fast, such strategy may suffer from some limitations. First, it requires the inversion of the nonlinearity mapping, which is not always straightforward. Additionally, any error produced by this preprocessing step is propagated into the unmixing step, which may drastically impact the unmixing accuracy. Conversely, unified approaches able to tackle the nonlinear unmixing problem directly are expected to be more robust.

The main classes of nonlinear models and corresponding nonlinear SU algorithms are described in the sequel.

3.1 Intimate Mixture Models

As stated before, a key assumption ensuring the validity of LMM is that the mixture processes should occur at a macroscopic scale. However, in some specific situations, interactions occur at a smaller scales, e.g., when the spatial scales involved in the mixtures are smaller than the path length followed by the photons. In this context, the materials are said to be intimately mixed [27]. An illustration of these interactions is represented in Fig. 4 (middle).

Such intimate mixture models (IMM) have been considered, e.g., for analyzing mixtures composed of sand or minerals [106], sometimes in laboratory [107]. Theoretical frameworks based on the RT theory have been built to accurately describe the intimate interactions. Probably the most popular approaches devoted to such mixtures are those introduced by Hapke [27]. Indeed, these models are mainly based on meaningful and interpretable parameters with physical significance, and can relate the spectral measurements to physical characteristics of the endmembers and corresponding abundances. Note that in this case, these abundances are associated with the relative mass fractions of the materials. For instance, Hapke [105] derives a nonlinear model to express the measured reflectances as a function of various parameters of interest, such as the mass fractions, the characteristics of the individual particles (density, size) and the single-scattering albedo. Conversely, Draine [108] and Shkuratov *et al.* [109] have proposed alternate and relevant approximating nonlinear mixing models. However, again, one limitation inherent to these models is that they also strongly depend on external parameters related to the acquisition, e.g., the perfect knowledge of the geometric positioning of the sensor. As a consequence, performing SU in such conditions can be difficult and the expected results can be irrelevant if such parameters are approximately known, in particular when tackling the SU problem in the challenging unsupervised scenario, i.e., when the spectral signatures of the materials are unknown and need to be jointly estimated with the fractional abundances. Some SU techniques dedicated to analyze remotely sensed images based on IMM or related alternatives have been exploited in the hyperspectral literature [110, 111]. In particular, a common approach to avoid the painful inversion of the reflectance measurements into the single scattering albedo consists of using neural-networks (NN) to learn this nonlinear function. This is the strategy followed by Guilfoyle *et al.* [110], for which several improvements have been proposed by Altmann *et al.* [112] to reduce the computationally intensive learning step. In these NN-based approaches, the endmembers should be assumed to be a priori known, and are required to train the NN. Other NN-based algorithms have been studied by Plaza *et al.* [113–116].

Besides, Broadwater and Banerjee [117–119] derived various kernel-based unmixing techniques that implicitly relied on the Hapke model. Even more recently, Close *et al.* [120] introduce a the multi-mixture pixel (MMP) model, which combines macroscopic and microscopic (i.e., intimate) mixtures

$$\mathbf{y}_p^{(\mathrm{MMP})} \approx \sum_{r=1}^{R} a_{r,p}\mathbf{m}_r + a_{R+1,p}\mathcal{R}\left(\sum_{r=1}^{R} f_{r,p}\mathbf{w}_r\right).$$

The first term is similar to the contribution encountered in LMM in (7) and comes from the macroscopic mixing process. The second one, considered as an additional endmember with abundance $a_{R+1,P}$, corresponds to an intimate mixture described as the average single-scattering albedo [105] expressed in the reflective domain, where $\mathbf{f}_p = [f_{1p}, \ldots, f_{Rp}]^T$ correspond to the microscopic proportions, $\{\mathbf{w}_r\}$ are the albedo-domain endmember signatures and $\mathcal{R}(\cdot)$ is the mapping function from the albedo domain to the reflectance domain. The same authors have introduced as well an unsupervised algorithm to perform MMP-based SU [120, 121].

3.2 Bilinear Mixing Models

Another assumption required to ensure that LMM is a relevant model to describe mixtures is related to the paths followed by the photons. Specifically, before reaching the hyperspectral sensor they must interact with only one material. However, when analyzing scenes remotely sensed over vegetated or urban areas are known to be subjected to more complex interactions that can not be properly taken into account by a simple LMM [97, 101–103, 122–125]. In these specific scenarios, due to differences in elevation between different materials (e.g., vegetation canopies vs the relatively flat soil surfaces), photons are submitted to multipath and scattering effects. An archetypal example of this kind of scene is shown in Fig. 4 (right). Similar interaction effects have been also encountered when analyzing urban scenes acquired over buildings areas, sometimes referred to as *urban canyon* [126–128]. Therefore, various attempts have been conducted to overcome the intrinsic limitations of the LMM. A large family of nonlinear models that have been proposed to analyze these so-called *multilayered* configurations consists of including powers of products of reflectance into the standard LMM, generally neglecting interactions of orders greater than two. The resulting models are known as the family of the bilinear mixing models (BMM) defined as [129]

$$\mathbf{y}_p^{(\mathrm{BMM})} \approx \sum_{r=1}^{R} a_{r,p}\mathbf{m}_r + \sum_{i=1}^{R-1}\sum_{j=i+1}^{R} b_{i,j,p}\mathbf{m}_i \odot \mathbf{m}_j \tag{14}$$

where $\odot$ stands for the termwise (Hadamard) product

$$\mathbf{m}_i \odot \mathbf{m}_j = \begin{pmatrix} m_{1,i} \\ \cdots \\ m_{L,i} \end{pmatrix} \odot \begin{pmatrix} m_{1,j} \\ \cdots \\ m_{L,j} \end{pmatrix} = \begin{pmatrix} m_{1,i}m_{1,j} \\ \cdots \\ m_{L,i}m_{L,j} \end{pmatrix}$$

In (14), the observed pixel is composed of a linear contribution similar to the LMM and an additive nonlinear term

$$\sum_{i=1}^{R-1}\sum_{j=i+1}^{R} b_{i,j,p}\mathbf{m}_i \odot \mathbf{m}_j \tag{15}$$

that models nonlinear interactions between the materials. The nonlinearity coefficients $\{b_{i,j,p}\}_{i,j}$ allow the amount of nonlinearity in the pth pixel to be adjusted between each pair of materials $\mathbf{m}_i$ and $\mathbf{m}_j$. Most of the various bilinear models of the literature mainly differ by the definition of these coefficients $b_{i,j,p}$ and the associated constraints they should satisfied. The most common models are described below.

Somers *et al.* [102] and Nascimento and Bioucas-Dias [130] propose to include the nonlinearity coefficients $\{b_{i,j,p}\}_{i,j}$ within the set of constraints (1), leading to

$$\mathbf{y}_p^{(\mathrm{NM})} \approx \sum_{r=1}^{R} a_{r,p}\mathbf{m}_r + \sum_{i=1}^{R-1}\sum_{j=i+1}^{R} b_{i,j,p}\mathbf{m}_i \odot \mathbf{m}_j \tag{16}$$

with

$$\begin{cases} a_{r,p} \geq 0, \quad \forall r, \quad \forall p \\ b_{i,j,p} \geq 0, \quad \forall r, \quad \forall i \neq j \\ \sum_{r=1}^{R} a_{r,p} + \sum_{i=1}^{R-1}\sum_{j=i+1}^{R} b_{i,j,p} = 1, \quad \forall p. \end{cases} \tag{17}$$

It is worthy to note that this so-called Nascimento's model (NM), also used by Raksuntorn and Du [131], reduces to the LMM when $b_{i,j,p} = 0, \forall i \neq j$. This is an interesting property since, as stated before, LMM is known to be an admissible first approximation of the actually involved physical processes. However, in a more general setting (i.e., $b_{i,j,p} \neq 0$), the linear abundance coefficients $\{a_{r,p}\}_{r=1}^{R}$ are not subject to the sum-to-one constraints defined in (1). Moreover, NM can be also interpreted as a specific LMM with additional virtual endmembers. Indeed, considering $\mathbf{m}_i \odot \mathbf{m}_j$ as a pure component spectral signature with corresponding abundance $b_{i,j,p}$, the model in (16) can be rewritten

$$\mathbf{y}_p \approx \sum_{s=1}^{\tilde{R}} \tilde{a}_{s,p}\tilde{\mathbf{m}}_s$$

with the non-negativity and sum-to-one constraints in (1) imposed on the set of new abundance coefficients $\tilde{a}_{s,p}$ and

$$\begin{cases} \tilde{a}_{s,p} \triangleq a_{r,p}, & \tilde{\mathbf{m}}_s \triangleq \mathbf{m}_r & s = 1,\ldots,R \\ \tilde{a}_{s,p} \triangleq b_{i,j,p}, & \tilde{\mathbf{m}}_s \triangleq \mathbf{m}_i \odot \mathbf{m}_j & s = R+1,\ldots,\tilde{R} \end{cases}$$

and $\tilde{R} = \frac{1}{2}R(R+1)$. Again, this formulation shows that NM reduces to a standard LMM when $\tilde{a}_{s,p} = 0$ for $s = R+1,\ldots,\tilde{R}$. Besides, Eq. (3.2) shows that, in the case where the endmember signatures $\mathbf{m}_1,\ldots\mathbf{m}_R$ are known, supervised SU, i.e., estimation of the abundance coefficients, can be conducted by an LMM-based inversion. More precisely, assuming an Euclidean discrepancy measure as in Eq. (9), the linear abundances $\mathbf{a}_p$ and nonlinear coefficients $b_{i,j,p}$ can be recovered by solving the constrained LS regression

$$\min_{\widetilde{\mathbf{a}}_p} \left\| \mathbf{y}_p - \widetilde{\mathbf{M}}\widetilde{\mathbf{a}}_p \right\|_2^2 \text{ s.t. } (1)$$

where $\widetilde{\mathbf{a}}_p \triangleq \left[\tilde{a}_{1,p},\ldots,\tilde{a}_{\widetilde{R},p}\right]^T$ and $\widetilde{\mathbf{M}} \triangleq \left[\widetilde{m}_1,\ldots,\widetilde{m}_{\widetilde{R}}\right]$. As a consequence, any LMM-based inversion algorithm introduced in Section 2.3.1, such as FCLS [51], can be employed to recover the abundances of the NM, given the matrix of real and virtual endmembers $\widetilde{\mathbf{M}}$. When NM-based SU should be conducted in a unsupervised framework, the endmembers signatures are unknown and need to be jointly estimated with the abundance and nonlinear coefficients. Specifically Gader *et al.* [132] introduced an unsupervised NM-based SU algorithm by implementing updating rules that generalize the SPICE algorithm introduced by Zare and Gader [133] for the LMM.

Conversely, Fan *et al.* [101] have defined the nonlinearity coefficients $b_{i,j,p}$ as the product of the abundances, $b_{i,j,p} \triangleq a_{i,p}a_{j,p}$, under the standard abundance constraints in (1), leading to the so-called Fan's Model (FM)

$$\mathbf{y}_p^{(\mathrm{FM})} \approx \sum_{r=1}^{R} a_{r,p}\mathbf{m}_r + \sum_{i=1}^{R-1}\sum_{j=i+1}^{R} a_{i,p}a_{j,p}\mathbf{m}_i \odot \mathbf{m}_j. \tag{18}$$

The main argument for relating the amount of nonlinear interactions (governed by $b_{i,j,p}$) to the amount of linear contribution (governed by $a_{i,p}$ and $a_{j,p}$) is straightforward: if the ith endmember is absent in the pth pixel, then $a_{i,p} = 0$ and there are no interactions between $\mathbf{m}_i$ and the other materials $\mathbf{m}_j$ ($j \neq i$). More generally, the quantity of nonlinear interactions in a given pixel between two materials is directly related to the quantity of each material present in that pixel: the more a given material is present in the pixel, the more nonlinear interactions may occur. However this bilinear model does not extend the LMM, which can restrict its applicability. In a supervised fashion, i.e., given the endmember signature matrix $\mathbf{M}$, the FM abundances can be estimated with the algorithm proposed by Fan *et al.* [101], which is based on a first-order Taylor series expansion of the nonlinearity. More recently, Eches ang Guillaume [134] have proposed an unsupervised FM-based SU algorithm, based on an extension of the NMF framework introduced in Section 2.4.

To cope with the issue raised by the fact FM does not generalize the LMM, the generalized bilinear model (GBM) proposed by Halimi *et al.* [135] weights the products of abundances $a_{i,p}a_{j,p}$ by additional free parameters $\gamma_{i,j,p} \in (0,1)$ that tune the amount of nonlinear interactions, leading to $b_{i,j,p} \triangleq \gamma_{i,j,p}a_{i,p}a_{j,p}$ and

$$\mathbf{y}_p^{(\mathrm{GBM})} \approx \sum_{r=1}^{R} a_{r,p}\mathbf{m}_r + \sum_{i=1}^{R-1}\sum_{j=i+1}^{R} \gamma_{i,j,p}a_{i,p}a_{j,p}\mathbf{m}_i \odot \mathbf{m}_j. \tag{19}$$

The interaction coefficient $\gamma_{i,j,p} \in (0,1)$ quantifies the nonlinear interaction between the spectral components $\mathbf{m}_i$ and $\mathbf{m}_j$. Having $\gamma_{i,j,p} > 0$ indicates that only constructive interactions are considered. This model has the same interesting characteristic as the FM: the amount of nonlinear interactions is governed by the presence of the endmembers that linearly interact. In particular, if an endmember is absent in a pixel, there is no nonlinear interaction coming from this endmember. However, it also has the nice properties of i) generalizing the LMM by enforcing $\gamma_{i,j,p} = 0$ ($\forall i,j$), similarly to NM but contrary to FM and ii) having the amount of nonlinear interactions to be proportional to the material abundances, similarly to FM but contrary to NM. Various algorithmic strategies have been proposed to solved the GBM-based SU in a supervised framework, i.e., when the endmember signatures $\mathbf{M}$ are known. Halimi *et al.* [135] have initially introduced a MCMC-based algorithm to approximate the Bayesian estimators associated with the GBM abundance and nonlinear parameters. To overcome the computational complexity imposed by MCMC, optimization techniques have been proposed. Following the strategy proposed by Fan *et al.* [101], Halimi *et al.* [136] propose to linearize the objective function via a first-order Taylor series expansion of the nonlinearity. Then, the FCLS algorithm proposed by Heinz and Chang [51] can be used to estimate the new parameter vector. An alternate scheme proposed by Halimi *et al.* [136] consists of resorting to a gradient descent method coupled with constrained line search procedure enforcing the constraints inherent to the mixing model to adjust the step-size parameter. Yokoya *et al.* [137] have proposed a semi-NMF framework to perform supervised GBM-based SU. When the endmember signatures also need to be recovered (i.e., in an unsupervised SU framework) Heylen and Scheunders [138] have adopted a geometrical point-of-view and conducted an integral analysis to compute geodesic distances on the nonlinear manifold induced by the GBM. This formulation opens the door to an EEA able to identify the simplex of maximum volume contained in the manifold defined by the GBM-mixed pixels, following the geometrical ideas detailed in Section 2.2.

3.3 Polynomial Mixing Models

All the bilinear mixing models introduced above only include between-component interactions $\mathbf{m}_i \odot \mathbf{m}_j$ with $i \neq j$ but no within-component interactions $\mathbf{m}_i \odot \mathbf{m}_i$. Meganem *et al.* [128] have derived a nonlinear mixing model based on RT theory to analyze a simple canyon-like urban scene, leading, after successive approximations and simplifying assumptions, to the following linear-quadratic mixing model (LQM)

$$\mathbf{y}_p \approx \sum_{r=1}^{R} a_{r,p}\mathbf{m}_r + \sum_{i=1}^{R}\sum_{j=i}^{R} b_{i,j,p}\mathbf{m}_i \odot \mathbf{m}_j \tag{20}$$

with the non-negativity and additivity constraints in (1) and $b_{i,j,p} \in (0,1)$. This model, also encountered in the blind source separation community [139], is similar to the general formulation of the bilinear models in Eq. (14), with the noticeable difference that the nonlinear contribution includes quadratic terms $\mathbf{m}_i \odot \mathbf{m}_i$. An NMF-like algorithm has been introduced by Meganem *et al.* [128] to perform unsupervised LQM-based SU.

Inspired by pioneering works in blind source separation by Taleb and Jutten [140], Altmann *et al.* [141] have proposed an approximating model able to describe a wide class of nonlinearities. More precisely, the pth observed pixel spectrum is defined as a nonlinear transformation $\mathbf{g}_p(\cdot)$ of a linear mixture of the endmember spectra

$$\mathbf{y}_p \approx \mathbf{g}_p\left(\sum_{r=1}^{R} a_{r,p}\mathbf{m}_r\right) \tag{21}$$

where the nonlinear function $\mathbf{g}_p$ is defined as a second order polynomial governed by a unique parameter b_p

$$\begin{aligned} \mathbf{g}_p : \ (0,1)^L & \rightarrow \mathbb{R}^L \\ \mathbf{x} & \mapsto \left[x_1 + b_p x_1^2, \ldots, x_L + b_p x_L^2\right]^T \end{aligned} \tag{22}$$

This polynomial post-nonlinear mixing model (PPNM) can be rewritten as

$$\mathbf{y}_p^{(\mathrm{PPNM})} \approx \mathbf{M}\mathbf{a}_p + b_p(\mathbf{M}\mathbf{a}_p) \odot (\mathbf{M}\mathbf{a}_p) \tag{23}$$

or

$$\mathbf{y}_p^{(\mathrm{PPNM})} \approx \mathbf{M}\mathbf{a}_p + b_p \sum_{i=1}^{R}\sum_{j=1}^{R} a_{i,p}a_{j,p}\mathbf{m}_i \odot \mathbf{m}_j. \tag{24}$$

Eq. (24) shows that PPNM includes bilinear terms $\mathbf{m}_i \odot \mathbf{m}_j$ $(j \neq i)$ similar to those involved in the wide class of BMM (i.e., NM, FM, and GBM), including quadratic terms $\mathbf{m}_i \odot \mathbf{m}_i$ encountered in the LQM (20), which may account for interactions between similar materials. The PPNM has demonstrated a great flexibility to model various nonlinearities not only for unmixing purposes [104, 141] but also to detect nonlinear mixtures in the analyzed scene [142]. This model has also the great advantage of having the amount of nonlinearity to be governed by a unique parameter b_p in each pixel, contrary to other BMM such as NM or GBM. When the endmember spectral signatures $\mathbf{M}$ are known, estimating the abundance vector $\mathbf{a}_p$ and the nonlinearity parameter b_p defining the PPNM can be performed following the various Bayesian and optimization algorithms proposed by Altmann *et al.* [141]. When these signatures are unknown, Altmann *et al.* [143] derives a unsupervised PPNM-based SU algorithm based on an MCMC method.

Note finally that other nonlinear models of higher order polynomials have been recently considered by Marinoni *et al.* [144] and Heylen and Scheunders [145].

3.4 Robust Linear Models

Although nonlinear mixing models can significantly improve the unmixing procedure in the presence of complex mixing processes, linear models often stand as a satisfactory trade-off between algorithm complexity and performance for most observed mixed spectra. Thus, recent methods have investigated robust LMMs generally defined as

$$\mathbf{y}_p^{(\mathrm{rLMM})} \approx \mathbf{M}\mathbf{a}_p + \mathbf{r}_p \tag{25}$$

where the additional term $\mathbf{r}_p$ represents any potential deviation from the classical LMM. This class of models for robust linear SU allows for a wide variety of additional deviations from the LMM to be handled, such as nonlinear effects, outliers or possible endmember variability (see Section 5). As illustrated in Sections 3.2 and 3.3, the term $\mathbf{r}_p$ can be specified to explicitly account for its dependency on the endmember spectra and/or abundances. This is the case when considering polynomial models (e.g., bilinear models and PPNM) or the residual component analysis (RCA) based model proposed by Altmann *et al.* [146], where $\mathbf{r}_p$ has been considered as a random vector described by Gaussian processes whose covariance matrices involve the endmember matrix $\mathbf{M}$. Such models will be further discussed in Section 3.5.

In this section, contrary to the previously mentioned models, the term $\mathbf{r}_p$ is assumed to be not explicitly parametrized by the endmember spectra or their respective proportions. Conversely, the set of terms $\{\mathbf{r}_p\}_p$ presents sparsity structures, i.e., contains many zero entries. Such assumption can be easily motivated by the fact that LMM is an admissible first approximation of the physical processes underlying the mixtures in both the spectral and the spatial domains. Févotte and Dobigeon [28] have proposed an NMF method to jointly estimate the endmembers and abundances, assuming that the potential non-zeros entries in $\mathbf{R} = [\mathbf{r}_1,\ldots,\mathbf{r}_P]$ appear only in a reduced number of spatial measurements. This is for instance the case when nonlinear mixtures or other anomalies occur in a few specific spatial locations (i.e., pixels) and produce nonlinear contributions in most corresponding wavelengths. Mathematically speaking, this robust LMM assumes that the energy of the residual terms $\|\mathbf{r}_p\|_2$ ($p = 1,\ldots,P$) is null for most pixels, i.e., the LMM holds, while some of these residual energies are non-zeros. In other words, most of the columns of the residual component matrix $\mathbf{R}$ are assumed to be null vectors (for spatial indexes p corresponding to pixels resulting from LMM). In such cases solving the unsupervised SU problem can be formulated as

$$\min_{\mathbf{M},\mathbf{A},\mathbf{R}} \sum_{p=1}^{P} D_p\left[\mathbf{y}_p|\mathbf{M}\mathbf{a}_p + \mathbf{r}_p\right] + \lambda\|\mathbf{R}\|_{2,1} \quad \text{s.t.}\,(1) \tag{26}$$

where $\|\mathbf{R}\|_{2,1} = \sum_{p=1}^{P}\|\mathbf{r}_p\|_2$ is the so-called $\ell_2 - \ell_1$ norm and λ is a regularization parameter which controls the structured sparsity of the solution.

Févotte and Dobigeon [28] have shown that the proposed approach seems to be sufficiently flexible to analyze hyperspectral data for which some measured pixel spectra result from various mixtures, e.g., BMM or MMP model described in Sections 3.1 and 3.2.

Similarly, Newstadt *et al.* [147] have considered spatial and spectral sparsity structures (i.e., blocks) in the anomaly/outlier matrix term $\mathbf{R}$, which allows deviations from the LMM to be accounted for when these deviations occur in specific spatial regions or spectral bands of the hyperspectral data. This is typically the case when actual outliers are present, but also when material spectral signatures present have significant variations in particular spectral ranges (e.g., due to natural variability of the materials). This approach has been extended by Altmann *et al.* [148] who have modeled the spectral and spatial dependencies through more flexible Markovian models. Both Newstadt *et al.* [147] and Altmann *et al.* [148] have proposed to solve the resulting (NP-hard) BSS problems involving ℓ_0-type penalizations using MCMC methods.

3.5 Nonparametric SU Techniques

SU appears even more challenging when no prior knowledge regarding the nonlinear process is available. To face with this issue, in a supervised scenario, Plaza *et al.* [149] and Li *et al.* [150] develop a classification-like scheme to solve the inversion step and estimate the abundance coefficients using support vector machines. In an unsupervised setting, when the endmember signatures $\mathbf{m}_1,\ldots,\mathbf{m}_R$ are not known, a geometrical-oriented SU can be conducted, based on graph-based approximate geodesic distances [151], or manifold learning techniques [152, 153]. Specifically, as stated by Dobigeon *et al.* [154], when the nonlinear effects are not too severe and the pure pixel assumption discussed in Section 2.2 is fulfilled, geometrical-based EEA still provide reasonable estimation of the endmember signatures.

Other relevant approaches lie on nonparametric methods based on reproducing kernels as developed by Chen *et al.* [155, 156], Li *et al.* [157], and Nguyen *et al.* [158]. More precisely, operating in reproducing kernel Hilbert spaces (RKHS) is a well-admitted framework in the machine learning and signal processing community. For instance, kernel-based methods have been widely considered for detection and classification in hyperspectral images [159]. Conversely, when tackling the SU problem nonlinear unmixing approaches exploiting RKHS formalism have received much less interest. Broadwater *et al.* [160] and Broadwater and Banerjee [117] have merely proposed to replace each inner product between endmember signatures involved in the target cost functions with a kernel function. This strategy has been motivated by its high efficiency when solving classification problems since it is well known that an appropriate nonlinear mapping generally increases the class separability [161]. It can be straightforwardly interpreted as a nonlinear mapping operating on each the endmember signature, while neglecting interactions between the materials. Unfortunately, such strategy is more difficult to be advocated when conducting SU. Indeed, nonlinear interactions involved in the mixing process are rarely uniquely dominated by material-wise distortions, but rather by nonlinear interactions between the components themselves, as discussed in the previous section. In a supervised setting, Chen *et al.* [155] have proposed to take the nonlinear interactions of the endmembers into account by exploiting a new kernel-based paradigm. More precisely, exploiting the mixing relation in Eq. (3), the authors have formulated the SU problem in Eq. (4) as its functional counterpart

$$\min_{f_{\mathbf{b}_p} \in \mathcal{H}} \left\| \mathbf{y}_p - f_{\mathbf{b}_p}(\mathbf{M}, \mathbf{a}_p) \right\|_2^2 + \lambda \left\| f_{\mathbf{b}_p} \right\|_{\mathcal{H}}^2 \tag{27}$$

where $\mathcal{H}$ stands for a given functional space, the function $f_{\mathbf{b}_p}$ in Eq. (3) is now parametrized by additional (nonlinear) parameters $\mathbf{b}_p$ and λ adjusts the trade-off between regularity of the function $f_{\mathbf{b}_p}$ and the data-fitting term. The RKHS $\mathcal{H}$ may drastically impact the quality of the SU task and should be carefully chosen to address various nonlinear interactions. Chen *et al.* [155] have proposed to define the function $f_{\mathbf{b}_p}$ as a combination of a linear contribution similar to LMM and a nonparametric term accounting for any nonlinear interactions

$$f_{\mathbf{b}_p}(\mathbf{M}, \mathbf{a}_p) = \mathbf{M}\mathbf{a}_p + f_{\mathrm{NLMM}}(\mathbf{M}) \tag{28}$$

where $f_{\mathrm{NLMM}}(\cdot)$ is a band-wise and material-wise real-valued function of an appropriate RKHS. In particular, when this latter is chosen as the space of 2-nd order polynomials, the proposed approach generalizes the PPNM introduced in Section 3.3. Moreover, due to the strong connections between RKHS and Gaussian process formalisms, the RCA-based model proposed by Altmann *et al.* [146] and further generalized by Altmann *et al.* [162] to account for the spatial dependencies between nonlinearity effects, can also be associated to a polynomial kernel. Other nonlinear interactions can be considered by choosing an alternate reproducing kernel, inducing a specific RKHS.

Another promising nonparametric strategy has been introduced by Altmann *et al.* [163] who considers a kernel-based unsupervised SU approach. It relies on a nonlinear dimensionality reduction through a Gaussian process latent variable model (GPLVM). In this work, a particular form of kernel have been considered to extend the generalized bilinear model in (19). The resulting Bayesian algorithm is fully unsupervised since the endmember signatures can be jointly estimated with the abundance coefficients, as a complementary result of a prediction step [163]. Another advantage of GPLVMs is their ability to accurately describe various nonlinearities. This strategy has been also considered by Nguyen *et al.* [158] to solve the so-called pre-image problem [164]. When addressing the SU context, GPLVMs assume that spectrally different pixels result from different abundance vectors and thus can be represented by different latent variables. Conversely, spectrally close pixels are expected to have similar abundance vectors and thus similar latent variables. Several approaches have been proposed to preserve similarities, including back-constraints [165], dynamical models [166] and locally linear embedding (LLE) [167].

4 EXPERIMENTS: UNMIXING EELS DATA

By far, SU has been mainly considered by the geoscience community to analyze hyperspectral images acquired over the Earth [19]. In a more limited way, unmixing algorithms have been also advocated to analyze planetological data, e.g., acquired over Mars [58, 87]. More recently, since the first attempt by Dobigeon and Brun [18], SU has been considered to analyze data acquired by scanning transmission electron microscope [168–170], in particular EELS data. Following this recent trending, this section demonstrates that SU can extract meaningful and relevant characteristics from EELS spectrum-images.

The hyperspectral dataset has been acquired on a multilayer consisting of altered layers of carbon and boron nitride deposited by ion beam assisted deposition (IBAD) [171]. The cross-sectional specimen has been prepared using tripod polishing and ion milling. The data have been acquired on an aberration-corrected Nion UltraSTEM microscope operated at 100 kV. The 250×39 pixels (0.078 nm/pixel) spectrum-image has been acquired using 0.05 s/pixel acquisition time, 0.3 eV/channel energy dispersion (1340 channels).

The data have been analyzed using a set of six representative SU methods, i.e., two linear and four nonlinear SU methods presented in Sections 2 and 3 of this chapter. More precisely, the first linear method, referred to as BLU, is the Bayesian unsupervised unmixing method proposed by Dobigeon *et al.* [85] and relying on an additive Gaussian noise model whose variance is constant across the observed pixels and wavelengths. The second linear method, referred to as BLU+, is also a Bayesian method relying on a Gaussian noise assumption and generalizes BLU by allowing the noise variance to change across wavelengths. This method is a special case of the robust linear SU proposed by Altmann *et al.* [148]. The first two nonlinear methods are based on the PPNM presented in Section 3. The first algorithm (referred to as SPPNM for supervised PPNM [141]) uses the spectral signatures estimated by BLU+ as endmembers and thus solves a supervised nonlinear SU problem. The second PPNM-based method, referred to as UPPNM (unsupervised PPNM), is the unsupervised SU method proposed by Altmann *et al.* [143]. The third method is the RCA-based algorithm from Altmann *et al.* [162] (GRCA for generalized RCA), which estimates the nonlinearities and abundances associated with the endmembers estimated by BLU+. One of the main advantages of these three nonlinear methods is the fact that the deviations from the LMM are characterized by these models using a single parameter (b_p for SPPNM and UPPNM and s_p^2 for GRCA) which tends to 0 when the nonlinear effects decrease and which can thus be used to quantify the levels of nonlinearity. The last three nonlinear methods rely on robust LMMs. Precisely, the fourth (RNMF) and fifth (RNMF-KL) methods are algorithms proposed by Févotte and Dobigeon [28] for unsupervised nonlinear unmixing and rely on Gaussian (squared-Euclidean distance) and Poisson noise (generalized Kullback-Leibler divergence) assumptions, respectively. Finally, the last method, referred to as RBLU for robust BLU, is the Bayesian algorithm proposed by Altmann *et al.* [148] based on a Markovian spatial/spectral anomaly model. The arbitrary parameters (if any) of each method have been tuned by cross-validation.

For the considered dataset, the number of endmembers has been chosen as $R = 6$ from preliminary runs. The corresponding endmember spectra estimated by the different algorithms are depicted in Fig. 5. The corresponding estimated abundance maps are depicted in Fig. 6. These two figures show that all methods generally estimate similar endmembers and abundances, although local differences can be observed.

The endmember #1 corresponds to the substrate. The other endmembers spectra characterize the constituents of the layers: the identified endmembers #2 and #6 correspond to a graphite-like carbon K-edge, the endmember #3 to a boron-like K-edge, and the endmembers #4 and #5 to interlayer compounds resulting from the complex apparatus geometry combining evaporation with ion bombardment. Thus endmembers #4 and #5 appear at the interface between nominal layers and could be related to intermediate compounds as non-stoichiometric BN_x and BC_x, respectively, as endmembers #4 and #5 exhibit changes in the fine structure of the B K edge at 180 eV related to different bonding of B compounds [71]. Indeed, a variety of bonding structures can be obtained by IBAD within the B-C-N system. The interest of SU in this case is clearly to reveal the existence of such compounds at a nanometer scale.

Fig. 7 depicts the average noise variance in each channel, estimated by the different methods. While the methods assuming the same noise level in all channels (i.e., BLU and RNMF) provide a single estimated variance, the other methods provide more informative results. Indeed they all show that the noise variance generally decreases when the energy dispersion increases. They also highlight specific channels of high data variability (two peaks around 200 and 300 eV).

Finally, Fig. 8 depicts the level of nonlinearity estimated by the nonlinear SU methods in the 250 × 39 pixels. Precisely, this level has been represented by $|b_p|$ for the PPNM-based methods, by the logarithm of s_p^2 for GRCA and by the norm of the deviation term $||\mathbf{r}_p||^2$ for the robust LMM-based methods. Although the results depend on the underlying mixing model, spatial structures in agreement with the abundances depicted in Fig. 6 can be identified. In particular, nonlinear effects seem to be identified at the boundaries between

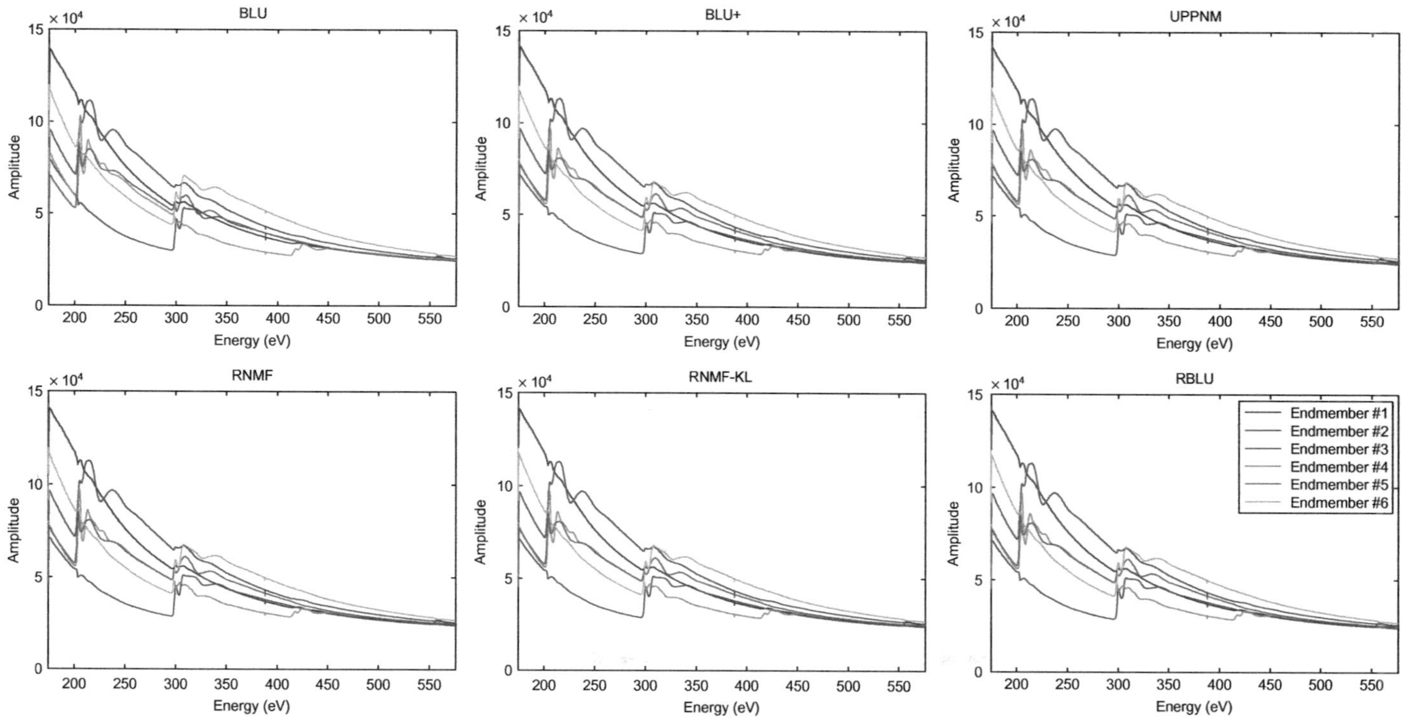

FIG. 5 Endmember spectral signatures extracted by various algorithms.

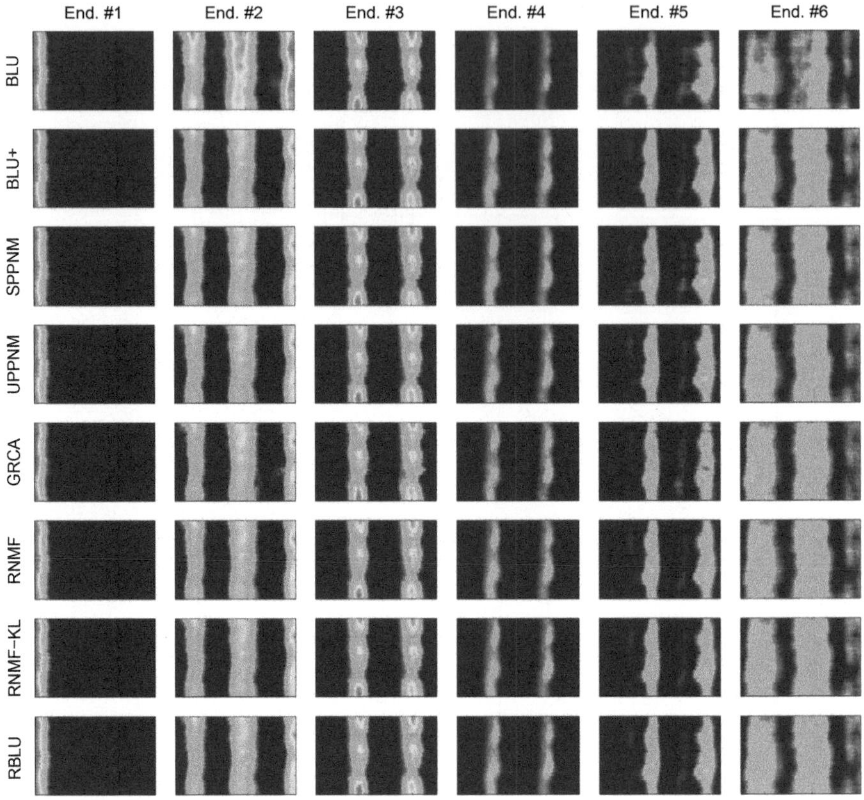

FIG. 6 Abundance maps estimated by various algorithms.

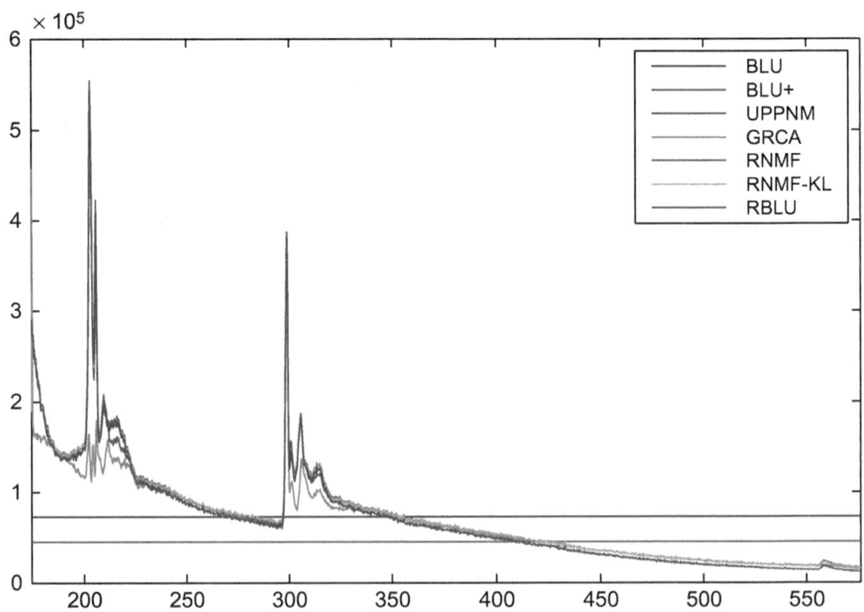

FIG. 7 Noise variances estimated by various algorithms.

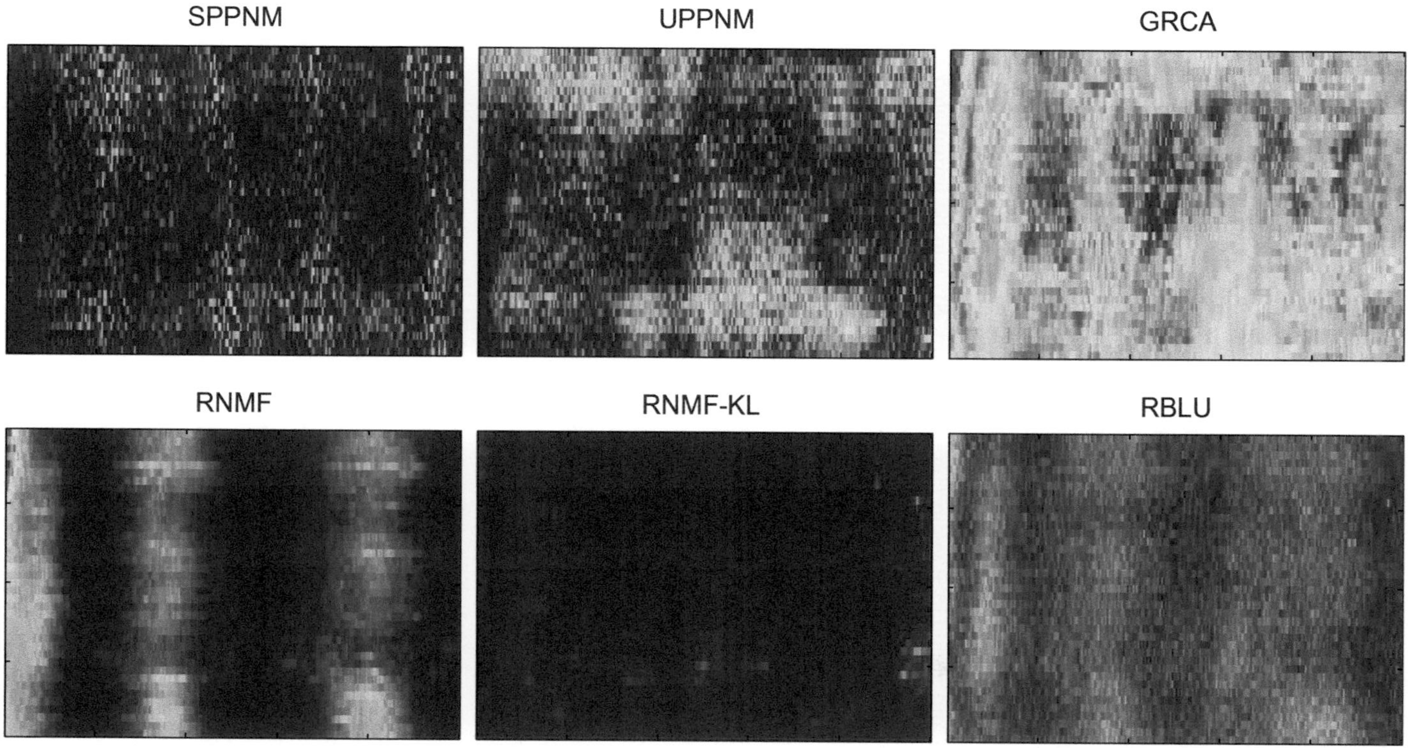

FIG. 8 Nonlinearity levels estimated by various algorithms.

the substrate and the graphite-like carbon K-edge and between the graphite-like carbon-and boron K-edges where complex interactions are expected to occur.

5 CONCLUSION

This chapter introduced spectral unmixing as a powerful analysis tool able to reveal latent and unobserved spectral and spatial structures in hyperspectral images acquired through various modalities, from long-range remote sensors to microscopy imagers. By identifying the spectral signatures of the main components present in the imaged scene while quantifying their respective spatial distributions over the scene, SU provides a compact, comprehensive and meaningful (i.e., physically interpretable) description of the whole set of measurements. In an unsupervised scenario, i.e., when both the endmember spectra and abundance vectors are unknown, SU can be formulated as a blind source separation problem. However, contrary to other popular methods of multivariate data analysis, such as principal or independent component analysis, the mathematical formulation underlying SU is driven by an acute will to model the physical interactions involved in the mixing process. This explains why SU provides latent parameters whose denominations (component spectral signatures and fractional abundances) directly refers to physical quantities of interest.

As a consequence, one key step when formulating SU consists of properly defining the mixture model that could explain the hyperspectral measurements. By far, the linear mixing model is the most commonly used since it is an admissible first approximation of the actual material interactions and generally provides relevant results. In this case, solving SU can be casted as a nonnegative matrix factorization problem, complemented by additional specific constraints. For more complex scenes, e.g., composed of multi-layered structures (vegetated or urban areas) or sand-like materials, other nonlinear mixing models need to be considered, at the price of a higher computational burden. To avoid the use of these nonlinear models when the linear mixing model is sufficient, detection-oriented techniques can be implemented to distinguish the linearly from the nonlinearly mixed pixels [142, 172, 173].

Akin to a large majority of BSS or NMF techniques, SU algorithms generally operates in a pixel-wise fashion, i.e., the mixing and unmixing processes are defined for each pixel independently. However some noticeable contributions have shown that exploiting contextual (e.g., spatial) information can improve the accuracy of the mixing model as well as the unmixing results. Indeed, firstly, when defining the model that governs the material interactions that occur in a given pixel, only components that are assumed to be present in this pixel are generally considered. However taking adjacency effects, i.e., additional interactions coming from spectral interference caused by atmospheric scattering, into account has been addressed in an unmixing context in Burazerovic *et al.* [174].

Moreover, when dealing with unmixing, most of the inversion or NMF-based SU algorithms work pixel-by-pixel. It is worthy to note that the Bayesian framework initially adopted by Dobigeon *et al.* [61, 85] to design SU algorithms offers the possibility of incorporating spatial dependencies between neighboring pixels through the use of Markov random fields. This strategy has been considered by Nishii and Ozaki [175] and Eches *et al.* [176, 177] to promote similar statistical characteristics of the abundance vectors associated with neighboring pixels. Conversely, such Markovian models have been also resorted by Altmann *et al.* [178] to improve the identification of the endmembers when performing semi-supervised SU. Deterministic counterparts of these Bayesian approaches have also shown interesting results when incorporating additional spatial information through an appropriate regularization. For instance Chen *et al.* [156] and Iordache *et al.* [179] have proposed to consider an ℓ_1-type regularizer form to promote piecewise-constant transitions in the fractional abundances. Legendre *et al.* [180] have defined an alternate regularizer to preserve the image edges.

Another major assumption commonly admitted when performing SU is the spectral invariance of the materials. More precisely, in common SU approaches, each component is assumed to be uniquely represented by a single spectral signature. However, spectral variability is known to be one of the major source of errors when conducting unmixing [181]. Various techniques have been proposed in the literature to mitigate this spectral variability. As stated by Somers *et al.* [33], two classes of SU methods have been mainly developed. The first one represents each endmember as a collection of spectral signatures, or bundles, representative of the corresponding materials [182]. Conversely, another class of methods represent the endmembers as statistical distributions [177, 183]. More recently, Thouvenin *et al.* [184] and Veganzones *et al.* [185] have proposed to derive parametric models to explicitly account for endmember variability over the pixels of a given image.

Finally, it is worthy to note that preliminary works have shown that SU can be implemented for a collection of hyperspectral images acquired over time, which opens the door to multitemporal SU [186, 187].

ACKNOWLEDGMENTS

N.D. and Y.A. would like to thank their close collaborators, Jean-Yves Tourneret, University of Toulouse; Steve McLaughlin and Abderrahim Halimi, Heriot-Watt University; Alfred O. Hero, University of Michigan; Cédric Févotte, University of Nice; and Jose M. Bioucas Dias, Universidade de Lisboa; for stimulating and fruitful discussions during the past years while facing with unmixing problems. N.B. would like to thank Virginie Serin (CEMES, Toulouse) and Ignacio Jimenez (ICMM-CSIC Madrid) for providing the C/BN sample and for comments on the results.

REFERENCES

[1] Schaepman ME, Ustin SL, Plaza A, Painter TH, Verrelst J, Liang S. Earth system science related imaging spectroscopy–an assessment. Remote Sens Environ 2009;113:123–37.

[2] Smith MO, Johnson PE, Adams JB. Quantitative determination of mineral types and abundances from reflectance spectra using principal component analysis. In: Proc Lunar and Planetary Sci Conf 1985;90:p. 797–904.

[3] Douté S, Schmitt B, Langevin Y, Bibring JP, Altieri F, Bellucci G, et al. South pole of mars: nature and composition of the icy terrains from mars express OMEGA observations. Planet Space Sci 2007;55:113–33.

[4] Bernard-Michel C, Douté S, Fauvel M, Gardes L, Girard S. Retrieval of mars surface physical properties from OMEGA hyperspectral images using regularized sliced inverse regression. J Geophys Res 2009;114.

[5] Gowen AA, O'Donnell CP, Cullen PJ, Downey G, Frias JM. Hyperspectral imaging : an emerging process analytical tool for food quality and safety control. Trends Food Sci Technol 2007;18(12):590–8.

[6] Berman M, Conner P, Whitbourn L, Coward D, Osborne B, Southan M. Classification of sound and stained wheat grains using visible and near infrared hyperspectral image analysis. J Near Infrared Spectrosc 2007;15(6):351–8.

[7] Mahest S, Manichavsagan A, Jayas D, Paliwall J, White N. Feasibiliy of near-infrared hyperspectral imaging to differentiate Canadian wheat classes. Biosyst Eng 2008;101(1):50–7.

[8] Akbari H, Kosugi Y, Kojima K, Tanaka N. Detection and analysis of the intestinal ischemia using visible and invisible hyperspectral imaging. IEEE Trans Biomed Eng 2010;57(8): 2011–7.

[9] Gendrin C, Roggo Y, Collet C. Pharmaceutical applications of vibrational chemical imaging and chemometrics: a review. J Pharm Biomed Anal 2008;48(3):533–53.

[10] Plaza J, Pérez R, Plaza A, Martínez P, Valencia D. Mapping oil spills on sea water using spectral mixture analysis of hyperspectral image data. In: Jensen JO, Thériault JM, editors. Proc. SPIE chemical and biological standoff detection III. 5995. SPIE; 2005. p. 79–86.

[11] Manolakis D, Siracusa C, Shaw G. Hyperspectral subpixel target detection using the linear mixing model. IEEE Trans Geosci Remote Sens 2001;39(7):1392–409.

[12] Stein D, Beaven S, Hoff L, Winter E, Schaum A, Stocker A. Anomaly detection from hyperspectral imagery. IEEE Signal Process Mag 2002;19(1):58–69.

[13] Jackson Q, Landgrebe DA. An adaptive method for combined covariance estimation and classification. IEEE Trans Geosci Remote Sens 2002;40(5):1082–7.

[14] Chang CI. Hyperspectral imaging: techniques for spectral detection and classification. New York: Kluwer; 2003.

[15] Rellier G, Descombes X, Falzon F, Zerubia J. Texture feature analysis using a Gauss-Markov model in hyperspectral image classification. IEEE Trans Geosci Remote Sens 2004;42(7):1543–51.

[16] Colliex C, Gloter A, Kociak M, March K, Stéfan O, Tencé M. The STEM multi-signal approach: learning the most from your nano-object. Microsc Anal 2012;6:33–42.

[17] Pennycook TJ, Oxley MP, Garcia-Barriocanal J, Bruno FY, Leon C, Santamaria J, et al. Seeing oxygen disorder in YSZ/SrTiO3 colossal ionic conductor heterostructures using EELS. Eur Phys J Appl Phys 2011;54(03).

[18] Dobigeon N, Brun N. Spectral mixture analysis of EELS spectrum-images. Ultramicroscopy 2012;120:25–34.

[19] Chang CI. Hyperspectral data exploitation: theory and applications. Hoboken, NJ: John Wiley & Sons; 2007.

[20] Henrot S, Soussen C, Brie D. Fast positive deconvolution of hyperspectral images. IEEE Trans Image Process 2013;22(2):828–33.

[21] Loncan L, Almeida LB, Bioucas-Dias JM, Briottet X, Chanussot J, Dobigeon N, et al. Hyperspectral pansharpening: a review. IEEE Geosci Remote Sens Mag 2015;3(3):27–46.

[22] Wei Q, Dobigeon N, Tourneret JY. Bayesian fusion of multi-band images. IEEE J Sel Topics Signal Process 2015;9(6):1117–27.

[23] Wei Q, Dobigeon N, Tourneret JY. Fast fusion of multi-band images based on solving a Sylvester equation. IEEE Trans Image Process 2015;24(11):4109–21.

[24] Keshava N, Mustard JF. Spectral unmixing. IEEE Signal Process Mag 2002;19(1):44–57.

[25] Bioucas-Dias JM, Plaza A, Dobigeon N, Parente M, Du Q, Gader P, et al. Hyperspectral unmixing overview: geometrical, statistical, and sparse regression-based approaches. IEEE J Sel Top Appl Earth Obs Remote Sens 2012;5(2):354–79.

[26] Somers B. Endmember library approaches to resolve spectral mixing problems in remotely sensed data: challenges, potential and applications. In: Ruckebusch C, editor. Resolving spectral mixtures–with application from ultrafast spectroscopy to super-resolution imaging. Oxford, UK: Elsevier; 2016.

[27] Hapke B. Theory of reflectance and emittance spectroscopy. Cambridge, UK: Cambridge Univ. Press; 1993.

[28] Févotte C, Dobigeon N. Nonlinear hyperspectral unmixing with robust nonnegative matrix factorization. IEEE Trans Image Process 2015;24(12):4810–9.

[29] Wang S, Wang N, Tao D, Zhang L, Du B. A K-L divergence constrained sparse NMF for hyperspectral signal unmixing. In: Proc. IEEE int. conf. security patt. anal. cybernetics (SPAC); 2014. p. 223–8.

[30] Kay SM. Fundamentals of statistical signal processing: estimation theory. Englewood Cliffs, NJ: Prentice Hall; 1993.

[31] Plaza A, Martinez P, Perez R, Plaza J. A quantitative and comparative analysis of endmember extraction algorithms from hyperspectral data. IEEE Trans Geosci Remote Sens 2004;42(3):650–63.

[32] Kent JT, Mardia KV. Spatial classification using fuzzy membership models. IEEE Trans Patt Anal Mach Intell 1988;10(5):659–71.

[33] Somers B, Asner GP, Tits L, Coppin P. Endmember variability in spectral mixture analysis: a review. Remote Sens Environ 2011;115(7):1603–16.

[34] Chang CI, Wang S. Constrained band selection for hyperspectral imagery. IEEE Trans Geosci Remote Sens 2006;44(6):1575–85.

[35] Jolliffe IT. Principal component analysis. New York: Springer-Verlag; 1986.

[36] Green AA, Berman M, Switzer P, Craig MD. A transformation for ordering multispectral data in terms of image quality with implications for noise removal. IEEE Trans Geosci Remote Sens 1988;26(1):65–74.

[37] Bioucas-Dias JM, Nascimento JMP. Hyperspectral subspace identification. IEEE Trans Geosci Remote Sens 2008;46(8):2435–45.

[38] Boardman JW. Kruse FA, Green RO. Mapping target signatures via partial unmixing of AVIRIS data. In: Green RO, editor. Summaries 5th annual JPL airborne earth science workshop. JPL Publication; 1995. p. 23–6.

[39] Winter ME. Fast autonomous spectral end-member determination in hyperspectral data. In: Proc. 13th int. conf. on applied geologic remote sensing, Vancouver. 2;1999. p. 337–44.

[40] Nascimento JMP, Bioucas-Dias JM. Vertex component analysis: a fast algorithm to unmix hyperspectral data. IEEE Trans Geosci Remote Sens 2005;43(4):898–910.

[41] Ma WK, Bioucas-Dias JM, Gader P, Chan TH, Gillis N, Plaza A, et al. Signal processing perspective on hyperspectral unmixing. IEEE Signal Process Mag 2013;31(1):67–81.

[42] Araújo U, Saldanha B, Galv ao R, T Yoneyama HC, Visani V. The successive projections algorithm for variable selection in spectroscopic multicomponent analysis. Chemometr Intell Lab Syst 2001;57:65–73.

[43] Chan TH, Ma WK, Ambikapathi A, Chi CY. A simplex volume maximization framework for hyperspectral endmember extraction. IEEE Trans Geosci Remote Sens 2011;49(11):4177–93.

[44] Wu CC, Chu S, Chang CI. Sequential N-FINDR algorithms. In: Proc. SPIE imaging spectrometry XIII. 7086; 2008.

[45] Craig MD. Minimum volume transforms for remotely sensed data. IEEE Trans Geosci Remote Sens 1994;32(3):542–52.

[46] Bowles JH, Palmadesso PJ, Antoniades JA, Baumback MM, Rickard LJ. Use of filter vectors and fast convex set methods in hyperspectral analysis. In: Strojnik M, Andresen BF, editors. Proc. SPIE infrared spaceborne remote sensing III, San Diego, CA, US. 2553. SPIE; 1995. p. 148–57.

[47] Li J, Bioucas-Dias JM. Minimum volume simplex analysis: a fast algorithm to unmix hyperspectral data. In: Proc. IEEE international conference on geoscience and remote sensing (IGARSS), Boston, USA. 3; 2008. p. 250–3.

[48] Chan TH, Ambikapathi A, Wing-KinMa, Chi CY. Robust affine set fitting and fast simplex volume max-min for hyperspectral endmember extraction. IEEE Trans Geosci Remote Sens 2013;51(7):3982–97.

[49] Chan TH, Ma WK, Ambikapathi AM, Chi CY. Chance-constrained robust minimum-volume enclosing simplex algorithm for hyperspectral unmixing. IEEE Trans Geosci Remote Sens 2011;49(11):4194–209.

[50] Bioucas-Dias JM. A variable splitting augmented Lagrangian approach to linear spectral unmixing. In: Proc. IEEE GRSS workshop hyperspectral image signal processing: evolution in remote sensing (WHISPERS), Grenoble, France; 2009.

[51] Heinz DC, Chang C. Fully constrained least-squares linear spectral mixture analysis method for material quantification in hyperspectral imagery. IEEE Trans Geosci Remote Sens 2001;29(3):529–45.

[52] Tu TM, Chen CH, Chang CI. A noise subspace projection approach to target signature detection and extraction in an unknown background for hyperspectral images. IEEE Trans Geosci Remote Sens 1998;36(1):171–81.

[53] Theys C, Dobigeon N, Tourneret JY, Lantéri H. Linear unmixing of hyperspectral images using a scaled gradient method. In: Proc. IEEE-SP workshop stat. and signal process. (SSP), Cardiff, UK; 2009. p. 729–32.

[54] Heylen R, Burazerovic D, Scheunders P. Fully constrained least squares spectral unmixing by simplex projection. IEEE Trans Geosci Remote Sens 2011;49(11):4112–22.

[55] Wei Q, Bioucas-Dias JM, Dobigeon N, Tourneret JY. Fast spectral unmixing based on Dykstra's alternating projection; 2015, http://arxiv.org/abs/1505.01740/. ArXiv preprint.

[56] Chouzenoux E, Legendre M, Moussaoui S, Idier J. Fast constrained least squares spectral unmixing using primal-dual interior point optimization. IEEE J Sel Top Appl Earth Obs Remote Sens 2014;7(1):59–69.

[57] Themelis K, Rontogiannis AA. A soft constrained MAP estimator for supervised hyperspectral signal unmixing. In: Proc. european signal process. conf. (EUSIPCO), Lausanne, Switzerland; 2008.

[58] Themelis KE, Schmidt F, Sykioti O, Rontogiannis AA, Koutroumbas KD, Daglis IA. On the unmixing of MEx/OMEGA hyperspectral data. Planet Space Sci 2012;68(1):34–41.

[59] Dobigeon N, Tourneret JY, Hero IIIAO. Bayesian linear unmixing of hyperspectral images corrupted by colored Gaussian noise with unknown covariance matrix. In: Proc. IEEE international conference on acoustics, speech and signal processing (ICASSP), Las Vegas, USA; 2008. p. 3433–6.

[60] Rodriguez-Yam G, Davis RA, Scharf L. A Bayesian model and Gibbs sampler for hyperspectral imaging. In: Proc. IEEE sensor array and multichannel signal processing workshop (SAM), Washington, DC; 2002. p. 105–9.

[61] Dobigeon N, Tourneret JY, Chang CI. Semi-supervised linear spectral unmixing using a hierarchical Bayesian model for hyperspectral imagery. IEEE Trans Signal Process 2008;56(7):2684–95.

[62] Mallat S, Zhang Z. Matching pursuits with time-frequency dictionaries. IEEE Trans Signal Process 1993;41(12):3397–415.

[63] Pati Y, Rezaiifar R, Krishnaprasad P. Orthogonal matching pursuit : recursive function approximation with application to wavelet decomposition. In: Proc. IEEE asilomar conf. signals, systems, computers; 1993. p. 40–4.

[64] Bioucas-Dias JM, Figueiredo MAT. Alternating direction algorithms for constrained sparse regression: application to hyperspectral unmixing. In: Proc. IEEE GRSS workshop hyperspectral image signal processing: evolution in remote sensing (WHISPERS). 1; 2010. p. 1–4.

[65] Iordache MD, Bioucas-Dias JM, Plaza A. Sparse unmixing of hyperspectral data. IEEE Trans Geosci Remote Sens 2011;49(6):2014–39.

[66] Themelis KE, Rontogiannis AA, Koutroumbas KD. A novel hierarchical Bayesian approach for sparse semisupervised hyperspectral unmixing. IEEE Trans Signal Process 2012;60(2):585–99.

[67] Févotte C. Bayesian audio source separation. In: Makino S, Lee TW, Sawada H, editors. Blind speech separation. Dordrecht, The Netherlands: Springer; 2007. p. 305–35.

[68] Févotte C, Bertin N, Durrieu JL. Nonnegative matrix factorization with the Itakura-Saito divergence. With application to music analysis. Neural Comput 2009;21(3):793–830.

[69] Bonnet N, Brun N, Colliex C. Extracting information from sequences of spatially resolved EELS spectra using multivariate statistical analysis. Ultramicroscopy 1999;77(3-4):97–112.

[70] Bosman M, Watanabe M, Alexander DTL, Keast VJ. Mapping chemical and bonding information using multivariate analysis of electron energy-loss spectrum images. Ultramicroscopy 2006;106(11-12):1024–32.

[71] Arenal R, de la Peña F. Stéphan O, Walls M, Tencé M, Loiseau A, et al. Extending the analysis of EELS spectrum-imaging data, from elemental to bond mapping in complex nanostructures. Ultramicroscopy 2008;109(1):32–8.

[72] Bayliss J, Gualtieri JA, Cromp R. Analyzing hyperspectral data with independent component analysis. In: Selander JM, editor. Proc. AIPR workshop exploiting new image sources and sensors. 3240. Washington, DC: SPIE; 1998. p. 133–43.

[73] Wang J, Chang CI. Applications of independent component analysis in endmember extraction and abundance quantification for hyperspectral imagery. IEEE Trans Geosci Remote Sens 2006;4(9):2601–16.

[74] Bonnet N, Nuzillard D. Independent component analysis: a new possibility for analysing series of electron energy loss spectra. Ultramicroscopy 2005;102(4):327–37.

[75] de la Peña F, Berger MH, Hochepied JF, Dynys F, Stephan O, Walls M. Mapping titanium and tin oxide phases using EELS: an application of independent component analysis. Ultramicroscopy 2011;111(2):169–76.

[76] Nascimento JMP, Bioucas-Dias JM. Does independent component analysis play a role in unmixing hyperspectral data? IEEE Trans Geosci Remote Sens 2005;43(1):175–87.

[77] Lee DD, Seung HS. Learning the parts of objects by non-negative matrix factorization. Nature 1999;401:788–91.

[78] Paatero P, Tapper U. Positive matrix factorization: a non-negative factor model with optimal utilization of error estimates of data values. Environmetrics 1994;5:111–26.

[79] Miao L, Qi H. Endmember extraction from highly mixed data using minimum volume constrained nonnegative matrix factorization. IEEE Trans Geosci Remote Sens 2007;45(3): 765–76.

[80] Huck A, Guillaume M. Robust hyperspectral data unmixing with spatial and spectral regularized NMF. In: Proc. IEEE GRSS workshop hyperspectral image signal processing: evolution in remote sensing (WHISPERS), Reykjavik, Iceland; 2010.

[81] Jia S, Qian Y. Constrained nonnegative matrix factorization for hyperspectral unmixing. IEEE Trans Geosci Remote Sens 2009;47(1):161–73.

[82] Qian Y, Jia S, Zhou J, Robles-Kelly A. Hyperspectral unmixing via $l_{1/2}$ sparsity-constrained nonnegative matrix factorization. IEEE Trans Geosci Remote Sens 2011;49(11):4282–97.

[83] Ammanouil R, Ferrari A, Richard C, Mary D. Blind and fully constrained unmixing of hyperspectral images. IEEE Trans Image Process 2014;23(12):5510–8.

[84] Moussaoui S, Brie D, Mohammad-Djafari A, Carteret C. Separation of non-negative mixture of non-negative sources using a Bayesian approach and MCMC sampling. IEEE Trans Signal Process 2006;54(11):4133–45.

[85] Dobigeon N, Moussaoui S, Coulon M, Tourneret JY, Hero AO. Joint Bayesian endmember extraction and linear unmixing for hyperspectral imagery. IEEE Trans Signal Process 2009;57(11):4355–68.

[86] Arngren M, Schmidt MN, Larsen J. Unmixing of hyperspectral images using Bayesian nonnegative matrix factorization with volume prior. J Signal Process Sys 2011;65(3): 479–96.

[87] Schmidt F, Moussaoui S, Dobigeon N. Material identification on martian hyperspectral images using Bayesian source separation. In: Proc. IEEE GRSS workshop hyperspectral image signal processing: evolution in remote sensing (WHISPERS); 2009. p. 1–4.

[88] Moussaoui S, Hauksdóttir H, Schmidt F, Jutten C, Chanussot J, Brie D, et al. On the decomposition of mars hyperspectral data by ICA and Bayesian positive source separation. Neurocomputing 2008;71(10-12).

[89] Carteret C, Dandeu A, Moussaoui S, Muhr H, Humbert B, Plasari E. Polymorphism studied by lattice phonon raman spectroscopy and statistical mixture analysis method. Application to calcium carbonate polymorphs during batch crystallization. Cryst Growth Des 2009;9(2): 807–12.

[90] Moussaoui S, Carteret C, Brie D, Mohammad-Djafari A. Bayesian analysis of spectral mixture data using Markov chain Monte Carlo methods. Chemometr Intell Lab Syst 2006;81(2): 137–48.

[91] Duchamp M, Lachmann M, Boothroyd CB, Kovács A, Haug FJ, Ballif C, et al. Compositional study of defects in microcrystalline silicon solar cells using spectral decomposition in the scanning transmission electron microscope. App Phys Lett 2013;102(133902).

[92] Bazot C, Dobigeon N, Tourneret JY, Zaas AK, Ginsburg GS, Hero AO. Unsupervised Bayesian linear unmixing of gene expression microarrays. BMC Bioinform 2013;14(99).

[93] Schmidt F, Guiheneuf M, Moussaoui S, Tréguier E, Schmidt A, Dobigeon N. Implementation strategies for hyperspectral unmixing using Bayesian source separation. IEEE Trans Geosci Remote Sens 2010;48(11):4003–13.

[94] Brie D. Bayesian positive source separation and factorization of spectral mixtures. In: Ruckebusch C, editor. Resolving spectral mixtures–with application from ultrafast spectroscopy to super-resolution imaging. Oxford, UK: Elsevier; 2016.
[95] Singer RB, McCord TB. Mars: Large scale mixing of bright and dark surface materials and implications for analysis of spectral reflectance. In: Proc. 10th lunar and planetary sci. conf., Houston; 1979. p. 1835–48.
[96] Clark RN, Roush TL. Reflectance spectroscopy: quantitative analysis techniques for remote sensing applications. J Geophys Res 1984;89(7):6329–40.
[97] Ray T, Murray B. Nonlinear spectral mixing in desert vegetation. Remote Sens Environ 1996;55(1):59–64.
[98] Combe JP, Launeau P, Carrère V, Despan D, Méléder V, Barillé L, et al. Mapping microphytobenthos biomass by non-linear inversion of visible-infrared hyperspectral images. Remote Sens Environ 2005;98(4):371–87.
[99] Liu W, Wu EY. Comparison of non-linear mixture models: sub-pixel classification. Remote Sens Environ 2005;18(2):145–54.
[100] Arai K. Nonlinear mixture model of mixed pixels in remote sensing satellite images based on Monte Carlo simulation. Adv Space Res 2008;41(11):1715–23.
[101] Fan W, Hu B, Miller J, Li M. Comparative study between a new nonlinear model and common linear model for analysing laboratory simulated-forest hyperspectral data. IJRS 2009;30(11):2951–62.
[102] Somers B, Cools K, Delalieux S, Stuckens J, der Zande DV, Verstraeten WW, et al. Nonlinear hyperspectral mixture analysis for tree cover estimates in orchards. Remote Sens Environ 2009;113:1183–93.
[103] Somers B, Tits L, Coppin P. Quantifying nonlinear spectral mixing in vegetated areas: computer simulation model validation and first results. IEEE J Sel Top Appl Earth Obs Remote Sens 2014;7(6):1956–65.
[104] Dobigeon N, Tits L, Somers B, Altmann Y, Coppin P. A comparison of nonlinear mixing models for vegetated areas using simulated and real hyperspectral data. IEEE J Sel Top Appl Earth Obs Remote Sens 2014;7(6):1869–78.
[105] Hapke BW. Bidirectional reflectance spectroscopy. I. Theory. J Geophys Res 1981;86(B4): 3039–54.
[106] Nash DB, Conel JE. Spectral reflectance systematics for mixtures of powdered hypersthene, labradorite, and ilmenite. J Geophys Res 1974;79(11):1615–21.
[107] Mustard JF, Pieters CM. Photometric phase functions of common geologic minerals and applications to quantitative analysis of mineral mixture reflectance spectra. J Geophys Res 1989;94(B10):13,619–34.
[108] Draine BT. The discrete-dipole approximation and its application to interstellar graphite grains. Astrophys J 1988;333:848–72.
[109] Shkuratov Y, Starukhina L, Hoffmann H, Arnold G. A model of spectral albedo of particulate surfaces: implication to optical properties of the Moon. Icarus 1999;137(2):235–46.
[110] Guilfoyle KJ, Althouse ML, Chang CI. A quantitative and comparative analysis of linear and nonlinear spectral mixture models using radial basis function neural networks. IEEE Trans Geosci Remote Sens 2001;39(8):2314–8.
[111] Nascimento JMP, Bioucas-Dias JM. Unmixing hyperspectral intimate mixtures. In: Bruzzone L, editor. Proc. SPIE image and signal processing for remote sensing XVI, Toulouse, France. 74830. SPIE; 2010. p. 78300C.
[112] Altmann Y, Dobigeon N, McLaughlin S, Tourneret JY. Nonlinear unmixing of hyperspectral images using radial basis functions and orthogonal least squares. In: Proc. IEEE

international conference on geoscience and remote sensing (IGARSS), Vancouver, Canada; 2011. p. 1151–4.

[113] Plaza J, Martinez P, Perez R, Plaza A. Nonlinear neural network mixture models for fractional abundance estimation in AVIRIS hyperspectral images. In: Proc. XIII NASA/Jet propulsion laboratory airborne earth science workshop, Pasadena, CA, USA; 2004. p. 1–12.

[114] Plaza J, Plaza A, Perez R, Martinez P. Automated generation of semi-labeled training samples for nonlinear neural network-based abundance estimation in hyperspectral data. In: Proc. IEEE international conference on geoscience and remote sensing (IGARSS); 2005. p. 345–50.

[115] Plaza J, Plaza A, Pérez R, Martinez P. Joint linear/nonlinear spectral unmixing of hyperspectral image data. In: Proc. IEEE international conference on geoscience and remote sensing (IGARSS); 2007. p. 4037–40.

[116] Plaza J, Plaza A, Perez R, Martinez P. On the use of small training sets for neural network-based characterization of mixed pixels in remotely sensed hyperspectral images. Phys Rev 2009;42(11):3032–45.

[117] Broadwater J, Banerjee A. A comparison of kernel functions for intimate mixture models. In: Proc. IEEE GRSS workshop hyperspectral image signal processing: evolution in remote sensing (WHISPERS); 2009. p. 1–4.

[118] Broadwater J, Banerjee A. A generalized kernel for areal and intimate mixtures. In: Proc. IEEE GRSS workshop hyperspectral image signal processing: evolution in remote sensing (WHISPERS); 2010. p. 1–4.

[119] Broadwater J, Banerjee A. Mapping intimate mixtures using an adaptive kernel-based technique. In: Proc. IEEE GRSS workshop hyperspectral image signal processing: evolution in remote sensing (WHISPERS), Lisbon, Portugal; 2011. p. 1–4.

[120] Close R, Gader P, Wilson J, Zare A. Using physics-based macroscopic and microscopic mixture models for hyperspectral pixel unmixing. In: Shen SS, Lewis PE, editors. Proc. SPIE algorithms and technologies for multispectral, hyperspectral, and ultraspectral imagery XVIII, Baltimore, Maryland, USA. 8390. SPIE; 2012. p. 83901L–83901L-13.

[121] Close R, Gader P, Zare A, Wilson J, Dranishnikov D. Endmember extraction using the physics-based multi-mixture pixel model. In: Shen SS, Lewis PE, editors. Proc. SPIE imaging spectrometry XVII, San Diego, California, USA. 8515. SPIE; 2012. p. 85150L–14.

[122] Borel CC, Gerstl SAW. Nonlinear spectral mixing model for vegetative and soil surfaces. Remote Sens Environ 1994;47(3):403–16.

[123] Zhang L, Li D, Tong Q, Zheng L. Study of the spectral mixture model of soil and vegetation in PoYang Lake area, China. Remote Sens Environ 1998;19:2077–84.

[124] Chen X, Vierling L. Spectral mixture analyses of hyperspectral data acquired using a tethered balloon. Remote Sens Environ 2006;103(3):338–50.

[125] Tits L, Delabastita W, Somers B, Farifteh J, Coppin P. First results of quantifying nonlinear mixing effects in heterogeneous forests: a modeling approach. In: Proc. IEEE international conference on geoscience and remote sensing (IGARSS); 2012. p. 7185–8.

[126] Huard P, Marion R. Study of non-linear mixing in hyperspectral imagery–a first attempt in the laboratory. In: Proc. IEEE GRSS workshop hyperspectral image signal processing: evolution in remote sensing (WHISPERS), Lisbon, Portugal; 2011. p. 1–4.

[127] Fontanilles G, Briottet X. A nonlinear unmixing method in the infrared domain. App Opt 2011;50(20):3666–77.

[128] Meganem I, Déliot P, Briottet X, Deville Y, Hosseini S. Linear-quadratic mixing model for reflectances in urban environments. IEEE Trans Geosci Remote Sens 2014;52(1):544–58.

[129] Altmann Y, Dobigeon N, Tourneret JY. Bilinear models for nonlinear unmixing of hyperspectral images. In: Proc. IEEE GRSS workshop hyperspectral image signal process: evolution in remote sensing (WHISPERS), Lisbon, Portugal; 2011. p. 1–4.

[130] Nascimento JMP, Bioucas-Dias JM. Nonlinear mixture model for hyperspectral unmixing. In: Bruzzone L, Notarnicola C, Posa F, editors. Proc. SPIE image and signal processing for remote sensing XV, Berlin, Germany. 7477. SPIE; 2009. p. 74770I.

[131] Raksuntorn N, Du Q. Nonlinear spectral mixture analysis for hyperspectral imagery in an unknown environment. IEEE Geosci Remote Sens Lett 2010;7(4):836–40.

[132] Gader P, Dranishnikov D, Zare A, Chanussot J. A sparsity promoting bilinear unmixing model. In: Proc. IEEE GRSS workshop hyperspectral image signal processing: evolution in remote sensing (WHISPERS), Shanghai, China; 2012.

[133] Zare A, Gader P. Sparsity promoting iterated constrained endmember detection in hyperspectral imagery. IEEE Geosci Remote Sens Lett 2007;4(3):446–50.

[134] Eches O, Guillaume M. A bilinear-bilinear non-negative matrix factorization method for hyperspectral unmixing. IEEE Geosci Remote Sens Lett 2014;11(4):778–82.

[135] Halimi A, Altmann Y, Dobigeon N, Tourneret JY. Nonlinear unmixing of hyperspectral images using a generalized bilinear model. IEEE Trans Geosci Remote Sens 2011;49(11): 4153–62.

[136] Halimi A, Altmann Y, Dobigeon N, Tourneret JY. Unmixing hyperspectral images using the generalized bilinear model. In: Proc. IEEE international conference on geoscience and remote sensing (IGARSS), Vancouver, Canada; 2011. p. 1886–9.

[137] Yokoya N, Chanussot J, Iwasaki A. Nonlinear unmixing of hyperspectral data using semi-nonnegative matrix factorization. IEEE Trans Geosci Remote Sens 2014;52(2):1430–7.

[138] Heylen R, Scheunders P. Calculation of geodesic distances in nonlinear mixing models: application to the generalized bilinear model. IEEE Geosci Remote Sens Lett 2012;9(4): 644–8.

[139] Duarte LT, Jutten C, Moussaoui S. Bayesian source separation of linear and linear-quadratic mixtures using truncated priors. J Signal Process Syst 2010;65(3):311–23.

[140] Taleb A, Jutten C. Source separation in post-nonlinear mixtures. IEEE Trans Signal Process 1999;47(10):2807–20.

[141] Altmann Y, Halimi A, Dobigeon N, Tourneret JY. Supervised nonlinear spectral unmixing using a post-nonlinear mixing model for hyperspectral imagery. IEEE Trans Image Process 2012;21(6):3017–25.

[142] Altmann Y, Dobigeon N, Tourneret JY. Nonlinearity detection in hyperspectral images using a polynomial post-nonlinear mixing model. IEEE Trans Image Process 2013;22(4): 1267–76.

[143] Altmann Y, Dobigeon N, Tourneret JY. Unsupervised post-nonlinear unmixing of hyperspectral images using a Hamiltonian Monte Carlo algorithm. IEEE Trans Image Process 2014;23(6):2663–75.

[144] Marinoni A, Plaza J, Plaza A, Gamba P. Nonlinear hyperspectral unmixing using nonlinearity order estimation and polytope decomposition. IEEE J Sel Top Appl Earth Obs Remote Sens 2015;8(6):2644–54.

[145] Heylen R, Scheunders P. A multilinear mixing model for nonlinear spectral unmixing. IEEE Trans Geosci Remote Sens 2016;54(1):240–51.

[146] Altmann Y, Dobigeon N, McLaughlin S, Tourneret JY. Residual component analysis of hyperspectral images–application to joint nonlinear unmixing and nonlinearity detection. IEEE Trans Image Process 2014;23(5):2148–58.

[147] Newstadt GE, Hero AO, Simmon J. Robust spectral unmixing for anomaly detection. In: Proc. IEEE-SP workshop stat. and signal process. (SSP), Gold Coast, Australia; 2014.

[148] Altmann Y, McLaughlin S, Hero AO. Robust linear spectral unmixing using anomaly detection. IEEE Trans Comput Imag 2015;1(2):74–85.

[149] Plaza J, Plaza A, Martinez P, Perez RM. Nonlinear mixture models for analyzing laboratory simulated-forest hyperspectral data. Proc SPIE Image Signal Process Remote Sens IX 2004;5238:480–7.

[150] Li PX, Bo W, Zhang L. Abundance estimation from hyperspectral image based on probabilistic outputs of multi-class support vector machines. In: Proc. IEEE international conference on geoscience and remote sensing (IGARSS); 2005. p. 4315–8.

[151] Heylen R, Burazerovic D, Scheunders P. Non-linear spectral unmixing by geodesic simplex volume maximization. IEEE J Sel Topics Signal Process 2011;5(3):534–42.

[152] Nguyen H, Richard C, Honeine P, Theys C. Hyperspectral image unmixing using manifold learning methods. Derivations and comparative tests. In: Proc. IEEE international conference on geoscience and remote sensing (IGARSS), Munich, Germany; 2012. p. 3086–9.

[153] Licciardi GA, Ceamanos X, Doute S, Chanussot J. Unsupervised nonlinear spectral unmixing by means of NLPCA applied to hyperspectral imagery. In: Proc. IEEE international conference on geoscience and remote sensing (IGARSS), Munich, Germany; 2012. p. 1369–72.

[154] Dobigeon N, Tourneret JY, Richard C, Bermudez JCM, McLaughlin S, Hero AO. Nonlinear unmixing of hyperspectral images: models and algorithms. IEEE Signal Process Mag 2014;31(1):89–94.

[155] Chen J, Richard C, Honeine P. Nonlinear unmixing of hyperspectral data based on a linear-mixture/nonlinear-fluctuation model. IEEE Trans Signal Process 2013;60(2):480–92.

[156] Chen J, Richard C, Honeine P. Nonlinear estimation of material abundances in hyperspectral images with ℓ_1-norm spatial regularization. IEEE Trans Geosci Remote Sens 2014;52(5):2654–65.

[157] Li X, Cui J, Zhao L. Blind nonlinear hyperspectral unmixing based on constrained kernel nonnegative matrix factorization. Signal Image Video Process 2012;1–13.

[158] Nguyen NH, Chen J, Richard C, Honeine P, Theys C. Supervised nonlinear unmixing of hyperspectral images using a pre-image methods. In: Mary D, Theys C, Aime C, editors. New concepts in imaging: optical and statistical models. EAS Publications Series, 59. Les Ulis, France: EDP Sciences; 2013. p. 417–37.

[159] Fauvel M, Tarabalka Y, Benediktsson JA, Chanussot J, Tilton JC. Advances in spectral-spatial classification of hyperspectral images. Proc IEEE 2013;101(3):652–75.

[160] Broadwater J, Chellappa R, Banerjee A, Burlina P. Kernel fully constrained least squares abundance estimates. In: Proc. IEEE international conference on geoscience and remote sensing (IGARSS); 2007. p. 4041–4.

[161] Fauvel M, Chanussot J, Benediktsson JA. Kernel principal component analysis for the classification of hyperspectral remote sensing data over urban areas. EURASIP J Adv Signal Process 2009. Article ID 783194.

[162] Altmann Y, Pereyra M, McLaughlin S. Bayesian nonlinear hyperspectral unmixing with spatial residual component analysis. IEEE Trans Comput Imag 2015;1(3):174–85.

[163] Altmann Y, Dobigeon N, McLaughlin S, Tourneret JY. Nonlinear spectral unmixing of hyperspectral images using Gaussian processes. IEEE Trans Signal Process 2013;61(10): 2442–53.

[164] Honeine P, Richard C. Preimage problem in kernel-based machine learning. IEEE Signal Process Mag 2011;28(2):77–88.

[165] Lawrence ND, Qui nonero-Candela J. Local distance preservation in the GP-LVM through back constraints. In: Proc. int. conf. machine learning (ICML), Pittsburgh, Pennsylvania. New York: ACM; 2006. p. 513–20.

[166] Wang JM, Fleet DJ, Hertzmann A. Gaussian process dynamical models for human motion. IEEE Trans Patt Anal Mach Intell 2008;30(2):283–98.

[167] Urtasun R, Fleet DJ, Lawrence ND. Modeling human locomotion with topologically constrained latent variable models. In: Proc. conf. human motion: understanding, modeling, capture and animation, Rio de Janeiro, Brazil. Berlin: Springer-Verlag; 2007. p. 104–18.

[168] Boothroyd CB, Moreno MS, Duchamp M, Kovács A, Monge N, Morales GM, et al. Atomic resolution imaging and spectroscopy of barium atoms and functional groups on graphene oxide. Ultramicroscopy 2014;145:66–73. Low-Voltage Electron Microscopy.

[169] Eljarrat A, López-Conesa L, López-Vidrier J, Hernández S, Garrido B, Magén C, et al. Retrieving the electronic properties of silicon nanocrystals embedded in a dielectric matrix by low-loss EELS. Nanoscale 2014;6:14971–83.

[170] Palacio I, Celis A, Nair MN, Gloter A, Zobelli A, Sicot M, et al. Atomic structure of epitaxial graphene sidewall nanoribbons: flat graphene, miniribbons, and the confinement gap. Nano Lett 2015;15(1):182–9.

[171] Torres R, Caretti I, Serin V, Brun N, Radnóczic G, Jiménez I. Reversed texture in nanometric carbon/boron nitride multilayers. Carbon 2014;74:374–8.

[172] Altmann Y, Dobigeon N, Tourneret JY, Bermudez JCM. A robust test for nonlinear mixture detection in hyperspectral images. In: Proc. IEEE international conference on acoustics, speech and signal processing (ICASSP), Vancouver, Canada; 2013. p. 2149–53.

[173] Imbiriba T, Bermudez JCM, Tourneret JY, Richard C. Detection of nonlinear mixtures using Gaussian processes: application to hyperspectral imaging. In: Proc. IEEE international conference on acoustics, speech and signal processing (ICASSP), Florence, Italy; 2014.

[174] Burazerovic D, Heylen R, Geens B, Sterckx S, Scheunders P. Detecting the adjacency effect in hyperspectral imagery with spectral unmixing techniques. IEEE J Sel Top Appl Earth Obs Remote Sens 2013;6(6). 1939–1404.

[175] Nishii R, Ozaki T. Contextual unmixing of geospatial data based on Markov random fields and conditional random fields. In: Proc. IEEE GRSS workshop hyperspectral image signal processing: evolution in remote sensing (WHISPERS), Grenoble, France; 2009.

[176] Eches O, Dobigeon N, Tourneret JY. Enhancing hyperspectral image unmixing with spatial correlations. IEEE Trans Geosci Remote Sens 2011;49(11):4239–47.

[177] Eches O, Dobigeon N, Mailhes C, Tourneret JY. Bayesian estimation of linear mixtures using the normal compositional model. Application to hyperspectral imagery. IEEE Trans Image Process 2010;19(6):1403–13.

[178] Altmann Y, Pereyra M, Bioucas-Dias JM. Collaborative sparse regression using spatially correlated supports–application to hyperspectral unmixing. IEEE Trans Image Process 2015;24(12):5800–11.

[179] Iordache MD, Bioucas-Dias JM, Plaza A. Total variation spatial regularization for sparse hyperspectral unmixing. IEEE Trans Geosci Remote Sens 2012;50(11):4484–502.

[180] Legendre M, Moussaoui S, Chouzenoux E, Idier J. Primal-dual interior-point optimization based on majorization-minimization for edge preserving spectral unmixing. In: Proc. IEEE int. conf. image process. (ICIP), Paris, France; 2014. p. 4161–5.

[181] Zare A, Ho K. Endmember variability in hyperspectral analysis: addressing spectral variability during spectral unmixing. IEEE Signal Process Mag 2014;31(1):95–104.

[182] Bateson CA, Asner GP, Wessman CA. Endmember bundles: a new approach to incorporating endmember variability into spectral mixture analysis. IEEE Trans Geosci Remote Sens 2000;38(2):1083–94.

[183] Halimi A, Dobigeon N, Tourneret JY. Unsupervised unmixing of hyperspectral images accounting for endmember variability. IEEE Trans Image Process 2015;24(12):4904–17.

[184] Thouvenin PA, Dobigeon N, Tourneret JY. Hyperspectral unmixing with spectral variability using a perturbed linear mixing model. IEEE Trans Signal Process 2016;64(2):525–38.

[185] Veganzones MA, Drumetz L, Tochon G, mura MD, Plaza A, Bioucas-Dias J, et al. A new extended linear mixing model to address spectral variability. In: Proc. IEEE GRSS workshop hyperspectral image signal processing: evolution in remote sensing (WHISPERS), Lausanne, Switzerland; 2014.

[186] Thouvenin PA, Dobigeon N, Tourneret JY. Online unmixing of multitemporal hyperspectral images accounting for spectral variability; 2015, http://arxiv.org/abs/1510.05893/. ArXiv preprint.

[187] Henrot S, Chanussot J, Jutten C. Dynamical spectral unmixing of multitemporal hyperspectral images; 2015, http://arxiv.org/abs/1510.04238/. ArXiv preprint.

Chapter 7

Independent Components Analysis: Theory and Applications

D. Jouan-Rimbaud Bouveresse and D.N. Rutledge[1]
UMR Ingénierie Procédés Aliments, AgroParisTech, INRA, Université Paris-Saclay, Massy, France
[1]*Corresponding author: e-mail: douglas.rutledge@agroparistech.fr*

Chapter Outline

1 INTRODUCTION

1.1 The History of Independent Components Analysis

For almost 30 years now, Independent Components Analysis (ICA) has been used to unmix complex signals. The unmixing of signals without using, or even having, any prior information on the source signals mixed together is

Data Handling in Science and Technology, Vol. 30. http://dx.doi.org/10.1016/B978-0-444-63638-6.00007-3
Copyright © 2016 Elsevier B.V. All rights reserved.

known as Blind Source Separation (BSS), and ICA is a particular procedure for performing BSS, which was developed in the field of signal processing. ICA is now used in different domains where signal treatment is involved, such as chemometrics (for the analysis of spectroscopic data, for example, as will be illustrated in Section 3), metabolomics, the medical field (for the detection and elimination of artifacts in electroencephalograms (EEGs), for example), telecommunications, etc. The main interest of ICA is that it aims to extract "source signals" and their mixing coefficients (proportions) from a set of observed signal mixtures, which means that the extracted information is more directly interpretable.

ICA was first introduced in 1986 by Jutten and Herault [1], who proposed a method to solve the BSS problem by assuming the independence of the source signals. In 1991, they proposed a theoretical approach [2], together with Comon [3]. In 1994, ICA was formalized by Comon [4]. Since then, different algorithms have been proposed to calculate ICA models, which differ in the way statistical independence is addressed, as will be explained in Section 2.

The objective of this chapter is to present ICA, and to briefly describe some algorithms used to compute ICA models, without entering into too many details into the higher-order statistics (HOS) involved. In his "Survey on Independent Component Analysis," Hyvärinen defines "ICA of a random vector [as] finding a linear transform $\mathbf{s}=\mathbf{Wx}$[1] so that the components $\mathbf{s_i}$ are as independent as possible, in the sense of maximizing some function $F(\mathbf{s_1}, \ldots, \mathbf{s_m})$ that measures independence" [5]. The function F, sometimes called "contrast function," carries the statistical information of independence on which the corresponding algorithm is based. This definition is equivalent to stating that the "ICA of a random vector $\mathbf{x}$ consists of estimating the following generative model for the data: $\mathbf{x}=\mathbf{As}$," where the vectors (components) $\mathbf{s}_i$ in the set of vectors $\mathbf{s}=(\mathbf{s}_1, \ldots, \mathbf{s}_n)^T$ are assumed to be independent, and $\mathbf{A}$ is a mixing matrix [5]. Several tutorials or introductions to ICA have already been published, among which one can cite Refs. [6–10], and it is not the objective in this chapter to imitate them. After a general presentation of ICA in this section, only a few of the most often-used algorithms will be briefly presented in Section 2, with particular focus on the Joint Approximate Diagonalization of Eigenmatrices (JADE) algorithm. In Section 3, some applications of ICA in chemometrics will be presented in order to show the potential of ICA in the analysis of a wide range of data types.

1. In this chapter, "standard" notations will be used: Matrices will be denoted by a bold, uppercase letter; vectors will be denoted by a bold lowercase letter; and scalar will be denoted by an italic (uppercase or lowercase) letter. The superscript "T" stands for vector or matrix transpose.

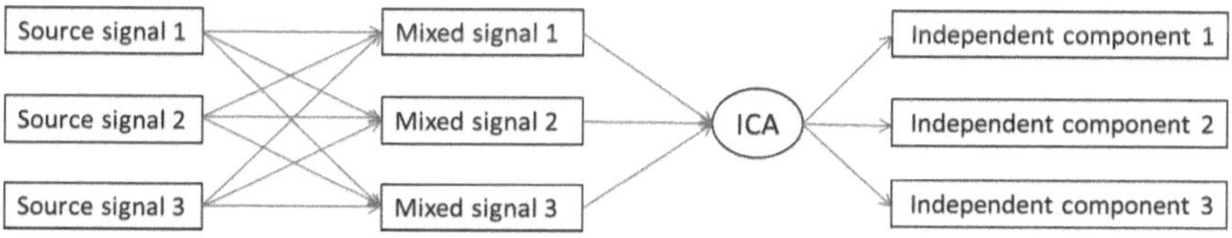

FIG. 1 Illustration of the "cocktail party problem," in the case where three microphones record the mixed signals obtained from three independent source signals.

1.2 The Cocktail Party Problem

In order to have an intuitive understanding of the concept of ICA, the example of the *cocktail party problem* (http:/research.ics.aalto.fi/ica/cocktail/cocktail_en.cgi) is often presented. Although this illustration is not stricto sensu an experimental scientific example, the application of ICA to this case enables a straightforward comprehension of the results ICA is able to deliver. This problem is illustrated in Fig. 1.

Imagine a room where a cocktail party is taking place. Different sound sources exist in this room: music is being played, a few groups of people are engaged in different conversations, a chairman is making announcements, noises from the street can be heard, etc. These sound sources are represented by "Source signals" 1, 2, and 3 in Fig. 1. From any given place in the room, one can perceive a sound signal which is a mixture of all these different sound source signals. These observed mixed sound signals are not the same at different locations in the room as they depend on the distances from each of the original source signals, as well as the relative intensity of each source. Now imagine a number of microphones, situated at different locations in the room, recording the different sound mixtures (mixed signals 1, 2, and 3 in Fig. 1). Provided there are as many microphones as original sources signals, ICA is able to decompose the signals obtained from all microphones, and to recover each original sound source (independent components (ICs) 1, 2, and 3 in Fig. 1), along with the proportions in which they are mixed in each observed signal mixtures. In many case studies, the number of mixed sources is greater than the number of independent signal sources, and this in fact facilitates the separation of the sources. A demonstration of the cocktail party problem can be found in http:/research.ics.aalto.fi/ica/cocktail/cocktail_en.cgi.

1.3 Principal Components Analysis vs Independent Components Analysis

ICA is often compared to Principal Components Analysis (PCA). The reason is that both methods aim to decompose a data matrix into two more informative matrices, one characterizing the individuals (rows) and the other the variables (columns), by calculating linear combinations of the original variables. In PCA, a small number of new variables, the principal components (PCs), also called factors or latent variables (LVs), are calculated from the original variables in such a way that they correspond to the largest variations in the data. Indeed, a data matrix **X** may be regarded as a set of points (the samples), distributed in a multidimensional space defined by the original variables, and the objective of PCA is to find orthogonal directions of maximum dispersion of the samples in that multidimensional space by calculating new directions (PCs) that extract the maximum of the variance of **X**. The hypothesis here is that variance is related to information in the data. The coordinates of the samples along the directions of the PCs are called the *scores*, and the contributions of the original variables to the definition of these directions are called the *loadings*. In ICA, however, a data matrix **X** is regarded as a set of observed signals (for example, a set of infrared spectra of mixtures), which are linear mixtures of source signals (for example, the infrared spectra of the different compounds present in the set of mixtures). The objective of ICA is to extract the source signals, the ICs, and the proportions in which they are mixed together in the signal mixtures. The distribution of intensities in signals containing information tends to be far from Gaussian, whereas noise and signal mixtures—because of the Central Limit Theorem—tend to be more Gaussian. Therefore, ICA algorithms seek to extract signal vectors that are as non-Gaussian as possible so as to have sources that are as independent as possible.

Because of this difference in the way the **X** matrix is viewed and decomposed, the ICs calculated by ICA tend to correspond to pure source signals, while for the PCA loadings vectors this is not necessarily the case. In most cases, PCA loadings vectors contain mixtures of source signals in different proportions.

In order to graphically illustrate the difference between ICA and PCA, a rank 2 data matrix was simulated. The simulated source signals and their histograms are presented in Fig. 2A while the corresponding proportions are presented in Fig. 2B. The two source signals are, respectively, a sinusoidal and a stairs-shaped signal. Concerning the proportions, two main directions can be distinguished, although they are included in a cloud of (noisy) data points. The resulting data matrix and its histogram are presented in Fig. 2C. This histogram is a little more difficult to read than standard one-variable histograms. Here each bar of the histogram itself consists of smaller bars each corresponding to a mixed signal. Hence each color corresponds to one mixed signal. As can be seen, all mixed signals have a similar histogram centered around 10. Therefore, although the original signals are far from Gaussian, their combinations are more Gaussian. This is an illustration of the Central Limit Theorem, which states that the sum of any independent random variables has a distribution which is "more Gaussian" than the distributions of the original independent variables. Therefore, non-Gaussian signals extracted from a set of mixture signals are likely to be more independent than the original Gaussian mixtures. This theorem is very important, as in most cases it

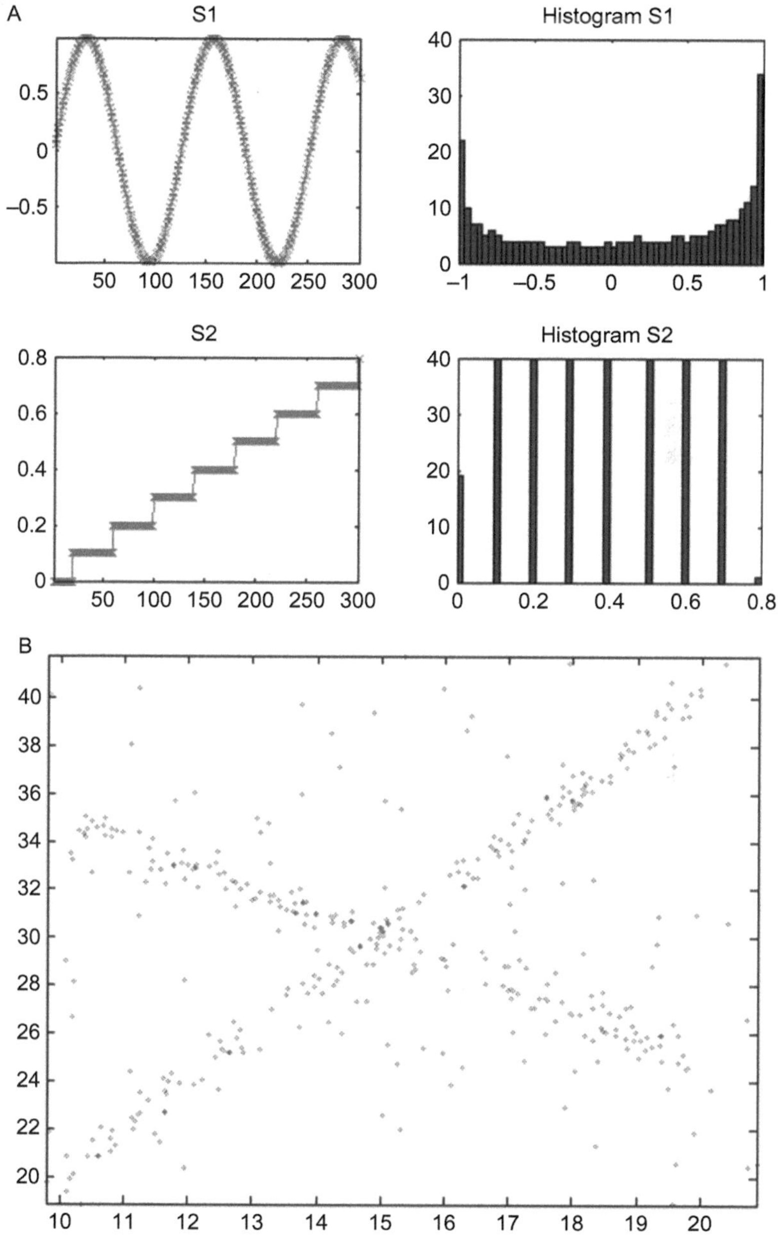

FIG. 2 (A) The two source signals and their (non-Gaussian) histograms; (B) the "proportions" of source signals 1 and 2;

(Continued)

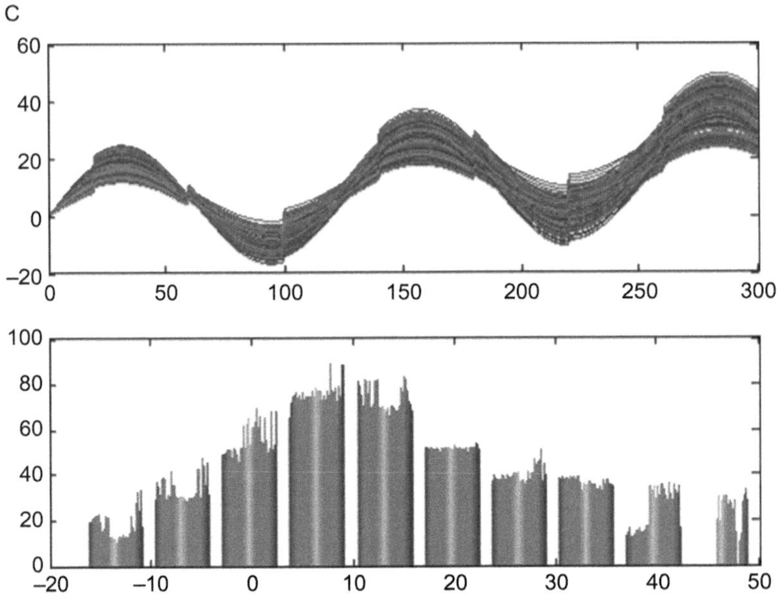

FIG. 2—CONT'D (C) the mixed signals and their histograms.

implies that when a signal has a Gaussian-looking histogram, it is very probable that it is a combination of pure signals or is just Gaussian noise, and can therefore not be a source signal, i.e., an IC. The opposite is true: if an extracted component is far from Gaussian, it is very probably an IC.

Therefore, in order to find ICs, one should calculate the least Gaussian signals possible.

PCA and ICA models were calculated for this data matrix, and the PCA scores and loadings are presented in Fig. 3A–C and the ICA proportions and sources in Fig. 4A–C.

PC scores (Fig. 3B) are not oriented in the same direction as the branches visible in the original data (Fig. 3A), but rather in the directions of the greatest dispersion of the samples, between the branches. As for the PC loadings (Fig. 3C), they correspond to the directions of dispersion of the samples that are oriented between the proportions of the original signals, and therefore they do not correspond to the original simulated signals, but to a combination of them (thus looking like "sinusoidal stairs"). On the other hand, ICA seeks to extract the source signals from the mixed signals and as can be seen in Fig. 4B, has in fact succeeded in recovering these original source signals. The IC proportions (or mixing coefficients, corresponding to *scores* in PCA), displayed in Fig. 4C, are therefore oriented exactly as the original data points (Fig. 2B), showing that the ICs really represent the two directions

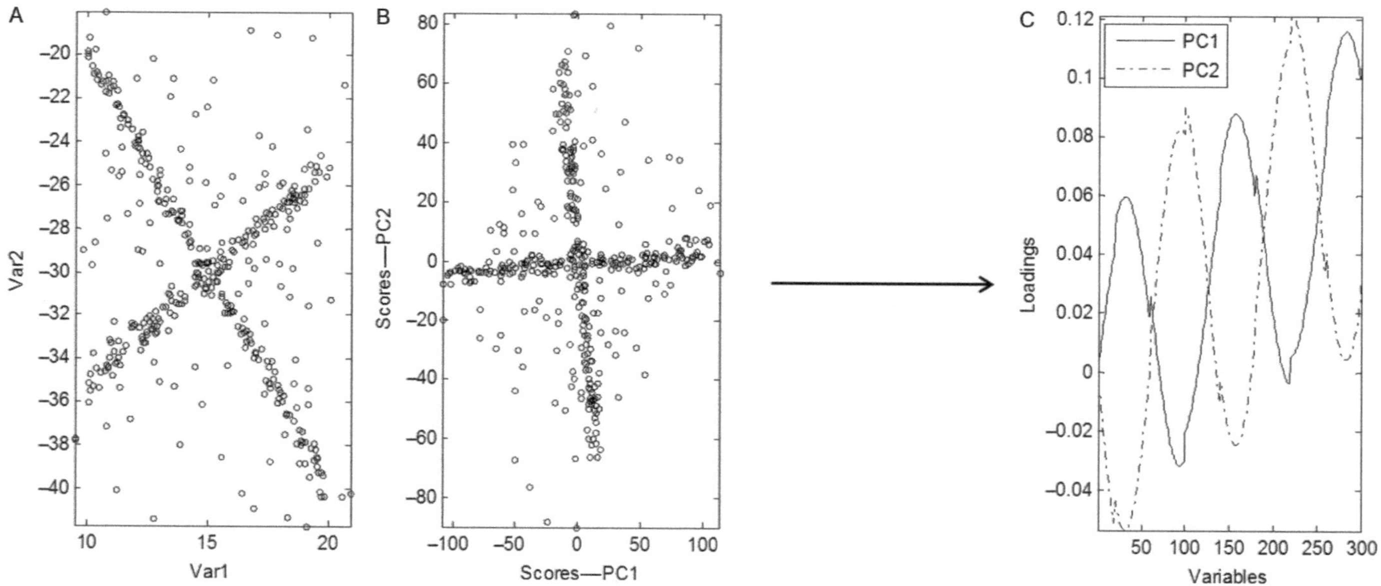

FIG. 3 Simulated data: (A) variable 2 vs variable 1; (B) PCA scores on PC2 vs PC1; (C) PCA loadings of PC1 and PC2 vs variable number.

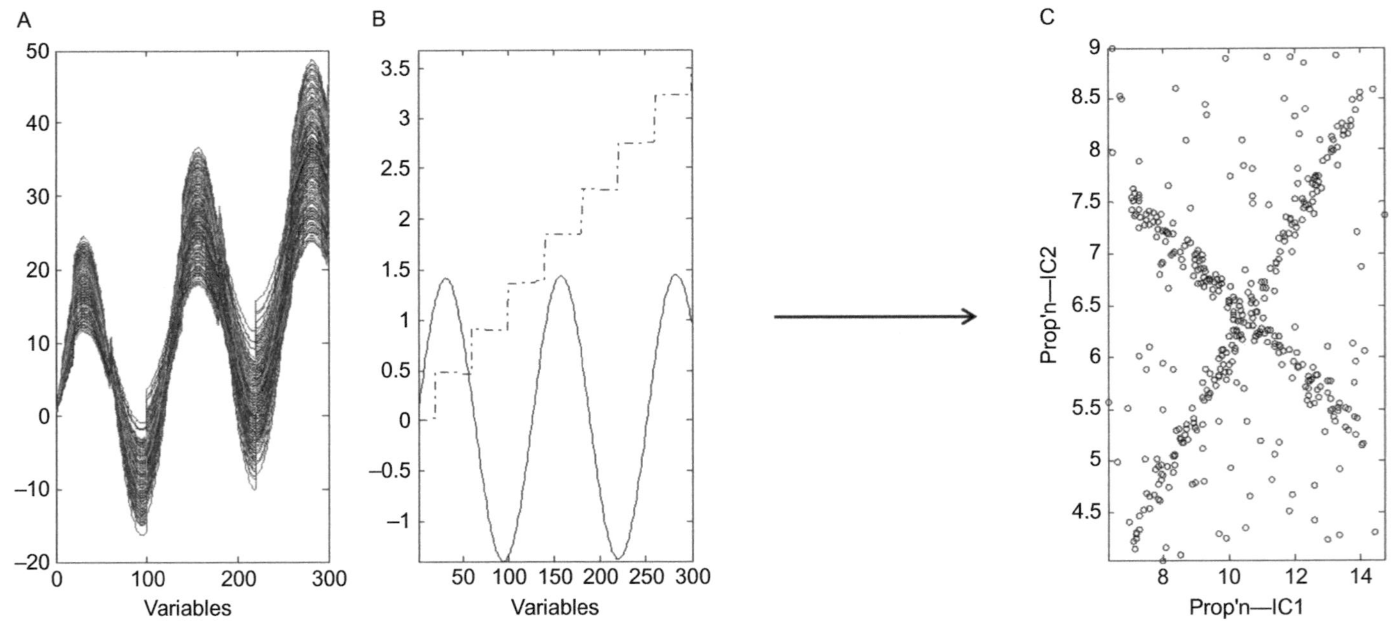

FIG. 4 Simulated data: (A) the mixed signals; (B) ICA source signals of IC1 and IC2 vs variable number; (C) ICA proportions of IC2 vs IC1.

(true proportions) in the original data set. Note that the scales of the IC mixing coefficients and signals are not the same as those of the original data.

Therefore, this simple simulated example shows the advantage of ICA over PCA when interpretation of the signals is at stake: PCA is based on the samples' dispersion, and looks for directions of maximal dispersion of samples, and the loadings are calculated to obtain these directions; whereas ICA looks for independent signals, and the proportions or "mixing coefficients" (similar to PCA "scores") are the projections of the samples onto these signals, not related to the dispersion of samples. It is noteworthy that in the original mixed data, even though the "stairs" structure is not clearly visible, only a few small sharp increases being visible, ICA is still able to recover the original stairs-shaped signal.

A comparison between PCA and ICA on real data was published by Jouan-Rimbaud Bouveresse and Rutledge [11], where the interpretation clearly shows that ICA extracts directly interpretable signals, while PCA extracts mixtures of these source signals.

2 THEORY

2.1 The Principles of ICA

ICA is a generative model, which calculates the source signals and the mixing coefficients that give the measured mixed signals. ICA searches for the most non-Gaussian, and therefore independent, components. As stated by Stone [12], "the more independent the extracted signals are, the more likely they are to be the required source signals." ICA is a "blind" method, since despite only knowing the mixed signals, both the sources signals and their mixing coefficients are estimated.

Independence is a statistical term with a strong meaning: two variables are independent when the modification of one cannot be used to predict the values of the other. It means the variables are not related in any way whatsoever. The notion of independence is stronger than that of orthogonality: two orthogonal vectors can be mathematically related, and thus be dependent. For example, in Fig. 5, the relationship $y=x^2$ is represented. Although the linear correlation between x and y is zero, the two variables are related, and are hence dependent. One can therefore conclude that although independence implies uncorrelatedness, the opposite is not true, uncorrelatedness does NOT imply independence.

In fact, if two vectors are uncorrelated, the linear correlation between them is zero: $r(x,y)=0$. This property is not sufficient for x and y to be independent. If x and y are independent, then whatever the nonzero integer values a and b: $r(x^a,y^b)=0$.[2] In the example presented just earlier, when $a=2$ and $b=1$, $r(x^2,y)=1$, so x and y are not independent.

2. This has to do with the fact that for two independent variables x and y, their joint probability density function is equal to the product of their marginal probability density functions.

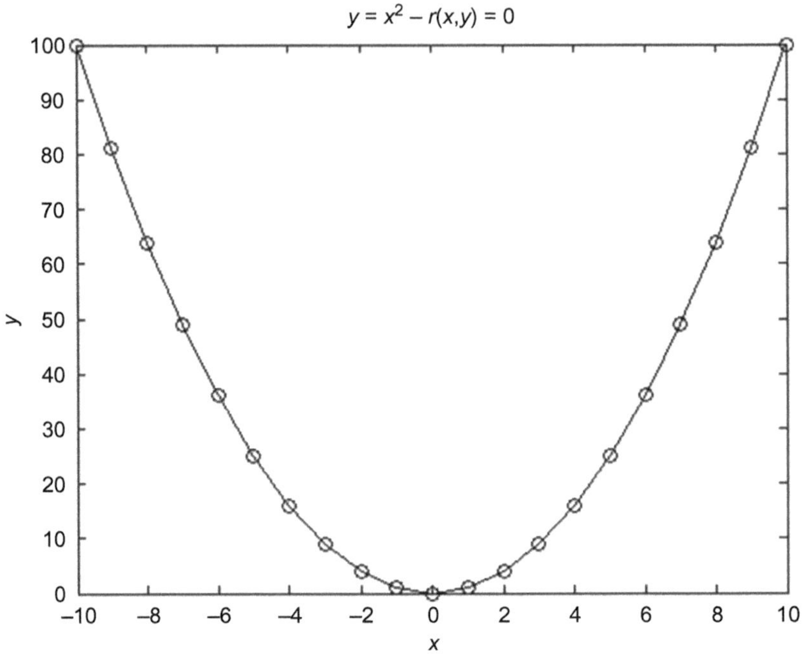

FIG. 5 Graphical representation of the function "$y=x^2$." Although uncorrelated, the variables y and x^2 are related, and therefore not independent.

In order to present ICA, let a signal be represented by the row vector $\mathbf{x}$ of dimension $1 \times p$. $\mathbf{x}$ can be written as the weighted sum of q pure source signals $\mathbf{s}_j$, ($j=1$ to q), the weights being given by a_j:

$$\mathbf{x} = \sum_{j=1}^{q} a_j \mathbf{s}_j \tag{1}$$

So for a set of n signals, each observed signal $\mathbf{x}_i$ ($i=1$ to n) can be given by:

$$\mathbf{x}_i = \sum_{j=1}^{q} a_{ij} \mathbf{s}_j \tag{2}$$

Eq. (2) implies that, although the source signals $\mathbf{s}_j$ are independent, the mixture signals, which are linear combinations of the same source signals, are not.

If the n signals are in the rows of an $\mathbf{X}$ matrix, Eq. (2) can be written in the matrix form:

$$\mathbf{X} = \mathbf{A} \cdot \mathbf{S} \tag{3}$$

$\mathbf{X}$ ($n \times p$) being the original data matrix of mixed sources, where n signals (in the rows of $\mathbf{X}$) are described by p variables (in the columns of $\mathbf{X}$); $\mathbf{A}$ ($n \times q$) is the mixing matrix, whose element a_{ij} describes the weight of source signal j in the mixed signal $\mathbf{x}_i$; and $\mathbf{S}$ $(q \times p)$ is the matrix of pure source signals (on the rows; q signals described by the original p variables).

The objective of ICA is to determine **A** and **S** knowing only **X**. Contrary to curve resolution methods such as Multivariate Curve Resolution (MCR) [13,14] (see Chapter 2), many ICA algorithms do not require constraints such as unimodality, nonnegativity, etc. to be applied during the calculation, nor any knowledge concerning the nature of the source signals or their proportions. It must be pointed out that, contrary to what is sometimes expected from ICA, most of the times its results are not comparable to those obtained by MCR. This is because while MCR aims to extract the signals of pure compounds, ICA extracts signals reflecting underlying independent *phenomena*, which may in fact be combinations of the signals of several pure compounds.

In order to find **A**, a demixing matrix **W** is calculated such that:

$$\mathbf{S} = \mathbf{W} \cdot \mathbf{X} \tag{4}$$

which implies that **W** is the (pseudo)inverse of **A**. After calculating **W**, Eq. (4) enables the calculation of the independent source signals, **S**, i.e., the ICs. **A** is then obtained from Eq. (5):

$$\mathbf{A} \cdot \mathbf{S} \cdot \mathbf{S}^{\mathrm{T}} \cdot \left(\mathbf{S} \cdot \mathbf{S}^{\mathrm{T}}\right)^{-1} = \mathbf{X} \cdot \mathbf{S}^{\mathrm{T}} \cdot \left(\mathbf{S} \cdot \mathbf{S}^{\mathrm{T}}\right)^{-1}$$

i.e.,

$$\mathbf{A} = \mathbf{X} \cdot \mathbf{S}^{\mathrm{T}} \cdot \left(\mathbf{S} \cdot \mathbf{S}^{\mathrm{T}}\right)^{-1} \tag{5}$$

Some characteristics of ICA models, which differ from other matrix decomposition methods, such as PCA or Partial Least Squares (PLS) regression, are worth mentioning here:

- ICA models are not nested: The ICs extracted from an ICA model with i ICs may be mathematically different from (although they may be similar to) the first i ICs extracted from models with more ICs (i.e., models with $j > i$ ICs).
- The extracted ICs are not ordered. Hence, if a data set is split into two representative subsets, which cover the same space as the original data set, the ICA models calculated from all three sets (the original set and its two subsets) might yield similar ICs, but extracted in different orders.
- The variance explained by the ICs cannot be calculated. Indeed, in most algorithms, the first step in the calculations is to "whiten" or "sphere" the data, which means that samples are equally dispersed along the directions of all the variables so that the influence of the dispersion, and therefore of the variance, is eliminated.

2.2 Data Preprocessing

Different algorithms exist to compute an ICA model. They differ in the parameters in the contrast function used to evaluate the independence and non-Gaussianity of a signal, and the procedure to optimize that contrast function. However, whatever the algorithm used, they all start with a preprocessing of the original data matrix of mixture signals, **X**. The two steps of preprocessing are row-*centering* and *whitening*:

- The effect of centering is to subtract the mean value from each spectrum, so that all variations are relative to its mean (see Figure 6), thus simplifying the ICA analysis (it will be seen later that the definitions of the different high-order statistical parameters are simpler when applied to centered vectors).
- Shlens writes that "whitening is an operation that removes all linear dependencies in a data set (i.e. second-order correlations) and normalizes the variance along all dimensions" [9]. The decorrelation is done by applying Principal Components Analysis to the row-centered **X** matrix. The normalisation is then done by giving equal "importance" to the resulting loadings vectors by setting all their variances to 1. In concrete terms, whitening is equivalent to using the PCA loadings matrix **P** after applying a Standard Normal Variates transform, rather than the original data matrix of mixture signals **X**.

The centering–whitening steps are graphically presented in Fig. 6.

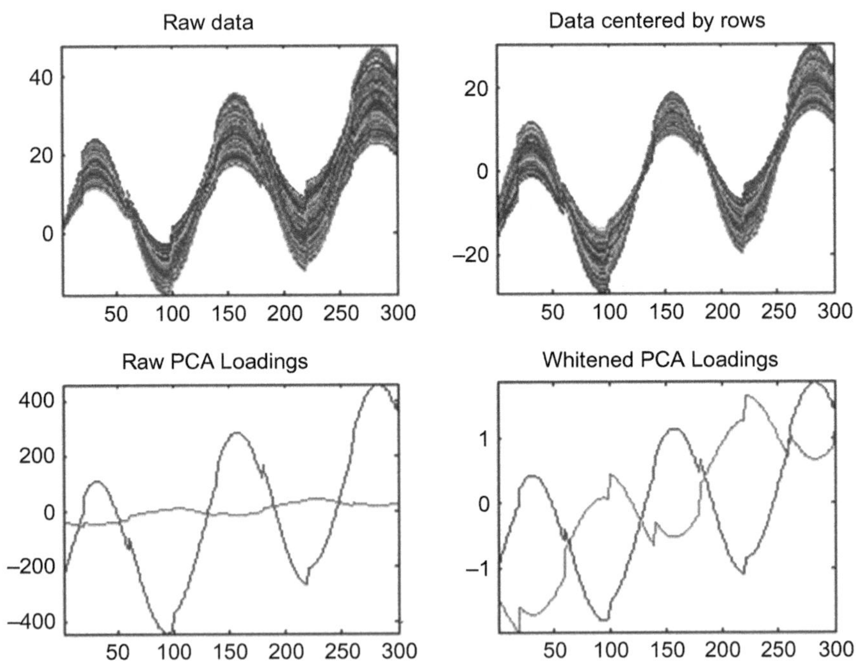

FIG. 6 (A) Original simulated signals; (B); Signals centered by rows; (C); Loadings of first two Principal Components; (D) Whitened PCA loadings.

2.3 The Different Algorithms

The different ICA algorithms can be sorted into two categories, based on their definition of independence: (i) algorithms maximizing the non-Gaussianity of the components or (ii) algorithms minimizing the mutual information. It is not the objective of this chapter to review exhaustively all algorithms. A review of a few of the most commonly encountered algorithms was published in 2008 [15]; more recently, an article presenting the comparison of not fewer than 30 algorithms [16] was also published, although no details were given concerning their algorithmic differences.

Among the most commonly used ICA algorithms, one can cite FastICA [17], JADE [18–20], Infomax ICA [21], Mean-Field ICA [22,23], Kernel ICA [24], the Mutual Information Least-dependent Component Analysis [25], the Stochastic Nonnegative Independent Component Analysis [26], RADICAL [27], etc. Before briefly presenting some of these algorithms, it appears necessary to include a few notions on HOS, which will help in understanding the philosophy of the different ICA algorithms.

2.3.1 *Definition of Independence and HOS*

The use of HOS in ICA makes sense as one looks for components as non-Gaussian as possible. Indeed, if a random variable follows a Gaussian distribution, all its moments and cumulants of order higher than 2 are null [28,29]. Hence, finding the ICs implies finding signals whose moments and cumulants of order higher than 2 are as far from 0 as possible.

Hence it is necessary to briefly introduce here the different notions of HOS useful in ICA.

2.3.1.1 Moments

The *j*th moment m_j of a variable x is equal to:

$$m_j = E\{x^j\} \tag{6}$$

where E is the expectation. If $j=1$, $m_1=\text{mean}(x)$.

The moments of a variable describe its probability density function, i.e., its distribution.

2.3.1.2 Central Moments

The *j*th central moment μ_j of a variable x is equal to the moment of the centered variable x, i.e.,

$$\mu_j = E\left\{(x-m_1)^j\right\} \tag{7}$$

Hence:

$\mu_1=0$ is the mean of x.

$\mu_2=\sigma^2$ is the variance of x.

$\mu_3 = E\{(x - m_1)^3\}$ is called the *skewness*, and it is a measure of the asymmetry of the distribution. Skewness can be negative, positive, or null in the case of a Gaussian distribution.

$\mu_4 = E\{(x - m_1)^4\}$ is the fourth moment of a variable x. It is related to its *kurtosis*, reflecting the flatness or pointedness of the distribution of values.

2.3.1.3 Kurtosis

For a centered variable x, the kurtosis, which is one measure of the (non-) Gaussianity of a distribution, is defined, as:

$$k(x) = E\{x^4\} - 3 \cdot \left[E\{x^2\}\right]^2 \tag{8}$$

For centered and whitened data, $E\{x^2\} = 1$, and hence, the kurtosis is equal to:

$$k(x) = E\{x^4\} - 3$$

If $k(x) = 0$, the distribution is Gaussian; if $k(x) > 0$, the distribution is said to be super-Gaussian, and the peak of the probability density function is very sharp; if $k(x) < 0$, the distribution is said to be sub-Gaussian, and the peak of the probability density function is rather flat.

Therefore, non-Gaussian components can be calculated by maximizing the absolute value of their kurtosis. The independence of the components can be maximized by simultaneously maximizing the kurtosis (maximum non-Gaussianity) of each one while at the same time minimizing their mutual kurtosis (minimum non-Gaussianity), which can be defined as a function of the fourth-order cumulant.

2.3.1.4 Cumulants

Cumulants can be compared to the covariances used in second-order statistics.

For centered variables, the first three cumulants are equal to their moments, i.e.,

$\kappa_1 = 0$.

$\kappa_2 = E\{x^2\}$ is the variance of x.

$\kappa_3 = E\{x^3\}$.

The fourth-order cumulant is defined as:

$$\kappa_4 = E\{x^4\} - 3\left[E\{x^2\}\right]^2 = k(x) \tag{9}$$

and therefore is equal to the kurtosis.

The fourth-order *auto*- and *cross*-cumulants of four vectors $\mathbf{v}_i$, $\mathbf{v}_j$, $\mathbf{v}_k$, and $\mathbf{v}_l$ are given by:

$$\begin{aligned}\kappa_4\{\mathbf{v}_i, \mathbf{v}_j, \mathbf{v}_k, \mathbf{v}_l\} = {} & E\{\mathbf{v}_i\mathbf{v}_j \cdot \mathbf{v}_k\mathbf{v}_l\} - E\{\mathbf{v}_i\mathbf{v}_j\} \cdot E\{\mathbf{v}_k\mathbf{v}_l\} \\ & - E\{\mathbf{v}_i\mathbf{v}_k\} \cdot E\{\mathbf{v}_j\mathbf{v}_l\} - E\{\mathbf{v}_i\mathbf{v}_l\} \cdot E\{\mathbf{v}_j\mathbf{v}_k\}\end{aligned} \tag{10}$$

From this general definition, one can see that the fourth-order *auto-*cumulant of a centered variable is equal to its kurtosis. The cumulants can be expressed in the form of a tensor (which can be regarded as a generalization of the notion of matrices to a higher order). A cumulant tensor is the generalization of the variance–covariance matrix, with *auto-cumulants* on its diagonal and *cross-cumulants* off the diagonal. Auto-cumulants characterize one vector (this is equivalent to the variance of a variable), while cross-cumulants characterize two vectors (equivalent of the covariance between two variables). Statistically, a set of independent vectors yields a "diagonal" cumulant tensor, with null off-diagonal elements and maximum auto-cumulants.

Although kurtosis is an appropriate estimation of the (non-)Gaussianity of a variable, it is very sensitive to outliers, and is therefore a nonrobust estimation of its (non-)Gaussianity. Another estimation of the (non-)Gaussianity of a variable, which is more robust and therefore preferred to kurtosis, is the negentropy.

2.3.1.5 Negentropy

As explained by Hyvärinen, "the entropy of a random variable can be interpreted as the degree of information that the observation of the variable gives. The more 'random', i.e. unpredictable and unstructured, the variable is, the larger its entropy. [...] A Gaussian variable has the largest entropy among all random variables of equal variance" [8]. Hence, entropy could also be a proper criterion to estimate the (non-)Gaussianity of a variable. It is defined, for a variable x, as:

$$H(x) = -\sum_i P(x = a_i)\log P(y = a_i) \tag{11}$$

and has therefore a negative value. One rather works with negentropy, which is greater than or equal to zero (in the Gaussian case), and which is defined as:

$$J(y) = H(y_{\text{Gauss}}) - H(y) \tag{12}$$

where y_{Gauss} is a Gaussian random variable with the same covariance matrix as y. The more "non-Gaussian" the variable, the higher the value of the negentropy. Therefore, when looking for ICs, one should try to maximize the negentropy of the components. However, the negentropy value is rather difficult to estimate in real cases, so one usually works with a simpler approximation. Hyvaärinen [8] presents several such approximations, given below:

$$J(y) = \frac{1}{12}E\{y^3\} + \frac{1}{48}\text{kurt}(y)^2 \tag{13a}$$

where y is a variable of zero mean and unit variance. However, this approximation relies on kurtosis, which is not a robust estimator. Another approximation can be used instead:

$$J(y) \approx k[E\{G(y)\} - E\{G(v)\}]^2 \tag{13b}$$

where k is a constant, G is a nonquadratic function, and v is a variable with mean 0 and unit variance. Hyvärinen proposes two "useful" [8] choices for G:

$$G_1(u) = \frac{1}{a_1} \log \cosh a_1 u \tag{13c}$$

and

$$G_2(u) = -\exp\left(\frac{-u^2}{2}\right) \tag{13d}$$

2.3.2 The FastICA Algorithm

This fixed-point algorithm was developed by Hyvärinen and coworkers, and has been presented in several books and articles (see, for example, Refs. [8,28]). This algorithm aims at maximizing an approximation of negentropy. If one considers the approximation of negentropy defined by Eq. (13b), one can say that in order to find an IC, one should maximize the negentropy of the searched source signal **s** (which according to Eq. 4 would be equal to $\mathbf{w}^T\mathbf{x}$), defined as:

$$J(\mathbf{w}^T\mathbf{x}) = \left[E\{G(\mathbf{w}^T\mathbf{X})\} - E\{G(v)\}\right]^2$$

where **w** is such that $E\{(\mathbf{w}^T\mathbf{x})^2\} = 1$.

The different steps of FastICA can be then summarized as follows:

- Choose an initial random "demixing" vector **w**.
- Maximize the non-Gaussianity of $\mathbf{w}^T\mathbf{x}$.
  (a) Define $\mathbf{w}^+ = E\{\mathbf{x}\ g(\mathbf{w}^T\mathbf{x})\} - E\{g'(\mathbf{w}^T\mathbf{x})\}\mathbf{w}$.
  (b) Set $\mathbf{w} = \mathbf{w}^+/\|\mathbf{w}^+\|$.
  (c) Check convergence of **w**. If it has not converged, go back to (a).
- After convergence, the data are deflated by $\mathbf{w}^T\mathbf{x}$ (i.e., decorrelation with respect to the previously estimated ICs) and the whole procedure is repeated to find the next demixing vector **w**.
- When all **w**'s have been computed, Eqs. (4) and (3), respectively, give **S** and **A**.

The first step of this algorithm is a random initiation of the **w** vector. This has consequences on the results obtained at each iteration of the algorithm, as in two successive runs of FastICA applied to the same data, different first estimates of **w** may lead to a different final solution. Another risk of random initialization is the possibility for the algorithm to converge to a local optimum, rather than the global optimum. The approximation of negentropy and the G function chosen are also critical parameters that can influence the output of FastICA.

2.3.3 Infomax ICA

Infomax ICA was proposed in 1995 by Bell and Sejnowski [21]. Infomax consists in finding the ICs by maximizing the joint entropy of the components. In their work, Bell and coworkers have tried to develop a method, based on Linsker's Infomax principle [30], in order to develop unsupervised learning rules for neural networks, and it turned out this could also solve the BSS problem. Infomax uses HOS in order to maximize the information, which is similar to minimizing the mutual information.

2.3.4 The JADE Algorithm

The JADE algorithm [18] was recently presented exhaustively, particularly in the context of signal treatment for chemometrics applications [19,20]. JADE consists in an eigenmatrix decomposition of a higher-order cumulant tensor. As stated by Hyvärinen, "the advantage of this approach is that it requires no knowledge of the probability densities of the independent components" [5]. The JADE algorithm includes different steps, which can be briefly summarized as follows: The **X** matrix is first transformed into a reduced set of PCA loadings that are then centered and whitened to have equal variances (whitening). The auto- and cross-cumulants of these loadings are then placed in a fourth-order tensor (hypercube) of dimensions $n \times n \times n \times n$ (where n is the number of loadings). The cumulants give a measure of the (non-)Gaussianity of the components. As explained earlier, the fourth-order tensor of loadings cumulants can be seen as a generalization of the second-order variance–covariance matrix. Independent vectors have null cross-cumulants and maximal auto-cumulants. Therefore, in a way analogous to PCA, where the eigenvalue decomposition results in diagonalization, JADE generalizes this methodology by diagonalizing the fourth-order cumulants tensor. To do this, the tensor is projected onto a set of orthogonal eigenmatrices, which are then diagonalized via a rotation matrix (using the Jacobi algorithm). This rotation matrix is applied to the whitening matrix used in the preprocessing step (corresponding to a scaled scores matrix), yielding the demixing matrix **W**. Eqs. (4) and (3) then give, respectively, the ICs and the mixing matrix.

This algorithm is deterministic, and so is not subject to convergence problems since random initial estimates are not required. As well, the calculations are very similar to the matrix rotations used in certain PCA algorithms.

As indicated by Westad and Kermit, the JADE algorithm "differs from both FastICA and Infomax in that it does not optimize a transform of the input data. Instead, the second and fourth order cumulants from the data is optimized" [31]. However, it presents the inconvenient that it is known to be quite slow when the studied data set is voluminous. This is mainly due to the first step, PCA, which slows down the execution of the algorithm when huge data sets are analyzed. Therefore, speeding up this step should enable to calculate most ICA models in a reasonable time. In order to overcome this problem when the analyzed data matrix is "large," "tall," or both "large and tall," the singular value decomposition often used when computing PCA models can be replaced, for example, by kernel methods [32,33] or (segmented-) PCT [34,35]. This solves JADE's most often mentioned drawback and so given its other advantages, makes it an algorithm of choice for the computation of ICA models.

Therefore, in the rest of this chapter, focus will be put on the JADE algorithm, which yields stable and satisfactory results.

2.4 Application of JADE-ICA on Simulated Data

Some applications of ICA on simulated data have already been presented in Refs. [19,20,36], and have shown that ICA can successfully recover the original source signals. Another simulation will be briefly presented here, in order to demonstrate the ability of ICA to recover the pure sources. In many real case studies, the source signals are unknown, and are estimated by ICA. In the case presented here, four IR spectra were downloaded from a NIST database (National Institute of Standards and Technology, www.nist.gov), and were manipulated so that they contained the same number of datapoints and had overlapping peaks (cf. Fig. 7A). These four source signals were mixed by calculating 100 weighted sums (the weights being chosen randomly and independently), and low level of Gaussian noise was added. The resulting mixed signals are presented in Fig. 7B.

An ICA model with 6-ICs was calculated on these data. Usually, the number of ICs has to be determined before computing the final model. In the present case, the "real" number of ICs was known to be 4, but up to 6 ICs were tested in order to show what the two supplementary signals look like in such a situation. The results are presented in Fig. 8.

A few conclusions can be drawn from these results:

- As expected, 4 ICs were significant, while the last 2 ICs seem to be related to Gaussian noise only. However, in real case studies, some late ICs, which a priori are not interesting, show some structure, and can render the choice of the number of ICs relatively tricky.
- Each IC corresponds to one of the original signals, except for the signs of IC2 and IC3. As in PCA, where the signs of the scores and loadings can be changed simultaneously without modifying the connection between them, in ICA the signs of the proportions and the source signals can also be changed. So as to have ICs of the same sign as the true source signals, their signs can be changed depending on the sign of either the highest proportion or the most intense signal point, simply by multiplying both the ICs and the corresponding proportions by -1.

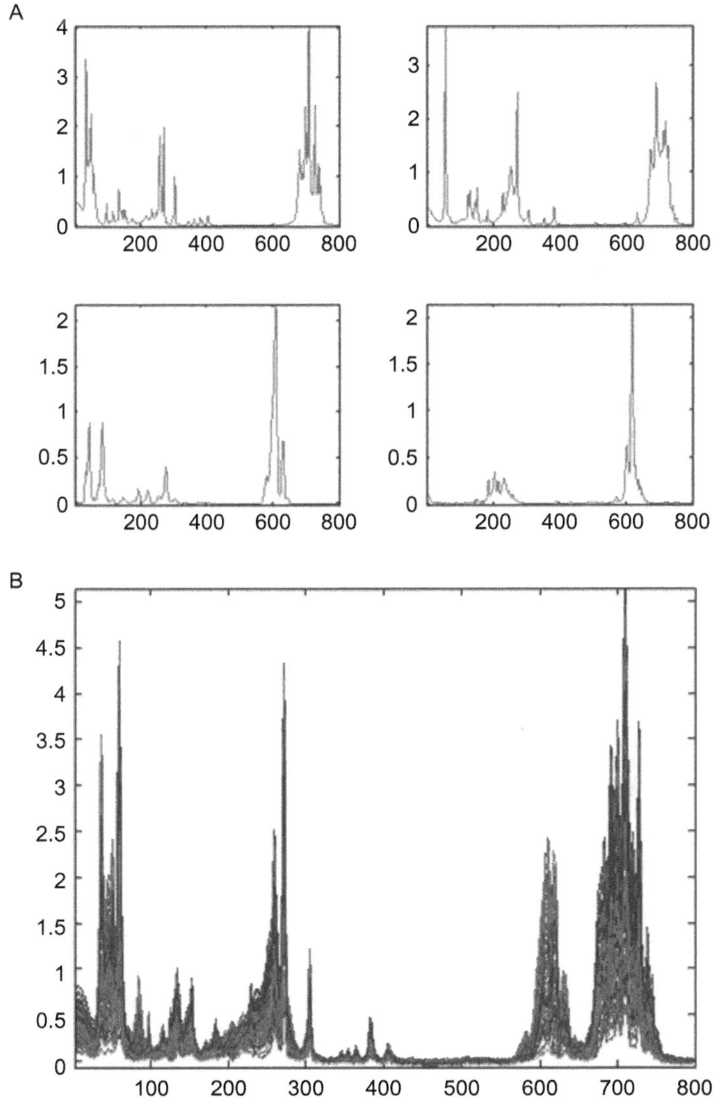

FIG. 7 Simulated data: (A) the four source signals; (B) the 100 mixed signals.

2.5 Determination of the Number of Components to Extract by ICA

The number of components to use in an ICA model is an important parameter, and when one has no idea about its value, time and effort should be invested to determine it. Indeed, if too few ICs are extracted, some of the significant

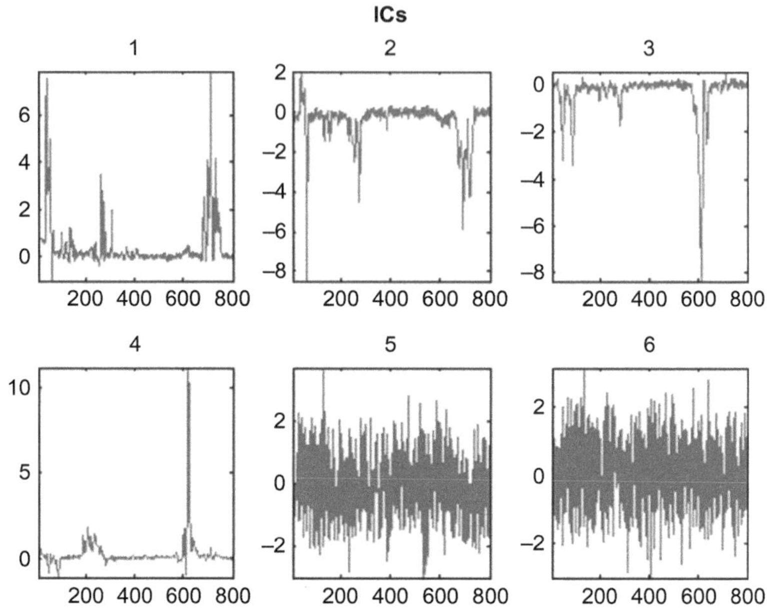

FIG. 8 Simulated data: the six estimated independent components.

components may remain in the residual matrix or be mixed together in the small number of extracted ICs; on the other hand, if too many ICs are extracted, some of the significant components might themselves be decomposed into "subcomponents" or the extracted ICs may be contaminated with noise. In both cases, this might lead to misinterpretation of the investigated underlying phenomena.

A first simple method which is sometimes used to determine the number of ICs to extract is based on the hypothesis that the number of ICs is equal to the number of significant sources of variation in a data matrix, **X**. This number would then be equal to the number of significant PCs which can be extracted from **X**. Hence a scree plot, representing the eigenvalues vs the PC index, can be drawn. The PCs being sorted in decreasing order of extracted variance, i.e., in decreasing order of eigenvalues, the scree plot presents a descending curve. The value at which the eigenvalues start to level off is often taken as the optimal number of PCs to work with, and hence as the optimal number of ICs to extract. Although very simple to implement, this method supposes that significant information is related to the variance, and that one does not wish to consider weak variations, which are assumed to be related to random noise. These two hypotheses are not always valid, and this method should only be used for a preliminary data analysis.

Few methods enable to find the optimal number of ICs. For example, the Amari index is defined as [37]:

$$\mathrm{AI} = \frac{1}{2N}\sum_{i,j=1}^{N}\left(\frac{|p_{ij}|}{\max_k(|p_{ik}|)} + \frac{|p_{ij}|}{\max_k(|p_{kj}|)}\right) - 1 \tag{14}$$

where $p_{ij}=(\hat{\mathbf{A}}^{-1}\mathbf{A})_{ij}$, the matrix **A** being the true mixing matrix and $\hat{\mathbf{A}}$ being the estimated mixing matrix; and N the number of samples (rows in **X**). The computation of the Amari index requires knowing the theoretical **A** matrix, which is normally, by definition, not the case. As to its interpretation, a small value for the Amari index is indicative of a satisfactory model. Indeed, the best possible model corresponds to $\hat{\mathbf{A}}=\mathbf{A}$, in which case AI would be equal to 0. Therefore, one can calculate the Amari index for models with from 1 to *ICmax* ICs, and plot it as a function of the number of ICs.

This procedure was applied to the simulated IR data, and the plot presented in Fig. 9 was obtained:

A minimum value for the Amari index is obtained with 4 ICs, which is the expected number.

However, although this method seems to deliver satisfactory results, it can be used only when the theoretical **A** matrix is known, which is not the case in most studies.

Similarly to what occurs with other methods relying on matrix decomposition, cross validation (CV) could be used with ICA. However, much caution should be taken when using CV in ICA, due to the fact that ICA is not nested, and that the order of the components is not determined: even if all deletion groups are representative of the whole data, nothing ensures that at each iteration of the CV, similar ICs will be extracted from the different subsets in the same order. To overcome this problem, a solution involving Procrustes rotation was proposed by Westad and Kermit [31]. However, whatever the type

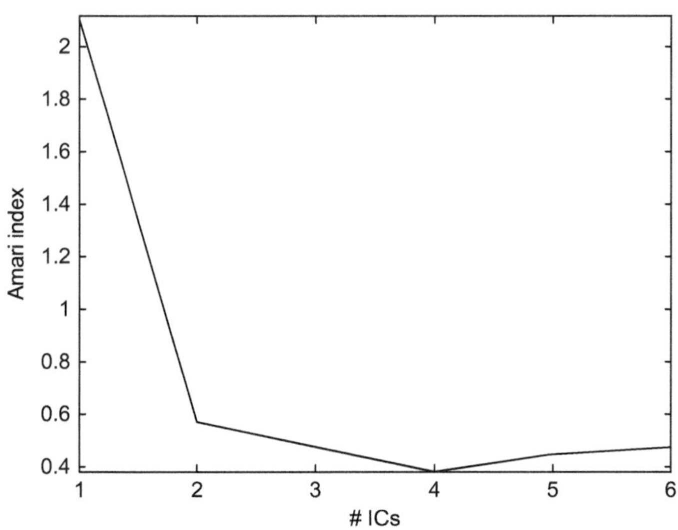

FIG. 9 The Amari index as a function of the number of ICs for the simulated IR data.

of model being considered (PLS, PCA, or ICA), CV is known to be an over-optimistic method for determining the optimal number of components in a model. It is also dependent on the investigated data (number of samples, representativity of the subsets), and is therefore not ideal.

Out of the few methods used to estimate the "optimal" number of ICs to work with, the methods based on the Durbin–Watson (DW) criterion [38], and the random ICA_by_Blocks method [36] are particularly interesting, and will be presented hereunder. Moreover, in cases where ICA factors are used in a multivariate (qualitative or quantitative) calibration model, the vectors obtained by ICA (proportions or sources) may be used in combination with another matrix (or vector) corresponding to the theoretical proportions or signals. In this case, specific methods have to be used for the determination of the optimal number of ICs.

2.5.1 The DW Criterion

The DW criterion is given by:

$$\mathrm{DW} = \frac{\sum_{i=2}^{n} (e_i - e_{i-1})^2}{\sum_{i=1}^{n} e_i^2} \tag{15}$$

where e_{i-1} and e_i are successive values in a series.

The DW criterion has been proposed as an estimation of the signal-to-noise ratio in the case of a *structured* signal [38]; it cannot be used in the case of unstructured signals, such as mass spectra. Its value is close to 0 when there is not much noise in the signal; it increases as noise increases, and its value gets close to 2 when there is (almost) only noise in the signal. In order to use this criterion as a validation method for ICA, the following procedure was proposed:

- ICA models with from 1 to *maxICs* ICs are computed.
- For each model:
  - An approximation of the original **X** matrix is estimated with Eq. (3). The matrix estimated with the *k*-ICs model will be denoted **X*k***.
  - The residual matrices **Rk** can be computed as: **Rk** = **X** − **Xk**.
  - For each row of **Rk**, corresponding to the residual signal, the DW criterion is calculated:
    - If DW tends to 0, it means there still remains some structure in the residual signal.
    - If DW gets larger, it means there is less structure and more noise in the residual signal.

When *maxICs* residual matrices have been calculated, an $n \times maxICs$ matrix of DW values is obtained, which can be represented graphically to choose the optimal number of ICs.

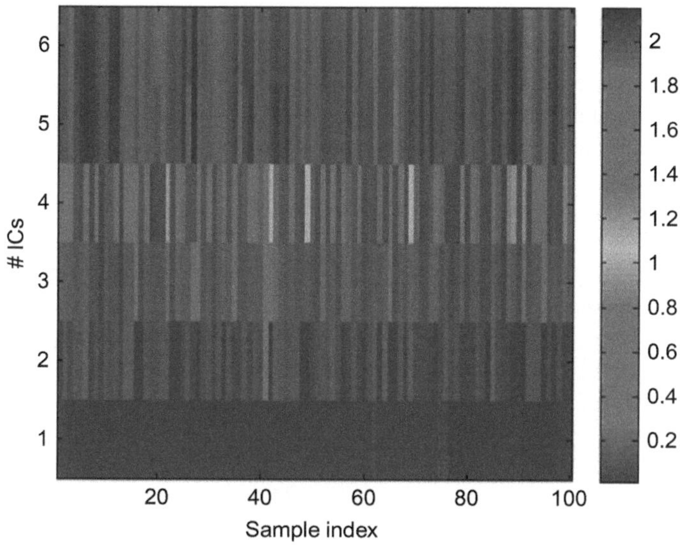

FIG. 10 Color plot of the Durbin–Watson criterion used to determine the optimal number of ICs.

As an illustration, the DW criterion was calculated for the simulated data presented in Section 2.4, and the graph presented in Fig. 10 was obtained:

In this figure, one can see the DW value for each reconstructed mixed signal (on the abscissa) with each of the tested ICA models (on the ordinates). The DW values represented for a given number of ICs correspond to the DW values of the signals BEFORE the model is constructed, i.e., they correspond to the residual signals from the previous model. For example, one can see that, with the 2-ICs model, all samples still have a low DW value: this means that the residuals from the 1-IC model are still structured (their signal-to-noise ratio is high), and therefore, at least 2 ICs are optimal. In this example, the optimal number of ICs is known to be 4. One can see that with up to 3 ICs, all residual signals have a low value for the DW criterion; therefore, one can conclude that there remains structure in the residual signals. With 5 and more ICs model, the DWs are high and so the residual signals are very noisy, and hence, it is concluded that all significant information has already been extracted in earlier models. With 4 ICs, most samples still have a low DW value, although for a few samples, a 3-ICs model would be sufficient. However, globally, it seems evident by looking at this plot that 4 ICs is the optimal number to extract, confirming the expected results.

2.5.2 *The "ICA_by_Blocks" Method*

This method was published in Ref. [36] and aims at determining the optimal number of ICs. In order to do so, the data set is split into two (or more, if necessary) representative subgroups (which are representative of the original data set). For each group, several ICA models are calculated, with from 1 to *max-ICs* ICs. For each model (i.e., each number of computed ICs), the correlations between all ICs extracted from one subgroup with all ICs extracted from the other subgroup are calculated. The idea underlying this method is that, as long as significant ICs are extracted, they should be extracted from both sets. However, because the order and the signs of the extracted ICs are undetermined, it is necessary, for a given model, to compare all ICs from one subgroup to all ICs of the other subgroup in order to check whether or not similar ICs are extracted in both sets. When noisy ICs start to be extracted, they are characteristic of the random noise present in their subgroup, and consequently, these ICs should have a low correlation with all ICs from the other group. Therefore, the optimal number of ICs is defined as the largest number of ICs for which all ICs of one subgroup are highly correlated with one IC from the other subgroup.

An illustration of ICA_by_Blocks on the simulated data is presented below. Models with 1 to 10 ICs are tested. In Fig. 11A, the correlations profiles of all models are tested. For each model, the *A* highest correlations are presented, *A* being the number of ICs included in the model. If all extracted ICs of a subgroup show a large correlation with an IC from the other subgroup, all correlations will be high. As soon as a noisy IC is extracted, the correlation decreases. One can see in Fig. 11A that this is the case from 5-ICs model on. One can then conclude that 4 is the optimal number of ICs to calculate in this model.

Fig. 11B is a simplified version of Fig. 11A: instead of plotting the correlation profile of each tested model, which can soon lead to an overcrowded plot if a large number of ICs is tested, only the lowest correlation value of each profile is plotted. As long as this value remains large, the extracted ICs are assumed to be significant. Although the same information is present in both plots, the plot presented in Fig. 11B is simpler to read than that of Fig. 11A.

2.5.3 *Random_ICA_by_Blocks*

The results obtained with ICA_by_Blocks are highly dependent on the distribution of the samples between the two (or more) blocks. Therefore, in order to avoid obtaining results that may not reflect the true structure of the data, a randomized version of ICA_by_Blocks can be applied: the ICA_by_Blocks is repeated a number of times on random distributions of the samples into the two subgroups, and all correlation results are plotted at the end, showing whether repeatable results can be obtained or not. For each run of ICA_by_Blocks, i.e., each random distribution of the samples into two subgroups, a profile similar to that of Fig. 11B is obtained. The gathering of all these profiles can be graphically displayed in different ways, as will be presented below by application of the so-called *Random_ICA_by_Blocks* to the simulated data. Twenty repetitions of ICA_by_Blocks were carried out, each time randomly splitting the samples into two subgroups, and calculating ICA models

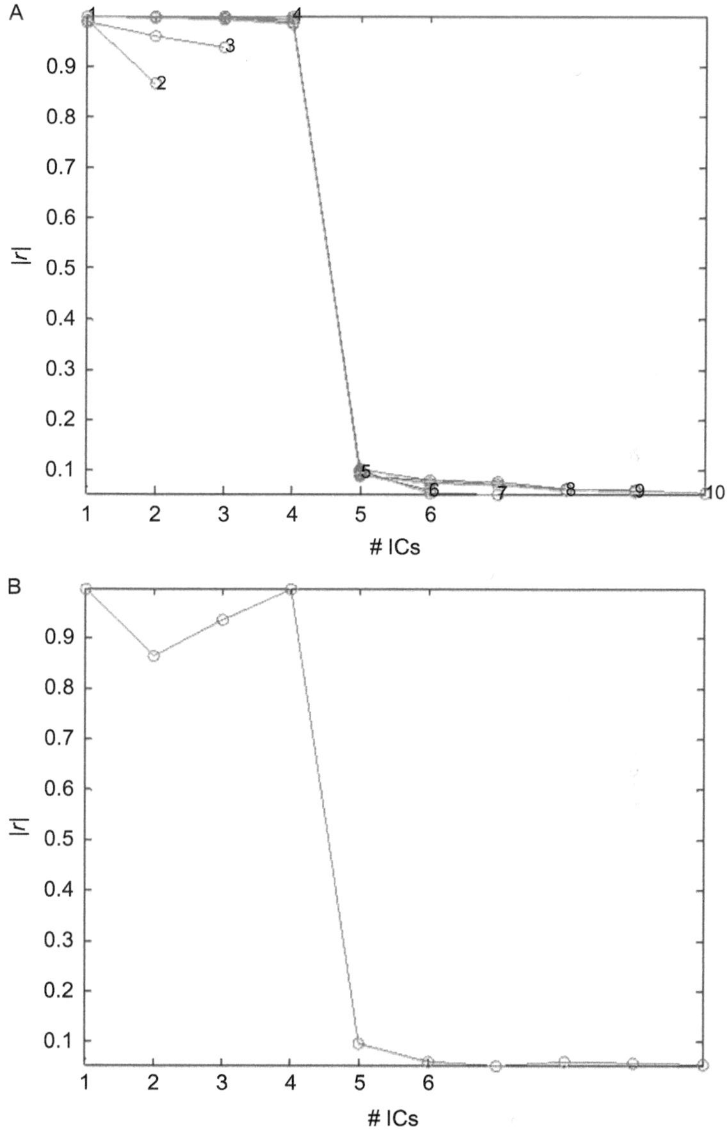

FIG. 11 ICA_by_Blocks applied to the simulated IR data. (A) Correlation profiles are plotted for each model (the number displayed at the end of each profile indicates the number of ICs in the corresponding model); (B) smallest correlations profile.

with from 1 to 10 ICs. The results are presented in Fig. 12A–C. The objective is to be able to distinguish as clearly as possible the different correlation profiles.

Looking at Fig. 12A, it can be seen that for 1-, 3-, and 4-ICs model, all absolute correlations between ICs extracted from each subgroup are always

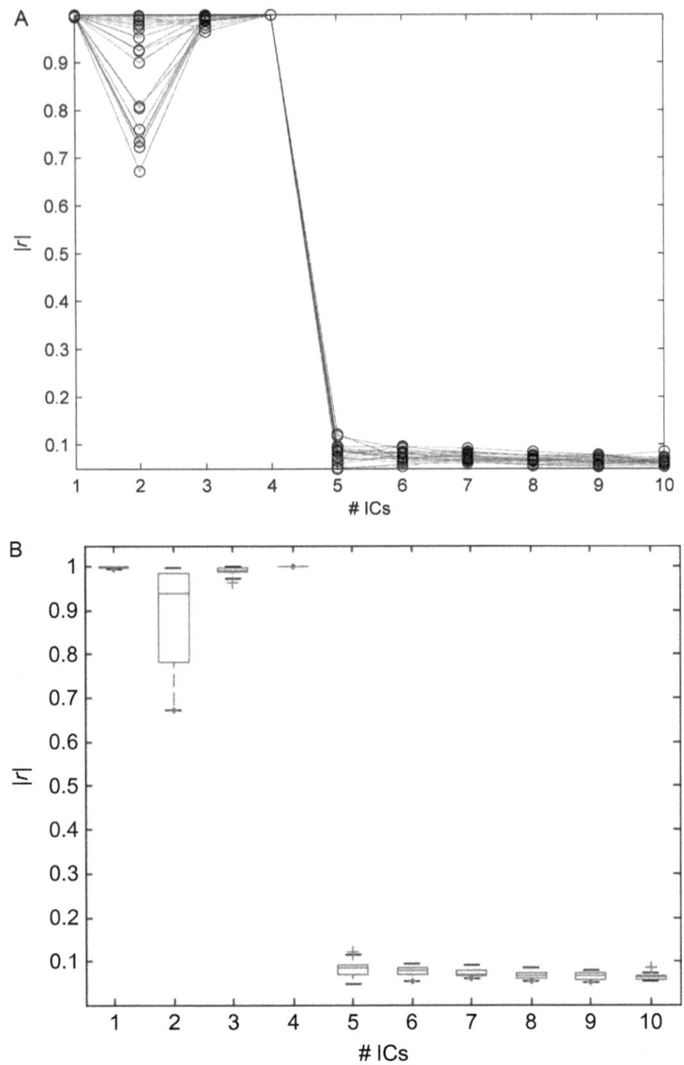

FIG. 12 (A) The absolute correlation profiles of the 20 runs of Random_ICA_by_Blocks; (B) boxplot of the profiles of the 20 runs of Random_ICA_by_Blocks;

close to 1, while with 2-ICs models depending on the distribution of the samples in the two subgroups, the correlations between the 2 ICs extracted from each subgroup are sometimes lower.[3] Still, the sharp decrease between the

3. If ICA_by_Blocks is only done once and "by chance" all the observed signals containing a particular IC are in only one of the blocks, then the correlation will be very low and too few ICs will be selected. This is avoided by doing ICA_by_Blocks several times with random distributions of the samples between the blocks.

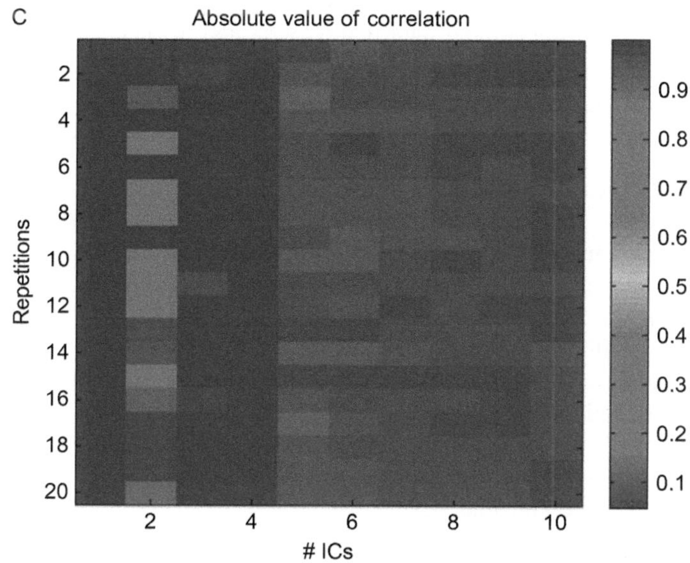

FIG. 12—CONT'D (C) scaled color image of profiles of the 20 runs of Random_ICA_by_Blocks.

4- and 5-IC models and the following plateau clearly indicate that the optimal number of ICs in this data set is 4. In bigger and more complex data sets, it is sometimes necessary to perform more than 20 repetitions. Moreover, the obtained profiles in real cases are rarely so similar with respect to one another, and therefore, the global figure obtained can be difficult to read. Therefore, Fig. 12B and C proposes alternative plots displaying the same information, either as a boxplot, where the three lines indicate the lower quartile, the median, and the upper quartile values and the "+" signs represent outlier values due to unrepresentative distributions of the samples between the blocks, or as a scaled color image display.

2.5.4 *RV_ICA_by_Blocks*

Instead of calculating the correlations between all pairs of IC vectors stemming from two models, another possibility is to calculate the so-called RV coefficient[4] between the two matrices of signals obtained. The RV coefficient is a common measure of the similarity of two matrices. It was introduced in 1973 by Escoufier [40,41]. The RV coefficient is calculated between two matrices, and can be interpreted as a coefficient of determination (R^2), ranging between 0 and 1. When the RV coefficient between two matrices is equal to 1, it means that the variance–covariance matrices of the two data tables are

4. "The '*R*' in RV refers to correlation coefficient [...]. The letter '*V*' refers to 'vector (random variables)'" [39].

similar, i.e., that their information content is similar. The RV coefficient between two column-centered data matrices $\mathbf{X}_1$ and $\mathbf{X}_2$ is defined as:

$$RV(\mathbf{X}_1, \mathbf{X}_2) = \frac{\text{trace}\left(X_1 X_1^T X_2 X_2^T\right)}{\sqrt{\text{trace}\left(X_1 X_1^T X_1 X_1^T\right)\text{trace}\left(X_2 X_2^T X_2 X_2^T\right)}} \tag{16}$$

In the RV_ICA_by_Blocks method, the data are split into two representative subsets, and several ICA models (with from 1 to *maxICs* ICs) are calculated. For each model, i.e., each number of ICs tested, the RV coefficient is calculated between the **S** matrices (matrices of extracted source signals) obtained for the two data sets. Therefore, a value of RV can be plotted vs the number of ICs included in the model.

Application of this variant of ICA_by_Blocks on the simulated data yields the "RV correlation" plot presented in Fig. 13.

As expected, the model with 4 ICs is found to be optimal.

Similarly to what was presented earlier, a "Random_RV_ICA_by_Blocks" method can also be considered.

2.5.5 *ICA_corr_Y*

If the true concentrations or the true signals of an analyte in the samples are known, this information can be used to choose the optimal number of ICs. The method presented here consists in computing the correlations between the proportions (**A** matrices) estimated using from 1 to *ICmax* ICA models and the known concentrations of the mixture constituents, or between the corresponding source signals (**S** matrices) and the known signals of the

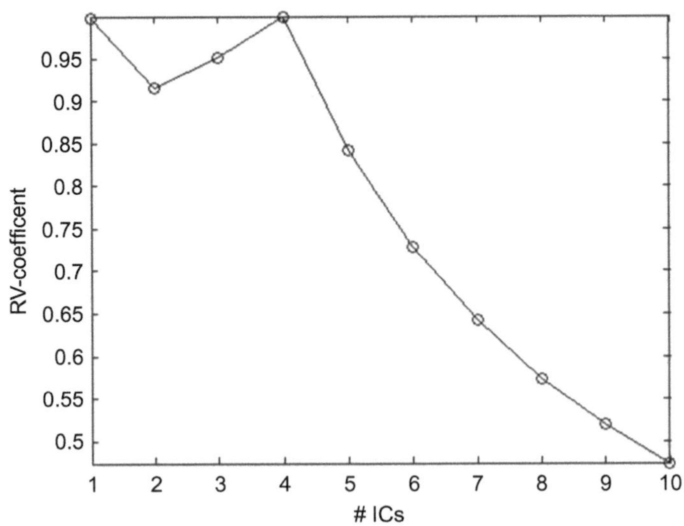

FIG. 13 RV_ICA_by_Blocks applied to the simulated IR data.

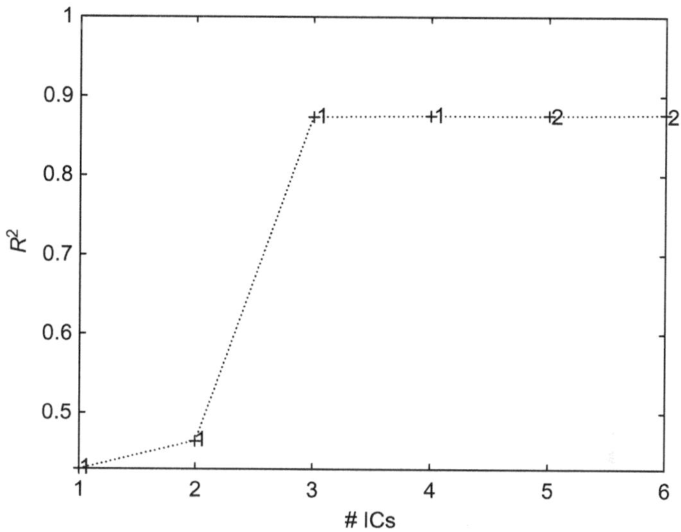

FIG. 14 ICA_corr_y: maximal correlation between the dependent vector **y** and the most correlated IC from each of the ICA models (calculated with from 1 to 6 ICs).

mixture constituents. The known matrix (of concentrations or of signals) is not used to influence the extraction of the ICs, but simply to determine the optimal number of ICs. The model giving the highest correlations with known concentrations or known signals is considered as corresponding to the optimal number of ICs [42]. ICA models not being nested, it is necessary, for each number of ICs tested, to calculate all correlations between known and estimated vectors. For each model, the maximal correlation is retained, as well as the index of the IC giving this maximal correlation.

As an example, a "concentration" **y** vector was created for the IR-simulated data, as being equal to the simulated mixing coefficients associated to the second pure source signal (see Fig. 7A), to which 5% random Gaussian noise has been added. The application of ICA_corr_Y to the **X** and **y** data yielded the plot presented in Fig. 14.

Models with from 1 to 6 ICs were tested. In each model, the coefficient of determination between **y** and each IC proportion was calculated, and the maximal value was retained, as well as the index of the IC with which that maximal value is associated (values indicated in the figure). One can then represent the maximal squared correlation coefficient (coefficient of determination) vs the number of ICs included in the model. In the example presented here, the maximal correlation was obtained between **y** and the first IC from a 3-ICs model.[5] Therefore, a model can be built to relate **y** to this IC. This relationship

5. Note that, since the ICs are not ordered, it came out as IC1 in this example, but it could also have had another index.

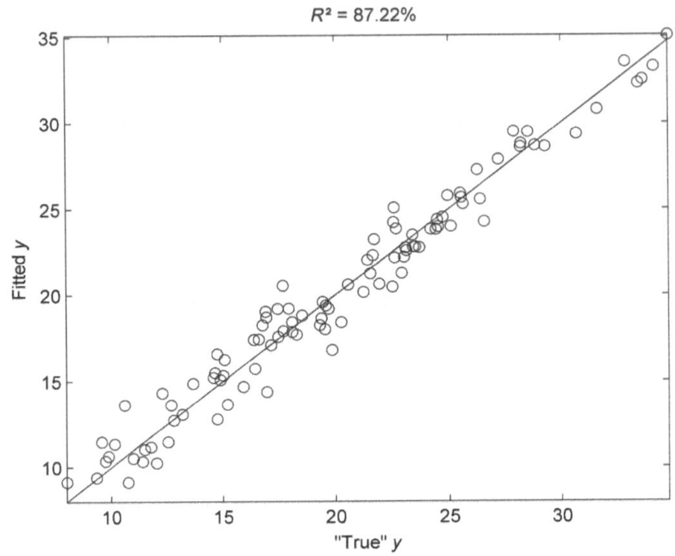

FIG. 15 MLR model between the most correlated IC and **y**.

is depicted in Fig. 15, and one can see a satisfactory relationship exists between **X** and **y** (the R^2 value of 0.87 corresponds to the value of R^2 which had been found by ICA_corr_y for the 3-ICs model). In order to evaluate the interpretability of this model, IC1 can be compared to the original source signal 2, whose mixing coefficients were used to generate **y**. Both vectors are represented in Fig. 16: clearly, IC1 resembles the simulated pure source signal, which explains why a satisfactory model could be obtained.

Therefore, although as underlined by Chen and Wang [43], "In practice, not all the m ICs can match the pure substances exactly, and therefore it cannot be taken for granted that the elements in the matrix [**A**] are concentrations. However like PCR, regression analysis can be performed between the concentration matrix **C** and [**A**]," in many cases good linear regressions can be had between one ICA proportions vector and the concentrations vector of the compound of interest. For cases when this is not possible, Chen and Wang proposed using Multivariate Linear Regression (MLR) between the matrix of proportions and the concentrations.

2.6 Regression on Independent Components Proportions (Scores)

MLR models are now often calculated using Principal Components Regression (PCR) [44] and PLS regression [45]. Both methods are based on LVs which means that instead of working with the original data, where the matrices are often "wide," i.e., the number of variables (greatly) exceeds the number

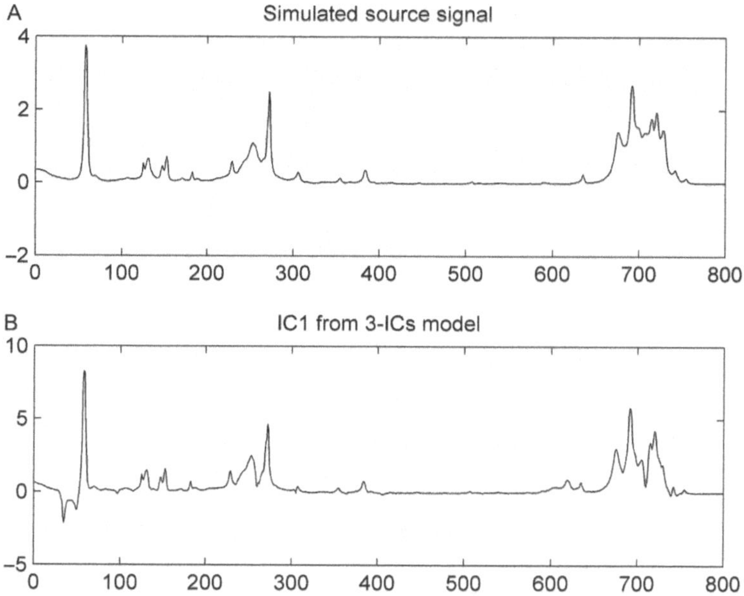

FIG. 16 (A) The second simulated source signal; (B) signal of IC1 extracted from the 3-ICs model.

of observations, the original data are transformed into LVs. The LVs are linear combinations of the original variables, with particular properties, such as the orthogonality, which then simplifies the second step of the modeling procedure, namely the multilinear regression of the response vector **y** onto the LVs. In PCR, the LVs are the PCs, and are obtained as linear combinations of the variables in **X**. As explained earlier, the first PC is oriented in the direction of maximal sample dispersion (variance) in **X**. Each following PC is orthogonal to the previously calculated PCs, and is oriented in the direction of largest dispersion of the residuals. In PLS, the LVs are oriented in the direction of maximal covariance between **X** and **y**, and are also orthogonal.

Similarly to what is done in PCR and PLS, Chen and Wang [43] demonstrated that ICA is capable of finding pure spectra from near-infrared spectroscopic data. They found that the proportions from ICA could be used in an MLR (giving Independent Components Regression, ICR) similarly to PCR. The interest of the ICR procedure was demonstrated again by Shao *et al.* [46] analyzing NIR spectra of tobacco.

Gustafsson [47] proposed two regression methods, where the LVs correspond to the proportions vectors extracted by ICA. In the first method, called Sorted Input Independent Components Regression, an ICA is applied to the **X** matrix, and a response vector is regressed onto the obtained proportions vectors via an MLR model (this method is equivalent to what is done in PCR). In the second method, called Sorted Input–Output Independent Components

Regression, the ICA model is calculated on a matrix which is the concatenation of the **X** and **y** matrices [**X**·**y**]. Therefore, the obtained ICs correspond to ICs in the {**X**‖**y**} space (where "‖" stands for row-wise concatenation). An MLR model is then built between the response variable and the proportions vectors (this method is in some ways similar to what is done in PLS, i.e., both **X** and **y** are taken into account for the calculation of the components).

In order to regress a vector **y** onto the ICA proportions extracted from **X**, which are in the columns of **A**, it is necessary to dispose of a method to determine the optimal number of ICs to use. This can be done using many of the methods presented earlier, but the most logical would consist in computing the correlations between the proportions (**A** matrices) estimated by from 1 to *ICmax* ICA models and the known concentrations of the mixture constituents. This can be done with the ICA_corr_Y method presented earlier.

2.7 Independent Components–Discriminant Analysis

In the preceding section, the use of ICA in a quantitative multivariate analysis method was presented. Similarly, it is possible to consider using ICA for qualitative multivariate analysis methods, aiming at discriminating samples into different groups, according to predefined characteristics.

Monakhova *et al.* [48] have shown that in favorable cases, nonsupervised analysis of data using ICA can discriminate between groups of samples as efficiently as supervised methods such as Linear Discriminant Analysis, Quadratic Discriminant Analysis [49], Soft Independent Modeling of Class Analogy [50], and Partial Least Squares–Discriminant Analysis (PLS-DA) [51,52]. This is possible in cases when different groups are characterized by either very different source signals or different proportions of the same source signals. In other situations, it may be necessary to orient the ICA calculations by introducing prior information about group membership. A multivariate method for the modelization of group membership is PLS-DA [51,52], in which a regression model is built between the **X** data matrix and a **Y** matrix, which is created with as many columns as there are groups, and whose elements are 0s and 1s, indicating the (non)membership of the corresponding sample to the corresponding group. For example, in Fig. 17A, the data for seven samples are assembled in **X**, each sample belonging to one of the three groups, represented by the three columns of the **Y** matrix: samples 1, 2, and 5 belong to group 1, samples 3 and 6 belong to group 2, and samples 4 and 7 belong to group 3. A PLS2 model is built to regress **Y** onto **X**, and for each new sample, a three-elements vector can be predicted, where two elements should have values close to 0 while the third should have a value close to 1, indicating to which group that sample belongs.

The method using ICA for discrimination, Independent Components–Discriminant Analysis (IC-DA), is based on the same idea, but is slightly different. As shown in Fig. 17B, the **X** and **Y** matrices are first concatenated

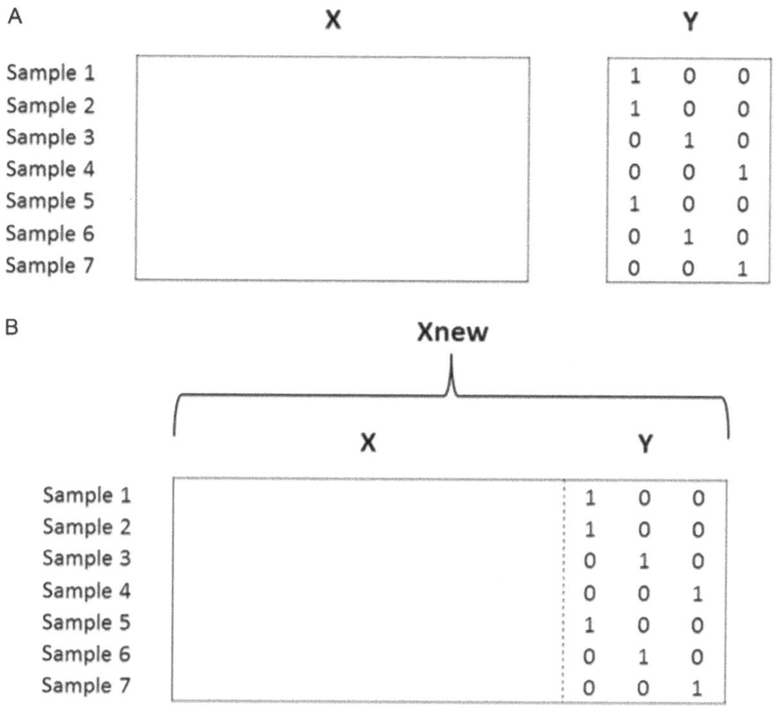

FIG. 17 Organization of data for (A) PLS-DA and (B) IC-DA.

row-wise into a single data table (**Xnew** in Fig. 17B) which is to be analyzed by ICA (IC-DA is therefore an oriented version of ICA). However, one must first ensure that the orders of magnitudes in both the **X** and **Y** matrices are similar. Indeed, if it is greater in **Y** then in **X**, the ICA will be mostly "influenced" by **Y**, so that the influence of the original **X** variables involved in the discrimination will not be significant; if it is smaller in **Y** than in **X**, then the method will not really be oriented, and the results will resemble those obtained by calculating an ICA model on **X** alone. In order to balance the orders of magnitude of the two matrices, the **Y** matrix is premultiplied by a factor depending on **X**, such as its median value. In order to determine the optimal number of ICs, a method specific to IC-DA is proposed (which is why it was not exposed earlier), which is based on the Wilks' λ coefficient [53], that characterizes the discrimination of the groups: the optimal number of ICs is the number that maximizes the value of the F of Fisher associated to Wilks' λ.

Once the optimal number of ICs is known, an ICA model of **Xnew** can be calculated. Examination of the proportions of the ICs may reveal some discrimination. If a discriminating IC is found, a group will be discriminated on a proportions plot, and the contribution of the **Y** variables on the IC plot will also indicate for which group(s) it is important. On this same IC plot, **X** variables with a large positive or negative value can then be interpreted as also being important variables for discrimination. However, the discrimination of samples that can be observed on these proportions plots have been obtained with proportions calculated from extracted signals containing a contribution from **Y** (according to Eq. 5).

Therefore, "new" proportions must be calculated, by first removing from the ICs the variables corresponding to the **Y** matrix. This results in reduced ICs of length equal to the number of variables in **X**. Examination of the new proportions plots will reveal whether a discrimination can still be observed, when it is only based on the effect of the signals containing the **X** variables.

When a discrimination is obtained with IC-DA, one must ensure that a similar discrimination could not have been obtained with random group assignments, so a validation step is performed at the end of the procedure: the rows of the scaled **Y** matrix are permuted before (the scaled) **Y** is concatenated to **X**, and the IC-DA procedure is repeated on this random data set. The Wilks' λ is calculated and saved, and the whole procedure is repeated a number of times (for example, 999 times), by ensuring that the true **Y**, if randomly generated, is not used. The distribution of the Fisher's F statistics associated with the obtained "random" λ's is represented graphically, and the F for the "true" λ, obtained from the non-randomized **Y**, is compared to this distribution. The farther away it is from the bulk of the random distribution, the more confidence one can have in the discriminant model.

When a new sample is available, it is projected onto the component signals according to Eq. (5), and its proportions are obtained. These can be compared to the proportions calculated with the "calibration samples," i.e., the samples which were used to calculate the parameters of the ICA model, and group membership can then be established.

An illustration of IC-DA and its properties will now be presented, based on the simulated infrared data presented in Fig. 7B. In order to create a data matrix with clusters to be separated, the four columns of values in Fisher's well-known Iris data set were used as the mixing coefficients [54]. This 150×4 data set represents four variables, the sepal lengths and widths and the petal lengths and widths, for the flowers of three different species of Iris (50 samples of each species), namely *Iris setosa* (group 1), *Iris versicolor* (group 2), and *Iris virginica* (group 3). The resulting simulated IR data, after addition of 5% random Gaussian noise, are presented in Fig. 18. A 150×3 groups matrix, **Y**, was constructed, indicating to which Iris species the corresponding sample belongs.

When applying IC-DA to the **Xnew** data set, obtained by concatenating the simulated Iris_IR matrix to the 150×3 groups matrix, an optimal model with 3 ICs is obtained, presented in Fig. 19. On IC2 and IC3, the three groups are clearly separated: IC2 separates group 1 from the other 2 (as indicated by

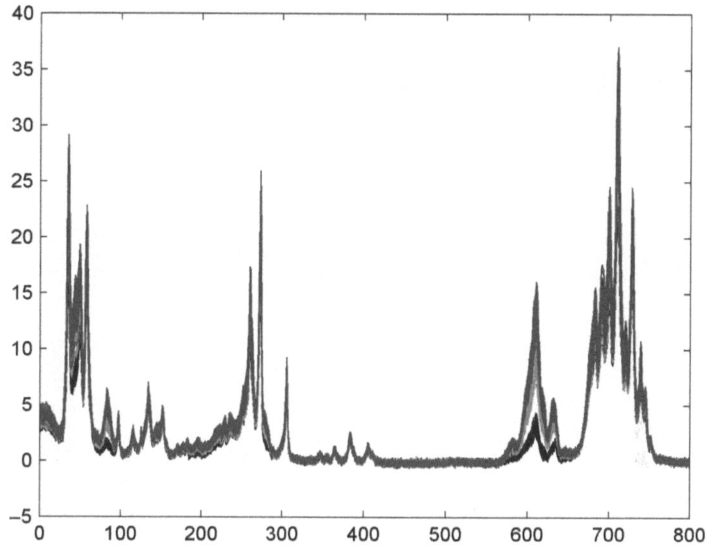

FIG. 18 The simulated Iris_IR data set.

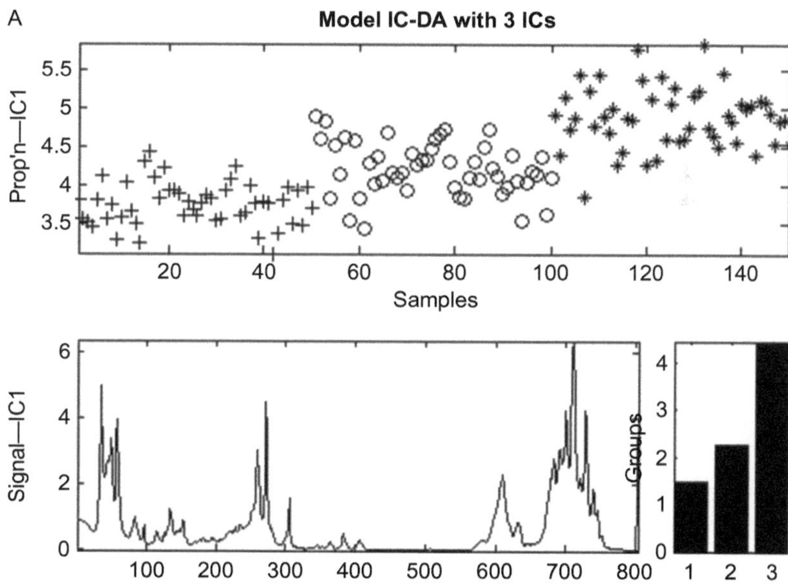

FIG. 19 IC-DA of the simulated IR_Iris data set: proportions and signals of (A) IC1,

(Continued)

the bar plot in the bottom part of Fig. 19B, showing that column 1 of the groups matrix has a higher contribution on IC2 than the other two columns), and IC3 opposes group 2 to group 3 as can also be seen in the IC3 bar plot. Looking at the corresponding source signals shows the IR parts of the spectra

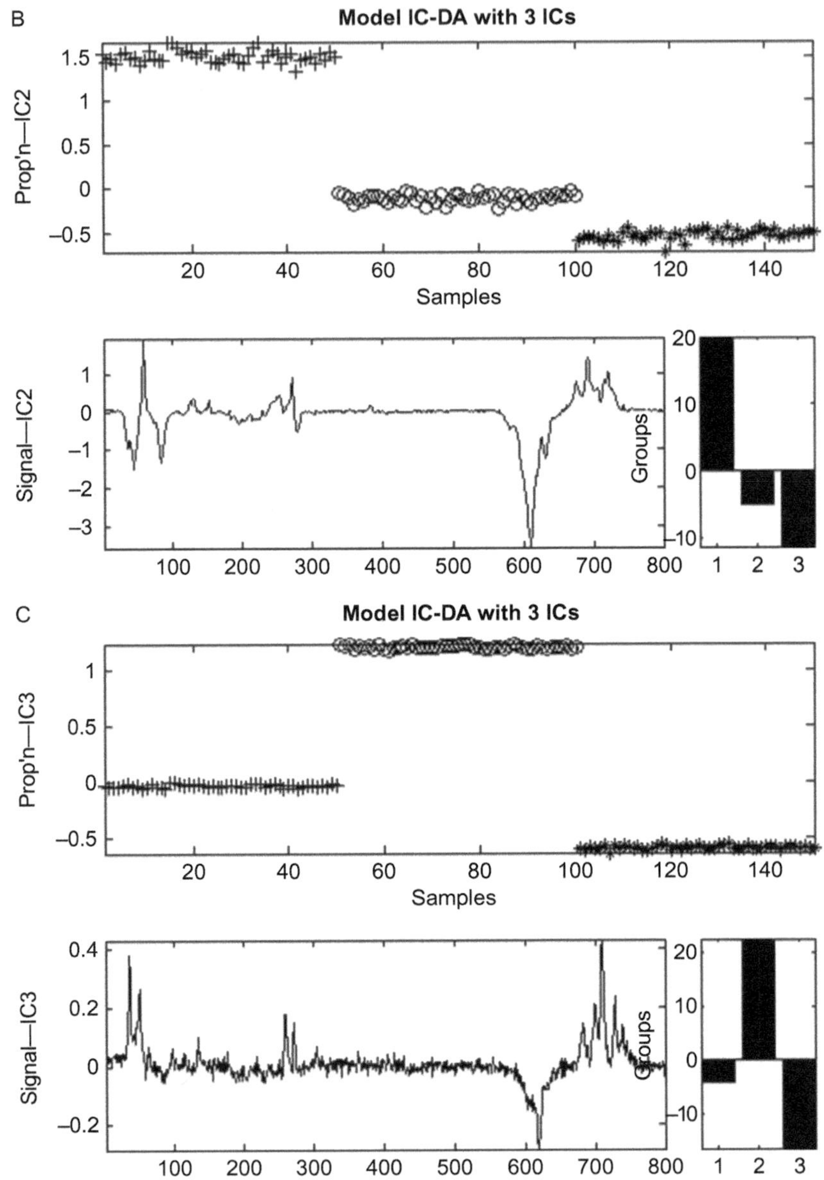

FIG. 19—CONT'D (B) IC2, and (C) IC3.

which discriminate groups 1 and 2, respectively. However, although a very satisfactory separation can be observed on the proportion plots, it must be remembered that these proportions were influenced by the contribution of the group membership matrix, which oriented the extraction of the ICs.

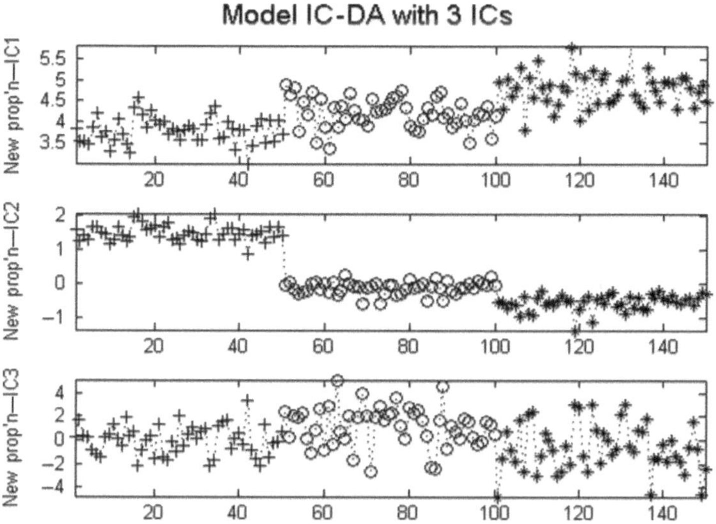

FIG. 20 "New" proportions of the IC-DA model, calculated without using the signals extracted from the group membership matrix.

Therefore, the last three values of each IC, corresponding to the groups contribution, were removed, and new proportions were recalculated, based on the IR variables only. These new proportions are presented in Fig. 20. It can be seen that group 1 is perfectly separated from the other two groups. Groups 2 and 3 are almost separated, but a slight overlap exists. This illustrates the fact that the proportions to consider should be those obtained without the groups contribution on the ICs, as keeping these contribution leads to overoptimistic results.

This discrimination must then be validated, by randomizing 999 times the assignment of the samples in the groups, and by recalculating each time an IC-DA model. The Wilks' λ coefficient can then be calculated on all so-called new proportions, and the distribution of these random λ values can be compared to the λ value calculated with the known distribution of the samples into groups by plotting the histogram of the corresponding Fisher's F values. These results are presented in Fig. 21.

It is clear from this figure that when the sample assignment in the three groups is correct, the resulting group separation is not due to random effect, but is based on "real" **X** data characteristics. Therefore, IC-DA is a suitable method for discrimination.

2.8 Comparison with Multiway Methods

The chemometrical analysis of multiway data requires the use of specific functions which have been developed recently for the decomposition of data

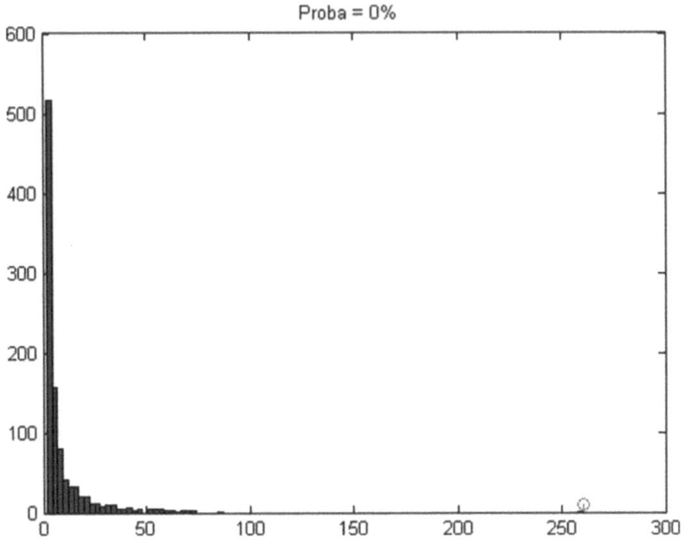

FIG. 21 Distribution of the F values for the "random" Wilks' λ coefficient values, and comparison with the "true" value.

arrays of dimensions greater than 2, that is, three-way cubes and n-way hypercubes. Typically, methods like Parafac [55] or Tucker [56] are applied. Normally, for such tri- (or more) linear data sets, it is advisable not to unfold the data with the aim of applying a "standard" multivariate analysis method such as PCA to the resulting matrix, because the trilinear structure in the data is lost, and so cannot be used to orient the decomposition toward a solution taking into account the inherently trilinear nature of the data. However, as will be seen in the example presented in Section 3.3, unfolding of multiway data, followed by an ICA analysis, and a refolding of the factors, can yield results which are similar to those which would have been obtained with a multiway method. This is because the Gaussianity/non-Gaussianity of the signals is not modified by the unfolding.

2.9 Comparison with MCR

Deconvolution methods, such as Multivariate Curve Resolution–Alternating Least-Squares (MCR-ALS) [13,14], are sometimes compared to ICA, as both these methods aim to solve a mixture problem and extract interpretable components, which should produce (an approximation of) the studied data. However, these two methods do not have the same objective, and are not based on the same assumptions, and so they do not optimize the same criteria. Indeed, as stated by Parastar *et al.* [57], while "MCR methods try to maximize the data variance explained by the different components under a set of constraints [...], ICA [...] tries to maximize the independence among the resolved components with a minimum loss of information." Therefore, there is no reason why they should yield the same results—MCR-ALS aims to extract the signals of individual components mixed together in the data, whereas ICA aims to extract source signals that characterize the different independent phenomena influencing the data values. Hence, care must be taken when comparing these two methods, as the obtained conclusions are data dependent.

Therefore, in ICA, extracting source signals does not mean that the obtained ICs correspond to the signals of the pure components, but they rather correspond to a signal characterizing a *phenomenon* modifying the acquired signals. In order to illustrate this difference, one can give the basic example of a chemical reaction, where a compound **A** reacts to yield a second compound **B**. It is then obvious that the appearance of compound **B** is related to disappearance of compound **A**. Hence, the signals of compounds **A** and **B**, which in many cases can be recovered separately by MCR, are not independent, and therefore, ICA will not extract two signals, but only one, which will be a mixture of a negative signal of **A** and positive signal of **B**.

In Ref. [57], Parastar *et al.* made an interesting comparison of MCR-ALS with different ICA algorithms. They conclude, among other things, that the comparison of ICA with MCR-ALS does not only depend on the data but also on the algorithm used for ICA, stating that "only MF-ICA can be strictly compared to MCR-ALS," because it allows nonnegativity constraints both on the mixing matrix and on the source signal matrix. However, the use of constraints has an impact on the independence of the extracted signals, which decreases. Moreover, "if the overlap of concentration profiles increases, the independence and the recoveries of the resolved spectra by ICA also become significantly affected." On the contrary, if there is (almost) no overlap in the concentration profiles, satisfactory solutions can be obtained with ICA, which can then extract signals that are similar to the real signals, and results comparable to those of MCR-ALS can be obtained. This will be illustrated with an application of ICA to Nuclear Magnetic Resonance (NMR) data of ternary mixtures in Section 3.2. At least, in accordance with what is explained earlier, the authors state that since MCR-ALS can include constraints (such as unimodality, closure), it is more flexible than ICA, and the authors recommend to use it instead of ICA for deconvolution problems.

2.10 An Unexpected Property of ICA

ICA is described as a method for signal processing, and the numerous applications found in the literature present the ICA of different types of signals. By signals, one tends to understand a structured and (auto)correlated array of measured data points. However, as is shown from the example later, working on structured arrays is not a necessary condition to apply ICA. The two source signals presented in Fig. 22A are linearly mixed to yield 100 mixed signals, to which 1% random Gaussian noise is added. An ICA model with

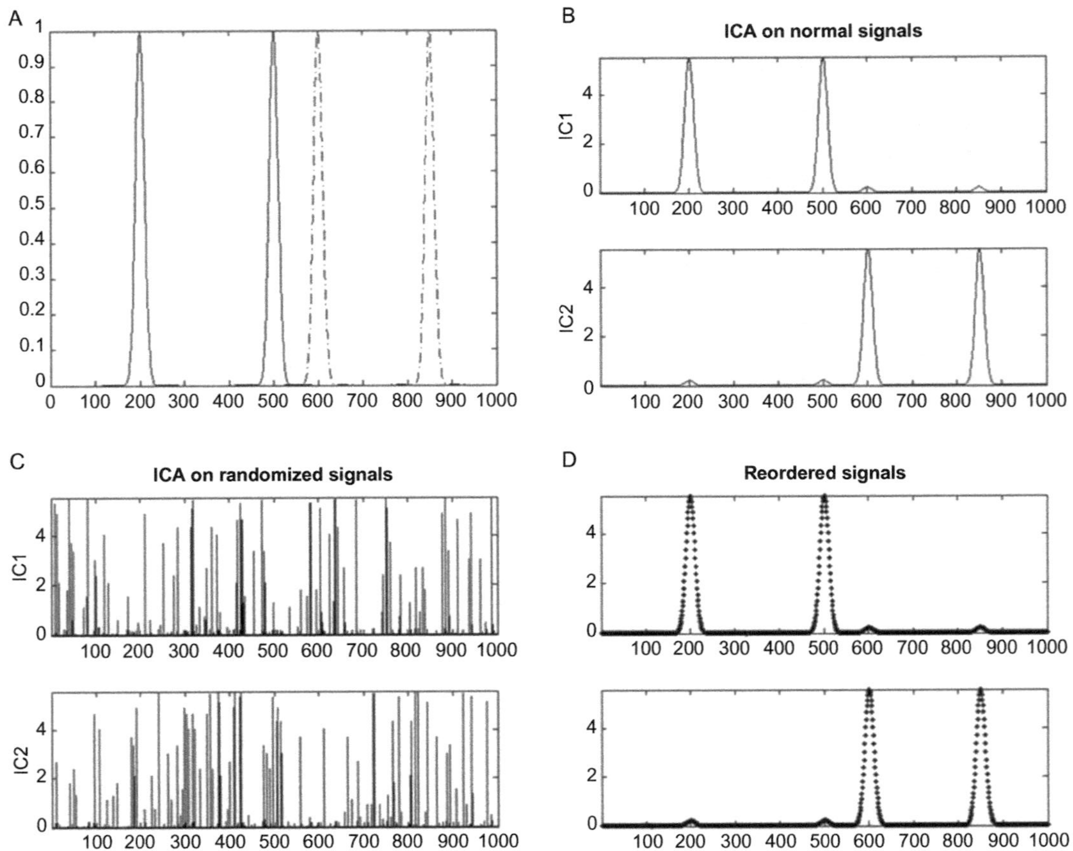

FIG. 22 (A) The two simulated source signals; (B) the ICs extracted from the original mixed signals matrix; (C) the ICs extracted from the randomized mixed signals matrix; and (D) reordered signals.

2 ICs is computed, and, as expected, the extracted ICs presented in Fig. 22B are highly correlated to the original source signals. If the variables of the 100 mixed signals are reorganized in random order, the structure and correlation in all signals will be lost. Nevertheless, if an ICA model with 2 ICs is then computed on this data set with randomized variables, and the extracted ICs are reordered according to the original indices of the signal, they exactly match the ICs extracted from the "unsorted" matrix of mixed signals (Fig. 22D). This result shows that ICA does not need the "signals" to be structured, but any type of data can be analyzed by ICA, and yield sensible results. This will be illustrated in Section 3.1. This is because ICA is based on the histograms of the intensities, which is more Gaussian for mixtures than for source signals, rather than on the signal structure.

3 SOME APPLICATIONS OF ICA

ICA was initially developed in the field of signal processing in telecommunications [2,4,8], but its use has now extended to a wide range of domains. A rapid general bibliographical research on ICA produces hundreds of different applications of ICA, in a lot of very diverse fields. Mostly, ICA is applied to separate significant components from signals but also to remove part of the noise, by eliminating the components carrying noise information. It is not the goal of this section to present an exhaustive review on all possible ICA applications but to illustrate the variety of case studies where ICA is encountered, a few of them will be cited below:

- Financial data analysis [58]
- Text analysis [59]
- Pattern recognition [60]
- In Astrophysics, ICA was used to separate "astrophysical components superposed in maps of the sky" [61].
- ICA is also much used in image analysis, to restore a blurred image [62] or to separate components from superimposed images [63].
- Westad used ICA on two sensory data sets [64], and "significance testing on each sensory attribute for the components gave valuable information about the relevance in interpreting the individual attributes and components." In this article, Westad showed that ICA is applicable to data vectors (sensory attributes) that are not what one normally considers as signals. He also proposed the idea of IC-PLSR—an extension of ICR [43], by making ICA an integral part of the PLS algorithm.
- ICA has also been used in process monitoring, for dimension reduction [65] and to facilitate the interpretation of the phenomena taking place [66,67].
- As well, ICA has been used in the petrochemical field [68,69], for example, to identify gasoline constituents from infrared spectra. In this study,

the authors used the JADE algorithm and showed that the correlation coefficients between the extracted ICA source signals and the true spectra were much higher than between PCA loadings and the true spectra.

- In the medical field, ICA has been found to be more efficient than PCA for the analysis of EEGs [70]. It has been used, for example, to decompose ictal EEGs into ictal and nonictal components [71], and hence helped to remove artifacts, for example, due to eyes or muscles movements from EEGs [72]. It was also used to remove artifacts from electrogastric signals and for biological signal separation [73] to improve their interpretability, or for the interpretation of ^{1}HMR spectra of tumors [74,75].
- In the pharmaceutical field, where nondestructive spectroscopic techniques are often used in combination with chemometric methods, ICA has been used to analyze data from Raman microscopy, and helped identify the active ingredients and excipients of a genuine tablet, and the constituents of a counterfeit tablet [76,77]. Moreover, the distribution of each constituent within the tablets could be determined.
- In analytical chemistry, a lot of applications of ICA can be found, for example, in Refs. [15,42,78–81]. Many of them are based on the multivariate analysis of spectroscopic data, measured on real samples. ICA is particularly suited to such data, as spectra can often be regarded as linear mixtures of "basis" spectra, which corresponds to the paradigm of ICA.

This list only presents a small sample of the numerous articles concerning the use of ICA. It shows that ICA is a method which can be applied whenever multivariate data analysis is required to extract information from data, or to remove noise from data. However, as indicated earlier, ICA can be applied to all types of data. Three concrete case studies will be presented later, involving the application of ICA to one-, two-, and three-dimensional data.

3.1 One-Dimensional Data

82 Gallo-roman ceramics samples, originating from three geographically neighboring French towns (Saintes, Soubran, and Petit-Niort), dated from between 25 BC and AD 250, had their mineral composition analyzed for 12 elements, namely SiO, AlO, CaO, KO, FeO, TiO, MnO, PO, MgO, Pb, Sr, and Zr. Rutledge *et al.* first applied PCA on the autoscaled data [82,83], and were able to separate almost all the Saintes site samples from the other two sites on the first PC, corresponding to 28% of the total variance. However, the other two sites, which are geographically very close to each other, and hence probably have a similar mineral composition, were overlapped in all score plots. ICA was applied to the 82×12 autoscaled data table (of so-called one-dimensional data, the 12 variables being discrete, so that one cannot really speak of sample "signals" with the usual definition of this term). A model with 3 ICs was constructed, and the corresponding proportions and

variables plots are presented in Fig. 23A, C, and E, respectively, along with the results obtained using PCA (in Fig. 23B, D, and F).

The objective here is to show that in the case of data sets where the variables do not constitute what one normally considers as a signal, ICA can still be used in the same way as PCA.

IC1 separates the "Saintes" samples from those from the other two towns. This is mainly due to large mineral contents in CaO, KO, FeO, MnO, MgO,

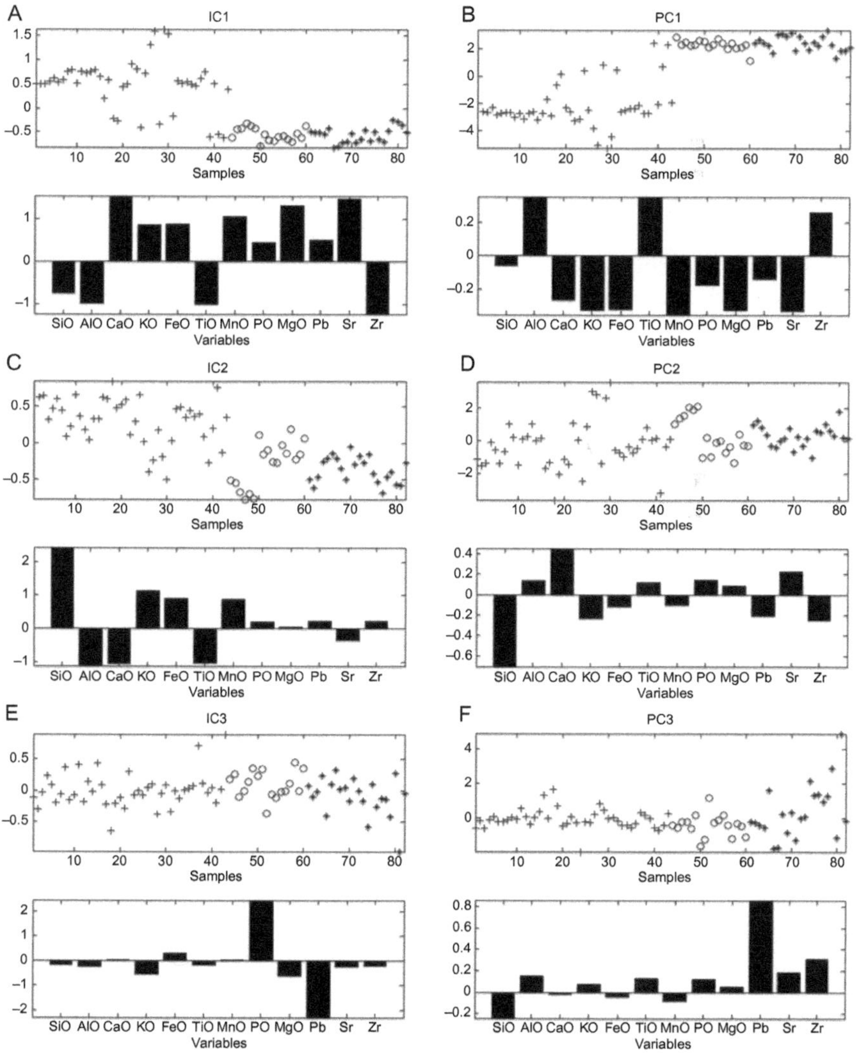

FIG. 23 ICA and PCA models on ceramics data: (A), (C), and (E): ICA proportions and variable plots; (B), (D), and (F) PCA scores and loadings plots.

and Sr, while the samples from Soubran and Petit-Niort have a relatively large mineral content in AlO, TiO, and Zr. These same conclusions were obtained from PCA (Fig. 23A and B). PC2 and IC2 (Fig. 23C and D) are also quite comparable, and also PC3 and IC3 to some extent (Fig. 23E and F). Therefore, although each sample is described by a few discrete variables, ICA can be applied to extract information from the data, similarly to what can be done with PCA.

3.2 Two-Dimensional Data

NMR data presented in Ref. [84] were downloaded from www.models.life.ku.dk/NMR_Mix_DoE. Two hundred and thirty-one samples of ternary mixtures of propanol, butanol, and pentanol, each with 21 concentrations varying between 0% and 100% (5% increments) and mixed according to a ternary mixture design, were analyzed by ^{1}H NMR on a Bruker Avance Ultra Shield 400 spectrometer (details of the measurements parameters can be found in Ref. [84]). PCA and MCR-ALS had been applied, and the authors showed that MCR-ALS was able to recover the NMR signals of each alcohol, together with their concentrations in the mixtures. ICA can also be applied to this sort of data set. An ICA model with 3 ICs was calculated, as three components were expected to be present in the signals acquired on this three-compound mixture design. Fig. 24A–C present the proportions plots, the IC signal plots, and the NMR spectra of each pure alcohol, and Fig. 24D presents the 3D proportions plot. Clearly, ICA has succeeded in recovering the pure NMR signal of each alcohol, as well as the ternary mixture design.

3.3 Three-Dimensional (and more) Data

An example will be given with a three-way data set, called *the Oil data*. Corn oils from two different suppliers were heated for 4 h at four different temperatures (120, 140, 170, and 190°C), either "as is" or after addition of two different antioxidants. The samples were analyzed by 3D fluorescence spectroscopy and a spectrum acquired every 15 min (excitation wavelengths: 280–500 nm, step 2 nm; emission wavelengths: 300–550 nm, step 2 nm). This resulted in a 3D data array (cube) of dimensions $320 \times 126 \times 111$. Each of the 320 matrices was unfolded resulting in a $320 \times 13{,}986$ matrix.

The optimal number of ICs to use in the model was determined using Random_ICA_by_Blocks (2 blocks, 20 repetitions, up to 15 factors tested). The final boxplot of correlations is presented in Fig. 25. Based on these results, nine source signals were extracted by ICA. At the same time, a nine-factor Parafac model was calculated using the N-way Toolbox [85]. Each IC signal, of length $1 \times 13{,}986$, was folded back to a 126×111 matrix, comparable to an excitation–emission matrix (EEM). As for the Parafac factors, the loadings on the second mode were multiplied by the loadings on the third

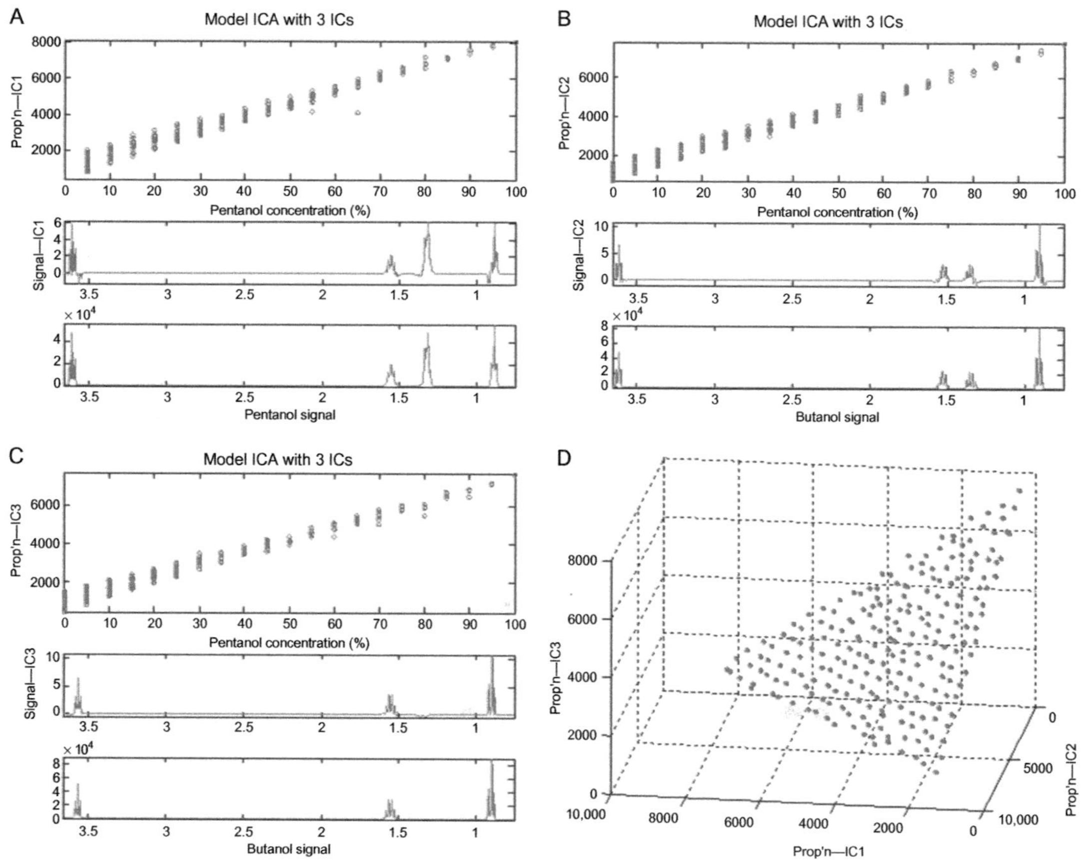

FIG. 24 ICA model with 3 ICs on the NMR data: (A)–(C): One-dimensional proportions and signals plots, on IC1 to IC3, and the corresponding NMR signal of each pure alcohol; (D) the three-dimensional proportions plot.

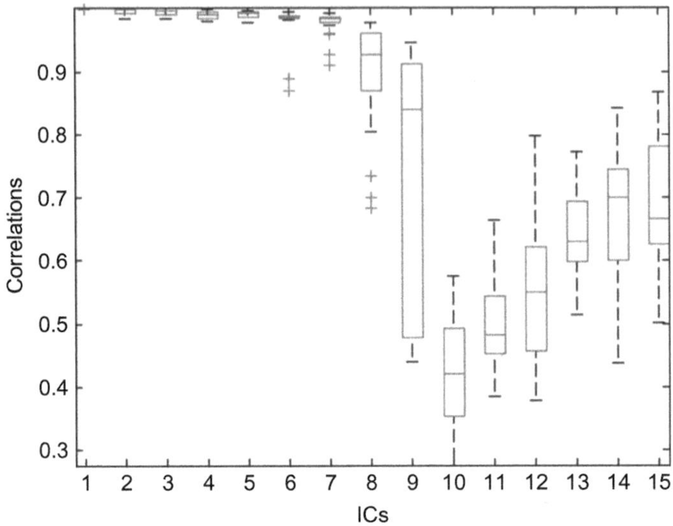

FIG. 25 Random_ICA_by_Blocks with 20 repetitions on the oil data. ICA signals correlation boxplot.

mode, in order to obtain an EEM-like surface. The comparison of the ICA and Parafac results revealed that similar factors were extracted in the two cases (although with different order, but one knows ordering of ICs is not an issue in ICA), as can be seen from Fig. 26.

Fig. 26A shows that most of the signals extracted by ICA are similar to those extracted by Parafac. The greatest discrepancies are for the pairs ICA/Parafac: 4/5, 7/6, and 8/9. This is not surprising since Parafac, using the Corcondia criterion, indicated that there were only four significant factors, whereas ICA, using the random ICA_by_Blocks methods, found 9 ICs. Fig. 26B shows a very close resemblance in the evolution of the ICA proportions and Parafac scores of the samples as a function of time and temperature for the two types of oils (one must sometimes multiply the signal by -1). The only major differences are for the pairs ICA/Parafac: 4/5, 7/6, and 8/9.

Therefore, in spite of not taking into account the trilinearity of the data by unfolding the EEM data cube into a matrix, ICA is able to produce results comparable to those of an inherently trilinear method, such as Parafac. This is because the unfolding has no influence on the distribution of the intensities. One advantage of applying ICA rather than Parafac lies in the way to optimize the number of factors to work with: the Corcondia criterion calculated on Parafac models does not always yield satisfactory results. CV has also been proposed to determine the number of Parafac factors, but this can lead to overfitted models. The different methods presented earlier for determining the optimal number of ICs usually give concordant, reasonable estimates.

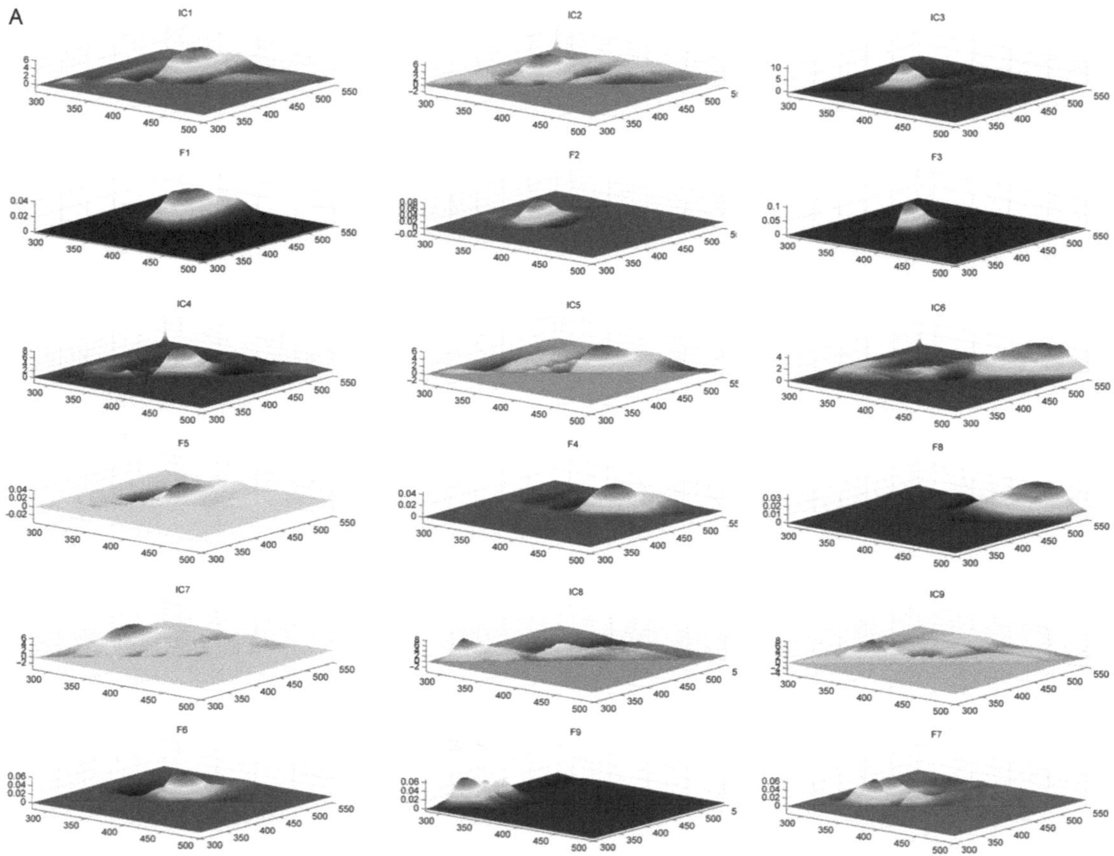

FIG. 26 (A) Signals extracted by ICA and surfaces reconstructed using factors extracted by Parafac;

(Continued)

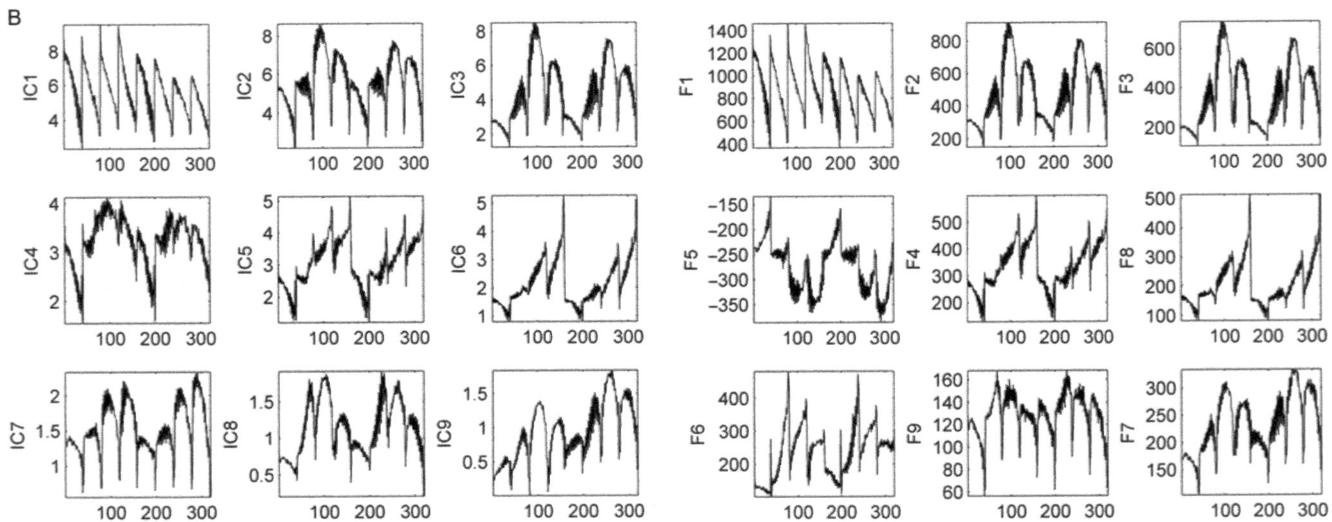

FIG. 26—CONT'D (B) proportions extracted by ICA and scores extracted by Parafac.

4 CONCLUSION

ICA is a powerful multivariate data analysis method, which has been gaining more and more interests in the past few years. A lot of different applications in diverse scientific domains exist. ICA can be used in quantitative regression problems, in discrimination problems, or can also be applied on unfolded multiway data. The differences between ICA and PCA, on the one hand, and between ICA and MCR, on the other hand, are important. They should always be kept in mind in order to avoid the confusion which sometimes arises during the interpretation of the results.

REFERENCES

[1] Herault J, Jutten C. Space or time adaptive signal processing by neural models. In: Denker JS, editor. Proceedings AIP conference: neural networks for computing. Woodbury, NY, USA: American Institute for Physics; 1986. p. 206–11. vol. 151.

[2] Jutten C, Herault J. Blind separation of sources, part I: an adaptive algorithm based on neuromimetic architecture. Signal Process 1991;24:1–10.

[3] Comon P, Jutten C, Herault J. Blind separation of sources, part II: problem statement. Signal Process 1991;24:11–20.

[4] Comon P. Independent component analysis: a new concept? Signal Process 1994;36:287–314.

[5] Hyvärinen A. A survey on independent component analysis, http://www.cmap.polytechnique.fr/~peyre/cours/x2003signal/ica_survey.pdf; 2003.

[6] De Lathauwer L, De Moor B, Vandewalle J. An introduction to independent component analysis. J Chemometr 2000;14:123–49.

[7] Hyvärinen A, Oja E. Independent component analysis: a tutorial, http://cis.legacy.ics.tkk.fi/aapo/papers/IJCNN99_tutorialweb/; 1999.

[8] Hyvärinen A, Oja E. Independent component analysis: algorithms and applications. Neural Netw 2000;13:411–30.

[9] Shlens J. A tutorial on independent component analysis; 2014. arXiv: 1404.2986v1.

[10] Stone JV. Independent component analysis: an introduction. Trends Cogn Sci 2002;6(2):59–64.

[11] Jouan-Rimbaud Bouveresse D, Rutledge DN. Introduction à l'Analyse en Composantes Indépendantes, et comparaison avec l'Analyse en Composantes Principales. Revue de la Société des Experts Chimistes de France 2012;976:35–41.

[12] Stone JV. Independent component analysis. A tutorial introduction. A Bradford book. Cambridge MA, USA: The MIT Press; 2004.

[13] Tauler R. Multivariate curve resolution applied to second order data. Chemometr Intell Lab Syst 1995;30:133–46.

[14] Tauler R, Smilde A, Kowalski B. Selectivity, local rank, three-way data analysis and ambiguity in multivariate curve resolution. J Chemometr 1995;9:31–58.

[15] Wang G, Ding Q, Hou Z. Independent component analysis and its applications in signal processing for analytical chemistry. Trends Anal Chem 2008;27(4):368–76.

[16] Al-Saegh A. Independent component analysis for separation of speech mixtures: a comparison among thirty algorithms. Iraq J Electr Electron Eng 2015;11(1):1–9.

[17] Hyvärinen A, Oja E. A fast fixed-point algorithm for independent component analysis. Neural Comput 1997;9:1483–92.

[18] Cardoso J-F, Souloumiac A. Blind beamforming for non-Gaussian signals. IEE Proc-F 1993;140(6):362–70.

[19] Rutledge DN, Jouan-Rimbaud Bouveresse D. Independent components analysis with the JADE algorithm. Trends Anal Chem 2013;50:22–32.

[20] Rutledge DN, Jouan-Rimbaud Bouveresse D. Corrigendum to "Independent components analysis with the JADE algorithm" Trends Anal Chem 2015;67:220.

[21] Bell AJ, Sejnowski TJ. An information maximization approach to blind separation and blind deconvolution. Neural Comput 1995;7:1129–59.

[22] Højen-Sørensen PA, Winther O, Hansen LK. Mean-field approaches to independent component analysis. Neural Comput 2002;14(4):889–918.

[23] Jalali-Heravi M, Parastar H, Sereshti H. Towards obtaining more information from gas chromatography-mass spectrometric data of essential oils: an overview of mean field independent component analysis. J Chromatogr A 2010;1217:4850–61.

[24] Bach FR, Jordan MI. Kernel independent component analysis. J Mach Learn Res 2002; 3:1–48.

[25] Stögbauer H, Kraskov A, Astakhov SA, Grassberger P. Least-dependent-component analysis based on mutual information. Phys Rev E 2004;70:066123-1–066123-17.

[26] Astakhov SA, Stögbauer H, Kraskov A, Grassberger P. Monte-Carlo algorithm for least dependent non-negative mixture decomposition. Anal Chem 2006;78:1620–7.

[27] Learned-Miller EG, Fisher III JW. ICA using spacings estimates of entropy. J Mach Learn Res 2003;4:1271–95.

[28] Hyvärinen A, Karhunen J, Oja E. Independent component analysis. New York: John Wiley and Sons; 2001.

[29] Khlaifi A. Estimation des sources de pollution atmosphérique par modélisation inverse; 2007. Thèse de Doctorat de l'Université Paris XII, Paris (France).

[30] Linsker R. Local synaptic learning rules suffice to maximize mutual information in a linear network. Neural Comput 1992;4:691–702.

[31] Westad F, Kermit M. Cross-validation and uncertainty estimates in independent component analysis. Anal Chim Acta 2003;490:341–54.

[32] Wu W, Massart DL, de Jong S. The kernel PCA algorithms for wide data. Part I: theory and algorithms. Chemometr Intell Lab Syst 1997;36:165–72.

[33] Wu W, Massart DL, de Jong S. Kernel-PCA algorithms for wide data. Part II: fast cross-validation and application in classification of NIR data. Chemometr Intell Lab Syst 1997;37:271–80.

[34] Barros AS, Rutledge DN. Principal components transform-partial least squares: a novel method to accelerate cross-validation in PLS regression. Chemometr Intell Lab Syst 2004;73:245–55.

[35] Barros AS, Rutledge DN. Segmented principal components transform-principal component analysis. Chemometr Intell Lab Syst 2005;78:125–37.

[36] Jouan-Rimbaud Bouveresse D, Moya-González A, Ammari F, Rutledge DN. Two novel methods for the determination of the number of components in independent components analysis models. Chemometr Intell Lab Syst 2012;112:24–32.

[37] Monakhova YB, Astakhov SA, Kraskov A, Mushtakova SP. Independent components in spectroscopic analysis of complex mixtures. Chemometr Intell Lab Syst 2010;103:108–15.

[38] Rutledge DN, Barros AS. The Durbin-Watson statistic as a morphological estimator of information content. Anal Chim Acta 2002;446:279–94.

[39] El Ghaziri A, Qannari EM. Measures of association between two datasets: application to sensory data. Food Qual Prefer 2015;40:116–24.

[40] Escoufier Y. Le traitement des variables vectorielles. Biometrics 1973;29:751–60.

[41] Robert P, Escoufier Y. A unifying tool for linear multivariate statistical methods: the Rv coefficient. Appl Stat 1976;25:257–65.

[42] Kassouf A, Ruellan A, Jouan-Rimbaud Bouveresse D, Rutledge DN, Domenek S, Maalouly J, et al. Attenuated total reflectance-mid infrared spectroscopy (ATR-MIR) coupled with independent components analysis (ICA): a fast method to determine plasticizers in polylactide (PLA). Talanta 2016;147:569–80.

[43] Chen J, Wang XZ. A new approach to near-infrared spectral data analysis using independent component analysis. J Chem Inf Comput Sci 2001;41:992–1001.

[44] Næs T, Martens H. Principal component regression in NIR analysis: viewpoints, background details and selection of components. J Chemometr 1988;2:155–67.

[45] Geladi P, Kowalski BR. Partial least-squares regression: a tutorial. Anal Chim Acta 1986;185:1–17.

[46] Shao X, Wang W, Houa Z, Cai W. A new regression method based on independent component analysis. Talanta 2006;69:676–80.

[47] Gustafsson MG. Independent component analysis yields chemically interpretable latent variables in multivariate regression. J Chem Inf Model 2005;45:1244–55.

[48] Monakhova YB, Rutledge DN, Roßmann A, Waiblinger H-U, Mahler M, Ilse M, et al. Determination of rice type by 1H NMR spectroscopy in combination with different chemometric tools. J Chemometr 2014;28:83–92.

[49] Vandeginste BGM, Massart DL, Buydens LMC, De Jong S, Lewi PJ, Smeyers-Verbeke J. Handbook of chemometrics and qualimetrics: part B. Data handling in science and technology, vol. 20B. Amsterdam, The Netherlands: Elsevier Science BV; 1998.

[50] Wold S. Pattern recognition by means of disjoint principal components models. Pattern Recogn 1976;8:127–39.

[51] Ballabio D, Consonni V. Classification tools in chemistry. Part 1: linear models. PLS-DA. Anal Methods 2013;5:3790–8.

[52] Barker M, Rayens W. Partial least squares for discrimination. J Chemometr 2003;17: 166–73.

[53] Krzanowski WJ. Principles of multivariate analysis. Oxford: Oxford University Press; 1988.

[54] Fisher RA. The use of multiple measurements in taxonomic problems. Ann Eugen 1936;7(2):179–88.

[55] Bro R. PARAFAC. Tutorial and applications. Chemometr Intell Lab Syst 1997;38:149–71.

[56] Andersson CA, Bro R. Improving the speed of multi-way algorithms: part I. Tucker3. Chemometr Intell Lab Syst 1998;42:93–103.

[57] Parastar H, Jalali-Heravi M, Tauler R. Is independent component analysis appropriate for multivariate resolution in analytical chemistry. Trends Anal Chem 2012;31:134–43.

[58] García-Ferrer A, González-Prieto E, Peña D. A conditionally heteroskedastic independent factor model with an application to financial stock returns. Int J Forecast 2012;28:70–93.

[59] Yokoi T, Yanagimoto H, Omatu S. Information filtering with extracted index words using ICA. Electron Commun Jpn 2009;92(2):1468–73.

[60] Yiang J, Gao X, Zhang D, Yang J-Y. Kernel ICA: an alternative formulation and its application to face recognition. Pattern Recogn 2005;38:1784–7.

[61] Maino D, Farusi A, Baccigalupi C, Perrotta F, Banday AJ, Bedini L, et al. All-sky astrophysical component separation with fast independent component analysis (FASTICA). Mon Not R Astron Soc 2002;334:53–68.

[62] Umeyama S. Blind deconvolution of blurred image by use of ICA. Electron Commun Jpn 2001;84(12):1–9.

[63] Meinecke FC, Harmeling S, Müller K-R. Inlier-based ICA with an application to superimposed images. Int J Imag Syst Technol 2005;15(1):48–55.

[64] Westad F. Independent component analysis and regression applied on sensory data. J Chemometr 2005;19:171–9.

[65] Li RF, Wang XZ. Dimension reduction of process dynamic trends using independent component analysis. Comput Chem Eng 2002;26(3):467–73.

[66] Albazzaz H, Wang XZ. Statistical process control charts for batch operations based on independent component analysis. Ind Eng Chem Res 2004;43:6731–41.

[67] Huang S-P, Chiu C-C. Process monitoring with ICA-based signal extraction technique and CART approach. Qual Reliab Eng Int 2009;25:631–43.

[68] Kardamakis AA, Mouchtaris A, Pasadakis N. Linear predictive spectral coding and independent component analysis in identifying gasoline constituents using infrared spectroscopy. Chemometr Intell Lab Syst 2007;89:51–8.

[69] Pasadakis N, Kardamakis AA. Identifying constituents in commercial gasoline using Fourier transform-infrared spectroscopy and independent component analysis. Anal Chim Acta 2006;578:250–5.

[70] Bugli C, Lambert P. Comparison between principal component analysis and independent component analysis in electroencephalograms modelling. Biom J 2007;49:312–27.

[71] Nam H, Yim T-G, Kee Han S, Oh J-B, Kun Lee S. Independent component analysis of ictal EEG in medial temporal lobe epilepsy. Epilepsia 2002;43(2):160–4.

[72] Urrestarazu E, Iriarte J, Alegre M, Valencia M, Viteri C, Artieda J. Independent component analysis removing artifacts in ictal recordings. Epilepsia 2004;45(9):1071–8.

[73] Hubka P, Rosík V, Ždiňák J, Tyšler M, Hulin I. Independent component analysis of electrogastrographic signals. Meas Sci Rev 2005;5(2):21–4.

[74] Hao J, Zou X, Wilson MP, Davies NP, Sun Y, Peet AC, et al. A comparative study of feature extraction and blind source separation of independent component analysis (ICA) on childhood brain tumour ^{1}H magnetic resonance spectra. NMR Biomed 2009;22:809–18.

[75] Szabo de Edenelyi F, Simonetti AW, Postma G, Huo R, Buydens LMC. Application of independent component analysis to ^{1}H MR spectroscopic imaging exams of brain tumours. Anal Chim Acta 2005;544:36–46.

[76] Boiret M, Rutledge DN, Gorretta N, Ginot Y-M, Roger J-M. Application of independent component analysis on Raman images of a pharmaceutical drug product: pure spectra determination and spatial distribution of constituents. J Pharm Biomed Anal 2014;90:78–84.

[77] Boiret M, Rutledge DN, Gorretta N, Ginot Y-M, Roger J-M. Utilisation de la microscopie Raman et des méthodes chimiométriques pour la détection de comprimés contrefaits. Spectra Anal 2014;298:72–8.

[78] Chuang Y-K, Chen S, Martin Lo Y, Tsai C-Y, Yang I-C, Chen Y-L, et al. Integration of independent component analysis with near infrared spectroscopy for rapid quantification of sugar content in wax jambu (*Syzygium samarangense Merrill & Perry*). J Food Drug Anal 2012;20(4):855–64.

[79] Chuang Y-K, Hu Y-P, Yang I-C, Delwiche SR, Martin Lo Y, Tsai C-Y, et al. Integration of independent component analysis with near infrared spectroscopy for evaluation of rice freshness. J Cereal Sci 2014;60:238–42.

[80] Masoum S, Seifi H, Ebrahimabadi EH. Characterization of volatile components in *Calligonum comosum* by coupling gas chromatography-mass spectrometry and mean field approach independent component analysis. Anal Methods 2013;5:4639–47.

[81] Rohart A, Jouan-Rimbaud Bouveresse D, Rutledge DN, Michon C. Spectrophotometric analysis of polysaccharide/milk protein interactions with methylene blue using independent components analysis. Food Hydrocoll 2015;43:769–76.

[82] Ducauze C. Chimie Analytique, analyse chimique et chimiométrie. Concepts, démarche et méthodes. Paris, France: Lavoisier Tec & Doc; 2014.

[83] Rutledge DN, Guesnier S, Lahanier C, Ducauze CJ. Classification of gallo-roman ceramics from the Bordeaux region by multidimensional statistical analysis of their chemical composition. Chemometr Intell Lab Syst 1993;19:111–7.

[84] Winning H, Larsen FH, Bro R, Engelsen SB. Quantitative analysis of NMR spectra with chemometrics. J Magn Reson 2008;190:26–32.

[85] Andersson CA, Bro R. The N-way toolbox for MATLAB, Chemometr Intell Lab Syst 2000;52(1):1–4. Downloaded from, www.models.life.ku.dk/source/nwaytoolbox.

Chapter 8

Bayesian Positive Source Separation for Spectral Mixture Analysis

D. Brie*[,1], S. Moussaoui[†], S. Miron*, C. Carteret[‡] and M. Dossot[‡]
**Centre de Recherche en Automatique de Nancy, Université de Lorraine, CNRS, Vandœuvre-lès-Nancy, France*
[†]*Ecole Centrale de Nantes, IRCCyN, UMR CNRS 6597, Nantes Cedex 3, France*
[‡]*Laboratoire de Chimie Physique et Microbiologie pour l'Environnement, Université de Lorraine, CNRS, Villers-lès-Nancy, France*
[1]*Corresponding author: david.brie@univ-lorraine.fr*

Chapter Outline

Data Handling in Science and Technology, Vol. 30. http://dx.doi.org/10.1016/B978-0-444-63638-6.00008-5
Copyright © 2016 Elsevier B.V. All rights reserved.

1 INTRODUCTION

Source separation is a statistical signal processing problem which arises in a class of applications where the observation model assumes that m measured signals, called *observations*, are mixtures of p nonobserved signals, called *sources* [1]. From n measurements of each observation signal, the linear instantaneous mixing model, in the noise-free case, can be represented by the following matrix formulation

$$\mathbf{X} = \mathbf{A}\,\mathbf{S}, \tag{1}$$

where $\mathbf{X}$ is the $(m \times n)$ data matrix, containing the m observation signals $\{x_{ik}; k=1,\ldots,n\}_{i=1}^{m}$ in its rows, $\mathbf{A}$ is a $(m \times p)$ matrix containing the mixing coefficients $\{a_{ij}; i=1,\ldots,m\}_{j=1}^{p}$ on its columns, and $\mathbf{S}$ is a $(p \times n)$ matrix containing the p source signals $\{s_{jk}; k=1,\ldots,n\}_{j=1}^{p}$ in its rows. In some applications, such as hyperspectral imaging and spectroscopy, due to their physical meaning, the source signals as well as the mixing coefficients are subject to nonnegativity constraints

$$\forall(i,j),\ \ a_{ij} \geq 0 \ \text{ and } \ \forall(j,k),\ \ s_{jk} \geq 0, \tag{2}$$

which will be denoted

$$\mathbf{A} \geq 0 \text{ and } \mathbf{S} \geq 0. \tag{3}$$

According to this model, the positive source separation goal is to find matrices $\mathbf{A}$ and $\mathbf{S}$ that jointly fulfill the mixing model (1) and satisfy the constraints (3). This problem is also called *nonnegative matrix factorization* (NMF). The first publications related to the NMF problem appeared in the field of linear algebra [2–5] where the authors concentrated their efforts to formulate conditions for the existence and the uniqueness of such a factorization. In the chemometrics community, this problem is termed as *multivariate curve resolution* [6], *mixture analysis* [7], and *factor analysis* [8], while in signal and image processing field, this problem is referred to as *positive blind source separation* [9]. In the sequel, the problem is referred to as spectral mixture analysis and we implicitly will assume a linear mixing. Nonlinear mixing may be encountered in a number of applications including hyperspectral image unmixing (see chapter 6 by Dobigeon *et al.*).

Regarding the decomposition algorithms, pioneering contributions of Refs. [10–12] proposed original algorithms to find an approximate factorization of a matrix in the case of spectroscopic data and noisy observations, by alternating nonnegative least squares estimation, in the first case, and penalized least squares estimation, in the second. More recently, Ref. [13] presented two fast algorithms dedicated to NMF. NMF and source separation with nonnegativity constraints have since remained an active research topic [9, 14–17].

However, before trying to effectively estimate the positive sources and the mixing coefficients, a key point is to answer some questions related to the model indeterminacies and to the uniqueness of the solution. Furthermore, when the solution is not unique, it is also useful to determine the range of admissible solutions. In that respect, a necessary and sufficient condition for the uniqueness of NMF has been formulated by Ref. [18] as a convex geometry problem. The most recent advances addressing the uniqueness of the NMF by the geometric approach can be found in Refs. [19–21]. It appears that the uniqueness relies on the sparsity of underlying latent variables. Also, Refs. [22, 23] have expressed some sufficient uniqueness conditions well adapted to specific applications, but they are not applicable for a general purpose application. In the particular case where only the source signals are constrained to be nonnegative and under the assumption of statistically mutually independent sources, the problem is called *nonnegative independent component analysis*. In this case, Ref. [24] has already provided an analysis of uniqueness conditions and proposed some separation algorithms [25].

In this chapter, we motivate the use of the proposed Bayesian approach by analyzing the rotational ambiguities of the NMF from a geometrical point of view. The proposed Bayesian Positive Source Separation (BPSS) [9] is particularly well adapted to difficult problems suffering from rotational ambiguities and with spectral sources having complex shapes. It consists in assigning Gamma distribution as an a priori model on pure component spectra and mixing coefficient. Three examples of successful applications of BPSS are given.

2 GEOMETRICAL INTERPRETATION OF THE NMF MODEL

2.1 Preliminaries

Definition 1. (Convex cone). A subset $\mathcal{C} \subseteq \mathbb{R}^n$ is a convex cone if

(C1) $\forall (v_1, v_2) \in \mathcal{C} \Rightarrow v_1 + v_2 \in \mathcal{C}$,
(C2) $\forall v \in \mathcal{C}, \lambda \in \mathbb{R}_+ \Rightarrow \lambda v \in \mathcal{C}$.

Definition 2. (Simplicial cone). The simplicial cone generated by the finite set of independent vectors $\{\mathbf{s}_1, \ldots, \mathbf{s}_p\} \in \mathbb{R}^n$ is the set

$$\mathcal{C}_\mathcal{S} = \{\mathbf{x} | \mathbf{x} = \sum_{j=1}^{p} a_j \mathbf{s}_j, a_j > 0\}. \tag{4}$$

The set of vectors $\{\mathbf{s}_1, \ldots, \mathbf{s}_p\}$ is termed as a frame of the simplicial cone and p represents the dimension of the cone (noted $p = \dim \mathcal{C}$). Gathering the vectors into the matrix $\mathbf{S} = \left[\mathbf{s}_1 \cdots \mathbf{s}_p\right]^{\mathrm{T}}$, we may write $\mathcal{C}_\mathcal{S} = \text{cone}(\mathbf{S})$.

Thus, the rows of the matrix $\mathbf{X} = \mathbf{A}\ \mathbf{S}$, with $\mathbf{A} \geq 0$, $\mathbf{S} \geq 0$, may be considered as vectors belonging to the simplicial cone $\mathcal{C}_\mathcal{S}$: $\mathbf{x}_i \in \mathcal{C}_\mathcal{S}$.

2.2 Chen's Necessary and Sufficient Uniqueness Condition

The decomposition of $\mathbf{X}=\mathbf{A}\,\mathbf{S}, \text{with}\mathbf{A}\geq 0, \mathbf{S}\geq 0,$ states that each row of $\mathbf{X}$ satisfies $\mathbf{x}_i \in \mathcal{C}_S = \text{cone}(\mathbf{S})$. Denoting $\mathcal{K}$ the convex hull of the data matrix $\mathbf{X}$, the necessary and sufficient condition for the uniqueness of the nonnegative factorization is given by the following theorem stated by Ref. [18].

Theorem 1. *The decomposition of* $\boldsymbol{X}$ *according to*

$$\boldsymbol{X}=\boldsymbol{A}\,\boldsymbol{S} \quad \text{with} \quad \boldsymbol{A}\geq 0, \boldsymbol{S}\geq 0, \tag{5}$$

is unique if and only if the simplicial cone $\mathcal{C}_S$, such as $\mathcal{K}\subset\mathcal{C}_S$, is unique.

The key point is that this theorem does not provide any numerical tool to check whether a particular matrix admits a unique nonnegative decomposition or not.

2.3 A Necessary Uniqueness Conditions

Consider the case of p source signals and let

$$\mathbf{A}=\begin{bmatrix}\mathbf{a}_1 & \cdots & \mathbf{a}_p\end{bmatrix} \quad \text{and} \quad \mathbf{S}=\begin{bmatrix}\mathbf{s}_1 & \cdots & \mathbf{s}_p\end{bmatrix}^{\mathrm{T}}, \tag{6}$$

where $\{\mathbf{a}_j\}_{j=1}^{p}$ are m-dimensional vectors and $\{\mathbf{s}_j\}_{j=1}^{p}$ are n-dimensional vectors. For convenience, we will note $\mathbf{a}_j(\ell)$ the ℓth element of vector $\mathbf{a}_j$ and similarly, $\mathbf{s}_j(k)$ the kth element of vector $\mathbf{s}_j$. We give here the necessary condition for having a unique NMF [19, 26] which links the uniqueness of the NMF model to the sparsity of the latent variables.

Proposition 1. *If the decomposition of* $\boldsymbol{X}$ *into* $\boldsymbol{A}$ *and* $\boldsymbol{S}$ *according to*

$$\boldsymbol{X}=\boldsymbol{A}\,\boldsymbol{S} \text{ with } \boldsymbol{A}\geq 0, \boldsymbol{S}\geq 0, \tag{7}$$

is unique, then the following conditions are satisfied

(A1) $\forall\, j_1 \neq j_2,\ \exists\, k$ such that $\mathbf{s}_{j_1}(k)=0, \mathbf{s}_{j_2}(k)>0,$
(A2) $\forall\, j_1 \neq j_2,\ \exists\, \ell$ such that $\mathbf{a}_{j_1}(\ell)=0, \mathbf{a}_{j_2}(\ell)>0.$

This proposition means that, to have a unique NMF, both the sources and the mixing coefficient should have at least a minimum number of zeros.

2.4 Convex Cone and Uniqueness

Now, following Ref. [27], we provide a geometrical interpretation of the NMF. For simplicity and without loss of generality, we restrain our illustrations to the case of three sources. Each row of data matrix $\mathbf{X}$ can be seen as a point in a n-dimensional space. As the data are nonnegative, every such point belongs to the positive orthant $\mathbb{R}^n_+$ of $\mathbb{R}^n$. The NMF model implies that all the rows of **X** are in the simplicial cone generated by the columns of **S**. Thus, the admissible solutions for $(\mathbf{s}_1, \mathbf{s}_2, \mathbf{s}_3)$ are confined in the three-dimensional subspace obtained by intersecting $\mathbb{R}^n_+$ with span(**S**). Let $\{\mathbf{e}_1, \mathbf{e}_2, \mathbf{e}_3\}$ be a canonical unitary basis of $\mathbb{R}^n_+ \cap \text{span}(\mathbf{S})$ (the vectors $\mathbf{e}_1, \mathbf{e}_2, \mathbf{e}_3$ are contained in the canonical planes of $\mathbb{R}^n$), then all the data lie inside $\mathcal{C}(\mathbf{s}_1, \mathbf{s}_2, \mathbf{s}_3)$ (Fig. 1).

Let us now consider a mapping (affine projection) of the three-dimensional space to the plane (1,1,1). By this mapping a simplicial cone of dimension p is mapped to a simplex of dimension $p - 1$. This affine projection is illustrated in Fig. 2 and can be seen as imposing a sum to one constraint to the mixing coefficient provided that the sources are normalized to one w.r.t. a given norm. In all what follows, we will consider that the data have been projected in this plane.

Now, we give a geometrical interpretation of the two necessary conditions for having a unique nonnegative factorization of the data matrix. In Fig. 3, the

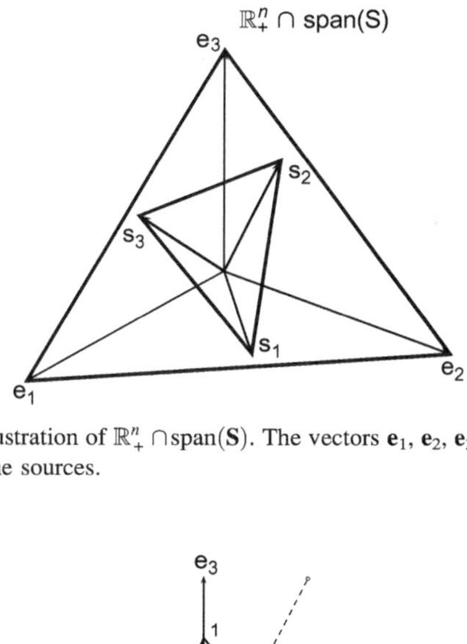

FIG. 1 Geometrical illustration of $\mathbb{R}^n_+ \cap \text{span}(\mathbf{S})$. The vectors $\mathbf{e}_1, \mathbf{e}_2, \mathbf{e}_3$ form a canonical unitary basis and $\mathbf{s}_1, \mathbf{s}_2, \mathbf{s}_3$ are the sources.

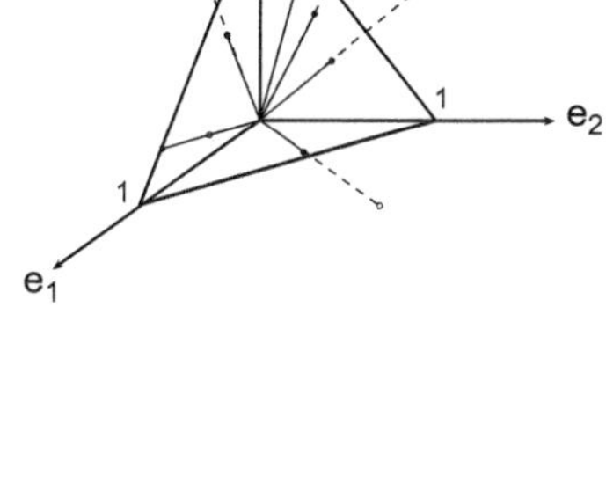

FIG. 2 Affine projection.

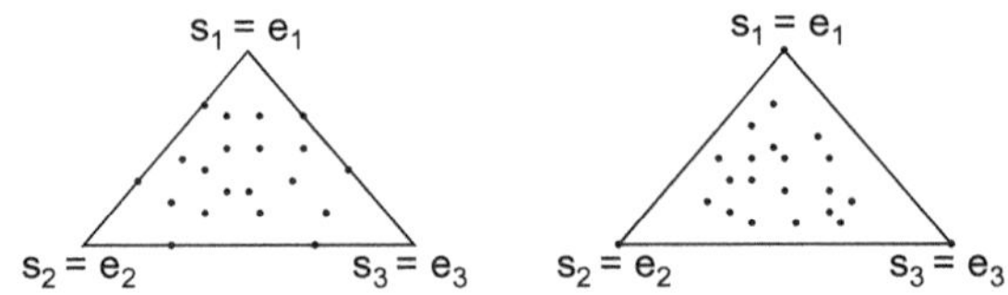

FIG. 3 Geometrical interpretation of two NMF uniqueness necessary conditions.

triangle is representing the simplex of dimension $p - 1$ and the dots correspond to the affine projection of the data. The case on the left hand side corresponds to the necessary condition of Ref. [28] which imposes that the sources should be equal to the vector of the canonical unitary basis and that there are at least $p - 1$ points on each facet of the simplex. The case depicted on the right hand side corresponds to the case in which the sources are still equal to the vector of the canonical unitary basis and the data should include each of the p pure source. This is a limit case of Donoho's sufficient condition obtained by moving points on the facets of the simplex toward the simplex vertices.

2.5 Reducing the Set of Admissible Solutions

Basically, the uniqueness of the NMF of the data matrix $\mathbf{X}$ is ensured if the (projected) data are sufficiently filling the simplex defined by $(\mathbf{e}_1,\cdots,\mathbf{e}_p)$. If that there are not enough data to fully cover it, then any simplicial cone including the data yields an admissible solution, the sources corresponding to the vertices of this simplicial cone. This is illustrated by Fig. 4, in which the white triangle is representing the true (unknown) solution for which a maximum number of points are on the facets of the convex cone. Any triangle including the white triangle defines an admissible solution and the domain of possible solutions is represented by the hatched area.

A fundamental question in source separation is the need to state further information, in addition to nonnegativity, allowing to define the best solution among the admissible ones. By incorporating additional assumptions or constraints, like unimodality and full additivity (closure) [29], the range of admissible solutions will be reduced to a smaller set. On the other hand, a penalized least squares or a Bayesian estimation method leads to one particular solution fulfilling at best the a priori distribution model.

We present, in the sequel, the main ideas of the approaches aiming at reducing the set of admissible solutions. These approaches can be roughly classified into four main categories:

- the geometrical approaches, among which the most representative is the minimum volume simplex approach [21, 30];
- the hard constraint approaches, such as unimodality and full additivity (closure) [29];

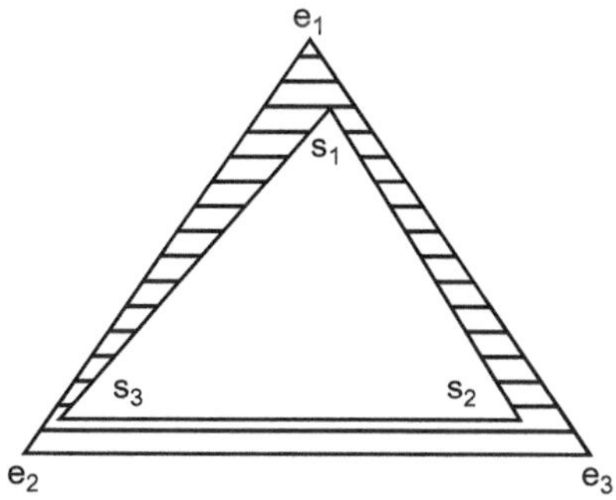

FIG. 4 Geometrical interpretation of the admissible solutions domain.

- the regularization (soft constraint) approaches which consist in formulating the problem as a compound criterion minimization problem. Among these approaches, representatives approaches are the positive matrix factorization (PMF) [11] and NNSC [14, 16]. The BPSS [9, 31] addresses the problem from a probabilistic point of view. This approach is detailed in the next section;
- the data augmentation approach which consists in using multiple measurement modalities. Representative approaches are Refs. [27, 32].

3 THE BAYESIAN POSITIVE SOURCE SEPARATION ALGORITHM

We turn our attention to spectral mixture analysis and we now consider the noisy NMF. Using matrix notations this model can be expressed as

$$\mathbf{X} = \mathbf{A}\,\mathbf{S} + \mathbf{E}, \tag{8}$$

where the row vectors of the $(m \times n)$ data matrix $\mathbf{X}$ contain the m measured spectra, and $\mathbf{A}$ is the $(m \times p)$ mixing matrix, with its column vectors representing the mixing coefficient of each pure component. $\mathbf{S}$ is a $(p \times n)$ matrix with its row vectors containing the p pure component spectra and $\mathbf{E}$ is the $(m \times n)$ additive noise matrix.

Spectral mixture analysis is an ill-posed inverse problem, since its solution is not unique. To obtain a physically relevant solution, any available knowledge on the data must be used or additional assumptions should be made. This inverse problem can be successfully addressed using Bayesian estimation theory, where the available information is encoded through the assignment of a priori distributions on the pure component spectra and concentration profiles. The formulation of source separation and factor analysis using Bayesian estimation theory has been suggested in Refs. [33–36] and used in signal and image processing problems [37, 38]. However, its application to the separation of spectral mixture data was originally addressed in Refs. [9, 39, 40].

Before going further, let us recall the Bayesian approach to mixture analysis which consists in assigning prior distributions on source spectra and mixing coefficients, to account for the available knowledge on the observed phenomenon. The probabilistic formulation of the problem enables the use of stochastic simulation tools to draw samples from the posterior distribution, which are used to evaluate various types of estimators. In this section, the BPSS algorithm [9] is presented, and its link with the *positive matrix factorization* method [11] is emphasized.

3.1 A Bayesian Approach to Spectral Mixture Analysis

The main idea of the Bayesian approach to mixture analysis is to encode the available knowledge on the pure spectra and concentration profiles through the assignment of prior distributions $p(\mathbf{S})$ and $p(\mathbf{A})$, which according to Bayes' theorem, and using the likelihood $p(\mathbf{X}|\mathbf{S}, \mathbf{A})$, leads to the a posteriori density

$$p(\mathbf{S},\mathbf{A}|\mathbf{X}) = p(\mathbf{X}|\mathbf{S},\mathbf{A}) \times p(\mathbf{S}) \times p(\mathbf{A}) \div p(\mathbf{X}), \tag{9}$$

where the independence between $\mathbf{A}$ and $\mathbf{S}$ is assumed. Since $p(\mathbf{X})$ is a normalization constant, one can write

$$p(\mathbf{S},\mathbf{A}|\mathbf{X}) \propto p(\mathbf{X}|\mathbf{S},\mathbf{A}) \times p(\mathbf{S}) \times p(\mathbf{A}). \tag{10}$$

This a posteriori density combines explicitly the a priori knowledge and assumptions on the pure component spectra and concentrations with the information coming from the measured data.

To get estimates of pure component spectra and concentrations, various Bayesian estimators [41] can be applied

- the *joint maximum a posteriori* (JMAP) estimates are obtained by

$$\left(\hat{\mathbf{S}}_{\mathrm{JMAP}}, \hat{\mathbf{A}}_{\mathrm{JMAP}}\right) = \arg\max_{\mathbf{S},\mathbf{A}} p(\mathbf{S},\mathbf{A}|\mathbf{X}). \tag{11}$$

This maximization is equivalent to a joint minimization of the corresponding criterion, defined by

$$J(\mathbf{S},\mathbf{A}) = -\log p(\mathbf{S},\mathbf{A}|\mathbf{X}), \tag{12}$$

and depends not only on the measured data but also on the prior distributions of pure spectra and concentration profiles

$$J(\mathbf{S},\mathbf{A}) = -\log p(\mathbf{X}|\mathbf{S},\mathbf{A}) - \log p(\mathbf{S}) - \log p(\mathbf{A}). \tag{13}$$

Each prior model leads to a particular regularization criterion. This can be seen as a penalty function design through a probabilistic formulation of the prior knowledge. This is known as a Bayesian interpretation of penalized least squares estimation methods.

- the *marginal maximum a posteriori* (MMAP) estimates are obtained by integrating (marginalizing) with respect to **S** or **A** and maximizing the resulting posterior marginal distribution $p(\mathbf{A}|\mathbf{X})$ or $p(\mathbf{S}|\mathbf{X})$.
- the *marginal posterior mean* (MPM) estimates are obtained from the mean of the marginal posterior distributions $p(\mathbf{A}|\mathbf{X})$ and $p(\mathbf{S}|\mathbf{X})$. In practice, the marginalization is generally performed using MCMC methods and the posterior means are calculated from the simulated Markov chains.

In the following, the proposed Bayesian inference for the analysis of spectral mixtures and its related estimation algorithm are presented. The mathematical computation details can be found in Ref. [9].

3.2 Noise Distribution and Likelihood

Each noise sequence $\{E_{(i,k)}, k=1,\ldots,n\}$ is assumed Gaussian, zero mean, independent and identically distributed with variance σ_i^2. Moreover, all the noise sequences are assumed mutually independent. Thus,

$$p(\mathbf{E}|\theta_1) = \prod_{i=1}^{m}\prod_{k=1}^{n} \mathcal{N}\left(E_{(i,k)}; 0, \sigma_i^2\right), \tag{14}$$

where $\theta_1 = \left[\sigma_1^2, \ldots, \sigma_m^2\right]^{\mathrm{T}}$ and $\mathcal{N}(x; \mu, v)$ denotes the normal distribution for the variable x with mean μ and variance v

$$\mathcal{N}(x;\, \mu, v) = \frac{1}{\sqrt{2\pi v}} \exp\left[-\frac{1}{2v}(x-\mu)^2\right]. \tag{15}$$

Using this noise distribution and the mixing model, the expression of the likelihood is obtained as

$$p(\mathbf{X}|\mathbf{S}, \mathbf{A}, \theta_1) \propto \prod_{i=1}^{m}\prod_{k=1}^{n} \left(\frac{1}{\sigma_i}\right)^n \exp\left[-\frac{1}{2\sigma_i^2}\left(X_{(i,k)} - [\mathbf{AS}]_{(i,k)}\right)^2\right]. \tag{16}$$

3.3 Prior Distributions of Pure Spectra and Mixing Coefficients

The pure component spectra are considered mutually independent and identically distributed, and each pure spectrum $\{S_{(j,k)}, k=1,\ldots,n\}$ is assumed distributed as Gamma distribution of parameters $(\alpha_j,\ \beta_j)$. These parameters are considered constant for each spectrum over all the spectral band but may differ from one pure spectrum to another. Thus, the prior distribution of the pure spectra is

$$p(\mathbf{S}|\theta_2) = \prod_{j=1}^{p}\prod_{k=1}^{n} \mathcal{G}(S_{(j,k)}; \alpha_j,\ \beta_j), \tag{17}$$

where $\mathcal{G}(z; \alpha, \beta)$ represents the Gamma density for the variable z with parameters α and β. The vector $\theta_2 = \left[\alpha_1, \ldots, \alpha_p, \beta_1, \ldots, \beta_p\right]^{\mathrm{T}}$ contains the parameters of all the prior distributions.

The Gamma density is expressed as

$$\mathcal{G}(z; \alpha, \beta) = \begin{cases} \dfrac{\beta^\alpha}{\Gamma(\alpha)} z^{\alpha-1} \exp[-\beta z], & \text{for } z \geq 0, \\ 0, & \text{for } z < 0, \end{cases} \tag{18}$$

where $\Gamma(\alpha)$ is the Gamma function. The choice of this distribution is motivated by the fact that:

- it takes into account explicitly the nonnegativity since $p(z < 0) = 0$;
- due to its two parameters α and β, this density can be used to fit various distributions and has the Khi-2 and exponential distributions as particular cases [42]. Fig. 5 illustrates three typical shapes of the Gamma density. Case (a) is well adapted to encode sparsity (concentration of the distribution in the neighborhood of zero), while case (b) corresponds, for example, to the case of a sparse pure spectrum superimposed on a background, typically consisting in the tail of a wide peak. Case (c) shows that the Gamma distribution can also lead to a noninformative positive prior.

Each jth column of the mixing matrix is also assumed distributed as a Gamma density of parameters (γ_j, δ_j). These parameters are considered constant for each concentration profile but may differ from one profile to another. The prior distribution of mixing coefficients is then given as

$$p(\mathbf{A}|\theta_3) = \prod_{i=1}^{m}\prod_{j=1}^{p} \mathcal{G}(A_{(i,j)};\ \gamma_j, \delta_j), \tag{19}$$

where $\theta_3 = \left[\gamma_1, \ldots, \gamma_p, \delta_1, \ldots, \delta_p\right]^{\mathrm{T}}$.

3.4 Posterior Density and Resulting Criterion

Noting $\theta = [\theta_1, \theta_2, \theta_3]$ and according to Eq. (10) one can deduce the expression of the posterior law

$$\begin{aligned} p(\mathbf{S}, \mathbf{A}|\mathbf{X}, \theta) \propto & \prod_{k=1}^{n}\prod_{i=1}^{m} \mathcal{N}(E_{(i,k)}; 0, \sigma_i^2) \\ & \times \prod_{k=1}^{n}\prod_{j=1}^{p} \mathcal{G}(S_{(j,k)}; \alpha_j, \beta_j) \times \prod_{i=1}^{m}\prod_{j=1}^{p} \mathcal{G}(A_{(i,j)}; \gamma_j, \delta_j). \end{aligned} \tag{20}$$

The corresponding criterion takes the form

$$J_1(\mathbf{S}, \mathbf{A}|\theta) = Q(\mathbf{S}, \mathbf{A}|\theta_1) + R_S(\mathbf{S}|\theta_2) + R_A(\mathbf{A}|\theta_3), \tag{21}$$

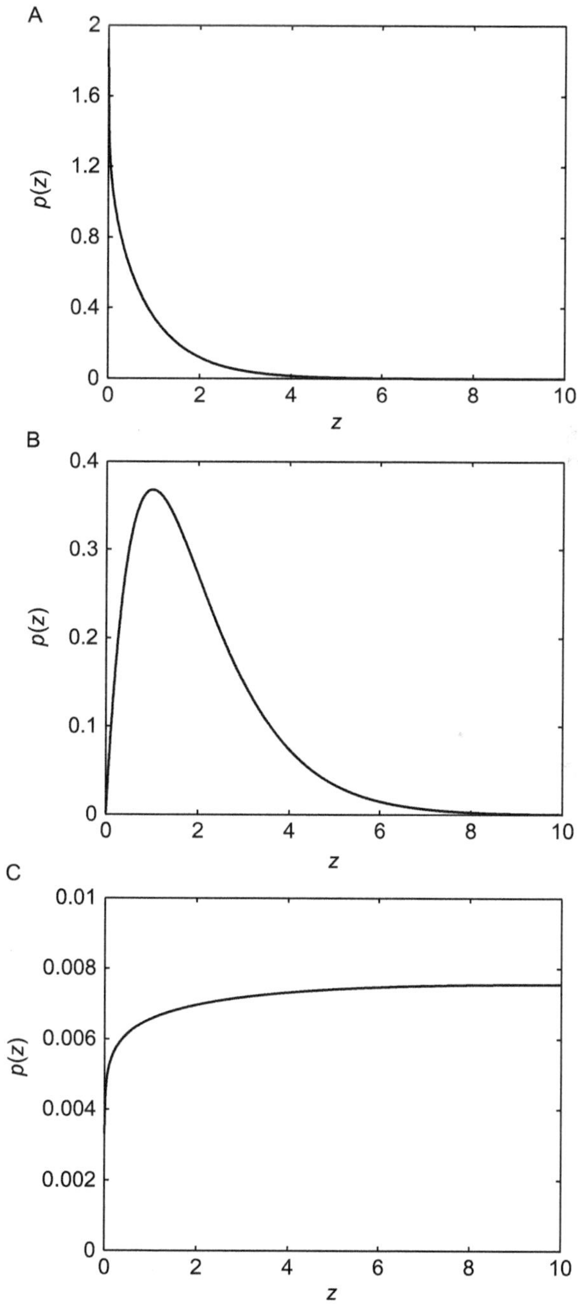

FIG. 5 Illustrations of the Gamma distribution shapes according to the values of its two parameters: (A) $\alpha = 0.9$, $\beta = 1$; (B) $\alpha = 2$, $\beta = 1$; and (C) $\alpha = 1.1$, $\beta = 0.1$.

where the terms Q, R_S, and R_A are given by

$$\begin{cases} Q(\mathbf{S},\mathbf{A}|\theta_1) = \sum_{i=1}^{m}\sum_{k=1}^{n}\frac{1}{2\sigma_i^2}\left(X_{(i,k)} - [\mathbf{AS}]_{(i,k)}\right)^2, \\ R_S(\mathbf{S}|\theta_2) = \sum_{j=1}^{p}\sum_{k=1}^{n}(1-\alpha_j)\log S_{(j,k)} + \sum_{j=1}^{p}\sum_{k=1}^{n}\beta_j S_{(j,k)}, \\ R_A(\mathbf{A}|\theta_3) = \sum_{j=1}^{p}\sum_{i=1}^{m}(1-\gamma_j)\log A_{(i,j)} + \sum_{j=1}^{p}\sum_{i=1}^{m}\delta_j A_{(i,j)}. \end{cases} \tag{22}$$

The first term Q can be seen as a data fitting measure, while the last two terms are regularization criteria that prevent negative values of **A** and **S**.

3.5 Bayesian Formulation of the PMF Criterion

In the PMF method [12], the minimized criterion is given by

$$\begin{aligned} J_2(\mathbf{S},\mathbf{A}|\theta) = & \sum_{i=1}^{m}\sum_{k=1}^{n}\frac{1}{2\sigma_{(i,k)}^2}\left(X_{(i,k)} - [\mathbf{AS}]_{(i,k)}\right)^2 \\ & -\alpha\sum_{j=1}^{p}\sum_{k=1}^{n}\log S_{(j,k)} + \beta\sum_{j=1}^{p}\sum_{k=1}^{n}S_{(j,k)}^2 \\ & -\gamma\sum_{j=1}^{p}\sum_{i=1}^{m}\log A_{(i,j)} + \delta\sum_{j=1}^{p}\sum_{i=1}^{m}A_{(i,j)}^2, \end{aligned} \tag{23}$$

where (α, β, γ, δ) are positive regularization parameters. The regularization terms (applied to both spectral sources and concentrations) are of the form

$$f(z) = -\alpha\log z + \beta z^2, \quad \text{with } z \geq 0 \tag{24}$$

and correspond to assign to the variable z a prior distribution defined as

$$p(z|\alpha,\beta) = K(\alpha,\beta)\exp[-f(z)] = K(\alpha,\beta)\, z^{\alpha}\exp\left[-\beta z^2\right] \quad \text{with } z \geq 0, \tag{25}$$

where $K(\alpha, \beta) = 2\beta^{(\alpha+1)/2}/\Gamma((\alpha + 1)/2)$. It can be noted that this distribution corresponds to assuming a Gamma prior distribution on z^2 with parameters $((\alpha + 1)/2, \beta)$. Fig. 6 shows three typical shapes of this distribution. One can particularly note a similarity with the Gamma distribution profiles shown in Fig. 5. To make a comparison between the PMF method and our proposed approach, the following discussion can be pointed out:

(i) The case $\alpha = 0$ corresponds to assigning an a priori truncated Gaussian distribution. Contrarily to the Gamma distribution, when $\alpha > 0$ this distribution is not appropriate to encode the sparsity since its mode is greater than zero.

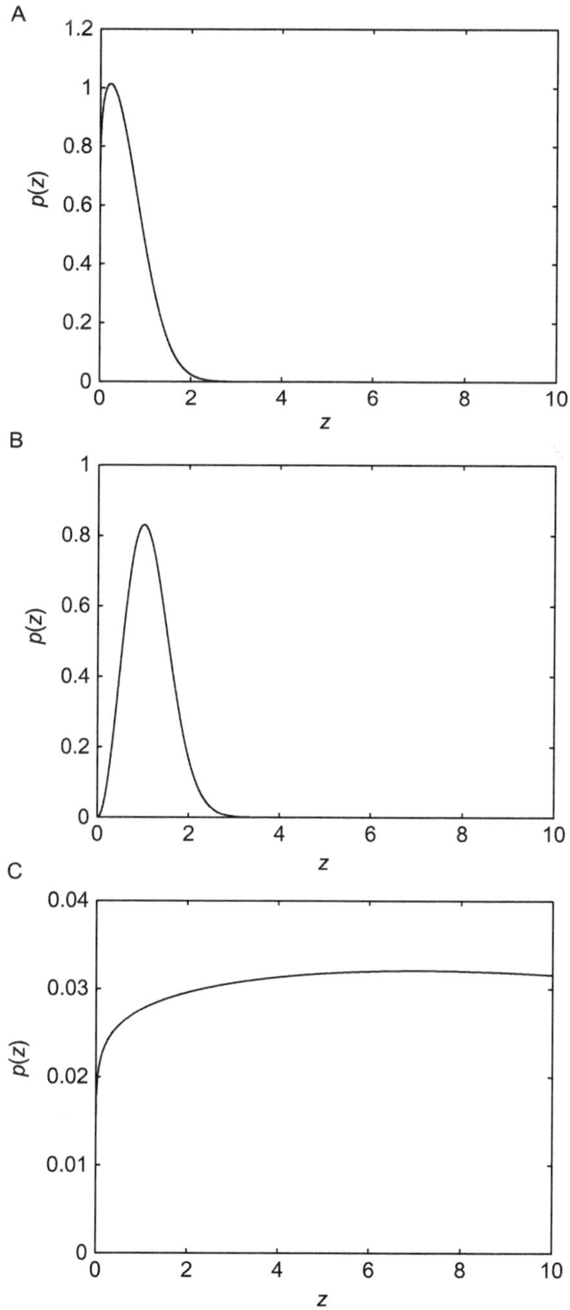

FIG. 6 Illustrations of the a priori distribution associated to the PMF criterion for (A) $\alpha = 0.1, \beta = 1$; (B) $\alpha = 2, \beta = 1$; and (C) $\alpha = 0.1, \beta = 0.001$.

(ii) In the PMF method, the values of the hyperparameters should be fixed a priori and are considered equal for all the pure spectra and for all the concentrations, which is nothing but assuming that all the pure spectra and concentration profiles have the same distribution. In the proposed inference these parameters may differ for different components and due to the probabilistic formulation, all the hyperparameters are estimated jointly with the pure component spectra and concentration profiles.

(iii) Due to the nonconvexity of the two criteria J_1 and J_2, when the hyperparameter estimation is added, deterministic optimization methods such as gradient or Newton–Raphson algorithms need either a good starting point or multiple random initializations in order to reach the global minimum.

To handle these optimization problems, the hyperparameters are included in the Bayesian model, and stochastic simulation tools are used to calculate the Bayesian estimators.

3.6 Inference of the Hyperparameters

In practice, the parameters of the Gamma distributions, as well as the noise variances are not known a priori. Thus, for an unsupervised processing, they must also be considered in the inference. Thus, by using Bayes' theorem and assigning appropriate a priori distributions to these hyperparameters, the whole a posteriori distribution, including the hyperparameters, is expressed as

$$p(\mathbf{S},\mathbf{A},\theta|\mathbf{X}) \propto p(\mathbf{S},\mathbf{A}|\mathbf{X},\theta) \times p(\theta). \tag{26}$$

The joint estimation of the pure spectra, the mixing coefficients, and the hyperparameters is then performed from this a posteriori distribution. However, this inference also needs the definition of appropriate priors for the hyperparameters, that is choosing a priori distributions allowing to set the range of admissible values of the hyperparameters, which is much easier than specifying a specific value of each parameter.

3.6.1 *Priors on the Noise Variances*

Each noise variance is positive; therefore, an appropriate distribution is to assign an a priori Gamma distribution to $1/\sigma_i^2$ with parameters $(2, \epsilon)$ where ϵ is set to a small value (10^{-3}). This distribution is chosen because it encodes the positivity and leads to simpler computations of the a posteriori distribution $p(\theta_1|\mathbf{S}, \mathbf{A}, \mathbf{X})$ [41].

3.6.2 *Priors on the Gamma Distribution Parameters*

The parameters of the Gamma distributions associated to the pure spectra and concentration profiles are assumed as a priori Gamma distributed with parameters $(2, \epsilon)$, where ϵ is set to a small value (10^{-3}). Similar to the noise variances, this distribution is chosen since it encodes the positivity of the parameters and leads to simpler computations of the a posteriori distributions $p(\theta_2|\mathbf{S})$ and $p(\theta_3|\mathbf{A})$. Note also that these prior distributions are not very informative (quasi-uniform shape) since their mode is $1/\epsilon$ and their variance is equal to $2/\epsilon^2$.

3.7 Estimation via Markov Chain Monte Carlo Methods

Markov Chain Monte Carlo (MCMC) is a class of stochastic simulation tools for generating random variables from univariate or multivariate probability distribution functions [43, 44]. These methods are extensively documented in the statistical literature (see Refs. [45–47] and the references therein). The main objective of MCMC methods is to simulate a stationary ergodic Markov chain whose samples asymptotically follow the posterior density $p(\mathbf{S},\mathbf{A},\theta|\mathbf{X})$. Estimates of the pure component spectra and the mixing coefficients are then calculated from the simulated Markov chains. In the sequel, the main steps of the MCMC algorithm for sampling the posterior distribution (26) are given.

3.7.1 *Posterior Density Simulation Using MCMC Methods*

To simulate $p(\mathbf{S},\mathbf{A},\theta|\mathbf{X})$, at each new iteration r of the algorithm, the main steps consist in simulating

1. the pure spectra $\mathbf{S}^{(r+1)}$ from

$$p\left(\mathbf{S}|\mathbf{X},\mathbf{A}^{(r)},\theta^{(r)}\right)\propto p\left(\mathbf{X}|\mathbf{S},\mathbf{A}^{(r)},\theta^{(r)}\right)\times p\left(\mathbf{S}|\theta^{(r)}\right);$$

2. the mixing coefficients $\mathbf{A}^{(r+1)}$ from

$$p\left(\mathbf{A}|\mathbf{X},\mathbf{S}^{(r+1)},\theta^{(r)}\right)\propto p\left(\mathbf{X}|\mathbf{S}^{(r+1)},\mathbf{A},\theta^{(r)}\right)\times p\left(\mathbf{A}|\theta^{(r)}\right);$$

3. the noise variances $\theta_1^{(r+1)}$ from

$$p\left(\theta_1|\mathbf{X},\mathbf{S}^{(r+1)},\mathbf{A}^{(r+1)}\right)\propto p\left(\mathbf{X}|\mathbf{S}^{(r+1)},\mathbf{A}^{(r+1)},\theta_1\right)\times p(\theta_1);$$

4. the source hyperparameters $\theta_2^{(r+1)}$ from

$$p\left(\theta_2|\mathbf{S}^{(r+1)}\right)\propto p\left(\mathbf{S}^{(r+1)}|\theta_2\right)\times p(\theta_2);$$

5. the mixing coefficient hyperparameters $\theta_3^{(r+1)}$ from

$$p\left(\theta_3|\mathbf{A}^{(r+1)}\right)\propto p\left(\mathbf{A}^{(r+1)}|\theta_3\right)\times p(\theta_3).$$

All the stochastic simulation steps including the expressions of the conditional posterior distributions and their simulation techniques are detailed in Ref. [9], where this method is termed as *Bayesian Positive Source Separation*.

3.7.2 The Marginal Posterior Mean Estimator

The Gibbs sampler [44] is iterated in such a way to simulate M samples $\left\{\mathbf{S}^{(r)}, \mathbf{A}^{(r)}, \theta^{(r)},\right\}_{r=1}^{M}$ from the posterior distribution (26). The first L samples corresponding to the burn-in of the Markov chains are discarded, and therefore, the Monte Carlo approximation for MPM estimation is achieved by

$$\hat{\mathbf{z}} \approx \frac{1}{M-L+1} \sum_{r=L+1}^{M} \mathbf{z}^{(r)}, \tag{27}$$

where $\mathbf{z} \in \{\mathbf{S}, \mathbf{A}, \theta\}$.

4 APPLICATIONS

In this section we present three successful applications of the BPSS algorithm in chemometrical analysis of spectral data obtained under varying conditions (type of experiments, type of spectroscopy). For each data set, we precise the experimental setup and highlight the key features making BPSS an effective method in the considered application.

4.1 Near Infrared Spectra of a Mixture of Alcanes

The goal of this experiment was to validate the proposed BPSS using actual near infrared (NIR) spectroscopy measurements. Three known chemical species (cyclopentane, cyclohexane, and n-pentane) are mixed experimentally with specified proportions. These species have been chosen for two main reasons. First, their available spectra in the NIR frequency band are highly overlapping and as a consequence are highly correlated. This precludes the use of standard nonnegative ICA methods [25] to achieve the separation. Second, being inert solvents, these species do not interact when they are mixed, guaranteeing that no new component appears. Thus, the number of sources as well as their concentrations in the mixtures is exactly known. In addition, their individual spectra can be (and have been) measured separately. Fig. 7 shows the pure spectra of the chemical species and their concentration profiles.

To assess the spatial correlation of the pure spectra, we calculated their empirical covariance matrix

$$\hat{\mathbf{R}}_{\mathrm{s}} = \begin{bmatrix} 1.00 & 0.76 & 0.84 \\ 0.76 & 1.00 & 0.86 \\ 0.84 & 0.86 & 1.00 \end{bmatrix}, \tag{28}$$

which shows that the available samples of the pure spectra present a high spectral correlation, so ICA methods will lead to an unacceptable solution.

Fig. 8 compares both the estimated and the true sources in the more challenging bands (3500–4500 and 5000–6000 cm^{-1}), where the peaks of the

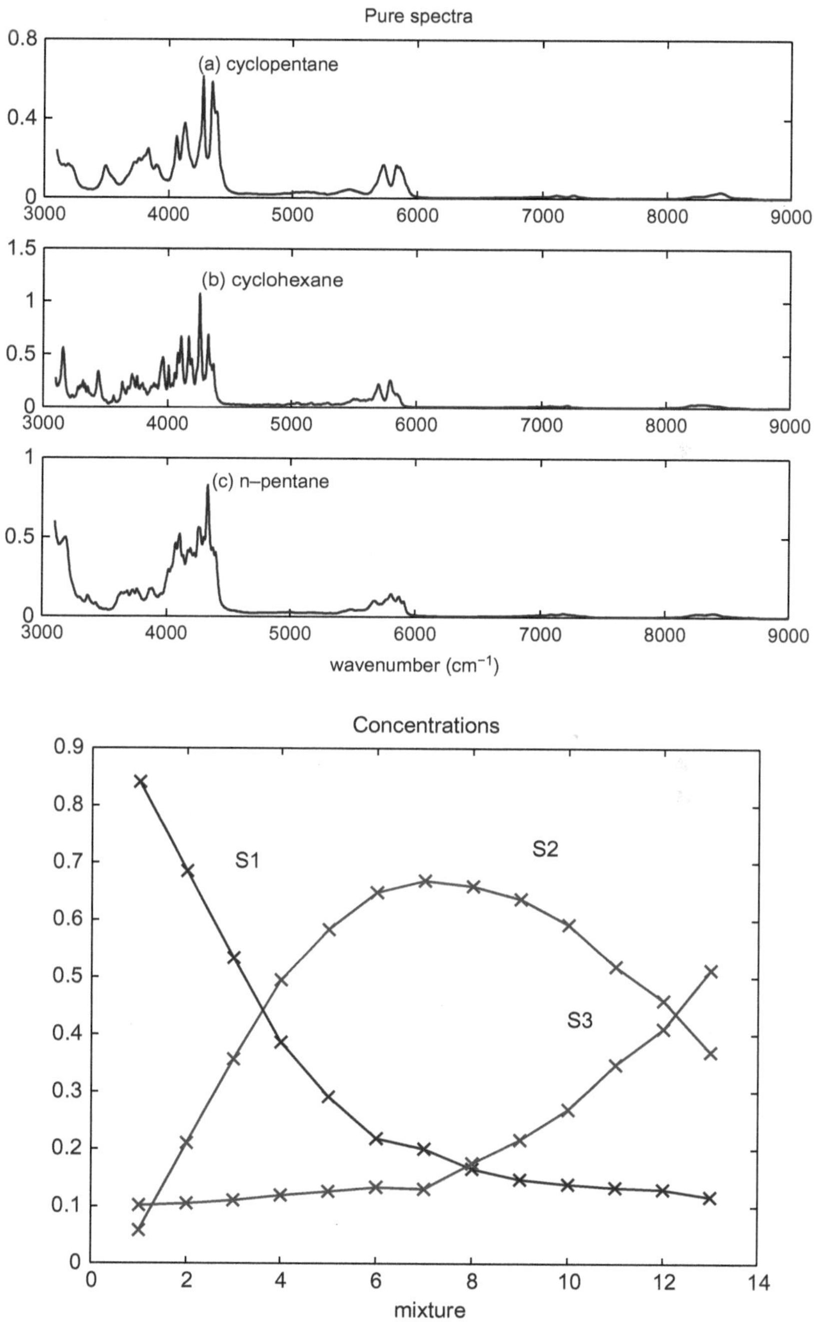

FIG. 7 Constituent spectra and concentration profiles.

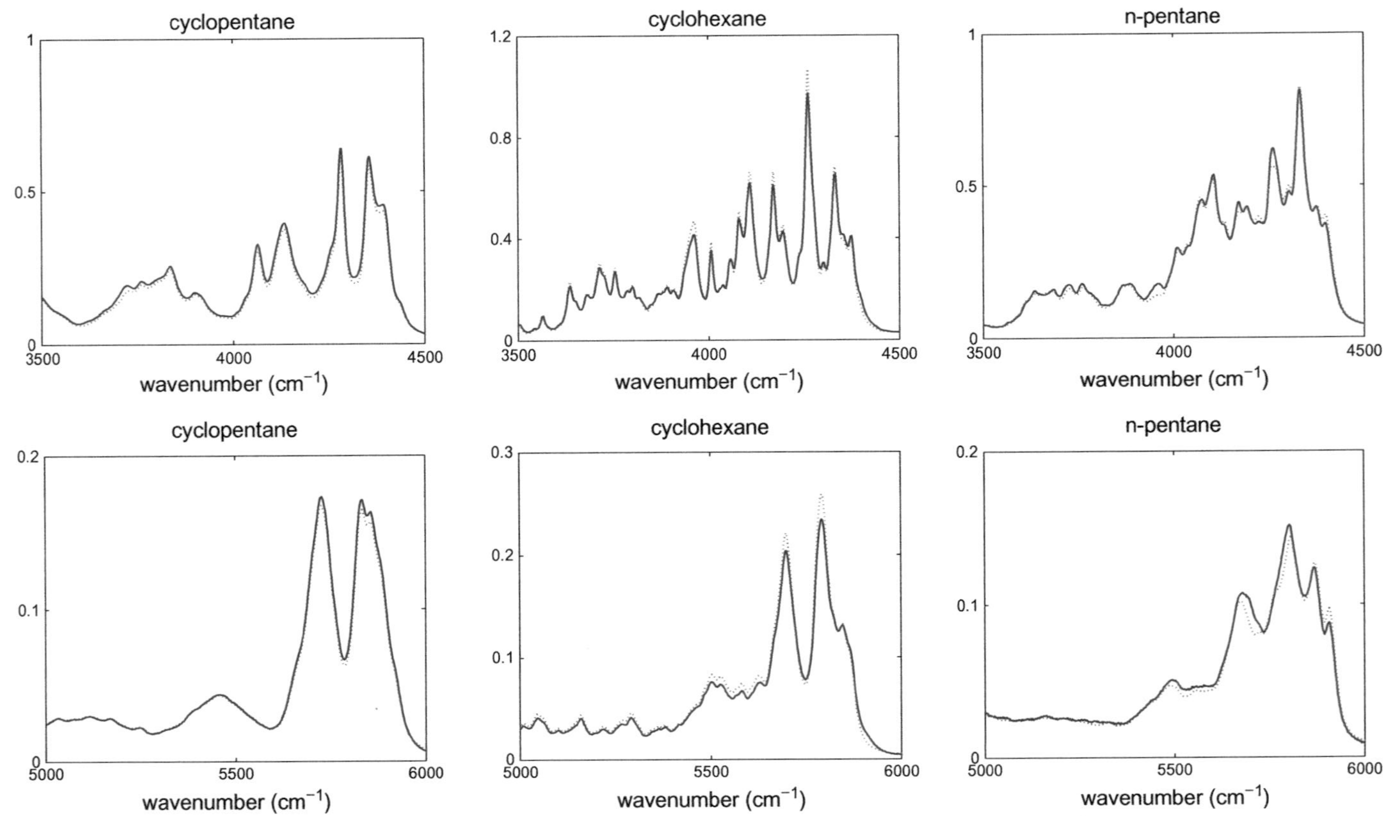

FIG. 8 Zoom of two subbands of the source spectra. The true spectra are shown in *dotted lines*.

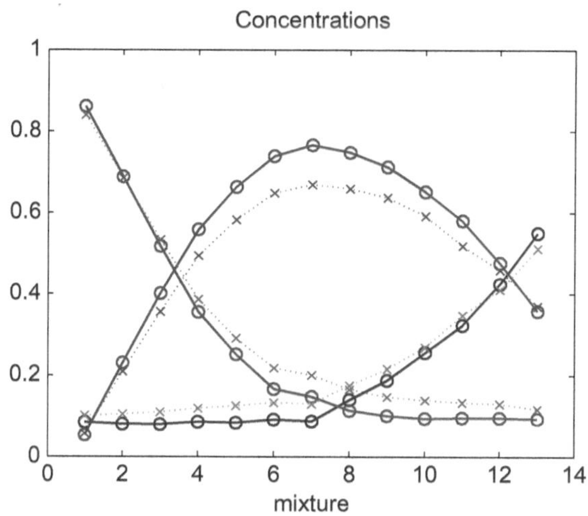

FIG. 9 Estimated (*continuous curve*) and true (*dotted curve*) concentration profiles.

different source spectra highly overlap, resulting in a significant cross-correlation. The source spectra are well reconstructed, which makes the identification of the components easier.

Concerning the estimation of the concentrations, Fig. 9 shows the similarity of the estimated mixing coefficients with the true concentration profiles; a small error can still be observed, but the results remain easily interpretable.

4.2 Raman Analysis of a Kerolite Sample

The Ni-rich mineral phases constituting the Ni ores in New Caledonia are dominated in most cases by kerolite–pimelite series, also called talc-like minerals. Despite their economic and geologic interests, the crystallographic structures of these minerals are not fully understood. In order to improve the knowledge of their crystallographic structure, a set of natural talc-like minerals of varying Ni/Mg ratio was selected. The aim of the study [48] was to investigate the structure of kerolite over the whole Mg–Ni substitution range by Raman spectroscopy. A particular focus was given at examining the behavior of the O–H stretching vibration bands with Ni content variations. In that respect, molecular Raman spectroscopy was combined with atomic analytical techniques (e.g., TEM-EDS, EMPA) to provide a correlation between chemical composition and structure of the kerolite. Thus, a set of 10 samples covering the Ni content between 0% and 100% was selected. Raman spectra were acquired on growing bands of polished sections previously analyzed by BSE-SEM and EMPA. After background subtraction [49] and normalization, the corrected Raman spectra were ordered as a function of the Ni–Mg substitution rate (SR) of the corresponding sample. Fig. 10A shows the

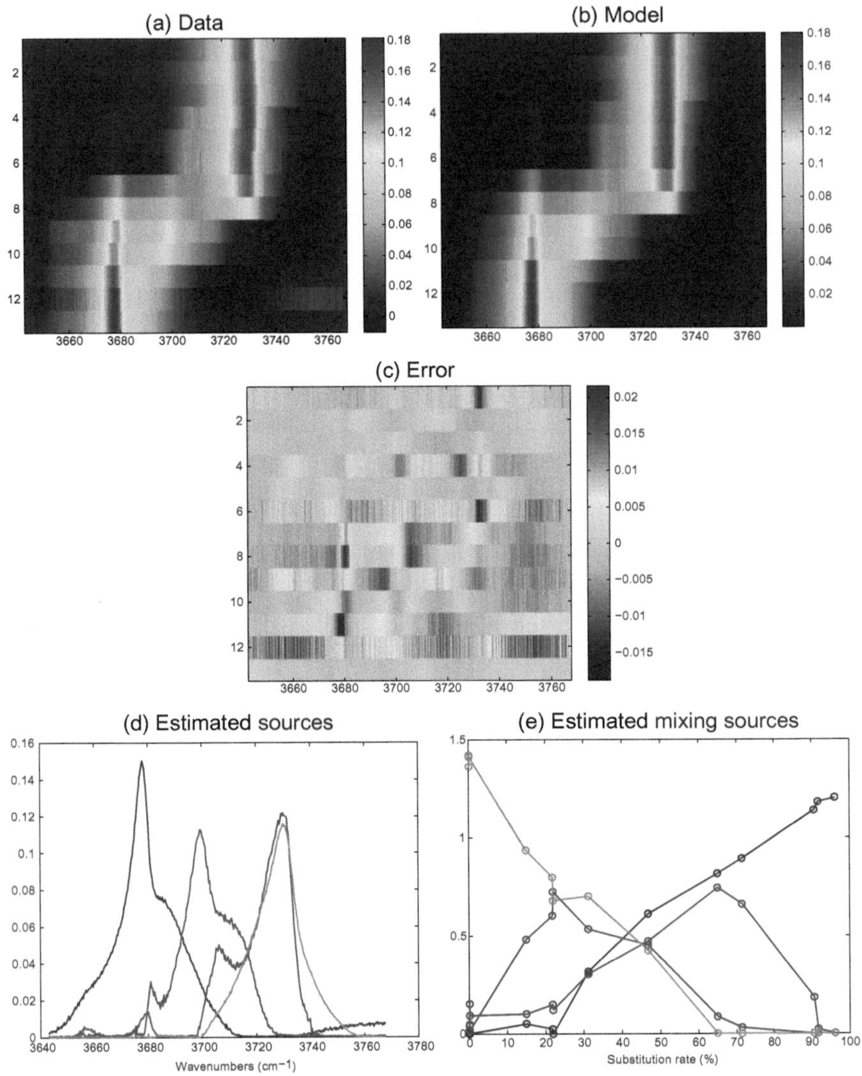

FIG. 10 Raman analysis of a kerolite sample using BPSS: (A) data, (B) model, (C) error, (D) estimated sources, and (E) estimated mixing coefficients.

corresponding data, the top and bottom spectra correspond, respectively, to the lower and higher SR. The set of Raman spectra was analyzed using the BPSS method. For physical reasons explained below, the rank of the decomposition was set to 4. After data processing, a similar map is drawn with the reconstructed model (Fig. 10B) resulting from the estimated sources and mixing coefficients shown in Fig. 10D and E, respectively. The difference between the two maps results in the error map (Fig. 10C). The errors are randomly distributed, symmetric to zero, and the error intensity maximum (0.02) is about 10% of the spectra intensity maximum (0.18). The contribution of each of these sources varies with SR. Source 1 is predominant at 0% of Ni and regularly decreases to be at zero at about 65% SR. Source 2 first increases to reach a maximum at 20% SR and then regularly decreases down to zero at 70% SR. Source 3 regularly increases from zero to reach a maximum of contribution in the signal at 70% SR and then decreases down to zero at 90% SR. Finally, source 4 is equal to zero up to 20% SR and then regularly increases to be predominant at 90% SR. The Raman spectra of the four sources (Fig. 10D) have peak maxima that move from high wavenumbers to lower wavenumbers with the increase of the Ni content. The shape of the bands is complex, with several local maxima except in source 1. Source 1 and source 2 maxima overlapped, but source 2 presents three other peaks.

The Raman spectra of Ni–Mg kerolite in the OH stretching vibration region are extremely complex. If four maxima are visible, no individual peak clearly appears in the band structure. The band shape strongly evolves with the chemistry of the mineral. The BPSS approach clearly distinguished four sources. Following the coherent and continuous evolution of the contribution of these sources with Ni content, each of them can be assigned to one environment of the OH groups, i.e., one arrangement of the Ni and Mg cations in the octahedral sites: (1) Mg_3; (2) Mg_2Ni; (3) $MgNi_2$; (4) Ni_3, the position of the OH bands shifting to lower wavenumbers when the Ni content increases. However, the second source corresponding to the Mg_2Ni arrangement which appears with the first Ni atoms cannot be fully differentiated from source 1 as the peak maximum of the two sources coincides. It might be due to a lack of samples in the range 1–8% NiO (1–15% SR) where the contribution of source 2 might range between 5% and 30%. Ideally, to better distinguish between source 1 and source 2, the set of samples should be completed by samples of Ni contents, in the range 1–15% SR. Unfortunately, the only samples we found in this range presented a different crystal chemistry, mainly a different Si/(Si + Mg + Ni) ratio (Fig. 2) and different Raman spectra, and thus were not included in this study. The sources are simultaneously present in the spectrum over a wide range of Ni content. For example, at 50% SR, each of the four sources contributes for about 25% to the Raman signal. Consequently, the distribution of Ni and Mg in the octahedral site is probably heterogeneous, with areas still free of Ni at relatively high Ni content, and Ni-rich areas in the first steps of Ni–Mg substitution. It validates one of the hypotheses stated in previous studies [50, 51] on the basis of EXAFS and IR studies of kerolite–pimelite series.

4.3 Polarized Raman Spectroscopy

This section is adapted from Ref. [27]. Raman scattering is a light–matter interaction process which reflects the molecular vibration properties of molecules and materials, thus characterizing the chemical composition of the analyzed sample [52, 53]. For materials presenting a regular atomic or molecular structure, a more accurate characterization of the sample can be achieved by using polarizers [54]. In particular, this is the case for crystals as their response to the polarized light excitation will reflect the crystallographic structure of the sample, motivating the development of polarized Raman spectroscopy. This is the type of problem addressed here since we are considering the polarized Raman spectroscopy of a rutile (TiO_2) crystal. It is worth mentioning that polarized Raman spectroscopy has other possible uses. For example, it can also be used for determining the molecular orientation distribution of polymeric materials [55].

The Raman measurements presented in this section were carried out in back scattering geometry with the same objective for excitation and collection of light. The confocal Raman spectrometer was equipped with a cooled CCD camera and the laser source was an ionized argon laser emitting at a wavelength $\lambda = 514.5$ nm. The analyzed crystal sample is fixed on a rotating stage as shown in Fig. 11. Two coordinate systems are used, one associated with the laboratory space-fixed coordinates (O, X, Y, Z) and another attached to the analyzed sample (O, x, y, z). The incident light is polarized such that the electric field arriving on the sample is oriented along the Y direction. The scattered light is analyzed by positioning an analyzer in front of the entrance slit of the spectrometer. The analyzer is oriented either along the

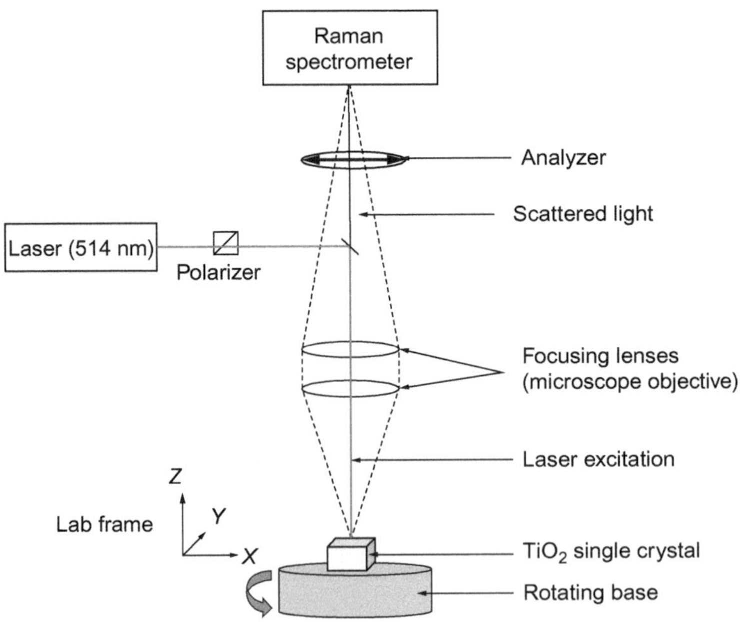

FIG. 11 Polarized Raman spectroscopy setup in backscattering geometry.

Y-axis (*parallel polarization*) or the X-axis (*crossed polarization*). Thus, the acquisition in one point of the sample yields a pair of spectra, one for the parallel polarization, indexed by Y, and another for the crossed polarization, indexed by X. The *rotational diversity* scheme consists in rotating the sample around the Z-axis (Fig. 11) with a fixed angular step (typically 10 degrees) and acquiring two polarized spectra for each step of the rotation. For the rotational diversity acquisition scheme, m polarized spectra are acquired for m different rotation angles ($\boldsymbol{\theta}_1, \ldots, \boldsymbol{\theta}_m$) of the analyzed sample. For this angular diversity data, the "sources" are represented by vibrational modes. Indeed, the vibrational modes are characterized by specific displacements of the atoms from their equilibrium position, which dictate the magnitude of the components of the Raman polarizability tensor. The change of polarized Raman intensity vs rotational angle, for a specific vibrational mode, will therefore be different from another one. Each mode in polarized Raman spectra will thus contribute as one source in the full spectrum.

Under the assumption of instantaneous linear mixture, the acquired data can be structured as two $m \times n$ matrices, corresponding to the two polarization orientations X and Y:

$$\mathbf{D}_X = \mathbf{A}_X \mathbf{S}_X + \mathbf{E}_X, \tag{29}$$

$$\mathbf{D}_Y = \mathbf{A}_Y \mathbf{S}_Y + \mathbf{E}_Y. \tag{30}$$

In Eqs. (29) and (30), matrices $\mathbf{E}_X$ and $\mathbf{E}_Y$ account for the additive noise on the sensors and the model errors.

If we further analyze the underlying physicochemical phenomenon generating the two data sets, the spectra of pure compounds are the same for the crossed and the parallel polarization [52, 53], since the vibrational modes are imposed by the structure of the crystal. This implies $\mathbf{S}_X = \mathbf{S}_Y = \mathbf{S}$, which is quite intuitive if we consider a geometrical point of view in which the crossed and parallel polarized spectra are projections of the same signal on two orthogonal axes. By injecting this information into Eqs. (29) and (30), we can write:

$$\begin{pmatrix} \mathbf{D}_X \\ \mathbf{D}_Y \end{pmatrix} = \begin{pmatrix} \mathbf{A}_X \\ \mathbf{A}_Y \end{pmatrix} \mathbf{S} + \begin{pmatrix} \mathbf{E}_X \\ \mathbf{E}_Y \end{pmatrix}. \tag{31}$$

Eq. (31) points out a bilinear model for the polarized spectra with rotational diversity considering both polarized spectra families jointly. Besides the fact that this is a more natural and compact representation of the data, the sample size is doubled in Eq. (33) compared to Eqs. (29) and (30) and the number of unknowns is lower; this should improve the accuracy of the estimated source parameters.

In order to simplify the presentation we use the following notations:

$$\mathbf{D} = \begin{pmatrix} \mathbf{D}_X \\ \mathbf{D}_Y \end{pmatrix}, \quad \mathbf{A} = \begin{pmatrix} \mathbf{A}_X \\ \mathbf{A}_Y \end{pmatrix}, \quad \mathbf{E} = \begin{pmatrix} \mathbf{E}_X \\ \mathbf{E}_Y \end{pmatrix}. \tag{32}$$

Eq. (31) can thus be rewritten in a more concise manner as:

$$\mathbf{D} = \mathbf{AS} + \mathbf{E}. \tag{33}$$

Given the nature of the data, the sources and the mixing coefficients are positive-valued, meaning that Eq. (33) expresses an NMF model. It should be noticed that stacking the data matrices $\mathbf{D}_X$ and $\mathbf{D}_Y$ into a bigger matrix $\mathbf{D}$ can also be seen as a data augmentation strategy. This kind of technique has already been proposed for diverse problems such as the analysis of multiple runs of gasoline blending processes [56]. Another example is the joint analysis of UV–visible spectra related to the complexation of the aluminum by caffeic acid and the titration of caffeic acid [32]. Actually, the benefit of matrix augmentation strategy is threefold: it allows to decrease estimation error uncertainties, may remove rank deficiency, and helps in reducing rotational ambiguities.

The approach was applied to a rutile TiO_2 crystal, as shown in Fig. 11. The crystallographic face (110) (Hermann–Mauguin international crystallographic symbols) is analyzed. The sample is rotated with respect to Z-axis only, meaning $\boldsymbol{\theta} = (0, 0, \chi)$. Fig. 12 presents the acquired polarized data for the parallel and crossed polarizations (matrices $\mathbf{D}_Y$ and $\mathbf{D}_X$). The data were acquired in a spectral range of 100–800 cm^{-1}, with an angular rotation step of 10 degrees between 0 and 190 degrees.

In the case of TiO_2, four Raman active modes denoted as A_{1g}, E_g, B_{1g}, and B_{2g} (Mulliken symbols for symmetry groups [52]) are expected from theory.

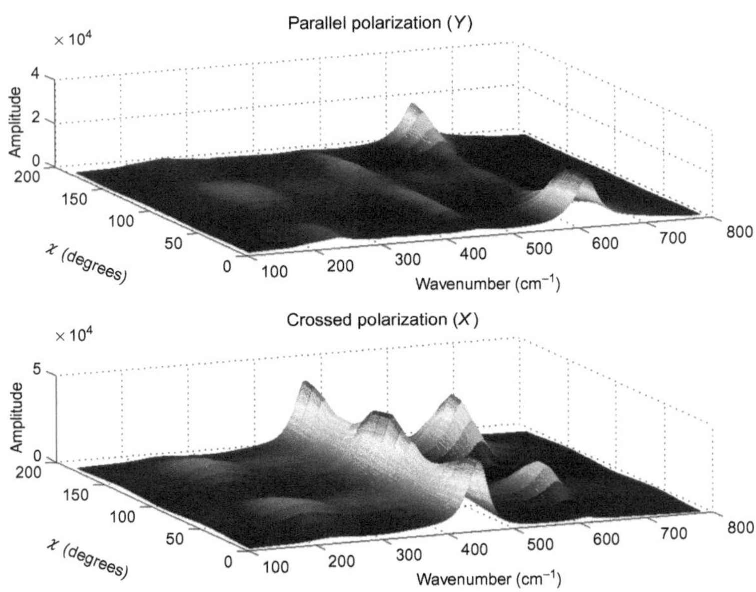

FIG. 12 Polarized Raman data vs rotational angle χ for rutile $TiO_2(110)$ single crystal.

However, the B_{2g} mode at 826 cm^{-1} is out of the spectral window used in this work (and anyway the B_{2g} has a very low Raman crossed section and is often not detected). The B_{1g} mode with the (110)-oriented crystal plane is inactive either in parallel or crossed polarizations. Consequently, one can expect two Raman active modes, i.e., two sources, in the data collected here. Nevertheless, three sources are necessary to properly describe the data, as indicated by the magnitude analysis of the singular values of data matrices $\mathbf{D}_X$, $\mathbf{D}_Y$, and $\mathbf{D}$. A theoretical explanation for the presence of this third source is provided below.

First, we illustrate the effect of the joint use of the crossed and parallel polarization data sets on the reduction of the NMF admissible solutions set. The NMF algorithm [13] was used to estimate the three source vectors and the corresponding mixing coefficients. The two data sets were processed separately and jointly and the results are presented in Fig. 13 for the source spectra and in Figs. 14 and 15 for the mixing coefficients. To evaluate the size of the admissible solutions set, we used 15 independent runs for each

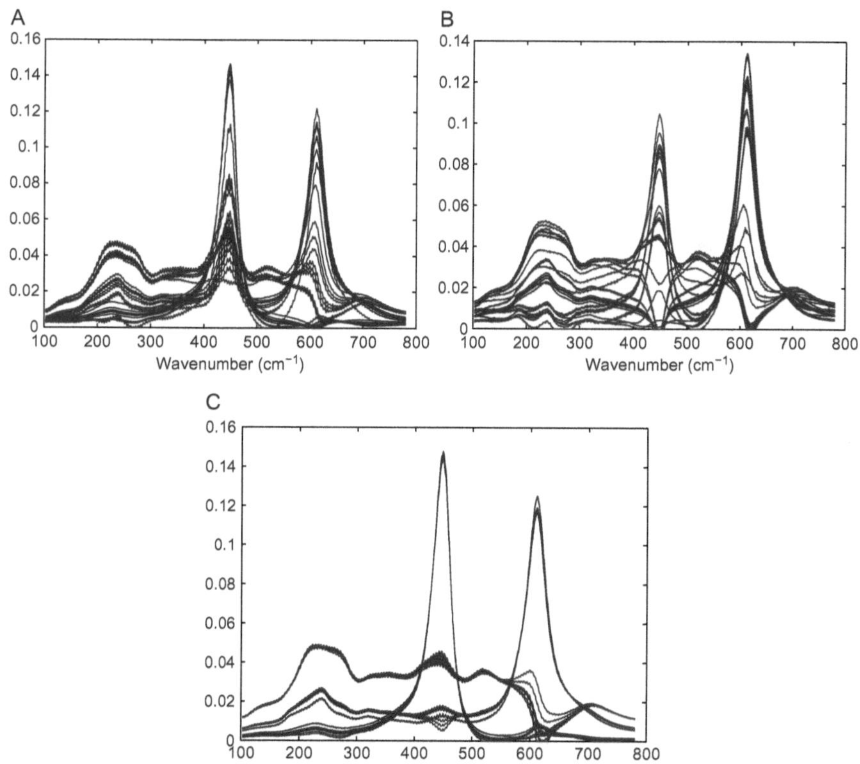

FIG. 13 Source spectra estimation by NMF (15 runs). (A) Crossed polarization (*X*). (B) Parallel polarization (*Y*). (C) Both polarization (*XY*).

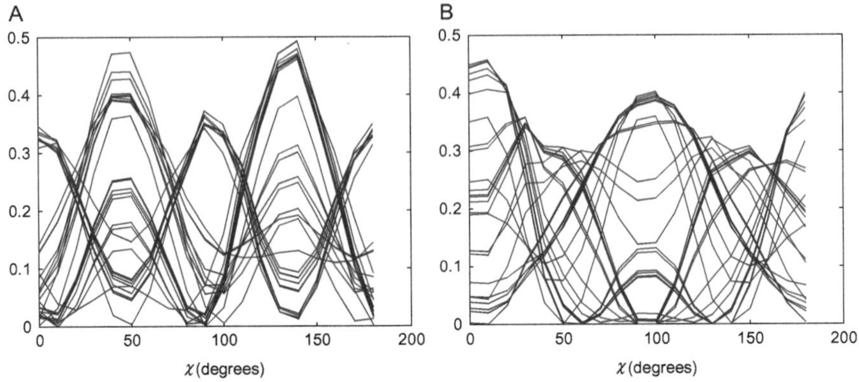

FIG. 14 Estimated coefficients by NMF for each polarization data set separately. (A) Polarization X. (B) Polarization Y.

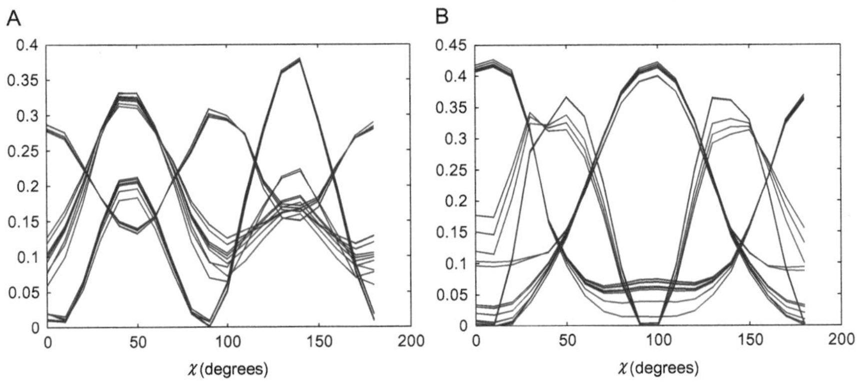

FIG. 15 Estimated coefficients by NMF for both polarization data sets jointly (15 runs). (A) Polarization X. (B) Polarization Y.

plot, with different random initial values for the matrices **A** and **S**. As one can see, by processing both polarization data sets jointly (Figs. 13C and 15) the admissible solution domain is largely reduced as compared to the case when only one polarization is used (Figs. 13A and B and 14). These results emphasize the utility of using both polarizations in Raman acquisitions.

Next, we illustrate the effect of the joint data processing on the estimation accuracy. First, BPSS was applied on the concatenated data as in Eq. (31) and then on each of the polarized data sets independently. The results for the joint processing of the polarizations are given in Fig. 16C for the source spectra, and in Fig. 18 for the mixing coefficients. Figs. 16A and B and 17 plot the estimated spectra and mixing coefficients, respectively, using only the parallel or the crossed polarization data. Three sources are identified: one (#1) with

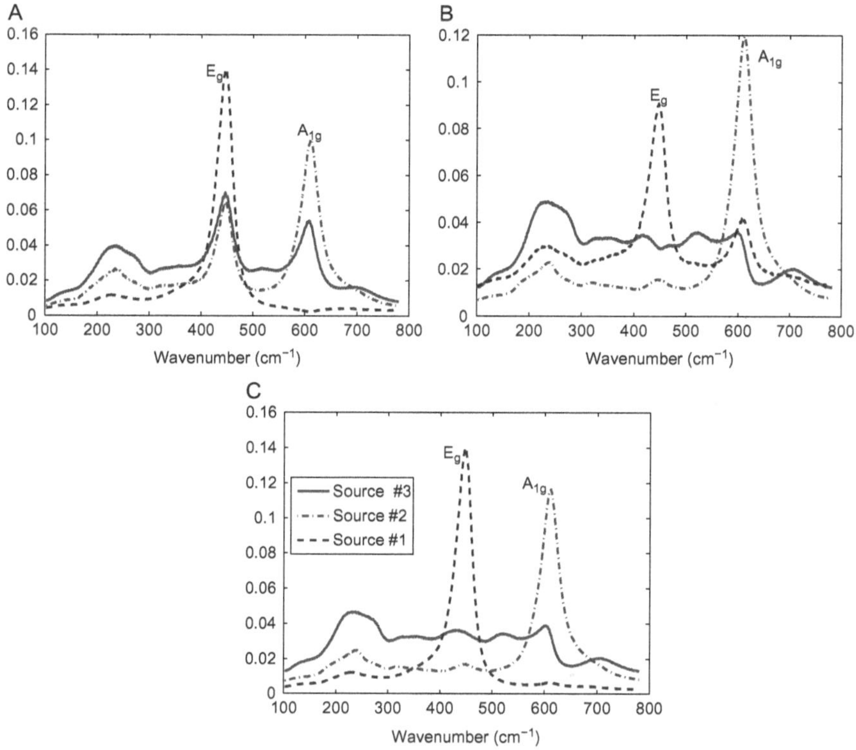

FIG. 16 Source spectra estimation by BPSS. (A) Crossed polarization (X). (B) Parallel polarization (Y). (C) Both polarization (XY).

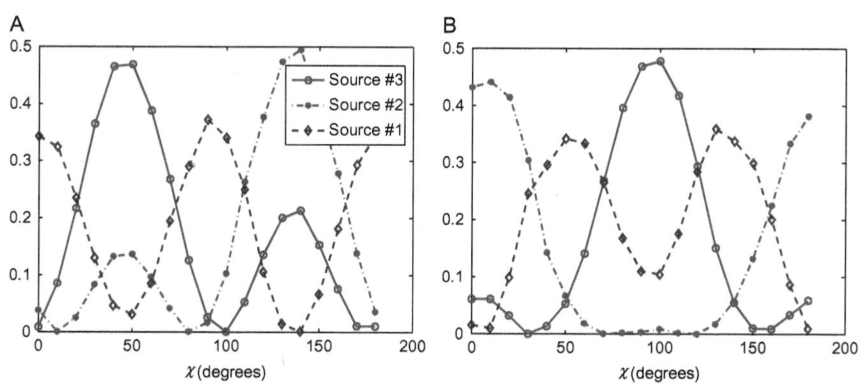

FIG. 17 Estimated coefficients by BPSS for each polarization data set separately. (A) Polarization X. (B) Polarization Y.

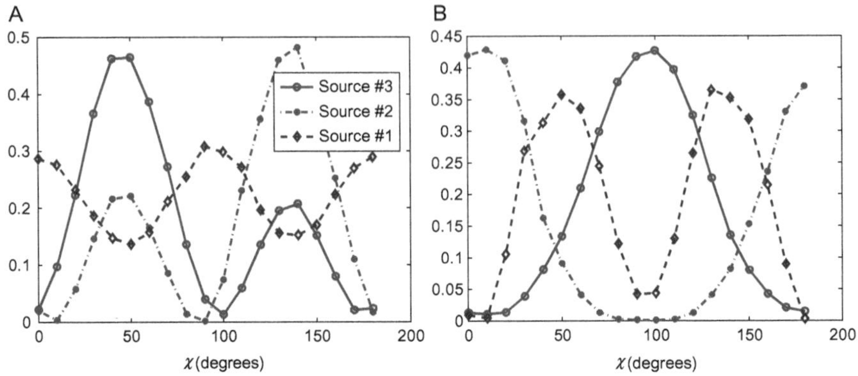

FIG. 18 Estimated coefficients by BPSS for both polarization data sets jointly. (A) Polarization *X*. (B) Polarization *Y*.

the main peak at 440 cm^{-1} corresponding physically to the vibrational mode E_g, another (#2) at 610 cm^{-1} corresponding to the mode A_{1g}, and a third one (#3) that is rather unexpected from the theoretical considerations exposed above. Indeed, one expects to find only two sources corresponding to the two Raman active modes A_{1g} and E_g. In fact, the spectral shape of source #3 corresponds to second-order Raman active mode and not to fundamental vibrational modes. This second-order mode has a structured shape (coming from the density of phonon states in rutile material) already observed and a complex behavior toward rotational diversity [57]. It is conjectured to be a harmonic or a combination band of lower frequency fundamental acoustic modes that are silent in Raman spectroscopy, but their combination becomes Raman active due to symmetry consideration and spectroscopic rules. Thus, by our approach we revealed the existence of this second-order mode, experimentally evidenced by Ref. [57], as a third source and retrieved its complex structure.

One can see that the use of the two polarizations jointly improves the results of the bilinear factorization. This is especially visible on the source spectra, where the polarization information allows a more accurate separation of source contributions. For example, in Fig. 16C, source #1 is clearly concentrated at a single wavenumber (440 cm^{-1}), while in Fig. 16B it presents an additional peak at the wavenumber corresponding to source #2. Similar remarks can be made for the two other sources in Fig. 16A.

The advantage of using the polarized data jointly is also visible on the mixing coefficients. For example, if we look at the crossed polarization data in Fig. 12 (bottom plot, near 440 cm^{-1}), one can see that the E_g mode, i.e., source #1, presents an offset. If both data sets are used, the aforementioned characteristic is successfully recovered (Fig. 18A), while this is not the case if the crossed polarization data are processed independently (Fig. 17A). Similar remarks can be made for the other mixing coefficients and polarization data set.

5 CONCLUSION

A Bayesian method for spectral mixture analysis, based on assigning Gamma distribution as an a priori model on the pure component spectra and mixing coefficients, was presented in this chapter. This approach allows to encode both the sparsity and the nonnegativity of the variables in the model. The estimation is achieved using an MCMC method, which allows to simulate samples from the joint a posteriori distribution. The use of this stochastic estimation technique precludes the convergence to local minima and yields a method working in an unsupervised framework. The only necessary parameters to tune are the number of components and the number of algorithm iterations. The approach is well adapted to difficult situations suffering from rotational ambiguities. Three applications are presented in which the method yields effective and interpretable results.

REFERENCES

[1] Jutten C, Hérault J. Blind separation of sources, part I: an adaptive algorithm based on neuromimetic architecture. Signal Process 1991;24:1–10.
[2] Markham TL. Factorizations of non-negative matrices. Proc Am Math Soc 1972;32:45–7.
[3] Campbell SI, Poole GD. Computing non-negative rank factorizations. Linear Algebra Appl 1981;35:175–82.
[4] Thomas LB. Rank factorizations of nonnegative matrices. SIAM Rev 1974;16:393–4.
[5] Cohen JE, Rothblum UG. Non-negative ranks decompositions and factorizations of non-negative matrices. Linear Algebra Appl 1993;190:149–68.
[6] Lawton WH, Sylvestre EA. Self-modeling curve resolution. Technometrics 1971;13:617–33.
[7] Windig W. Mixture analysis of spectral data by multivariate methods. Chemometr Intell Lab Syst 1990;9:7–30.
[8] Malinowski ER. Factor analysis in chemistry. 3rd ed. New-York: John Wiley & Sons; 2000.
[9] Moussaoui S, Brie D, Mohammad-Djafari A, Carteret C. Separation of non-negative mixture of non-negative sources using a Bayesian approach and MCMC sampling. IEEE Trans Signal Process 2006;54(11):4133–45.
[10] Tauler R, Kowalski B, Fleming S. Multivariate curve resolution applied to spectral data from multiple runs of an industrial process. Anal Chem 1993;65:2040–7.
[11] Paatero P, Tapper U. Positive matrix factorization: a non-negative factor model with optimal utilization of error estimates of data values. Environmetrics 1994;5:111–26.
[12] Paatero P. Least squares formulation of robust non-negative factor analysis. Chemometr Intell Lab Syst 1997;37:23–35.
[13] Lee DD, Seung HS. Learning the parts of objects by non-negative matrix factorization. Nature 1999;401(6755):788–91.
[14] Hoyer PO. Non-negative sparse coding. In: Proceedings of IEEE workshop on neural networks for signal processing (NNSP'2002); 2002. p. 557–65.
[15] Sajda P, Du S, Brown TR, Stoyanova R, Shungu DC, Mao X, et al. Nonnegative matrix factorization for rapid recovery of constituent spectra in magnetic resonance chemical shift imaging of the brain. IEEE Trans Med Imaging 2004;23(12):1453–65.
[16] Hoyer PO. Nonnegative matrix factorization with sparseness constraints. J Mach Learn Res 2004;5:1457–69.

[17] Pascual-Montano A, Carazo JM, Kochi K, Pascual-Marqui R. Nonsmooth nonnegative matrix factorization. IEEE Trans Pattern Anal Mach Intell 2006;28(3):403–15.
[18] Chen JC. Nonnegative rank factorisation of nonnegative matrices. Linear Algebra Appl 1984;62:207–17.
[19] Huang K, Sidiropoulos N, Swami A. Non-negative matrix factorization revisited: uniqueness and algorithm for symmetric decomposition. IEEE Trans Signal Process 2014;62(1):211–24.
[20] Gillis N, Glineur F. On the geometric interpretation of the nonnegative rank. Linear Algebra Appl 2012;437(11):2685–712.
[21] Ouedraogo WSB, Souloumiac A, Jaidane M, Jutten C. Non-negative blind source separation algorithm based on minimum aperture simplicial cone. IEEE Trans Signal Process 2014;62(2):376–89.
[22] Park ES, Spiegelman CH, Henry RC. Bilinear estimation of pollution source profiles and amounts by using multivariate receptor models. Environmetrics 2002;13:775–98.
[23] Smilde AK, Hoefsloot HCJ, Kiers HAL, Bijlsma S, Boelens HFM. Sufficient condition for unique solutions within a certain class of curve resolution models. J Chemom 2001;15: 405–11.
[24] Plumbley MD. Conditions for nonnegative independent component analysis. IEEE Signal Process Lett 2002;9(6):177–80.
[25] Plumbley MD. Algorithms for non-negative independent component analysis. IEEE Trans Neural Netw 2003;14(3):534–43.
[26] Moussaoui S, Brie D, Idier J. Non-negative source separation: range of admissible solutions and conditions for the uniqueness of the solution. In: Proceedings of IEEE international conference on acoustics, speech, and signal processing (ICASSP'2005). Philadelphia, PA; 2005.
[27] Miron S, Dossot M, Carteret C, Margueron S, Brie D. Joint processing of the parallel and crossed polarized Raman spectra and uniqueness in blind nonnegative source separation. Chemometr Intell Lab Syst 2011;105(1):7–18.
[28] Donoho D, Stodden V. When does non-negative matrix factorization give a correct decomposition into parts? In: Advances in neural information processing systems vol. 16. Cambridge, MA: MIT Press; 2003.
[29] Bro R, Sidropoulos ND. Least squares algorithms under unimodality and non-negativity constraints. J Chemom 1998;12:223–47.
[30] Chan TH, Chi CY, Huang YM, Ma WK. A convex analysis-based minimum-volume enclosing simplex algorithm for hyperspectral unmixing. IEEE Trans Signal Process 2009;57(11):4418–32.
[31] Moussaoui S, Carteret C, Brie D, Mohammad-Djafari A. Bayesian analysis of spectral mixture data using Markov chain Monte Carlo methods. Chemometr Intell Lab Syst 2006;81(2):137–48.
[32] Ruckebusch C, De Juan A, Duponchel L, Huvenne JP. Matrix augmentation for breaking rank-deficiency: a case study. Chemometr Intell Lab Syst 2006;80(2):209–14.
[33] Roberts SJ. Independent component analysis: source assessment and separation, a Bayesian approach. IEEE Proc Vision Image Signal Process 1998;145(3):149–54.
[34] Knuth K. A Bayesian approach to source separation. In: Cardoso JF, Jutten C, Loubaton P, editors. Proceedings of international workshop on independent component analysis and signal separation (ICA'99). Aussios, France; 1999, p. 283–8.
[35] Mohammad-Djafari A. A Bayesian approach to source separation. AIP Proc 1999;567:221–44.
[36] Rowe DB. Multivariate Bayesian statistics: models for source separation and signal unmixing. Boca Raton, FL: CRC Press; 2003.

[37] Snoussi H, Idier J. Bayesian blind separation of generalized hyperbolic processes in noisy and underdeterminate mixtures. IEEE Trans Signal Process 2006;54(9):3257–69.

[38] Févotte C, Godsill SJ. A Bayesian approach for blind separation of sparse sources. IEEE Trans Audio Speech Lang Process 2006;14(6):2174–88.

[39] Ochs MF, Stoyanova RS, Arias-Mendoza F, Brown TR. A new method for spectral decomposition using a bilinear Bayesian approach. J Magn Reson 1999;137:161–76.

[40] Miskin J, MacKay D. Ensemble learning for blind source separation. In: Roberts S, Everson R, editors. Independent component analysis: principles and practice. Cambridge: Cambridge University Press; 2001. p. 209–33.

[41] Robert CP. The Bayesian choice. 2nd ed. New-York: Springer-Verlag; 2001.

[42] Stuart A, Ord JK. Kendall's advanced theory of statistics. Distribution theory, vol. 1. 6th ed. New-York: John Wiley & Sons; 1994.

[43] Hastings WK. Monte Carlo sampling methods using Markov chains and their applications. Biometrika 1970;57(1):97–109.

[44] Geman S, Geman D. Stochastic relaxation, Gibbs distributions and the Bayesian restoration of images. IEEE Trans Pattern Anal Mach Intell 1984;6(6):721–41.

[45] Gilks WR, Richardson S, Spiegehalter DJ. Markov chain Monte Carlo in practice. London: Chapman & Hall; 1999.

[46] Robert CP. Monte Carlo statistical methods. New-York: Springer-Verlag; 1999.

[47] Chen MH, Shao QM, Ibrahim JG. Monte Carlo methods in Bayesian computation. New York: Springer-Verlag; 2000.

[48] Cathelineau M, Caumon MC, Massei F, Brie D, Harlaux M. Raman spectra of Ni-Mg kerolite: effect of Ni-Mg substitution on O-H stretching vibrations. J Raman Spectrosc 2015;46(10):933–40.

[49] Mazet V, Carteret C, Brie D, Idier J, Humbert B. Background removal from spectra by designing and minimising a non-quadratic cost function. Chemometr Intell Lab Syst 2005;76:121–33.

[50] Manceau A, Calas G. Heterogeneous distribution of nickel in hydrous silicates from New Caledonia ore deposits. Am Mineral 1985;70(5–6):549–58.

[51] Gérard P, Herbillon J. Infrared studies of Ni-bearing clay minerals of the kerolite-pimelite series. Clays Clay Miner 1983;31(2):143–51.

[52] Long DA. The Raman effect: a unified treatment of the theory of Raman scattering by molecules. New-York: John Wiley & Sons; 2002.

[53] Turrell G. Infrared and Raman spectra of crystals. New York: Academic Press; 1972.

[54] Jiménez C, Caroff T, Bartasyte A, Margueron S, Abrutis A, Chaix-Pluchery O, et al. Raman study of CeO_2 texture as a buffer layer in the $CeO_2/La_2zR_2O_7/Ni$ architecture for coated conductors. Appl Spectrosc 2009;63(4):401–6.

[55] Tanaka M, Young RJ. Review: polarised Raman spectroscopy for the study of molecular orientation distributions in polymers. J Mater Sci 2006;41(3):963–91.

[56] Jaumot J, Menezes JC, Tauler R. Quality assessment of the results obtained by multivariate curve resolution analysis of multiple runs of gasoline blending processes. J Chemom 2006;20(1–2):54–67.

[57] Porto SPS, Fleury PA, Damen TC. Raman spectra of TiO_2, MgF_2, ZnF_2, FeF_2, and MnF_2. Phys Rev 1967;154(2):522.

Chapter 9

Multivariate Curve Resolution of Wavelet Compressed Data

P. de B. Harrington[1]
Clippinger Laboratories, Center for Intelligent Chemical Instrumentation, Ohio University, Athens, OH, United States
[1]*Corresponding author: e-mail: Peter.Harrington@OHIO.edu*

Chapter Outline

1 INTRODUCTION

Multivariate curve resolution (MCR) is an established method for exploiting differences that occur in a set of mixture spectra to resolve independent components of the mixture. To avoid confusion with independent component analysis, for this case, independence is not used as its strict statistical definition. Instead, it refers to distinct components of a mixture whose measurements vary differently within a set of observations. For example, the concentration of two chemical compounds may enter a spectrometer at different times. Several other examples will be given later in Section 1, but as a basic principle for MCR, the measurementor spectrum or spectrum of each component must differ from the others, the measurements of the components must change with respect to concentration across a set of observations, and these changes in concentration must also differ from each other.

MCR can be used to perform virtual chromatography and in some cases eliminate time-consuming separation stages. My first involvement with MCR arose when we were developing algorithms for ion mobility spectrometry (IMS). IMS uses atmospheric ionization of the analytes and anytime you have an ionization process under atmospheric conditions you have a mixture. Under the simplest of circumstances, the mixture comprises the reactant and analyte ions, and the task for MCR is to separate the analytical peaks from

Data Handling in Science and Technology, Vol. 30. http://dx.doi.org/10.1016/B978-0-444-63638-6.00009-7
Copyright © 2016 Elsevier B.V. All rights reserved.

the background of reactant ions [1]. Two examples of MCR as a form of virtual chromatography are the modeling of ions based on the clear-down rate of the instrument [2]. The clear-down rate is the return of an instrument to its baseline response; many measurements often ignore this information. The other application was using the profiles obtained from thermal desorption that also were previously ignored by IMS researchers [3–5].

Interest in data compression of analytical data often goes in cycles, as improvements in computational capacity, and analytical measurement speed and resolution continually leapfrog each other. The timely importance of data compression is attributed to several factors. First, the design of small analytical sensors makes them portable, so the accompanying computational storage device must also be small, consume low amounts of power, and have a high capacity for logging the data. The alternative is to use wireless transmission of data, but wireless transmission often is bandwidth limited so for either logging or transmission, data compression may be required. My interest in data compression was renewed when we were interested in the wireless transmission of data from a handheld IMS sensor [6]. Later, with a project sponsored by the US Army, we developed compression algorithms for IMS spectrometers mounted in drones. At that time, they were referred to as unmanned aerial vehicles [7]. Other requirements for data compression are that with advances in instrument resolution and speed, analytical instrumentation can generate overwhelming amounts of data at faster rates. Some measurements can easily exceed the storage and computational capacity of modern workstations. Examples are high-resolution mass spectrometry coupled with high-resolution chromatography or hyperspectral imaging. Both techniques may produce vast quantities of data that may overwhelm the capabilities of modern computers, so compression can greatly simplify modeling.

Fortunately, data compression not only makes the data size smaller, but for many cases, it also can make the data better by removing noise. Often, high-frequency noise components are removed, but low-frequency baseline drift can be removed as part of the data compression as well. The simpler approaches for reducing the data size are slowing the data acquisition rate down, discarding signals below an intensity threshold, or monitoring specific variables, all of which discard information that is available from the instrument. Because most compression algorithms implicitly have a signal averaging step, better quality data can be achieved by maximizing the number of digital data points per analog waveform by increasing the digital sampling rate, and then compressing the oversampled measurement. This approach should be a general principle of analytical chemistry.

A note on terminology, dimensionality will be used to refer to the number of elements for each way of multiway data. For example, if there are 100 measurements for 50 observations, there are two ways, but the dimensionality of the observations would be 50, and the measurements would be 100.

The dimensionality of the object or sample would be 5000. The term *mode* or *modality* will be used for the model or method applied to the data.

Fig. 1 gives the scheme for the basic strategy. For the example, a series of spectra are acquired with respect to time and are arranged as the rows of the data matrix. The spectral way is compressed first to produce a reduced number of measurements. The second step compresses the time way, to yield a reduced number of rows. These steps also could be accomplished at the same time, by using a two-dimensional wavelet compression, but the results would be the same. The compressed data are then modeled using MCR that will decompose the data into two matrices that characterize the changes in time referred to as the concentration profiles and the corresponding spectral features. These two matrices are in the compressed representation, so the final step is to apply an inverse transform so the concentration profiles and spectra are returned to their native representations.

Two-way compression is most common. In other cases such as hyperspectral imaging, there would be three ways (two spatial ways and a third spectral way). For some cases, the hyperspectral image may be unfolded from a three-way representation to a two-way representation, although this transformation is made for convenience and is unnecessary. However, this added way would just require one more wavelet compression.

Three approaches will be used with this overall scheme. The first is linear wavelet compression using orthogonal wavelets (e.g., Daubechies, symlet,

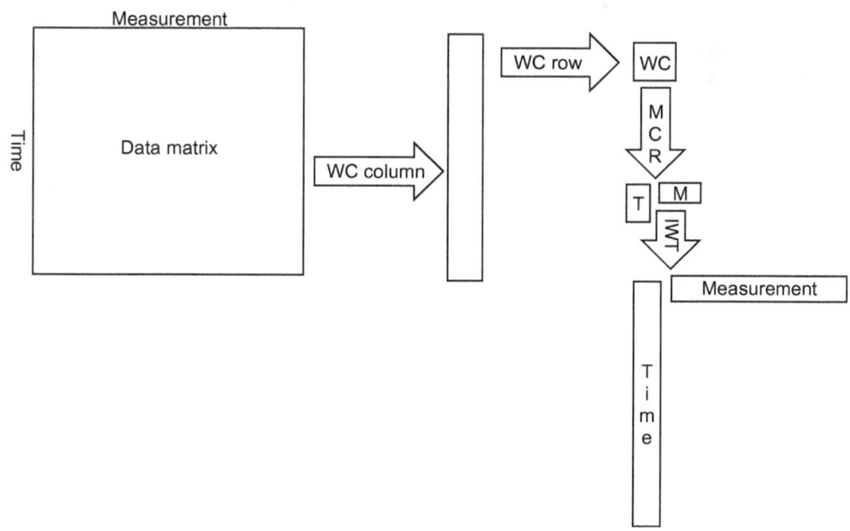

FIG. 1 The basic scheme for which a data matrix of spectra as rows collected with respect to time is compressed in two ways, first the measurement and then the time way, the compressed matrix is subjected to MCR, and then the inverse wavelet transform is used to convert the model back to its native representation and size.

Haar, etc.), the second is the use of biorthogonal wavelets (e.g., Villasenor) for which the inverse transform may be skipped and the compressed model directly interpreted, and the third is the use of nonlinear wavelet compression to yield even greater compression. Other compression methods such as Fourier compression can also be used [8]. Orthogonal wavelets have the property that the smooth and detail filters are orthogonal to each other and orthogonal to themselves when shifted by multiples of two points. They will form an orthogonal basis so that the wavelet coefficients can be obtained by multiplication. The biorthogonal wavelet requires two filters to maintain orthogonality instead of a single filter and hence the name biorthogonal. By sacrificing some orthogonality symmetric biorthogonal wavelet filters may be obtained, so the wavelet coefficients will not be shifted during compression and thus amenable to direct interpretation without the inverse transform to the uncompressed representation.

2 THEORY

MCR is a decomposition method, but unlike principal components, the factors should model independently varying components of the mixture. In other words, for any MCR method to work, the mixture component profiles must not be correlated. The initial method was simple-to-use interactive self-modeling analysis (SIMPLISMA) [9]. SIMPLISMA seeks to find independently varying concentration profiles by finding pure variables. Ideally, a pure variable is specific to a component, in that the variable is not a significant variable among the other components. The pure variables are identified by satisfying two criteria. First, the variable must be changing with respect to the observations or time. Typically, the relative standard deviation may be used, but a damping factor α is used to remove very small variances that may be below the noise level.

A data matrix $\boldsymbol{X}$ comprises n measurements as the columns and m observations as the rows. Each measurement j is evaluated for its purity p_j:

$$p_j = \frac{\sigma_j}{\bar{x}_j + \alpha} \times w_j \tag{1}$$

for which p_j is the purity for measurement j, σ_j is the standard deviation of the measurement, $\bar{x}_j$ is the average for measurement j that is obtained from $\boldsymbol{X}$, and w_j is a weight that measures the mutual independence compared to the other measurements in the model. Measurements are selected sequentially by the largest purity value.

The weight w_j measures the independence of the candidate pure variable j, by evaluating the determinant comprised of correlation coefficients in Eq. (2)

$$w_j = \begin{vmatrix} \rho_{11} & \rho_{12} & \rho_{1j} \\ \rho_{21} & \rho_{22} & \rho_{2j} \\ \rho_{j1} & \rho_{j2} & \rho_{jj} \end{vmatrix} \tag{2}$$

for which the correlation coefficient ρ_{1j} is determined in Eq. (3):

$$\rho_{1j} = \frac{\sum_{i=1}^{m} x_{i1} x_{ij}}{\sqrt{\sum_{i=1}^{m} (x_{i1} + \alpha)^2} \sqrt{\sum_{i=1}^{m} (x_{ij} + \alpha)^2}} \tag{3}$$

Once the number of pure variables has been selected an iterative set of regressions are used until convergence is achieved. The data matrix $\boldsymbol{X}$ is regressed onto columns of the data matrix $\boldsymbol{C}$ corresponding to the pure variables, and then the spectral profiles $\boldsymbol{S}$ are used to reestimate the concentration profiles $\boldsymbol{C}$:

$$\boldsymbol{S} = \boldsymbol{D}^{\mathrm{T}} \boldsymbol{C} \left(\boldsymbol{C}^{\mathrm{T}} \boldsymbol{C} \right)^{-1} \tag{4}$$

$$\boldsymbol{C} = \boldsymbol{D} \boldsymbol{S} \left(\boldsymbol{S}^{\mathrm{T}} \boldsymbol{S} \right)^{-1} \tag{5}$$

Our version of SIMPLISMA incorporates alternating least squares (ALS) without any constraints, so Eqs. (4) and (5) are iteratively applied until convergence is achieved. An important note is that the concentration profiles do not have units of concentration; however, if the instrument gives a linear response, then the profiles will be proportional to concentration and the concentrations can be achieved by dividing each profile by its corresponding analytical sensitivity ($\partial x/\partial c$).

Selection of the number of components is an important parameter for modeling. For MCR, the simplest approach is to measure the noise from the baseline response and then add the components successively until a candidate concentration profile has a maximum that exceeds the noise level [10]. Then, the component is added to the model and then a new candidate component is modeled. A second approach uses the ratio of the purity values to the first purity. This approach accounts for the independence of the candidate component. Components are added until the purity drops below a specified threshold [11]. Determination of the components automatically is important for real-time MCR [11–13]. Sometimes, peak shifting will occur during the measurement and peak shifts will result in additional components, and the spectral components will appear as first-derivative peaks. Usually, peak shift conveys useful chemical information, so negative peaks are important; nevertheless sometimes it is desirable to remove negative features from the model. If peak shifting and shape are important, then the error-based approach is better for determining the number of components. Otherwise, the ratio of purity values is the preferred method.

Negative features are removed by using ALS [14] with nonnegativity constraints [15]. This calculation simply continues the iterative regressions of Eqs. (4) and (5), except that a least-squares inverse is calculated that is constrained by nonnegativity. In MATLAB, the function is *lsqnonneg*. A more pedestrian and less rigorous approach is simply to set any values of the concentration and spectral profile that are negative to zero during the iteration.

Wavelets provide a flexible and convenient method for data compression [16]. They will be briefly described here, and further details on wavelets and their applications in chemistry can be found in one book [17] and several tutorials [18–20]. Like the Fourier transform (FT), the wavelet transform (WT) relies on orthogonal bases to encode frequency information. The FT uses continuous waves, while the wavelet is a short wave packet, so it provides local support. In other words, it also contains phase information. The wavelets act as digital filters and come in pairs that comprise a smoothing filter and a detail filter. The smooth wavelet coefficients encode the low-frequency portion of the data window, and the detail coefficients encode the high-frequency portion of the data. The transform is accomplished by convolution of the window of wavelet filter points with the analytical data.

For orthogonal wavelets, a set of quadrature mirror filters (QMFs) are used. Without getting too technical, the detail coefficients are derived by reversing the smooth filter points and then alternating the sign of the points. Also, because the filters are orthogonal in matrix form, the inverse transform can simply be performed by transposing the wavelet matrix. QMFs are generated in even numbers of points with the simplest filter having only two points. The number of vanishing moments is the number of QMF points divided by 2.

In Eq. (6) the four smooth filter points $\boldsymbol{c}$ are in the odd-numbered rows and the four detail filter points $\boldsymbol{c}$ are immediately below them in the even-numbered rows. Then in the subsequent rows, the pair of filters is shifted by two columns. The bottom two rows of the matrix wrap around so as to model a periodic function. The smooth $\boldsymbol{s}$ and detail $\boldsymbol{d}$ coefficients result from multiplication of the filter matrix with signal $\boldsymbol{x}_i$ and can be separated. Each multiplication of the filter results in half the number of smooth and detail coefficients and is referred to as a level:

$$\begin{bmatrix} c_1 & c_2 & c_3 & c_4 & 0 & 0 & 0 & 0 \\ c_4 & -c_3 & c_2 & -c_1 & 0 & 0 & 0 & 0 \\ 0 & 0 & c_1 & c_2 & c_3 & c_4 & 0 & 0 \\ 0 & 0 & c_4 & -c_3 & c_2 & -c_1 & 0 & 0 \\ 0 & 0 & 0 & 0 & c_1 & c_2 & c_3 & c_4 \\ 0 & 0 & 0 & 0 & c_4 & -c_3 & c_2 & -c_1 \\ c_3 & c_4 & 0 & 0 & 0 & 0 & c_1 & c_2 \\ c_2 & -c_1 & 0 & 0 & 0 & 0 & c_4 & -c_3 \end{bmatrix} \begin{bmatrix} x_1 \\ x_2 \\ x_3 \\ x_4 \\ x_5 \\ x_6 \\ x_7 \\ x_8 \end{bmatrix} = \begin{bmatrix} s_1 \\ d_1 \\ s_2 \\ d_2 \\ s_3 \\ d_3 \\ s_4 \\ d_4 \end{bmatrix} \rightarrow \begin{bmatrix} s_1 \\ s_2 \\ s_3 \\ s_4 \end{bmatrix}, \begin{bmatrix} d_1 \\ d_2 \\ d_3 \\ d_4 \end{bmatrix} \quad (6)$$

Each convolution (multiplication of the wavelet matrix) of the wavelet filter pair will divide the data into two sets of wavelet coefficients each equal to half the size of the data and is referred to as a level. The number of levels to completely transform the data is the base 2 logarithm of the number of data points n. Hence, just like the fast FT, the size of the data is usually dyadic (i.e., a power of 2), which may be accomplished by truncating the data or padding the data with zeros or the last point in the data array so that it appears to be continuous. The linear compressions pad with the last point of the data array, while the nonlinear compressions pad with zeros.

Because the wavelet coefficients are orthogonal, one may apply the inverse transform to reconstruct the data using the transpose of the wavelet matrix. The multiplication has been transposed and is given by Eq. (7):

$$\begin{bmatrix} s_1 \\ s_2 \\ s_3 \\ s_4 \end{bmatrix} \rightarrow \begin{bmatrix} s_1 \\ 0 \\ s_2 \\ 0 \\ s_3 \\ 0 \\ s_4 \\ 0 \end{bmatrix}^{\mathrm{T}} \begin{bmatrix} c_1 & c_2 & c_3 & c_4 & 0 & 0 & 0 & 0 \\ c_4 & -c_3 & c_2 & -c_1 & 0 & 0 & 0 & 0 \\ 0 & 0 & c_1 & c_2 & c_3 & c_4 & 0 & 0 \\ 0 & 0 & c_4 & -c_3 & c_2 & -c_1 & 0 & 0 \\ 0 & 0 & 0 & 0 & c_1 & c_2 & c_3 & c_4 \\ 0 & 0 & 0 & 0 & c_4 & -c_3 & c_2 & -c_1 \\ c_3 & c_4 & 0 & 0 & 0 & 0 & c_1 & c_2 \\ c_2 & -c_1 & 0 & 0 & 0 & 0 & c_4 & -c_3 \end{bmatrix} = \begin{bmatrix} \hat{x}_1 \\ \hat{x}_2 \\ \hat{x}_3 \\ \hat{x}_4 \\ \hat{x}_5 \\ \hat{x}_6 \\ \hat{x}_7 \\ \hat{x}_8 \end{bmatrix} \tag{7}$$

Note that for the reconstruction shown earlier, the detail coefficients can be eliminated from the wavelet matrix so that it only comprises the smooth coefficients (i.e., odd rows) which will speed up the reconstruction.

For the linear compression as illustrated by Eqs. (6) and (7), the pyramid algorithm is used [16]. The pyramid algorithm will apply successive WTs to each set of smooth coefficients and discard the detail coefficients that encode the high-frequency information.

In Fig. 2, linear wavelet compression is applied to a matrix-assisted laser desorption/ionization-mass spectrum (MALDI-MS) of a bacterial sample. Each application or level of the WT yields half the number of smooth and detail wavelet coefficients. The smooth coefficients resemble the original data with half the number of points or resolution. MCR is amenable to these smooth coefficients as long as the important analytical information is encoded. If the data are overly compressed, resolution is lost, and separate peaks may become blended together which may deleteriously affect the modeling. See Fig. 2 for the 7 and 10 wavelet filter applications. After 10 wavelet filter applications, the data have been reduced to 2^{-10} of its original size. Usually, when features appear in the detail coefficients that correspond to features in the smooth coefficients, the data are being overcompressed in that analytical signals are being removed by the pyramid algorithm.

One may determine the degree of compression by visual examination of the coefficients or reconstructed spectra. Because wavelet compression also removes noise, the greater the compression, the greater the reconstruction error. However, one may also compare the average spectrum with the average of the reconstructed spectrum, which will be less affected by noise. Even better is to compare the reconstructed spectra or measurements from modeling with those from the model built with the original data, but this metric implies that the data size is small enough that it can be modeled directly [8,21].

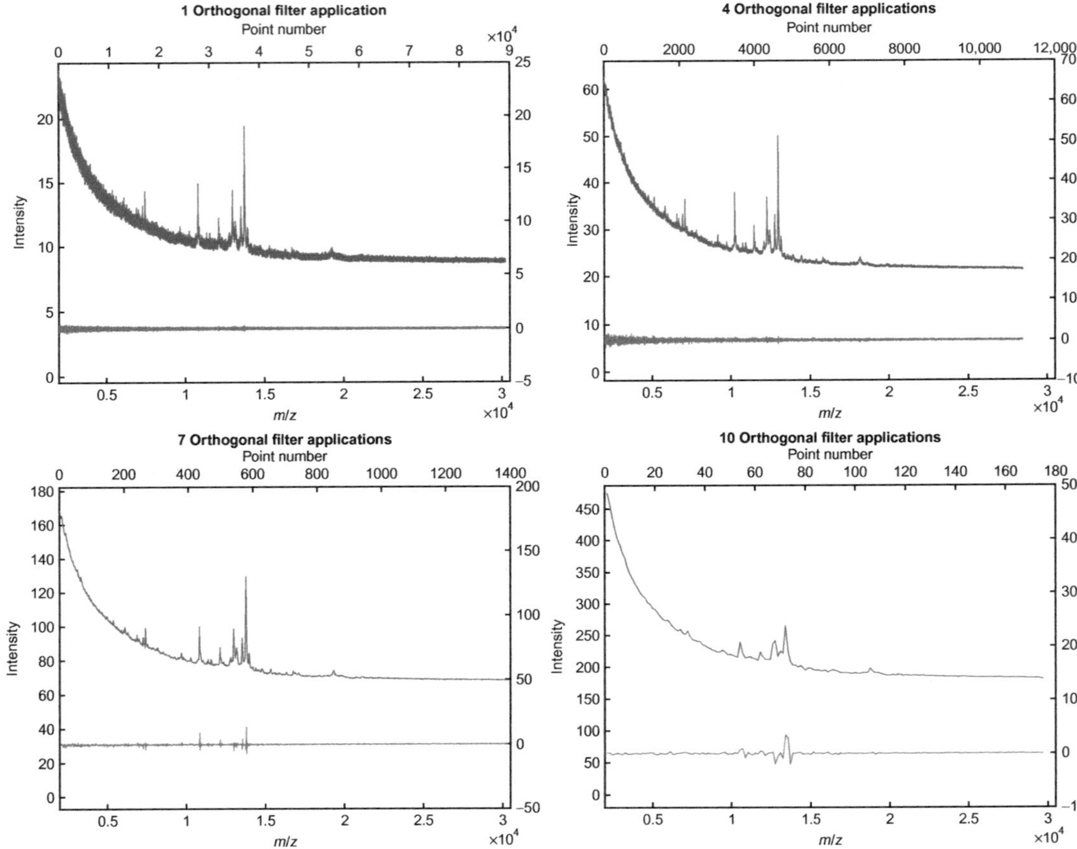

FIG. 2 *Escherichia coli* MALDI-MS smooth (*blue*; *dark gray* in the print version) and detail (*orange*; *gray* in the print version) coefficients after various numbers of orthogonal wavelet filter applications. The left ordinate corresponds to the smooth coefficients and the right ordinate corresponds to the detail coefficients. Each application will divide the data into equal numbers of smooth and detail coefficients. To save space, the figure comprises the results after 1, 4, 7, and 10 filter applications. The Daubechies filter was used with 10 filter points.

The problem is very similar to determining the number of principal components; if too many are selected, then noise will be embedded in the model, and too few components result in the loss of signal. Unfortunately just like PCA, there is no simple criterion for selecting the proper degree or level of compression.

Orthogonal wavelet coefficients are not symmetric, and the coefficients may shift the signal with respect to the decimated resolution elements. However, by applying the inverse transforms, the peak positions will be shifted back to the correct resolution element. Also, the wavelet coefficient will resemble the shape of the wavelet and not the peak. Reconstruction by applying the inverse transform will also return the signal to its original peak shape. Biorthogonal wavelets have the advantage that the wavelet coefficients will be at the decimated resolution element and can be directly interpreted. Reconstruction will restore the peak shape in the same way as for the orthogonal wavelets. However, the biorthogonal filter points are not QMFs and instead two separate digital filters must be used. Otherwise, they are similar to the orthogonal wavelet filters.

The pyramid algorithm was applied to the same MALDI-MS spectrum as with the orthogonal wavelet coefficients, but for this case a biorthogonal wavelet filter pair was used. For this case, the Villasenor 5 filter was used that has a QMF with three points and a dual QMF with seven points. These filters differ in that they have odd numbers of points.

Fig. 3 is virtually identical to Fig. 2 except for subtle differences in peak shape and their locations. The advantage of the biorthogonal wavelet compression is that the compressed data may be modeled, and the models are amenable to interpretation without reconstruction back to the original representation, which can be enabling for very large data sets.

The pyramid algorithm works well for smooth or continuous data. For data for which the signal is contained in higher frequency components (e.g., sharp peaks) the nonlinear wavelet compression may be suitable. The nonlinear WT will completely transform the data so that all that remains is a single smooth wavelet coefficient and all the other data points are the detail coefficients. For this case, the detail coefficients $\boldsymbol{d}$ are retained. The detail coefficients resemble first-derivative spectra, so they will have negative features as Fourier coefficients. These transformed coefficients may be directly modeled by MCR, but one must be careful regarding nonnegativity constraints. If the peaks or features have diverse shapes or widths, it may be apropos to model in the frequency domain using the detail coefficients. If the peaks all have similar shapes and widths, they will have a similar frequency response, so the detail coefficients can be used to reconstruct a low-resolution representation of the data that comprise the smooth coefficients, and nonnegativity constraints will no longer be problematic.

The nonlinear compression first performs a complete WT and then stores the wavelet coefficients of greatest magnitude and their locations. In this

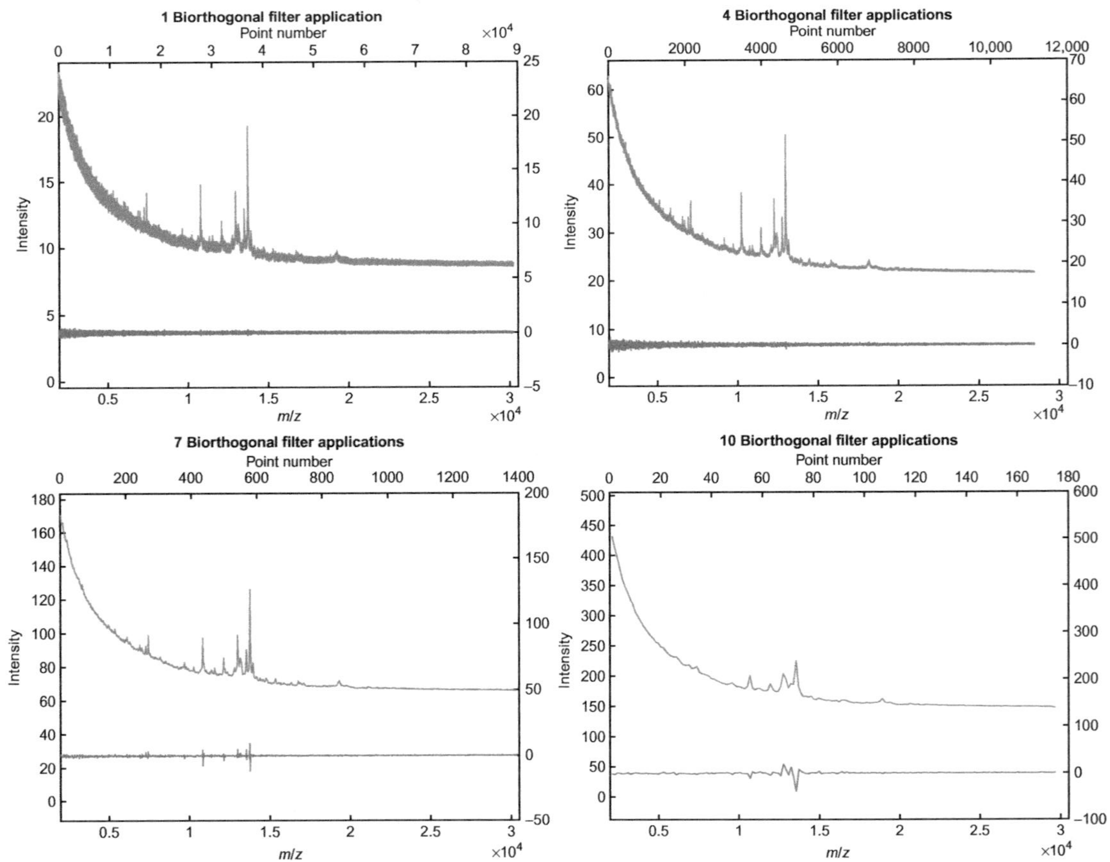

FIG. 3 *Escherichia coli* MALDI-MS smooth (*blue*; *dark gray* in the print version) and detail (*orange*; *gray* in the print version) coefficients after various numbers of biorthogonal wavelet filter applications. The left ordinate corresponds to the smooth coefficients and the right ordinate corresponds to the detail coefficients. Each application will divide the data into equal numbers of smooth and detail coefficients. To save space, the figure comprises the results after 1, 4, 7, and 10 filter applications. The Villasenor filter biorthogonal was used with 3 and 7 filter points.

approach high- and low-frequency information may be stored so long as it is important (i.e., the coefficient or contribution is large). Before MCR, the data are reconstructed to the smooth coefficients, but it can be a partial reconstruction to generate a smaller low-resolution version of the data. Then, after constructing the model, the components of the model may be appropriately reconstructed in their original representation.

In Fig. 4, one can see that the nonlinear compression is more efficient than the pyramid algorithm. The nonlinear compression was accomplished so that the number of coefficients would correspond with those obtained from the pyramid algorithm used for Figs. 2 and 3. The spectral features were attenuated with linear compression when compressed to 175 data points; however, nonlinear compression preserved the spectral features. The reason is that for nonlinear compression only the large coefficients are stored, so in regions of the data without features the coefficients are discarded, and space is saved. Conversely, the pyramid algorithm uniformly samples the entire range of the data by retaining the smooth coefficients regardless of their size. Therefore, when the signal is sparse in that the data range is large and there are only a few narrow spectral features, nonlinear compression is more efficient than linear compression.

3 EXPERIMENTAL DETAILS

The ion mobility spectrometer was a Graseby (Graseby Ionics, Ltd., Watford, UK) chemical agent monitor (CAM) with the acetone dopant cartridge removed. The ionization chemistry is based on water from the humidity in the air. This instrument is handheld and has a silicone membrane that the analytes must perfuse through. The purpose of the membrane is to keep the instrument dry so that the ionization efficiency and sensitivity are improved. However, an unexpected advantage is that analytes will perfuse through the membrane at different rates and effectively provide a separation.

The instrument removes the analytes after sampling by sorptive scrubbing using a bed of molecular sieves. This clean-up process that allows the instrument to return to baseline response presents a second method of separation by the selective sorption rates of different analytes onto the molecular sieves. A schematic of the instrument is given here [2].

In the experiment, a vial containing a 1:1 (*v*/*v*) mixture of dicyclohexylamine (DCHA) and diethylmethylamine (DEMP) was sampled by placing the tube close to the inlet of the instrument and then removing it (aka, sniffing). The experimental details may be found elsewhere [21,22]. For this measurement, 1560 spectra were collected each with 1500 data points using an 80 kHz sampling rate in a time span of 2.2 min. All measurements were made in positive ion mode.

For SIMPLISMA, the damping factor α that attenuates the influence of small variances is set to 5% of the maximum average spectral intensity for

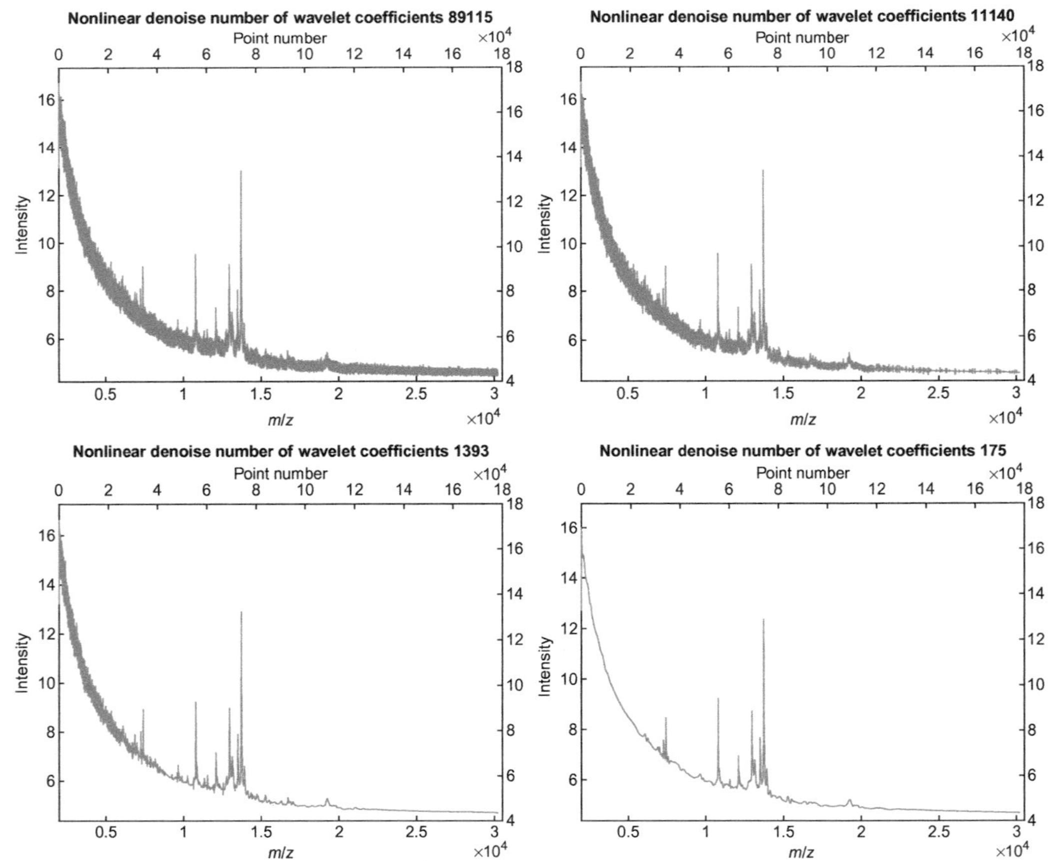

FIG. 4 *Escherichia coli* MALDI-MS smooth (*blue*; *dark gray* in the print version) coefficients after various numbers of nonlinear orthogonal wavelet filter applications. Each application will divide the data into equal numbers of smooth and detail coefficients. To save space, the figure comprises the results after 1, 4, 7, and 10 filter applications. The nonlinear compression used with the Daubechies filter of 10 points.

the data set. This value is fairly standard and is the default value for my research group. The number of components was determined automatically using the ratio of purity value method. The ratios were converted to their negative natural logarithms and the threshold value was set to 10, which also is a default value. The number of components for all SIMPLISMA calculations was automatically determined to be 4 using the previously state criterion. The home built ALS function used the MATLAB function from the Optimization Toolbox *lsqnonneg* to perform all regressions. The wavelet core functions were obtained from the freeware Wavelab Toolbox version 8.50 (http://www-stat.stanford.edu/~wavelab). All calculations were completed in MATLAB 8.5 with the Optimization Toolbox 7.1 (MathWorks, Natick, MA) on a Core i7 860 CPU operating at 2.8 GHz with 16 GB of RAM and Windows 8.1 64-bit Enterprise Edition.

4 DISCUSSION OF RESULTS

In Fig. 5, the left side is a photograph of the CAM sniffing a vial that is a 1:1 (v:v) mixture of DCHA and DEMP. On the right side, the average spectrum has three prominent peaks. The first peak at 6.2 ms is the water reactant ion peak (RIP). The peaks at 7.7 and 10.2 ms, respectively, are the DEMP monomer and dimer ions. At 8.8 ms, there is a small peak that could be easily overlooked that belongs to DCHA.

The disadvantage of using an average spectrum when dynamic processes are involved is the loss of all temporal information. However for many practitioners of analytical chemistry, reducing the data into a single average measurement is the standard operating procedure. As will be seen, the historical approach should be abandoned, because of the loss of temporal information. Any field measurement made outside the controlled environment of the laboratory may also contain dynamic information as an instrument scans a surface or as the wind blows chemical vapor plumes toward the inlet of the instrument.

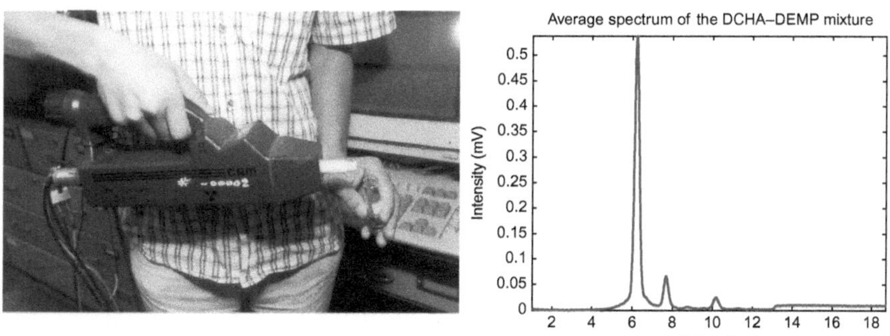

FIG. 5 *Left*: A CAM sniffing a vial of a DCHA and DEMP mixture. *Right*: The average spectrum from the measurement.

Two other commonly used display methods are given in Fig. 6 that are the surface plot and the heat map. It is difficult to visualize small features in both of these renderings. For heat maps if the contour levels are not carefully controlled, smaller features may be lost in the background.

MCR models decompose the data into two perspectives or renderings that allow smaller features to be seen. The results may be viewed as two two-dimensional figures that are linear and can be interpreted quantitatively unlike the surface in Fig. 6. Because independently varying components from the model can be plotted, smaller features are apparent that may be lost in surface plots or heat maps.

SIMPLISMA was applied first to select pure variables and determine the number of components. Then ALS was applied using the concentration profiles from SIMPLISMA to initiate the algorithm.

Fig. 7 gives the SIMPLISMA concentration profiles and spectra extracted from the DCHA–DEMP set, which richly displays the chemistry involved in the dynamic response of the IMS instruments positive ion response of the vial

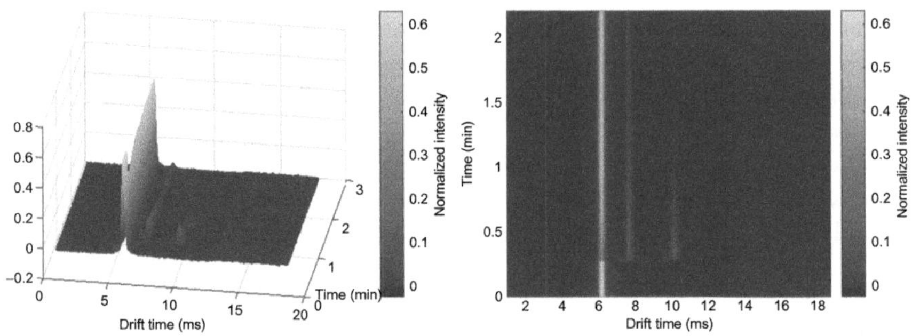

FIG. 6 The DCHA–DEMP IMS measurement displayed as a surface plot (*right*) and a heat map (*left*).

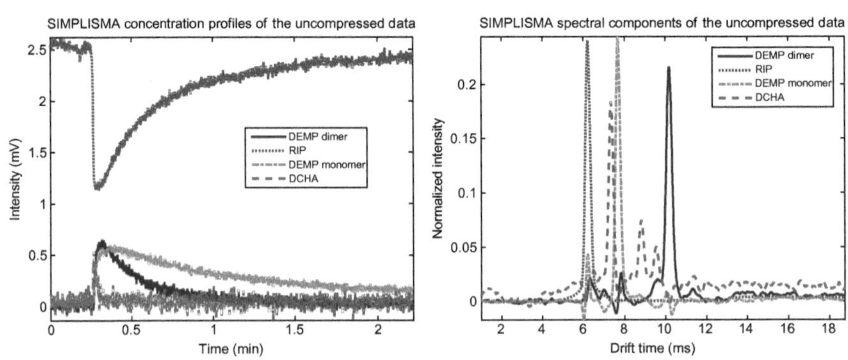

FIG. 7 SIMPLISMA concentration profiles and spectra for the CAM sniff of a mixture of DCHA and DEMP.

sniff of the DCHA–DEMP mixture. The RIP is a cluster ion of water molecules bound to a proton. This peak is the baseline response of the instrument because there is always moisture present in ambient air. The peak will transfer charge to the analytes in the form of proton attachment to the analytes. Thus, the RIP decreases when the vial is sampled as the analytes enter the instrument, and the charge is transferred from the RIP to the analytes.

Competitive charge transfer is very important for ionization processes. For this case, positive ions are measured, and a more a basic analyte can remove the charge from less basic analytes and render them neutral and thereby undetectable by the spectrometer. In this case, DEMP has a greater proton affinity than DCHA and suppresses the DCHA signal by competitive charge transfer. However, DCHA is more efficient at perfusion through the silicon membrane of the instrument and is detected initially in the concentration profile before being quenched by DEMP. One can see the DCHA initially in the concentration profile, and then as the DEMP is detected, it disappears in the concentration profile and the spectra. The next peak in the concentration profile to arise is the DEMP dimer peak. Then, the more persistent peak in the measurement is the DEMP monomer peak.

For the spectra in Fig. 7, some negative dips between peaks are apparent. There are also some negative features in the concentration profiles, but as a reminder, a properly corrected baseline should have an equal number of negative and positive features or deviations about the baseline so that the average or sum of the baseline points will be zero. Many baseline correction algorithms are incorrect so that all the baseline points are positive which can introduce bias and deleteriously affect subsequent data analysis.

The ALS approach is implicit in the SIMPLISMA algorithm so that when ALS is used to refine the SIMPLISMA results, it has implemented nonnegativity constraints for the concentration profiles and spectra. The ALS results are given in Fig. 8.

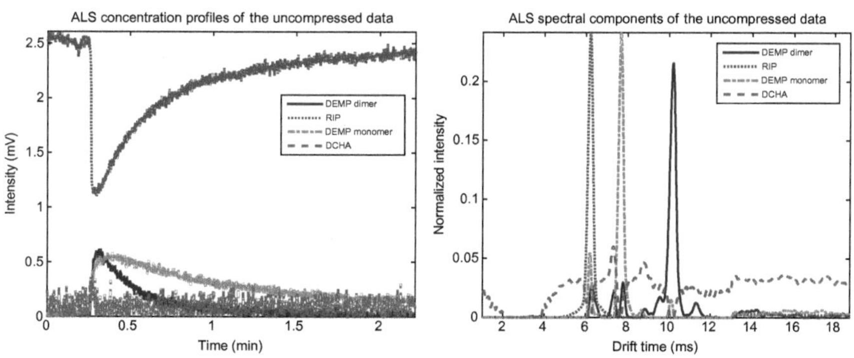

FIG. 8 The ALS results implementing nonnegativity constraints to the SIMPLISMA results from the positive IMS sniff of the DCHA–DEMP mixture.

In Fig. 8 one can see that all the negative features have been removed from the MCR model. Furthermore, the DCHA response has been attenuated in the concentration profiles, and the peaks have been degraded in the spectral profiles.

Orthogonal, biorthogonal, and nonlinear wavelet compressions will be coupled to MCR and applied to the DCHA–DEMP measurements. For the first case, the Daubechies wavelet filter with 12 QMF points was chosen (six vanishing moments). Although there are many filters and several options for filter size, for the most part, they all function similarly. As long as the data are not overly compressed so that important variances are removed, the choice of filter and filter size is not critical.

The IMS data were compressed by a factor of 16 for both the spectral and concentration ways so that the data matrix that was 1560×1500 before compression and is 99×90 after compression or reduced to 4‰ of its original size. The compressed data matrix was then subjected to SIMPLISMA using the same set of parameters as for the uncompressed matrix.

For Fig. 9, the key difference between modeling the uncompressed and compressed data is the removal of the high-frequency noise components. For a DCHA peak, a negative artifact peak appears at 7.9 ms. The compressed data were then submitted to ALS, and then the obtained profiles were back-transformed to their original representation.

ALS fails to accurately model the DCHA component from the uncompressed data in Fig. 8. However, the implicit signal-to-noise improvement of the WT allows the DCHA component to be accurately modeled in Fig. 10. Therefore, not only does the compression improve the computational efficiency by removing high-frequency noise from the data, but it also improves the efficacy of the model by improving the signal-to-noise ratio.

Biorthogonal wavelets are a less popular but useful variant of the wavelets. They work about the same when applied to the data using the same

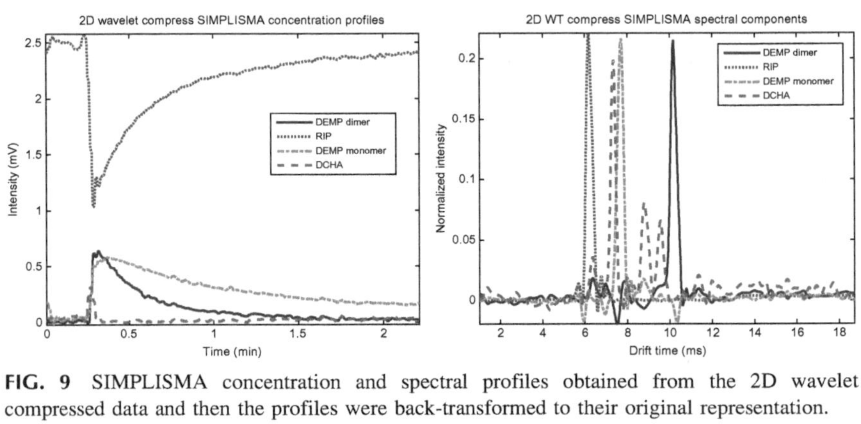

FIG. 9 SIMPLISMA concentration and spectral profiles obtained from the 2D wavelet compressed data and then the profiles were back-transformed to their original representation.

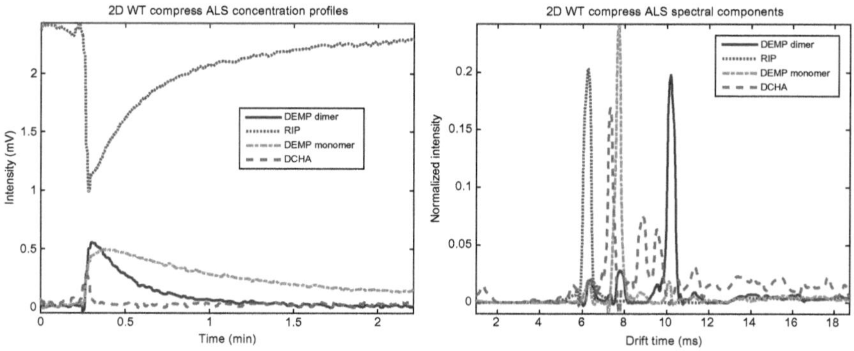

FIG. 10 The reconstructed ALS profiles from the 2D wavelet compressed data.

conditions. They have an advantage that the wavelet coefficients are not shifted, so if you are unconcerned about peak shape, you can skip the inverse transform step to reconstruct the data. Two-dimensional biorthogonal wavelets were used to compress gas chromatography–mass spectrometry data so that the two-way retention time × spectral images would be amenable to pattern recognition [23].

Fig. 11 compares the decimated DEMP component from ALS of the uncompressed data set with the ALS DEMP component from the two-way compressed data using the orthogonal Daubechies coefficients and the biorthogonal Villasenor coefficients. One can see that the biorthogonal coefficients preserve the feature location, while the orthogonal coefficients will shift it. Both sets of wavelet coefficients were divided by four because they were compressed with four levels of the wavelet filter so that the intensity scale will correspond to the uncompressed component. However, if the peak shape is unimportant, the components from the biorthogonal wavelet coefficients are good approximations. Thus, the inverse transform of the model can be avoided, and the biorthogonal coefficients can be used directly. Typically, the wavelet coefficient features will have the shape of the wavelet, so when picking a wavelet, it is useful to find one that resembles the analytical peak shape if the reconstruction step shall be bypassed.

When it is important to model high-frequency components, the nonlinear WT can be used [7,10,24]. In this case, one models the wavelet coefficients above a given threshold. Higher compression rates can be achieved by neglecting uninformative regions of the measurement and storing only the wavelet coefficients above a certain magnitude threshold.

For this data set, 1124 wavelet coefficients with the greatest magnitude that were within the bounds defined by the pyramid algorithm used before for the 4 × 4 level compressions. The four row- and column-level compression was used earlier and is used for the nonlinear compression to reconstruct a low-resolution data set that is modeled by the MCR methods. This procedure

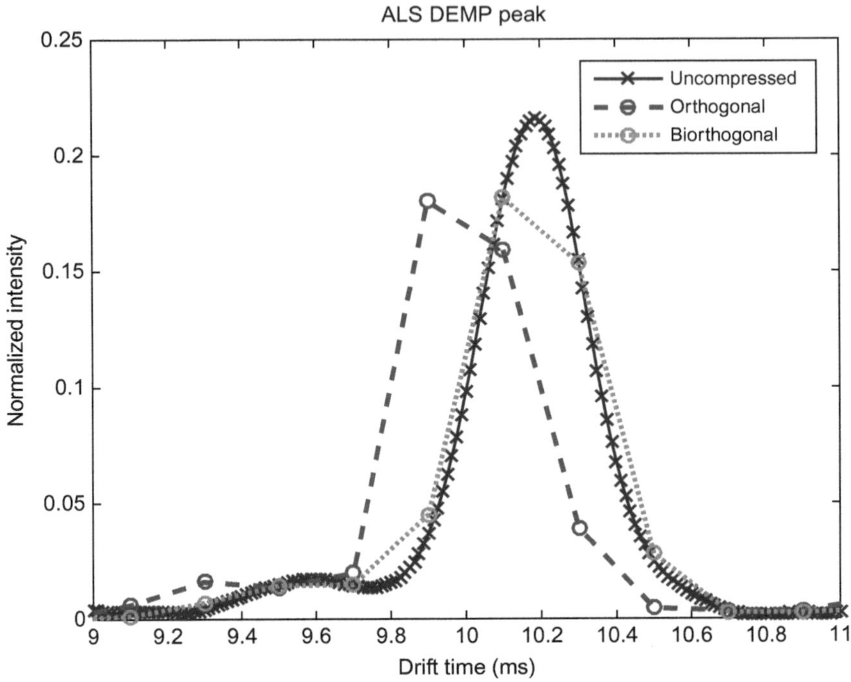

FIG. 11 A comparison of the uncompressed DEMP dimer ALS component, the component extracted from the orthogonal Daubechies coefficients, and the component extracted from the biorthogonal Villasenor coefficients.

compressed the data to 50 ppm of the original matrix size. There is also a requirement to store a control vector that contains the index of where the wavelet coefficients belong in the reconstructed data matrix. There are many schemes to reduce the size of this vector, but if a simple integer array is used, it will not add much to the computational memory.

The SIMPLISMA concentration profiles and spectra were obtained from a low-resolution reconstruction of the compressed data in Fig. 12. For nonlinear compression, the data are completely transformed to the detail coefficients and one smooth coefficient, the coefficients above a given intensity threshold are stored. These coefficients are placed back into a smaller matrix 128×128 and the back-transformed to generate a low-resolution reconstruction of the data. These low-resolution data are subjected to SIMPLISMA and ALS. Then the model profiles are back-transformed to their original size and representation.

In Fig. 12, for the RIP component, the Gibbs effect can be observed by the ringing pattern at the beginning and ending of the concentration profile and some small ringing around the base of the peak in the spectral profile. Wavelets just like the FT require the signal to be periodic, and when discontinuities

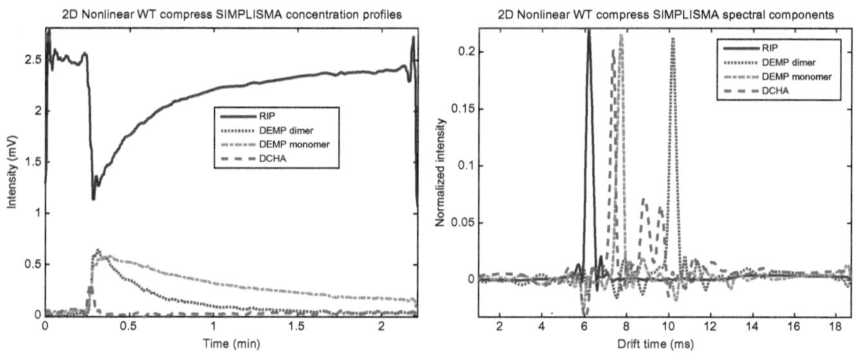

FIG. 12 The SIMPLISMA concentration and spectral profiles from the nonlinear wavelet compressed data to 0.005% of its original size.

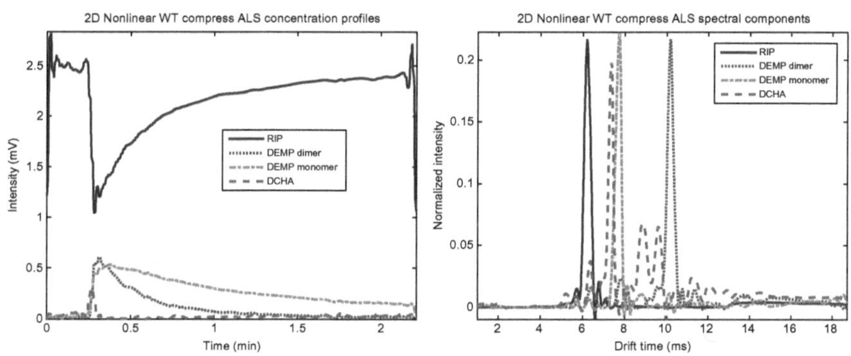

FIG. 13 The ALS concentration and spectral profiles from the nonlinear wavelet compressed data to 0.005% of its original size.

are observed this effect is manifested in the reconstructed data. As mentioned before, wavelets require the data to be of a dyadic length and this effect arises from the zero filling that occurs at the end of the concentration profile for the padding step. Note that it does not appear during the linear compression. Typically one can remove some of the points at either end of the reconstructed concentration profile to eliminate this effect, especially if measurements are acquired before the sample introduction and data are acquired until the instrument has returned to its baseline response.

As in all the cases before, the concentration profile from SIMPLISMA was used to initiate the ALS algorithm with nonnegativity constraints. The ALS model from the low-resolution reconstruction from the 2D nonlinear wavelet is given in Fig. 13. When comparing the profiles for SIMPLISMA and ALS, the ALS has fewer negative artifacts by removing some of the Gibbs effects especially in the spectral profiles around the bases of the peaks.

TABLE 1 The Ratios of the Negative Logarithms of Purity Ratios for SIMPLISMA with Respect to the Different Data Compressions

Component #	Uncompressed	Orthogonal	Biorthogonal	Nonlinear
1	0	0	0	0
2	1.91	1.96	1.96	0.6
3	3.96	4.10	4.03	3.93
4	7.48	7.45	8.50	7.51
5	11.8	13.9	14.3	13.3

As a final note, the ratio of the purity values also improves with wavelet compression, thereby making it easier to determine the correct number of components. The ratios are reported as their negative natural logarithms in Table 1. The threshold that was used to determine the number of components was a value of 10, so all ratios below 10 were retained and used in the SIMPLISMA and ALS models.

For the nonlinear compression, the component numbers for the RIP and DEMP dimer have been reversed from the other models. It is for this reason that the second component's purity ratio differs between the nonlinear wavelet compression and the other SIMPLISMA results.

5 CONCLUSION

Wavelet compression offers an approach that allows one to reduce the size of the data while at the same time improving its quality through the removal of high-frequency noise components. Data can easily be reduced below 1% of its original size. By modeling the linear, compressed, smooth coefficients directly and then using the inverse transform on the model saves computation time and space but also improves the quality of the model by eliminating noise. Once the model is obtained, there is no need to reconstruct the entire data set. Accurate SIMPLISMA and ALS models were obtained from reduced data sizes of 4‰ using linear and 50 ppm using nonlinear compressions. The number of components may be automatically determined by SIMPLISMA, and SIMPLISMA provides good initial estimates for ALS. Wavelet compression improves the contrast between the signal and noise components when determining the number of components by purity ratio. The reduction in the size of the data set and improvement in signal-to-noise ratio achieved through wavelet compression allow MCR methods to be applied faster on smaller computer devices and concomitantly improve the MCR model quality.

ACKNOWLEDGMENTS

Xinyi Wang is thanked for her helpful comments. Cyril Ruckebusch is thanked for his helpful suggestions and meticulous corrections. The US Army ECBC is thanked for providing the IMS instrument and funding this research. The Ohio University Center for Intelligent Chemical Instrumentation is thanked for its continual support.

REFERENCES

[1] Harrington PD, Reese ES, Rauch PJ, Hu LJ, Davis DM. Interactive self-modeling mixture analysis of ion mobility spectra. Appl Spectrosc 1997;51:808–16.
[2] Rauch PJ, Harrington PD, Davis DM. Chemometric resolution of mixture components by cleardown rates. Anal Chem 1998;70:716–23.
[3] Buxton TL, Harrington PD. Rapid multivariate curve resolution applied to identification of explosives by ion mobility spectrometry. Anal Chim Acta 2001;434:269–82.
[4] Reese ES, Harrington PD. The analysis of methamphetamine hydrochloride by thermal desorption ion mobility spectrometry and SIMPLISMA. J Forensic Sci 1999;44:68–76.
[5] Shaw LA, Harrington PD. Seeing through the smoke with dynamic data analysis—detection of methamphetamine in forensic samples contaminated with nicotine. Spectroscopy 2000; 15:40.
[6] Slagel SL, Harrington PD, Schmittauer P. Pulling the plug on the data cable: wireless transmission of chemical sensor data using a cordless phone. Spectroscopy 1999;14:42–66.
[7] Cao L, Harrington PD, Harden CS, Mchugh VM, Thomas MA. Nonlinear wavelet compression of ion mobility spectra from ion mobility spectrometers mounted in an unmanned aerial vehicle. Anal Chem 2004;76:1069–77.
[8] Harrington PD, Rauch PJ, Cai CS. Multivariate curve resolution of wavelet and Fourier compressed spectra. Anal Chem 2001;73:3247–56.
[9] Windig W, Guilment J. Interactive self-modeling mixture analysis. Anal Chem 1991;63: 1425–32.
[10] Cao LB, Harrington PD, Liu JD. SIMPLISMA and ALS applied to two-way nonlinear wavelet compressed ion mobility spectra of chemical warfare agent simulants. Anal Chem 2005;77:2575–86.
[11] Chen G, Harrington PB. Real-time interactive self-modeling mixture analysis. Appl Spectrosc 2001;55:621–9.
[12] Chen GX, Harrington PB. Real-time two-dimensional wavelet compression and its application to real-time modeling of ion mobility data. Anal Chim Acta 2003;490:59–69.
[13] Rauch PJ, Harrington PDB, Davis DM. Near real-time self-modeling mixture analysis. Chemom Intell Lab Syst 1997;39:175–85.
[14] Young F, De Leeuw J, Takane Y. Regression with qualitative and quantitative variables: an alternating least squares method with optimal scaling features. Psychometrika 1976;41: 505–29.
[15] Tauler R, Kowalski B, Fleming S. Multivariate curve resolution applied to spectral data from multiple runs of an industrial-process. Anal Chem 1993;65:2040–7.
[16] Mallat S. A theory for multi-resolution signal decomposition: the wavelet representation. IEEE Trans Pattern Anal Mach Intell 1989;11:674–93.
[17] Walczak B, editor. Wavelets in chemistry. Amsterdam: Elsevier; 2000.
[18] Alsberg BK, Woodward AM, Kell DB. An introduction to wavelet transforms for chemometricians: a time-frequency approach. Chemom Intell Lab Syst 1997;37:215–39.

[19] Leung AKM, Chau FT, Gao JB. A review on applications of wavelet transform techniques in chemical analysis: 1989–1997. Chemom Intell Lab Syst 1998;43:165–84.

[20] Walczak B, Massart DL. Wavelets—something for analytical chemistry. Trends Analyt Chem 1997;16:451–63.

[21] Urbas AA, Harrington PB. Two-dimensional wavelet compression of ion mobility spectra. Anal Chim Acta 2001;446:393–412.

[22] Harrington PB, Urbas A. Strategies for smarter sensors. Int J Ion Mobil Spectrom 2001;4:26–30.

[23] Sun XB, Zimmermann CM, Jackson GP, Bunker CE, Harrington PB. Classification of jet fuels by fuzzy rule-building expert systems applied to three-way data by fast gas chromatography-fast scanning quadrupole ion trap mass spectrometry. Talanta 2011;83: 1260–8.

[24] Cao LB, Harrington PD, Liu C. Two-dimensional nonlinear wavelet compression of ion mobility spectra of chemical warfare agent simulants. Anal Chem 2004;76:2859–68.

Chapter 10

Chemometric Resolution of Complex Higher Order Chromatographic Data with Spectral Detection

D.K. Pinkerton*, K.M. Pierce† and R.E. Synovec*,[1]
*University of Washington, Seattle, WA, United States
†Seattle Pacific University, Seattle, WA, United States
[1]Corresponding author: e-mail: synovec@chem.washington.edu

Chapter Outline

1 INTRODUCTION

One-dimensional (1D) chromatography and two-dimensional (2D) chromatography offer well-established instrumental platforms that produce separations of chemical compounds in complex samples in a variety of fields and applications. The most commonly implemented chromatographic instruments are 1D gas chromatography (GC), 1D liquid chromatography (LC), comprehensive 2D gas chromatography (GC × GC), and comprehensive 2D liquid chromatography (LC × LC) [1–4]. Coupling or "hyphenating" chromatographic instruments to spectroscopic detectors like multiwavelength absorbance diode array detection (DAD), mass spectrometry (MS), and time-of-flight mass spectrometry (TOFMS) results in powerful instruments

Data Handling in Science and Technology, Vol. 30. http://dx.doi.org/10.1016/B978-0-444-63638-6.00010-3
Copyright © 2016 Elsevier B.V. All rights reserved.

capable of producing informative data that facilitate quantitative results with high chemical selectivity [5,6]. While there are other chromatographic instruments and methods coupled with various detectors, this chapter focuses on GC–MS, LC–DAD, GC × GC–TOFMS, and LC × LC–DAD, since the majority of research that employs chemometric resolution is performed with these systems.

Concurrent with these chromatographic instruments producing very large and information-rich data (typically 50–500 MB per sample injection onto the instrument), there are significant challenges in working with multiple chromatograms of this size, especially when analyzing complex samples. Different chemicals that possess similar chemical functional groups often have similar spectroscopic and chromatographic properties that make it difficult for analysts to physically resolve, identify, and quantify the chemical analytes in a complex mixture. It is the accurate and precise identification and quantification of the chemical analytes in complex mixtures that is paramount to converting the experimental data into useful information.

When physical resolution by chromatography is not fully achieved, analysts often turn to chemometrics to achieve mathematical resolution, to unmix the signal contributions from overlapped analytes. To illustrate mathematical resolution, Fig. 1 (left) shows the observed signals for a simulated unresolved two-component mixture, LC–DAD (top) and GC × GC–TOFMS (bottom), wherein each chemical analyte is represented in the chromatographic domain(s) by a Gaussian elution profile with total signal proportional to concentration. Many chemometric resolution techniques are available which,

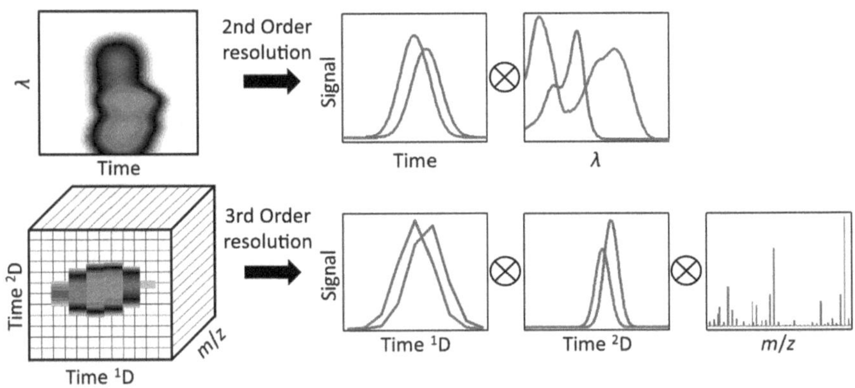

FIG. 1 Observed sample profiles of chromatographically overlapped two-component mixture (*left*) with LC–DAD (*top*) and GC × GC–TOFMS (*bottom*). Mathematically resolved component profiles (*right*) after application of second-order (*top*) and third-order (*bottom*) chemometric resolution. LC–DAD data decomposed into chromatographic elution profiles, signal vs time, and UV–vis absorbance spectra, absorbance vs wavelength (λ). GC × GC–TOFMS data decomposed into column one elution profiles, signal vs time ^{1}D, column two elution profiles, signal vs time ^{2}D, and time-of-flight mass spectra, signal vs m/z.

when operated under proper conditions and given proper information, may be able to extract the pure component elution profiles and the pure component spectral profiles, as shown in Fig. 1 (right). The mathematically resolved, pure component profiles for each analyte in the sample can then be individually examined and quantified. Fig. 1 (top) is an example of resolving second-order data, while Fig. 1 (bottom) is an example of resolving third-order data. A 1D chromatography instrument coupled with a univariate detector (e.g., GC-FID) would produce first-order data. Even higher dimensions of data can be achieved when multiple second-order or multiple third-order chromatographic runs are simultaneously analyzed, thus adding a sample dimension to the data. The order of the data determines which chemometrics can be applied to resolve overlapping analyte peaks and different orders of data provide different advantages, sometimes referred to as the second-order advantage, bilinear data, or third-order advantage, trilinear data [7]. In general, when aiming to achieve mathematical resolution of overlapping analyte peaks, the advantage is that higher order data require a user to input less information than lower order data. For example, chemometric resolution methods for first-order data (e.g., classical least squares (CLS) analysis) require the user to input a pure reference spectrum for each component in the mixture. An alternative approach to using CLS for first-order data would be to simultaneously analyze multiple samples to artificially increase the order of the data so higher order chemometrics can be used. Samples with second- and third-order advantages involve freeing the analyst from providing pure reference spectra (a big advantage for the analysis of unknown samples) and freeing the analyst from providing more than one chromatogram to analyze any component in the sample (a big advantage in the midst of analyte peak retention time shifting and other sources of uncontrollable and chemically irrelevant variations among chromatograms).

The ability to mathematically resolve analytes within a single sample significantly relaxes the requirement for sample-to-sample retention time precision, minimizing the need to implement retention time alignment as a data preprocessing step because replicate sample injections to the instrument can be analyzed independently. However, we should be clear that producing higher order data does not necessarily ensure that the data are amenable to analysis by higher order chemometrics or that the user can make use of the second- or third-order advantages. Producing data that can take full advantage of these chemometric methods strongly depends on the balance between the instrumental design, how the user operates an instrument, and the specific capabilities of the algorithm(s) implemented within the chemometric methods.

While there are specific advantages and disadvantages to each chemometric method, which will be discussed in more detail in later sections of this chapter, mathematical resolution generally requires at least some selectivity that exists between analyte components in each data dimension. The ability of a data set to be decomposed into individual analyte component vectors also generally requires the data to be bilinear for second-order data or trilinear for third-order data, meaning each of the data dimensions are linearly independent from one another and the pixel-level data are composed of unique, reproducible, and analyte concentration-dependent signals at consistent retention times from each independent and unique analyte source that is present. To be more precise, contributions to the signal are also provided by baseline and indeterminate noise sources. Linear dependence between dimensions or inconsistencies in retention times can be removed by reducing the order of the data, but this also means that the user may not be able to fully utilize the advantages of the true dimensionality of the data. Accordingly, this chapter will first describe the instrumental designs and instrumental operation techniques that produce good quality high-order data. Then the chapter will describe several chemometric methods that provide mathematical resolution for ND data, where $N > 1$, specifically, multivariate curve resolution-alternating least squares (MCR-ALS) for two-way data, generalized rank annihilation method (GRAM) for two-way data, and parallel factor analysis methods (PARAFAC and PARAFAC2) for three-way data [8–16].

2 INSTRUMENT DESIGN AND UTILIZATION TO PRODUCE HIGH-ORDER CHEMICAL SEPARATIONS DATA

Producing ND chromatographic data that engender second-order and third-order advantages strongly depends on the quality of the instrument design and utilization. It is important to note that when discussing and comparing GC–MS, LC–DAD, GC × GC–TOFMS, and LC × LC–DAD, or other related chromatographic instruments coupled with spectral detection, the instrumentation and experimental challenges are generally different, but the data analysis challenges for applying chemometric resolution methods are very similar.

GC–MS, LC–DAD, GC × GC–TOFMS, or LC × LC–DAD should chromatographically separate as many analytes as possible with the shortest separation run time possible. However, sometimes it is not feasible to chromatographically resolve all the analytes, especially if a complex sample contains hundreds to thousands of chemical compounds, and high-throughput applications demand fast chromatographic acquisition rates (high experimental duty cycle). Therefore, the key goal for the analyst is to strike a balance between too little chromatographic resolution (time saving) and total chromatographic resolution (time consuming), while collecting appropriate data that can be readily analyzed and transformed into useful information to answer a specific analytical question of interest.

2.1 Goals for Two-Way Data: GC–MS and LC–DAD

The instrumental goals for 1D chromatography with spectral detection such as GC–MS and LC–DAD are to maximize selectivity, provide physical separation of analyte peaks, and keep the run time fast to improve the experimental duty cycle, so narrow chromatographic peaks are preferred. Narrower peaks not only improve separation peak capacity, they also focus the analyte signal, improving sensitivity, and signal-to-noise ratio (*S/N*). However, chemometric algorithms require some degree of selectivity between overlapped peaks when chromatographic resolution is not achieved, so each peak should be wide enough and sampled with a data acquisition rate high enough to produce at least ~10 to 20 spectral scans for each peak's elution profile, and provide enough chemical selectivity for chemometric algorithms to appropriately resolve analyte peaks. This requirement for 10–20 scans per peak width at base does not mesh well with simultaneously trying to obtain narrow peaks, especially for quadrupole mass spectrometers (qMS) which are commonly used in GC–MS. A typical qMS with unit mass resolution can collect up to 2–4 scans/s over a 40–600 *m/z* range per scan. Therefore, analyte peaks in GC–qMS need to have a width of ~4 to 6 s to achieve ~10 to 20 scans/peak. In contrast, TOFMS can collect 500 full spectra/s over the same 40–600 *m/z* range, although most work with TOFMS is performed at 100 spectra/s. Therefore, at the 100 spectra/s data acquisition rate, GC–TOFMS provides ~10 to 20 spectra/peak for peaks as narrow as ~100 to 300 ms wide. This fast scanning rate is why the TOFMS is ideally suited for detecting the second chromatographic dimension of peaks in GC × GC–TOFMS.

When qMS is used in GC–MS, then collection of a sufficient number of spectra across the chromatographic peak width is especially important because of a spectral skew imposed by this detector. Spectral skew occurs because the qMS detector does not scan all *m/z* simultaneously, and within a single observed spectrum, later *m/z* can be measured at a different analyte concentration than earlier *m/z* because the Gaussian-shaped peak rapidly swept through the detector between observations, exposing each observation to a slightly different analyte concentration. This causes the collection of spectra in a GC–MS chromatogram to be not only a function of retention time between spectra (totally expected, since this is chromatography after all) but also a function of retention time *within* each spectrum (potentially troubling to a chromatographer who wanted to collect bilinear data). This skewing effect is most noticeable when the peak signal is changing rapidly relative to the detector scan rate, such as the case with narrow chromatographic peaks. The TOFMS does not have this spectral skew because all *m/z* within each spectrum are analyzed at the same time, rather than scanned one-by-one while time passes. When the data are skewed and the analyte spectrum varies across the peak elution profile, the GC–MS data are not strictly bilinear. Fortunately, skewed GC–qMS data can be corrected by applying an unskewing algorithm that corrects the signal intensity for each mass based on the measured intensity and the time needed to go from the start of the scan to each mass [17].

Application of mathematical resolution to LC–DAD data has the same peak sampling and spectral acquisition rate requirements as GC–MS data. A typical diode array detector offers a scan rate up to ~100 scans/s, but is commonly implemented at 40 scans/s over a UV–vis wavelength range of 200–700 nm [6,18]. The fast scan rates for DAD means that peaks only need to be ~0.5 s wide to achieve 10–20 spectra/peak, but the performance of 1D LC typically produces peak widths that range from 5 s to 1 min. A 5 s wide peak detected with a DAD spectral scan rate of 40 scans/s results in 200 spectra/peak, well beyond the minimum requirements for mathematical resolution. Indeed, the fast spectral scan rate of the DAD makes it ideally suited for the second chromatographic dimension of peaks in LC × LC–DAD. Achieving an adequate number of spectral scans across the chromatographic peaks with LC–DAD is less of a concern than in GC–MS, but UV–vis DAD is inherently less chemically selective than MS. Decreased detector selectivity generally requires additional chromatographic selectivity in order to resolve overlapped analyte peaks within a single chromatographic run if the analyst does not have a priori knowledge of the analyte spectrum or elution profile. Alternatively, this challenge is often handled by stacking multiple 2D chromatograms and analyzing the data as a single three-way array [19,20].

Combining replicates of 1D chromatography with spectral detection for simultaneous analysis can significantly reduce analysis time and has the potential to produce trilinear data that are amenable for third-order chemometric methods, but there are some common nonideal features in the data that often need to be addressed. Among these nonideal features are differences in baseline signals and misalignment of peaks between samples. When either of these issues is present in the data, they can cause significant challenges (and shortcomings) in applying third-order chemometrics. Proper tuning of the detector and running samples close together temporally can reduce these issues. However, even with great care in the instrumental analysis, these problems can persist. Because of this, replicates of two-way chromatograms can be fused end-to-end in the time domain into a longer two-way data set (spectral dimension × concatenated retention time dimension) rather than "stacked" as a three-way array (spectral dimension × retention time dimension × sample dimension). Fig. 2 (top) provides a graphical representation of the rearrangement schemes for replicates of 1D chromatography with spectral detection. Reducing the data dimensionality in this manner can remove the deviations from trilinearity that are caused by analyte peak retention time shifts. The analyst also has the option to use peak alignment software to make the data approximately trilinear and amenable for third-order chemometrics [21–23].

2.2 Goals for Three-Way Data: GC × GC–TOFMS and LC × LC–DAD

Comprehensive 2D chromatographic instruments, GC × GC–TOFMS and LC × LC–DAD, produce a third-order chromatogram for each sample that is injected into the instrument. Before discussing the specific instrumental/experimental considerations that are required to produce sufficiently trilinear chromatograms, it is necessary to describe the general design of the

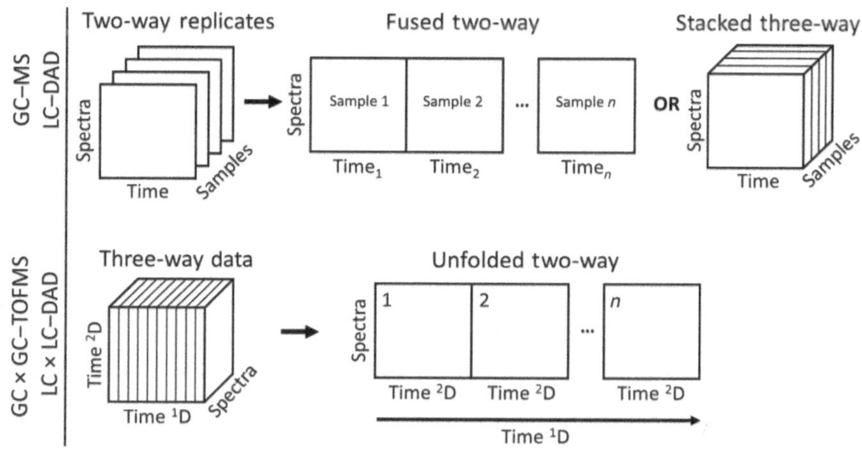

FIG. 2 Graphical representation of data structures (*left*) for replicates of 1D chromatography with spectral detection (*top*) and single run of 2D chromatography with spectral detection (*bottom*), with data rearrangement schemes (*right*). Replicates of two-way data (*top*) are shown fused together end-to-end in the time dimension to create a single two-way array (spectral dimension × concatenated retention time dimension), where Time_n is the time dimension of the nth sample, and stacked along the sample dimension to create a three-way array (spectral dimension × retention time dimension × sample dimension). Three-way data (*bottom*) can be unfolded along the time ^{1}D dimension (modulation dimension) such that each of the n modulations are concatenated, each modulation is as a two-way slab with dimensions spectra × time ^{2}D from the original three-way array.

instruments. In comprehensive 2D chromatography, two columns are connected in series with a modulator between them to move eluate from one column to the next, and a spectral detector is positioned at the end of the second column [4,18]. The two columns should have stationary phases that produce separations that are "complementary" and not overly correlated, in order to more fully utilize the two-dimensional peak capacity offered by comprehensive 2D chromatography. Peak capacity, at unit chromatographic resolution, is simply the separation run time in a given separation dimension divided by the average analyte peak width at base. Hence, the ideal peak capacity of a comprehensive 2D separation is the product of the peak capacities for the first column (^{1}D) and second column (^{2}D) separation dimensions, far greater than the peak capacity offered by 1D chromatography. Peak capacity is a figure-of-merit that roughly indicates the amount of chemical information that can be gleaned from D and 2D chromatographic separations. The modulator serves as an automated sample injection device that interfaces the two columns, placed in series, in the 2D chromatographic instrument. Analytes are initially separated along the ^{1}D separation dimension, with the modulator periodically injecting eluate from this ^{1}D column onto the ^{2}D separation dimension using a constant duty cycle. The ^{2}D separations occur in real-time while preserving chromatographic resolution that was achieved on ^{1}D. GC × GC–TOFMS commonly uses a thermal modulator which is capable of periodically focusing the ^{1}D eluate and improving S/N. LC × LC–DAD commonly uses an 8 or 10 port switching valve as the modulator. The time length of the ^{2}D separation is referred to as the modulation period (P_M), and it is necessarily short (on the order of one second to several seconds) relative to the ^{1}D separation time (on the order of minutes) so that each ^{1}D peak can be sampled by the modulator and separated by ^{2}D an appropriate number of times. Since the detector is positioned at the end of the second column and the second column serially followed the first column, the number of spectra detected and collected along the ^{2}D chromatographic peak profile is determined directly from the second column peak width at base ($^{2}W_b$) relative to the detector spectral scan rate. In contrast, detector sampling of the ^{1}D chromatographic peak profile is dictated by the modulation ratio (M_R). The M_R of a comprehensive 2D chromatographic peak is defined as the first column peak width at base ($^{1}W_b$) divided by P_M ($M_R = {}^{1}W_b/P_M$) [24].

For comprehensive 2D chromatography with spectral detection, generally, a M_R of ~2 to 4 is capable of producing a trilinear chromatogram with sufficient chemical selectivity in all three dimensions, concurrent with sufficient data density to quantitatively define the ^{1}D and ^{2}D peaks, and all the while reasonably utilizing the high peak capacity offered by comprehensive 2D chromatography. Smaller M_R values do result in noticeable broadening of ^{1}D peaks due to "undersampling" of the analyte peaks [25–27]. Larger M_R values are better at defining the ^{1}D peak and its shape, avoiding broadening, and better at detecting chemically selective signals among overlapping peaks. However, the trade-off for larger M_R is reduced P_M, which results in reduced peak capacity in ^{2}D, which in turn may cause a reduced overall 2D peak capacity. For these reasons, $M_R \sim 2$–4 is preferred to balance peak capacity utilization, while appropriately sampling peaks in both dimensions. Analysts should strive to design a comprehensive 2D separation by optimizing the peak capacity of the ^{1}D separation first, and then designing the ^{2}D separation to have the appropriate P_M taking into account $^{1}W_b$ to provide a proper M_R. The resulting ^{2}D peak capacity at this P_M will provide a true improvement to the overall comprehensive 2D peak capacity. These concepts are discussed in the context of specific GC × GC–TOFMS and LC × LC–DAD instrumental operation parameters in the following paragraphs.

When a ^{1}D separation in GC × GC–TOFMS is sufficiently optimized and $^{1}W_b$ is kept small ($^{1}W_b \sim 4$–8 s), a relatively fast modulation period ($P_M \sim 1$–2 s) is required to achieve a desirable $M_R \sim 4$. Peak widths of $^{2}W_b$ on the ^{2}D separation range from ~40 to 400 ms, so the 500 spectra/s provided by the TOFMS is ideally suited for this instrumental platform, readily providing the ~10 to 20 spectra/peak, and a ^{2}D peak capacity range of ~10 to 25. An alternative GC × GC–TOFMS method is commonly practiced, with the goal of running a longer modulation period ($P_M \sim 5$–8 s) to improve the selectivity of the ^{2}D separation (i.e., to increase the ^{2}D peak capacity). Achieving the same M_R with this longer P_M would then require that the ^{1}D separation be run suboptimally in order to produce wide enough peaks

($^1W_b \sim 20$–30 s). Ironically, the overall comprehensive 2D separation peak capacity is compromised by this second approach. Indeed, for all comprehensive 2D separations, the analyst should strive to initially optimize the ^{1}D separation, so as not to reduce separation performance. If the ^{1}D separation is basically optimized, then the use of a long P_M translates into an "undersampling" of the ^{1}D analyte peaks, resulting in a $M_R < 2$, which negates the ability to produce chemically selective third-order data for chemometric analysis. To this point, a long P_M has been shown to be problematic for chemometric analysis, causing significant ^{2}D retention time shifts, defined as Δ^2t_R, between modulations of an analyte, due to the necessary use of temperature programming in GC × GC [28]. The term trilinearity deviation ratio (TDR) was defined as a metric to study this issue. The TDR is the ^{2}D retention time shift normalized relative to the peak width ($\text{TDR} = \Delta^2t_R/^2W_b$) and provides a quantitative metric for the trilinearity of the data that describe the performance of trilinear chemometrics. When TDR = 0, it indicates that the data are strictly trilinear and no shifting has occurred along the ^{2}D separations from one peaklet to another for a given analyte modulated from the ^{1}D separation, as shown in Fig. 3 (top). Herein, we refer to each peak detected in the ^{2}D separation dimension as a peaklet, because groups of the peaklets originate from each ^{1}D dimension analyte peak. Note that the M_R is not necessarily equal to the number of peaklets observed for a given ^{1}D chromatographic peak. While it is still possible to use trilinear chemometric methods when TDR > 0, large TDR values increase the quantitative errors associated with those modeling techniques. If third-order chemometric models are to be used, then the data must be sufficiently trilinear (TDR ~ 0–0.05 in the initial study); often the use of a long $P_M \sim 5$–8 s coupled with fast temperature programs in

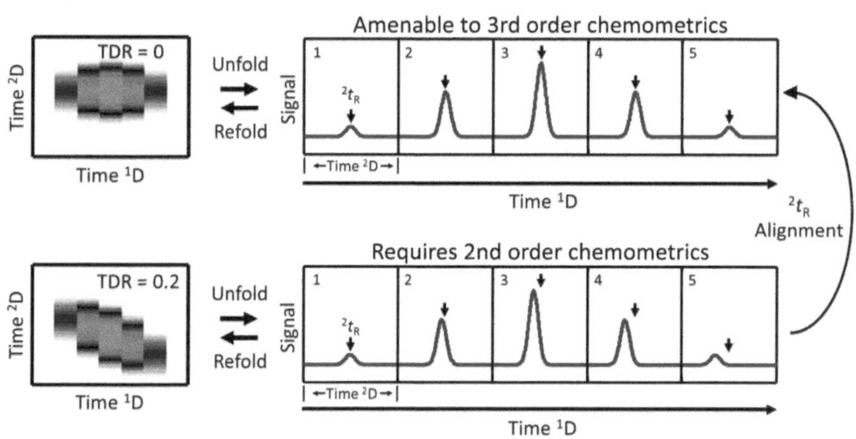

FIG. 3 2D chromatogram of one analyte for a single *m/z* or wavelength taken from GC × GC–TOFMS or LC × LC–DAD sample. Trilinear data (TDR = 0) (*top*) and nontrilinear data (TDR = 0.2) (*bottom*) are shown folded (*left*) and unfolded (*right*). Numbered windows in unfolded data represent individual modulations. *Arrows* in windows of unfolded data indicate the ^{2}D retention time (^2t_R) of the first modulated ^{2}D peaklet.

GC × GC results in data that are not amenable for analysis with third-order chemometrics [28]. Under these circumstances, the data must be rearranged prior to chemometric analysis. Generally, this is accomplished by reducing the order of the data to apply second-order chemometrics, unfolding the three-way data matrix (time ^{2}D dimension × time ^{1}D dimension × spectral dimension) by concatenating each of the n modulations, along the time ^{2}D dimension, to produce a two-way matrix (spectral dimension × n modulations of time ^{2}D dimension). Fig. 2 (bottom) is a graphical depiction of the native three-way data structure with modulations depicted as slabs (spectra × time ^{2}D) along the time ^{1}D dimension, and the unfolded two-way data with n modulations connected end to end. Alternatively, localized peak retention time alignment can be applied, if third-order chemometrics are to be used. To illustrate localized peak retention time alignment, Fig. 3 (bottom) shows a nontrilinear chromatogram, TDR = 0.2, that would require retention time alignment prior to analysis by third-order chemometrics.

The necessary data considerations for applying mathematical resolution to LC × LC–DAD are generally very similar to those of GC × GC–TOFMS; but unlike GC × GC–TOFMS, the solutions to these challenges often come from changing how the LC × LC–DAD instrument functions, rather than varying the separation parameters. Overall, there is significantly less variability in how analysts operate LC × LC–DAD instruments than compared to GC × GC–TOFMS. Typical LC × LC experiments have a total ^{1}D run time of 30–60 min and produce ^{1}D peaks with a $^{1}W_{b} \sim 30$ s to 1 min. Unfortunately, the LC × LC valve modulation system does not provide analyte focusing to improve the S/N, which is a benefit provided by the GC × GC–TOFMS thermal modulator. Also, the P_{M} selection in state-of-the-art LC × LC–DAD is generally restricted to a relatively narrow range ($P_{M} \sim 15$–30 s), with ^{2}D peak widths $^{2}W_{b} \sim 0.5$–2.5 s, resulting in a ^{2}D peak capacity range of ~10 to 20. The instrumental constraints of LC × LC–DAD, particularly the relatively long P_{M} necessary to produce adequate selectivity on the ^{2}D separation dimension, coupled with the peak widths on ^{1}D, result in $M_{R} \sim 1$–3, which in turn impacts applicability of some chemometric methods, so extra care may be required to ensure that the data are amenable to third-order chemometric analysis for a single sample run. Recent LC × LC instrumental developments have helped to reduce the need for ^{2}D peak alignment or the use of lower order chemometric resolution methods, namely the introduction of a synchronized dual valve modulation system, typically implemented as a single 8 or 10 port valve, and independently programmable ^{2}D separation gradients [18,21,29,30].

3 CHEMOMETRIC RESOLUTION METHODS

The chemometric resolution methods discussed in this chapter have at least one design property in common: they all assume that the observed signal profile of a sample mixture is a linear combination of the signal profiles of the pure components in the mixture, and chemically irrelevant sources of variations in observed signals are minimized. This is often a reasonable assumption for chromatography detectors, especially for absorbance spectroscopy when the spectra of unresolved components are additive according to the Beer–Lambert Law, or when ion counts provided by mass spectrometers are additive. However, chemometric resolution methods have unique benefits and tradeoffs in terms of the amount of information a user must provide and the limitations imposed by uncontrollable, chemically irrelevant variations in the data. For example, aside from requiring the user to input the one-way observed mixed spectrum of the original sample mixture ($\mathbf{x}$) as a vector, CLS also requires the user to input the pure spectrum of each component in the mixture ($\mathbf{S}$) as a matrix. Then CLS uses the model

$$\mathbf{x} = \mathbf{c}\mathbf{S}^{\mathrm{T}} + \mathbf{e} \tag{1}$$

to compute the concentrations of the components in the mixture ($\mathbf{c}$) vector by applying the pseudoinverse of $\mathbf{S}$ and minimizing the least squares error of fit ($\mathbf{e}$) vector between $\mathbf{x}$ and the $\mathbf{c}\mathbf{S}^{\mathrm{T}}$ model. CLS outputs the resolved quantitative spectrum for each component in the mixture, but the user needs to know the identity of the components in the mixture before CLS can successfully resolve one-way data. To resolve the overlapping signals of a two-way sample profile using MCR-ALS, the user needs to only input the number of components in the mixture (the identity of those components can remain unknown, if necessary); this is an example of a second-order advantage. The second dimension of chemical selectivity provides chemically relevant information to the MCR-ALS algorithm that was unavailable when CLS was given only one-way data. An analogous additional advantage is provided by third-order data for PARAFAC algorithms wherein fewer user inputs are required to achieve mathematical resolution of overlapped analyte peaks due to the additional data dimension contributing additional chemically selective information. Please note that for equations that deal with mathematically manipulating ND chromatograms in this chapter, lower case letters represent scalars, superscript T means the former variable is transposed, bold lower case letters represent vectors, bold uppercase letters represent matrices, and underlined bold uppercase letters represent 3D arrays.

3.1 MCR-ALS

In terms of two-way chromatographic data, the chemometric model for MCR-ALS is

$$\mathbf{D} = \mathbf{C}\mathbf{S}^{\mathrm{T}} + \mathbf{E} \tag{2}$$

where $\mathbf{D}$ represents a given observed two-way chromatogram with retention times in the rows dimension and wavelengths (or ion m/z) in the columns dimension, $\mathbf{C}$ contains pure predicted chromatograms for the components, $\mathbf{S}^{\mathrm{T}}$ contains pure predicted spectra for the components, and $\mathbf{E}$ is the minimized error of residuals between the predicted and observed two-way chromatogram. We refer the reader to Chapter 2 of this book and to excellent references that thoroughly describe MCR-ALS [12,20,31,32,51]. In brief, while using appropriate constraints, initial estimated pure chromatograms and pure spectra are multiplied to yield the $\mathbf{CS}^{\mathrm{T}}$ model, tested for convergence, then **C** and **S** values are alternately and repeatedly iterated, multiplied, and again tested for convergence, until the convergence criterion is finally met. MCR-ALS outputs a pure two-way chromatogram for each component present in **D**. To apply MCR-ALS, the user does not have to know the identity of the components in the mixture. Instead, the user inputs initial values for **C** and **S**, and the user chooses appropriate constraints (such as unimodality, nonnegativity, or local rank) and an appropriate convergence criterion (such as threshold number of iterations, minimal value of **E**, or an improvement threshold for lack of fit between the predicted model and **D**). Initial values for **C** and **S** can be provided by evolving factor analysis, simple-to-use self-modeling analysis (SIMPLISMA, which is described in Chapter 3 of this book) orthogonal projection approach, or key set factor analysis, all of which, in general, try to select the purest rows or columns present in the original data as initial estimates [12].

As mentioned previously, MCR-ALS requires that the two-way input data are bilinear. In GC–MS and LC–DAD applications, if samples or replicates compose a third dimension, users can append matrices to remove the effects of retention time variations. Users can unfold the three-way array along t_{R} dimensions to produce a suitable two-way matrix, as in Figs. 2 and 3, and then the resolved component profiles that are produced can be refolded into the original three-way form [32]. In GC–MS and LC–DAD applications, if samples or replicates compose a fourth dimension and t_{R} variations are problematic, then unfolding along the t_{R} dimension and augmenting samples or replicates into a single dimension can remove the problematic effects of analyte peak retention time shifts between chromatograms.

3.2 GRAM

Like MCR-ALS, GRAM is also applicable to resolving overlapping peaks in two-way chromatographic data. Unlike MCR-ALS, GRAM is not iterative, meaning it *does not* require the user to input convergence criteria (or iteration stopping point). GRAM outputs a final unique set of quantitative chromatograms for the resolved components in a given sample mixture chromatogram (**M**). However, GRAM *does* require the user to input a standard mixture chromatogram (**N**). In brief, GRAM models **M** and **N** as

$$\mathbf{M} = \mathbf{X}\mathbf{C}_{\mathbf{M}}\mathbf{Y}^{\mathrm{T}} \tag{3}$$

and

$$\mathbf{N} = \mathbf{X}\mathbf{C}_{\mathbf{N}}\mathbf{Y}^{\mathrm{T}} \tag{4}$$

where **X** is a matrix containing the pure component profiles in the first chromatographic dimension, **Y** is a matrix containing the pure component profiles in the second chromatographic dimension, and $\mathbf{C_N}$ is a diagonal matrix containing the concentration of *k* known components in the standard sample. Thus, given **N**, **M**, $\mathbf{C_N}$, and *k* (*k* can be objectively and algorithmically estimated when *k* is initially unknown), it is possible to calculate the resolved pure chromatographic profiles of the components in the unknown sample mixture (**X** and **Y**) as well as the component's concentrations relative to the standard ($\mathbf{C_M}$). We refer the reader to excellent references that thoroughly describe GRAM [13,14,33–39]. In brief, GRAM mathematically resolves the components of the sample mixture, given the following requirements: the unknown sample must contain relative analyte component concentrations that differ from the standard (for example, if the two analytes of interest that overlap in the unknown sample have concentration ratio = 2:1, the standard must contain the two analytes at concentration ratio ≠ 2:1), the overlapped peaks must have some chromatographic and spectral resolution on the respective dimensions, and the analyte retention times and peak shapes (shapes, not magnitudes) must be equal between the unknown sample and the standard (the data must be bilinear). The unknown sample can contain interference peaks that are absent in the standard and vice versa. The key is that the user augments **M** and **N** into a single matrix so that all components that are common to both **M** and **N** are modeled. Thus, GRAM is initiated by either augmenting or summing **M** and **N** into a single matrix. Then the GRAM algorithm goes on as follows: it performs singular value decomposition (SVD) on that single matrix, the SVD decomposition matrices are truncated according to *k* components, an eigenvalue problem is solved for the truncated matrices yielding eigenvectors and eigenvalues. The eigenvectors and decomposition matrices from SVD are used to determine the pure component profiles (**X** and **Y**), and the eigenvalues and $\mathbf{C_N}$ are used to determine the concentrations of the unknown sample components ($\mathbf{C_M}$) relative to the standard mixture.

3.3 PARAFAC

As described earlier, GC × GC–TOFMS and LC × LC–DAD instruments produce trilinear data wherein the signal from an analyte is essentially the outer product of three fixed vectors: the first separation dimension profile (^{1}D), the second separation dimension profile (^{2}D), and the spectrum. PARAFAC can mathematically resolve a given trilinear structure by modeling each component as the outer product of its three fixed vectors, while simultaneously filtering out nontrilinear signals such as noise contributions to signals. The trilinear chemometric model for PARAFAC is

$$\underline{\mathbf{R}} = \sum_{(\text{for}\, i=1\, \text{to}\, n)} \mathbf{x}_i \otimes \mathbf{y}_i \otimes \mathbf{z}_i + \underline{\mathbf{E}} \tag{5}$$

wherein $\underline{\mathbf{R}}$ represents a given observed three-way chromatogram, $\mathbf{x}_i$ is the pure predicted ^{2}D chromatogram for the ith component, $\mathbf{y}_i$ is the pure predicted ^{1}D chromatogram for the ith component, $\mathbf{z}_i$ is the pure predicted spectral profile of the ith component, $\underline{\mathbf{E}}$ is the minimized error of residuals between the predicted and observed chromatograms, and n is the total number of factors in the chromatographic subregion (factors are the sum of number of analyte components and number of sources of noise or baseline contributions to the total signal). We refer the reader to excellent references that thoroughly describe PARAFAC [6,8,28,40–50] and an enhanced version called PARAFAC2 [9,15]. In brief, PARAFAC mathematically resolves overlapping components of a 2D chromatographic subregion while filtering out nontrilinear noise contributions (as long as there is not an excessive number of factors in the model) by using alternating least squares to iteratively converge upon the $\mathbf{x}_i$, $\mathbf{y}_i$, and $\mathbf{z}_i$ values that minimize $\underline{\mathbf{E}}$.

In theory, PARAFAC performs well given the following requirements: the unknown sample array must be sufficiently trilinear (TDR $\leq$ 0.05), it must contain chemically selective information for each component on two of the three dimensions, the user must choose appropriate initialization values (common choices are random values or values provided by generalized eigenvalue decomposition of the raw data), and the user must choose stopping criterion (some form of minimizing $\underline{\mathbf{E}}$ such as reaching a small threshold value for relative change in fit between two successive models). It is beneficial if a small maximum number of components are known to be present in the three-way data. But in the absence of this a priori knowledge, convergence of the algorithm, and objective determination of the number of components present can be achieved by building models of increasing number of factors and automatically detecting the number of factors at which overfitting occurs that number minus one factor is the optimal model. Optional constraints such as nonnegative signals and unimodality in peak shape are often reasonably acceptable assumptions that can be input into a PARAFAC algorithm to facilitate efficient convergence to a final solution, as well, but these are not requirements. As mentioned previously, PARAFAC requires trilinear data. But to be clear, PARAFAC is robust against some ^2t_R shifting, and it is still useful for data with some deviations from trilinearity [28]. With GC × GC–TOFMS and LC × LC–DAD data, if ^2t_R variations are problematic, use appropriate alignment techniques before submitting the data to PARAFAC. If samples or replicates compose a fourth dimension in the data, then unfold the ^2t_R dimension to reduce the data from four-way to three-way and align all the chromatograms to a target before doing PARAFAC. It is rare to apply PARAFAC to two-way GC–MS data (GC–MS or LC–DAD), but if samples or replicates compose a third dimension, then align all of the chromatograms to a target chromatogram before submitting the three-way data to PARAFAC. Another excellent chemometric option to consider is PARAFAC2, which is an advanced version of PARAFAC that provides as many $\mathbf{x}_{i,k}$ vectors as there are modulations in the three-way data, where $\mathbf{x}_{i,k}$ is the pure predicted ^{2}D chromatogram for the ith component in the kth modulation. PARAFAC2 is robust against ^2t_R variations within a comprehensive 2D chromatographic run because PARAFAC2 is built to assume the covariance matrix of each second column elution profile is constant (e.g., the cross product matrix of $\mathbf{X}_k$, specifically $\mathbf{X}_k^T\mathbf{X}_k$, is constant for each component as a function of retention time). This condition is met when Δ^2t_R is constant between pairs of $\mathbf{X}_k$ matrices, and there are no changes in the shape of each component peak profile.

4 APPLICATIONS AND CONCLUSIONS

Skilled researchers have developed and utilized MCR-ALS, GRAM, and PARAFAC algorithms to mathematically resolve complex samples and quantify interesting analytes from environmental samples, industrial materials, fuels, and biological samples. Figs. 4 and 5 show two examples from the literature of chromatographic data before and after quantitative mathematical

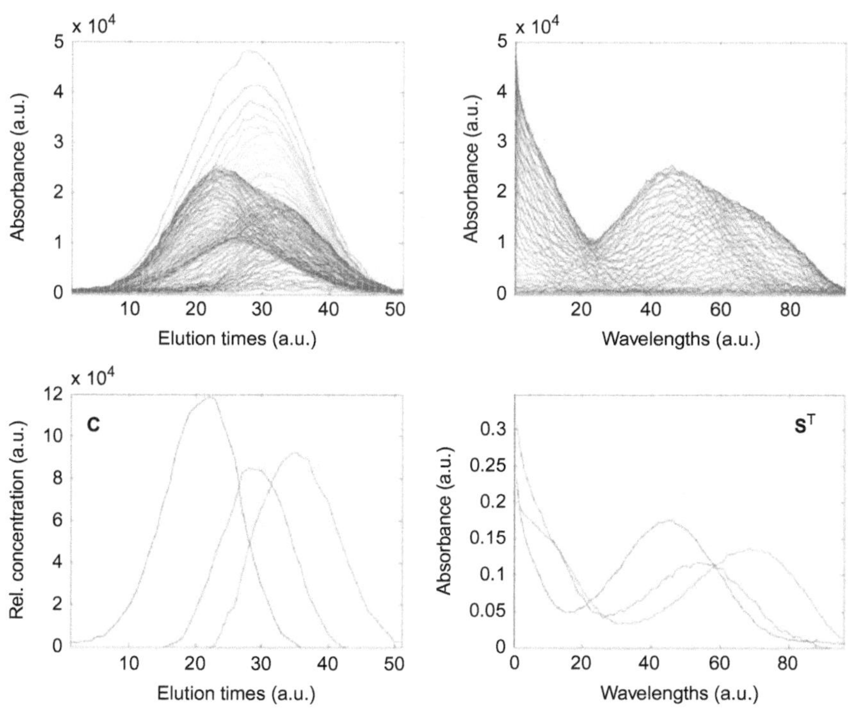

FIG. 4 The top plots are views of a subregion of an HPLC-DAD chromatogram (**D**) containing unresolved analytes. Matrix **D** was submitted to MCR-ALS, yielding the mathematically resolved pure chromatograms (**C**) and pure spectra (**S**) of three components. *This figure was taken with permission from Juan Ad, Jaumot J, Tauler R. Multivariate curve resolution (MCR). Solving the mixture analysis problem. Anal Methods 2014;6:4964–4976.*

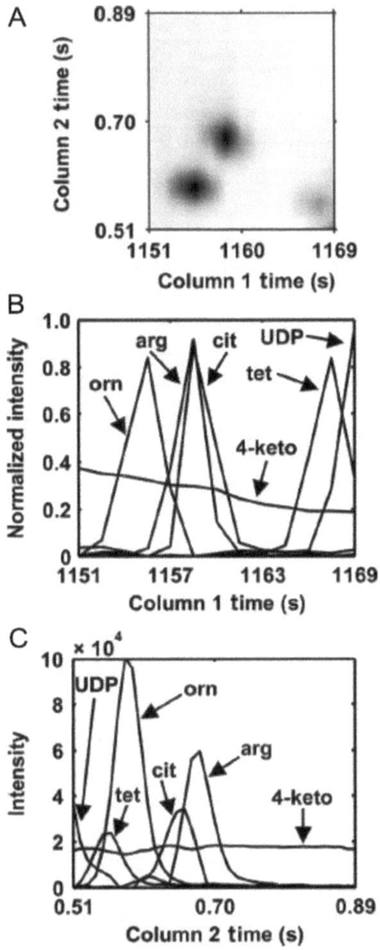

FIG. 5 The *top plot* (A) is a grayscale TIC image of a subregion of a GC × GC–TOFMS chromatogram of a derivatized metabolite sample. The *middle plot* (B) shows the PARAFAC loadings on the first separation dimension, revealing six mathematically resolved sample components were present in the subregion: ornithine (orn), citric acid (cit), 4-ketoglucose (4-keto), arginine (arg), tetradecanoic acid (tet), and a UDP glucose fragment (UDP). The *bottom plot* (C) shows the PARAFAC loadings on the second separation dimension for the same components labeled in plot (B). The PARAFAC results also calculated a pure mass spectrum for each sample component (not shown). *This figure was taken with permission from Hoggard JC, Synovec RE. Automated resolution of nontarget analyte signals in GC × GC–TOFMS data using parallel factor analysis. Anal Chem 2008;80:6677–6688.*

resolution of analytes of interest in complex sample mixtures. The key benefit in each case is the mathematical resolution, via chemometrics, of the analyte peaks that otherwise were only partially resolved by the chromatographic separation dimension(s). The spectral dimension adds significant chemical selectivity and added power to the chemometric methods to facilitate the mathematical resolution. The chemometrics enhances the detection, identification, and quantification of the analyte peaks that otherwise would not be readily analyzed by more conventional data analysis methods. Hence, using chemometric resolution methods substantially broadens the scope of these instrumental platforms to provide a more in-depth and confident analysis of ever increasing complex samples. All of these chemometric methods seem to provide a satisfactory level of accuracy and precision in applications, if the data input for analysis meets specific standards of quality. Future challenges involve the general acceptance, more common implementation, and automation of chemometric software with the relatively large data sets that are produced with GC–MS, LC–DAD, GC × GC–TOFMS and LC × LC–DAD, and in particular for the comprehensive 2D separation methods. Currently, most applications of chemometric resolution requires the analyst to be an expert in the chemometrics field, so most software that is commonly applied is provided by the instrument manufacturer, using algorithms that are not in the public domain and accordingly are not traditionally labeled as "chemometric" software, although sharing the same goal to convert chemical data into chemical information.

REFERENCES

[1] Bushey MM, Jorgenson JW. Automated instrumentation for comprehensive two-dimensional high-performance liquid chromatography of proteins. Anal Chem 1990;62:161–7.

[2] Carr PW, Stoll DR, Wang X. Perspectives on recent advances in the speed of high-performance liquid chromatography. Anal Chem 2011;83:1890–900.

[3] Giddings JC. Two-dimensional separations: concept and promise. Anal Chem 1984;56:1258–70.

[4] Liu Z, Phillips JB. Comprehensive two-dimensional gas chromatography using an on-column thermal modulator interface. J Chromatogr Sci 1991;29:227–31.

[5] Pierce KM, Hoggard JC, Hope JL, Rainey PM, Hoofnagle AN, Jack RM, et al. Fisher ratio method applied to third-order separation data to identify significant chemical components of metabolite extracts. Anal Chem 2006;78:5068–75.

[6] Porter SEG, Stoll DR, Rutan SC, Carr PW, Cohen JD. Analysis of four-way two-dimensional liquid chromatography-diode array data: application to metabolomics. Anal Chem 2006;78:5559–69.

[7] Booksh KS, Kowalski BR. Theory of analytical chemistry. Anal Chem 1994;66:782A–91A.

[8] Bro R. PARAFAC. Tutorial and applications. Chemometr Intell Lab Syst 1997;38:149–71.

[9] Bro R, Andersson CA, Kiers HAL. PARAFAC2—part II. Modeling chromatographic data with retention time shifts. J Chemometr 1999;13:295–309.

[10] Hoggard JC, Synovec RE. Parallel factor analysis (PARAFAC) of target analytes in GC × GC-TOFMS data: automated selection of a model with an appropriate number of factors. Anal Chem 2007;79:1611–9.

[11] Jalali-Heravi M, Parastar H, Kamalzadeh M, Tauler R, Jaumot J. MCRC software: a tool for chemometric analysis of two-way chromatographic data. Chemometr Intell Lab Syst 2010;104:155–71.

[12] Juan Ad, Jaumot J, Tauler R. Multivariate curve resolution (MCR). Solving the mixture analysis problem. Anal Methods 2014;6:4964–76.

[13] Li S, Hamilton JC, Gemperline PJ. Generalized rank annihilation method using similarity transformations. Anal Chem 1992;64:599–607.

[14] Sanchez EE, Kowalski BR. Generalized rank annihilation factor analysis. Anal Chem 1986;58:496–9.

[15] Skov T, Hoggard JC, Bro R, Synovec RE. Handling within run retention time shifts in two-dimensional chromatography data using shift correction and modeling. J Chromatogr A 2009;1216:4020–9.

[16] Tauler R. Multivariate curve resolution applied to second order data. InCINC'94 selected papers from the first international chemometrics internet conference. Chemometr Intell Lab Syst 1995;30:133–46.

[17] Pool WG, de Leeuw JW, van de Graaf B. A rapid routine to correct for skewing in gas chromatography/mass spectrometry. J Mass Spectrom 1996;31:213–5.

[18] Cook DW, Rutan SC, Stoll DR, Carr PW. Two dimensional assisted liquid chromatography—a chemometric approach to improve accuracy and precision of quantitation in liquid chromatography using 2D separation, dual detectors, and multivariate curve resolution. Anal Chim Acta 2015;859:87–95.

[19] de Juan A, Tauler R. Comparison of three-way resolution methods for non-trilinear chemical data sets. J Chemometr 2001;15:749–71.

[20] Parastar H, Tauler R. Multivariate curve resolution of hyphenated and multidimensional chromatographic measurements: a new insight to address current chromatographic challenges. Anal Chem 2014;86:286–97.

[21] Bailey HP, Rutan SC, Carr PW. Factors that affect quantification of diode array data in comprehensive two-dimensional liquid chromatography using chemometric data analysis. J Chromatogr A 2011;1218:8411–22.

[22] Johnson KJ, Wright BW, Jarman KH, Synovec RE. A high-speed peak matching algorithm for retention time alignment of gas chromatographic data for chemometric analysis. J Chromatogr A 2003;996:141–55.

[23] Nadeau JS, Wilson RB, Hoggard JC, Wright BW, Synovec RE. Study of the interdependency of the data sampling ratio with retention time alignment and principal component analysis for gas chromatography. J Chromatogr A 2011;1218:9091–101.

[24] Khummueng W, Harynuk J, Marriott PJ. Modulation ratio in comprehensive two-dimensional gas chromatography. Anal Chem 2006;78:4578–87.

[25] Davis JM, Stoll DR, Carr PW. Effect of first-dimension undersampling on effective peak capacity in comprehensive two-dimensional separations. Anal Chem 2008;80:461–73.

[26] Murphy RE, Schure MR, Foley JP. Effect of sampling rate on resolution in comprehensive two-dimensional liquid chromatography. Anal Chem 1998;70:1585–94.

[27] Seeley JV. Theoretical study of incomplete sampling of the first dimension in comprehensive two-dimensional chromatography. J Chromatogr A 2002;962:21–7.

[28] Pinkerton David K, Parsons Brendon A, Anderson Todd J, Synovec RE. Trilinearity deviation ratio: a new metric for chemometric analysis of comprehensive two-dimensional gas chromatography time-of-flight mass spectrometry data. Anal Chim Acta 2015;871:66–76.

[29] Stoll DR, Cohen JD, Carr PW. Fast, comprehensive online two-dimensional high performance liquid chromatography through the use of high temperature ultra-fast gradient elution reversed-phase liquid chromatography. J Chromatogr A 2006;1122:123–37.

[30] Thekkudan DF, Rutan SC, Carr PW. A study of the precision and accuracy of peak quantification in comprehensive two-dimensional liquid chromatography in time. J Chromatogr A 2010;1217:4313–27.

[31] Kuligowski J, Quintás G, Tauler R, Lendl B, de la Guardia M. Background correction and multivariate curve resolution of online liquid chromatography with infrared spectrometric detection. Anal Chem 2011;83:4855–62.

[32] Parastar H, Radović JR, Jalali-Heravi M, Diez S, Bayona JM, Tauler R. Resolution and quantification of complex mixtures of polycyclic aromatic hydrocarbons in heavy fuel oil sample by means of GC × GC-TOFMS combined to multivariate curve resolution. Anal Chem 2011;83:9289–97.

[33] Bruckner CA, Prazen BJ, Synovec RE. Comprehensive two-dimensional high-speed gas chromatography with chemometric analysis. Anal Chem 1998;70:2796–804.

[34] Faber NM, Buydens LMC, Kateman G. Generalized rank annihilation method. I: derivation of eigenvalue problems. J Chemometr 1994;8:147–54.

[35] Moler CB, Stewart GW. An algorithm for generalized matrix eigenvalue problems. SIAM J Numer Anal 1973;10:241–56.

[36] Poe RB, Rutan SC. Effects of resolution, peak ratio and sampling frequency in diode-array fluorescence detection in liquid chromatography. Anal Chim Acta 1993;283:845–53.

[37] Prazen BJ, Bruckner CA, Synovec RE, Kowalski BR. Second-order chemometric standardization for high-speed hyphenated gas chromatography: analysis of GC/MS and comprehensive GC × GC data. J Microcolumn Sep 1999;11:97–107.

[38] Prazen BJ, Synovec RE, Kowalski BR. Standardization of second-order chromatographic/spectroscopic data for optimum chemical analysis. Anal Chem 1998;70:218–25.

[39] Sanchez E, Ramos LS, Kowalski BR. Generalized rank annihilation method: I. Application to liquid chromatography-diode array ultraviolet detection data. J Chromatogr A 1987; 385:151–64.

[40] Carroll JD, Chang J-J. Analysis of individual differences in multidimensional scaling via an N-way generalization of "Eckart-Young" decomposition. Psychometrika 1970;35:283–318.

[41] Comas E, Gimeno RA, Ferré J, Marcé RM, Borrull F, Rius FX. Quantification from highly drifted and overlapped chromatographic peaks using second-order calibration methods. J Chromatogr A 2004;1035:195–202.

[42] Harshman RA. University microfilms. Michigan: Ann Arbor; 1970.

[43] Hoggard JC, Siegler WC, Synovec RE. Toward automated peak resolution in complete GC × GC-TOFMS chromatograms by PARAFAC. J Chemometr 2009;23:421–31.

[44] Hoggard JC, Synovec RE. Automated resolution of nontarget analyte signals in GC × GC-TOFMS data using parallel factor analysis. Anal Chem 2008;80:6677–88.

[45] Hoggard JC, Wahl JH, Synovec RE, Mong GM, Fraga CG. Impurity profiling of a chemical weapon precursor for possible forensic signatures by comprehensive two-dimensional gas chromatography/mass spectrometry and chemometrics. Anal Chem 2010;82:689–98.

[46] Oca ML, Ortiz MC, Herrero A, Sarabia LA. Optimization of a GC/MS procedure that uses parallel factor analysis for the determination of bisphenols and their diglycidyl ethers after migration from polycarbonate tableware. Talanta 2013;106:266–80.

[47] Rutledge DN, Bouveresse DJ-R. Multi-way analysis of outer product arrays using PARAFAC. Chemometr Intell Lab Syst 2007;85:170–8.

[48] Vosough M, Bayat M, Salemi A. Matrix-free analysis of aflatoxins in pistachio nuts using parallel factor modeling of liquid chromatography diode-array detection data. Anal Chim Acta 2010;663:11–8.

[49] Yang S, Nadeau JS, Humston-Fulmer EM, Hoggard JC, Lidstrom ME, Synovec RE. Gas chromatography–mass spectrometry with chemometric analysis for determining 12C and

13C labeled contributions in metabolomics and 13C flux analysis. J Chromatogr A 2012;1240:156–64.

[50] Yang S, Sadilek M, Synovec RE, Lidstrom ME. Liquid chromatography–tandem quadrupole mass spectrometry and comprehensive two-dimensional gas chromatography–time-of-flight mass spectrometry measurement of targeted metabolites of *Methylobacterium extorquens* AM1 grown on two different carbon sources. J Chromatogr A 2009;1216:3280–9.

[51] Bailey HP, Rutan SC. Chemometric resolution and quantification of four-way data arising from comprehensive 2D-LC-DAD analysis of human urine. Chemometr Intell Lab Syst 2011;106:131–41.

Chapter 11

Multivariate Curve Resolution of (Ultra)Fast Photoinduced Process Spectroscopy Data

O. Devos*[,1], S. Aloïse*, M. Sliwa*, R. Métivier[†], J.-P. Placial[†] and C. Ruckebusch*

*Université de Lille, Sciences et Technologies, LASIR CNRS, LILLE, France

[†]PPSM, ENS Cachan, CNRS, Cachan Cedex, France

[1]Corresponding author: e-mail: olivier.devos@univ-lille1.fr

Chapter Outline

1 INTRODUCTION

Understanding fast and ultrafast reactions that govern the processes induced by light relies on the spectrodynamic study of different types of photoactive systems, such as photosynthetic systems and light-harvesting complexes [1–3], photoconductive materials [4–6], fluorescent proteins [4], and photochromic molecules [7–9]. In order to study these processes, time-resolved spectroscopy (TRS) is the best suited technique. In general, TRS can be applicable in a time range covering many orders of magnitude, from slow processes occurring in the nanosecond to millisecond time domain to ultrafast processes occurring in the femtosecond–picosecond time domain. The study

Data Handling in Science and Technology, Vol. 30. http://dx.doi.org/10.1016/B978-0-444-63638-6.00011-5
Copyright © 2016 Elsevier B.V. All rights reserved.

of the photoreactivity and dynamic quenching of triplet excited states investigated by flash photolysis is an example of a "slow" process [10,11]. On the other side are "ultrafast" processes, such as *cis–trans* isomerization and excited state intramolecular proton transfer (ESIPT) that can be investigated by femtosecond transient absorption spectroscopy [12]. If the process under study is slow (>ms), conventional spectrophotometric techniques can be used. However, for ultrafast processes, specific pump–probe optical techniques are required to register transient absorption spectra. In this case, the light generated by a pulsed laser promotes ("pumps") a population of molecules (around 10%) from the ground state to an excited electronic state. This latter can be measured ("probed") after a certain time delay. The data obtained are difference spectroscopy data (difference absorbances) that are usually arranged in a matrix where one mode corresponds to spectral variables (e.g., wavelength) and the other to time. The investigation of these two-way data enables to study the mechanistics and kinetics of the photoinduced processes occurring in the excited states and to characterize intermediates and photoproducts.

Different approaches can be used to deal with TRS data [13–15]. Among these approaches, it is commonly admitted to distinguish between methods based on a model (parametric model or hard model) and model-free approaches (soft modeling). For parametric methods, kinetic traces are fitted with a sum of exponential components, as in global analysis [15,16]. The amplitudes associated with the exponential decays can then be tentatively interpreted to derive spectral information. For hard modeling, as in target analysis [15], a photophysical or photochemical kinetic model is assumed, from which a set of ordinary differential equations (ODEs) can be derived. The concentration of each species is obtained by first integrating the ODEs and then solving a nonlinear least squares problem to obtain rate constants [17]. This approach is the most widespread in TRS. However, when too little information is available to preclude a mechanistic model, an alternative can be provided by soft-modeling approaches such as multivariate curve resolution (MCR). Multivariate curve resolution by alternating least squares (MCR-ALS) [18,19] was initially developed to deal with process spectroscopy data. When applied to TRS data MCR-ALS provides additive pure contributions, to which are associated time-dependent concentration profiles and spectra. This enables to investigate photoinduced processes and, potentially, to infer a reaction mechanism. This mechanism can be written as a model and translated into kinetic constraints, as in hard–soft multivariate curve resolution (HS-MCR) [20,21].

In this chapter, we will focus on the application of MCR-ALS for the analysis of TRS data investigating fast and ultrafast photoinduced processes. We will cover aspects ranging from data preprocessing to kinetic modeling. Three examples will be developed in order to illustrate these different aspects and to show the potential of an MCR approach in the field. The first data deal with the photophysics of benzophenone (BP) and revisit one of the most important issues concerning carbonyl compounds photochemistry. The second example investigates the photochemistry of a diarylethene and provides insights into the kinetics and photochromic quantum yields of the interconnected photoreactions involved in this system. The last example considers the photochromism of salicylidene aniline (SA) in solution. Particular attention is paid to ESIPT involved at femtosecond timescale, which is of interest from a fundamental point of view and for applications.

2 PHOTOINDUCED PROCESSES

Two types of photoinduced processes may be distinguished according to classical textbooks: photophysical processes, on the one hand, and photochemical processes, on the other hand [11,22]. Upon light excitation, a molecule undergoes either a relaxation, back to the ground state, or a reaction to form new photoproducts. If, on overall, the structure of the molecule does not undergo structural change, this is referred to as photophysical process. Otherwise, if the interaction with light results in a chemical reaction, this is photochemistry. In Fig. 1, we propose a schematic representation of the main elementary photoinduced reactions that occur in the excited states. These processes cover different timescales ranging from second to femtosecond.

2.1 Photophysical Processes

Photophysical processes define transitions that occur between two electronic states without any structural change of the molecular entity. Among photophysical processes (see Fig. 1), absorption is a very fast transition

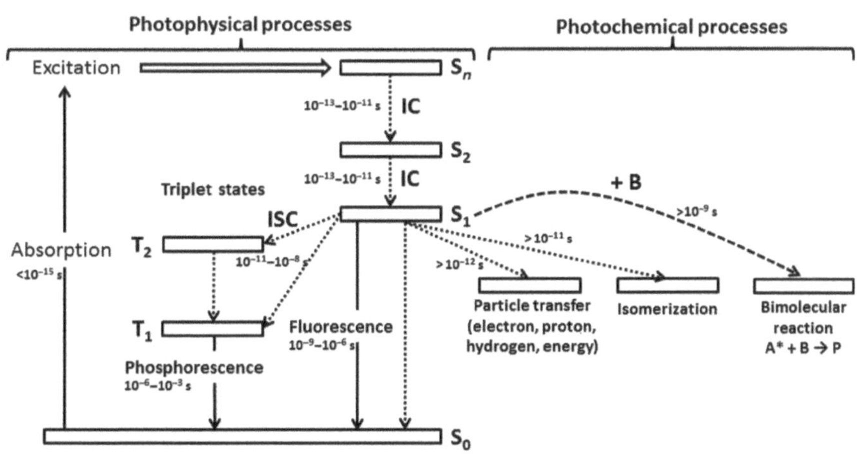

FIG. 1 Schematic representation of the photophysical and photochemical processes and their associated characteristic times. Radiative (resp. nonradiative) transitions are marked with *plain* (resp. *dashed*) *arrows*. *IC*, internal conversion; *ISC*, intersystem crossing.

($<10^{-15}$ s) that populates an excited singlet state S_n. Absorption usually occurs in the UV–visible domain and involves electronic states of the molecules. Ground state (S_0) recovery is obtained through the relaxation of the excess of energy and may involve different processes. Internal conversion (IC) and intersystem crossing (ISC) are nonradiative transitions that occur in the picosecond and nanosecond timescales, respectively. The former involves transitions between two electronic states of identical multiplicity (singlet to singlet or triplet to triplet), whereas, for the latter, different multiplicities are involved. At a longer timescale, a way for a molecule to return to the ground state is spontaneous emission of radiation, which can happen through fluorescence or phosphorescence. Fluorescence can occur from a singlet excited state to a singlet ground state, typically from S_1 to S_0 with characteristic times ranging from 10^{-9} to 10^{-6} s. Phosphorescence, on the other hand, involves transitions between a triplet excited state and a singlet ground state, typically T_1 and S_0, and is a much slower process.

2.2 Photochemical Processes

Unlike photophysical processes, photochemical processes are initiated from a photoexcited state (generally S_1). These processes lead to structural changes that can be followed in the femtosecond to the millisecond time domain.

Three types of elementary processes can be distinguished: (i) particle transfer such as proton transfer, (ii) isomerization (i.e., *cis–trans* isomerization), and (iii) bimolecular reactions leading to the formation of a photoproduct. It should be noted that bimolecular reactions are slower than reactions of type (i) or (ii) since they are limited by diffusion (μs timescale).

3 TIME-RESOLVED SPECTROSCOPY

The study of photoactive compounds, i.e., compounds for which the reaction/transformation is activated by light, involves a set of spectroscopic techniques for probing and characterizing the photoproducts, transient species, and short-lived excited states. Whatever the timescale considered, these techniques have in common that the reaction is first induced by light, most often laser light, and then investigated taking absorbance spectra at successive time delays.

3.1 Time-Resolved Absorption Spectroscopy

3.1.1 *Steady-State Spectroscopy*

When reactions on the millisecond timescale are investigated, absorbance variations can be monitored using conventional steady-state spectrophotometers. This approach often involves working under continuous irradiation (Hg/Xe lamp, for example, with filters). Typical applications can be the study of photostationary states or the evaluation of the quantum yield of a photoreaction.

3.1.2 Laser Flash Photolysis

The measurement of phenomena occurring in the nanosecond to microsecond range is generally performed using laser flash photolysis [22]. After excitation of the sample with a nanosecond pulsed laser, a polychromatic beam (e.g., from a Xe continuous arc lamp) is used to probe the sample. The transmitted light is then separated using a spectrograph, and spectra are measured as a function of time. In this time domain triplet decay, back electron transfer and back reaction of photoproduct are observed.

3.1.3 Pump–Probe Transient Absorption Spectroscopy

To study ultrafast reactions occurring in the femtosecond–picosecond time domain, pump–probe optical techniques are required [23–25]. Pump–probe spectroscopy uses two femtosecond laser pulses. The first is a monochromatic pump that is used to trigger the reaction by promoting molecules to the excited states. The second is a weak polychromatic probe pulse that goes through the same volume of the sample but with a certain time delay with respect to the pump. As the delay between the pump and the probe can be tuned using an optical delay line, the evolution of the processes in the excited states and the formation of transient species can be monitored measuring difference absorption as a function of time. The time resolution (hundred of femtosecond) mainly depends on the duration of the pump. Femtosecond transient absorption spectroscopy allows measuring triplet formation, monomolecular reactions (isomerization, electron and proton transfer, etc.), and photophysical relaxation.

3.2 Specificity of Ultrafast TRS

In pump–probe transient absorption spectroscopy, due to the nature of the measurement, some specific issues have to be considered for data analysis, as detailed in the following sections.

3.2.1 Difference Spectroscopy

In TRS, only a limited number of molecules are excited (typically about 10%). Therefore, difference spectroscopy is required to emphasize absorption variation. ΔA values correspond to the difference observed between the absorption of the sample in the presence of excitation and the one observed in the absence of excitation at a given time delay.

An example of difference femtosecond transient absorption spectra is provided in Fig. 2. Raw data are shown together with the stationary absorbance spectrum and the fluorescence spectrum of the molecule under study.

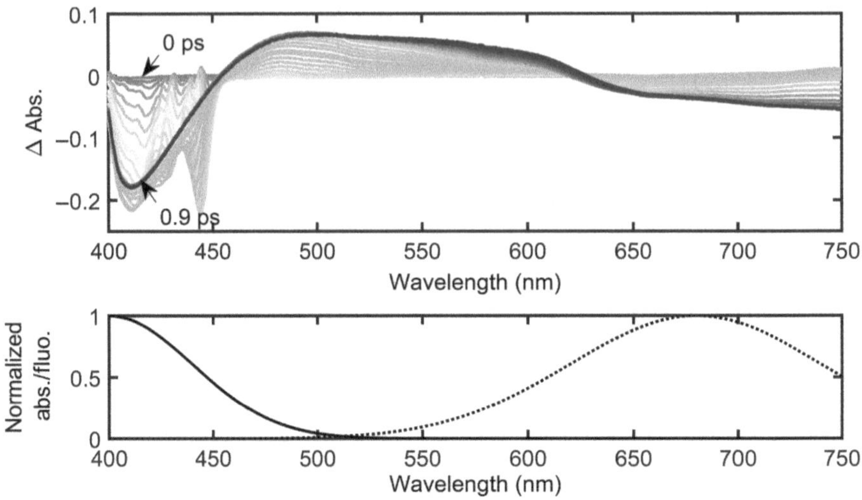

FIG. 2 Femtosecond transient absorption spectra (0–0.9 ps from *red* (*gray* in the print version) to *blue* (*dark gray* in the print version)) of 2-(1-pyridinio) benzimidazolate (SBPa) in acetonitrile following laser excitation at 390 nm. The stationary absorbance (*plain line*) and fluorescence spectra (*dotted line*) of the SBPa ground state are presented in the *bottom*.

Difference absorbance ΔA takes positive value at wavelengths where the new species absorb. On the other hand, ΔA can also be negative. This may happen in two different situations. The first one is when ground state bleaching, i.e., depopulation, is observed. Bleaching is an instantaneous signal which corresponds to the promotion of molecules from the ground state to the excited state. The bleaching component corresponds roughly to the negative image of the absorption spectrum of the studied molecule (e.g., range 400–500 nm in Fig. 2). The second situation is when stimulated emission is observed. Stimulated emission is a relaxation from the excited state to the ground state accompanied by an emission of light. Stimulated emission is generally observed in the same spectral range as steady-state fluorescence (e.g., range 500–750 nm in Fig. 2). In addition to these contributions, at short timescale, coherent artifact signals can be observed. These signals will be discussed further in Section 3.2.2.

3.2.2 Coherent Artifacts

The use of ultrafast laser pulses to trigger photoreactions often results in the observation of coherent optical phenomena [26,27]. They are observed when the pump and probe pulses overlap and they generate some "artifacts" such as two-photon absorption, stimulated Raman amplification (SRA), and cross-phase modulation. SRA is a scattering signal of the solvent that shows very characteristic spectrokinetic features. These signals are observed only during a few hundreds of femtoseconds, as their kinetic behavior is similar to the instrumental response function (IRF). Their typical spectral signature consists of a series of intense and sharp negative signals (see Fig. 2 at 445 nm). These artifacts are solvent specific and can be extremely dependent on the experimental conditions. This means that, in practice, SRA signals are hardly perfectly reproducible from pure solvent to solvent/solute experiments. Therefore, solvent subtraction is seldomly possible and dedicated data preprocessing methods have to be considered, as will be discussed in Section 4.1.

3.2.3 Convolution with the IRF

In ultrafast transient absorption spectroscopy, as time resolution is usually a few hundreds of femtoseconds, any faster or on the same timescale occurring transient phenomena are affected by the convolution with the IRF. A visualization of the IRF signal is given in Fig. 3. The kinetic traces of the IRF can be modeled with Gaussian shapes. In Fig. 3, a time resolution of about 150 fs corresponds to the full-width at half-maximum (FWHM) of the Gaussian. More information about how the IRF can be taken into account in the resolution will be detailed in Section 4.3.2.

3.2.4 Group Velocity Dispersion

As a result of the use of a polychromatic probe pulse, femtosecond transient absorption spectra are distorted by group velocity dispersion (GVD). GVD reflects that the velocity of light in a dispersive medium is wavelength dependent (it is always so, but it can be neglected at usual timescales). As a consequence, the time zero of the reaction is shifted toward positive time delays at longer wavelength. This phenomenon can be observed in Fig. 3 where a delay of 0.06 ps is observed between the time 0 at 425 and the 1 at 600 nm (this zero

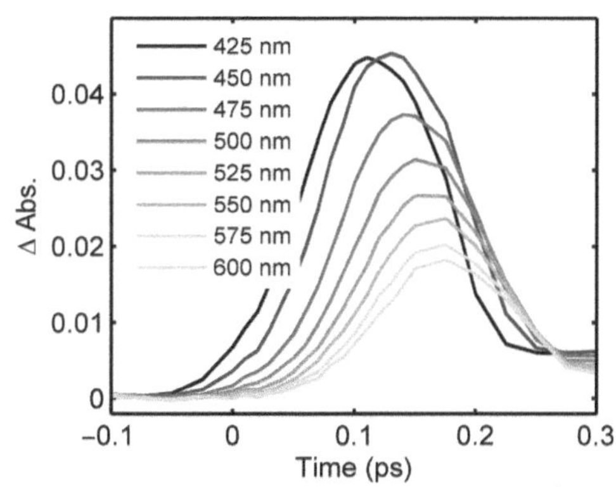

FIG. 3 IRF pump–probe cross-correlation signal (convolution of the pump and probe laser pulses) obtained by measuring two-photon absorption in a thin BK7 microscope coverslip at irradiation wavelength of 390 nm.

is taken at the maximum of the cross-correlation signal). GVD can be corrected by preprocessing (see Section 4.1).

4 MCR OF TRS DATA

Assuming the Beer–Lambert law, TRS data can be described as a superposition of the spectroscopic properties of the transient species weighted by the values of their time-dependent concentration profiles. These data can be arranged in a data matrix **D** of dimensions $m \times n$. Each row of **D** corresponds to a transient absorption spectrum registered at a certain time delay and each column is associated to the variation of the absorbance at a particular wavelength. Thus, the experimental data **D** ($m \times n$) can be described as in Eq. (1):

$$\mathbf{D} = \mathbf{C} \cdot \mathbf{S}^{\mathrm{T}} + \mathbf{E} = \mathbf{c}_1 \cdot \mathbf{s}_1{}^{\mathrm{T}} + \mathbf{c}_2 \cdot \mathbf{s}_2{}^{\mathrm{T}} + \ldots + \mathbf{c}_k \cdot \mathbf{s}_k{}^{\mathrm{T}} + \mathbf{E} \quad (1)$$

where the concentration matrix **C** ($m \times k$) ideally contains the time-dependent concentration profiles of the k pure contributions in the system at m time delays and the matrix $\mathbf{S}^{\mathrm{T}}$ ($k \times n$) the corresponding spectral contributions at n wavelengths. The error matrix **E** ($m \times n$) contains the residuals. In the case of difference spectroscopy, the spectra in **S** are difference spectra.

To resolve **D** into pure contributions, MCR-ALS (see Chapter 2) is the method of choice in spectroscopy. The main steps to be considered for an MCR analysis of TRS data are presented in Fig. 4.

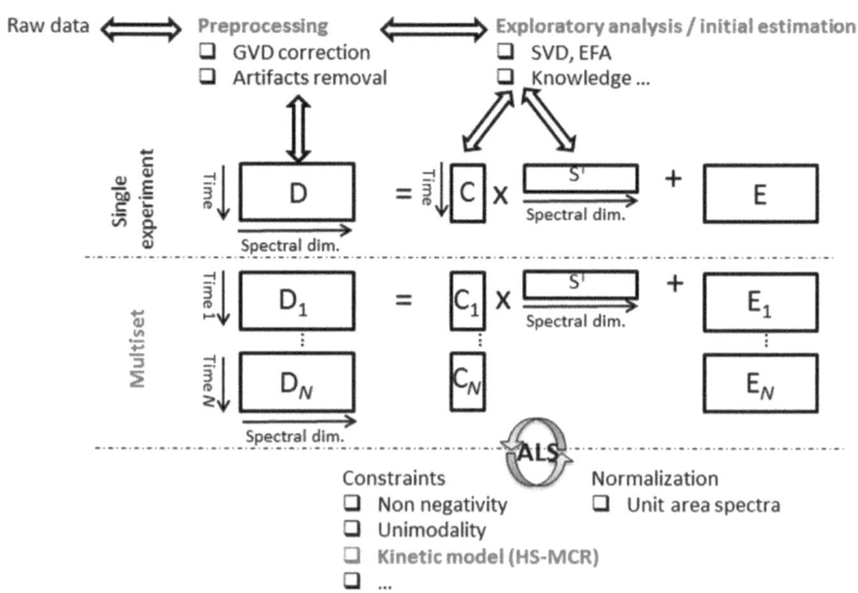

FIG. 4 Schematic representation of MCR applied to TRS data. In this chapter, particular attention is paid to the points in bold marked in *red* (*light gray* in the print version).

4.1 Preprocessing Time-Resolved Spectra

As introduced in Section 3.2, ultrafast artifacts are phenomena that hamper the application of methods based on a bilinear factor decomposition, such as MCR-ALS. For this reason, data preprocessing is required. To illustrate this point, the effect of GVD is visualized in Fig. 5A, where it can be observed that the detection of the spectral components at long wavelengths is delayed compared to ones at shorter wavelengths. Roughly, GVD preprocessing can be understood as "resetting initial time delays to zero" for all the wavelengths λ. For this purpose, the delay $\tau(\lambda)$ induced by GVD can be modeled using a low-order polynomial or an exponential function [28], as in Eq. (2),

$$\tau(\lambda) = A\left(1 - \exp\left(\frac{\lambda - \lambda_{\text{pump}}}{B}\right)\right) \tag{2}$$

where A and B are the parameters to optimize. Once the model is fitted (see Fig. 5A, black line), the data are then corrected by interpolating at a new time position $t_{\text{corr}} = t - \tau$. As a result, the signals at different wavelengths "start" at the same position in time, as shown in Fig. 5B.

Data should also be preprocessed to handle coherent artifacts that strongly distort spectrokinetic data at ultrashort timescale. This is not trivial due to the (nonlinear) nature of the physical phenomena involved. Among these artifact signals, SRA is a common issue and leads to the presence of narrow negative peaks superimposed on smooth UV–visible absorption spectra. Several ways were proposed in the literature to cope with these signals [26,29]. In the example presented in Section 5.3 we will focus on approaches based on multi-set MCR-ALS. However, it should be noted that signal smoothing with penalized asymmetric least squares is another possible approach [30,31] (Chapter 14). The principle is to smooth spectra using a weighted least squares approach. Let $\mathbf{y}$ (m wavelengths) be the raw spectrum affected by SRA and $\mathbf{z}$ the smooth spectrum estimated. The cost function S to minimize is given by Eq. (3):

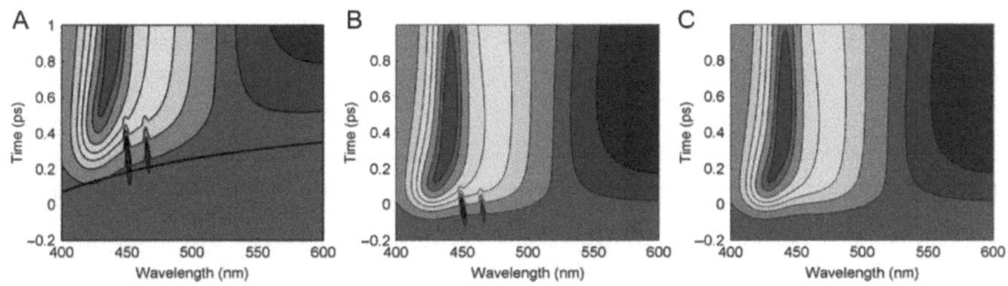

FIG. 5 Preprocessing femtosecond transient spectra (A) raw data affected by GVD (*black curve*) and SRA artifact (sharp negative signals around 450 nm), (B) GVD corrected data, and (C) data smoothed with penalized asymmetric least squares.

$$S = \sum_{i=1}^{m} w_i (y_i - z_i)^2 + \lambda \sum_{i=1}^{m} (\Delta^d z_i) \tag{3}$$

where w_i represents the weights for each data points, $\Delta^d z_i$ the differences between successive points of order d, and λ the trade-off parameter to balance between signal reproduction and signal smoothness. Extension of this method to two-dimensional data was later developed on the basis of tensor products of P-splines [30]. An example of the kind of results that can be obtained is provided in Fig. 5C.

4.2 MCR Basics

Here only a short description of MCR-ALS focusing on the practical aspects related to TRS data is presented.

The first point is to estimate the number of species (k) that contribute to the variation observed in the data. This is often addressed by performing singular value decomposition (SVD) or principal component analysis [32]. The second step aims at obtaining initial estimates of either $\mathbf{C}$ or $\mathbf{S}^{\mathrm{T}}$. For TRS data, the profiles in $\mathbf{C}$ can be first estimated by evolving factor analysis (EFA) [33–35]. The practical aspects related to SVD and EFA will be illustrated in the example in Section 5.1. MCR-ALS is then performed, alternatively fitting $\mathbf{C}$ and $\mathbf{S}^{\mathrm{T}}$ to the experimental data $\mathbf{D}$ in the least squares sense. The residuals are obtained computing the difference between the data $\mathbf{D}$ and the reproduced data $\mathbf{C}\cdot\mathbf{S}^{\mathrm{T}}$. At each iterative cycle of the optimization, constraints can be applied to the profiles in $\mathbf{C}$ and/or $\mathbf{S}^{\mathrm{T}}$. The quality of the decomposition is assessed from the lack of fit (lof, in %) between the experimental data matrix $\mathbf{D}$ and the data reproduced from the product $\mathbf{C}\cdot\mathbf{S}^{\mathrm{T}}$, defined in Eq. (4)

$$\mathrm{lof}(\%) = 100 \times \sqrt{\frac{\sum \left(d_{i,j}^{*} - d_{i,j} \right)^2}{\sum d_{i,j}^2}} \tag{4}$$

where d_{ij} is one element of the experimental matrix and d_{ij}* is the analogous element of the reproduced data. Convergence is achieved when the relative difference in lack of fit between two consecutive iterations goes below a threshold or calculation stops when the number of iterations exceeds a predefined value.

MCR-ALS can be performed on a single data set but can be extended to multiset (multiple data sets) combining data coming from different experiments or different techniques (data fusion) in order to provide a complete and accurate description of the chemical system under study. Multisets often correspond to experiments monitoring the same chemical system (as shown in Fig. 4) through processes developed in different experimental conditions and, as a consequence, showing different but complementary behaviors.

Multiset MCR-ALS was applied in photochemistry to comprehensively describe a complex reaction system having parallel pathways [36].

4.3 Hard–Soft Multivariate Curve Resolution

Hard–soft multivariate curve resolution method (HS-MCR) is an evolution of the soft-modeling algorithm including a kinetic hard-modeling constraint. It combines the benefits of both model-free and hard-modeling approaches [20,21]. The implementation of a kinetic constraint can be particularly relevant to confirm a reaction mechanism and estimate kinetic rates, but it has to be tailored to match the specificities of ultrafast TRS data.

We refer to Chapter 2 of this book for a more detailed description, but the principle of the application of a kinetic constraint in HS-MCR can be summarized as follows. At each iteration, the concentration profiles in **C** (calculated by least squares) are fitted by the hard-model profiles $\mathbf{C}_{\text{fit}}$, and those profiles update the soft-modeled ones. The fitting step is performed using a Newton–Gauss Levenberg–Marquardt-based algorithm, which aims to minimize the value of sum of squares *ssq* defined in Eq. (5)

$$ssq = \|\mathbf{C}_{\text{fit}} - \mathbf{C}\| \tag{5}$$

It should be noted that not all the profiles in **C** have to be fitted. Along the same idea, in multiset experiments, the batches can also obey different hard models [21] and model-based and model-free experiments can be analyzed together [37]. By experience this constraint can be applied successfully when soft MCR-ALS results with classical constraints already show concentration profiles with a clear kinetic behavior. In the following, we would like to emphasize two developments that have been realized in order to address some issues specific of ultrafast TRS data.

4.3.1 Incorporation of Quantum Yields in HS-MCR

Most photochemical processes involve not only sequential steps but parallel competitive reactions (branching). For illustration, we consider in Fig. 6 two parallel reactions where the excited state precursor **A** reacts to give two different photoproducts **B** or **C**. The formation quantum yield of the photoproduct **B** ($\phi_{A\to B}$) can be formulated as in Eq. (6):

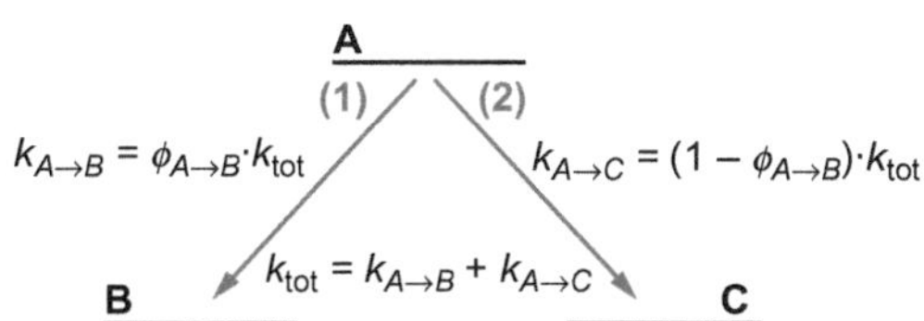

FIG. 6 Two parallel pathways with a branching ratio defined by the quantum yield $\phi_{A\to B}$: photochemical reactions leading to (1) the formation of the photoproduct **B** and (2) the formation of **C**.

$$\phi_{A\to B} = \frac{[\mathbf{B}]}{[\mathbf{B}]+[\mathbf{C}]} = \frac{k_{A\to B}}{k_{A\to B}+k_{A\to C}} = \frac{k_{A\to B}}{k_{\text{tot}}} \tag{6}$$

where $k_{A\to B}$ and $k_{A\to C}$ correspond to the kinetic formation rate of **B** and **C**, respectively, and k_{tot} to the global decay rate of **C**.

Without additional knowledge, the complete resolution of the simple system described in Fig. 6 cannot be obtained since a single kinetic profile describes two reactions (reaction (1) to form **B** and reaction (2) to form **C**). Reliable estimation of the rate constant and associated kinetics is only possible if the quantum yield is known. In this case, it can be incorporated in HS-MCR rewriting the process as a unique reaction modifying the stoichiometric coefficients as in Eq. (7):

$$\mathbf{A} \xrightarrow{k_{\text{tot}}} \phi_{A\to B}\cdot\mathbf{B} + (1-\phi_{A\to B})\cdot\mathbf{C} \tag{7}$$

We refer to Ref. [36] and the case study provided in Section 5.3 for further discussion.

4.3.2 IRF Convolution

Another aspect related to the application of HS-MCR to ultrafast TRS data is convolution of the kinetics with the IRF [38,39]. The IRF function is described as a Gaussian $irf_\sigma(t)$, where σ is the parameter describing the time resolution of the measurement. The kinetic profile convoluted by the IRF ($\mathbf{C}_{\text{fit}}$) is obtained from the pure undistorted kinetic profile $\mathbf{C}_{\text{kin}}$, as in Eq. (8):

$$\mathbf{C}_{\text{fit}} = \mathbf{C}_{\text{kin}} * irf_\sigma \text{ with } irf_\sigma(t) = \exp\left(-\frac{t^2}{2\cdot\sigma^2}\right) \tag{8}$$

where the sign * denotes the operation of convolution.

It should be noted that the parameter σ can be fixed if known a priori or optimized.

5 APPLICATIONS

5.1 Resolving Controversy about BP Photophysics

Understanding the photophysics of BP in its excited states is still an important issue when dealing with the photochemistry of carbonyl compounds. It is well established experimentally that the excited state $\mathbf{S}_1(n,\pi^*)$ relaxes very rapidly through ISC to the lowest excited triplet state $\mathbf{T}_1(n,\pi^*)$ [40]. However, a controversy existed regarding the exact photophysical mechanism of the relaxation of $\mathbf{S}_1$, with two possible pathways (see Fig. 7A): (1) a direct $\mathbf{S}_1(n,\pi^*)\rightarrow\mathbf{T}_1(n,\pi^*)$ ISC [41] or (2) an indirect mechanism involving a $\mathbf{T}_2(\pi,\pi^*)$ intermediate [11]. To further discuss this controversy, MCR-ALS was applied to transient absorption spectroscopy data [42].

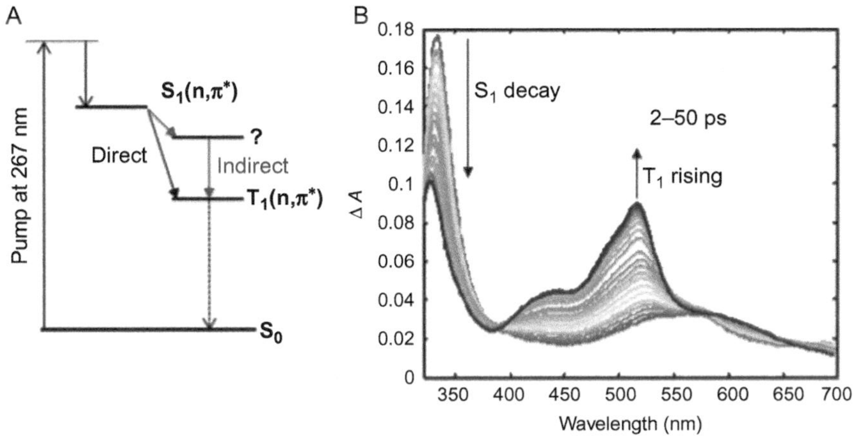

FIG. 7 (A) Direct vs indirect mechanism for the relaxation of the photoexcited state $\mathbf{S_1}$ of BP and (B) transient absorption spectra (*red* (*light gray* in the print version) to *blue* (*dark gray* in the print version), 2–50 ps time window) of BP at 5×10^{-4} M in acetonitrile following laser excitation at 267 nm.

The experimental data allowing to track in "real-time" $\mathbf{S_1}(n,\pi^*) \rightarrow \mathbf{T_1}(n,\pi^*)$ were obtained by transient absorption spectroscopy of BP in acetonitrile after excitation at 267 nm for the time window 2–50 ps. It should be noted that shorter times were not considered in the analysis as they are affected by the relaxation of Franck–Condon excited states. It should also be noted that the transition from the triplet state to the ground state is not observed since it occurs at a longer timescale (microseconds). Preprocessed spectra are shown in Fig. 7B and qualitative description of these data is straightforward. The decay of the transient absorption band peaking at 330 nm corresponding to $\mathbf{S_1}(n,\pi^*)$ is concomitant with the growth of the transient absorption bands peaking at 525 nm and corresponding to $\mathbf{T_1}(n,\pi^*)$. Obviously, basic analysis of the data would consist of a monoexponential fit of the kinetic traces taken at these two absorption maxima. Such an approach returns characteristic times of ~10 ps in both cases. However, at other wavelengths, the fitting results are more puzzling with characteristic times ranging from 10 to 17 ps. To address the possible presence of an intermediate species, MCR soft-modeling approach was performed.

SVD was applied first in order to estimate the number of significant contributions. It can be observed from the results provided in Fig. 8A that at least three contributions are required. The second step consisted of applying EFA to construct initial estimates of the concentration profiles in **C**, assuming that the process is sequential. The results are shown in Fig. 8B and the initial concentration profiles in Fig. 8C. However, the choice between three or more contributions can be questioned, and it is more than often so. In these situations, it is advised to perform several MCR-ALS resolutions with a different number of components and compare the different results obtained. Apart from

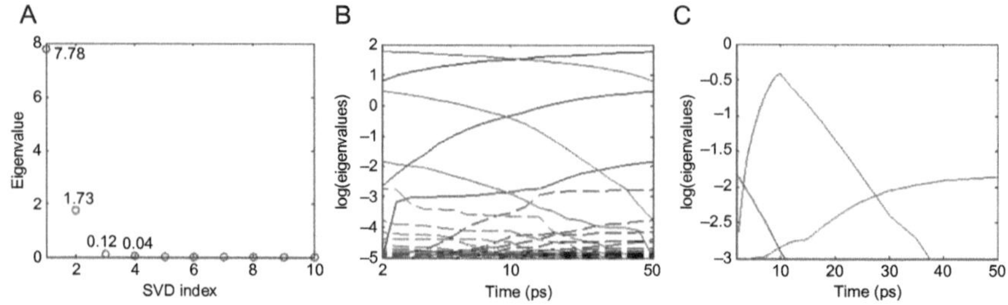

FIG. 8 Exploratory analysis of the BP TRS data (A) plot of the first 10 eigenvalues calculated using SVD, (B) results of forward (*black line*) and backward (*red line* (*gray line* in the print version)) EFA, the plain lines corresponds to significant contributions, and (C) initial concentration profiles obtained from EFA.

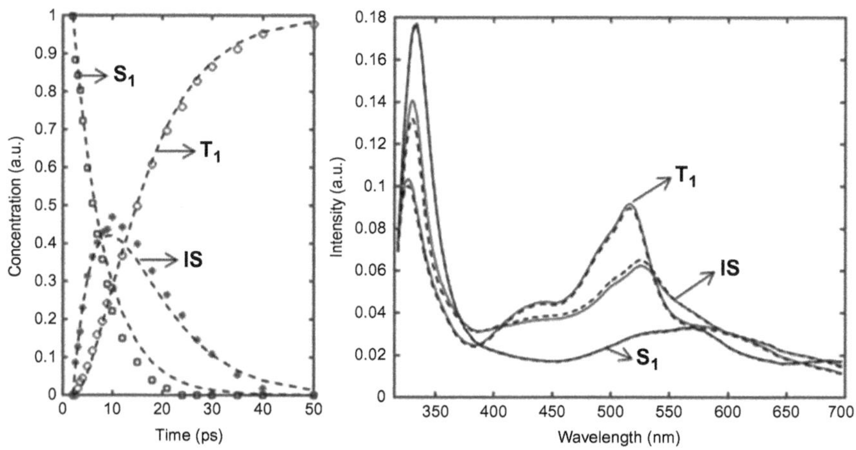

FIG. 9 Kinetic profiles and spectra obtained from MCR-ALS (markers and colored lines, respectively) and from HS-MCR analysis (*dotted black lines* in both cases) using the model $\mathbf{S_1} \rightarrow \mathbf{IS} \rightarrow \mathbf{T_1}$ for BP excited at 267 nm in acetonitrile. The color code (different gray shades in the print version) is the following: *blue* (*dark gray* in the print version) for $\mathbf{S_1}$; *green* (*gray* in the print version) for **IS**; *red* (*light gray* in the print version) for $\mathbf{T_1}$.

the numerical estimators (e.g., lack of fit) the chemical interpretability of the solution should significantly drive the choice of the appropriate model. In the situation at hand, by applying MCR-ALS with 2, 3, or 4 components, and looking both at the residuals and the interpretability, a three-component model was retained [42,43].

The MCR-ALS results are presented in Fig. 9. The initial spectra ($\lambda_{max} \approx 330$ and 570 nm) and the final spectra ($\lambda_{max} \approx 325$ and 525 nm) can be assigned to $\mathbf{S_1}(n,\pi^*)$ and $\mathbf{T_1}(n,\pi^*)$, respectively. The corresponding concentration profiles are relevant, as the disappearance of $\mathbf{S_1}(n,\pi^*)$ (blue squares (light gray in the print version)) matches the appearance of $T_1(n,\pi^*)$ (red circles (dark gray in the print version)). Furthermore an intermediate contribution (IS) is extracted. These profiles also clearly show an evolution compatible with the ones corresponding to sequential first-order reactions. To support the soft-modeling results, an HS-MCR model was applied (see Section 4.3 for further details on HS-MCR). A kinetic constraint that describes a two-step sequential model ($\mathbf{S_1} \rightarrow \mathbf{IS} \rightarrow \mathbf{T_1}$) was considered. Initial estimates of the kinetic profiles were those obtained from soft-modeling MCR-ALS. The HS-MCR results are presented in Fig. 9 (dotted black lines). The spectra obtained with HS-MCR are very similar to the previous ones, but now the time-dependent concentration profiles are forced to match a sequential first-order reaction model. It is important to point out that this model, which includes an intermediate species, gives better lack of fit and residuals than the direct one-step model ($\mathbf{S_1} \rightarrow \mathbf{T_1}$). The rate constant obtained for $\mathbf{S_1} \rightarrow \mathbf{IS}$ is $0.154 \pm 0.002\ \mathrm{ps}^{-1}$, whereas the one for $\mathbf{IS} \rightarrow \mathbf{T_1}$ is $0.093 \pm 0.001\ \mathrm{ps}^{-1}$. Other experiments using neighboring molecules, different solvents, and excitation wavelengths have been performed in order to confirm that **IS** is chemically relevant [42]. Finally, this "species" was interpreted as a mixture of two states, $\mathbf{T_2}(\pi,\pi^*)$ and a hot $\mathbf{T_1}(n,\pi^*)$, and, recently, theoretical calculations have confirmed this interpretation [44].

5.2 A Photochromic Study of CMTE

In this section, we investigate the photochromic study of *cis*-1,2-dicyano-1,2-bis(2,4,5-trimethyl-3-thienyl)ethene, abbreviated CMTE, in solution. This molecule is a typical example of diarylethene whose photochemistry has been extensively studied for more than 25 years [7,45–47]. CMTE was the first diarylethene to show stable and reversible photocyclization [48]. The CMTE molecule in its open *cis* isomer (**A** in Scheme 1) possesses three adjacent double bonds that permit a reversible photocyclization reaction leading to a closed-ring form (**B** in Scheme 1) corresponding to a 1,3-cyclohexadiene-like molecular configuration. In addition to this reaction, reversible photoisomerization of the open *cis* isomer (**A**) to the open *trans* isomer (**C** in Scheme 1) can occur and compete with the photocyclization. Despite the various systems and applications for which CMTE has been used [47,49–51], the kinetics and photochromic quantum yields of the interconnected photoreactions involved in this system remain mostly unknown [46,52].

We propose to investigate the spectral and dynamic properties of these three isomers that are engaged in several photochemical reactions under continuous irradiation at different irradiation wavelengths. Indeed, both **A** and **C** forms of CMTE are yellowish, since they absorb light mostly in the UV and blue region of the visible spectrum (up to 430 nm) [53]. The **B** isomer of CMTE shows a strong absorption band in the visible region, between 450 and 600 nm, providing its intense red color. Then, three different wavelengths can be used for irradiation: 365, 405, and 546 nm. The **A** and **C** isomers are

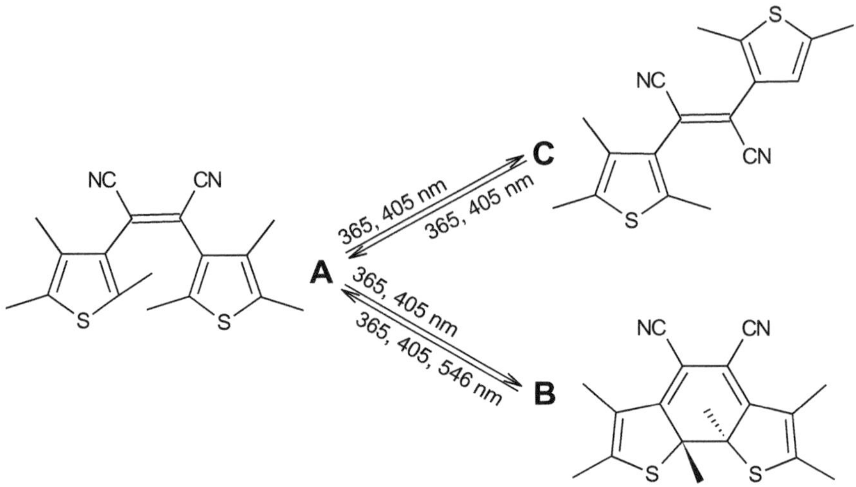

SCHEME 1 Photochromic reaction of CMTE showing the reversible photocyclization and the *cis/trans* photoisomerization. **A**, **B**, and **C** correspond to the open *cis* isomer, closed isomer, and open *trans* isomer, respectively.

sensitive to 365 and 405 nm, whereas **B** can be irradiated using 365, 405, and 546 nm. Photocyclization of CMTE is generally reported only at 405 nm. At this wavelength [46], both the reversible photocyclization (**A** → **B** and **B** → **A**) and *trans* ↔ *cis* photoisomerization (**A** → **C** and **C** → **A**) occur (see Scheme 1). At 365 nm both **A** ↔ **B** and **A** ↔ **C** reversible reactions occur, but the formation of a new photoproduct **D** is suspected. At 546 nm, only **B** absorbs light and can undergo the **B** → **A** ring opening reaction to form the open *cis*-isomer **A**. The aims of this study are precisely to characterize these interconnected photochromic reactions and obtain information (spectra, time evolution) on the three isomers **A**, **B**, and **C** but also on the photoproduct **D** produced under 365 nm irradiation.

For this purpose, the photochromic reactions of CMTE were induced in situ by an irradiation with a Hg/Xe lamp (200 W, Hamamatsu Lightningcure LC6) equipped with narrow-band interference filters of appropriate wavelength. The transmitted light from a Xe source (white probe light) was recorded on a CCD camera through a polychromator. UV–visible absorption spectra were acquired from 260 to 625 nm every 0.7 s. For each experiment, a known concentration of the pure isomer **A**, **B**, or **C** is introduced in a cuvette and irradiated with 365, 405, or 546 nm (see Table 1). A comprehensive description of the system is given by the seven data sets described in Table 1 and shown in Fig. 10: two data sets are available for **A** and for **C**, using 365 and 405 nm irradiation light and three data sets are available for **B**, using 365, 405, and 546 nm irradiation light. These data were analyzed with a multiset MCR-ALS approach to unravel the time-dependent profiles and spectra of the photoproducts involved.

TABLE 1 Initial States and Irradiation Wavelengths of the Seven Experiments

Name of the experiment	Initial state (conc. in mol L^{-1})			Irradiation wavelength (nm)
	A	B	C	
B546	—	5.05×10^{-5}	—	546
A405	6.2×10^{-5}	—	—	405
B405	—	6.6×10^{-5}	—	405
C405	—	—	1.1×10^{-4}	405
A365	6.7×10^{-5}	—	—	365
B365	—	7.4×10^{-5}	—	365
C365	—	—	1.2×10^{-4}	365

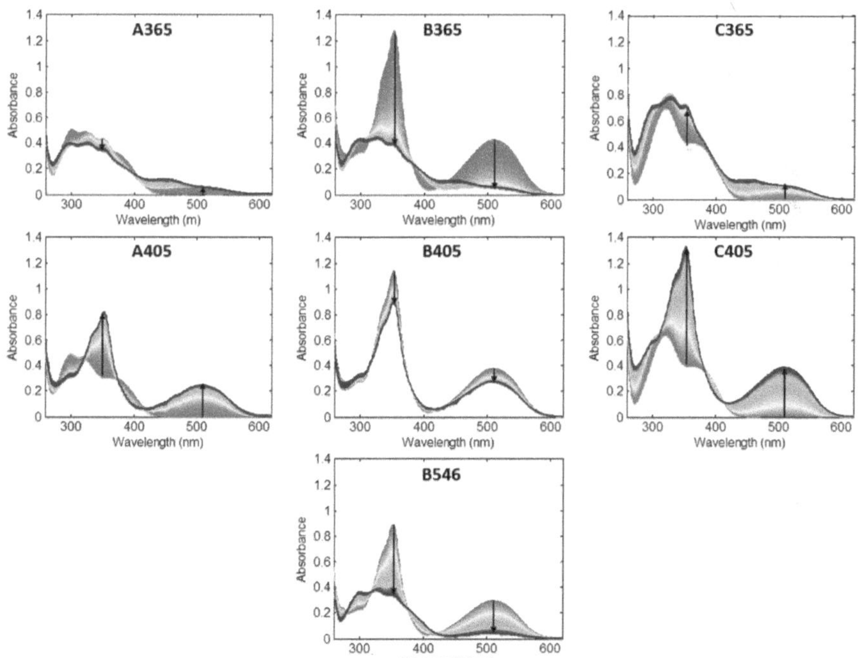

FIG. 10 UV–visible absorption spectra of the CMTE isomers (called **A**, **B**, and **C**) at different times for irradiation wavelengths of 365, 405, and 546 nm (experiment abbreviations and experimental conditions are presented in Table 1). UV–visible absorption spectra are recorded every 0.7 s. The *red* (*light gray* in the print version) spectra correspond to data acquired at the beginning of the process and the *blue* (*dark gray* in the print version) ones to the last spectra of the series.

When exposed to the same wavelength (365 or 405 nm) the final spectra are quite similar whatever the starting point, i.e., starting from isomers **A**, **B**, or **C**. This indicates that a photostationary state is reached (or at least a state evolving very slowly in the case of 365 nm irradiation). When exposed at 405 nm light, the final photostationary state is mostly composed of the closed form **B**, but minor contributions of **A** and **C** forms are certainly present. After irradiation at 365 nm, it is worth noting that the final spectra show a well-observable absorption band centered at 450 nm which corresponds to none of the pure isomers **A**, **B**, or **C**, as depicted by the first spectra of the different data sets. This observation is a clear indication that a specific photoproduct **D** is yielded under irradiation at 365 nm, in addition to the isomers **A**, **B**, and **C**. Finally when the closed form **B** is irradiated at 546 nm only specific bands of **A** appear, which is in accordance with the fact that only the **B**→**A** reaction is possible.

Direct application of MCR-ALS on the data sets at 365 nm, where the simultaneous presence of the four components **A**, **B**, **C**, and **D** is expected, did not allow unmixing of their respective contributions. Additional information was thus added in the form of the three data sets obtained starting from **A**, **B**, and **C** under continuous irradiation at 405 nm (data A405, B405, and C405). For these data, the one thing that is clear is that **D** is absent, and only three components are expected. For the same reason, the data set B546, where the forms **C** and **D** are absent, was considered in order to obtain a reliable description of the forms **A** and **B**. Applying this strategy, we end up with a multiset MCR-ALS performed on the data obtained by appending columnwise all the data sets described earlier, as shown in Fig. 11. Particular attention was paid to the application of constraints and to the specification of the correspondence between the different species and the different data sets (in Fig. 11 concentration profiles in gray mark the absence of a species). It should

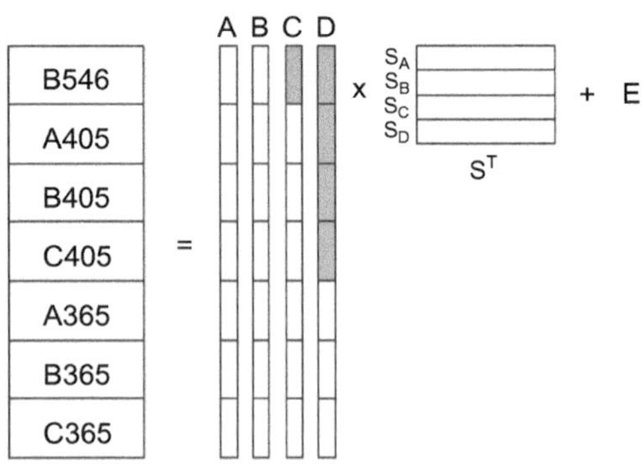

FIG. 11 Multiset MCR analysis of the seven data sets of CMTE under irradiation at 365, 405, and 546 nm. **A** corresponds to the open *cis* form, **B** to the closed form of CMTE, **C** to the open *trans* form, and **D** is a photoproduct that can be produced only at 365 nm. The *gray color* marks the absence of compounds in a given data set.

be added that nonnegativity for the concentration and spectra, as well as closure of the concentration, was applied as constraints. In addition, the known initial concentrations of each pure species (see Table 1) were used.

The time-dependent concentration profiles and pure spectra of the four CMTE forms obtained with MCR-ALS are provided in Fig. 12. For all the data sets, one component is always highly predominant at time zero indicating that only very little mixing occurs. First, the extracted spectra for **A**, **B**, and **C** (Figure 12, bottom) are quite realistic, since they are similar to the known spectra of the isolated species, obtained by chromatographic methods. Furthermore, the concentration profiles obtained for the individual components (see Fig. 12, top) are in agreement with Scheme 1. It illustrates how the global MCR-ALS analysis can provide a comprehensive mechanism of the different steps involved by the irradiation experiments. First, when **B** is irradiated at 546 nm, a clear single interconversion **B** → **A** is observed. Second, the sequential and interconnected **C** ↔ **A** ↔ **B** photoreactions are highlighted under 405 nm irradiation. As a typical example, when **C** is illuminated at 405 nm (experiment C405), the disappearance of **C** and the concomitant increase of **A** appear first, then followed by a decrease of both **A** and **C** in favor of **B** which rapidly increases to become the major species in the photostationary state. The data sets under irradiation at 365 nm are more complicated. As a general scheme, the first step involves the interconversion between **A**, **B**,

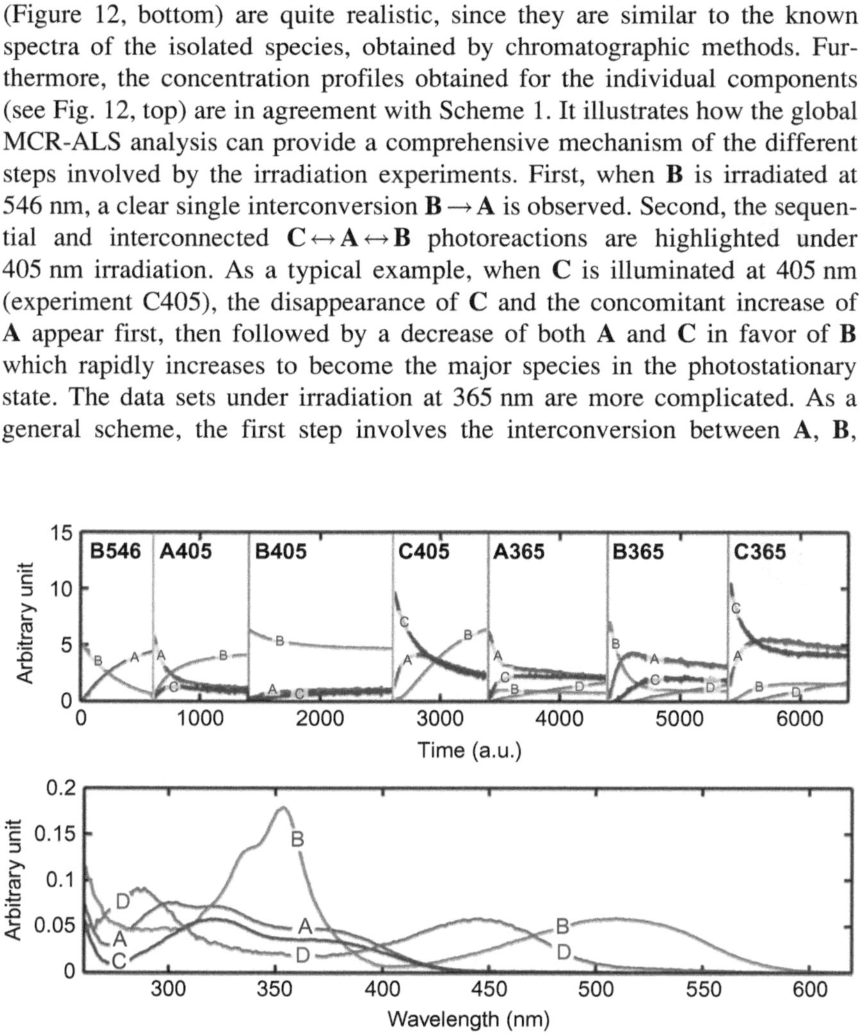

FIG. 12 Time-dependent concentrations and spectra obtained from MCR-ALS analysis with nonnegativity and closure constraints for irradiation at 546, 405, and 365 nm. **A** (open *cis* form) is represented in *green* (*gray* in the print version), **B** (close form) in *red* (*light gray* in the print version), **C** (open *trans* form) in *blue* (*dark gray* in the print version), and **D** (new photoproduct) in *violet* (*light gray* in the print version).

and **C**, but contrary to the 405 nm case, the **B** isomer does not reach a high level, since the photoproduct **D** appears in a second step, with a concentration profile showing a linear increase at longer irradiation times. This behavior is in agreement with a **C**$\leftrightarrow$**A**$\leftrightarrow$**B** equilibrium, followed by a slow **B**$\rightarrow$**D** irreversible photoreaction involving oxygen, confirmed by an NMR study [53]. Furthermore, the spectrum of **D** obtained with MCR-ALS is very similar to the one obtained after purification and separation of this species [53], with a large band around 450–475 nm and another one around 270 nm.

Overall, the use of multiset MCR-ALS, in combination with the incorporation of a priori knowledge on the presence or absence of the three photochromic forms (**A**, **B**, and **C**) and one photoproduct (**D**) for the different irradiation wavelengths, has allowed to extract chemically relevant spectra and concentration profiles of these four different species of CMTE. Meaningful information can be obtained from the MCR-ALS results. First, the spectra of the pure species are well compatible with the corresponding experimental ones, obtained from isolated forms of **A**, **B**, and **C**. Second, the time-dependent concentration profiles provide an excellent description of the kinetic photoreactions of the sequential transformations between the three isomers **C**$\leftrightarrow$**A**$\leftrightarrow$**B** and the photoproduct **B**$\rightarrow$**D**. In particular, at 365 nm, the photokinetic data can be well reproduced with a unique photoproduct **D**, and reliable spectral information could be obtained for the signature of this species **D**. As a conclusion, the photochemical reactivity of CMTE provides a representative example of the useful contribution of MCR-ALS analysis techniques to unravel both spectroscopic and kinetic descriptions of complex systems involving multiple reactions.

5.3 Clarification of the Photochromism of Anils by HS-MCR

In this study, the photochromism [54] of SA in solution is considered. This mechanism involves, first, a fast (<100 fs) ESIPT [12,55]. This process is of interest from a fundamental point of view (excited state proton transfer is a common reaction observed in nature) as well as for fast optical switching applications [56]. The overall photodynamics of SA generally accepted is presented in Fig. 13. Under ultrashort irradiation, an ESIPT occurs from the noncolored stable enol form to give a *cis*-keto form, which isomerizes to the final metastable (few milliseconds in solution) colored *trans*-keto photoproduct [12]. Nevertheless the mechanism of the formation of the photoproduct is still unclear and questions remain regarding (i) the existence of several excited species after the proton transfer (different *cis*-keto* isomers) and (ii) the nature of the precursor species of the final *trans*-keto photoproduct. Therefore, HS-MCR was used to investigate the kinetics and spectra of the transient species involved in the photochemistry of SA.

The transient absorption spectra of SA in acetonitrile at excitation wavelength 384 nm (data set $\mathbf{D}_{SA}$) were taken in the spectral range 400–700 nm

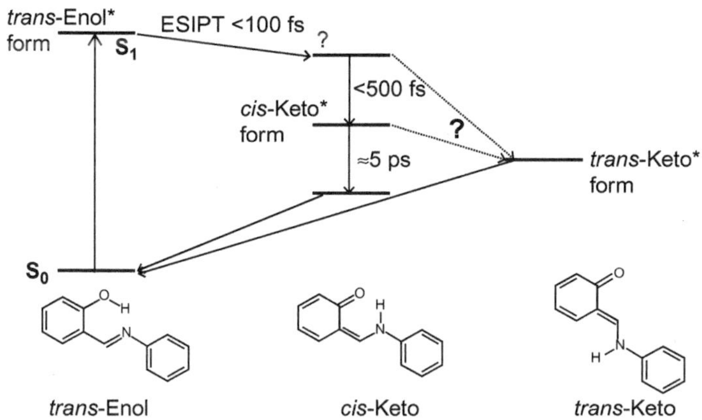

FIG. 13 Overall photoinduced processes for SA. (*) is used to identify species in the excited states. Question marks correspond to species and transitions that remain controversial in the literature.

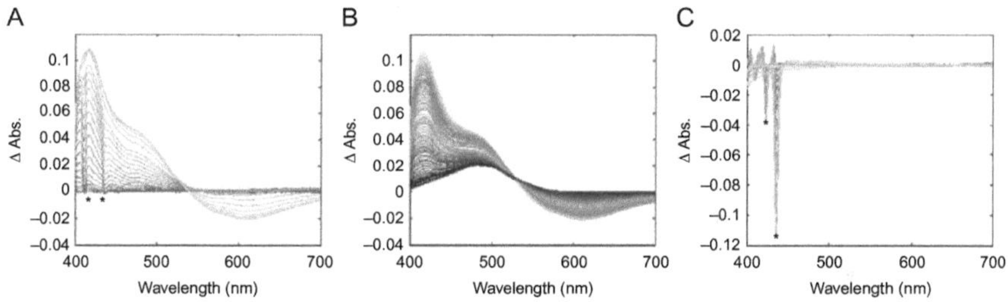

FIG. 14 Transient absorption spectra of SA in acetonitrile following laser excitation at 384 nm observed (A) between −200 and 300 fs (B) and between 300 fs and 50 ps (data set $\mathbf{D}_{SA}$) and (C) of acetonitrile (data set $\mathbf{D}_{SRA}$) between −200 and 75 fs. The negative peaks at 412 and 435 nm indicated by * correspond to SRA signals.

and for 132 time delays in the range −0.2 to 50 ps. A more comprehensive description of the setup and the experimental data can be found in reference [38]. The IRF (150 fs FWHM) was estimated from the SRA signals of acetonitrile (Fig. 14C). The spectra obtained after GVD preprocessing are shown in Fig. 14A and B. Narrow negative peaks appear between 412 and 435 nm at ultrashort timescale (below 300 fs). These peaks are mainly attributed to SRA, which hinders the spectrokinetic features of SA at short time delay (<300 fs), a time range during which the ESIPT occurs. To overcome this issue a multiset MCR-ALS approach was taken. For this purpose, SRA was measured in acetonitrile providing the data set $\mathbf{D}_{SRA}$. The data are shown in Fig. 14C.

Exploratory data analysis was first performed independently for the two data sets. SVD was also performed on the augmented data [$\mathbf{D}_{SRA}$; $\mathbf{D}_{SA}$]. Overall, seven contributions were found relevant: three to describe the variance of the SRA data and four associated to SA data. MCR-ALS was then applied on the augmented data. Nonnegativity and unimodality were applied on the profiles in **C**. Correspondence of the different contributions in the two experiments was set to indicate that the three SRA contributions are shared between the two matrices, while the four other contributions are related to the kinetic processes observed in $\mathbf{D}_{SA}$. The results obtained for time-dependent concentration profiles and spectra are shown in Fig. 15. The contributions observed in Fig. 15B show narrow peaks characteristic of SRA signals. The evolution of the time-dependent profiles in Fig. 15C shows four consecutive species, which will be called **A**, **B**, **C**, and **D** following their emerging sequence. The first transient species (**A**) appears at ultrashort timescale and presents a wide absorption band with a maximum around 420 nm. As can be observed, some narrow SRA contributions are still present. This is a consequence of the convolution of the kinetics with the IRF. The resulting "mixing" effect cannot be fully accounted for in a soft-modeling approach. To the contributions **B** and **C** correspond transient spectra with a large absorption

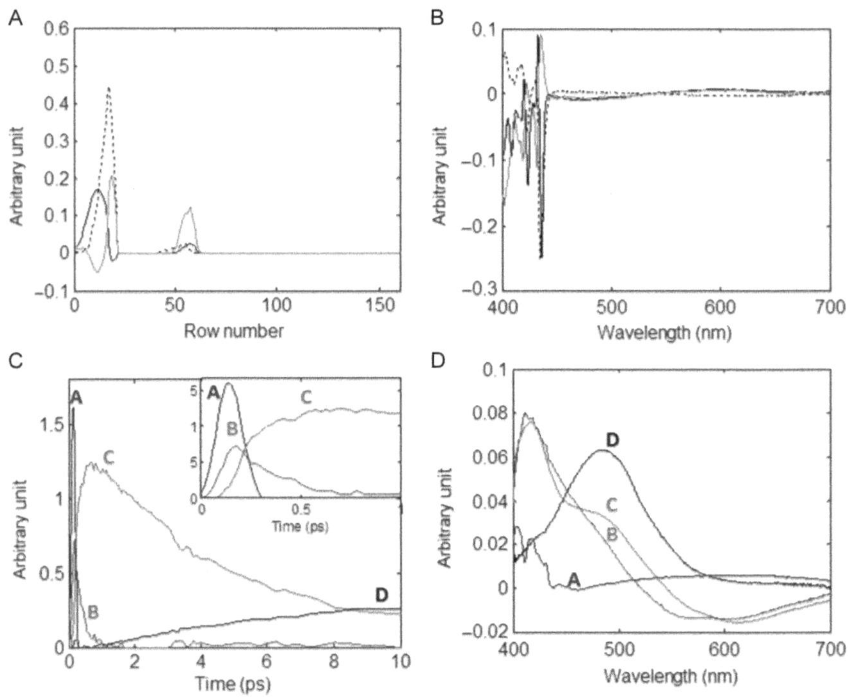

FIG. 15 MCR-ALS time-dependent concentration profiles and spectra obtained for: (A) and (B) the three artifact contributions (represented in *black*, *gray*, and *dotted line*) and (C) and (D) the contributions **A**, **B**, **C**, and **D** associated to the photoinduced processes (respectively, in *blue* (*dark gray* in the print version), *green* (*gray* in the print version), *red* (*light gray* in the print version), and *black*).

band showing a maximum near 420 nm with a shoulder between 460 and 550 nm, and a negative band with a maximum around 600 nm, respectively. This latter is related to stimulated emission. The contributions **B** and **C** can be associated with the excited *cis*-keto form of SA with vibrational relaxation. For species **D**, a large absorption band is observed with a maximum at 480 nm corresponding to the photochromic *trans*-keto form of SA.

From the shape of the time-dependent concentration profiles and the knowledge of the photochemistry of SA, several sequential kinetic models could be tentatively applied and the simplest model in Eq. (9) was finally chosen on the basis of the residuals and in terms of interpretability:

$$(\mathbf{T}) \xrightarrow{k_1} \mathbf{B} \xrightarrow{k_2} \mathbf{C} \xrightarrow{\Phi k_3} \mathbf{D} \tag{9}$$

where **T** is a transparent contribution, k_3 is the kinetic decay rate of the **C** species, and Φ corresponds to the formation quantum yield of the *trans*-keto photo product **D** (see Section 4.3.1). The first contribution **A** was not included into the kinetic model since this species appears and disappears faster than the time resolution of the experiment (it was left outside of the hard model). Furthermore, the quantum yield for the formation of **D** was fixed to 0.14, as estimated from another experiment [12]. The parallel reaction (see process (2), Fig. 6) corresponds to the formation of a colorless species (*cis*-keto) that will return to the enol form in its ground state [12]. This branching was taken into account in the HS-MCR resolution for which the results shown in Fig. 16 were obtained. The hard-modeled fitted profiles were convoluted by the IRF as described in Section 4.3.2. The spectra of **B**, **C**, and **D** are similar to the ones obtained using MCR-ALS. The rate constants associated to the kinetics are $k_1 = 12.2 \pm 0.2\ \text{ps}^{-1}$, $k_2 = 3.08 \pm 0.03\ \text{ps}^{-1}$, and $k_3 = 0.17 \pm 0.01\ \text{ps}^{-1}$, to which can be associated the characteristic times $\tau_1 = 82 \pm 1$ fs, $\tau_2 = 324 \pm 3$ fs, and $\tau_3 = 5.9 \pm 0.1$ ps.

The characteristic time 82 ± 1 fs obtained for τ_1 could be attributed to the ESIPT occurring after the photoexcitation of the enol form of SA. The determination of this characteristic time below the instrumental time resolution was possible through an HS-MCR approach taking into account IRF

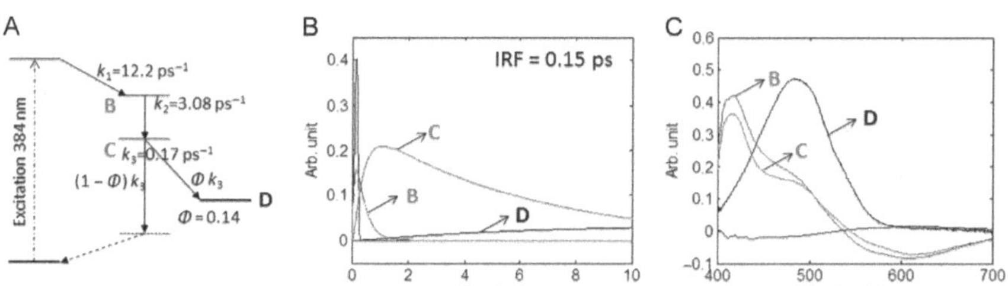

FIG. 16 (A) General photomechanism of SA with kinetic rates values provided by HS-MCR (B) kinetics and (C) spectra of the four contributions involved.

convolution. The results obtained tend to confirm (1) the presence of two *cis*-keto forms: a vibrationally hot *cis*-keto* form which relaxes to the cold form (spectra with the more pronounced shoulder) and (2) that the cold *cis*-keto* is the main precursor of the *trans*-keto photoproduct. Finally the photodynamical scheme for SA (see Fig. 16A) can be drawn with a cascade of first-order kinetic process which is in agreement with published results [12].

6 CONCLUDING REMARKS

We have illustrated the potential of MCR for the study of photoinduced spectrodynamic processes observed in time-resolved absorption spectroscopy. We have highlighted the main advantages of MCR-ALS approaches, including multiset analysis and the possibility to incorporate tailored kinetic models as additional constraints in so-called hybrid hard and soft MCR approaches. These approaches applied to the study of the photophysics of BP allowed the tentative assignment of an intermediate species, a result that was later confirmed by theoretical calculations. Another representative example of the usefulness of MCR-ALS analysis techniques to unravel both spectroscopic and kinetic descriptions of complex systems involving multiple reactions was provided by the study of the photochemical reactivity of a photochromic diarylethene. Here, reliable spectral information about a new photoproduct could be obtained, bridging the gap with some results obtained in an NMR study. At last, the overall photodynamics of SA were investigated, providing insights into the involvement of two precursor species in the formation of the main photoproduct.

REFERENCES

[1] Kosumi D, Maruta S, Horibe T, Fujii R, Sugisaki M, Cogdell RJ, et al. Ultrafast energy-transfer pathway in a purple-bacterial photosynthetic core antenna, as revealed by femtosecond time-resolved spectroscopy. Angew Chem Int Ed Engl 2011;50:1097–100.

[2] Mezzetti A. Light-induced infrared difference spectroscopy in the investigation of light harvesting complexes. Molecules 2015;20:12229.

[3] Stahl AD, Di Donato M, Van Stokkum I, Van Grondelle R, Groot ML. A femtosecond visible/visible and visible/mid-infrared transient absorption study of the light harvesting complex II. Biophys J 2009;97:3215–23.

[4] Katayama T, Ishibashi Y, Morii Y, Ley C, Brazard J, Lacombat F, et al. Ultrafast delocalization of cationic states in poly(N-vinylcarbazole) solid leading to carrier photogeneration. Phys Chem Chem Phys 2010;12:4560–3.

[5] Law KY. Organic photoconductive materials: recent trends and developments. Chem Rev 1993;93:449–86.

[6] Wallentin J, Anttu N, Asoli D, Huffman M, Aberg I, Magnusson MH, et al. InP nanowire array solar cells achieving 13.8% efficiency by exceeding the ray optics limit. Science 2013;339:1057–60.

[7] Irie M, Fukaminato T, Matsuda K, Kobatake S. Photochromism of diarylethene molecules and crystals: memories, switches, and actuators. Chem Rev 2014;114:12174–277.

[8] Ishibashi Y, Fujiwara M, Umesato T, Saito H, Kobatake S, Irie M, et al. Cyclization reaction dynamics of a photochromic diarylethene derivative as revealed by femtosecond to microsecond time-resolved spectroscopy. J Phys Chem C 2011;115:4265–72.
[9] Miyasaka H, Satoh Y, Ishibashi Y, Ito S, Nagasawa Y, Taniguchi S, et al. Ultrafast photodissociation dynamics of a hexaarylbiimidazole derivative with pyrenyl groups: dispersive reaction from femtosecond to 10 Ns time regions. J Am Chem Soc 2009;131:7256–63.
[10] Das PK, Encinas MV, Scaiano JC. Laser flash photolysis study of the reactions of carbonyl triplets with phenols and photochemistry of p-hydroxypropiophenone. J Am Chem Soc 1981;103:4154–62.
[11] Turro NJ, Ramamurthy V, Scaiano JC. Modern molecular photochemistry of organic molecules. Sausalito, California: University Science Book; 2010.
[12] Sliwa M, Mouton N, Ruckebusch C, Poisson L, Idrissi A, Aloise S, et al. Investigation of ultrafast photoinduced processes for salicylidene aniline in solution and gas phase: toward a general photo-dynamical scheme. Photochem Photobiol Sci 2010;9:661–9.
[13] Bonneau R, Wirz J, Zuberbuhler AD. Methods for the analysis of transient absorbance data. Pure Appl Chem 1997;69:979–92.
[14] Ruckebusch C, Duponchel L, Sombret B, Huvenne JP, Saurina J. Time-resolved step-scan FT-IR spectroscopy: focus on multivariate curve resolution. J Chem Inf Comput Sci 2003;43:1966–73.
[15] Van Stokkum IHM, Larsen DS, Van Grondelle R. Global and target analysis of time-resolved spectra. Biochim Biophys Acta Bioenerg 2004;1657:82–104.
[16] Puxty G, Maeder M, Hungerbühler K. Tutorial on the fitting of kinetics models to multivariate spectroscopic measurements with non-linear least-squares regression. Chemom Intell Lab Syst 2006;81:149–64.
[17] Maeder M, Zuberbuehler AD. Nonlinear least-squares fitting of multivariate absorption data. Anal Chem 1990;62:2220–4.
[18] De Juan A, Tauler R. Multivariate curve resolution (Mcr) from 2000: progress in concepts and applications. Crit Rev Anal Chem 2006;36:163–76.
[19] Tauler R. Multivariate curve resolution applied to second order data. Chemom Intell Lab Syst 1995;30:133–46.
[20] De Juan A, Maeder M, Martinez M, Tauler R. Combining hard- and soft-modelling to solve kinetic problems. Chemom Intell Lab Syst 2000;54:123–41.
[21] De Juan A, Maeder M, Martínez M, Tauler R. Application of a novel resolution approach combining soft- and hard-modelling features to investigate temperature-dependent kinetic processes. Anal Chim Acta 2001;442:337–50.
[22] Klán P, Wirz J. Photochemistry of organic compounds: from concepts to practice. Chichester: Wiley-Blackwell; 2009.
[23] Diels JC, Rudolph W. Ultrashort laser pulse phenomena. Second edition, Academic Press, Oxford: Elsevier Science; 2006.
[24] Rulliere C. Femtosecond laser pulses: principles and experiments. Berlin: Springer; 2007.
[25] Zewail AH. Femtochemistry: recent progress in studies of dynamics and control of reactions and their transition states. J Phys Chem 1996;100:12701–24.
[26] Ernsting NP, Kovalenko SA, Senyushkina T, Saam J, Farztdinov V. Wave-packet-assisted decomposition of femtosecond transient ultraviolet-visible absorption spectra: application to excited-state intramolecular proton transfer in solution. J Phys Chem A 2001; 105:3443–53.
[27] Lorenc M, Ziolek M, Naskrecki R, Karolczak J, Kubicki J, Maciejewski A. Artifacts in femtosecond transient absorption spectroscopy. Appl Phys B Lasers Opt 2002;74:19–27.

[28] Nakayama T, Amijima Y, Ibuki K, Hamanoue K. Construction of a subpicosecond double-beam laser photolysis system utilizing a femtosecond Ti:sapphire oscillator and three Ti:sapphire amplifiers (a regenerative amplifier and two double passed linear amplifiers), and measurements of the transient absorption spectra by a pump-probe method. Rev Sci Instrum 1997;68:4364–71.

[29] Ruckebusch C, Sliwa M, Réhault J, Naumov P, Huvenne JP, Buntinx G. Hybrid hard- and soft-modelling applied to analyze ultrafast processes by femtosecond transient absorption spectroscopy: study of the photochromism of salicylidene anilines. Anal Chim Acta 2009;642:228–34.

[30] De Rooi JJ, Devos O, Sliwa M, Ruckebusch C, Eilers PHC. Mixture models for two-dimensional baseline correction, applied to artifact elimination in time-resolved spectroscopy. Anal Chim Acta 2013;771:7–13.

[31] Devos O, Mouton N, Sliwa M, Ruckebusch C. Baseline correction methods to deal with artifacts in femtosecond transient absorption spectroscopy. Anal Chim Acta 2011;705:64–71.

[32] Malinowski ER. Factor analysis in chemistry. 3rd ed. New York: Wiley; 2002.

[33] Gampp H, Maeder M, Meyer CJ, Zuberbuehler AD. Quantification of a known component in an unknown mixture. Anal Chim Acta 1987;193:287–93.

[34] Keller HR, Massart DL. Evolving factor analysis. Chemom Intell Lab Syst 1991;12:209–24.

[35] Maeder M. Evolving factor analysis for the resolution of overlapping chromatographic peaks. Anal Chem 1987;59:527–30.

[36] Mouton N, Devos O, Sliwa M, De Juan A, Ruckebusch C. Multivariate curve resolution—alternating least squares applied to the investigation of ultrafast competitive photoreactions. Anal Chim Acta 2013;788:8–16.

[37] Muñoz G, De Juan A. pH- and time-dependent hemoglobin transitions: a case study for process modelling. Anal Chim Acta 2007;595:198–208.

[38] Mouton N, De Juan A, Sliwa M, Ruckebusch C. Hybrid hard- and soft-modeling approach for the resolution of convoluted femtosecond spectrokinetic data. Chemom Intell Lab Syst 2011;105:74–82.

[39] Mouton N, Sliwa M, Buntinx G, Ruckebusch C. Deconvolution of femtosecond time-resolved spectroscopy data in multivariate curve resolution. Application to the characterization of ultrafast photo-induced intramolecular proton transfer. J Chemom 2010;24:424–33.

[40] Yabumoto S, Sato S, Hamaguchi H-O. Vibrational and electronic infrared absorption spectra of benzophenone in the lowest excited triplet state. Chem Phys Lett 2005;416:100–3.

[41] El-Sayed MA, Leyerle R. Low field Zeeman effect and the mechanism of the S1 → T1 non-radiative process. J Chem Phys 1975;62:1579–80.

[42] Aloïse S, Ruckebusch C, Blanchet L, Réhault J, Buntinx G, Huvenne J-P. The benzophenone S1(N,Π*) → T1(N,Π*) states intersystem crossing reinvestigated by ultrafast absorption spectroscopy and multivariate curve resolution. J Phys Chem A 2008;112:224–31.

[43] Ruckebusch C, Aloïse S, Blanchet L, Huvenne JP, Buntinx G. Reliable multivariate curve resolution of femtosecond transient absorption spectra. Chemom Intell Lab Syst 2008;91:17–27.

[44] Sergentu D-C, Maurice R, Havenith RWA, Broer R, Roca-Sanjuan D. Computational determination of the dominant triplet population mechanism in photoexcited benzophenone. Phys Chem Chem Phys 2014;16:25393–403.

[45] Kawai T, Koshido T, Yoshino K. Optical and dielectric properties of photochromic dye in amorphous state and its application. Appl Phys Lett 1995;67:795–7.

[46] Spangenberg A, Piedras Perez JA, Patra A, Piard J, Brosseau A, Metivier R, et al. Probing photochromic properties by correlation of UV-visible and infra-red absorption

spectroscopy: a case study with cis-1,2-dicyano-1,2-bis(2,4,5-trimethyl-3-thienyl)ethene. Photochem Photobiol Sci 2010;9:188–93.

[47] Takashi K, Tsuyoshi K, Katsumi Y. Novel photomemory effects in an amorphous photochromic dye. Jpn J Appl Phys 1995;34:L389.

[48] Irie M, Mohri M. Thermally irreversible photochromic systems. Reversible photocyclization of diarylethene derivatives. J Org Chem 1988;53:803–8.

[49] Chen Q, Hiraga T, Men L, Tominaga J, Atoda N. Optical properties and application of photochromic diarylethene. Mol Cryst Liq Cryst Sci Technol Sect A Mol Cryst Liq Cryst 2000;345:21–6.

[50] Mizokuro T, Mochizuki H, Yamamoto N, Horiuchi S, Tanigaki N, Hiraga T, et al. A formation of organic rewritable optical memory media using the vapor transportation method. J Photopolym Sci Technol 2003;16:195–8.

[51] Murase S, Teramoto M, Furukawa H, Miyashita Y, Horie K. Photochemically induced fluorescence control with intermolecular energy transfer from a fluorescent dye to a photochromic diarylethene in a polymer film. Macromolecules 2003;36:964–6.

[52] Ishitobi H, Sekkat Z, Irie M, Kawata S. The photoorientation movement of a diarylethene-type chromophore. J Am Chem Soc 2000;122:12802–5.

[53] Erko FG, Berthet J, Patra A, Guillot R, Nakatani K, Métivier R, et al. Spectral, conformational and photochemical analyses of photochromic dithienylethene: cis-1,2-dicyano-1,2-bis(2,4,5-trimethyl-3-thienyl)ethene revisited. Eur J Org Chem 2013;2013:7809–14.

[54] Dürr H, Bouas-Laurent H. Photochromism. Molecules and systems. Amsterdam: Elsevier Science; 2003.

[55] Ziolek M, Kubicki J, Maciejewski A, Naskrecki R, Grabowska A. An ultrafast excited state intramolecular proton transfer (ESPIT) and photochromism of salicylideneaniline (SA) and its "double" analogue salicylaldehyde azine (SAA). A controversial case. Phys Chem Chem Phys 2004;6:4682–9.

[56] Sliwa M, Letard S, Malfant I, Nierlich M, Lacroix PG, Asahi T, et al. Design, synthesis, structural and nonlinear optical properties of photochromic crystals: toward reversible molecular switches. Chem Mater 2005;17:4727–35.

Chapter 12

Experimental and Data Analytical Approaches to Automating Multivariate Curve Resolution in the Analysis of Hyperspectral Images

D.M. Haaland*[,1], H.D.T. Jones[†] and J.A. Timlin[‡]
*Spectral Resolutions, Albuquerque, NM, United States
[†]HyperImage Solutions, Rio Rancho, NM, United States
[‡]Sandia National Laboratories, Albuquerque, NM, United States
[1]Corresponding author: e-mail: dmhaaland@comcast.net

Chapter Outline

1 INTRODUCTION

Multivariate curve resolution (MCR) is a widely used approach for identifying spectral pure components and their relative concentrations from hyperspectral images [1–10]. Chapters 2 and 3 present the basics of the MCR algorithm and

Data Handling in Science and Technology, Vol. 30. http://dx.doi.org/10.1016/B978-0-444-63638-6.00012-7
Copyright © 2016 Elsevier B.V. All rights reserved.

standard approaches used in extracting pure spectra and their relative concentrations from spectral data. This chapter focuses on the use of advanced techniques to improve the unmixing of component spectra obtained from hyperspectral images and to generate spatial components (images) from the MCR-determined relative concentrations in complex samples, such as biological samples. Biological samples present additional challenges for MCR analysis due to their high propensity for spectral overlap and spatial colocalization. Using multiple examples from real and realistically simulated hyperspectral images, we will demonstrate that a careful characterization of the instrument response and noise along with intelligent experimental designs can lead to automating image and spectral preprocessing, minimizing user input, and enhancing the accuracy of the extraction of pure spectral components and concentrations from one or more hyperspectral images. Though the examples used to demonstrate the approaches here are from our hyperspectral confocal fluorescence microscope [11], which uses an electron multiplying charge-coupled device (EMCCD) detector [12], the artifacts and noise, as well as the approaches to combat them are relevant to other spectral imaging microscopes, including commercially available instruments.

2 THEORY OF MCR ANALYSES

Chapters 2 and 3 have detailed the MCR methods; therefore, the methods used in this work will be described in brief. MCR is based upon constrained alternating least squares methods, which use as their basis the classical least squares (CLS) methods that have been detailed extensively in the past [13–15]. CLS methods make the assumption that (1) the spectra are linear combinations of the individual pure-component spectra of all those spectrally active components in the sample and (2) the amounts of each of the pure-component spectra in each spectrum are proportional to the concentrations of those components in the sample. In what follows, matrices are represented with bold upper case letters, vectors are represented as bold lower case letters, the transpose of matrices and vectors is represented by a superscript T, scalars are represented by lower case italics letters, and a superscript + is used to represent the pseudoinverse of the matrix. The CLS model for an $n \times p$ data set, **D**, consisting of n spectra each with p spectral channels can be expressed as in Eq. (1):

$$\mathbf{D} = \mathbf{C}\mathbf{S}^{\mathrm{T}} + \mathbf{E} \tag{1}$$

where **C** is an $n \times m$ matrix of component concentrations with m being the number of independently varying emitting species in the image, **S** is a $p \times m$ matrix of the pure-component spectra, and **E** is an $n \times p$ matrix consisting of noise and model error. Since we often do not have accurate a priori information about the numbers of components in a sample, we must first empirically determine the number of independently varying spectral components in the spectral image. We can accomplish this goal by first performing a principal component analysis (PCA) on the spectra in the image [15–17]. A Scree plot consisting of the semilog of the PCA eigenvalues vs numbers of eigenvectors is then used to determine the number of independent spectral components in the sample based on the location of the bend within the plot after which remaining eigenvalues are relatively small and similar value [18,19]. The Scree plot can be assessed manually, but we use methods previously developed for selecting numbers of factors to include in multivariate calibrations to automatically determine this number [20].

Once the number of components is determined, the initial estimates of pure spectra in **S** for the MCR analysis are required. These initial estimates can simply be sets of uniform random numbers varying between 0 and 1 if no a priori information about the component spectra is known (or normally distributed random numbers varying between −1 and 1 if the pure-component spectrum is not constrained to be nonnegative). If one or more components are known or suspected, then digital representations of these components can and should be used as starting spectra for initiating the MCR analysis. The least squares solution for the CLS model in Eq. (1) is

$$\hat{\mathbf{C}} = \mathbf{D}\left(\mathbf{S}^{\mathrm{T}}\right)^{+} \tag{2}$$

where the ^symbol above **C** indicates the least squares estimate of **C**. Once $\hat{\mathbf{C}}$ has been obtained, the least squares estimate for $\mathbf{S}^{\mathrm{T}}$ can be obtained from

$$\hat{\mathbf{S}}^{\mathrm{T}} = \hat{\mathbf{C}}^{+}\mathbf{D} \tag{3}$$

However, there are an infinite number of possible solutions to Eqs. (2) and (3) due to rotational ambiguity in the least squares solution [21]. Nonnegative constraints can be employed in an alternating least squares algorithm to restrict the range of solutions since real spectral species have nonnegative intensities and nonnegative concentrations. Rigorous least squares methods are employed in these constrained alternating least squares methods to assure that the analysis converges during the iterative computations [22–24].

Typical hyperspectral images from our microscope have >40,000 spectra/image, and an image data set may contain 12–45 hyperspectral images. In order to increase computational speed given the significant numbers of spectra present in these hyperspectral image data sets, we have employed a variety of techniques. First, a PCA compression of the data is performed and a set of compressed results in the form of scores and loading vectors are retained. A judicious selection of the order of matrix operations assures that the intermediate matrix dimensions remain as small as possible and the matrix computations are performed as fast as possible. Also, computationally efficient methods based on the rigorous least squares methods developed by Bro and De Jong [22] have been improved to dramatically reduce the number of computations required [23]. These enhanced approaches to the MCR computations allow the MCR analysis to be completed in approximately the time required to collect the spectral images even when nearly a million spectra of 512 wavelengths are included in the analyses.

Additionally, the range of solutions is further restricted by employing equality constraints when a spectral component or a portion of the spectral component or its concentrations are known. The method of direct elimination is used in order to maintain rigorous least squares solutions to the equality constrained solution [24]. Details of practical application of full equality constraints for spectral offsets and the use of partial equality constraints to minimize mixing between pure spectral components will be described in later sections. For some MCR models, it may be appropriate to remove the nonnegativity constraint for either the spectra or the concentrations. The basis for this removal will be explained in more detail later in this chapter. In addition to nonnegativity constraints on pure spectra and concentrations, equality constraints can be applied to concentrations, e.g., if one knows that certain pixels consist of a single component or if one or more components are known to be absent from certain pixels. Also, monotonic and unimodal constraints can be applied especially in cases of pure spectra of fluorescence species since their pure emission spectra are generally known to be unimodal. For a detailed description of various constraints, see de Juan and Tauler and references therein [25].

3 APPROACHES TO IMPROVE THE PROBABILITY OF SUCCESSFUL MCR SOLUTIONS

As mentioned previously, rotational ambiguity allows a range of possible solutions to the MCR analysis. Please refer to Chapters 4 and 5 for comprehensive coverage of the rotational ambiguity issue. Additionally, noise, background, instrument artifacts, spectral/spatial overlap, and sparse signals also contribute to reduced success in developing accurate MCR models from hyperspectral images of biological samples. In this section we detail both experimental and preprocessing approaches that have demonstrated improved MCR results. The goal here is to improve the identification of pure spectra and reduce the range of possible solutions using methods beyond the standard nonnegativity constraints placed on the pure spectra and their relative concentrations.

3.1 Experimental Approaches for Successful MCR Analysis

Several experimental approaches can be used to reduce rotational ambiguity in the analyses. Any experimental process that decreases the complexity of the data or increases the variation in the data can serve to shrink the range of possible MCR solutions. One experimental approach that has demonstrated success in our laboratory is to include multiple images in the analysis such that the samples that are imaged cover a range of different experimental conditions or variables. These composite data sets, also known as multiset data, serve to decrease component correlations and force the components to become more orthogonal. Please refer to Chapter 2 for additional discussion of multiset data. Samples with reduced numbers of components or containing components at the extremes of their concentration ranges are helpful. For example, in the case of MCR analysis of high spectrally and spatially overlapped photosynthetic components in cyanobacteria, different mutant strains of genetically modified bacteria with various photosynthetic species absent from the cells have resulted in accurate MCR models even in these challenging samples [3,26].

In situations where genetic modification is prohibitive, a relatively simplistic experimental approach to take with fluorescence images is to obtain hyperspectral images of the samples as they are photobleached. Since each spectral component photobleaches at a different rate, the spectral variance of the samples increases and the components become more orthogonal as the samples are photobleached. Similarly, some emitting species are known to be temperature dependent, so their emission variance increases with temperature changes or when temperature gradients are introduced. In living cells, the presence or concentrations of some species are dependent on growth conditions or environmental factors that can be controlled or manipulated to introduce orthogonality into the system.

Finally, improved MCR results can be obtained by using known information. For example, one or more of the spectral components may be known to be present or absent in a particular sample. In this case, the MCR analysis can be initiated with measured or library spectra of those components known to be present to assist the algorithm in convergence. Though possible, it is not generally helpful to equality constrain the spectra of known components since environmental conditions in the sample (such as temperature, pH, interactions with other components and solvents, and intensity of light exposure) may change the spectra in the sample relative to the spectra obtained from the library or an isolated component. If a component or components are known to be absent from an image or a portion of the image data set, the concentrations of those components can be equality constrained to be zero for those pixels where the component(s) are absent.

3.2 Automated Preprocessing strategies for Improving MCR Analysis

The following is a condensed discussion of the data and preprocessing methods presented in Jones *et al.* [18]. The preprocessing techniques described later are general and can be applied to any hyperspectral image data collected with a CCD-type detector. The preprocessing is applied in two stages. Stage 1 preprocessing identifies and removes known spectral artifacts including spikes due to cosmic rays, a large detector offset, and the structured-dark-noise components present in the detector response. Stage 2 preprocessing defines and limits spatial and/or spectral regions of interest to be included in the MCR analyses. By preprocessing the data in an automated and reproducible fashion to remove artifacts and select the pixels with the most information, the potential for successful extraction of correct pure spectra components and their concentrations during MCR analyses is increased. Automated data preprocessing also improves detection limits and increases the accuracy of the quantitation of the extracted components (see Sections 4.2 and 4.3).

3.2.1 Stage 1 Preprocessing: Spectral Artifacts and Noise in Hyperspectral Images

An important approach to improving the success of an MCR analysis is to identify any artifacts in the data and to consider all sources of spectral noise that serve to limit the analyses. To accomplish this goal, one must carefully characterize the instrumentation to fully understand artifacts and noise sources in the data. Artifacts include data spikes due to cosmic rays striking the detector, detector offsets, and structured noise present in array detectors.

Cosmic spikes can be relatively rare in fluorescence data due to the short integration times necessary (0.24 or 0.48 ms), but when they are present, they can greatly distort the MCR analyses or any other spectral model development. The spectral variance due to cosmic ray spikes is so large that each spike often needs a separate MCR factor to account for its presence. We have developed automated, rapid software to despike the hyperspectral image data by identifying spectra with spikes present based on their variance and then replacing the entire spectrum with a spatially adjacent pixel that is free of spikes [18]. Since less than 0.01% of the image pixel spectra contain cosmic spike artifacts, this approach can be justified. Additional algorithms for removing cosmic spikes have been demonstrated and could be employed for data that have a higher prevalence of cosmic events or situations when replacing the entire spectrum is not acceptable [27–29].

Offset and structured-noise components are removed after despiking has been performed on both the sample and dark current spectral images. The common practice of direct dark image subtraction has the disadvantage that the noise in the subtracted result is inflated by the square root of 2, and it also makes the assumption that both the shape and intensities of the structured noise are the same for the dark and sample pixels. Modeling the structured noise, however, only assumes that the shape of the structured noise is constant, and importantly this modeling allows for the different amounts of structured noise on a per pixel basis that we find in real data.

Efficient and automated removal of both of the offset and structured-noise components is facilitated by the presence of a 30 pixel dark spectral region of the sample image where the emission signal from the sample is zero. This dark region is present in our system due to the presence of a rugate filter that is used to assure that the laser radiation does not strike the detector. Most fluorescence imaging spectrometers will have some type of similar filter to assure that light from the laser does not impinge on the detector. If absent from a given system, a similar type optical filter can be added to the system to assure that a set of dark spectral channels are present in the detector. An additional significant advantage of this method of removing structured noise and offset is that it is not necessary to collect the same number of dark images as contained in the hyperspectral sample image. The ability to use a single dark image in the modeling and removal of structured noise and offset reduces the data collection time considerably when multiple confocal images are needed in a 3D image or in a time-resolved image. As will be seen later in this chapter, these dark channels not only allow automated offset and structured-noise removal from the data, but they can also be used to automatically generate a spatial mask (Section 3.2.2) and facilitate the MCR analyses (Sections 4.2.2 and 4.3.4).

Characterizing the noise present in the data is important to both understand the limitations in the data and develop strategies to mitigate the detrimental effects of noise on the MCR results. The MCR algorithm is based upon least squares methods that assume that the noise is independent and identically distributed (i.i.d. noise) meaning that the noise has constant variance throughout the spectrum and the noise is uncorrelated from data point to data point. In reality, most real systems experience heteroscedastic noise, i.e., the variance of the noise changes throughout the spectrum, and the noise is often correlated. The correlated error may simply be a result of baseline variations or high-frequency correlations due to the nature of the array detectors that are often used in hyperspectral imaging. In modern array detectors, read noise is present, and in the many array detectors used in our lab, this read noise contains high-frequency correlated errors. A careful examination of the dark image with no light impinging on the detector can yield a measure of the read noise present in the detector. A PCA analysis of this dark image can reveal if correlated errors are present in the read noise. Sections 4.2 and 4.3 present demonstrations of the high-frequency correlated errors present due to the EMCCD detector (see also Figs. 1D and 5). Unfortunately, these correlated errors are extremely difficult if not impossible to model, and therefore, these sources of correlated noise reduce sensitivity to minor components. Fortunately, the read noise is uncorrelated between spatial pixels, so spatial compression can be used to reduce the correlated read noise and improve the detection and extraction of minor spectral components with MCR (see Section 4.3.3).

In addition to read noise, Poisson noise is present in hyperspectral fluorescence image data. Poisson noise, or shot noise, is due to photon fluctuations and cannot be avoided if the sensitivity of the detector is sufficient to not be dominated by other noise sources. Poisson noise is almost always observable with modern, highly sensitive detectors used in spectral imaging systems, and it can become the dominant noise source when emission intensities are

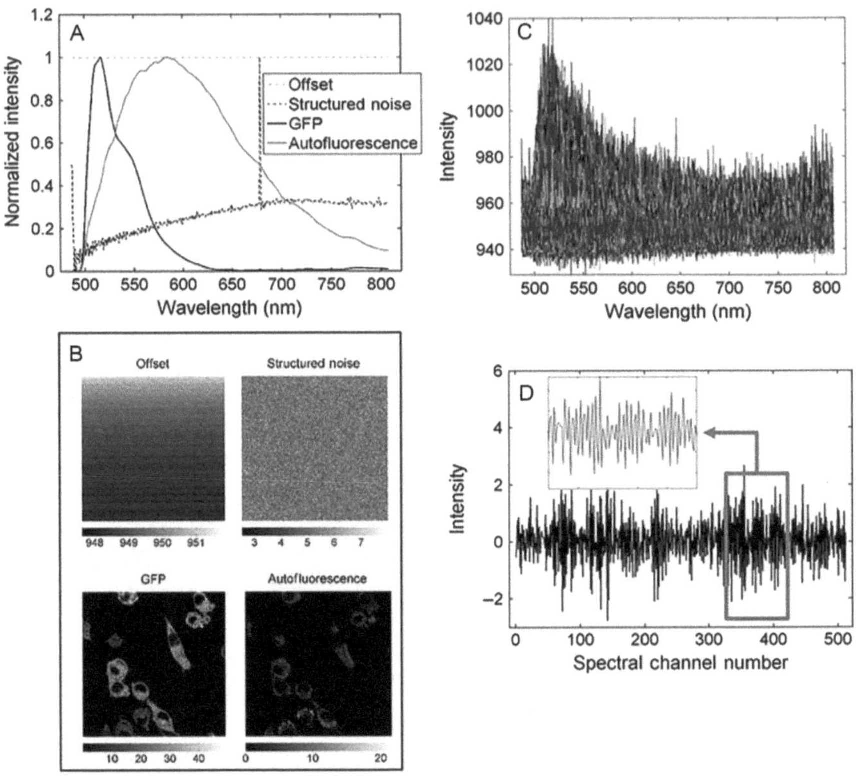

FIG. 1 Creating a realistic simulation image data set of macrophage cells containing GFP-labeled RelA protein. (A) The pure spectral components and (B) concentrations used to reconstruct the spectral data presented in (C). This data set was based on a real sample image. The spectra contain the realistic shapes and magnitudes of the major spectral artifacts and noise components present in our hyperspectral microscope, including a correlated noise source shown in (D). *From Jones HDT, Haaland DM, Sinclair MB, Melgaard DK, Collins AM, Timlin JA. Preprocessing strategies to improve MCR analyses of hyperspectral images. J Chemometr Intell Lab Syst 2012;117:149–58, Fig. 1, p. 152.*

large. The variance of Poisson noise increases linearly with light intensity until the detector becomes saturated. Poisson noise in spectral imaging microscopes can be characterized with the use of broadband fluorescence slides and neutral density filters (see Ref. [30] for the details of the methods to characterize the Poisson noise). Although the Poisson noise variance is directly proportional to the signal intensity, there exists a multiplicative noise inflation factor of ~3 in systems equipped with an EMCCD detector due to the electron multiplying nature of EMCCD detectors [12,18]. The presence of Poisson noise can have significant implications for the MCR analysis of images because the MCR algorithm assumes that the noise variance is constant throughout the spectrum. Since the MCR algorithm minimizes the total sum of squared errors in all the data, regions of high intensity serve to reduce the sensitivity to low-intensity components even if their pure spectra are spectrally or spatially isolated from the major spectral components.

One method to improve the MCR analysis in the presence of large variations in Poisson noise is to make use of weighted least squares analyses. Unfortunately, for large numbers of spatial and spectral pixels, full-weighted least squares methods are often computationally too time consuming to be practical. A computationally efficient alternative method is to scale the data in an attempt to equalize the noise throughout the data set [19,30–33]. Data scaling is an approximation to weighted least squares analysis and is accomplished by premultiplying the spectral image data by the diagonal matrix consisting of the mean image intensity (square root of the mean image variance) and postmultiplying the image data by the diagonal matrix consisting of the mean spectrum intensity (square root of the mean spectral variance). These scaling methods are computationally very fast and therefore are practical for MCR analysis of large hyperspectral images. For a more complete discussion of these scaling methods used to accommodate both heteroscedastic noise and read noise in the data, see Ref. [30]. Because these scaling methods are approximate methods, they are only helpful in certain cases, specifically when large variations in data relative to the mean spectra and mean image are reduced. Spatial compression, which is readily achieved by averaging numbers of adjacent spatial pixels, reduces variations in the data so the mean image becomes a better representation of the data. Therefore, spatial compression often precedes scaling whenever scaling is to be employed. Once pure spectra are obtained from the MCR analysis of compressed and scaled data, the resulting scaled pure spectra are unscaled and then used in a CLS analysis of the uncompressed data to obtain component concentrations at the full spatial resolution of the original data.

In addition to artifacts, drift of the instrument often causes the wavelength calibration of the instrument to vary between spectral images collected at different times. If the wavelength calibration of the instrument varies from image to image and these combined images are analyzed in a composite image data set, then MCR results can be greatly degraded. In the same manner, when pure spectra extracted from one image are used in a CLS analysis of subsequent images, wavelength shifts between images will lead to errors in concentration estimates. Therefore, the wavelength axis of the spectrometer should be periodically monitored with the use of a calibration lamp whose sharp emission peaks are well known (e.g., a low-pressure Kr–Ar lamp or similar lamp with known, well-resolved emission peaks). These emission peaks can be used to correct the wavelength axes between images and minimize degradation of the MCR results due to wavelength errors.

3.2.2 Stage 2 Preprocessing: Removing Uninformative Data

Unlike satellite imagery and nonimaging process analytical spectroscopy data, many biological images are information sparse; i.e., only a fraction of the pixels contain spectral features above the noise. The presence of these uninformative pixels during the MCR analysis only adds correlated and uncorrelated read noise to the data, and therefore, their presence in the analysis can significantly degrade the MCR results. In those cases where the uninformative pixels are a large fraction of the total number of pixels, the correlated read noise present in these pixels can actually overwhelm the signal of sparse and weakly present spectral features. In fact, we find in some cases of very sparse and weak signals that PCA compression can be dominated by the ubiquitous presence of correlated read noise in every spatial pixel, and the pixels with signals of interest are not present in the compressed PCA data even when large numbers of PCA factors are included in the compression. This loss of signal in the PCA-compressed data is the consequence of the spectral variance of the correlated read noise being greater than the spectral signal of interest. To increase the probability that the pixels with signal of interest are present in the PCA-compressed image data, we advocate using large numbers of PCA factors in PCA compression of the data when performing MCR (64 PCA factors work well for most experiments from a hyperspectral microscope with 512 spectral pixels). This recommendation is in contrast to the standard practice of using the same number of PCA factors as the number of independently varying spectral components discovered by the Scree plot obtained from the PCA analysis of the spectral images (see Section 4.3 for an example of the effect of weak, sparse signals where even 64 factors are not sufficient for the signal of interest to be present in the PCA-compressed data). The use of 64 PCA factors also allows our algorithm for selecting the optimal number of factors to define the noise floor in the image data and to apply statistics to determine when a factor rises above the noise.

In order to mitigate the detrimental effects of background pixels on the MCR results, spatial pixel masks that limit the MCR analyses to only those pixels whose spectral signal is significantly above the noise are employed. Masks are generated by the use of the dark pixels on the CCD, rendered dark by the use of a laser-blocking notch filter. The variance of the read noise in the dark pixels is representative of the variance of the read noise throughout all detector channels. Therefore, the mean and the standard deviation (sd) of the noise in 19 dark channels are determined. Then spectral pixels in each spectrum of the image with the highest intensity are identified, and the mean intensity of the 19 spectral channels centered on the peak intensity channel is determined. If the mean intensity of this most intense spectral region is 3 sd above the mean intensity of the dark spectral region for this spatial pixel, then the spatial mask is given a value of 1 and the pixel is included in the analysis. If the mean intensity is less than 3 sd above the mean intensity of the dark region, then the mask is given a value of 0 and the pixel is excluded from the analysis.

The use of this automatically generated mask has multiple advantages: (1) It serves to reduce the overall amount of correlated read noise present in the data being analyzed, (2) it improves the computation speed by reducing the total numbers of pixels in the MCR analysis, and (3) it increases the cases where the noise weighting described in Section 3.2.1 can be productively applied since it makes the mean spectrum and mean image scaling matrices more appropriate scale factors for the remaining pixels included in the MCR analysis.

4 COMPARISON OF STANDARD AND AUTOMATED PREPROCESSING APPROACHES TO MCR ANALYSES OF REALISTICALLY SIMULATED AND REAL HYPERSPECTRAL CONFOCAL FLUORESCENCE IMAGES

4.1 Image Acquisition Details

The hyperspectral image data presented in this chapter were collected with a custom hyperspectral confocal fluorescence microscope that has been described in detail elsewhere [11]. Briefly, we use a 488-nm solid-state laser to excite the fluorescence in the samples being imaged. The laser light is focused onto the sample with the use of a $60\times$ oil immersion plan apochromat (NA 1.4) objective to achieve diffraction-limited spatial resolutions measured at $250\times250\times600$ nm [11]. The fluorescence emission is then dispersed with a custom imaging prism spectrometer and detected with a backside thinned EMCCD detector. As configured, this system obtains 512 emission wavelengths from $\sim$485 to $\sim$800 nm at a spectral resolution varying from 1 nm at the shortest wavelength to 3 nm at the longest wavelength. An image is generated in one spatial dimension by scanning the laser with a rapidly scanning galvanomic-driven mirror, while the second dimension is obtained by continuously moving the sample under the beam with the use of a high-precision linear motor-driven XY stage. A rapid detector readout mode allows up to 8300 spectra/s to be obtained. In those cases where a 3D image is to be obtained, the third *z*-dimension is obtained by lowering the sample with a piezoelectric stage. However, the examples presented in this chapter are either single XY hyperspectral confocal images or a series of time-resolved two-dimensional hyperspectral confocal images.

Prior to imaging, a single XY dark image was obtained with the same collection parameters as the sample image. For our detector, the noise characteristics of the dark image remained relatively unchanged for approximately 30 min. Therefore, dark images were collected periodically throughout the experiment at approximately 30 min intervals to ensure sample images have a representative dark image available for background correction.

4.2 Realistically Simulated Hyperspectral Images of Macrophage Cells

4.2.1 Generating the Simulation Data

In order to test our automated preprocessing methods, we constructed a realistically simulated hyperspectral image based upon a measured hyperspectral image of mouse macrophage cells in order to follow the generation and location of the RelA protein, which is important to initiate cell signaling response following bacterial infection. Specifically, RAW264.7 mouse macrophage cells expressing green fluorescent protein (GFP)-RelA were grown on coverslips and placed in an imaging chamber. A 100×100 μm area of the cells was imaged using a $60\times$ oil immersion lens with an individual spatial pixel exposure time of 0.24 ms (spatial resolution 0.48 μm, details given elsewhere [18]).

An MCR solution for this hyperspectral image, which we believe to be the most accurate MCR representation of the cell image, was used as the basis for the simulated spectral image. The simulated hyperspectral image data were reconstructed from the spectral components and concentrations derived from the MCR result for this image (see Fig. 1A for pure-component spectra of the labeled GFP component and the endogenous autofluorescence component and see Fig. 1B for their respective relative concentration maps). Realistic levels of offset and noise consisting of Poisson distributed, structured, and correlated read noises were then added to the spectral data as described in detail in Ref. [18]. Briefly, the MCR-estimated pure spectra of GFP and autofluorescence shown in Fig. 1A were multiplied by their respective MCR-estimated relative concentration maps (208 rows $\times$ 204 columns). The resulting 42,432 pixel spectral image data now represent the known truth for this simulated image. Realistic noise was added to this noiseless representation of the hyperspectral image in the following manner. First, Poisson noise is generated for each wavelength in each spatial pixel in the image. Since we know that the Poisson noise variance for our system is proportional to 3.87 times the emission signal present at each wavelength of the simulated image, this Poisson-distributed noise with the same dimensions as the data are readily estimated and added to the simulated spectral intensities. The factor of 3.87 is the gain factor or noise inflation factor for the EMCCD detector used in these measurements and was empirically measured using the method described in Ref. [30]. Next, a dark image was added to this simulated hyperspectral image containing Poisson noise to provide realistic levels of the offset, structured noise, and correlated read noise to create the final realistically simulated image. The offset and structured-noise components obtained from an MCR analysis of a dark current spectral image are shown in Fig. 1A, and their relative concentrations are presented in Fig. 1B. Fig. 1C shows all the final simulated spectra with the realistic noise present. Fig. 1D exhibits the type of correlated error that would be present in a single spectrum. Because this correlated error structure varies from spectrum to spectrum, it is very difficult, if not impossible, to model.

4.2.2 Comparisons of Different Levels of Preprocessing and Subsequent MCR Results

Fig. 1 demonstrates the offset and structured component characteristics of our hyperspectral imaging system that were obtained from an MCR analysis of the dark current that was added to the simulated data. A concentration image of the offset component normalized to unit peak intensity is shown in Fig. 1B. Note that the magnitude of the offset changes in a uniform fashion across the image and that its intensity is high (~950 counts). Fig. 1A also presents the typical shape of the structured-noise component present in our EMCCD detector when the exposure time is 0.24 ms/spectrum. The spectral shape of this structured-noise component depends on exposure time used in collecting the spectral image. Importantly, Fig. 1A illustrates that there is significant structure contained in the first 30 dark channels. Fortunately, the shape of the structured-noise component is relatively constant over periods of ~30 min and the use of only the first 20 of the 30 dark channels is adequate to determine the amount of this structured noise in the entire spectrum for each spatial pixel. Thus, we can make use of the dark region at the beginning of the spectrum (no signal present except read noise of the EMCCD) to estimate the amount of offset and structured noise in a hyperspectral image obtained from a sample. Since no light is impinging on these first 20 detector channels, there is no Poisson noise present that would degrade the offset and structured-noise component estimates derived from these dark channels for each image pixel.

MCR applied to the Stage 1 and 2 preprocessed data that contains only useful fluorescence information void of structured noise and artifacts has a much greater chance of successfully obtaining the true pure-component spectra and their relative concentrations that represent biological variation. The advantages of this automated preprocessing technique can be visualized by comparing the results obtained from an MCR analysis of the simulated hyperspectral image described earlier when no preprocessing is performed, when the offset and structured noise are modeled during the MCR analysis rather than removed by preprocessing, and when fully automated Stage 1 and 2 preprocessing methods are applied. The MCR pure spectra and relative concentration maps obtained with these three approaches are presented in the three panels in Fig. 2. Panel A shows the MCR results when the raw data are analyzed. The semilog plot of the eigenvalues vs number of PCA factors (see the Scree plot at the bottom of panel A) shows that there are three independently varying spectral components present in this hyperspectral image. Therefore, the MCR algorithm was initiated with three sets of uniform random numbers for the number of spectral components detected. The spectral components resulting from the MCR analysis of these raw data are presented at the top of panel A. Component 3 is clearly the GFP component known to be present in the data (compare with Fig. 1A). Component 1 is somewhat similar to the autofluorescence component, but is both noisy and has an incorrect short-wavelength feature which is most likely due to mixing with the GFP component. By comparing the MCR concentration images with the true concentration maps of the simulated data, the concentration map of Component 2 most closely resembles that of the offset. The structured noise does not appear in panel A as an identifiable separate spectral component. In general, it would

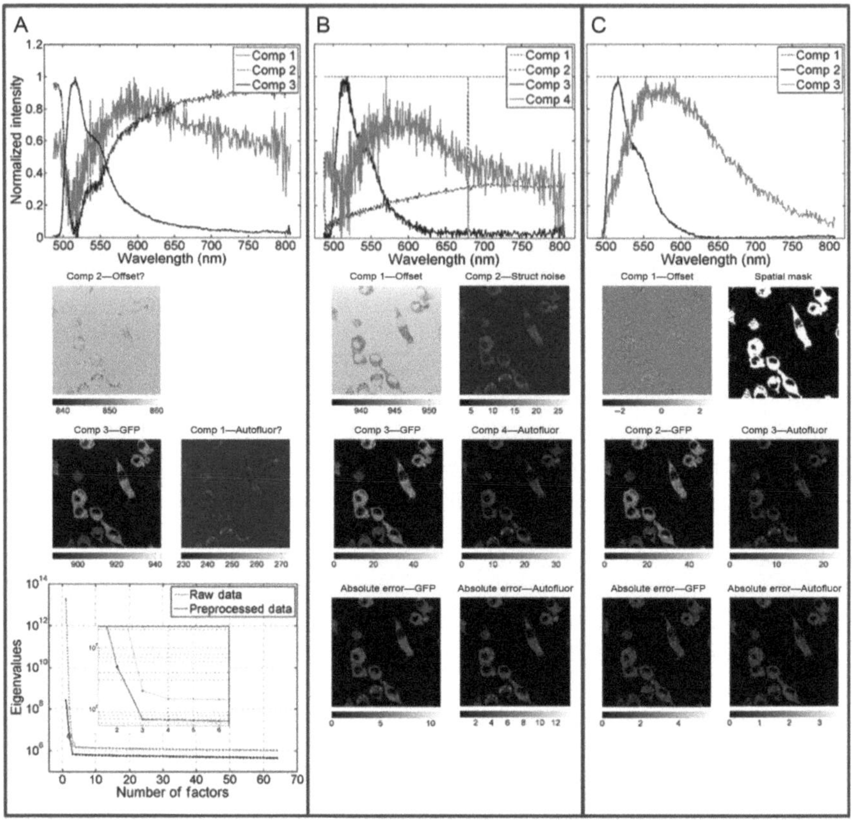

FIG. 2 MCR analysis results from the simulated macrophage cell data. Each panel shows the MCR pure spectral components and their corresponding concentrations. Contained within (A) is a Scree plot used to estimate the number of components for each analysis. Panels (B) and (C) contain the absolute error for the GFP and autofluorescence components. The panels show MCR results for: (A) no preprocessing, (B) no preprocessing with an attempt to model and remove the offset and structured-noise component during MCR analysis, and (C) Stage 1 and 2 preprocessing methodology (spatial mask included). *From Jones HDT, Haaland DM, Sinclair MB, Melgaard DK, Collins AM, Timlin JA. Preprocessing strategies to improve mcr analyses of hyperspectral images. J Chemometr Intell Lab Syst 2012;117:149–58, Fig. 5, p. 155.*

have been difficult to interpret these results without the benefit of the known pure spectra and their true relative concentration maps.

Panel B of Fig. 2 shows the MCR results when the offset and structured-noise components known to be present in the data are explicitly added to the MCR analysis and are spectrally equality constrained. The structured-noise spectral shape is readily obtained from the MCR analysis of the dark image used in creating the simulated data, so this is the best-case scenario for this direct modeling case. In this MCR analysis, the approach used is to estimate the amounts of offset and structured noise present in the simulated data while discovering the other spectrally varying components in the image. The MCR analysis is initiated with an offset, the structured noise, and two sets of uniform random numbers. The top of panel B shows the two equality constrained offset and structured noise along with the two spectral components discovered in the MCR analysis. Component 3 is again clearly the more intense GFP component, while Component 4 is more similar to the autofluorescence component added to the simulated data and is less mixed with the GFP component than before. However, the autofluorescence component is more mixed with the structured-noise component as evidenced by the negative-going spectral features at ~490 and ~680 nm that are characteristic of the structured noise present in our detector. More evidence of the mixing of the offset and structured components with GFP and autofluorescence is found when comparing the MCR concentration maps with the known concentrations in the simulated data. Both the offset and structured-noise components show strong presence of the cellular structures in their concentration maps in contrast to the true concentration maps for these two components in Fig. 1B where no cellular structure is present. The source of this mixing can be explained by the more intense GFP and autofluorescence components adding Poisson noise variance to spectra that increases linearly with the total fluorescence intensity. The high-noise regions of the spectra degrade the concentration estimates for the minor components independent of where they appear in the spectra since the MCR analysis is based upon constrained least squares methods that minimize the total sum of squared spectral errors. Unfortunately, this mixing also degrades the concentration errors of the GFP and autofluorescence components of interest. The absolute concentration errors of the two fluorescence components when directly modeling offset in structured noise during the MCR analysis are presented in the bottom two concentration maps of panel B in Fig. 2.

Panel C shows the MCR results of automatically preprocessed data with Stage 1 removal of the offset and structured noise and Stage 2 preprocessing with a spatial mask applied to retain only those pixels with spectral features that are greater than 3 sd above the mean signal level of the dark channels 5 through 23 of the detector. The spatial mask is shown in the upper right concentration map of panel C where the white pixels represent less than 20% of the original pixels. The Scree plot at the bottom of panel A also shows that the noise floor is reduced by the Stage 1 and 2 preprocessing. It is important to note that a small amount of offset (±a few detector counts) is still present in the fully preprocessed data since our offset correction using only 20 dark channels does not perfectly remove all of the offset. Therefore, the offset correction is further improved during MCR modeling by routinely adding and equality constraining an offset. When this offset component is added, the nonnegativity concentration constraint for this component is removed during MCR analyses since the remaining offset can be either positive or negative in magnitude. The MCR results can be improved even further by setting the first 20 spectral channels to zero and equality constraining these 20 zeroed spectral channels for all nonoffset spectral components since we know there is no signal present in these 20 dark channels due to the presence of a blocking filter. This procedure helps prevent the baseline fluorescence levels from drifting up and mixing with the offset and the other spectral components. Using this approach, the MCR analysis is initiated with an equality constrained offset and two sets of uniform random numbers with the first 20 numbers set to zero and equality constrained. The resulting pure spectra are shown in panel C along with the concentration maps of the offset, the GFP and autofluorescence along with the respective concentration maps. The pure spectra are close replicas of the components that were used to generate the simulated data; only they have greater noise present than the true spectral components. The bottom two concentration maps in panel C are the absolute concentration errors for both fluorescence components. The concentration errors are about a factor of two lower for GFP relative to panel B and are about a factor four improved for the autofluorescence component. It is important to note that the results in panels B and C both involve the modeling of the offset and structured noise. However, the modeling presented in panel C models these components before the MCR is performed using the first 20 dark channels of the sample spectral image, and therefore, their modeling is not negatively impacted by the Poisson noise present in the rest of the spectra. This improvement is in spite of the fact that the modeling makes use of only 20 of the 512 spectral channels present in each spatial pixel.

The earlier example makes use of an automatically generated spatial mask. It is also helpful, in some cases, to make use of a spectral mask. Jones *et al.* [18] demonstrates an example where the use of a spectral mask makes a vast improvement in the MCR concentration estimates. In this case, cells are imaged from an algae sample with photosynthetic pigments present. The long-wavelength fluorescence from the chlorophyll pigments is approximately 400 times as intense as the shorter-wavelength carotenoid fluorescence (see figure 6A in Ref. [18] showing the mean spectrum obtained from the spectral image). Even though the carotenoid and photosynthetic pigments are spectrally separated, an MCR analysis that includes the entire spectral region fails to detect the carotenoid component since the Poison noise from the chlorophyll fluorescent is more than an order of magnitude larger than the carotenoid spectral signal. Since MCR methods minimize the sum of spectral errors throughout the entire spectrum, the presence of the large Poisson noise overwhelms the weaker carotenoid signal even though it is spectrally isolated from the chlorophyll fluorescence. The solution to correctly extract and quantify the carotenoid fluorescence component required performing MCR analysis on only the shorter wavelengths characterizing the carotenoid fluorescence. This procedure allowed the weak carotenoid component to be extracted and its concentration map to be quantified (see figure 6B for the carotenoid component below 604 nm and figure 6F for the corresponding carotenoid concentration map in Ref. [18]). In cases where strong and weak fluorophores overlap, weighting or scaling of the data would be required in order to make the noise more uniform across the entire spectrum. The methods to weight the data were discussed earlier in this chapter and are detailed elsewhere [19,30,31].

4.3 Hyperspectral Confocal Fluorescence Images of Immune Cells Labeled with Quantum Dots

4.3.1 Experimental Details

For this example, we use an image data set comprised of a series of time-resolved hyperspectral images of quantum dot (QD)-labeled rat basophilic leukemia (RBL-2H3) cells during cellular activation. The details of the experimental parameters and the measurements are given elsewhere [34]. Briefly, QDs coated with the IgE protein were exposed to RBL cells for the purpose of studying the kinetics of FcεRI receptor cell signaling. QDs were used in this study since they have narrow, bright emission spectra, their emission frequency is dependent on their size, and they do not photobleach [35]. The QD-IgE binds the FcεRI receptor signaling molecules in the cell making it possible to follow the spatial location of multiple signal molecules in the cell as a function of time over relatively long periods. We used five different sized QDs (emission centers nominally at 565, 585, 605, 625, and 655 nm) so that five different emission labels could be monitored simultaneously as a function of time. Since each batch of QDs had a distribution of sizes and shapes, each individual QD's emission maximum could vary about these nominal peak wavelengths and the width of some of the emission bands could also vary slightly [34,36].

Cells with IgE-labeled QDs incorporated into them were placed on 15 mm coverslips. The coverslips were mounted on 25 × 75 mm glass slides equipped with a 2 mm rubber spacer. At the start of the imaging of a single RBL cell, the solution surrounding the samples was injected with a dilute solution of DNP-BSA cell signal activating compound and the activation process was followed for 240 s (60 time-resolved hyperspectral images) [11,34]. Specifically, 60 images of one 104 × 100 pixel field of view were obtained at 0.25 frames/s in order to follow the kinetics of cell activation.

4.3.2 Conventional MCR Analysis of the Cells with QD Labels

The initial MCR analysis followed the more standard approach taken by most chemometricians. However, the 60-frame kinetic hyperspectral image and its associated dark image were first despiked using the approach described in Jones *et al.* [18] in order to remove detrimental cosmic ray spikes. The dark signal was removed from the sample image with the use of the method described in Ref. [30]. A PCA analysis of the resulting despiked, dark-corrected image was then obtained in order to determine the number of independently varying spectral components in the image. As previously discussed a break in the Scree plot can be used to determine the number of independently varying spectral components in the image. In this case, the break in the plot is at three eigenvectors, indicating only three independently varying components (see Fig. 3). In previous studies, we have found that a varying spectral offset arising from the EMCCD detector is usually one of the spectral components present in the hyperspectral image data. Therefore, this PCA result would suggest that only two spectrally independent emitting components are present in this cellular image data labeled with five distinct classes of QD labels. A careful examination of individual spectral pixels in the various frames of the hyperspectral image clearly shows the existence of all five classes of QD labels in the cell (see Fig. 4 for spectra obtained from individual image pixels from each of the five classes of QDs and the corresponding mean image from a single time slice). Why does the PCA only indicate two independently varying QD components when there is experimental evidence for all 5?

A careful examination of the PCA loading vectors and the raw data leads to the following explanations: (1) the QD signal in any given pixel is very weak, at or just above the noise level, (2) only about 0.5% of the image pixels contain QD spectral features, and (3) the noise present in every spatial pixel of the image contains highly correlated spectral noise. We observe that the PCA analysis is dominated by the two strongest emitting QDs (605 and 655 nm QDs). The remaining QDs are too low in intensity and present in <0.5% of the pixels, so their signals in the PCA-compressed data are swamped by the correlated noise that is present in every image pixel, and their spectra are not present in any of the PCA eigenvectors retained for the analysis, even when we retain as many as 64 factors during the PCA compression. If we had not examined the raw spectral data in detail, we might have incorrectly

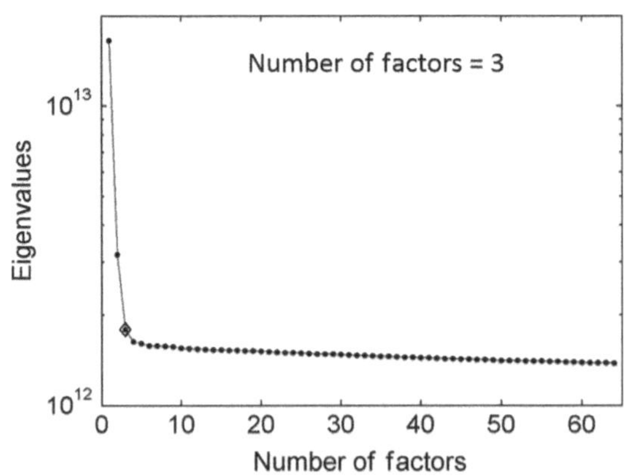

FIG. 3 PCA Scree plot used to estimate the number of independently varying spectral components in the RBL time-resolved image with QD labeling. Image data were despiked and dark image subtracted with the use of 60 dark images with the same collection parameters as the sample image.

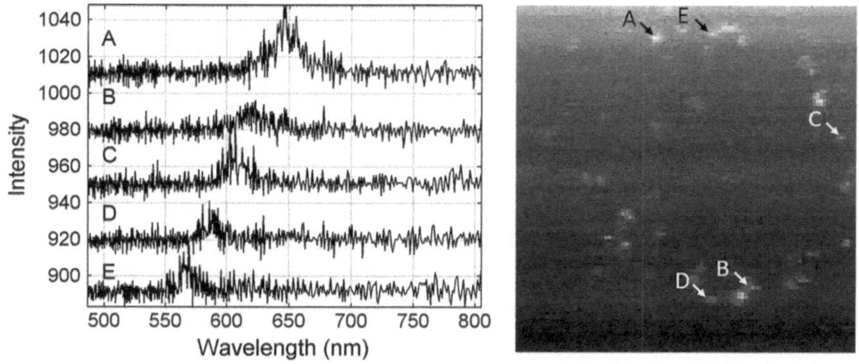

FIG. 4 *Left panel*: Spectra obtained from individual image pixels from QDs nominally at 655 nm (A), 625 nm (B), 605 nm (C), 585 nm (D), and 565 nm (E). *Right panel*: Mean spectral image obtained from the raw data showing the location of the pixels selected for the display in the *left panel*.

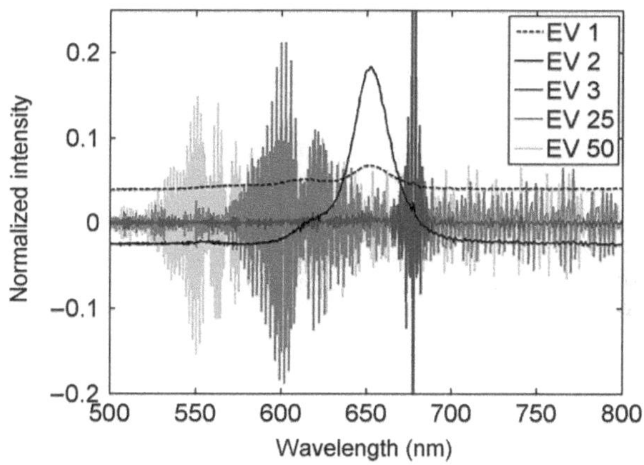

FIG. 5 Selected eigenvectors from the PCA analysis of the sample image with standard preprocessing.

concluded from the PCA-compressed data that the weaker QDs were too low in intensity to be detected by our hyperspectral imaging system. This result reinforces our belief that the raw spectral images should always be carefully examined before any chemometric analysis is performed on the data.

Fig. 5 shows the first 3 PCA loading vectors of the time-resolved hyperspectral image along with the 25th and 50th loading vectors. Only the first two factors are associated with the QD spectra. The third is associated primarily with Poisson noise and the structured noise present in our specific detector, while all the remaining loading vectors (in addition to the 25th and 50th loading vectors) are primarily high-frequency correlated noise present in the detector.

4.3.3 Spatial Compression to Mitigate Effects of Correlated Noise in MCR Analyses

Correlated detector noise, like that shown in this example, is common in all CCD detectors used for spectroscopy and spectral imaging studies, whether electron multiplying or conventional design is employed. There are several approaches that could be used to minimize the detrimental effects of the correlated detector noise. The data could be compressed in the wavelength dimension by averaging successive wavelength channels to minimize the effects of high-frequency noise or in the spatial dimension by averaging spatial pixels in both the x and y dimensions for each image. We empirically find that the spatial compression is more effective than the spectral compression at minimizing the detrimental effects of correlated noise. Spatial compression is effective since the correlated spectral error is random in the spatial dimensions, while the QD spectra are not randomly distributed in the spatial dimension. In addition, due to the large numbers of image pixels, the spatial dimension can be compressed more extensively than the spectral dimension. If the columns are spatially compressed by a factor of 10 and the rows by a factor of 4 (i.e., converting a 100×104 image to a 10×26 image), a very different set of PCA loading vectors result and the eigenvalue plot indicates that seven independently varying sources of spectral variation are present in the compressed hyperspectral image data (see Fig. 6). The PCA loading vectors show that six are associated with the QDs and that correlated spectral noise does not appear until the seventh loading vector.

An MCR analysis of these spatially compressed data was initiated with an offset and six sets of random normally distributed numbers between 0 and 1.

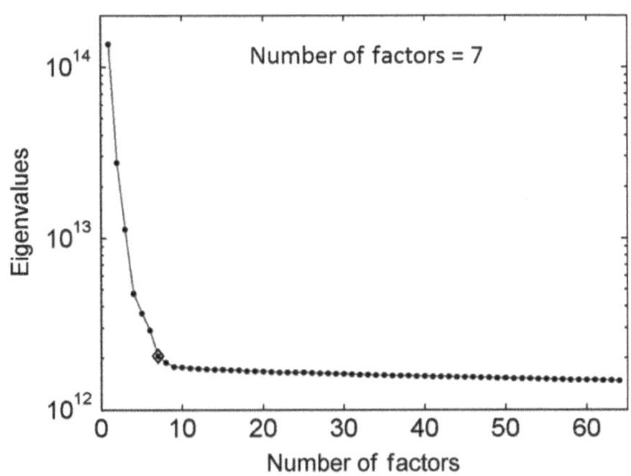

FIG. 6 PCA Scree plot used to estimate the number of independently varying spectral components in the RBL time-resolved image with QD labeling after spatial compression by a factor of 40. Image data were despiked and dark image subtracted with the use of 60 dark images with the same collection parameters as the sample image followed by 10×4 spatial compression.

The concentrations of the spectral offset were not constrained to be nonnegative since the offset is expected to be centered about zero. In addition, the first 20 data points in each nonoffset pure spectrum are set to zero and equality constrained in order to prevent mixing of these spectral components with each other and the offset. This MCR analysis results in six apparent QD spectra (see Fig. 7 upper panel). The most intense QD at 655 nm appears as two separate QD pure spectra centered near the expected nominal 655 nm wavelength. Since we expect each individual QD to vary in frequency about its nominal wavelength due to a small size distribution for each class of QDs, this result can be considered to be reasonable. That is, a linear combination of the two closely spaced pure QD spectra could be used to represent the range of variation in spectral position of individual QDs. Although modeling this

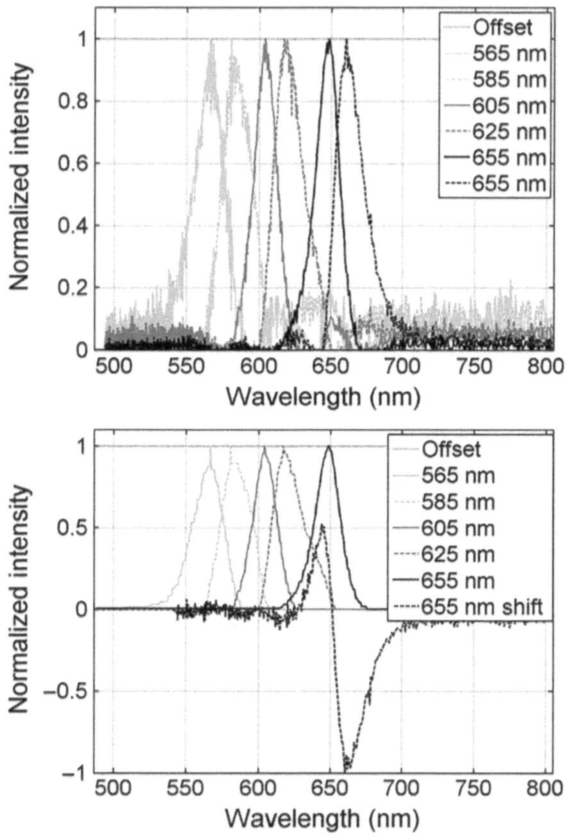

FIG. 7 MCR-estimated pure spectra scaled to unit peak height from analysis of the sample image with standard preprocessing followed by spatial compression by a factor of 40. *Upper panel*: A nonnegativity constraint was applied to all factors except the spectral offset. *Lower panel*: A nonnegativity constraint was applied to all factors except the spectral offset and the 655 nm shift component. See text for additional MCR conditions applied.

wavelength shift as the linear combination of two pure QD spectra is possible, we have shown previously [6] that it is often preferable to model wavelength shifts in a component with the use of a spectrum centered at the average center wavelength and a first derivative of that spectrum. When using this representation of the MCR results, the concentration of the 655 nm class of QDs would be indicated by the concentration of the nonnegative, centered QD spectrum, and the magnitude or concentration of the first derivative of the individual QD spectrum would represent the amount of spectral shift, either to higher or to lower wavelengths, from the centered QD spectrum depending on the sign of the derivative spectrum's concentration. In our opinion, this representation of the results is much easier to interpret than trying to understand the concentration maps of two separate QDs centered about the nominal 655 nm average QD peak wavelength. This same strategy was later successfully applied in process spectroscopy to deal with deviation from the ideal bilinear model [37]. The MCR analysis was reinitiated starting with the 5 QD spectra from the first MCR analysis along with a spectrum of normally distributed random numbers varying from -1 to 1 (again setting the first 20 spectral channels of this derivative component to 0 and equality constraining these first 20 spectral channels to avoid mixing with the offset and the 5 nonnegative constrained QD spectral components). Both the spectrum and the concentrations of this sixth component are left unconstrained and are allowed to go negative (except for the first 20 zeroed spectral channels). The pure spectra derived from this MCR analysis are shown in Fig. 7 (lower panel). Note that it is probable that we do not detect additional shift components for the other four QDs because the intensity of the associated QD derivative spectra is likely below the correlated noise level of the spatially compressed image data for these weakly emitting QDs, and therefore, derivative components for these four classes of QDs are not detected. In order to achieve concentration maps of the QDs at the full spatial resolution of the original data, the offset, the five QD pure spectra, and the derivative of the 655 nm QD spectrum are used in a CLS prediction analysis of the preprocessed spectral data at full (uncompressed) spatial resolution. Note that this CLS analysis must be applied to the preprocessed hyperspectral image spectra rather than the PCA-reconstructed spectra in order for all the QD concentrations to be properly predicted in these time-resolved image data. Some of the full resolution concentration maps associated with the QDs are displayed in Ref. [38] and the publication's associated supplemental material.

4.3.4 Automated Preprocessing of Cell Hyperspectral Images with QD Labels

As satisfying as this final MCR analysis appears to be, a spectroscopist with knowledge of the spectral properties coupled with multiple iterations of the analysis approach was necessary to obtain this final model. The question remains, could this analysis be further automated? Will the preprocessing methods described earlier in this chapter be advantageous for this image where the QD intensities are so low and infrequent that the correlated errors obscure much of the QD information upon performing a PC analysis? Therefore, we applied our automated preprocessing methods to the original time-resolved hyperspectral image data. After Stage 1 preprocessing, Stage 2 preprocessing was performed to select a spatial pixel mask to restrict the analysis pixels to only those pixels where the S/N of the brightest 19 contiguous pixels was 3 sd above the noise present in the 19 dark channels of the spectral data (channels 5–23). The pixel mask indicates only 3458 pixel spectra of the 624,000 pixels in the image have spectral features more than 3 sd above the noise level; i.e., only 0.55% of the spectral pixels have signal 3 sd above the noise. When PCA is applied to these 3458 pixels, the semilog plot of the eigenvalues vs number of loading vectors indicates nine independently varying spectral components including the spectral offset (see Fig. 8). Clearly, when the PCA analysis is restricted to only those image pixels with spectral signal above the noise, the correlated noise of the detector no longer obscures the QD spectral signals even for uncompressed data. In fact, this restricted set of data allows even greater sensitivity in observing independently varying spectral components than was possible with the previous unmasked data with 40 × spatial compression applied, since two additional spectrally independent factors are found above the noise.

The MCR analysis of these automatically preprocessed data was initiated with an offset and 8 sets of randomly distributed numbers with the first 20 spectral points set to zero and equality constrained. The concentration of the spectral offset is unconstrained, and the other eight spectral components are constrained to be nonnegative. The 565, 585, and 625 nm QDs are readily apparent in the resulting MCR-estimated pure-component spectra (see Fig. 9 upper panel). The 605 nm QD has two spectral features splitting the 605 nm

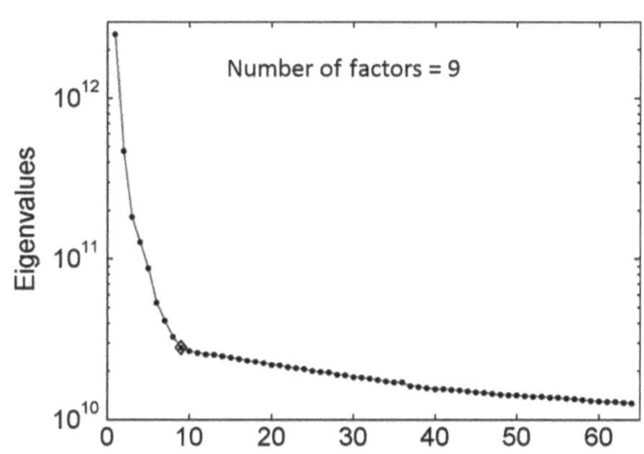

FIG. 8 PCA Scree plot used to estimate the number of independently varying spectral components in the RBL time-resolved image with QD labeling after Stage 1 and 2 preprocessing.

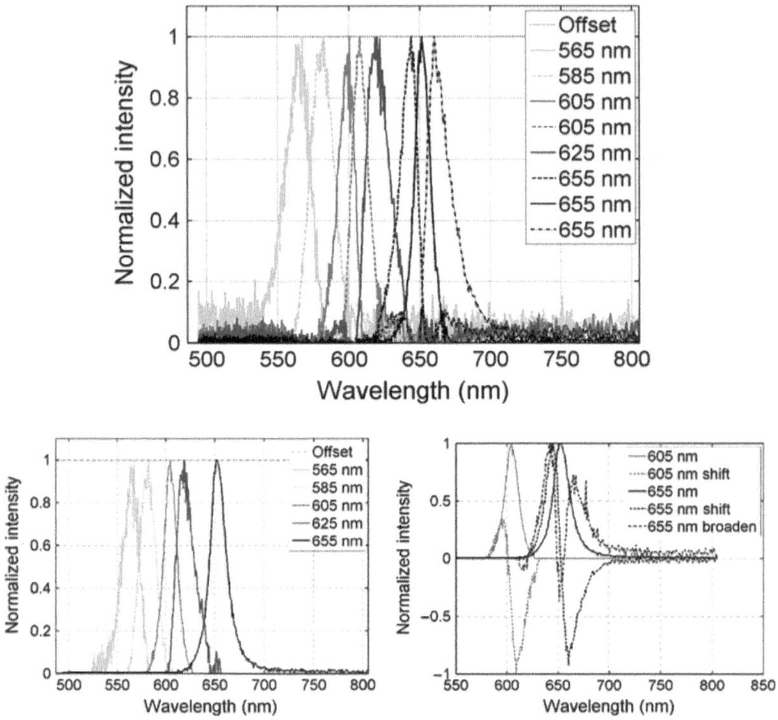

FIG. 9 MCR-estimated pure spectra scaled to unit peak height from analysis of the sample image with full automated preprocessing methods applied including a spatial mask limiting the pixels to those whose spectral signal was >3 sd above the noise. *Upper panel*: A nonnegativity constraint was applied to all factors except the spectral offset. *Lower panels*: A nonnegativity constraint was applied to all factors except the spectral offset and the 605 nm shift, 655 nm shift, and 655 broadening components. *Lower left*: The QD pure spectra derived from the nonnegatively constrained pure components from the five classes of QDs present in the cells. *Lower right*: Only the 605 and 655 nm QD pure spectra are shown along with their first derivatives and the second derivative of the 655 nm QD. See text for additional MCR conditions applied.

wavelength at 600 and 608 nm. As described earlier and in Haaland *et al.* [6], since we know we only introduced one class of 605 nm QDs into the cells, these two features arise from the wavelength shift due to the size distribution of these nominally 605 nm QDs. This 605 nm QD is the second brightest class of QD in the cell, and it is reasonable that its intensity is sufficient for the size-dependent shift to be observed when the analysis is restricted to only those 0.5% of pixels with QDs present. The 655 nm QD is by far the most intense class of QD in the cell, and the MCR result indicated three spectral features centered about the 655 nm wavelength, i.e., with peaks at 643, 652, and 661 nm. As seen in Andrews *et al.* [34], if there is sufficient intensity in the QD spectral features, then three spectral components are possible, which account for the average QD and spectral shifts and spectral broadening between QDs of a single class of QDs. These three spectral features can be represented as a linear combination of three nonnegative spectral features or as a single nonnegative feature representing the mean QD spectrum, a first derivative feature representing spectral shifts between QDs and a second derivative representing broadening or narrowing of individual QD spectra relative to the mean QD spectrum. Again the interpretation of the spectral components with first and second derivative spectral features is more straightforward, and therefore, we relax the nonnegativity constraints for the spectra and concentrations of three of the eight nonoffset spectral components to allow first and second derivative spectral features to be discovered. The first 20 spectral points of these 8 spectral components are again set to zero and equality constrained to avoid mixing with the spectral offset and each other. In addition, to avoid small amounts of mixing between components, we applied unimodal constraints to this MCR analysis since we know that the QD emission spectra are unimodal. The resulting pure spectra components derived from this MCR analysis are shown in the two lower panels of Fig. 9.

The MCR results presented in Figs. 8 and 9 clearly show the advantages of automated preprocessing. Specifically the automated pixel selection method greatly simplifies the analyses and provides increased sensitivity and more complete results than possible with standard methods applied to difficult hyperspectral image data. The conventional methods of treating the data fail to find more than two of the five classes of QDs in the RBL cell that was imaged. Even with spatial compression applied, loss of sensitivity to independently varying spectral components was observed relative to our automated data preprocessing methods.

5 CONCLUSIONS

MCR has been shown to be a highly effective tool in the extraction of pure spectra and their relative concentration images from hyperspectral images. MCR has often suffered from the requirement of significant user input in order to achieve correct and meaningful pure-component spectra and concentration images. In this chapter, we have introduced automated preprocessing methods that can be applied to the raw hyperspectral images in order to greatly reduce or even eliminate the need for user input and to achieve more accurate MCR results with greater sensitivity to minor spectral components. Our automated preprocessing methods, which can be applied to any hyperspectral imaging data acquired with 1D or 2D EMCCD or CCD detectors, bring us much closer to the ultimate goal of routine hyperspectral image analysis without user input.

We have introduced the use of a dark spectral region to achieve the majority of our preprocessing goals of removing offsets, structured noise, and uninformative pixels from the data. The presence of a dark spectral region also allows us to reduce mixing between spectral components during the MCR analysis to further improve the results. These improved MCR results are achieved through our standard automated process of setting to zero and equality constraining the dark region of the initial pure-component spectra that are used to initiate the MCR analysis.

The use of our automated preprocessing and improved MCR approaches has been compared in this work to more conventional MCR approaches to image analysis with the use of realistically simulated data obtained from hyperspectral fluorescence confocal images of cells tagged with GFP. These simulated data dramatically demonstrate and quantify the improvements possible when the methods we propose are used. We have also demonstrated significant improvements in the analysis of real hyperspectral images of QD-labeled cells that failed to yield appropriate results with standard MCR approaches. These examples have demonstrated the greater accuracy and sensitivity of our automated preprocessing methods and improved MCR approaches during the analysis of spectral images.

ACKNOWLEDGMENTS

The authors gratefully acknowledge Michael Sinclair for the maintenance and use of the hyperspectral confocal fluorescence instrument used in this study, Nicholas Andrews and Diane Lidke for the QD-labeled RBL cells, and Ryan Davis for acquisition of the QD images. Additional gratitude is given to David Melgaard, Mark Van Benthem, Michael Keenan, Greg Poulter, and Omar Garcia for their assistance in the development of the original MCR software. This work was funded in part by Sandia National Laboratories Laboratory Directed Research and Development program (original data collection, algorithm development and analysis, as well as support to J.A.T. for her role during the compilation and writing of this chapter).

REFERENCES

[1] Boiret M, De Juan A, Gorretta N, Ginot Y-M, Roger J-M. Distribution of a low dose compound within pharmaceutical tablet by using multivariate curve resolution on Raman hyperspectral images. J Pharm Biomed Anal 2015;103:35–43.

[2] Budevska BO, Sum ST, Jones TJ. Application of multivariate curve resolution for analysis of FT-IR microspectroscopic images of in situ plant tissue. Appl Spectrosc 2003;57:124–31.

[3] Collins AM, Liberton M, Garcia OF, Jones HDT, Pakrasi HB, Timlin JA. Photosynthetic pigment localization and thylakoid membrane morphology are altered in *synechocystis* 6803 phycobilisome mutants. Plant Physiol 2012;158:1600–9.

[4] Felten J, Hall H, Jaumot J, Tauler R, De Juan A, Gorzsás A. Vibrational spectroscopic image analysis of biological material using multivariate curve resolution–alternating least squares (MCR-ALS). Nat Protoc 2015;10:217–40.

[5] Gallagher NB, Shaver JM, Martin EB, Morris J, Wise BM, Windig W. Curve resolution for multivariate images with applications to TOF-SIMS and Raman. Chemom Intell Lab Syst 2004;73:105–17.

[6] Haaland DM, Jones HDT, Van Benthem MH, Sinclair MB, Melgaard DL, Stork CL, *et al.* Hyperspectral confocal fluorescence imaging: exploring alternative multivariate curve resolution approaches. Appl Spectrosc 2009;63:271–9.

[7] Sutherland V, Timlin JA, Nieman LT, Guzowski JF, Chawla MK, Roysam B, *et al.* Advanced imaging of multiple mRNAs in brain tissue using a custom hyperspectral imager and multivariate curve resolution. J Neurosci Methods 2007;160:144–8.
[8] Vajna B, Farkas I, Farkas A, Pataki H, Nagy Z, Madarász J, *et al.* Characterization of drug–cyclodextrin formulations using Raman mapping and multivariate curve resolution. J Pharm Biomed Anal 2011;56:38–44.
[9] Zhang D, Wang P, Slipchenko MN, Ben-Amotz D, Weiner AM, Cheng J-X. Quantitative vibrational imaging by hyperspectral stimulated Raman scattering microscopy and multivariate curve resolution analysis. Anal Chem 2012;85:98–106.
[10] Zhang X, Tauler R. Application of multivariate curve resolution alternating least squares (MCR-ALS) to remote sensing hyperspectral imaging. Anal Chim Acta 2013;762:25–38.
[11] Sinclair MB, Haaland DM, Timlin JA, Jones HDT. Hyperspectral confocal microscope. Appl Opt 2006;45:3283–91.
[12] Robbins MS, Hadwen BJ. The noise performance of electron multiplying charge-coupled devices. IEEE Trans Electron Devices 2003;50:1227–32.
[13] Haaland DM, Easterling RG. Application of new least-squares methods for the quantitative infrared analysis of multicomponent samples. Appl Spectrosc 1982;36:665–73.
[14] Haaland DM, Easterling RG, Vopicka DA. Multivariate least-squares methods applied to the quantitative spectral analysis of multicomponent samples. Appl Spectrosc 1985;39:73–83.
[15] Martens H, Naes T. Multivariate calibration. Chichester: Wiley; 1991.
[16] Geladi P, Grahn H. Multivariate image analysis. Chichester: John Wiley and Sons; 1996.
[17] Joliffe IT. Principal component analysis. New York: Springer-Verlag; 2002.
[18] Jones HDT, Haaland DM, Sinclair MB, Melgaard DK, Collins AM, Timlin JA. Preprocessing strategies to improve MCR analyses of hyperspectral images. J Chemometr Intell Lab Syst 2012;117:149–58.
[19] Keenan MR, Kotula PG. Accounting for Poisson noise in the multivariate analysis of TOF-SIMS spectrum images. Surf Interface Anal 2004;36:203–12.
[20] Haaland DM, Thomas EV. Partial least squares methods for spectral analyses. 1. Relation to other quantitative calibration methods and the extraction of qualitative information. Anal Chem 1988;60:1193–202.
[21] Tauler R, Smilde A, Kowalski B. Selectivity, local rank, three-way data analysis and ambiguity in multivariate curve resolution. J Chemometr 1995;9:31–58.
[22] Bro R, De Jong S. A fast non-negativity-constrained least squares algorithm. J Chemometr 1997;11:393–401.
[23] Van Benthem MH, Keenan MR. Fast algorithm for the solution of large scale non-negativity constrained least squares problems. J Chemometr 2004;18:441–50.
[24] Van Benthem MH, Keenan MR, Haaland DM. Application of equality constraints on variables during alternating least squares procedures. J Chemometr 2002;16:613–22.
[25] de Juan A, Tauler R. Chemometrics applied to unravel multicomponent processes and mixtures: revisiting latest trends in multivariate resolution. Anal Chim Acta 2003;500: 195–210.
[26] Vermaas WFJ, Timlin JA, Jones HDT, Sinclair MB, Nieman LT, Hamad S, *et al.* In vivo hyperspectral confocal fluorescence imaging to determine pigment localization and distribution in cyanobacterial cells. Proc Natl Acad Sci USA 2008;105:4050–5.
[27] Zhang D, Ben-Amotz D. Removal of cosmic spikes from hyperspectral images using a hybrid upper-bound spectrum method. Appl Spectrosc 2002;56:91–8.
[28] Zhang D, Hanna JD, Ben-Amotz D. Single scan cosmic spike removal using the upper bound spectrum method. Appl Spectrosc 2003;57:1303–5.

[29] Zhao J. Image curvature correction and cosmic removal for high-throughput dispersive Raman spectroscopy. Appl Spectrosc 2003;57:1368–75.

[30] Jones HDT, Haaland DM, Sinclair MB, Melgaard DK, Van Benthem MH, Pedroso MC. Weighting hyperspectral image data for improved multivariate curve resolution results. J Chemometr 2008;22:482–90.

[31] Keenan MR, Kotula PG. Optimal scaling of TOF-SIMS spectrum-images prior to multivariate statistical analysis. Appl Surf Sci 2004;231:240–4.

[32] Ohlhausen JA, Keenan MR, Kotula PG, Peebles DE. Multivariate statistical analysis of time-of-flight secondary ion mass spectrometry images using AXSIA. Appl Surf Sci 2004;231–232:230–4.

[33] Ruppert D, Wand MP. Multivariate locally weighted least squares regression. Ann Stat 1994;22:1346–70.

[34] Andrews NL, Pfeiffer JR, Martinez AM, Haaland DM, Davis RW, Kawakami T, *et al.* Small, mobile FcεRI aggregates are signaling competent. Immunity 2009;31:469–79.

[35] Jovin TM. Quantum dots finally come of age. Nat Biotechnol 2003;21:32–3.

[36] Peng X, Manna L, Yang W, Wickham J, Scher E, Kadavanich A, *et al.* Shape control of CdSe nanocrystals. Nature 2000;404:59–61.

[37] Debus B, Sliwa M, Miyasaka H, Abe J, Ruckebusch C. Multivariate curve resolution—alternating least squares to cope with deviations from data bilinearity in ultrafast time-resolved spectroscopy. Chemom Intell Lab Syst 2013;128:101–10.

[38] Lidke DS, Andrews NL, Pfeiffer JR, Jones HDT, Sinclair MB, Haaland DM, *et al.* Exploring membrane protein dynamics by multicolor single quantum dot imaging using wide field, TIRF, and hyperspectral microscopy. In: Colloidal quantum dots for biomedical applications II, 2007 San Jose, CA, USA. SPIE, 64480Y-8; 2007.

Chapter 13

Multiresolution Analysis and Chemometrics for Pattern Enhancement and Resolution in Spectral Signals and Images

M. Li Vigni and M. Cocchi[1]
University of Modena and Reggio Emilia, Modena, Italy
[1]*Corresponding author: e-mail: marina.cocchi@unimore.it*

Chapter Outline

1 INTRODUCTION

Nowadays the paradigm of data-driven discovery is becoming more and more adopted in scientific research. In many fields of chemistry and more generally of life and material sciences this results in a change in the methodology used to approach the study of complex systems. From the analysis of carefully selected subsystems on the basis of some a priori knowledge, in order to collect specific data, concerning a few classes of constituents/properties, scientists moved to a more holistic view, where the whole phenomenon/process (and the many steps involved) is characterized. This means that

Data Handling in Science and Technology, Vol. 30. http://dx.doi.org/10.1016/B978-0-444-63638-6.00013-9
Copyright © 2016 Elsevier B.V. All rights reserved.

system/sample characterization is kept as extended as possible and information is extracted by data analysis a posteriori.

The natural choice to achieve complete physical/chemical characterization of a system/sample as a whole, though implicitly, is acquiring a fingerprint through one or more instrumental analysis techniques, going from chromatographic–spectrometric series, through hyperspectral imaging and higher dimensional spectroscopic techniques, to batteries of sensors. When these techniques, most of which are also nondestructive and fast, are used solely to implement routine analysis and/or continuous monitoring, it could be of secondary importance to decipher the specific constituents responsible for the observed time trends or differentiating sample categories or systems status. However, data-driven discovery implies the need for a deep understanding of the system under observation and an increase in the knowledge of the same. To this aim the use of data analysis technique capable of resolving (also if partially) the signals/images in terms of specific molecular/functional groups contributions is necessary.

There are several different methodologies to achieve this aim, most of which are covered by this book. These can be grouped into two general categories: a first one is represented by techniques achieving resolution by decomposition techniques using criteria that explicitly or implicitly recover pure component profiles, such as multivariate curve resolution, MCR [1,2]; independent component analysis, ICA [3–5]; blind source separation [6,7]; sparse-based models [8,9]; and filter-based methods [10,11]. A second category, which is dealt with in this chapter, is made by approaches based on the assumption that chemical information retrieval can be accomplished a posteriori by filtering the outcome of fingerprinting techniques on the basis of frequency contributions that reflect the different sources contributing to the observed signals/images.

In fact, deciphering a signal or image fingerprint of complex matrix samples requires taking into account that the information pertinent to the problem is mixed with many uninformative sources of variations that may overwhelm the specific response of chemical constituents. An example is given by near infrared spectroscopy, where relevant information is usually hidden by the presence of a strong background, since bands are poorly resolved and physical effects have high influence and sum up to absorptions bands related to specific chemical compounds. At the same time different types and levels of noise add almost to any signal/image acquired on complex matrix samples. From the data analysis point of view, these issues can be dealt with adequate preprocessing and/or feature selection methodologies.

In the framework of filtering methods, the use of a data transform approach which shifts the data analysis step to a different domain may allow the unwanted sources of variation to be separated if they correspond to much lower or higher frequency response with respect to the one given by the source which we are interested in depicting. For example, white noise is typically a high-frequency contribution, while background is a very low frequency one; thus, they can usually be well separated from each other, and from relevant information, by some of the most popular currently applied methods, above all Fourier transform and wavelet transform (WT) [12–15].

The aim of this chapter is to illustrate the advantages that integration of WT with multivariate data analysis in multiresolution analysis can offer to signal and image analysis to recover the most salient features in order to enhance information content with respect to chemically and physically relevant phenomena. In particular, we will focus on the possibilities of simultaneously accomplish denoising and feature enhancement/selection. The latter is relevant in the context of multivariate image analysis, where a salient aspect is the knowledge boost given by the simultaneous analysis of textural variability (the spatial structure of pixels) and optical variation properties (the "third" dimension, that is the channels acquired for each pixel). Moving from grayscale to RGB and to multispectral images, the degree of information increases from a simple evaluation of texture homogeneity to a pixel-scale differentiation in chemical composition. This allows tackling several problems, such as the detection of product defects in quality control [16,17], the monitoring of changes in process behavior and its feedback control [18–20], and the prediction of product properties (by using multivariate image regression (MIR) methods) [21,22].

Different applicative contexts are presented in this chapter, ranging from feature selection applied to spectroscopic signals in classification and regression tasks, to multiresolution of multivariate image analysis with special attention to quality monitoring, fault detection, and classification.

The chapter is organized into three methodological sections: First one briefly recalling the salient features of WT and its application in signal/image processing, a second one dealing with algorithms proposed by our research group for signal data analysis, and a third one illustrating our proposal of coupling WT with MIA approach for image analysis. Following these remarks about the methods, four cases of study covering applications in the food and materials sciences are described to highlight the practical potentialities of the approach to different scientific challenges.

2 METHODS

2.1 Wavelet Transform

The main aspects of WT focusing on discrete wavelet transform (DWT) and multiresolution of signals will be recalled here, while extension of DWT to image is outlined in Section 2.3.

In general, a transform method allows an alternative representation of the data. In other words, the data domain is changed by mathematically expanding the raw signal as a combination of a set of basis functions.

The coefficients of the combination are the transform coefficients and can be used as alternative variables instead of the original collected ones. In order to guarantee that the transform coefficients are informative (hence the domain change is effective), the basis functions must have a shape which is properly linked to the characteristic features of the signal.

To analyze a signal in terms of frequency contents the Fourier transform, which goes from the original domain (generally referred to as time domain) to the frequency domain, is well suited when the phenomena contributing to the signal are periodic or extend to the whole domain. However, it lacks effectiveness to detect local phenomena, or more in general signals which present complex shapes (this is indeed a very common situation in chemistry), where, e.g., a contribution is present in a small-signal range or a high-frequency perturbation is present and these features are superimposed on a background of low-frequency components which persist in a wide signal range. In fact, FT assumes that the features extracted, of different frequencies, are present all along the original signal domain (e.g., time). Therefore, when frequency changes are localized (nonstationary) in the original domain, FT loses the ability of finding where/when that change happens [14]. To recover this information windowed (or short time) FT has been proposed where FT is applied to windows of the raw signal; still there are limitations due to the fixed window size. The WT overcomes this issue thanks to the properties of the basis functions. First of all, they have limited duration in time [14,15,23–28]. In addition, by applying scaling (stretching or compressing the wavelet function, i.e., mother wavelet) and translation operations (position on the time/original signal domain axis) on the basis function it is possible to map the signal as a function both of the original domain and of the frequency domain (also named scale domain). As a result, WT achieves good time localization in the high-frequency regions and good frequency localization in the low-frequency region; in other words the basis functions change shape according to the region they occupy. In continuous WT [12,15] scale and translation parameters change continuously. In this chapter, we will focus instead on the description and use of the discrete formulation of WT (DWT) as implemented through Mallat's pyramid algorithm [26] or fast wavelet transform (FWT). FWT works by recursively applying two filters, which identify the wavelet function (several wavelet families are available from the literature and implemented in dedicated software, such as Daubechies, Symlets, etc.), to the signal; an example is given in Fig. 1.

The low-pass filter maintains the low-frequency contributions in the so-called approximations vector (cA1), and the high-pass filter stores the high-frequency contributions in the details vector (cD1). Thus, the information contained in the signal is completely preserved after the decomposition, but it is split into cA1 and cD1, each one having length approximately equal to one-half of the original signal; i.e., downsampling (or decimation) of the wavelet coefficients is operated after filtering (Fig. 2A).

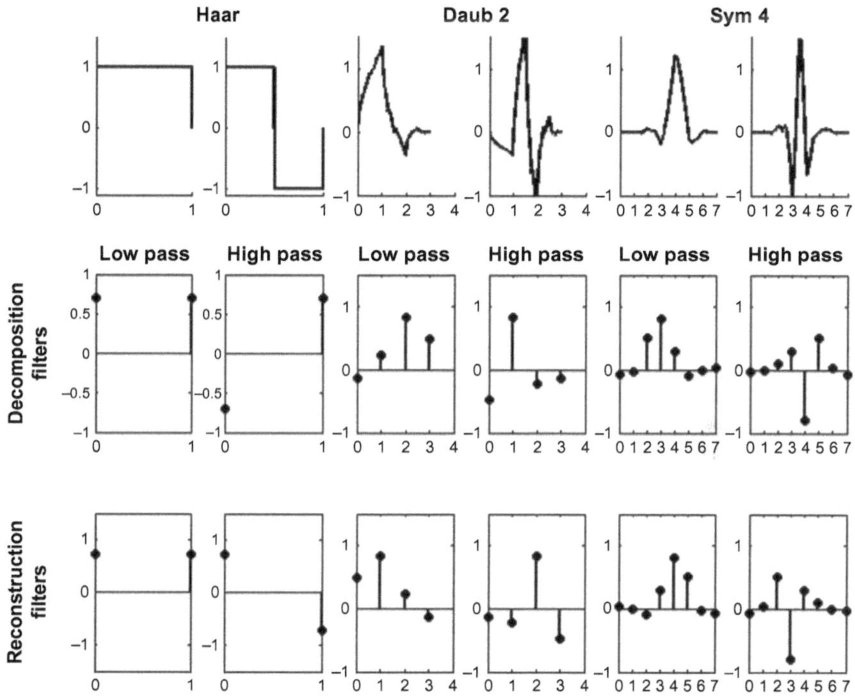

FIG. 1 Example of wavelet filters. Scaling function (*left*) and wavelet function (*right*) for each given filter (*top*) and corresponding decomposition (*middle*) and reconstruction filters (*bottom*).

The decomposition goes on iteratively by applying the same two filters to the previous approximation vector cAn, which is split into cAn + 1 and cDn + 1 (Fig. 2B). The maximum possible decomposition level (or scale) is reached when the approximations and details vectors cannot be further halved. There is also an implementation which does not operate downsampling by upsampling the filters, known as undecimated or stationary (SWT) wavelet transform [29,30], as reported in Fig. 2C.

The signal decomposition by WT is also called multiresolution analysis because at each scale (level) a more coarse resolution of the signal is depicted in the approximations vector, while the sharp features (high-frequency contribution) are stored in the details vectors.

A different implementation of WT is the wavelet packet transform (WPT) [15], which works similar to DWT, the only difference being that at each step details vectors are also decomposed, as represented in Fig. 2D, Thus, while with DWT at each decomposition level the signal can be represented by only one orthonormal basis (i.e., at level n the ensemble of vectors cAn, cDn, cDn − 1, cDn − 2, …, cD1, as shown in Fig. 3B), when using WPT the decomposition is redundant, so that one has to choose the best

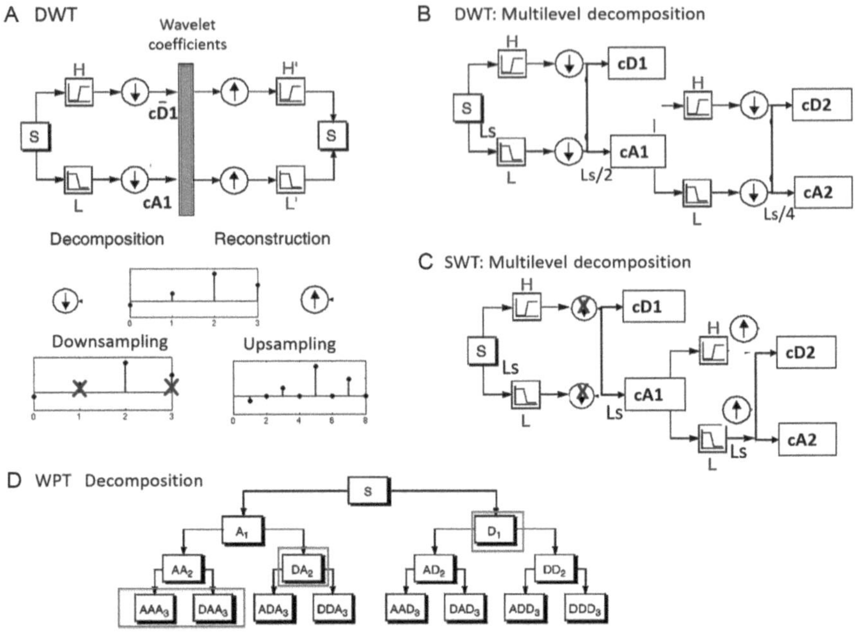

FIG. 2 Representation of wavelet transform decomposition schemes. (A) 1D discrete wavelet transform (DWT) for one-level decomposition and reconstruction. *H* and *L* stand for high-pass and low-pass filters, respectively. *Down-arrow* indicates a downsampling (or decimation) operation and *up-arrow* upsampling, as shown for an example coefficients vector, just below; (B) 1D discrete wavelet transform for a two-level decomposition; (C) stationary wavelet transform decomposition scheme for a two-level decomposition; (D) wavelet packet transform decomposition tree, the *gray rectangles* frame the blocks belonging to one of the possible complete basis.

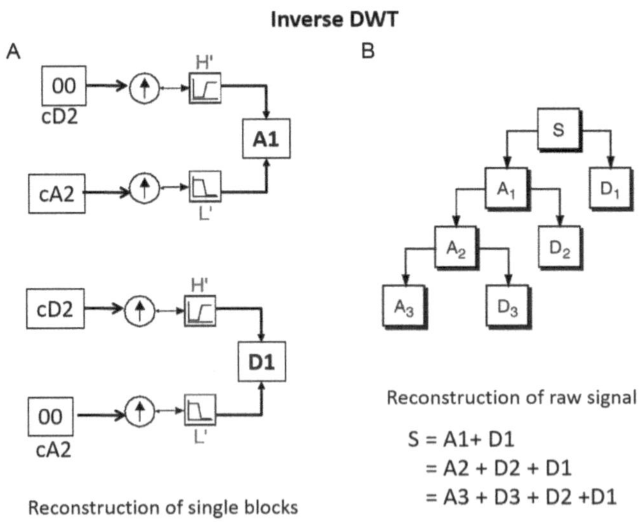

FIG. 3 Representation of the inverse discrete wavelet transform (i-DWT) application: (A) when each decomposition block is reconstructed separately; (B) to obtain a complete reconstruction of the original signal.

combination of blocks, i.e., approximations or details vectors at a given node, giving a complete orthonormal basis, as shown in Fig. 1D. In this case, among all the possible basis, the one that will ensure that the features of interest are well separated in different blocks, and decoupled from noise, has to be selected.

Lastly, adaptive wavelets have been developed: instead of using wavelet basis functions from given families, the filter is designed in an iterative procedure which optimizes a particular criterion based on the data; i.e., wavelet mother function is adapted to the specific application of the multiresolution analysis [31].

Independently of the specific implementations of WT, the representation in the wavelet domain offers the possibility to take into account not only the single intensity values of the signal but also peak widths, slopes of particular regions, degree of smoothness, and many other shape aspects which can be useful in classification and regression tasks. Moreover, removing baseline effects and reducing noise are easier in the WT domain than in the original one, since there is a good possibility that a single decomposition block, which can be completely removed from the analysis, captures these uninformative effects.

2.2 Feature Selection in Wavelet Domain

As any other transform, WT has its inverse, which brings back the decomposed signal to the original domain. Wavelet analysis of a signal consists of filtering and downsampling, and the wavelet reconstruction process consists of upsampling (the coefficients vectors) and filtering as schematically shown in Fig. 2A. The inverse DWT uses a couple of complementary filters, also called reconstruction filter (Fig. 1), which have to be consistent to the decomposition ones to compensate distortion due to downsampling in the decomposition phase. The decomposition and reconstruction filters form a system which is called quadrature mirror filters [28]. Approximations and details themselves can be reconstructed separately by considering their coefficients vector and inserting zeros for the coefficients of the blocks which are not to be considered in the reconstruction, as shown in Fig. 3A. In multilevel (multiscale) decomposition there can be different ways of recovering the signal, though in general the original signal is obtained by using An and Dn to D1, as highlighted in Fig. 3B.

If all wavelet coefficients are kept, decomposition followed by reconstruction brings back to the original signal. However, when a single approximation or details at a given level/scale are reconstructed, this may help isolating and inspecting specific signal/image feature as shown in Fig. 4.

Obviously if the aim is to remove some specific contribution, e.g., stochastic noise and/or background, this can be accomplished by selectively removing (zeroing) some wavelet coefficients at given levels/scales in specific

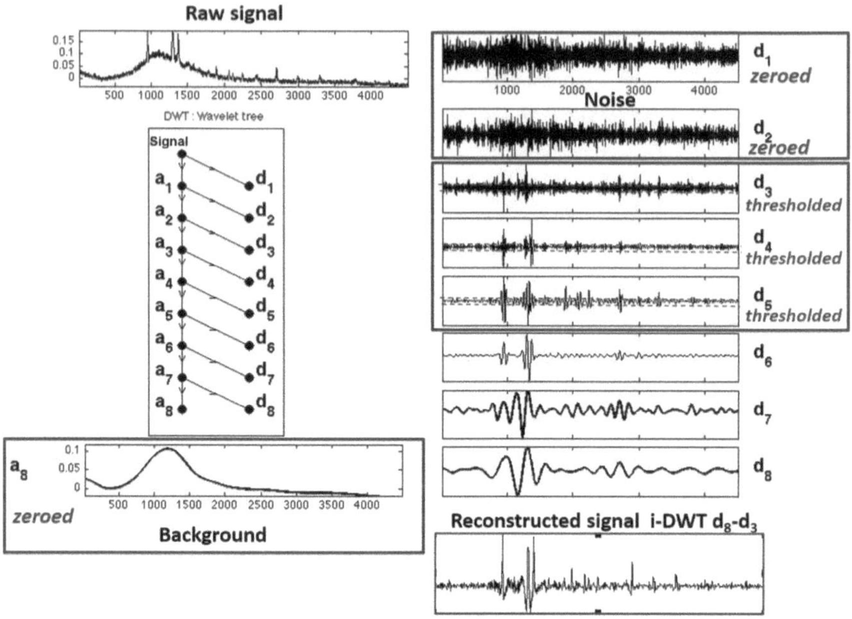

FIG. 4 Example of DWT decomposition of an XRDP signal for denoising and background filtering. After decomposition level 8 is reached, each block is either zeroed (Approximation 8 and Details at levels 1 and 2) or thresholded (Details at levels 3, 4, and 5) so that only the relevant features are highlighted as shown in the reconstructed signal, where the contribution of the amorphous phase (depicted by a_8) and most of noise have been removed.

approximations or details vectors and/or all the coefficients of some blocks (e.g., zeroing all coefficient in first-level details vector usually filter most of white noise) before applying the inverse transform for reconstruction.

More generally, wavelet coefficients obtained from DWT decomposition of a data set of signals can be seen as a set of "new" variables with respect to the original intensities in the specific raw signal domain. These alternative variables can be used for any multivariate data analysis method. In particular, feature selection can be done in the WT domain, i.e., selecting wavelet coefficients, and this will reflect in using only part of the information present in the original data. In this context denoising can be seen as a special case of feature selection.

Obviously, from the modeling point of view, reconstruction in the original domain is not necessary; e.g., discrimination or calibration models can be obtained by using wavelet coefficients. However, reconstruction becomes particularly useful for the interpretation of the models.

In the beginning, most of the applications of WT in the area of chemistry [14,15,32–34] have been focused on denoising, i.e., how to develop the most suitable criteria for wavelet coefficient thresholding in order to remove noise from signal/images and at the same time obtain the least distortion in the reconstructed signal/image. Later on, applications of WT in the feature selection context progressively increased [35], different approaches have been proposed operating either feature selection scale/level-wise [36] or coefficients-wise [37–39], coupling with orthogonal signal correction [40], genetic algorithm [41,42], or Bayesian methods for feature selection [43].

In the case of wavelet coefficients-wise selection the criteria employed in thresholding also differ, e.g., basing on the evaluation of partial least squares (PLS) weights [37] or of PLS regression coefficients [12,38,39], percentage of variance retained [40], or increasing of model performance when the selected coefficients are used [41].

We have contributed to two algorithms [44–46] aimed at building parsimonious classification and regression models by operating feature selection in WT domain, which will be described in this chapter. WPTER (wavelet packet transform for efficient pattern recognition of signals) [45,47–49] is a WPT-based algorithm for the selection of wavelet coefficients to build optimal classification models. WILMA-GA (wavelet interface to linear modeling analysis—genetic algorithm) [50], a modification of our WILMA [44,51–54] algorithm, is a DWT-based algorithm for the selection of wavelet coefficients coupled to genetic algorithm [55] to build optimal calibration models.

Since a key aspect in dealing with WT is the setting of the relevant parameters, such as the choice of the best wavelet filters, the padding criterion [28], and the identification of the optimal level of decomposition (or of the optimal basis for WPT), we have focused on two main targets:

- **(i)** developing fully automated algorithms programmed to iterate over the different parameters with the aim of restricting the a priori choices and to limit as much as possible the user intervention in the phase of model development;
- **(ii)** operating feature selection coefficient-wise, i.e., removing single coefficients instead of whole approximations or details vectors, with the aim of achieving at the same time compression and denoising, in this way avoiding the need of a separate signal preprocessing step;
- **(iii)** paying attention to presentation/interpretation of results.

All algorithms are written in Matlab© and use some of the routines implemented in the Wavelet Toolbox for Matlab© [56].

2.2.1 WPTER Algorithm

WPTER is a supervised classification tool; Fig. 5 reports a schematic view of the algorithm.

To simplify the following description, from now on we will name a specific approximations or details vector at a given scale/level a block and use the term level (although maybe the term "scale" is more used in the literature) to specify the decomposition level, numbering them starting from the first

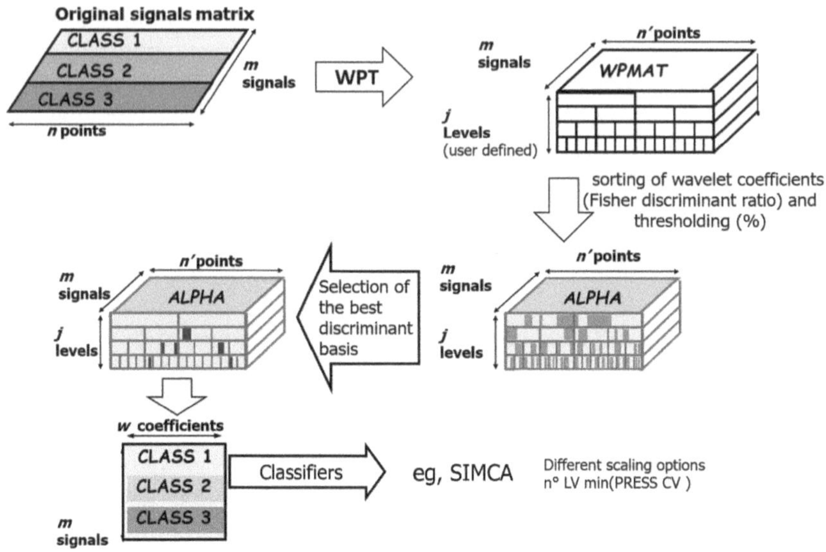

FIG. 5 Scheme of the algorithm for WPTER.

decomposition step. In this way, as an example, the coarse representation of the signal will belong to the approximations of the last decomposition level.

Feature selection is accomplished in two steps (the order of the steps depends on the criterion adopted for the best discriminant basis selection, as it will be detailed further on):

(1) univariate pruning of the most discriminant coefficients is operated in each block;
(2) a best discriminant basis is sought.

The final set of selected wavelet coefficients, after steps 1 and 2, are a set of descriptors which can be fed to any classification methods. WPTER is natively equipped with two classification methods: SIMCA [57–59] and a criterion called percentage of assignation (PA) which is described in Ref. [45]. Since the latter is not used in the applications presented here, it will not be further detailed.

Finally, by using the inverse wavelet transform (iWT), the selected features can be projected back into the original domain for model interpretation. For easiness of presentation, the main procedural steps are described later with distinct subtitles.

2.2.1.1 Signal Decomposition and Choice of the Best Discriminant Basis

The algorithm decomposes each signal of length n (constituting a row of an $m \times n$ data set) into the WPT tree until reaching a decomposition level, j, specified by the user. In this way an array (***WPMAT***) of dimension $m \times n' \times j$ is obtained which contains the decomposition of each one of the m considered signals in the wavelet packet (WP) domain. The dimension $n' \geq n$ is due to the fact that the value depends on the decomposition level and on the length of the filters; in other words downsampling brings to a length which is an exact half only in the case Haar filter is used, while with other filters $n' = (n + \text{length of filter} - 1)/2$.

The discrimination capability of each wavelet coefficient is estimated according to the Fisher discriminant ratio [60]; in this way a matrix (***ALPHA***) structured as a WPT tree is obtained containing the value of the discrimination capability of the wavelet coefficient in the corresponding position in the tree.

In the WPTER algorithm for best basis identification, on user choice, it is possible to use the classical entropy-based methods [61] on matrix ***ALPHA***, or our alternative approach, referred to as classification ability (CA) criterion.

The classical entropy-based methods, implemented in the Wavelet Toolbox for Matlab©, search for the best basis by minimizing the chosen entropy function, i.e., only those blocks whose entropy is lower than the sum of the entropies of the two blocks immediately below inside the tree are retained. Then, for the blocks belonging to the best basis, feature selection is performed by hard thresholding, i.e., keeping, in the decomposition array ***WPMAT***, only a fixed percentage of the wavelet coefficients having the highest discriminant ratio values, while all the other coefficients are set to zero.

When using the CA criterion, first hard thresholding is applied, the percentage of coefficients to retain can be fixed by the user (equal for all the blocks) or calculated automatically retaining only those coefficients whose Fisher ratio is significant at 95% confidence level according to an F-test. Then, for each block in the thresholded ***WPMAT***, the Euclidean distances, d_{ik}, between each possible couple of the m signals are calculated over the retained wavelet coefficients and normalized by dividing by the maximum value. The normalized distance vector is referred to as $\boldsymbol{d}'$. The classification ability for the considered block, CA, is defined according to Eq. (1):

$$\text{CA}(\text{of a block}) = \sum_{q=1}^{\lambda} d''_q \tag{1}$$

where λ is the number of elements in the distance vector; d''_q is defined on the basis of the class membership of the element in $\boldsymbol{d}'$:

$$\begin{aligned} d''_{ik} &= d'_{ik} && \text{if } i \text{ and } k \text{ belong to the same class} \\ d''_{ik} &= 1 - d'_{ik} && \text{if } i \text{ and } k \text{ belong to different classes} \end{aligned}$$

Thus, CA assumes low values for those blocks giving good class separation. Therefore, the best basis, named classification ability basis, which is not forced to be a complete basis, can be identified as the one constituted by the orthogonal blocks giving the overall lowest CA value, and the wavelet coefficients herein contained will then be the selected features.

It must be underlined that the CA parameter, contrary to entropy, does not satisfy an additive-type property. Therefore, the actual procedure [46] consists in sorting all the blocks of the best complete orthogonal basis in ascending order, calculating cumulative CA values by adding one block at time, until reaching a complete basis or a CA value greater than the mean of the CA values of the single blocks. At this point, the combination of blocks giving the minimum cumulative CA value is considered as the final best basis choice.

Once feature selection is attained, by the steps described earlier, the wavelet coefficients retained are used as input for the preferred classifier; in our case SIMCA is used and the number of components for each class is set according to minimum cross-validation classification error.

2.2.1.2 Choice of Best Performing WPTER Cycle

Model is validated by means of an external test set, decomposing it according to the selected wavelet basis and retaining the same wavelet coefficients selected in the calibration phase; hence, projection on the classification model according to the wavelet coefficient values will give predicted class membership.

The possibility of using different wavelet filters leads to different decompositions of a signal into the corresponding wavelet domain, so that the choice of an appropriate wavelet is a crucial point. To this respect, it must be underlined that in many cases, when dealing with WT, wavelets are chosen a priori on the basis of empirical evaluations, e.g., looking at the decomposition of the mean spectrum or at the shape of the signals. In WPTER, several wavelet filters are tested, and also other parameters are defined by the user, like the maximum level of decomposition, the method for coefficients preselection (different fixed percentages or automatic F-based selection), and the method for the identification of the best basis (classical entropy-based techniques or CA). Thus, many trials can be performed, corresponding to different values for the user-defined parameters, and this is made automatically by WPTER, cycling over all the possible combinations of the chosen parameter values.

For each cycle of calculation, the efficiency of the resulting classification model is evaluated, e.g., by the plots of Sensitivity (SENS) and Specificity (SPEC) values of the corresponding SIMCA models calculated on the selected wavelet coefficients. A monitoring set, i.e., an external test set, can be used to choose the best cycle if a further test set is available for validation.

2.2.1.3 Signal Reconstruction and Model Interpretation

For interpretative purposes, the selected wavelet coefficients for each signal can be reconstructed by using the iWT and plotted separately. Also, for each class, the *mean reconstructed signal*, i.e., the signal consisting of the mean values of the corresponding reconstructed signals, can be plotted.

The obtained plots represent the projection of the selected wavelet coefficients into the original domain, highlighting the salient signal regions for classification and, most important, giving information about the scales (frequencies) at which the features of interest are located. Therefore, if a given region of a reconstructed signal looks "jagged," it means that in the correspondence to this region high-frequency (low-scale) features are responsible for classification, while a more "smooth" behavior is observed in the region of the reconstructed signal where low-frequency (high-scale) features are meaningful. Obviously, the regions of the original signals in which there are no discriminant features correspond, in the mean reconstructed signals, to regions where the coefficients are equal to zero.

However, since the reconstructed signals are obtained using only the few selected wavelet coefficients, the reconstructed signals look generally very different from the original ones. Thus, for interpretative purposes, the corresponding *mean original signal* of each class is also plotted, highlighting the regions corresponding to selected features.

2.2.2 WILMA-GA Algorithm

We proposed the WILMA algorithm [44], based on DWT, for accomplishing feature selection in wavelet domain in the context of multivariate calibration/regression. WILMA has also been coupled with genetic algorithm [55] (WILMA+GA) to improve the search of wavelet coefficients.

A scheme of the key steps of WILMA is given in Fig. 6 and a brief description follows.

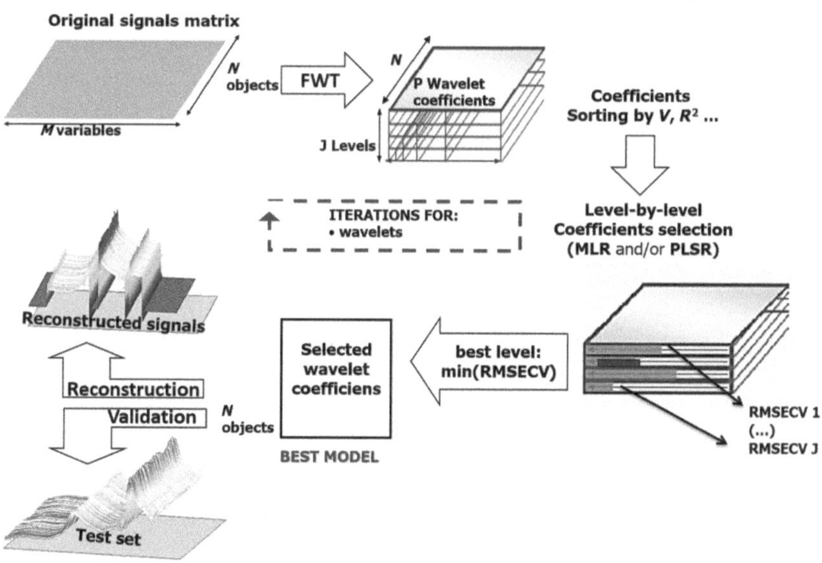

FIG. 6 Scheme of the algorithm for WILMA.

The first step consists of decomposing each signal of the data set into the wavelet domain by DWT until the maximum allowed decomposition level. Then, for each decomposition level, all the obtained wavelet coefficients (i.e., including approximations and details vectors) are ranked according to one of these criteria, considering the set of calibration signals: variance, squared values of the correlation coefficient or squared covariance with the response variable, or to the Spearman's rank coefficient. Then, for each level, the optimal number of wavelet coefficients is iteratively selected. This is performed by including increasing numbers of wavelet coefficients, until attaining the minimum root mean square error estimated in cross-validation (RMSECV). Leave-one-out, venetian blinds and random subsets criteria are implemented for cross-validation. The regression techniques implemented are either multilinear regression, and in this case, pruning of highly intercorrelated wavelet coefficients is first applied, or PLS regression.

Finally, the optimal decomposition level is chosen as the one giving the lowest RMSECV. Model is then validated by means of an external test set.

The wavelet coefficients selected with this procedure, for interpretative purposes, are reconstructed into the original domain by the inverse DWT (i-DWT); this allows highlighting the salient signal regions to model the response variables.

Even if with WILMA quite good results are obtained, the selection procedure is guided by the prior univariate ranking of the wavelet coefficients. Therefore, it was improved by employing a more efficient search method, like GA. This brought us to develop WILMA-GA, where the GA procedure for feature selection developed by Leardi [55] has been implemented for the wavelet coefficients selection. The key steps of WILMA-GA are reported in Fig. 7.

The first step as in WILMA is the decomposition of the signals matrix into the wavelet domain by DWT up to the maximum decomposition level. Then the GA is implemented as follows:

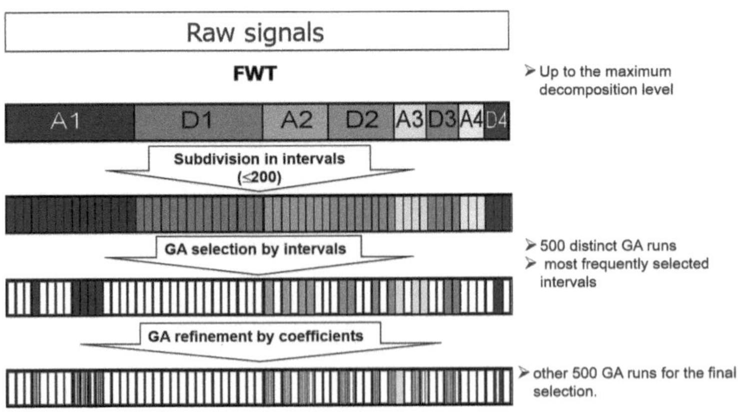

FIG. 7 Schematic view of the GA implementation to WILMA for coefficients selection.

1. All the wavelet coefficients from a redundant representation (both approximations and details for each decomposition level are considered) are searchable (it keeps track of the block and position to which they belong).
2. A first step of GA selection considers intervals of coefficients instead of single coefficients. Thus, a whole interval will be considered as "included" or "excluded" according to the 1/0 codification in chromosomes.
3. This will allow both pruning of wavelet coefficients belonging to each decomposition block (either an approximations or a details vector) and selection of the most promising blocks.
4. Taking into account the nature of WT decomposition scheme, i.e., coefficients belonging to higher level of decomposition cover wider regions of the signal when reconstructed in the original domain, each block is divided into a different number of intervals, which is progressively increased going from the maximum level of decomposition to the first level. Blocks belonging to the same decomposition level are divided in the same number of intervals.
5. Retaining all decomposition blocks for all levels allows considering all possible levels/scales combinations during GA selection in the wavelet domain, avoiding any a priori preselection or ranking of coefficients.
6. In order to avoid overfitting, which can be a common, yet serious issue when GA operates on a huge number of variables, the total number of intervals is kept to a maximum of about 200 and each interval corresponds to a variable in/out in GA selection.
7. Then, according to Leardi's, 500 independent GA runs are calculated and the most frequently selected intervals are retained, up to a global number of maximum 250 wavelet coefficients.
8. At this point a second GA step is accomplished, where a further selection of coefficients is done, starting from the coefficients belonging to the previously selected intervals, this time considering single coefficients as a variable in/out in GA selection. Other 500 GA runs are performed for the final selection.
9. A stepwise procedure is then adopted at this point to choose the final model; i.e., the unique coefficients selected over the 500 runs are sorted according to their frequency of selection and PLS models are calculated sequentially by adding a coefficient at a time, and then the most parsimonious model among those attaining an RMSECV value not significantly different from the absolute RMSECV minimum is selected.

As many WILMA-GA runs as wavelet filters to be tested are automatically performed. The obtained models are then ranked according to the RMSECV and the choice is left to the user in order to accomplish the best compromise between the lowest RMSECV and most parsimonious model.

If the selected wavelet coefficients come from a nonredundant basis (i.e., the selection does not contain coefficients from Ai-1, when coefficients from

Ai and/or Di are selected), it is possible to operate i-DWT to reconstruct the selected coefficients in the original domain for interpretative purposes. However, each decomposition block where one or more coefficients have been selected can always be reconstructed separately. Moreover, the sign of the regression coefficients in the finally selected PLS model can be taken into account to distinctly reconstruct positive and negative variables (i.e., wavelet coefficients) to ease the interpretation.

With respect to the WILMA algorithm the implementation of GA search allows to avoid any a priori preselection or ranking of the wavelet coefficients. Furthermore, it is not needed to cycle through levels. Moreover, the WILMA-GA approach, taking advantage of the preliminary GA selection of intervals, can deal with a huge number of variables.

2.3 WT in the Context Multivariate Image Analysis

The most frequent use of WT in image analysis, as for signals, is for denoising, deblurring, and compression. In MIA context [62–65] most of the work is related to feature extraction. Extension of DWT to image decomposition, i.e., 2D-DWT, is accomplished by applying the same monodimensional wavelet filters recursively on the two image directions: rows and columns, following an analogous scheme to the Mallat algorithm. This is illustrated in Fig. 8: for each channel *ch*, the low-pass and high-pass filters are first operated row-wise on the image and then, after downsampling (if SWT is used downsampling is skipped and filters are upsampled as illustrated in Section 2.1) of the coefficients in each of the resulting blocks, column-wise.

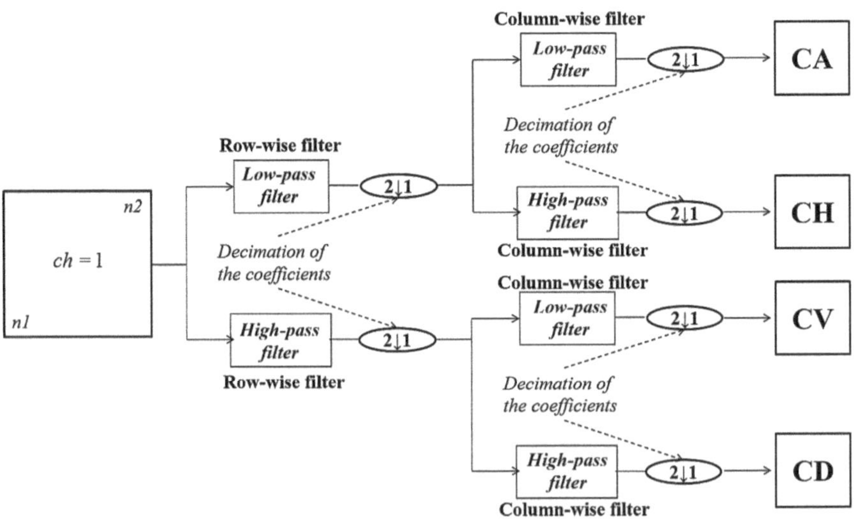

FIG. 8 Representation of the 2D discrete wavelet transform for one level of decomposition.

Considering all their possible combinations, this gives four decomposition blocks: Approximations (low-pass + low-pass), Horizontal details (low + high), Vertical details (high + low), and Diagonal details (high + high). The same procedure is recursively applied to the Approximations coming from the previous level until the maximum decomposition level, *L*, compatible with image size is reached.

In this section, we will focus on feature extraction/enhancement in wavelet domain and illustrate how it can be integrated to MIA, starting with recalling the basic of MIA approach.

From its beginning, as proposed by Bharati and MacGregor [66,67], the MIA approach consists of a framework (Fig. 9A) that can be summarized in two main steps:

I. Feature extraction (or enhancement) phase: the image (preprocessed, if necessary) is treated so that the texture information carried out by the pixels is made clearer.

II. Feature reduction (or analysis) phase: a suitable multivariate analysis method is applied, e.g., PCA, PLS, partial least squares discriminant analysis (PLSDA) on the feature matrix obtained after the first phase.

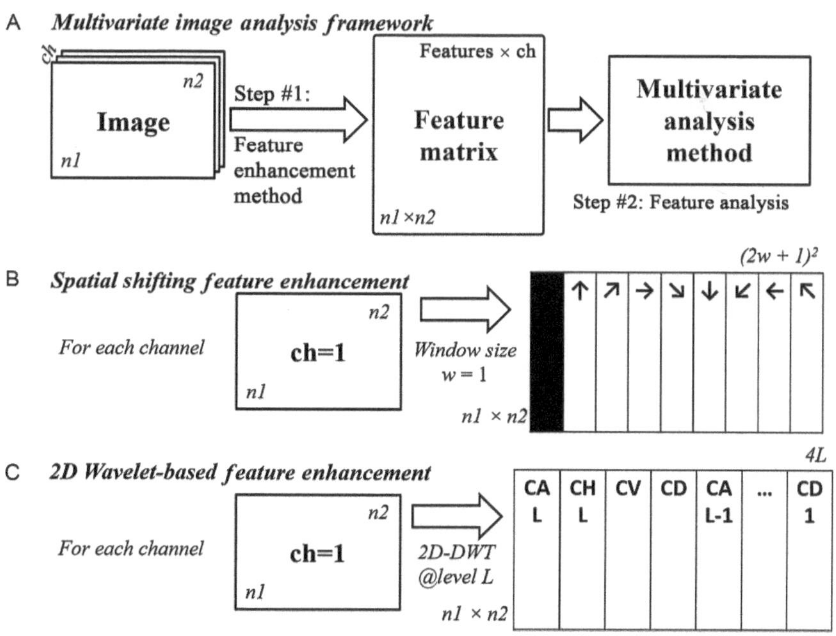

FIG. 9 The general multivariate image analysis framework (A) and schematic view of the feature matrix assembly for (B) Bharati and MacGregor spatial shifting feature enhancement and (C) 2D wavelet-based feature enhancement.

The first step can strongly influence the outcome of the following analysis. When choosing the feature enhancement method, it must be taken into account that it is equally important to preserve both the information given by the channels (to this aim a simple unfolding of the image pixel-wise could be sufficient) and the information pertaining to the correlation among neighboring pixels, i.e., the texture. Most of all, both sources of variability (texture and channel-based properties) have to be easily assessed by the following analysis step (whose method choice depends on the specific application).

Several approaches to feature extraction have been proposed and discussed [17,62,64,67], among which it is worth mentioning, together with the wavelet-based which will be considered in this chapter, the first relevant implementation by Bharati and MacGregor [62]. The spatial shifting feature enhancement, summarized in Fig. 9B, consists in capturing, for each channel *ch*, the pixel proximity correlation by means of a spatial shifting of neighboring pixels with respect to each pixel of the original image, according to a selected window aperture parameter, w. In practice, starting from a pixel element of the image $p_{i,j}$, a row vector is created by adding the intensity value of the channel corresponding to the closest surrounding pixels, and this is done for all the pixels as reported in Fig. 9B for a single-channel image. In the more general case, a total feature matrix is obtained which has as many rows as the number of pixels $I = n_1 \times n_2$, and as many columns as the number of channels *ch* times the number of spatial shifted pixels, which is given by $(2w + 1)^2$. Regardless of the number of channels, for a window parameter $w = 1$ (the closest neighbors) the feature matrix is $9 \times ch$, and when moving to $w = 2$ (the closest neighbors and the next surrounding layer), the feature matrix is $25 \times ch$. This implies a fast increase of the number of variables considered in the analysis.

Other feature extraction methods proposed in the course of the years are based on the application of a transformation of the image, again for each channel, in a different domain, such as the Fourier domain (via the 2D-FT) or the wavelet domain [17,62,67,68]. In most of the cases the feature extracted in wavelet domain is compressed to global descriptors [67,69], for a given decomposition level and block, by using, e.g., the Frobenius norm (energy) or the standard deviation of the coefficients; in this way the orientation information is only implicitly maintained by means of the level-block combination at which it is computed. On the contrary, working at pixel level [22,62], by using the wavelet coefficients extracted, keeps unaltered local spatial information.

We recently propose an approach [70,71] in this framework that considers each decomposition block per channel as an image. Our 2D wavelet-based feature enhancement approach is shown in Fig. 9C. The four decomposition blocks obtained from each level of decomposition (one Approximation block, CA, and three Details blocks, CH, CV, and CD) are independently reconstructed by means of the inverse 2D–DWT so that their dimensions are the same of the original image. This leads to a set of $4 \times L$ images for each channel *ch*, which can be unfolded and column-wise merged to obtain a total feature matrix which has as many rows as the number of pixels $I = n_1 \times n_2$ and as many columns as $4 \times L \times ch$. If we compare this column dimension to the one obtained with the Spatial Shifting approach, which is $(2w + 1)^2 \times ch$, it might appear that there is little benefit in terms of reduction of the feature matrix dimensions. By considering that the two approaches lead to the same number of feature descriptors if $L = \text{round}\left[(w + ½)^2\right]$, this results, e.g., a window parameter of $w=2$, in a decomposition level $L=7$. Due to the nature of the wavelet filters and how they interact with the image surface in the deconvolution, such a decomposition level (if allowed by the nature of the chosen wavelet) usually leads to the evaluation of correlations and high distance relationships among pixels to an extent which is superior to the use of a moving window of fixed size.

Once the 2D WT up to a given level is applied for each channel of an image, all of the resulting pseudo-images are then unfolded pixel-wise and (when made of compatible dimensions) mid-level data fused to a feature matrix which is used for the feature analysis phase. In this way, all possible relationships among blocks, decomposition levels, and channels are assessed in a single multivariate analysis step. The requirement of having pseudo-images of compatible dimensions depends on the considered WT. DWT requires that wavelet filters are passed at each cycle over downsampled versions of the image, thus leading to coefficients matrices which have a different dimension at each decomposition level. Therefore, if all the resulting blocks from all decomposition levels are to be fused in a single feature matrix, an additional reconstruction step in the original pixel dimension by using the i-DWT is mandatory. The use of SWT, on the contrary, does not require the decimation of the coefficients since it is the filter which is upsampled at each cycle over decomposition levels. In this way, each decomposition block preserves the dimensions of the original image, thus making the feature matrix immediately ready for the feature analysis step.

The main advantage can be found in a general reduction of the dimensions of the feature matrix when compared to the spatial shifting approach, while preserving the texture information and, possibly, extending the evaluation of pixel correlation to a larger scale, which would be practically infeasible by requiring a w-value much higher than what is normally used. Also, referring to wavelets, several filters, levels, and also transformation schemes (for example, 2D-DWT or 2D-SWT) can be adopted, allowing more possibilities in terms of investigation of the pixel correlation structure.

3 APPLICATIONS

3.1 WILMA + GA PLS for the Calibration of Flour Leavening Compounds

In the context of an industrial bread-making leavening phase, an important aspect is to monitor in the most precise possible way the correct progression of dough rising, so that cases of under- or overleavening are avoided, which could result in a failure of compliance for the final product. In this application, the flour-to-dough transformation process was studied from a chemical point of view by quantifying the amount of acids and sugars, which are formed or consumed during the leavening process. A wavelet-based calibration method, WILMA+GA PLS, was used to obtain calibration models for the concentration of these species and insight on the spectral regions which could be selected for an online implementation.

Two carboxylic acids were determined and quantified in doughs by means of gas chromatography, using a free induction decay detector (GC-FID), that is malic and succinic acids. As far as sugars and polyalcohols are concerned, six species were determined: glucose, fructose, sucrose, maltose, glycerol, and inositol, which are identified as either leavening starters (sucrose, glucose, fructose, and maltose, which are nutrients for the yeast) or leavening products (glycerol and inositol). For details about sample preparation and the chromatographic method for the determination of sugars and acids, attention is addressed to Refs. [53,54,72] from which the methods have been derived. Here, only the calibration model for glycerol will be presented to illustrate the advantages of the adopted approach.

3.1.1 Data Set

Sixteen batches of dough—indicated with letters from A to P—were sampled during 8 days of production. Each day, two batches were picked up from the production line in six different points before the baking phase, for a total of 96 samples. The production process consists of two separate leavening phases: three points represent the first and longer leavening step, and the other three points represent the second leavening, which is shorter and includes a second mixing of the first dough with new ingredients. Each sample was analyzed by means of GC-FID to determine sugars and acids, and by means of near infrared spectroscopy (NIRS) by acquiring the spectra at-line, directly on the fresh product taken from the production line.

3.1.2 Results and Discussion

Exploratory analysis showed a common variability caught by both NIRS and GC variables: therefore, the multivariate calibration models were computed, with the aim of exploiting the clear advantages NIRS has on chromatographic techniques, in terms of speed (some minutes instead of some hours) and sample manipulation (analysis "as is" instead of using reagents of known risk for the operator). The multivariate calibration was conducted using both the classical PLS and WILMA-GA-PLS. Seven wavelets, belonging to three different families, were tested: Daubechies (db1, db3, and db5), Symlet (sym4, sym5, and sym6), and Coiflet (coif4).

The calibration models were validated by means of both cross-validation and external validation. As for PLS models, the cross-validation was conducted by using the random subsets procedure, with 8 cancelation groups and 20 iterations; the optimal LVs number for each model was chosen as the minimum RMSECV. As far as WILMA-GA-PLS models are concerned, the number of LVs is chosen as the minimum RMSECV in cross-validation, according to a venetian blinds procedure, with six split groups. In order to operate an external validation, the data set was split into a training set of 66 samples (11 batches) and a test set of 30 samples (five batches) for both PLS and WILMA-GA.

The obtained calibration model for glycerol was satisfactory, both in terms of cross-validation and external prediction ability. As for WILMA+GA, the best cycle of calculation, among the seven wavelet filters tested for signal decomposition, was chosen as the most parsimonious one among the cycles statistically equivalent in terms of RMSECV values, and measured vs predicted values are reported in Fig. 10. The results in Table 1 demonstrate that generally feature selection leads to an improvement of the predictive ability of the model, particularly in terms of RMSEP values.

In addition, the WILMA+GA model presents three more advantages. While PLS needs the preprocessing of the signals matrix (in this case Savitsky–Golay filtering and second derivative with a 21 points width window and a second-order polynomial), WILMA+GA does not need any

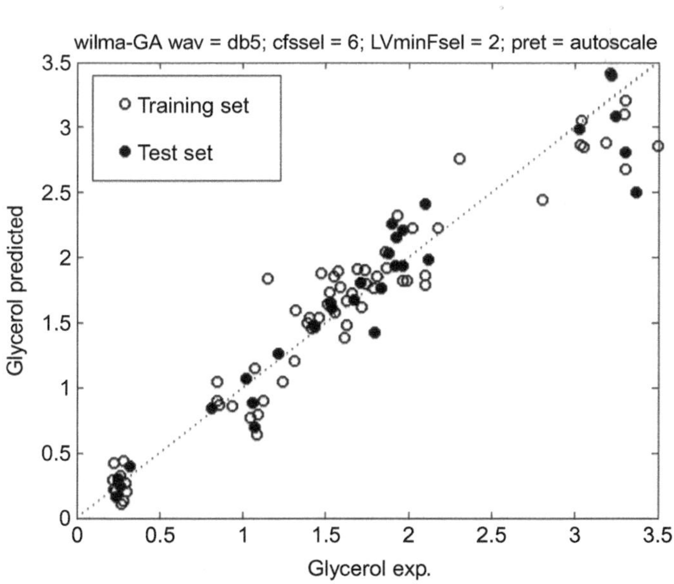

FIG. 10 Glycerol calibration by means of WILMA-GA PLS: measured vs predicted plot, *empty circles,* training set; *filled circles,* test set.

TABLE 1 Performances of the Calibration Models for Glycerol

Calibration method	Preprocess/ pretreatment	# Var	# LVs	R^2 CV	R^2 test	RMSECV	RMSEP
PLS1	SG+second der. (21 2 2)/ mncn	1336	6	0.89	0.89	0.30	0.31
WILMA-GA (db5)	None/auto	6	2	0.92	0.93	0.25	0.24

preprocessing, because the feature selection in the wavelet domain removes baseline effects, noise, and overlapped bands. Second, the model obtained by feature selection is definitely more parsimonious. Finally, if some spectral regions of particular influence for the calibration are to be selected, for e.g., sensor customization for online use, the pseudo-regression coefficients of PLS models are useful only if a clear predominance of a specific region is present, which is typically the case for the calibration of macrocomponents of the samples, e.g., water or protein content. On the contrary, the selected wavelet coefficients can be interpreted to identify more precisely the spectral regions which were selected as most important for the regression, especially when they are present at lower concentration and their contribution result hidden by broader bands due to macroconstituents.

The WILMA+GA PLS results for glycerol are shown in Fig. 11 where the three successive decomposition levels of the details selected by the model are represented. When reconstructed to the original spectral domain through the i-DWT it is possible to notice that the selected region globally covers the spectral wavenumbers between 5000 and 6500 cm^{-1}. In particular, a certain predominance in the regression can be ascribed to absorptions in two regions: at 5686 and 5817 cm^{-1}, which can be attributed to the first overtone of the stretching of C–H bond in the CH_2 group, and at 5516–5570 cm^{-1}, where several combination modes are reported, and in particular the second overtone of the O–H bond stretching/C–O stretching combination. These regions are particularly coherent with the molecular structure of glycerol and were not easily observable in the raw signals, i.e., smooth shoulders; the main drawback is however related to the fact that a clear discussion of the sign of the regression coefficients is lost in this representation. However, by looking at the sign of the regression coefficients for each selected wavelet coefficient in the PLS model, it could be observed that the only one with positive sign belongs to the details at level 4 (D4 block) and when reconstructing the signal based only on this wavelet coefficient it points to the shoulder at about 5940 cm^{-1}.

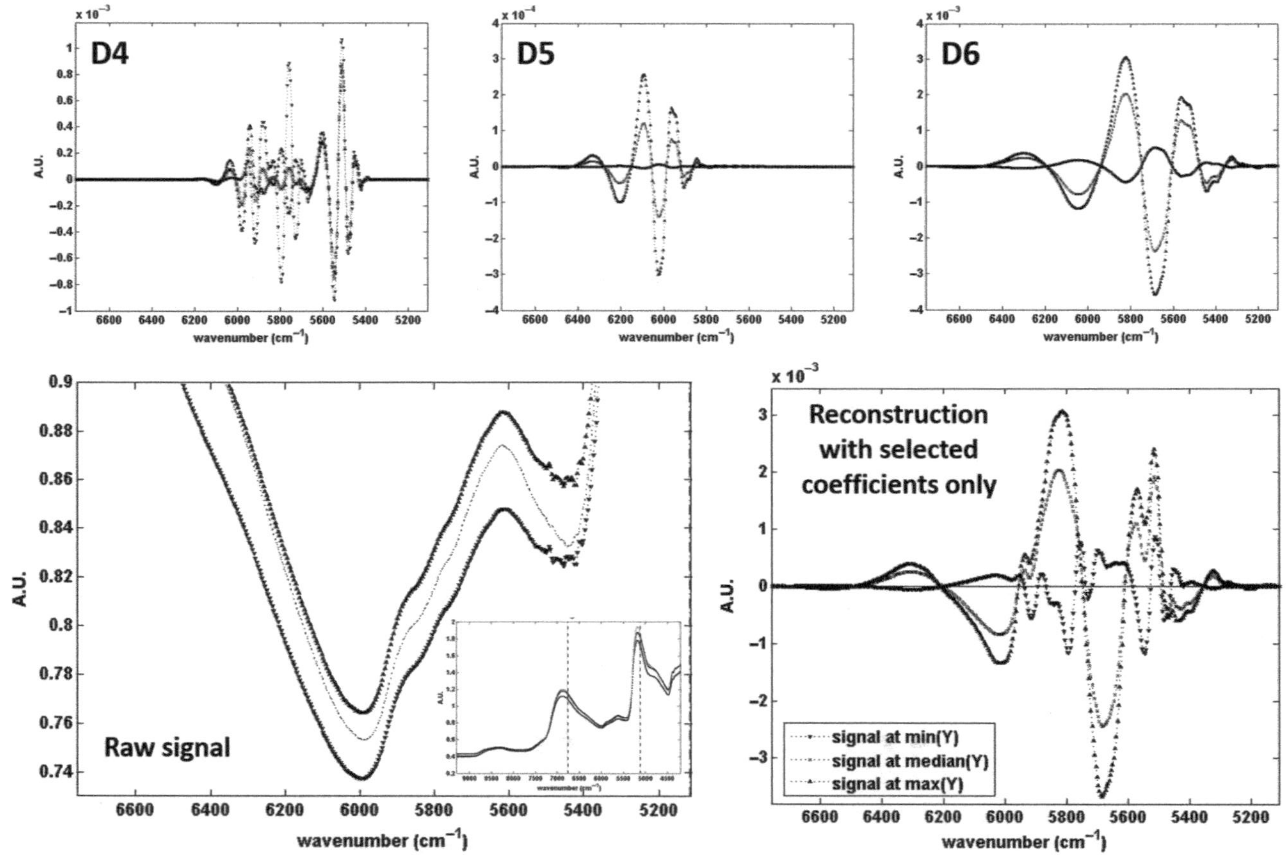

FIG. 11 Glycerol calibration by means of WILMA-GA PLS: indication of the GA selected wavelet coefficients for blocks D4, D5, and D6 (*top row*) and reconstruction of the selected NIR spectral region (*bottom right*) compared to the raw data variability in the same spectral region (*bottom left*).

As a final remark, the possibility of calibrating these species adds up to the ability of determining several features of the process with a single measurement dealt with online, at different points of the process line. With this kind of approach, for instance, leavening could be monitored and controlled to determine whether the dough has undergone an optimal rising or the conditions were too extreme and resulted in an accelerated or delayed process.

3.2 Classification of Animal Feed by WPTER

In this application, the discrimination of animal feed of different origins by NIR scanning microscopy [73,74] has been assessed with the aim of discriminating among meat bone meal (MBM) of different animal species and MBM-free feeding stuffs. The importance of this discrimination lays in the regulations which prohibit the use of animal feed for the feeding of cattle (e.g., bovines), due to the well-known cases of the late 1990s of mad cow disease (bovine spongiform encephalopathy) [75,76]. It is also important to be able to recognize feedstuff samples contaminated by MBM and, possibly, the percentage of contamination.

3.2.1 Data Set

NIR spectra of sediment fragments were collected in diffuse reflectance mode by an Autoimage microscope connected to a Perkin Elmer FT-NIR spectrometer. The microscope aperture (magnification: $10\times$) was of 50×50 μm, and spectra were recorded in the 6000–4000 cm^{-1} range, with an 8 cm^{-1} resolution, by accumulating 16 scans. The data set consisted of samples of commercial foodstuff (negative), and MBM of different animal species. These data were split randomly in a training set for constructing the model and a test set for validation. From the available maps, the following number of spectra was considered (test set dimensions are reported in parenthesis): Mammal pure samples (bovine and swine): 671 (135); Fowl: 369 (65); Fish: 331 (65); MBM free (negative): 736 (125). After model computation and validation, the predictive performance was evaluated on three commercial foodstuff spiked with MBM of mixed animal origin: Pos01 (spiked at 0.1%, 169 spectra); Pos03 (spiked at 0.3%, 141 spectra); and Pos05 (spiked at 0.5%, 144 spectra). Additionally, some reference MBM samples were considered, belonging to the following classes: Mammal (21 spectra), Fowl (15 spectra), and Fish (7 spectra).

3.2.2 WPTER–SIMCA Results and Discussion

In order to address the aim of this study, class modeling is well suited, since the objective is to assess the presence of any percentage of MBM in animal feed (negative or not) and possibly recognize the species (can belong to one of mammal, fowl, fish); thus SIMCA was used as a classification method.

From application of the WPTER algorithm, we expected on one hand to improve the classification performance, handling nonrelevant spectral variability in the NIR data, such as scattering, and on the other hand to gain interpretation about the most significant signal regions to identify the presence of MBM. This can be useful to suggest further development of specific device using a reduced signal range.

For comparative purposes the analysis was performed both on the whole signal after preprocessing of the spectra (Savitsky–Golay first derivative and mean centering), and by using the WPTER algorithm, the selected wavelet coefficients were then input to SIMCA. WPTER calculations have been carried on the raw signals, by taking into account the following parameters: type of wavelet (db1, db2, db5, db10, sym4, sym5, sym6, sym7, coif1, coif3, coif4, and coif5), decomposition level (up to maximum, that is 6), and different percentage of wavelet coefficient thresholding (1%, 0.5%, and 0.1%). The best performing cycle has been chosen on the basis of global (considering all four classes) sensitivity and specificity for the training set and corresponds to the use of coif1 wavelet at decomposition level 4. This results in the final selection of two wavelet coefficients. In Fig. 12, we compare the scores plot for the first two components of an exploratory PCA on the whole preprocessed signals and the scatter plot of the two selected coefficients; it can be observed that the distinction among the classes is clearly improved.

Table 2 reports the comparison in terms of sensitivity and specificity toward each category of SIMCA and WPTER-SIMCA results. Sensitivity values are on the diagonal terms, highlighted in gray, while specificity values are reported as extradiagonal entries and have to be read from top to down direction; e.g., going down Mammal class column the specificity of mammal category toward Fowl, Fish, and MBM-free categories, respectively, may be found. It is possible to notice by considering the results on the validation set that generally both SIMCA on the whole signals and WPTER-SIMCA perform in a good way, with possibly a higher sensitivity of WPTER-SIMCA for the detection of MBM-free samples. If we consider the capability of both models in prediction, that is on the external sets, Table 3 reports the results in terms of percentage of signals assigned to the different categories for the foodstuffs spiked with mixed MBM at known percentages and for the pure MBM reference spectra projected onto the model. In the first case it is important to notice that spiking quantity is known, but not with which of the three MBM classes. Both classification models recognize that there is MBM in the samples, but only the model based on WPTER-SIMCA shows a decreasing percentage of spectra assignation to the MBM-free class with the increase of spiking, which is mostly related to an increasing percentage of Mammal and Fowl. As far as pure MBM reference spectra are concerned, for both models the MBM-free class correctly rejects all the MBM samples. WPTER-SIMCA model shows a higher sensitivity, especially by not confounding Fowl with Fish: this has some commercial relevance, since Fish is

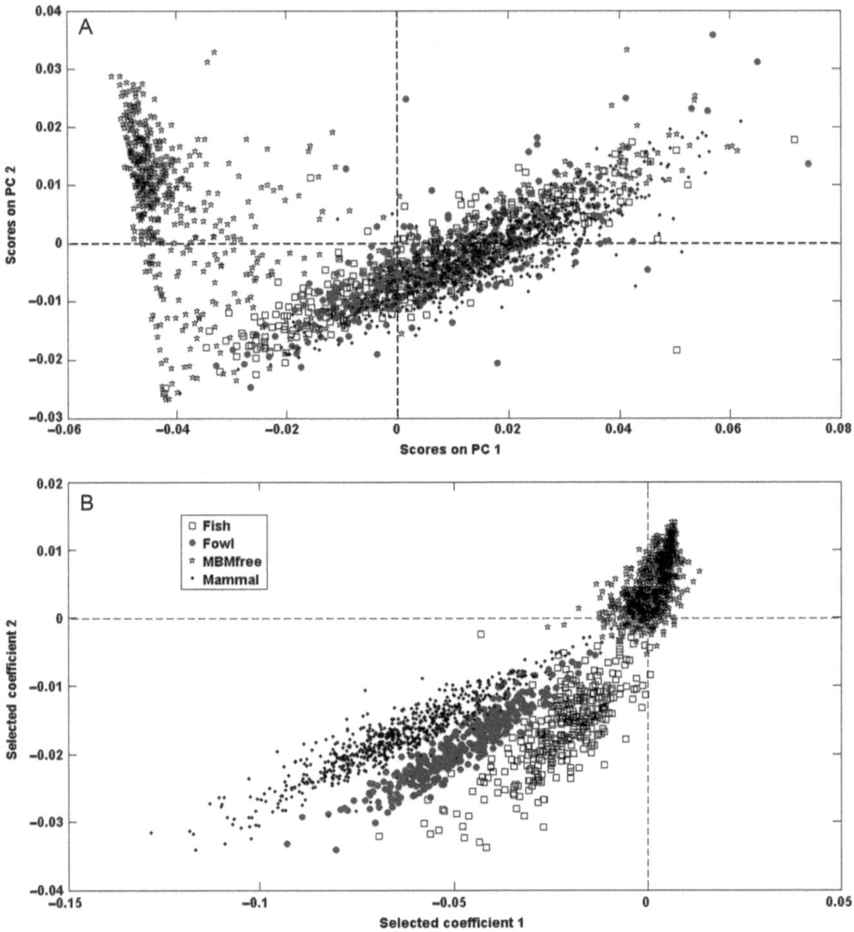

FIG. 12 PCA scores plot of the whole preprocessed signals (A) and distribution of the samples according to the two coefficients selected by WPTER (B).

actually allowed to some percentage in foodstuffs for the alimentation of non-ruminant animals.

Fig. 13 reports the average reconstructed signals for each category and a representation of the signal region is considered by the two selected detail coefficients, that is a portion of the signal in the 4200–5200 cm^{-1} range. Two main portions are reconstructed: the one at lower wavenumbers mainly discriminates MBM-free samples from MBM, while the other at higher wave-numbers is characterized by the fact that the different animal species show different intensity values and variability. In particular, these regions are mostly related to contributions which can be attributed to, respectively, fat content and protein–water interactions.

TABLE 2 Results Comparison of Whole-Signal SIMCA and WPTER-SIMCA for the Four Classes

			Mammal		Fowl		Fish		MBM free/negative	
Classes	SIMCA	LV	Train	Test	Train	Test	Train	Test	Train	Test
Mammal	Whole signal	6	90	93	100	100	99	99	100	100
	WPTER Two selected coefficients	1	92	94	97	98	99	98	100	99
Fowl	Whole signal	8	100	99	83	92	92	100	100	100
	WPTER Two selected coefficients	1	95	95	87	91	94	95	100	100
Fish	Whole signal	8	100	100	100	100	86	99	100	100
	WPTER Two selected coefficients	1	100	100	97	97	90	97	98	100
MBM free/negative	Whole signal	7	100	100	100	100	99	99	87	89
	WPTER Two selected coefficients	1	100	100	100	100	99	97	96	93

Values on the diagonal (highlighted in *gray*) indicate % Sensitivity; extradiagonal values indicate % Specificity toward each class (read column-wise).

TABLE 3 Results Comparison of Whole-Signal SIMCA and WPTER-SIMCA for the External Test Set Samples

MBM-spiked samples		Mammal	Fowl	Fish	MBM free/negative
WPTER sel cfs	Pos 0.1%	4	4	5	85
	Pos 0.3%	13	14	21	49
	Pos 0.5%	18	28	13	34
Whole signal	Pos 0.1%	2	0	67	9
	Pos 0.3%	2	2	35	22
	Pos 0.5%	4	1	24	15
MBM pure samples		**Mammal**	**Fowl**	**Fish**	**MBM free/negative**
WPTER sel cfs	Mammal	91	5	0	0
	Fowl	7	93	0	0
	Fish	0	0	100	0
Whole signal	Mammal	81	0	0	0
	Fowl	0	73	13	0
	Fish	0	0	86	0

For MBM-spiked samples, values indicate the percentage of signals assigned to a class. For MBM pure samples, % Sensitivity (diagonal terms) and % Specificity (extradiagonal terms) are reported.

3.3 Wavelet-Based Multivariate Image Analysis for the Evaluation of Bread Surface Defectiveness

Bread crust and crumb are the principal responsible of organoleptic characteristics of bakery products. Bread crumb appearance and properties have been studied in several circumstances [77–79], due to the fact that its texture, porosity, and resistance to staling are a direct consequence of the most important phases of the bread-making process. The interest on bread external surface is less widespread, frequently addressed to the identification of chemical compounds which are generated during the baking process from Maillard reactions [80,81]. However, in an industrial bread-making context, bread quality is first evaluated in terms of the visual aspect of the crust of bread loaves. In the application discussed here, an industrial production of "bun"-type bread for fast food restaurants, loaf dimensions, crust color, and homogeneity are parameters which are important and measured online for each bread loaf. This potentially gives feedback control on recipe formulation and parameters such as leavening and baking conditions to be modified when defects on bread crust arise.

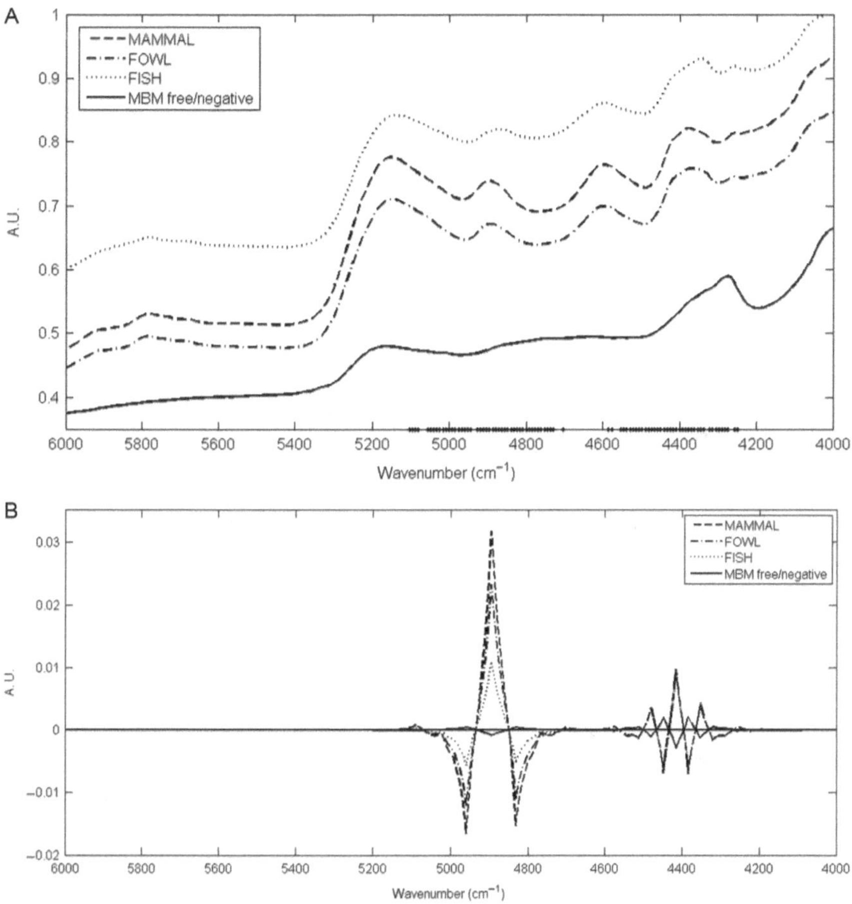

FIG. 13 Representation for each class of the (A) average NIR signal and (B) of the reconstructed signal on the basis of the two features selected by WPTER.

The study presented here considers a collection of bread loaves that are cataloged as in target by production managers, and loaves that present surface defects, such as dark spots, blisters, and pale areas. These defects arise from different causes, some of which not perfectly known, and are sometimes difficult to be detected by common vision systems, such as RGB-based online cameras. MIA of multispectral images can improve the acquisition of information on bread quality, combining spectral and textural information. Compared to RGB, the addition of some extra wavelengths (or channels) in the spectral regions close to visible range (i.e., ultra violet and/or near infrared) potentially brings more chances to visualize the defects in a clearer way. Since UV and NIR respond also to chemical changes, these channels can show localized variability on the surface connected to a different composition, thus better highlighting defects which can be due to, e.g., an abnormal presence or transformation of a given ingredient in a spot-size area. Also, the MIA framework suits well this case for online product quality control: by defining the latent space of well-manufactured loaves, it is possible to monitor abnormal behavior and detect where possible defects are located, thus rising a quick alarm to discard the faulty product and correct the process.

3.3.1 Image Acquisition and Pretreatment with 2D WT

Multichannel images of selected bread loaves were acquired by means of a Videometerlab 2 (Videometer A/S Hørsholm—Denmark), courtesy of the Faculty of Life Sciences, University of Copenhagen (Denmark). The instrument is able to record an 18-channel HD image of the whole sample, considering the visible range (from 430 to 700 nm, 10 channels) and the short-wavelength NIR range (from 850 to 970 nm at ~20 nm resolution, 8 channels), thus giving both color and chemical information together with the textural one.

In this application, MIA framework with 2D-DWT for feature enhancement has been used, following the approach described in Section 2.3. A set of seven normal images, representative of target compliance for product quality, and seventeen defective images, showing several different faults, ranging from macroscopic fissures and dark stains to pale areas and small blisters difficult to spot have been considered and processed according to the workflow presented in Fig. 14.

The preprocessing step of image compression and background removal provided a significant improvement in the results when compared to direct wavelet decomposition and analysis of the raw image. The ability to capture into Approximation and Detail blocks different aspects of pixel variability is in this case exploited by isolating in the Approximation coefficients (sym6 wavelet, sixth level of decomposition) the structure of background

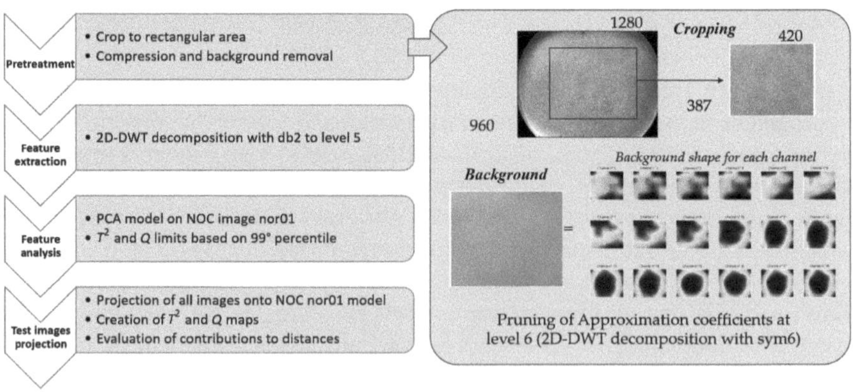

FIG. 14 The MIA framework followed in the application. *On the right*: details about pretreatment step with image compression and background removal.

illumination, which differs from visible to NIR channels and follows the rounded shape of the bread loaf. Also, together with coefficient decimation to get rid of the undesired variability due to lighting, a compression effect can be obtained, thus making the feature matrix easier to handle from a computational point of view.

3.3.2 2D-DWT-MIA Analysis: Results and Discussion

The Feature Extraction phase was performed on each image for every channel separately; a Daubechies 5 wavelet was used up to decomposition level 5. In this way, a data matrix of dimension $I_{\text{pixels}} \times 360$ $(5 \times 4 \times 18)$ is obtained (MIA with a window size of 2 would result in 450 features, taking into account a lesser degree of details). Since the problem can be defined as a multivariate quality monitoring chart strategy, where bread test images are compared to "normal" bread crust images, the multivariate analysis phase was based on PCA. The reference PCA model for nondefective image has been calculated by considering the feature matrix obtained by applying the decomposition on an image of a representative acceptable crust sample, as described before (Fig. 9C; Section 2.3). Then, distance-based statistics that is Q residuals and Hotelling's T^2 were computed. Critical limits for each of the two statistics were computed on the basis of the 99th percentile. The PCA reference model was calculated by considering each of the seven normal images as a reference; however, since similar findings were obtained in all case, here the results based on the single nor01 image are presented.

For all projected images, Q and T^2 distance values were computed for each pixel: this information can be used in a twofold way. The total percentage of pixels which exceed the critical limit can be provided, as reported in Table 4, to obtain a global index of how significant is the deviation of the test image from the normality described by the reference model. It is possible to notice that a clear detection of faulty images can be obtained, with relevant percentages of pixels above the critical limits for both distances for all the defective crust images. Normal labeled images show either one or both the percentages below the confidence limit, with two significant cases of higher deviation, that is nor04 and nor09. The second way to use the information extracted from the model can help in explaining these anomalies in particular and, more generally, why an image has high percentages of pixels outside the limits. The Q and T^2 values can be refolded to the original pixel $\times$ pixel domain to obtain a grayscale distance map, which gives a direct distinction between normal regions of the crust (white) and defective ones (gray to black). This is shown for two example images, nor04 and def13, in Figs. 15 and 16, respectively. The use of grayscale maps gives a great enhancement of the detection of commonly hard to identify defects; moreover, the distance parameter (in the example T^2) can be expressed in terms of its contributions in terms of the original variables, i.e., the spectral channels. Therefore, the spectral

TABLE 4 Out of Model Pixel Percentages for T^2 and Q Distances for Projected Images

	% Beyond T^2 (99th perct)	% Beyond Q (99th perct)
nor02	0.5	0.7
nor04	1.3	2.2
nor05	0.5	0.6
nor08	0.9	1.5
nor09	1	2
nor10	0.8	1.7
def01	12	9.6
def02	7.2	6.2
def03	4.3	4.3
def04	3.3	3.2
def05	9.2	8.5
def06	5.1	3.8
def07	4.6	4.3
def08	4.9	3.5
def09	7.4	5.2
def10	4.6	4.1
def11	6.6	5.6
def12	20	19
def13	7.1	6.8
def14	16	12
def15	12	7.6
def16	13.9	11.7
def17	13	9.9

information on pixels of a defective area can be calculated and shown as contribution plots, which could help to identify, in a qualitative way, the chemical nature of some of these defects in order to better evaluate the product surface. As an example, both def13 and nor04 show some areas of small pale

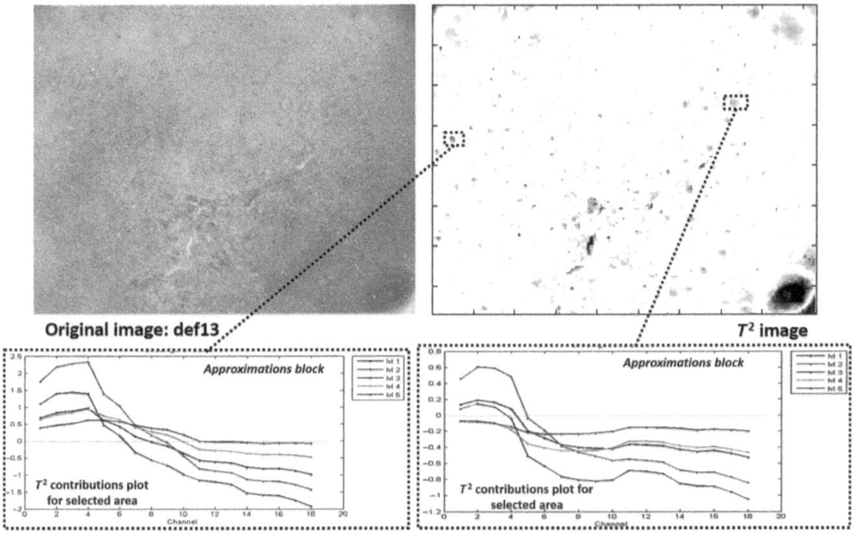

FIG. 15 Results of the projection of image nor04. *Top left*: the original appearance. *Top right*: T^2 values refolded in the original pixel × pixel dimensions. *Bottom*: contribution plots for the *two highlighted spots* in T^2 image: the different curves refer to the channel profile at each level of decomposition.

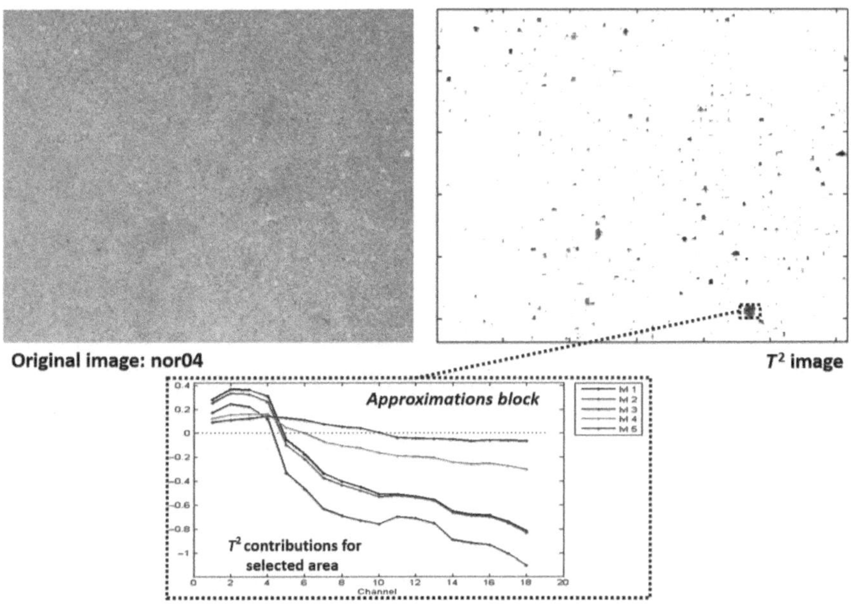

FIG. 16 Results of the projection of image def13. *Top left*: the original independent components appearance. *Top right*: T^2 values refolded in the original pixel × pixel dimensions. *Bottom*: contribution plots for the two highlighted spots in T^2 image: the different curves refer to the channel profile at each level of decomposition.

blistering: from the NIR channels which appear as significant from the contribution plots a tentative explanation of the origin of the defect can be given in terms of the presence of fat (e.g., olive oil) on the surface of the dough, which leads to lack of uniformity in the structure of the crust during leavening and then baking.

3.4 Wavelet-Based Multivariate Image Analysis of TEM Images for the Characterization of Innovative Nanomaterials

This application of the WT to the Feature Extraction phase of the MIA framework regards the elaboration of transmission electron microscopy (TEM) images of nanostructured glasses, that is glass materials functionalized by the addition of metal oxides, in the specific case made by TiO_2 nanoparticles doped with Ce (0.5% mol) [82,83]. The nature of TEM images is quite simple in practice, being grayscale (single-channel) images, therefore vectors of intensities for each of the unfolded pixels. However, the texture domain can give several important information, depending on the scale of magnification of the sample: at a >10 nm scale they can be used to evaluate particle dimensions and their distribution, so that it is possible to monitor the obtained material in terms of how well-defined the size distribution of the nanoparticles is. At a lower scale, ($\sim$2 nm), the magnification is able to give information about the presence of dominions of different crystalline phases (and/or the presence of amorphous phases), which rises other targets such as the possible to isolate the crystal planes and measure the planar dimensions which allow characterizing the material from a chemical point of view. Also, combining the information carried by the texture structure at this scale with a proper multivariate analysis method, the "resolution" of different overlapping domains might be achieved, thus facilitating the attribution and measurement processes which are usually left to the experience of the operator.

When dealing with grayscale images, several methods have been implemented to obtain information, which are usually based on segmentation. This usually requires correction approaches which eliminate the illumination and background variability prior to the application of a proper threshold which operates a binary conversion of the image. The two steps are critical whenever the illumination/background effect cannot be regularized to a known, analytical profile to be filtered, and when the differences in terms of levels of gray are very small. Thresholding, in fact, fixes a gray value below which all is considered as "not an object," thus leading to a clear separation of the features to be quantified, in terms of, e.g., dimensions or area, at the price of a huge approximation if the borders are, for example, not particularly defined. Also, the final results of segmentation are usually global indicators, such as the number of objects, the distribution curve of their areas or dimensions, or a global index of homogeneity of the surface. The application of the MIA framework to a single-channel, grayscale image can be used to obtain the same enhanced information that can be extracted from an RGB or multichannel image, that is an evaluation of the pixel domain in terms of texture and feature variability.

3.4.1 Considerations on MIA Framework Setup

In this application, the analysis is performed on a single TEM image, as shown in Fig. 17, in which it is possible to observe two domains corresponding to mixed crystalline planes, upon an amorphous background. The MIA framework is applied according to the usual two-step procedure of Feature Extraction and Feature Analysis: in this case, Feature Extraction is wavelet based, using the 2D-SWT, so that no reconstruction step is needed to compose the feature matrix. More precisely, a Daubechies 5 wavelet was used for the 2D-SWT at decomposition level 6.

The choice of the multivariate method for the feature analysis step is here target driven. The goal of this analysis is enhancing the image so that the crystal planes are well separated from the amorphous background and the illumination effects, and possibly different planes—hence different crystalline phases—are resolved into different components. PCA is used in this case for explorative purposes, but for its nature it highly relies on the (multi-) Gaussian characteristics of the sources of variability, which contribute to the whole data matrix, and operates on the minimization of data covariance (and, as a consequence, maximization of variance). Therefore, the resulting components are for sure orthogonal vectors which give a complete description

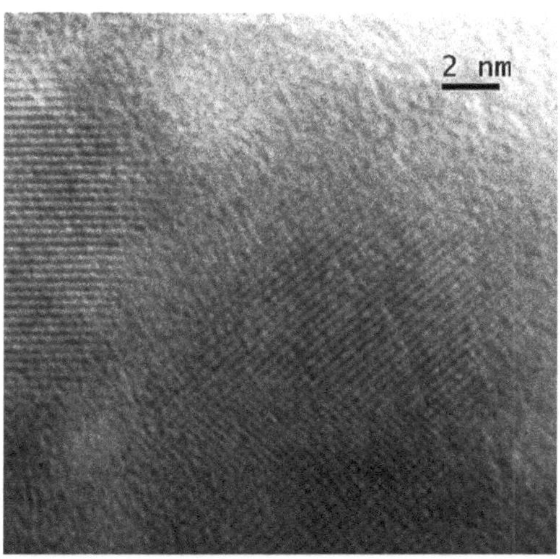

FIG. 17 TEM image of a nanoglass sample.

of the highest sources of variance in the data set, but they might not be a good "resolution" of the different components which concur to generate the measured image. In this case, for example, the amorphous phase contributes mostly as a huge source of white noise, and the different crystalline planes can be seen as independent sources measured simultaneously, which differ in terms of spatial orientation and interplanar distance. In these conditions, ICA might give a better interpretation of the images, in terms of isolation of different contributions in separate components. In brief, ICA aims at minimizing higher order statistics (e.g., skewness) of the data matrix, minimizing the mutual information and hence giving, as a result, components which act as independent sources of variability. Their contribution to the whole signal, in a way which is similar to what scores are, is given by mixing values, i.e., how much of the detected independent signal or component is present in each sample (in this case, each pixel). Exactly like scores can be reshaped in the original pixel dimensions of the image leading to score images, the mixing matrix values can be rearranged to obtain mixing images.

3.4.2 Results and Discussion

The results of PCA and ICA in terms of score images and mixing images are reported, respectively, in Figs. 18 and 19. Since the two techniques relay on different statistical features of the data, the results are quite different. PCA (preprocessing: autoscale) detects two directions of maximum variance (PC1 and PC2), which appear in terms of score images as a mix of different contributions (basically, background variability in dark vs regular shapes of plane crests in white), regardless of their nature and orientation. The following components capture variance which is more related to illumination effects (PC3) and amorphous structure (PC4), but again there is a significant degree of mixing with the regular structures of the planes which can still be seen in the score images.

When looking at ICA mixing images (ICs cannot be ranked in terms of importance; therefore the numbering is purely arbitrary), it is possible to notice that IC1 and IC2 describe almost uniquely two different features present in the image, i.e., the illumination effect and the amorphous structure, respectively. The other two ICs take into account the crystalline planes: in particular, IC3 isolates in a clear way the region on the left, which corresponds to a well-defined crystalline domain; IC4 shows a small degree of overlapping with this region, but mainly isolates the other, larger, crystalline domain.

ICA weights (Fig. 20) for IC3 and IC4 show which decomposition blocks of the wavelet decomposition are seen as a combination to form the two independent components, that is Horizontal and Diagonal details at decomposition level 4 and Diagonal details at level 3 for IC3 and all Horizontal details for the first four level of decomposition, together with Vertical and Diagonal details for level 1.

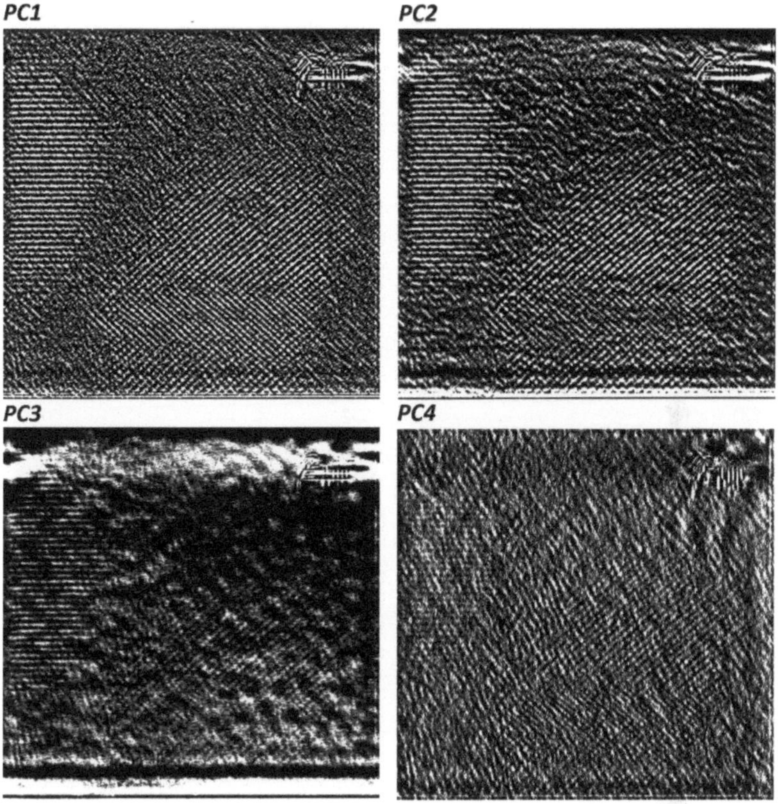

FIG. 18 Principal component analysis. Scores images for the first four principal components.

In conclusion, ICA is able to separate the two crystalline domains into two different ICs, which helps in the evaluation of features such as the distance between planes. This gives a good insight in terms of nanomaterials properties: for example, the interplane distance calculated for the crystal domain isolated by IC3 is 0.314 nm, which is in good accordance to the value reported in the literature for TiO_2 anatase plane distance [84]. Instead, the crystal domain isolated by IC4 shows a distance between planes which ranges from 0.311 nm to 0.261 nm: this could be connected to the doping effect of cerium ions in the anatase structure.

4 REMARKS

In this chapter we have looked at the thematic of resolving signal components from a different perspective, that is gathering salient features and pattern in spectral signals and images as a result of multivariate analysis of all possible "pictures" obtained at different resolutions from spectral and/or imaging data.

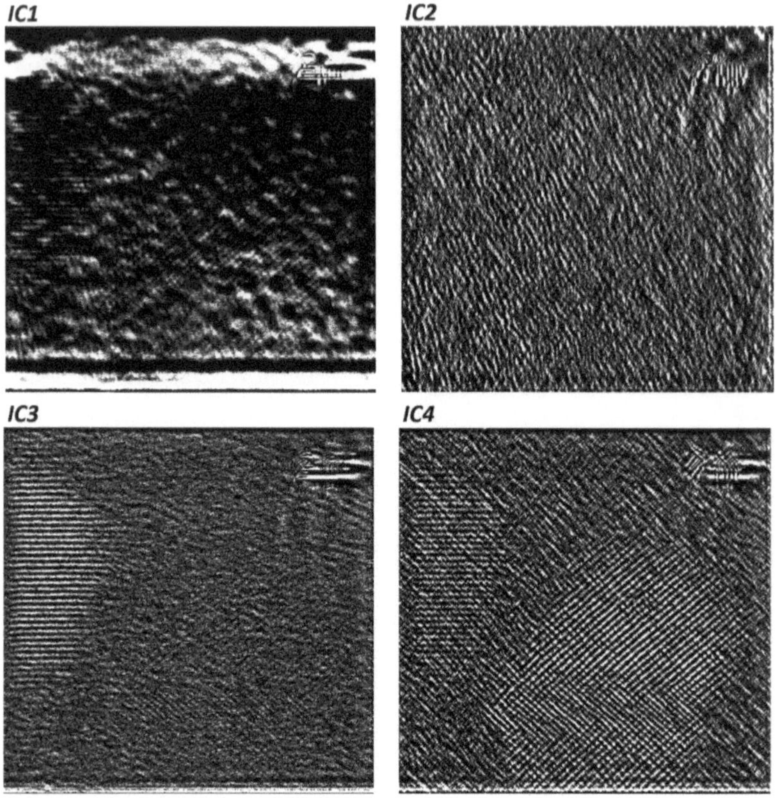

FIG. 19 Independent component analysis. Mixing images for the four IC components.

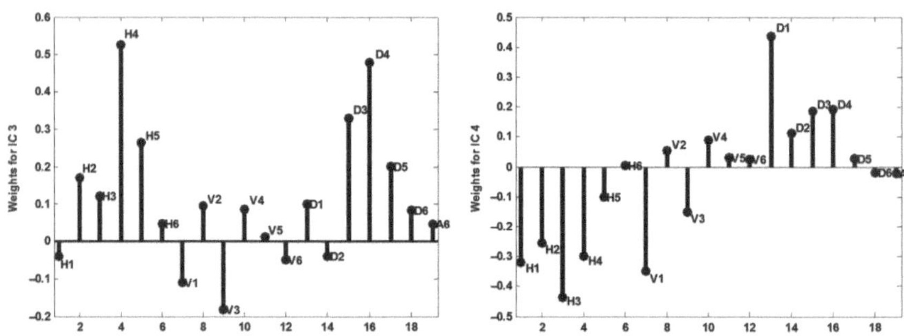

FIG. 20 Independent component analysis: weights for independent components 3 and 4.

To this aim feature selection and feature enhancement algorithms in wavelet domain have been applied, benefiting from the properties of WT of capturing different aspects of signal/image in disjoint subspaces, i.e., profiting of dual time frequency localization. The fact that each wavelet coefficient pertaining to a given block (approximations, hence smooth aspects, or details, hence sharp aspects) and scale/level (different degree of coarseness) is so information rich allowing to depict a specific shape/texture characteristic located in a specific region when reconstructed in the original domain makes it possible to translate the resolution issue in a feature selection/ isolation issue.

The applications presented furnish a panoramic of several possible fields that can benefit from such approach.

ACKNOWLEDGMENTS

We would like to acknowledge the contributions of Dr. G. Foca and Prof. A. Ulrici to Applications 3.1 concerning both methods and experimental parts. East Balt S.r.l. for sampling and financial support to Applications 3.1 and 3.3. Dr. D. Pavino and Istituto Zooprofilattico Sperimentale del Piemonte, Liguria e Valle d'Aosta are acknowledged for experimental work and financial support concerning Applications 3.2. Regarding Applications 3.3 Prof. R. Bro and F. Van Der Berg are acknowledged for making the employed instrumentation available. Prof. G. Malavasi and G. Lusvardi are acknowledged for experimental work and interpretative support concerning Applications 3.4.

REFERENCES

[1] de Juan A, Tauler R. Multivariate curve resolution (MCR) from 2000: progress in concepts and applications. Crit Rev Anal Chem 2006;36(3–4):163–76.

[2] Ruckebusch C, Blanchet L. Multivariate curve resolution: a review of advanced and tailored applications and challenges. Anal Chim Acta 2013;765:28–36.

[3] Hyvärinen A, Oja E. Independent component analysis: algorithms and applications. Neural Netw 2000;13(4–5):411–30.

[4] Mishra P, et al. Application of independent components analysis with the JADE algorithm and NIR hyperspectral imaging for revealing food adulteration. J Food Eng 2016;168:7–15.

[5] Westad F, Kermit M. 2.14—independent component analysis. In: Brown SD, Tauler R, Walczak B, editors. Comprehensive chemometrics; Oxford (UK): Elsevier; 2009. p. 227–48.

[6] Comon P, Jutten C, editors. Handbook of blind source separation. Oxford (UK): Elsevier Press; 2010.

[7] Moussaoui S, et al. Separation of non-negative mixture of non-negative sources using a Bayesian approach and MCMC sampling. IEEE Trans Signal Process 2006;54(11):4133–45.

[8] Filzmoser P, Gschwandtner M, Todorov V. Review of sparse methods in regression and classification with application to chemometrics. J Chemometr 2012;26(3–4):42–51.

[9] Friedman J, Hastie T, Tibshirani R. Sparse inverse covariance estimation with the graphical lasso. Biostatistics 2008;9(3):432–41.

[10] Saeys Y, Inza I, Larrañaga P. A review of feature selection techniques in bioinformatics. Bioinformatics 2007;23(19):2507–17.

[11] Yu L, Liu H. Feature selection for high-dimensional data: a fast correlation-based filter solution. Proceedings of the Twentieth International Conference on Machine Learning (ICML-2003), Washington DC, 2003. p. 1–8.

[12] Ghasemi JB, Heidari Z, Jabbari A. Toward a continuous wavelet transform-based search method for feature selection for classification of spectroscopic data. Chemom Intell Lab Syst 2013;127:185–94.
[13] Hoang VD. Wavelet-based spectral analysis. TrAC Trends Anal Chem 2014;62:144–53.
[14] Reis MS, Saraiva PM, Bakshi BR. 2.03—Denoising and signal-to-noise ratio enhancement: wavelet transform and Fourier transform. In: Brown SD, Tauler R, Walczak B, editors. Comprehensive chemometrics; Elsevier, Oxford (UK); 2009. p. 25–55.
[15] Walczak B, editor. Wavelets in chemistry. Oxford (UK): Elsevier Press; 2000.
[16] Antonelli A, et al. Automated evaluation of food colour by means of multivariate image analysis coupled to a wavelet-based classification algorithm. Anal Chim Acta 2004;515(1):3–13.
[17] Facco P, Masiero A, Beghi A. Advances on multivariate image analysis for product quality monitoring. J Process Control 2013;23(1):89–98.
[18] Bharati MH, MacGregor JF, Tropper W. Softwood lumber grading through on-line multivariate image analysis techniques. Ind Eng Chem Res 2003;42(21):5345–53.
[19] Boldrini B, et al. Hyperspectral imaging: a review of best practice, performance and pitfalls for in-line and on-line applications. J Near Infrared Spectrosc 2012;20(5):483–508.
[20] Kourti T. Process analytical technology beyond real-time analyzers: the role of multivariate analysis. Crit Rev Anal Chem 2006;36(3–4):257–78.
[21] Elmasry G, et al. Principles and applications of hyperspectral imaging in quality evaluation of agro-food products: a review. Crit Rev Food Sci Nutr 2012;52(11):999–1023.
[22] Liu JJ, MacGregor JF. Estimation and monitoring of product aesthetics: application to manufacturing of "engineered stone" countertops. Mach Vis Appl 2006;16(6):374–83.
[23] Coifman RR, Meyer Y, Wickerhauser MV. Wavelet analysis and signal processing. In: Misiti M, Misiti Y, Oppenheim G, Poggi JM, editors. Wavelets and their applications; London (UK): ISTE Ltd; 1992. p. 153–78.
[24] Daubechies I. Ten lectures on wavelets. Philadelphia (USA): Society for Industrial and Applied Mathematics; 1992.
[25] Fryzlewicz P. Wavelet methods. Wiley Interdiscip Rev Comput Stat 2010;2(6):654–67.
[26] Mallat S, editor. A wavelet tour of signal processing. London, UK: Academic Press; 1998.
[27] Shao X-G, Leung AK-M, Chau F-T. Wavelet: a new trend in chemistry. Acc Chem Res 2003;36(4):276–83.
[28] Strang G, Nguyen T, editors. Wavelets and filter banks. Wellesley, MA: Wellesley-Cambridge Press; 1996.
[29] Coifman RR, Donoho DL. Translation-invariant de-noising. In: Antoniadis A, Oppenheim G, editors. Wavelets and statistics. New York: Springer Verlag; 1995. p. 125–30.
[30] Juneau P-M, Garnier A, Duchesne C. The undecimated wavelet transform–multivariate image analysis (UWT-MIA) for simultaneous extraction of spectral and spatial information. Chemom Intell Lab Syst 2015;142:304–18.
[31] Mallet Y, et al. Classification using adaptive wavelets for feature extraction. IEEE Trans Pattern Anal Mach Intell 1997;19(10):1058–66.
[32] Alsberg BK, et al. Wavelet denoising of infrared spectra. Analyst 1997;122:645–52.
[33] Bos M, Hoogendam E. Wavelet transform for the evaluation of peak intensities in flow-injection analysis. Anal Chim Acta 1992;267(1):73–80.
[34] Mittermayr CR, et al. Wavelet denoising of Gaussian peaks: a comparative study. Chemom Intell Lab Syst 1996;34:187–202.
[35] Liu Y, Brown SD. Wavelet multiscale regression from the perspective of data fusion: new conceptual approaches. Anal Bioanal Chem 2004;380(3):445–52.

[36] Alsberg BK. Parsimonious multiscale classification models. J Chemometr 2000;14(5–6):529–39.
[37] Alsberg BK, et al. Variable selection in wavelet regression models. Anal Chim Acta 1998;368(1–2):29–44.
[38] Cai W, Li Y, Shao X. A variable selection method based on uninformative variable elimination for multivariate calibration of near-infrared spectra. Chemom Intell Lab Syst 2008;90(2):188–94.
[39] Jouan-Rimbaud D, et al. Application of wavelet transform to extract the relevant component from spectral data for multivariate calibration. Anal Chem 1997;69(21):4317–23.
[40] Eriksson L, et al. Orthogonal signal correction, wavelet analysis, and multivariate calibration of complicated process fluorescence data. Anal Chim Acta 2000;420(2):181–95.
[41] Depczynski U, et al. Quantitative analysis of near infrared spectra by wavelet coefficient regression using a genetic algorithm. Chemom Intell Lab Syst 1999;47:179–87.
[42] Van Dijck G, Van Hulle MM. Genetic algorithm for informative basis function selection from the wavelet packet decomposition with application to corrosion identification using acoustic emission. Chemom Intell Lab Syst 2011;107(2):318–32.
[43] Vannucci MT, Shab N, Brown PJ. NIR and mass spectra classification: Bayesian methods for wavelet-based feature selection. Chemom Intell Lab Syst 2005;77:139–48.
[44] Cocchi M, Seeber R, Ulrici A. Multivariate calibration of analytical signals by WILMA (wavelet interface to linear modelling analysis). J Chemometr 2003;17:512–27.
[45] Cocchi M, Seeber R, Ulrici A. WPTER: wavelet packet transform for efficient pattern recognition of signals. Chemom Intell Lab Syst 2001;57(2):97–119.
[46] Ulrici A, et al. Multivariate analysis of analytical signals to decipher relevant chemical information. In: Tassi L, Colombini MP, editors. New trends in analytical, environmental and cultural heritage chemistry. Trivandum, India: Research Signpost; 2008. p. 77–136.
[47] Cocchi M, et al. Application of a wavelet-based algorithm on HS-SPME/GC signals for the classification of balsamic vinegars. Chemom Intell Lab Syst 2004;71(2):129–40.
[48] Cocchi M, Durante C, Marchetti A, et al. Characterization and discrimination of different aged "Aceto Balsamico Tradizionale di Modena" products by head space mass spectrometry and chemometrics. Anal Chim Acta 2007;589(1):96–104.
[49] Foca G, et al. Different feature selection strategies in the wavelet domain applied to NIR-based quality classification models of bread wheat flours. Chemom Intell Lab Syst 2009;99(2):91–100.
[50] Cocchi M, Durante C, Foca G, et al. Efficient variables selection in multivariate analysis of signals by coupling fast wavelet transform and genetic algorithms. In: Proceedings of VI Colloquium Chemiometricum Mediterraneum. Saint-Maximin-la-Sainte-Baume, France; 2007.
[51] Ferrari E, et al. Adulteration of the anthocyanin content of red wines: perspectives for authentication by Fourier transform-near infrared and 1H NMR spectroscopies. Anal Chim Acta 2011;701(2):139–51.
[52] Palacios-Santander JM, et al. Multicomponent analysis in the wavelet domain of highly overlapped electrochemical signals: resolution of quaternary mixtures of chlorophenols using a peg-modified Sonogel–Carbon electrode. Chemom Intell Lab Syst 2008;91(2):110–20.
[53] Cocchi M, et al. Durum wheat adulteration detection by NIR spectroscopy multivariate calibration. Talanta 2006;68(5):1505–11.
[54] Cocchi M, et al. Simultaneous determination of sugars and organic acids in aged vinegars and chemometric data analysis. Talanta 2006;69(5):1166–75.

[55] Leardi R, Seasholtz MB, Pell RJ. Variable selection for multivariate calibration using a genetic algorithm: prediction of additive concentrations in polymer films from Fourier transform-infrared spectral data. Anal Chim Acta 2002;461:189–200.
[56] Misiti M, et al. Wavelet toolboxTM user's guide. Natick, MA: MathWorks; 2008.
[57] Bevilacqua M, et al. Classification and class-modelling. In: Marini F, editor. Chemometrics in food chemistry. Amsterdam: Elsevier Science Ltd; 2013. p. 215–28.
[58] Durante C, Bro R, Cocchi M. A classification tool for N-way array based on SIMCA methodology. Chemom Intell Lab Syst 2011;106(1):73–85.
[59] Wold S. Pattern recognition by means of disjoint principal components models. Pattern Recogn 1976;8(3):127–39.
[60] Fisher R. The use of multiple measurements in taxonomic problems. Ann Eugen 1936;7(2):179–88.
[61] Coifman RR, Wickerhauser MV. Entropy-based algorithms for best basis selection. IEEE Trans Inf Theory 1992;38(2):713–8.
[62] Bharati MH, Liu JJ, MacGregor JF. Image texture analysis: methods and comparisons. Chemom Intell Lab Syst 2004;72(1):57–71.
[63] Hackmack K, et al. Multi-scale classification of disease using structural MRI and wavelet transform. Neuroimage 2012;62(1):48–58.
[64] Liu JJ, MacGregor JF. On the extraction of spectral and spatial information from images. Chemom Intell Lab Syst 2007;85(1):119–30.
[65] Nattkemper TW. Multivariate image analysis in biomedicine. J Biomed Inform 2004;37(5):380–91.
[66] Duchesne C, Liu JJ, MacGregor JF. Multivariate image analysis in the process industries: a review. Chemom Intell Lab Syst 2012;117:116–28.
[67] Prats-Montalbán JM, de Juan A, Ferrer A. Multivariate image analysis: a review with applications. Chemom Intell Lab Syst 2011;107(1):1–23.
[68] Ganesan R, Das TK, Venkataraman V. Wavelet-based multiscale statistical process monitoring: a literature review. IIE Trans 2004;36(9):787–806.
[69] Reis MS, Bauer A. Wavelet texture analysis of on-line acquired images for paper formation assessment and monitoring. Chemom Intell Lab Syst 2009;95(2):129–37.
[70] Li Vigni M, Cocchi M, Prats-Montalban JM, Ferrer-Riquelme A. Coupling 2D-wavelet decomposition and multivariate image analysis (2D WT-MIA); submitted for publication.
[71] Prats-Montalbán JM, Cocchi M, Ferrer A. N-way modeling for wavelet filter determination in multivariate image analysis. J Chemometr 2015;29(6):379–88.
[72] Cocchi M, et al. Determination of carboxylic acids in vinegars and in aceto balsamico tradizionale di modena by HPLC and GC methods. J Agric Food Chem 2002;50(19):5255–61.
[73] Bhargava R, Levin IW. Fourier transform infrared imaging: theory and practice. Anal Chem 2001;73(21):5157–67.
[74] Lewis EN, Schoppelrei JW, Lee E, Kidder LH. NIR chemical imaging as a process analytical tool. In: Bakeev KA, editor. Process Analytical Technology. Hoboken NJ (USA): Blackwell Publishing Ltd; 2005. p. 187–225.
[75] Harman JL, Silva CJ. Bovine spongiform encephalopathy. J Am Vet Med Assoc 2009;234(1):59–72.
[76] Wilesmith JW, et al. Temporal aspects of the epidemic of bovine spongiform encephalopathy in Great Britain: holding-associated risk factors for the disease. Vet Rec 2000;147:319–25.
[77] Cocchi M, et al. Use of multivariate analysis of MIR spectra to study bread staling. Ann Chim 2005;1:1–10.

[78] Mondal A, Datta AK. Bread baking—a review. J Food Eng 2008;86(4):465–74.
[79] Scanlon MG, Zghal MC. Bread properties and crumb structure. Food Res Int 2001;34(10):841–64.
[80] Purlis E, Salvadori VO. Modelling the browning of bread during baking. Food Res Int 2009;42(7):865–70.
[81] Vanin FM, Lucas T, Trystram G. Crust formation and its role during bread baking. Trends Food Sci Technol 2009;20(8):333–43.
[82] Cocchi M, et al. Wavelet-enhanced multivariate image analysis for the food and materials sciences. In: V IASIM Conference in Spectral Imaging. Rome, Italy; 2014.
[83] Li Vigni M *et al.* Multivariate analysis of TEM images for the characterization of innovative nano-materials. In: Atti del XXV Congresso Nazionale della Società Chimica Italiana. Arcavacata di Rende (CS). Italy: University of Calabria; 2014. p. 202.
[84] Chen X, Mao SS. Titanium dioxide nanomaterials: synthesis, properties, modifications and applications. Chem Rev 2007;107(7):2891–959.

Chapter 14

A Smoothness Constraint in Multivariate Curve Resolution-Alternating Least Squares of Spectroscopy Data

S. Hugelier, O. Devos and C. Ruckebusch[1]
Université de Lille, Sciences et Technologies, LASIR CNRS, LILLE, France
[1]*Corresponding author: e-mail: cyril.ruckebusch@univ-lille1.fr*

Chapter Outline

1 INTRODUCTION

Evaluating the smoothness of signals and images is of great interest in a variety of applications and can provide useful signal characteristics to consider in the resolution of process spectroscopy and hyperspectral images data. By smoothness of a signal, an image, or a shape, we mean information of relatively low-frequency character. In contrast, when evaluating the smooth part of a signal (or image), the complementary can be considered to be the rough part and is related to the high-frequency character. The idea to work with the smoothness/roughness of a signal comes from an early work of Eilers and Marx [1], which was later adapted and presented as a way of "unimodal

Data Handling in Science and Technology, Vol. 30. http://dx.doi.org/10.1016/B978-0-444-63638-6.00014-0
Copyright © 2016 Elsevier B.V. All rights reserved.

smoothing" [2]. In the original work of Eilers and Marx [1], penalized B-splines, also called P-splines, were introduced as a flexible way to smooth signals. The flexibility of the method lies in different facts: the B-spline basis can be of different orders (first order, second order, etc.), the number of splines used for the fit is variable (the more splines, the more flexible the fitting can be), and the penalty term on the difference between adjacent spline coefficients can be tuned (the more important the penalty term, the more smooth the signal will be). At that time, data smoothing with splines was very trendy in statistics as a way to reduce noise, but additional advantages of using this method result from the fact that a discrete signal is converted into a continuous representation from which missing data can be interpolated and trends can be estimated [3,4]. Eilers [2] then used this method to conceive the principle of "unimodal smoothing," which allowed unimodality to be extended to multiple peaks of the same signal, as there is no consideration anymore of a single monotone left and right part [5,6], but a smoothly changing signal. Combining this last characteristic with the spatial frequency character of a signal, we can reproduce the smooth and rough parts of a signal. Low-frequency signals indeed change in a gradual (or smooth) way and can thus be separated from high-frequency signals (abrupt changes).

Modern spectroscopic techniques allow investigating complex samples and produce data such as time-resolved spectra and hyperspectral images (further abbreviated by HSI). Now, because we almost always work with spectral mixtures—sometimes of high chemical and physical complexity-separation (unmixing) of the pure components has to be provided afterwards, using dedicated chemometric techniques. Among self-modeling curve resolution methods [7], multivariate curve resolution-alternating least squares (MCR-ALS) [8] is a major player. There is no need to extensively introduce the principles of MCR-ALS in this chapter, as information can be found in Refs. [9–11] and Chapter 2. However, a basic understanding of the main principle of MCR-ALS and its limitations should be considered.

The usefulness of MCR-ALS applied to spectroscopic data lies in the fact that it follows an approximation of the Beer–Lambert law. This translates into a bilinear model of pure signal contributions, as provided in Eq. (1):

$$\mathbf{D} = \mathbf{C}\mathbf{S}^{\mathbf{T}} + \mathbf{E} \tag{1}$$

where the data matrix $\mathbf{D}$ $(m \times n)$ contains the spectra observed in a multicomponent system and the data matrices $\mathbf{C}$ $(m \times k)$ and $\mathbf{S}^{\mathbf{T}}$ $(k \times n)$ are dyads, containing the concentration profiles or spectral profiles of the k different components present in the mixture, respectively. The matrix $\mathbf{E}$ $(m \times n)$ expresses the residuals or error contribution of the model. For HSI data, this model can still be used, as the Beer–Lambert law remains valid for the image pixels. The only difference with process data is that HSI data should be unfolded first, from the original spatial–spectral dimensions to a two-way data matrix [10,12,13]. The problem described in Eq. (1) is an instance of an "inverse problem" and is impossible to resolve uniquely. Indeed, in MCR-ALS, Eq. (1) can be decomposed in a lot of possible solutions with the same quality of fit [14]. This concept is called ambiguity and is the source of the uncertainty in the solutions (see Refs. [8,10] and Chapter 4). The basis of limiting the extent of the ambiguity of the solutions is by adding mathematical conditions that the solutions should fulfill to the optimization process. By using these so-called constraints, it is possible to reduce the number of possible solutions for $\mathbf{C}$ and $\mathbf{S}^{\mathrm{T}}$ in a considerable way, and constraints are in fact essential to obtain meaningful solutions for the problem. The constraints are mostly derived from the physical nature of the system (e.g., only positive values should be expected in the concentration profiles, and, in many spectroscopic applications, also in the spectral profiles) or from prior knowledge available on the system (e.g., a closure constraint on the concentration profiles for a reaction-based system). For process analysis or applications in spectroscopy, a palette of soft, hard, and hard–soft constraints is available [6,9–11,15–18]. Many of them can be, but not exclusively, applied on the concentration profiles $\mathbf{C}$. However, during the analysis of hyperspectral images, the number of applicable constraints for concentration distribution maps is sparse and existing constraints mainly affect individual pixels or groups of adjacent pixels [19,20], due to the unfolding of the hyperspectral matrix cube to a matrix $\mathbf{D}$. Recently, an approach was proposed [12] that allows extending the number of available constraints [21,22], enabling the potential application of any global spatial constraints on the component distribution maps. This encompasses constraints coming from image processing approaches (e.g., segmentation, filtering, deblurring). The smoothness constraint is also among those constraints.

Within the chapter, we focus on this constraint that uses the smooth (or, complementary, rough) character of a signal shape (albeit one-dimensional or two-dimensional) as information that can help in the MCR-ALS analysis of spectral and hyperspectral data. The algorithm that the constraint uses is explained in detail, supported by simulated data sets of low complexity to fully illustrate the concept, followed by showing several case studies of experimental data to highlight the potential and strengths of it.

2 SMOOTHING SIGNALS AND IMAGES

Sometimes, from experience or from theory, one can know certain properties of a signal. Straightforward properties are, for example, non-negativity for concentration profiles or unimodality for some spectral profiles. However, a less evident property is that a signal can be smooth, which is related to the low-frequency character of the signal, as explained in Section 1. This section elaborates more on the P-splines algorithm used to accommodate for this smoothness.

2.1 Smooth vs Rough Information

The motivation behind the development of a smoothness constraint is that the smooth (or rough) character of a signal can be used within the MCR-ALS analysis to help resolve the data matrix into the unmixed contributions of the individual components. To illustrate this in a comprehensive way, an example is shown in Fig. 1. This figure shows in blue (gray in the print version) the (normalized) Fourier transform infrared (FTIR) absorption signal of a two-component mixture within the range of 2500–4000 cm^{-1}. One of the components is related to water, that gives the large OH stretching vibration band covering almost the whole wavenumber range, while the second component is related to the CH stretching vibration that provides the narrow peaks in the range of 2750–3100 cm^{-1}. As shown in Fig. 1, the smooth character of the original signal (OH contributions) can be fitted with P-splines, providing the signal indicated with the black asterisk. What remains can be regarded as rougher signal contributions (CH contributions, indicated by the full black line). By knowing this type of information, it can be applied as a constraint during the MCR-ALS decomposition.

2.2 Smoothing with Splines

We start with the example shown in Fig. 2. The simulated signal is the sum of three Gaussian peaks, on which noise was added. In this example, it should be

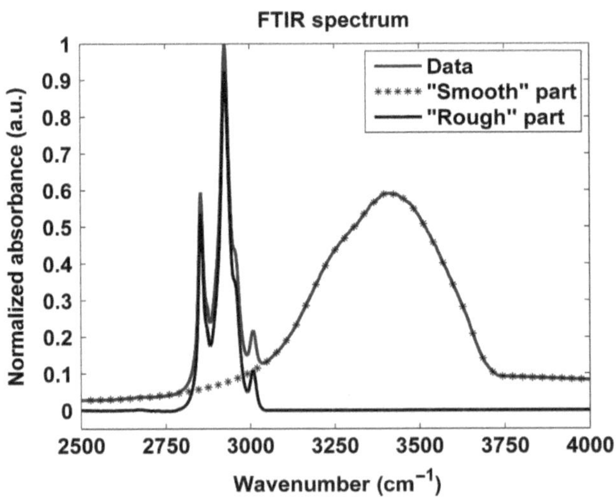

FIG. 1 Normalized FTIR absorption spectrum of a two-component mixture within the range of 2500–4000 cm^{-1}. The data (*blue* (*dark gray* in the print version) *curve*) contain a component with a smooth character (indicated with the *black asterisks*) and a component with a rough character (indicated with a *full black line*), which can be used as information during the MCR-ALS decomposition.

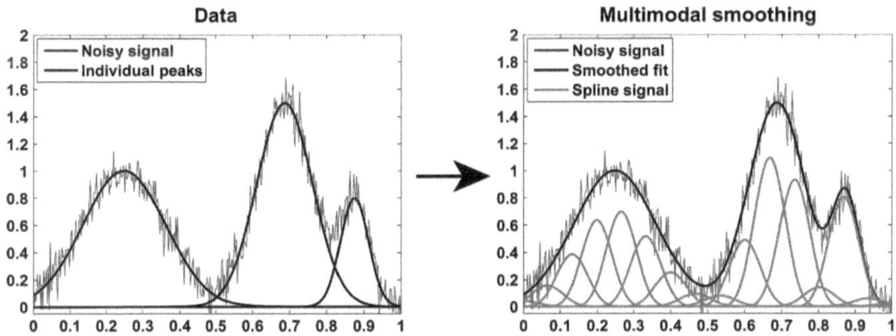

FIG. 2 Using P-splines for multimodal smoothing.

noted that the main variation of the signal associated to the presence of three Gaussian peaks is smooth, but the conventional signal unimodality constraint would not hold. By using P-splines to smooth the signal, "multimodal fitting" under noisy circumstances can be performed. The smoothed fit (black signal in the second panel) is the sum of all the individual spline signals (indicated in red (gray in the print version) in the second panel).

Splines are curves, constructed from joining polynomial "pieces" (also called segments) in a smooth way and the positions on the x-axis where the different segments come together are called "knots". Different orders of splines can be considered (e.g., linear splines are first order, cubic splines are third order, etc.), but cubic splines are usually applied in practice. B-splines can be used to provide a smooth approximation of a signal, which can be done by computing a set of B-splines that would minimize the loss function given in Eq. (2), and thus smoothing can be reduced to linear regression.

$$S = \|\mathbf{y} - \mathbf{B}\boldsymbol{\alpha}\|^2 \tag{2}$$

where $\mathbf{B}$ is the B-spline basis and $\boldsymbol{\alpha}$ the vector of spline coefficients. The explicit solution to this equation is given in Eq. (3):

$$\hat{\boldsymbol{\alpha}} = \left(\mathbf{B}'\mathbf{B}\right)^{-1}\mathbf{B}'\mathbf{y} \tag{3}$$

A possible way to control the amount of smoothing is thus determined by the size of the B-spline basis (and thus also by the number of knots). However, a more elegant way to control the smoothness of the fit can be considered. By adding a penalty to the loss function presented in Eq. (2), a smooth result can also be obtained. The new loss function is then shown in Eq. (4) and its solution in Eq. (5):

$$S = \|\mathbf{y} - \mathbf{B}\boldsymbol{\alpha}\|^2 + \lambda\|\boldsymbol{\Delta}_\mathbf{d}\boldsymbol{\alpha}\|^2 \tag{4}$$

$$\hat{\boldsymbol{\alpha}} = \left(\mathbf{B}'\mathbf{B} + \lambda\boldsymbol{\Delta}_\mathbf{d}'\boldsymbol{\Delta}_\mathbf{d}\right)^{-1}\mathbf{B}'\mathbf{y} \tag{5}$$

in which $\boldsymbol{\Delta}_\mathbf{d}$ is called the difference operator and is of order d [23]. Note that when using a penalized loss function, one would refer to P-splines rather than B-splines. Fig. 3A and B shows the influence of the number of knots used to smooth a signal and it should be clear that when 20 B-splines are used for the smoothing, the fit appears to be "rougher", because the fit is more flexible. On the other hand, when a high penalty parameter λ is used (Fig. 3C) instead of the low penalty in combination with 20 B-splines, this also leads to a smooth result.

An important aspect that has not been taken into account in the loss function presented in Eq. (4) is the fact that the contribution of the signal that is unwanted (and thus needs to be removed) can be either positive or negative with respect to the original signal. Fig. 4 shows two examples in which the rough component to be removed has positive (Fig. 4A) or negative (Fig. 4B) contributions. To accommodate for this and distinguish between the smooth baseline and the signal part of an (unwanted) peak, asymmetry is encoded using a weight matrix $\mathbf{W}$ in the least squares approach. The elements of the matrix $\mathbf{W}$, w_i, can take two values at point i: $w_i = p$ (when positive differences are observed between the smoothed and original signal) or $w_i = 1 - p$ (in the other case). The parameter p is called the asymmetry parameter.

The complete loss function is then summarized in Eq. (6):

$$S = (\mathbf{y} - \mathbf{B}\boldsymbol{\alpha})'\mathbf{W}(\mathbf{y} - \mathbf{B}\boldsymbol{\alpha}) + \lambda\|\boldsymbol{\Delta}\boldsymbol{\alpha}\|^2 \tag{6}$$

For which the solution is

$$\hat{\boldsymbol{\alpha}} = \left(\mathbf{B}'\mathbf{W}\mathbf{B} + \lambda\boldsymbol{\Delta}_\mathbf{d}'\boldsymbol{\Delta}_\mathbf{d}\right)^{-1}\mathbf{B}'\mathbf{W}\mathbf{y} \tag{7}$$

The fitting routine is repeated until convergence, in which the weights are iteratively updated and optimized. This ensures an optimal fit for the smooth signal and maximum removal of the unwanted contributions.

2.3 Smoothing Images with Tensor Products of Splines

The algorithm can be extended to the fitting of images, for which the observation is a matrix $\mathbf{Y}$ ($n \times m$). The rows have coordinates $i = 1, \ldots, n$ and the columns have coordinates $j = 1, \ldots, m$. The surface of interest, represented by $\mathbf{M}$, can be estimated with tensor products of P-splines [24]. This tensor product is formed from two bases of splines: B ($n \times n_x$) as the basis in row direction and $\breve{B}$ ($m \times n_y$) as the basis in column direction. The smooth surface $\mathbf{M}$ can then be written as

$$\mathbf{M} = \mathbf{B}\mathbf{A}\breve{\mathbf{B}}' \tag{8}$$

where $\mathbf{A}$ ($B_1 \times B_2$) represents the matrix of coefficients.

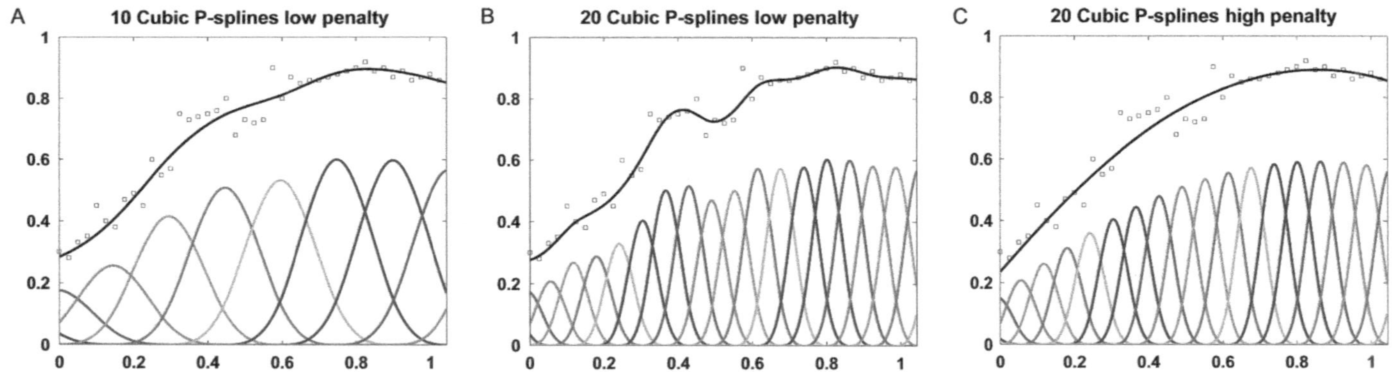

FIG. 3 Scatterplot data fitted by a smooth curve by using (A) 10 basis splines and low penalty, (B) 20 basis splines and low penalty, and (C) 20 basis splines and high penalty. The individual B-splines (with the correct coefficients) are shown, as well as their sum representing the fit (*thick black line*).

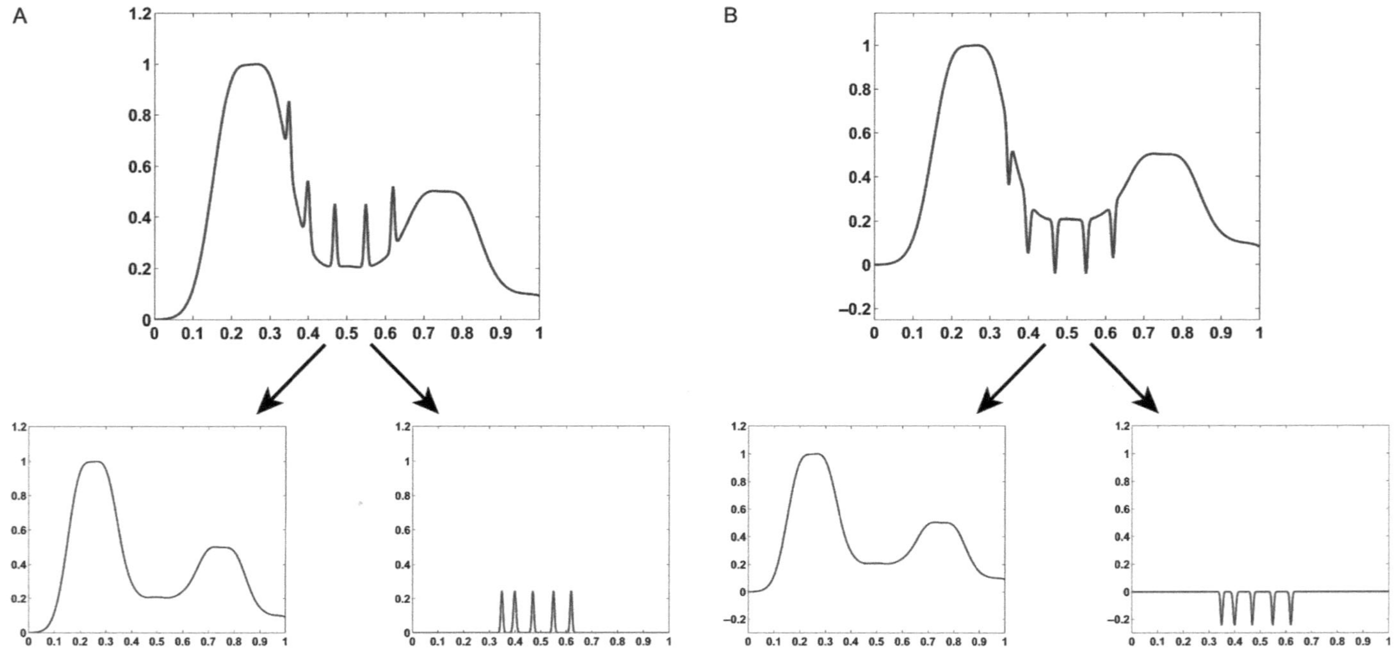

FIG. 4 The removal of unwanted peaks from a signal, composed of a *smooth part* and a *rough part*, which can either be (A) positive or (B) negative with respect to the original signal.

As previously for one-dimensional signals, a penalty term is introduced to the least squares loss function, just as in Eq. (4) to tune the smoothness of the surface:

$$\text{Pen} = \lambda \|\mathbf{\Delta A}\|_{\text{F}} + \breve{\lambda} \|\breve{\mathbf{\Delta}}'\mathbf{A}\|_{\text{F}} \tag{9}$$

with $\mathbf{\Delta}$ and $\breve{\mathbf{\Delta}}$ the differencing operators, $\|.\|_F$ the Frobenius norm, and λ and $\breve{\lambda}$ the smoothing parameters for the row-wise and column-wise penalty, respectively.

Of course, asymmetry should be encoded to this least squares approach too, to remove the unwanted signals. This is done again by using a weight matrix $\mathbf{W}$, of which its elements, w_{ij}, are assigned to the different data pixels ij. These weights w_{ij} can also take two values at a given point ij: $w_{ij}=p$ (when positive differences are observed between the smoothed and original surface) or $w_{ij}=1-p$ (in the other case). Again, the parameter p is called the asymmetry parameter.

2.4 Parameter Settings

Besides the order of the splines, which is usually set to 3 as this gave the best results in general, there are several other parameters to be considered. A first choice that has to be made is the number of splines, n. This value should be kept as low as the data allows, in order to discriminate between the smooth and rough parts of the data and to ensure a maximum smoothness of the fit. For one-dimensional signals, only one value is needed, whereas for images, a value should be provided for each of the two directions. A second choice to be made concerns the penalty term (see Eq. 4 and Eq. 9 for one-dimensional signals and two-dimensional signals, respectively), as the parameters λ (and $\breve{\lambda}$) can be tuned to control the smoothness of the fit. The values are typically between 10^{-8} and 10^{8}. The last parameter that should be provided is the asymmetry parameter, p, which ranges between 0 and 1. To remove positive peaks from the smooth surface, the value must be close to 0 and for negative peaks, it should be close to 1. To accommodate for intermediate situations, with both positive and negative contributions, values in between can be used.

3 SMOOTHNESS CONSTRAINTS IN MCR-ALS

As mentioned in Section 1, the motivation behind the development of a smoothing constraint is that the information can be used within the MCR-ALS analysis to help resolve the data matrix into the unmixed contributions of the individual components. Within the MCR-ALS algorithm, the smooth part would be fitted and can be used to impose a shape constraint to a component profile. Conversely, the complementary part of that fit can be used to force a high-frequency signal. The previous section showed that using

P-splines is a possibility to fit a multimodal signal, providing a useful complement to the unimodality constraint, which can only be used in limited cases (mainly on concentration profiles in a process reaction). It is indeed so that we often deal with data that are not multimodal, but still smooth (e.g., UV–VIS spectra), for which the smoothness constraint can be of use.

3.1 Application to Process Spectroscopy Data

3.1.1 Implementation

The implementation of the smoothness constraint in MCR-ALS is quite straightforward. It is applied directly after the application of the non-negativity constraint within the algorithm and can be used on both concentration profiles and spectral profiles, as shown in Fig. 5. As can be seen from this figure, the constraint is applied at each least squares step to $\mathbf{C}$ and/or $\mathbf{S}^{\mathrm{T}}$ on the constituent profiles. After application of the constraint, the resulting profile is then re-entered in the ALS loop as the constrained concentration profile, $\hat{\mathbf{c}}_{\mathbf{i}}$, or as the constrained spectral profile $\hat{\mathbf{s}}_{\mathbf{i}}^{\mathrm{T}}$. Fig. 5 illustrates the application of the smoothness constraint (by exploiting the low-frequency character of a signal)

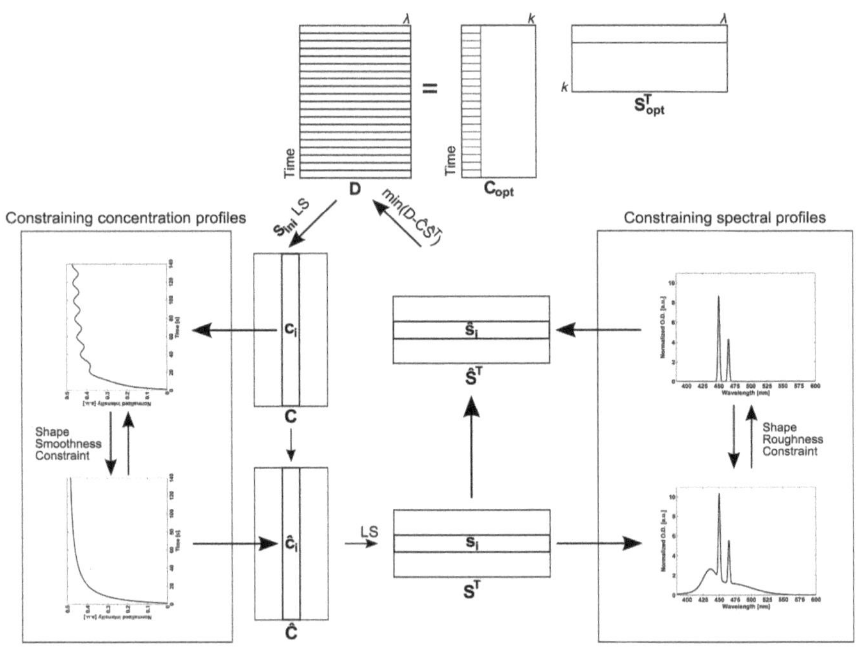

FIG. 5 A schematic of the smoothness constraint applied in the MCR-ALS framework on process data. In this schematic, high-frequency information present in the concentration profile is removed by using the smoothness constraint so that only the low-frequency information is left. Additionally, a priori information of the high-frequency character of the spectral profile is used to eliminate other (low-frequency) contributions from its signal.

on a concentration profile on the one hand and the exploitation of the high-frequency character of a spectral profile to enforce this information on the other hand. Note that the constraint can be applied on individual or multiple components within the same MCR-ALS analysis and can be combined with any other constraints that hold for the data, such as non-negativity or closure.

On top of the conventional settings described in Section 2.3, one should consider every component of the mixture individually during the MCR-ALS analysis, and thus parameters have to be tuned independently. We advise to determine useful parameters in advance, outside the ALS procedure, by testing them on the signals of interest obtained from MCR-ALS. However, finding adequate parameters can be a trial and error procedure, but, when inappropriate parameter settings are used, convergence will not be obtained during the MCR-ALS resolution procedure with the smoothness constraint. We would also like to refer to other publications [1,22,23] for more information about the fitting procedure and parameter estimation.

3.1.2 Simulated Example: Process Data

To show the usefulness of the constraint on process data, a two-component reaction has been simulated resulting in the profiles shown in Fig. 6A (the black curve represents component A, while the blue (dark gray in the print version) curve represents component B). Associated to these concentration profiles are the spectral profiles, which are a simple Gaussian profile for each of the components. Now, adding to the system an "oscillating" component, C (red (gray in the print version) curve in Fig. 6A), for which, obviously, unimodality does not hold, complicates the resolution significantly. Noise was added to the final data. Let us now solve this problem by using MCR-ALS with a non-negativity constraint. The result obtained is shown in Fig. 6B (LOF: 8.87%), and we can remark different things. The concentration profile of component B is mixed with the concentration profile of component C, and none of the spectral profiles seem to correspond to what has been simulated. Less obvious, looking at the concentration profile of component A, we can see that it seems a bit off from what has been simulated (reaction appears to be slower and less complete). Now, if we take a close look at the spectral profiles, you can notice that there are no selective data points for component A, as it overlaps at lower wavelengths with component C and at higher wavelengths with component B. Additionally, there is also a small overlap between the spectral profiles of component B and component C.

To improve the solutions obtained, we can, for example, add unimodality to the non-negativity constraint as we know that all the spectral profiles are Gaussian, and additionally, we can also constrain the concentration profiles of components A and B as they are expected to be unimodal. The results are presented in Fig. 6C (unimodality tolerance: 1.05; LOF: 8.98%), and they appear to be improved, as can be noticed from the spectral profiles and the concentration profile for component B that is now correct.

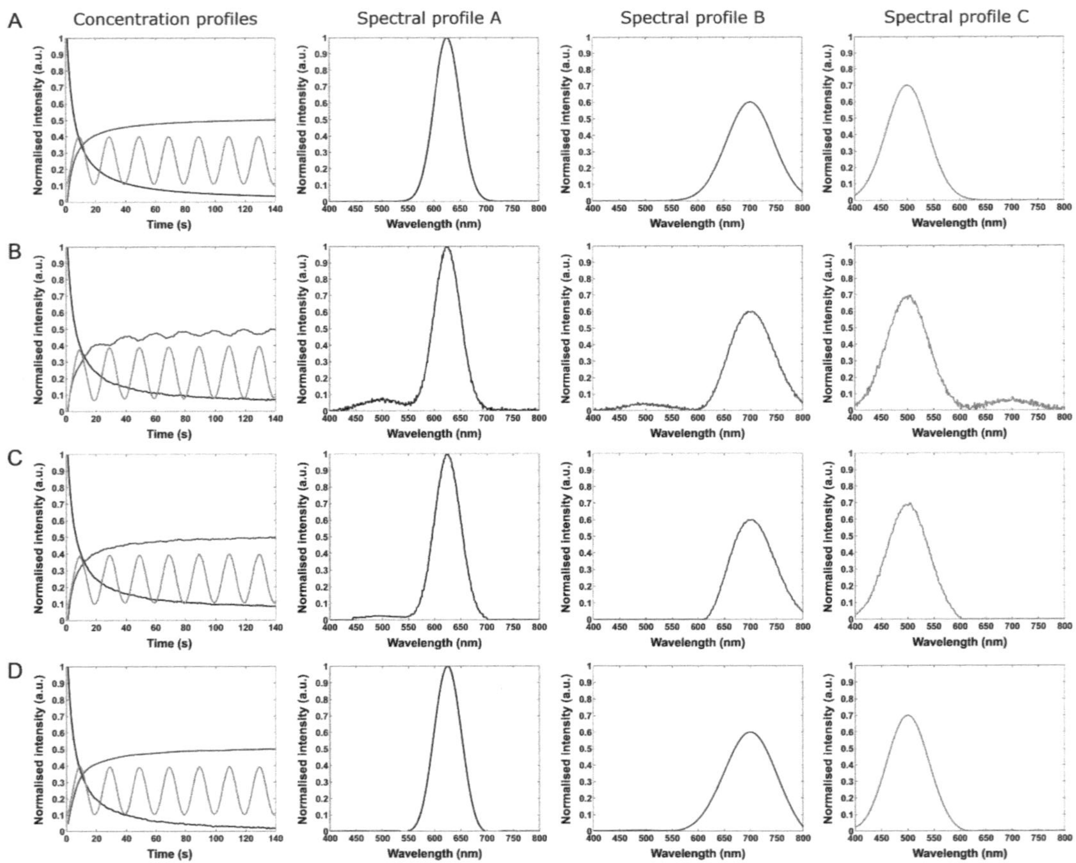

FIG. 6 Concentration profiles and spectral profiles of the simulated process data (three-component system: two reactants and an interfering component). (A) Simulated concentration profiles and spectral profiles (LOF: 9.40%); MCR-ALS results obtained using (B) non-negativity (LOF: 8.87%); (C) non-negativity and unimodality of concentration profile B and spectral profiles (LOF: 8.98%); (D) non-negativity and smoothness of concentration profile B and spectral profiles (LOF: 9.25%).

Unfortunately, we do not solve the system entirely as the concentration profile for component A did not change. In addition, application of unimodality resulted in some mild distortion in the spectral profile of component B, which seems to be cut off in quite a drastic way, making it not a Gaussian profile anymore.

Let us now use the smoothness constraint instead of unimodality. By doing so, we force the low-frequency character of the spectral profiles. Parameters used on the spectral profile of component B were 40, 100, and 0.2 for n, λ, and p, respectively. Additionally, we also applied the constraint on the other spectral profiles (with parameters 40, 1, and 0.1) and on the concentration profile of component B (with parameters 25, 0.1, and 0.5). Doing so allows us to show the viability of the constraint and demonstrate its flexibility as no unimodality is needed anymore. The results can then be seen in Fig. 6D (LOF: 9.25%) and correspond much better to the simulation. By using a relatively high penalty parameter for the spectral profile of component B, the profile is forced to change smoothly. Additionally, this example is perfect to illustrate the removal of high-frequency contributions from a profile where only low-frequency contributions are expected (concentration profile of component B). By using a low number of splines and putting the asymmetry factor at 0.5, the P-spline fit of the profile could never accommodate for the high-frequency fluctuations present in this profile and therefore assigned them to another component (component C).

By using the smoothness constraint in this apparent simple example, we managed to solve it while all other soft-modeling constraints failed. We showed the flexibility of the constraint by using it on a concentration profile to eliminate odd contributions with a high frequency and by using it on spectral profiles to smoothly change their evolution along the spectral direction.

3.2 Constraining Smoothness in HSI Data

3.2.1 Implementation

Now, dealing with HSI data is slightly different than dealing with process data as the data come under the form of a data cube. However, if the data cube is unfolded to a data matrix **D**, MCR-ALS can still be used as the Beer–Lambert law holds for every pixel in the image. As a result, the spatial–spectral data cube is transformed into a data matrix in which every line represents a pixel, as shown in Fig. 7. When the MCR-ALS solutions are found, the matrix $\mathbf{C_{opt}}$ is then refolded in the same way as it was unfolded to obtain the component distribution maps for each component. However, having to do this unfolding step to obtain the matrix **D** of course has consequences for the spatial structure present in our image, as seen in Fig. 7. The pixels indicated in gray are neighbors from one another in the data cube, but this is no longer the case when the data is unfolded. So, in order to apply a global spatial constraint

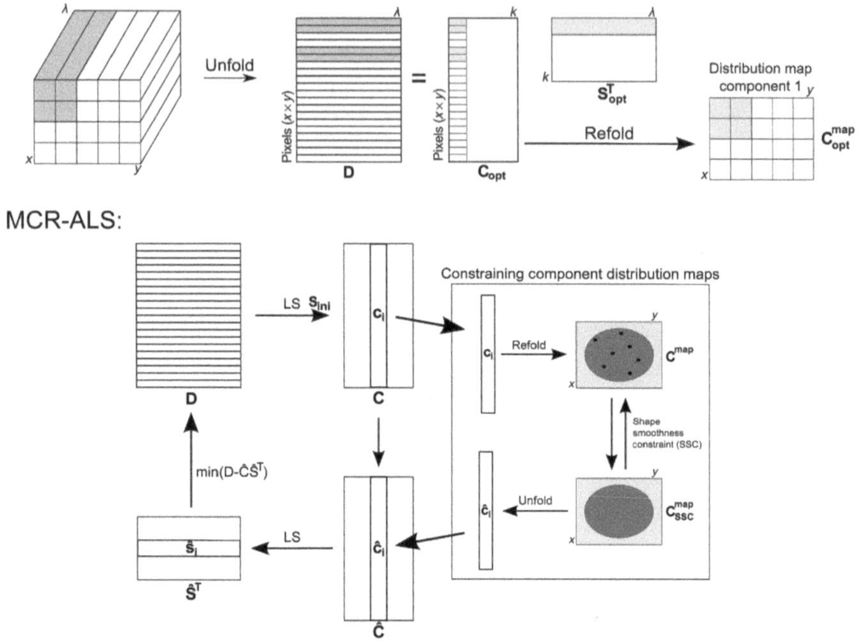

FIG. 7 A schematic representation of constraining smoothness for hyperspectral imaging (HSI) data. A HSI data cube is unfolded into the two-way matrix **D** (top) which is then iteratively optimized in the MCR-ALS framework (bottom). It shows that artifacts with a high-frequency character can be removed from the component distribution map constraining smoothness. These odd contributions are then assigned to other component (or the residuals).

on a component distribution maps (e.g., smoothness/roughness of the image shape characteristics), this spatial information has to be recovered first. This is done by refolding the individual concentration profile $\mathbf{c_i}$ at each least squares step, as indicated by $\mathbf{C^{map}}$ in Fig. 7. After application of the constraint, and removal of the unwanted information, $\mathbf{C^{map}_{SSC}}$ is unfolded again and re-entered into the ALS loop as before. The figure shows the application of the shape smoothness constraint to remove high-frequency information present in the distribution map, but, most obviously, the removal of low-frequency information can be done in a similar way (see example in Section 3.2.2). Note that this figure does not show any constraining of the spectral profiles, but this is of course done in the same way as discussed in Section 3.1.1.

The parameters can be determined in a similar way as described in Section 3.1.1, but the number of splines used in every direction for a single component (n_x and n_y) does not necessarily have to be the same, which might be useful for HSI data of which the two spatial directions are not equal in size. Additionally, in the same way, the penalty parameters (λ and $\check{\lambda}$) can be tuned separately too for each spatial direction.

3.2.2 Simulated Example: Hyperspectral Imaging Data

The simulated data correspond to a three-component system with different spectral and spatial object features. Two low-frequency spatial signals (Gaussian shape) are present, which are both centered on the image, whereas a third component corresponds to higher frequency signals. For this component, the objects are small and scattered across the image. A visual three-dimensional representation of the sum of the spatial contributions of the three components is given in Fig. 8, and the corresponding individual distribution maps are shown in Fig. 9A. The spectral profiles of the different components consist of many Gaussian peaks of different amplitudes and widths, and there is strong overlap of the peaks between the different components. Homoscedastic noise, corresponding to 1.5% of the maximum amplitude of the data, was added to the data, which corresponds to a lack of fit of 12.60%.

In a first step, MCR-ALS analysis is performed with only the application of non-negativity. As can be seen from the results in Fig. 9B (LOF: 12.56%), the problem cannot be resolved. It can simply be explained because the criteria of Manne's theory [25], for the resolution of chemical mixtures, are not met. Indeed, regions of rank 1–3 can be observed in the image: the rank is 1 at pixel positions where only the very low-frequency component (component 3) is present, it is rank 2 at every pixel positions where the high-frequency component (component 2) is absent, and rank 3 at all other positions. This means that the pure spectrum of component 1 can be readily obtained from these pixels (theorem 2). The two other spectral profiles cannot be obtained as there is no pixel position where they appear selectively.

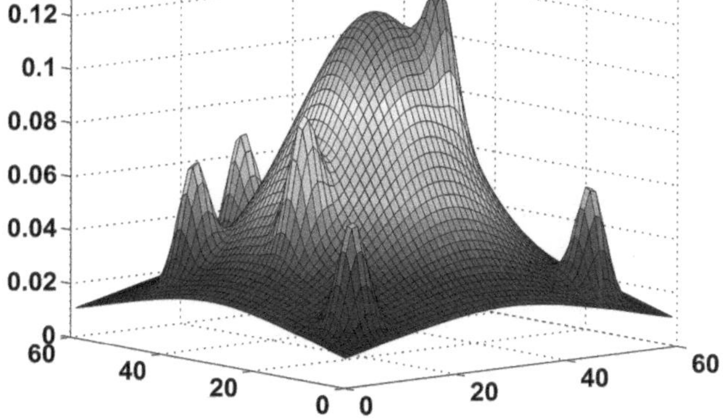

FIG. 8 Three-dimensional representation of the sum of the distribution maps of the three-component system. A very low-frequency background, a low-frequency component, and a high-frequency component can be observed.

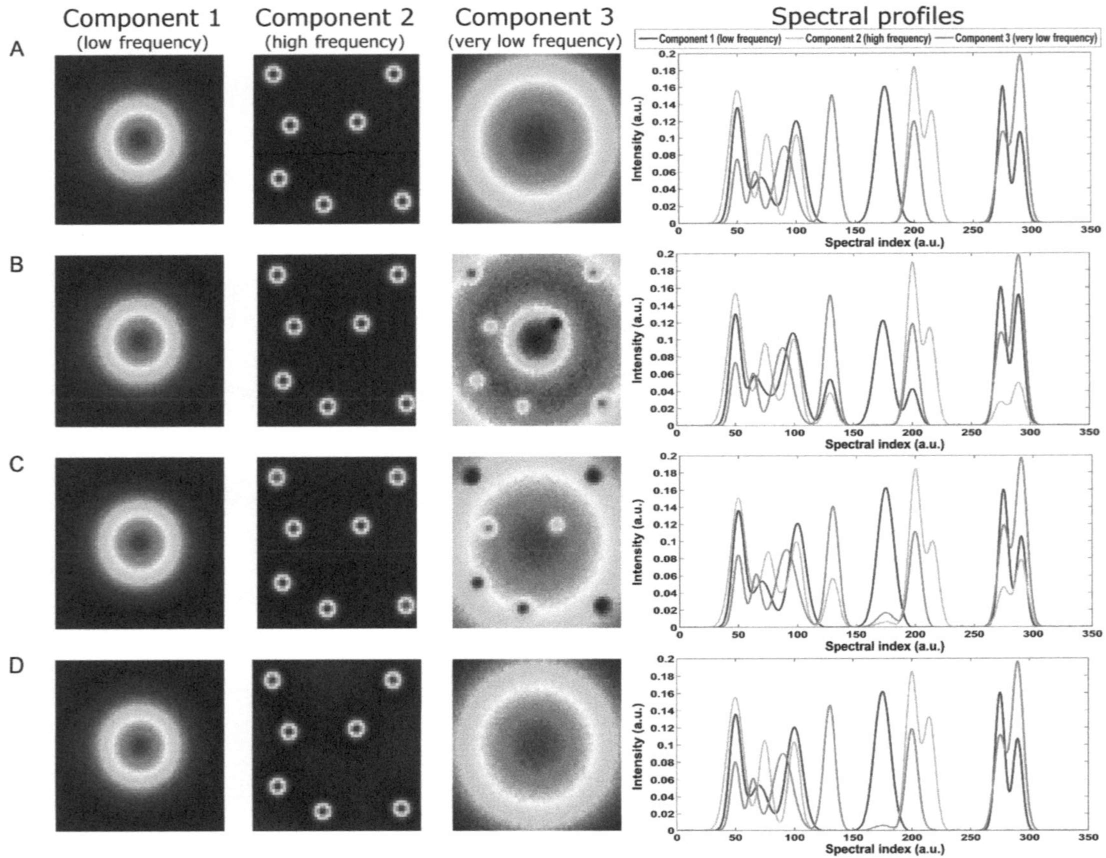

FIG. 9 Distribution maps (intensity coded from *blue* (*dark gray* in the print version) to *red* (*gray* in the print version)) and spectral profiles for the simulated data of the three-component system. (A) Simulated distribution maps and spectral profiles; MCR-ALS results obtained using (B) non-negativity (LOF: 12.56%); (C) non-negativity and smoothness of component 1 (LOF: 12.86%); (D) non-negativity and roughness of component 2 (LOF: 12.72%).

Additionally, due to the fact that the other components do not appear outside the concentration windows of the third component, the concentration profile of this component cannot be obtained (theorem 1). Now, getting back to the results shows that the distribution map of component 3 is not having the simulated bell-shaped curve. Severe component mixing can be observed for this distribution map, resulting in a lower intensity in the center of the map (i.e., retention of component 3 in the distribution map of component 1) as well as several lower intensity spots (i.e., retention of component 3 in the distribution map of component 2). Using the local rank constraint will slightly improve the results obtained in Fig. 9B, but for the same reasons, it cannot ensure the recovery of all the pure compound profiles (results not shown).

This example depicts a typical situation where additional spatial information about some global features of the shape of the object(s) in the image could help improving resolution. MCR-ALS analysis was thus carried out with the smoothness constraint applied to the distribution map of component 1 in a first occasion. The parameters used were the following: 15, 0.01, and 0.01 for $n_{x,y}$, λ and $\check{\lambda}$, and p, respectively. By applying the smoothness constraint on the distribution map of component 1, this component and component 3 are better separated, and the bell-shaped curve of component 3 was restored. However, the resolution was still not complete as the distribution between component 2 and component 3 was not completely correct. It should be clear from Fig. 9C (LOF: 12.86%) that several lower intensity spots are still present in the component distribution map of component 3, making resolution incomplete, also confirmed in the spectral profiles of these components. For this reason, the rough characteristics of the component distribution map of component 2 were constrained and added to the previous constraints. The parameters used here are 15, 10, and 0.1 for $n_{x,y}$, λ and $\check{\lambda}$, and p, respectively. A note that has to be made here is that even though the rough character of the component distribution map is constrained, the algorithm smooths the entire component distribution map and the residuals are then used for further ALS calculations. This is the reason why the value for the penalty parameters is relatively high, as it is desirable to remove as much low-frequency character of the object as possible, so that only the high-frequency character of the object remains. The results are shown in Fig. 9D (LOF: 12.72%). They reveal that the distribution maps of the different components could all be recovered successfully, apart from some noise, which can be solved by also using the smoothness constraint on the component distribution map of the third component (results not shown). Additionally, the spectral profiles of the different components correspond almost perfectly to the real simulated profiles, and thus it was shown that by using the smoothness constraint, a better resolution of hyperspectral data set could be achieved.

4 CASE STUDIES

4.1 Ultrafast Time-Resolved Absorption Spectroscopy of Salicylidene Aniline

Pump-probe transient absorption spectroscopy allows investigating ultrafast photochemical processes (at femtosecond–picosecond timescale) (see Refs. [17,26] and Chapter 11). The data, $\mathbf{X_1}$, used in this example are shown in Fig. 10A. Spectral artifacts can affect this type of data, which translates into the observation of narrow "negative" peaks in the range of 400–450 nm. To correct for these artifacts (related to stimulated Raman amplification, SRA), a second measurement on the pure solvent, $\mathbf{X_2}$, was performed to observe them independently (Fig. 10B). However, they are not perfectly reproducible from solvent to solution measurement.

The MCR-ALS analysis was applied on a multiset data [$\mathbf{X_1}$; $\mathbf{X_2}$]. Initial estimates were obtained by using a method of pure variable detection [27] and seven components were considered (four true components and three artifact components). During the resolution with MCR-ALS, the non-negativity constraint was only applied on the concentration profiles of the components in $\mathbf{X_1}$, while no non-negativity was used on data set $\mathbf{X_2}$. Unimodality was also applied, but only on the components present in the reaction, and not on the artifacts, in both $\mathbf{X_1}$ and $\mathbf{X_2}$. The results of the MCR-ALS analysis are then presented in Fig. 11A (LOF: 2.85%). The spectral profiles of species B and species C are both associated to excited forms of salicylidene aniline (SA) [26]. However, they are, despite the multiset approach, affected by some spectral features attributed to artifacts. Species A may correspond to the absorption of the excited enol form of SA [17,26] but no information about that species is known. What is important for this species is its concentration profile: it is mixed with the SRA signal artifacts due to convolution effects, which means that complete resolution of this profile is difficult.

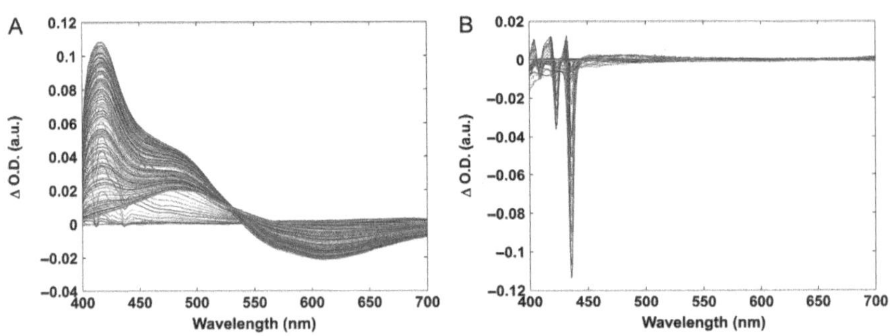

FIG. 10 Data of salicylidene aniline in acetonitrile, acquired by femtosecond transient absorption spectroscopy. (A) The sample data for a timescale of −0.2 to 50 ps and (B) the solvent data (acetonitrile) for a timescale of −0.2 to 0.75 ps. See Ref. [17] for further details.

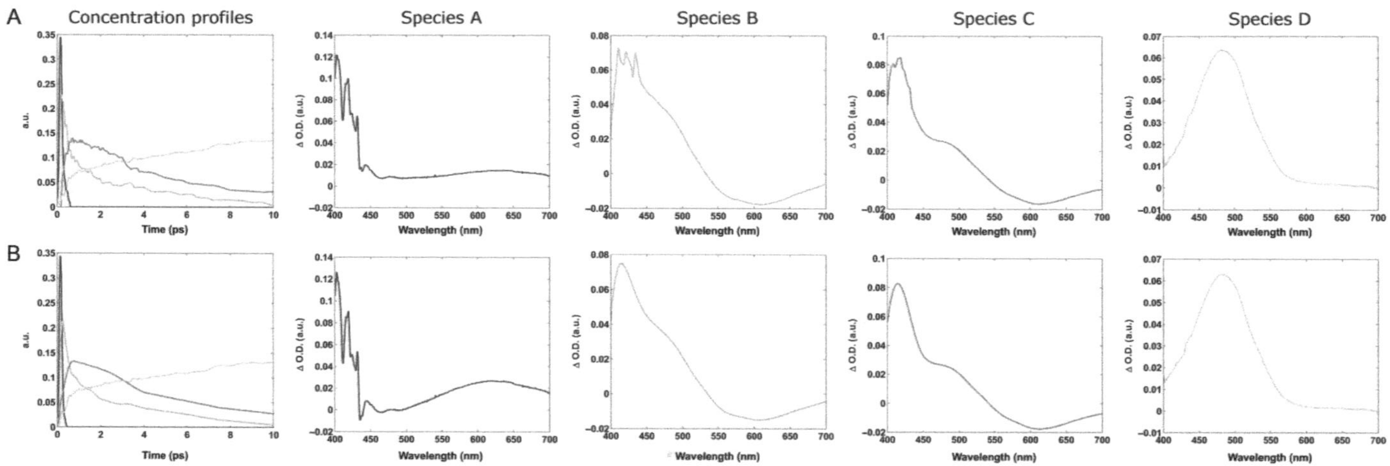

FIG. 11 Comparison of the concentration profiles and spectral profiles of the components of the experimental transient absorption spectrokinetic data. (A) Augmented data set resolution with non-negativity and unimodality (LOF: 2.85%) and (B) augmented data set resolution with non-negativity, unimodality, and smoothness (LOF: 3.37%).

Additional information is required for full resolution of the data. The smoothness constraint could be added to the ones previously listed. It was applied on the spectral profile of species B to ensure a smooth shape of the signal by eliminating the artifacts. The same has been done for species C. Applied parameters were 45, 1, and 0.7 for n_B, λ_B, and p_B, respectively, and 45, 1, and 0.7 for n_C, λ_C, and p_C, respectively. Additionally, the constraint was also used on the concentration profiles of these same components with parameters 100, 1, and 0.2 for component B and 45, 1, and 0.5 for component C. The results are shown in Fig. 11B (LOF: 3.37%). One can notice the differences with the previous case as there are no artifacts present anymore in the spectral profiles of species B and species C, whereas the spectral profiles for species A and species D did not change significantly. Additionally, the concentration profile of species A changed in an important matter, as the decay of this species is now faster, which could be interpreted as a better unmixing. The other concentration profiles did not change a lot, apart from their smoothness. The reason why the constraint was also used on the concentration profiles was because even though the results were significantly improved by only applying the constraint on the spectral profiles, the lack of fit and reproduction of the original data set were even better when the constraint was also used on the concentration profiles.

This transient absorption spectroscopy example shows that the smoothness constraint can be used on complex real data as a way of enforcing properties about spectral shapes. It is also important to mention that in the example taken no other constraint (apart from non-negativity on some of the components) can be used. An additional advantage of the constraint is that it is a soft constraint, and thus small deviations from this constraint are allowed [6], whereas this is not possible for hard–soft constraints.

4.2 Hyperspectral Imaging of an Oil-in-Vinegar Emulsion

This hyperspectral data set was acquired by using FTIR spectroscopy. It represents a sample that does not contain highly structured patterns and is used to illustrate that the smoothness constraint can also be used on practical HSI data. The FTIR spectra of a mayonnaise sample—taken as a representative case study of an emulsion—were collected over a frequency range of 1100–3850 cm^{-1} at room temperature (24°C) by using a Bruker Vertex 70 FTIR spectrometer, equipped with a Bruker Hyperion 3000 IR microscope. The instrument uses an extended focal plane array detector of mercury cadmium telluride type and a KBr beam splitter. The spectra were recorded with a resolution of 4 cm^{-1} using 512 scans, with a 36× objective. The acquired image is 64 pixels by 64 pixels. The mean image and the corresponding spectra (for the purpose of analysis, the spectral range was limited between 2950 and 3850 cm^{-1}) can be found in Fig. 12.

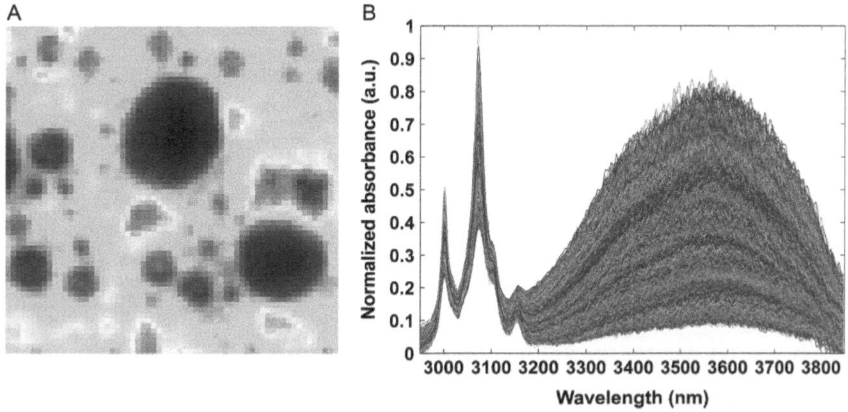

FIG. 12 FTIR HSI of mayonnaise (64 pixels by 64 pixels). In (A) the mean image of the sample (intensity coded from *blue* (*dark gray* in the print version) to *red* (*gray* in the print version)) and in (B) the spectra obtained for every pixel for a range of 2950–3850 cm^{-1}.

Initial data investigation revealed the presence of two main components. Mayonnaise is indeed classified as a permanent emulsion of an oily phase and a water phase, which is created when a shear force is applied so that the oily phase is dispersed in the water phase as droplets that remain throughout this water phase [28,29]. However, when looking at this mean image in Fig. 12A, this is not quite clear. The mean image shows everywhere a mixture of the two major components, including in low-intensity regions (a wrong impression could be gotten by the observation of the map, as the total intensity over the OH spectral region is much higher than the one for the CH contributions).

MCR-ALS analysis can be used to distinguish the different components from one another. In a first step, the analysis is done with non-negativity for all profiles as a constraint (Fig. 13A; LOF: 5.64%). The spectral profiles show that a first component can be assigned to an aqueous phase and the second component is assigned to an oily phase. It is clear that everywhere the aqueous phase is absent, a big contribution can be found for the oily phase. However, the spectral profile for the aqueous component reveals a contribution for the CH vibrations around 3000–3100 cm^{-1}, translating that the two components are not completely unmixed.

To improve resolution, the smoothness constraint was used on the component distribution map of the oily phase. The constraint is applied using following parameters: 35, 0.001, and 0.2 for $n_{x,y}$, λ and $\check{\lambda}$, and p, respectively. Results are shown in Fig. 13B (LOF: 5.65%). As this spatial constraint is applied, one can notice the similarity with the results previously shown. However, the change in the component distribution map is, as expected, more

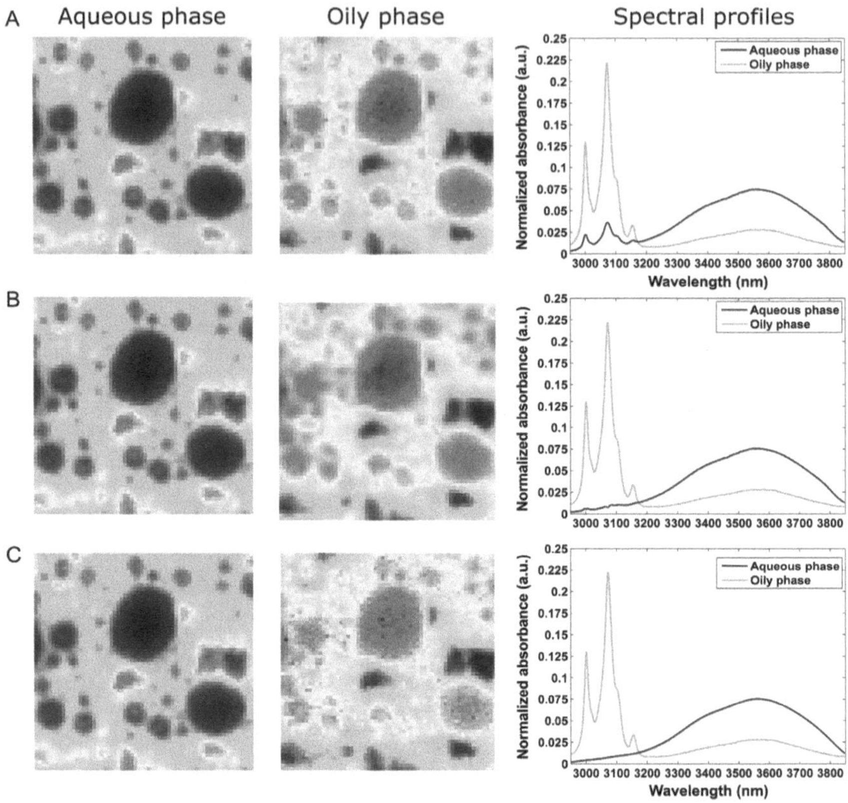

FIG. 13 Distribution maps (intensity coded from *blue* (*dark gray* in the print version) to *red* (*gray* in the print version)) and spectral profiles for experimental data of mayonnaise, obtained by FTIR imaging for a spectral range of 2950–3850 cm^{-1}. MCR-ALS results obtained using (A) non-negativity on all profiles (LOF: 5.64%); (B) non-negativity and smoothness on the component distribution map of the oily phase (LOF: 5.65%); and (C) non-negativity and smoothness on the spectral profile of the aqueous phase (LOF: 5.66%).

gradually compared to the previous results. This smoother profile lead to a change of the spectral profile of the aqueous phase, in which the CH contribution at 3000–3100 cm^{-1} has been removed.

Let us now show the flexibility of the smoothness constraint by comparing the results obtained applying the smoothness constraint on the spectrum of the water phase to the previous ones where the constraint was used on the component distribution map of the oily phase. The parameters used were 35, 1, and 0.4 for n, λ, and p, respectively, and the results are shown in Fig. 13C (LOF: 5.66%). As can be seen here, the CH contributions are removed and the spectral profile of the aqueous component appears to be quite smooth. Looking at the concentration distribution maps, not much difference can be seen with the two other analyses, apart from some noisy appearance of the component distribution map of the oily phase.

This case study has shown that the application of the smoothness constraint, on both a component distribution map and spectral profile of experimental HSI data. Further investigations are required, but we demonstrate a feasible way to enforce a certain smoothness of global spatial information onto this data and helps unraveling components in HSI data.

5 CONCLUDING REMARKS

In this book chapter, we elucidate on a new global constraint, for use in MCR-ALS, which uses signal characteristics to enforce smoothness onto that signal. By using a combination of parameters, associated to the use of P-splines, the main pattern (whether it should be smooth or rough) can be separated from the unwanted signal, by assigning the odd contributions to other components or to the residuals. The constraint can be applied on both one-dimensional, i.e., concentration profiles and spectra, and two-dimensional signals, i.e., chemical images, to aid in the unraveling of the individual components in a mixture, in a highly flexible way.

We have illustrated on seemingly easy simulated systems of process data and HSI data that the constraint is viable and can work complementary to conventional constraints, providing solutions in situations where conventional constraints fail. Furthermore, we successfully demonstrated its potential on two real case studies dealing with time-resolved spectroscopy data and hyperspectral images, respectively.

ACKNOWLEDGMENTS

We would like to thank Isabelle De Waele (LASIR, Université de Lille—Sciences et Technologies) and Patrizia Firmani (Università di Roma La Sapienza) for their assistance in acquisition of the hyperspectral imaging data.

REFERENCES

[1] Eilers PHC, Marx BD. Flexible smoothing with B-splines and penalties. Stat Sci 1996;11:89–121.
[2] Eilers PHC. Unimodal smoothing. J Chemometr 2005;19:317–28.
[3] Silverman BW. Some aspects of the spline smoothing approach to nonparametric regression curve fitting (with discussion). J R Stat Soc Ser B 1985;47:1–52.
[4] Wand MP, Jones MC. Kernel smoothing. London: Chapman and Hall; 1993. 224 p.
[5] Bro R, Sidiropoulos ND. Least squares algorithms under unimodality and non-negativity constraints. J Chemometr 1998;12:223–47.
[6] Gemperline PJ, Cash E. Advantages of soft versus hard constraints in self-modeling curve resolution problems. Alternating least squares with penalty functions. Anal Chem 2003;75:4236–43.
[7] Lawton WE, Sylvestre EA. Self-modeling curve resolution. Technometrics 1971;13:617–33.
[8] Tauler R, Smilde A, Kowalski B. Selectivity, local rank, three-way data analysis and ambiguity in multivariate curve resolution. J Chemometr 1995;9:31–58.

[9] De Juan A, Tauler R. Multivariate curve resolution (MCR) from 2000: progress in concepts and applications. Crit Rev Anal Chem 2006;36:163–76.
[10] De Juan A, Jaumot J, Tauler R. Multivariate curve resolution (MCR). Solving the mixture analysis problem. Anal Methods 2014;6:4964–76.
[11] Ruckebusch C, Blanchet L. Multivariate curve resolution: a review of advanced and tailored applications and challenges. Anal Chim Acta 2013;765:28–36.
[12] Hugelier S, Devos O, Ruckebusch C. On the implementation of spatial constraints in multivariate curve resolution alternating least squares for hyperspectral image analysis. J Chemometr 2015;29:557–61.
[13] Prats-Montalbán JM, de Juan A, Ferrer A. Multivariate image analysis: a review with application. Chemometr Intell Lab Syst 2011;107:1–23.
[14] Tauler R, Maeder M. Two-way data analysis: multivariate curve resolution—error in curve resolution. In: Brown S, Tauler R, Walczak B, editors. Comprehensive chemometrics, vol. 2. Oxford: Elsevier; 2009. p. 345–63.
[15] Blanchet L, Ruckebusch R, Huvenne J-P, de Juan A. Hybrid hard- and soft-modelling applied to difference spectra. Chemometr Intell Lab Syst 2007;89:26–35.
[16] De Juan A, Maeder M, Martinez M, Tauler R. Combining hard- and soft-modelling to solve kinetic problems. Chemometr Intell Lab Syst 2000;54:123–41.
[17] Mouton N, de Juan A, Sliwa M, Ruckebusch R. Hybrid hard- and softmodeling approach for the resolution of convoluted femtosecond spectrokinetic data. Chemometr Intell Lab Syst 2011;105:74–82.
[18] Tauler R, Maeder M, de Juan A. Two-way extended curve resolution. In: Brown S, Tauler R, Walczak B, editors. Comprehensive chemometrics, vol. 2. Oxford: Elsevier; 2009. p. 473–505.
[19] De Juan A, Maeder M, Hancewicz T, Tauler R. Use of local rank-based spatial information for resolution of spectroscopic images. J Chemometr 2008;22:291–8.
[20] Hancewicz TM, Wang JH. Discriminant image resolution: a novel multivariate image analysis method utilizing a spatial classification constraint in addition to bilinear non-negativity. Chemometr Intell Lab Syst 2005;77:18–31.
[21] Aram P, Shen L, Pugh JA, Vaidyanathan S, Kadirkamanathan V. An efficient TOF-SIMS image analysis with spatial correlation and alternating non-negativity-constrained least squares. Bioinformatics 2015;31:753–60.
[22] Hugelier S, Devos O, Ruckebusch C. Constraining shape smoothness in multivariate curve resolution–alternating least squares. J Chemometr 2015;29:448–56.
[23] De Rooi JJ, Eilers PHC. Mixture models for baseline estimation. Chemometr Intell Lab Syst 2012;117:56–60.
[24] Eilers PHC, Currie IB, Durbán M. Fast and compact smoothing on large multidimensional grids. Comput Stat Data Anal 2006;50:61–76.
[25] Manne R. On the resolution problem in hyphenated chromatography. Chemometr Intell Lab Syst 1995;27:89–94.
[26] Sliwa M, Mouton N, Ruckebusch C, Poisson L, Idrissi A, Aloise S, et al. Investigation of ultrafast photoinduced processes for salicylidene aniline in solution and gas phase: toward a general photo-dynamical scheme. Photochem Photobiol Sci 2010;9:661–9.
[27] Windig W, Guilment J. Interactive self-modeling mixture analysis. Anal Chem 1991;63:1425–32.
[28] Chippie AL, Jamieson PR, Golt CM, Hsu C-H, Lo YM. Quantitative analysis of fat and moisture in mayonnaise using Fourier transform infrared spectrometer. Int J Food Prop 2002;5:655–65.
[29] van Nieuwenhuyzen W, Szuhaj BF. Effects of lecithins and proteins on the stability of emulsions. Fett/Lipid 1998;100:282–91.

Chapter 15

Super-Resolution in Vibrational Spectroscopy: From Multiple Low-Resolution Images to High-Resolution Images

M. Offroy and L. Duponchel[1]

LASIR CNRS UMR 8516, Université Lille 1, Sciences et Technologies, Villeneuve d'Ascq Cedex, France

[1]*Corresponding author: e-mail: ludovic.duponchel@univ-lille1.fr*

Chapter Outline

Data Handling in Science and Technology, Vol. 30. http://dx.doi.org/10.1016/B978-0-444-63638-6.00015-2
Copyright © 2016 Elsevier B.V. All rights reserved.

1 INTRODUCTION

Nowadays, a growing interest in nanoscience is observed in many research fields like physics, chemistry, and biology. Spectroscopic instrumentation must be therefore efficient and request improvements to address the submicrometric analysis challenges. Vibrational spectroscopy is a valuable tool to analyze complex sample and provides significant molecular information. Beyond macroscopic characterizations, the coupling of spectrometers with microscopes makes possible map generation that represents the spatial distribution of chemical components from samples. Considering conventional spectroscopic imaging, the microscope first focuses photons on specific area of the sample we want to analyze and second collects nonabsorbed or scattered photons. A motorized stage is placed on the microscope to enable movement in the three directions of the space of the sample surface. Hyperspectral acquisition system can be thus categorized into two groups [1,2] (Fig. 1). The first group is composed of global imaging systems. It corresponds to a global illumination of the sample area. Spectroscopic images are acquired by a multichannel detector. The second group is mapping imaging system. It scans subareas of the sample point by point or line by line. A single channel

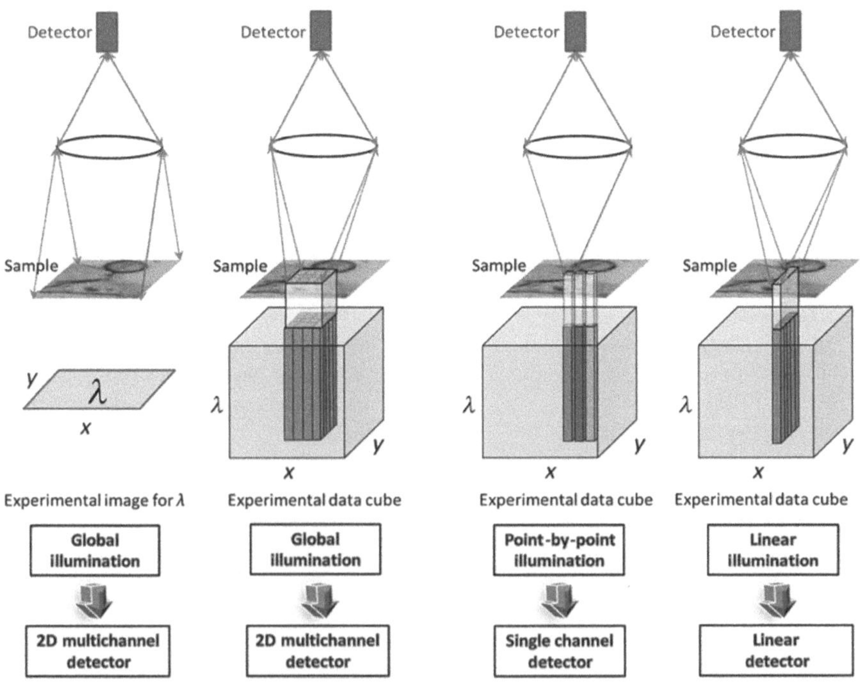

FIG. 1 Spectral imaging acquisition systems.

detector or a linear one is used for collecting the data. A motorized stage is used to move the sample.

In general, the most attractive methods are the ones generating a spectral data cube because monochromatic images are often biased due to the lack of wavelength selectivity. Beyond that, a point-by-point acquisition gives the best spatial resolution, but it is in general very slow. On the contrary, a 2D multichannel detector acquires the data cube in one shot, but the spatial resolution is often lower. All instruments later used in this chapter will generate an experimental data cube of size $x \times y \times \lambda$, where x and y are the coordinates of the microzone in the sample plane and λ is the spectral variable (Fig. 2). Hyperspectral imaging purpose is then to generate a chemical image from the acquired data cube. If a specific wavelength can be found for the compound of interest, a signal integration procedure is used to generate the corresponding chemical map, i.e., tacking a slice from the data cube. Otherwise chemometric tools such as MCR-ALS (see Chapter 2) have to be used when wavelength specificity is not observed. Significant spectral and structural molecular information is deduced which gives the opportunity to understand complex chemical samples, systems, and processes.

Even if hyperspectral images are powerful analytical measurements, far-field spectroscopic imaging instruments are not always adapted to the micrometric or submicrometric sample size. Indeed, the spatial resolution limit is first and foremost dictated by the photon wavelength due to diffraction limit. Considering the spectral domain of vibrational spectroscopy (i.e., 2.5–25 μm), spatial resolution becomes a real constrain when micron- or submicron-sized samples are analyzed.

Two approaches have emerged to push this resolution limit back. The first approach focuses on instrumental development with the so-called near-field

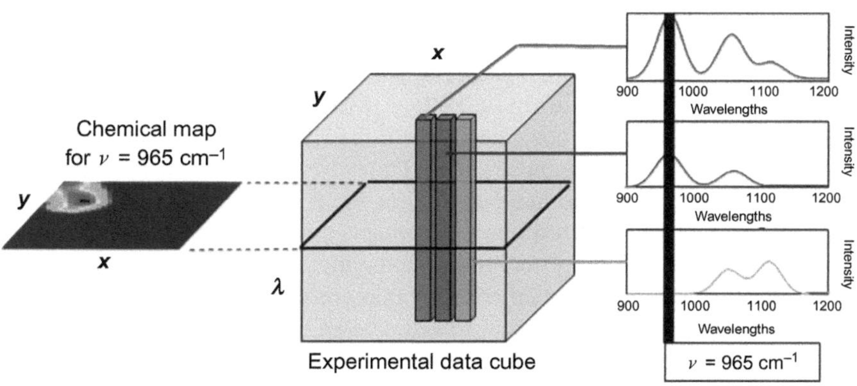

FIG. 2 A mapping procedure is used to acquire the experimental data cube (x pixels $\times$ y pixels $\times$ λ). As a consequence, each pixel is a microzone of analysis for which we have a spectrum. If a specific wavelength is known (i.e., 965 cm^{-1} in the presented example), it is then possible to integrate the signal for each pixel to generate its corresponding chemical map.

spectroscopy. However, this approach is complex to set up for both instrumental and environmental reasons [3]. The second approach is an algorithmic one and will be detailed in this chapter. Indeed, we will introduce super-resolution, a data processing concept in order to overcome the instrumental limits in vibrational spectroscopic imaging. The main idea of super-resolution is to combine the information of a set of images slightly shifted from each other to generate one higher-resolution image. The first part in the chapter will expose the concept of the super-resolution. We will next explain how to adapt the super-resolution to vibrational spectroscopic imaging. A last part will demonstrate the potential of the approach in order to surmount instrumental limits on several complex samples explored with mid-infrared (MIR), near-infrared (NIR), and Raman spectroscopy.

2 THE SUPER-RESOLUTION CONCEPT

High-resolution (HR) image generation is a key step for many imaging applications. Indeed, the more details of an observed object you have, the better the representation of the reality. In the 1970s many optical systems have been developed with multichannel detectors, such as the well-known CCD (Charge-Couple Device) [4]. However, these detectors have shown quickly their spatial resolution limits. Indeed, the first way to increase the spatial resolution was to reduce the pixel size. Unfortunately the shot noise was increased. At this moment, it was demonstrated that 40 μm^2 was the optimal pixel size. The second idea was to increase the chip size, but as a consequence, the cost was really increased.

In the 1980s the signal processing community showed the possibility to overcome the limitations of these optical systems. The image processing concept called super-resolution was born [5,6]. The main idea of super-resolution is the fusion of several low-resolution (LR) images of the same object observed from "different angles" (different point of view) in order to obtain one higher-resolution image. The first results were published by Tsai and Huang [5]. The task was to find a system of equations between the Discrete Fourier Transform (DFT) coefficients from the LR images and the HR image sampling with the Continuous Fourier Transform (CFT). The system had a solution deduced from the inverse DFT in the frequency domain. For more than 20 years, different algorithmic approaches have been developed in the frequency or the multiple sampling domains, including Bayesian theory [7,8]. Most of these recent methods take into account the optical instrumental response and the scene movement to estimate an HR image. Super-resolution becomes then a real alternative particularly when instrumental setup improvement is not available or even possible. Indeed, existing imaging systems can still be used. Moreover, it is sometimes the only way to increase the spatial resolution when a better instrumentation is not available. Super-resolution is a real research field since the early 2000s. Many domains have applied it with

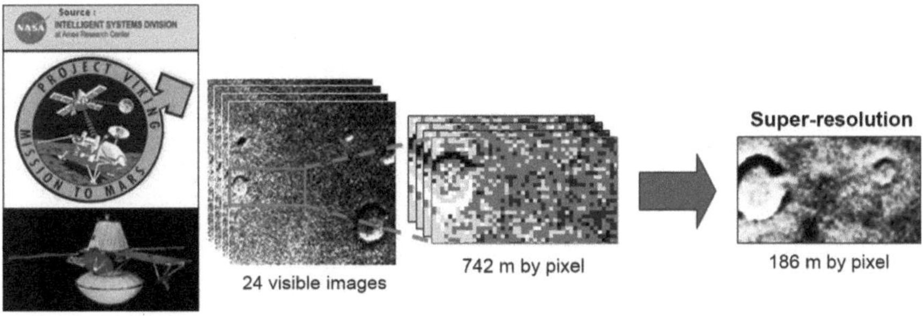

FIG. 3 The super-resolution method applied on 24 visible images of MARS. *Adapted from Lee H. Maximum entropy and Bayesian methods. Kluwer Academic Publishers; 1996. p. 305, Fig. 5.*

success, like, signal video processing, medical imaging, tomography, X-ray imaging, forensics, remote sensing, astronomy, or the military applications.

At first glance, the super-resolution algorithms improve the visual quality of an image by increasing the number of pixels. But in practice, this approach uncovers the information hidden upon acquisitions by increasing the spatial resolution. Fig. 3 presents the super-resolution concept applied on data acquired during the NASA Viking's program [9]. One of the goals of this mission was to take pictures of MARS surface. The super-resolution image is deduced from 24 LR images acquired by the space probe Viking Orbiter I. It is a very interesting example because one of the two craters observed on the super-resolution image is absolutely not observed on LR ones. With this example, it is clear that the super-resolution concept gives an opportunity to break the spatial resolution limits of an optical system.

2.1 The Analytical Model Used in the Super-Resolution Concept

In order to understand the super-resolution concept, it is now necessary to introduce the image deformation/degradation model which makes the link between the HR image and an LR image. In fact, we consider an LR image as a noisy, undersampled, blurred, and warped version of the original HR image (Fig. 4). The aim of super-resolution is then to find an HR image with the fusion of the LR ones, given estimates of noise, undersampling, blur, and shifts.

Given n LR gray scale images denoted $\{\mathbf{C}_k\}_{k=1}^{n}$. These LR images (with low pixel density) are defined by $m_1 \times m_2$ pixels. They can be considered as worse and different representations of a single image of much higher-resolution $\mathbf{C}_{\mathrm{HR}}$ defined by $l_1 \times l_2$ pixels, where $l_1 > m_1$ and $l_2 > m_2$ for $1 \leq k \leq n$. The super-resolution concept considers that each LR image $\mathbf{C}_k$ is the result of a particular geometric warping (translations in x and y directions, rotations, or both), linear space-invariant blurring, and uniform rational decimation (undersampling) performed on the ideal HR image $\mathbf{C}_{\mathrm{HR}}$. In general,

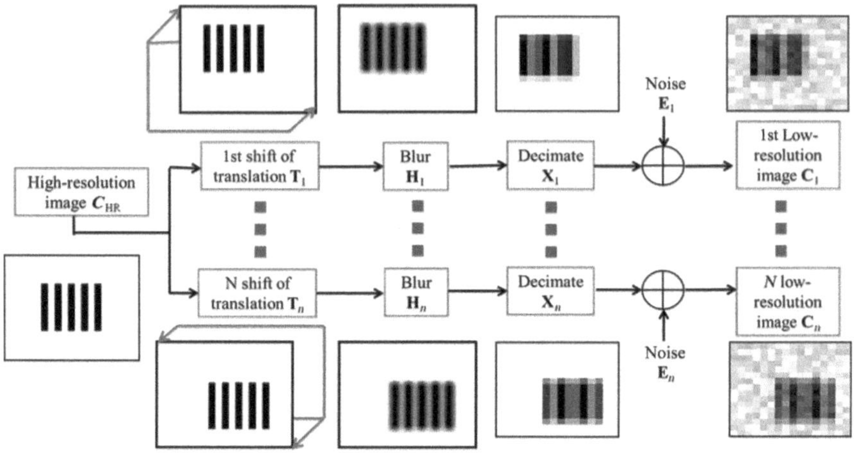

FIG. 4 Representation of the analytical model used in the super-resolution where LR images are represented as noisy, undersampled, blurred, and warped versions of the original HR image.

we also consider that LR images are corrupted with an additive Gaussian noise. Considering Eq. (1), it is possible to propose an analytical model in order to express the steps previously described:

$$\mathbf{C}_k = \mathbf{X}_k \mathbf{H}_k \mathbf{T}_k \mathbf{C}_{\text{HR}} + \mathbf{E}_k \tag{1}$$

The matrix $\mathbf{T}_k$-sized $[l_1 l_2 \times l_1 l_2]$ corresponds to the geometric warping between $\mathbf{C}_{\text{HR}}$ and $\mathbf{C}_k$ images. The matrix $\mathbf{H}_k$-sized $[l_1 l_2 \times l_1 l_2]$ defines the blur of the optical system also called Point Spread Function (PSF). The matrix $\mathbf{X}_k$-sized $[m_1 m_2 \times l_1 l_2]$ is the decimation resulting in $\mathbf{C}_k$. It corresponds to the reduction of the number of observed pixels in the measured image. The additive Gaussian noise observed in the k-th image is described with $\mathbf{E}_k$. Considering a generalization of this model applied to all n available LR images, we obtain the system of equations:

$$\begin{bmatrix} \mathbf{C}_1 \\ \vdots \\ \mathbf{C}_k \end{bmatrix} = \begin{bmatrix} \mathbf{X}_1 \mathbf{H}_1 \mathbf{T}_1 \\ \vdots \\ \mathbf{X}_n \mathbf{H}_n \mathbf{T}_n \end{bmatrix} C_{HR} + \begin{bmatrix} \mathbf{E}_1 \\ \vdots \\ \mathbf{E}_n \end{bmatrix} \tag{2}$$

As a consequence, the super-resolution is an inverse problem since $\mathbf{C}_{\text{HR}}$ will be found from the known set of LR maps $\{\mathbf{C}_k\}_{k=1}^{n}$ and estimations of $\mathbf{X}_k$, $\mathbf{H}_k$, and $\mathbf{T}_k$ matrices for all $k = 1, \ldots, n$. In general, this mathematical problem is hard to resolve for two main reasons. First, it is an ill-posed problem because the number of LR images is often much smaller than the dimensions of $\mathbf{C}_{\text{HR}}$. Second, it is an ill-conditioned problem because $\mathbf{X}_k \mathbf{H}_k \mathbf{T}_k$ product is difficult to invert. It is an inherent numerical instability of such inverse problem. As a result, small variations on $\{\mathbf{C}_k\}_{k=1}^{n}$ induce large errors on $\mathbf{C}_{\text{HR}}$.

Nevertheless, it is possible to simplify the analytical model of the super-resolution concept when spectroscopic imaging is considered.

2.2 The Super-Resolution Model in Far-Field Imaging Spectroscopy

Whatever the detection mode in far-field imaging, the instrumental setup is often a spectrometer coupled to a microscope with a motorized stage (Fig. 5).

For spectroscopic experiments, $\mathbf{T}_k$ matrices are just defined by the shift between LR chemical maps $\{\mathbf{C}_k\}_{k=1}^{n}$, with respect to one of them taken as a reference. These shifts are translations movement from the *xyz* motorized stage between two acquisitions. The main idea of the super-resolution concept is that these shifts must be lower than the pixel size of the LR image. Contrary to major applications of super-resolution, where the scene is not under control and algorithms have to be used in order to estimate relatives moves, we have the advantage here of knowing precisely the translation vector from the piezo stage. Moreover, $\mathbf{H}_k$ can be considered constant during all the experiments. Indeed, all LR images are obtained with the same optical system and the same experimental conditions (time exposure, spectral resolution, etc.). We make then the hypothesis that the PSF remains constant all over the time experiment. Generally, this matrix is estimated from a spatial resolution evaluation, but even a rough guess of it, like Gaussian filter, can be used to resolve the inverse problem. The matrix $\mathbf{X}_k$ depends only on the decimation ratio between the HR image we want to obtain and the LR images, i.e., the ratio between the

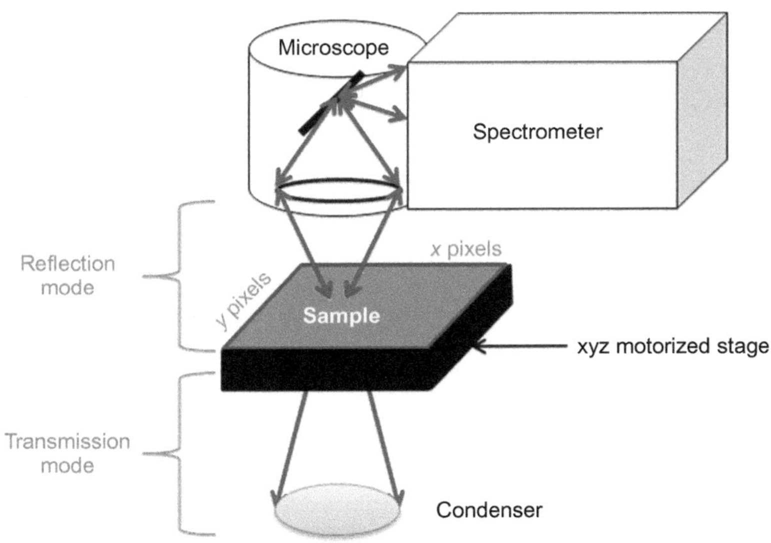

FIG. 5 Instrumental setup for far-field imaging spectroscopy.

numbers of pixels of the two. Indeed, l_1 and l_2 are user-defined parameters that always satisfy the following rule of thumb:

$$l_1 l_2 \leq n m_1 m_2 \tag{3}$$

Given $l_1 = km_1$ and $l = km_2$, we must select the highest k-value to satisfy Eq. (3). As a consequence, Eq. (1) can be rewritten as follows:

$$\mathbf{C}_k = \mathbf{XHT}_k \mathbf{C}_{\mathrm{HR}} + \mathbf{E}_k \tag{4}$$

From Eq. (4), it is now possible to retrieve the HR image $\mathbf{C}_{\mathrm{HR}}$ with a classical restoration method such as the Maximum A Posteriori (MAP) estimation [10]. This methodology comes from Bayesian statistics. MAP is particularly well suited for solving inverse problem. It is closely related to Maximum Likelihood (ML) method, but employs an optimization procedure, which incorporates prior information (Eq. 5).

$$\mathbf{C}_{\mathrm{HR}} = \arg\min_{n} \sum_{k=1}^{n} \|\mathbf{XHT}_k \mathbf{C}_{\mathrm{HR}} - \mathbf{C}_k\|_p^p + \gamma A\{\mathbf{C}_{\mathrm{HR}}\} \tag{5}$$

A regularization term $A\{\mathbf{C}_{\mathrm{HR}}\}$ is necessary in order to find a stable solution of $\mathbf{C}_{\mathrm{HR}}$ (Eq. 6).

$$A\{\mathbf{C}_{\mathrm{HR}}\} = \|\nabla \mathbf{C}_{\mathrm{HR}}\|_1 \tag{6}$$

With the idea of preserving edges in the reconstruction of $\mathbf{C}_{\mathrm{HR}}$, the regularization term called Total Variation was chosen for all experiments presented in this chapter [5,6,11].

3 CRITERIA TO MEASURE THE SPATIAL RESOLUTION IN IMAGING SPECTROSCOPY

In order to estimate the potential of the super-resolution concept, it is necessary to define and measure spatial resolution. Spatial resolution is defined as the smallest observable distance between two separate objects. There are numerous experimental methods to estimate it [12,13]. However, the purpose of this part is to propose a technique able to objectively and quantitatively estimate the intrinsic spatial resolution of imaging systems and ultimately the improvement brought by the super-resolution. In general, the two methods used to measure spatial resolution are the Rayleigh criterion and the step-edge [14]. These two methods require spectroscopic measurements on an ideal sample called target which is considered as an ideal sample because it is spatially and chemically well defined. Indeed, the target sample is a substrate where patterns are uniformly distributed and well defined with different sizes (Fig. 6). Many target samples have been developed depending on the experiment and the instrument [15]. When developing a new target sample,

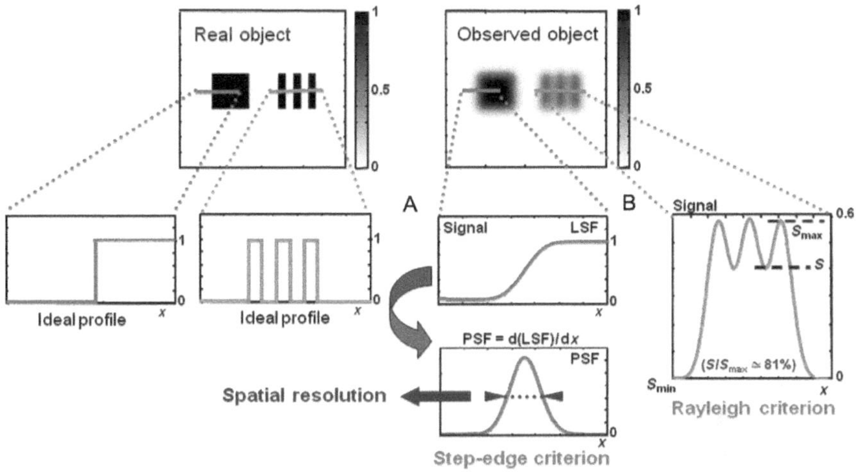

FIG. 6 Spatial resolution measurement on a "target" sample. (A) Step-edge method and (B) Rayleigh criterion.

the main idea is to select materials in order to obtain the best contrast in the image between the substrate and patterns. For the Rayleigh criterion, spectral acquisitions are performed in a selected direction though well-defined consecutive bars (three or more). Considering the signal profile along perpendicular axis of the bars, if the signal value (S) between two consecutive peaks is lower than $8/\pi^2$ (i.e., 81%) of the intensity of the maximum (S_{max}), then the three "bars" are considered as resolved. The spatial resolution is then defined by the smaller resolved pattern group present on the target sample. For the step-edge method, the main idea is to observe a step on the target sample. Considering the step-edge as an input to an optical system, the signal along the perpendicular direction to the edge is defined as the Line Spread Function (LSF) with respect to position. The derivative of the LSF is the PSF in the considered direction. It is the response of the system to a point source observation. The full width at half maximum (FWHM) of the PSF is used as an estimation of spatial resolution. This approach is very attractive since it allows in a single experiment to determine directly the spatial resolution. Moreover, it can be easily used both on LR and super-resolution images.

The next parts of the chapter will demonstrate the potential of the super-resolution concept in vibrational spectroscopy and, more particularly, in MIR, NIR, and Raman imaging in order to push back the spatial and chemical resolution of far-field instruments. Three different case studies will be successively developed by defining for each the instrumentation used and the target sample for the estimation of the system's intrinsic spatial resolution. Finally real samples will be analyzed and the potential of super-resolution demonstrated.

4 CASE STUDY #1: SUPER-RESOLUTION CONCEPT IN MIR SPECTROSCOPY [16]

Fourier transformed infrared (FTIR) spectroscopic imaging is a valuable tool to obtain information about molecular distribution of various compounds in complex samples for many industrial or scientific fields. FTIR spectrometers are thus coupled with microscopes to enable the spectral analysis of successive microzones of the considered sample. Modern infrared spectrometers often use Focal Plane Array (FPA) detectors which are able to deliver a rapid spatial and chemical information. As a consequence, this technique can quickly provide chemical maps for the compounds of interest.

4.1 FTIR Instrument

The aim of this part is to define the experimental setup used for this case study. An equinox 55 spectrometer (Bruker, Ettlingen, Germany) coupled with a Hyperion 3000 microscope equipped with a 64×64 Mercury–Cadmium–Telluride Focal Plane Array (MCT-FPA) detector is used to acquire IR reflectance spectra between 3900 and 900 cm^{-1} with a 32 cm^{-1} spectral resolution. The number of scans is 5 corresponding to a total acquisition time of few seconds. Using a conventional signal integration procedure on reflectance spectra at a specific wavenumber (of the compound of interest), it is then possible to obtain a 64×64 pixels image. The surface analyzed by one FPA measurement is then defined by a square of 270 μm × 270 μm due to the use of 15× Cassegrain objective (NA = 0.45). Thus the pixel size on the raw chemical image is 4.22 μm × 4.22 μm. Between two FPA measurements, it is possible to move the sample with an XY motorized stage with a 0.1 μm repeatability in position.

The spatial resolution of a microscope at a selected wavelength is ideally determined by the diffraction limit. When a monochromatic radiation point source is passed through the microscope, the well-known Airy pattern is observed at the beam focus. The central circular area, called the Airy disc, has thus a radius r given by:

$$r = 0.61 \times \frac{\lambda}{\text{NA}} \tag{7}$$

λ is the wavelength of photons and NA is the numerical aperture of the microscope. Considering the previous equation, two objects are completely resolved if they are separated by $2r$ and hardly resolved for r (Rayleigh criterion of spatial resolution). Thus the diffraction limit and the spatial resolution can be defined by more or less the wavelength of the photons when conventional Cassegrain objectives are used in the air, i.e., from 2.5 to 25 μm for MIR imaging. Unfortunately, this relation is rarely valid in practice. First, the FPA detector is not a point and has a finite size. In this case, a pixel detector becomes a sort of aperture. Therefore when the pixel size is higher than the wavelength, the system is said to be pixel size limited [2]. In these conditions, the spatial resolution is limited by the pixel size and less by the diffraction phenomenon. Second, a low axial spatial resolution (in z direction, i.e., depth of the sample) can degrade the achievable optimal lateral spatial resolution. Third, the optical setup of the microscope itself has a direct influence on the spatial resolution such as the number of aperture in the optical pathway [17,18]. For all the above reasons, it is often impossible to find and use a theoretical model or formula in order to evaluate the spatial resolution of a microspectrometer particularly when FPA detectors are used. Therefore a direct measurement of the spatial resolution with a target sample is preferred when considering a specific instrumental setup.

4.2 The Target Sample

In order to evaluate the spatial resolution before and after applying super-resolution concept, all spatial resolution measurements are done on the well-known 1951 USAF resolution target. This chart is a resolution test pattern conformed to the MIL-STD-150A standard, set by US Air Force in 1951. The target sample has a chrome metal coating pattern on soda-lime glass. It is an appropriated sample for infrared imaging experiments because contrasted images are obtained due to the high reflectivity of metal and the high absorbance of glass. The patterns consist of groups of three bars from large to small dimensions. More precisely, it consists of nine groups divided into five layers of patterns (Fig. 7). The largest groups form the first layer located on the outer sides. The smaller layers follow the same patterns but are

FIG. 7 The 1951 USAF resolution target used for all infrared spectroscopic measurements (*LP*, line pairs/mm).

progressively smaller toward the center. Each group consists of six elements, numbered from 1 to 6 (except for group 9). Within the same layer, the odd-numbered groups appear contiguously from 1 to 6 from the upper right corner. The first element of the even-numbered groups is at the lower right of the layer, with the remaining elements from 2 to 6, at the left. As described in Fig. 7, the scales and dimensions of the bars depend on the spatial frequency noted LP, i.e., the number of line pairs per mm. For a given group g and element e, the spatial frequency is determined by Eq. (8).

$$LP = 2^{\left[g + \left(\frac{e-1}{6}\right)\right]} \tag{8}$$

Table 1 shows the evolution of the spatial frequency and the line width (LW) for a given element. As it can be noticed, it is even possible (if necessary) to analyze patterns with subwavelength dimensions from high group numbers.

4.3 The Intrinsic Spatial Resolution Evaluation with a Target Sample

The first objective of this part is to evaluate the benefits and drawbacks of the *step-edge* and the *Rayleigh criterion* methods. Fig. 8 illustrates the application of these two methods on an infrared image of the target sample generated by integration at 3600 cm^{-1} (wavelength of 2.77 μm) from the spectral data cube in order to evaluate the spatial resolution along the x direction. Both A and B white arrows in Fig. 8 indicate areas where the step-edge method is applied, while the black ones indicate areas where the Rayleigh's criterion is used from element 1 group 5 to element 5 group 6. On Rayleigh's criterion bars profiles, the green dotted lines represent the limit of $8/\pi^2$ for the ratio equal to S_{max}/S previously explained. It can be observed without ambiguity that patterns are resolved from the element 1 group 5 to element 3 group 6 because the signal minima are always below this limit. As the element 5 group 6 is not resolved, the effective resolution is then between these elements and the element 5 group 6.

Considering the spatial characteristics of these elements (Table 1), the spatial resolution would be between 9.84 μm ($2 \times LW_{G6E5}$) and 12.40 μm ($2 \times LW_{G6E3}$). We can observe that this method is not so accurate and rather difficult to use. Indeed, it is necessary to analyze many different patterns in order to estimate an interval of spatial resolution. Furthermore, considering smaller patterns, the bar profiles are defined by a lower number of points. As a consequence, it becomes really difficult to observe real minima and maxima on signal profiles. Therefore the real S_{max}/S ratio cannot be estimated and the comparison with the $8/\pi^2$ threshold is biased. It is even more difficult to appreciate this ratio because the C contrast (Eq. 9) always decreases for smaller patterns.

TABLE 1 Evolution of the Spatial Frequency (LP) and the Line Width (LW) for a Given Element in a Specific Group for the USAF Test Chart

	LP (mm)	LW (μm)	LP (mm)	LW (μm)	LP (mm)	LW (μm)	LP (mm)	LW (μm)	LP (mm)	LW (μm)
	Group number									
Element	*0*		*1*		*2*		*3*		*4*	
1	1.00	500.00	2.00	250.00	4.00	125.00	8.00	62.50	16.00	31.25
2	1.12	445.45	2.24	222.72	4.49	111.36	8.98	55.68	17.96	27.84
3	1.26	396.85	2.52	198.43	5.04	99.21	10.08	49.61	20.16	24.80
4	1.41	353.55	2.83	176.78	5.66	88.39	11.31	44.19	22.63	22.10
5	1.59	314.98	3.17	157.49	6.35	78.75	12.70	39.37	25.40	19.69
6	1.78	280.62	3.56	140.31	7.13	70.15	14.25	35.08	28.51	17.54
	5		*6*		*7*		*8*		*9*	
1	32.00	15.63	64.00	7.81	128.00	3.91	256.00	1.95	512.00	0.98
2	35.92	13.92	71.84	6.96	143.68	3.48	287.35	1.74	574.70	0.87
3	40.32	12.40	80.63	6.20	161.27	3.10	322.54	1.55	645.08	0.78
4	45.25	11.05	90.51	5.52	181.02	2.76	362.04	1.38	—	—
5	50.80	9.84	101.59	4.92	203.19	2.46	406.37	1.23	—	—
6	57.02	8.77	114.04	4.38	228.07	2.19	456.14	1.10	—	—

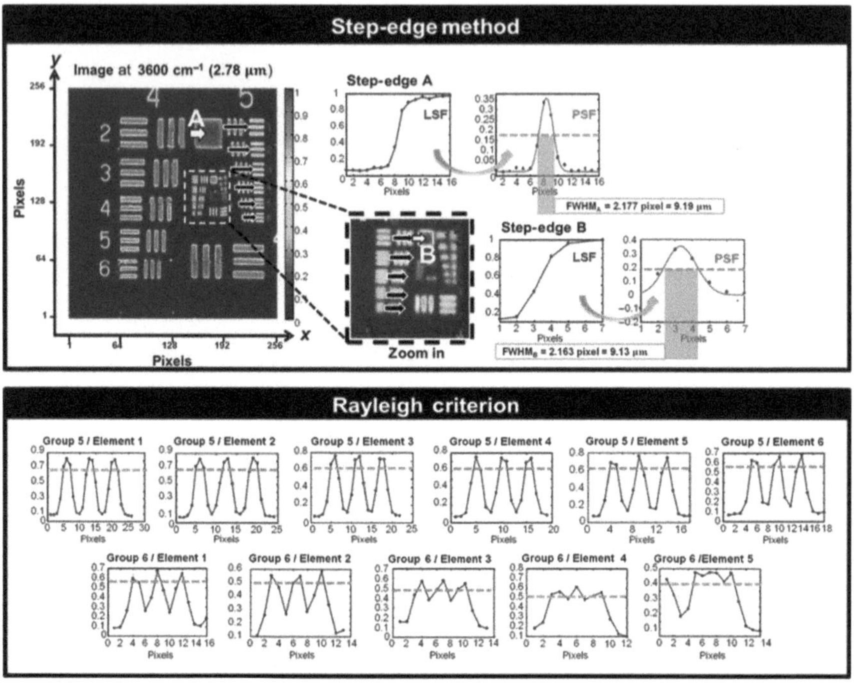

FIG. 8 The intrinsic spatial resolution measured by the "step-edge" and the Rayleigh criteria with the "target" sample at a wavelength equal to 3600 cm^{-1}.

$$C = \frac{S_{\max} - S_{\min}}{S_{\max} + S_{\min}} \tag{9}$$

The last problem is the number of FPA measurements necessary for the analysis of all bar profiles. In our case, three FPA acquisitions were necessary in order to observe all profiles from element 1 group 5 to element 5 group 6 due to a restricted field of view (FOV) of the FPA, i.e., 270 μm × 270 μm. If we consider now the step-edge method, the derivative of the step-edge profile in area A is obtained from only one line of a single FPA acquisition. It allows, after a Gaussian fit, to find a resolution of 9.19 μm corresponding to the FWHM of PSF. The same procedure applied on the step-edge of area B (square with smaller dimensions) gives a consistent result of 9.13 μm. It is obvious that the precision of this method is related to the number of experimental points on the LSF. When possible, a higher number of points should be used in order to obtain a better Gaussian fit and a more accurate estimation of the spatial resolution. In this example, the step-edge profile in area A is then preferred. Moreover, a single FPA measurement can give several estimation of the spatial resolution by the analysis of successive and parallel LSF profiles from the same area. It is then possible to propose a standard deviation for the evaluation of the spatial resolution. In conclusion, the step-edge method is the

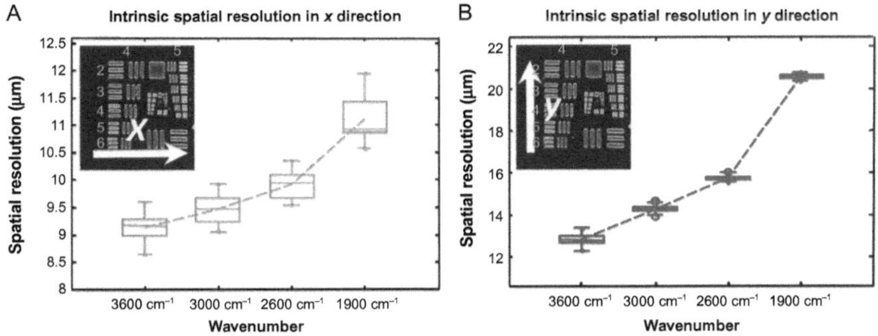

FIG. 9 The intrinsic spatial resolution of the spectroscopic system.

most appropriate for effective quantitative estimation of the spatial resolution. Fig. 9 presents measurements of spatial resolution from raw chemical images of the target sample at different wavelength in the *x* and *y* directions. We are characterizing here the intrinsic spatial resolution of the system. For each wavenumber, 17 different LSF profiles were used in order to evaluate the dispersion of the spatial resolution. A classical box plot representation is used for which the bottom and the top of the box represent the 25th and 75th percentile, respectively, the bold band near the middle of the box is the median, the ends of the whiskers are the lowest spatial resolution still within 1.5 IQR (interquartile range) of the lower quartile, and the highest spatial resolution still within 1.5 IQR of the upper quartile. Circles represent outliers when present. It can be observed that for a given wavenumber the spatial resolution dispersion is rather low, indicating the robustness of the step-edge method. As we already know, spatial resolution is wavelength dependent, i.e., the higher the wavenumber the better the spatial resolution.

Nevertheless it does not follow Eq. (7) because the system is not diffraction limited but pixel size limited. Another very important point is the observation of different spatial resolutions in the *x* and *y* directions when comparing Fig. 9A and B. In fact, this situation is due to the instrumental setup using a half mirror in order to direct the infrared light through the objective. In this condition, the full numerical aperture of the objective is only used in the parallel direction to the half mirror (*x* direction). In the perpendicular direction (*y* direction), the numerical aperture is lower and the spatial resolution is decreased. In order to obtain the same resolution in *x* and *y* directions, it would be necessary to use a beam splitter. Nevertheless, in this case and considering the reflection mode, only 25% of the light would finally reach the detector compared to the 50% obtained for the half mirror setup. From a more general point of view, evaluating spatial resolution on target sample is a very important task because it is impossible to detect spatial resolution differences for different directions when we analyze real samples.

4.4 The Super-Resolution Concept Evaluation with the Target Sample

Considering previous results, the step-edge method will be used in this part to demonstrate the improvement of the spatial resolution with the super-resolution concept. As a first step, raw data cubes are successively acquired by moving the sample with a subpixel shift (lower than the pixel size of the FPA) in the *x* and *y* directions between each FPA measurement. Thus a multiple of 1 μm (lower than 4.22 μm) is used as a shift. LR images are then generated from signal integration at different wavelengths. Fig. 10 presents an example of the super-resolution concept applied on 16 raw FPA images at 3600 cm^{-1}.

This global number of FPA images was optimized in order to obtain the best spatial resolution within a reasonable time for all the spectroscopic analysis (less than 1 min). Looking at this example, the super-resolution image presents a 50% improvement in spatial resolution beyond the increase of pixel density compared with LR images. The step-edge method applied on this new super-resolution image confirms a spatial resolution equal to 6.09 μm compared with 9.13 μm for the initial spatial resolution at 3600 cm^{-1}. It demonstrates that considering the super-resolution method as an oversampling technique is a counterintuitive idea. Indeed, it is true that there are more points on the LSF profile of the super-resolution image, but its shape is really

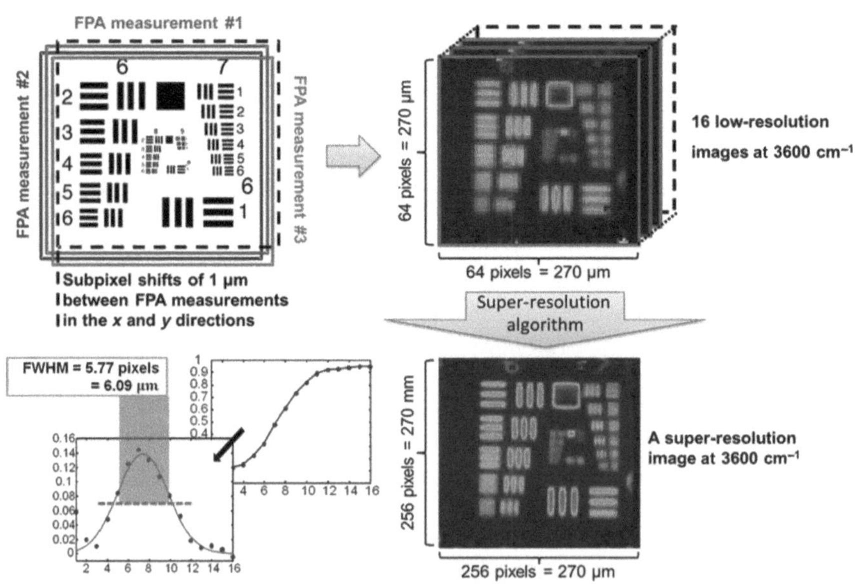

FIG. 10 The super-resolution concept applied on 16 raw images of the target sample at 3600 cm^{-1}.

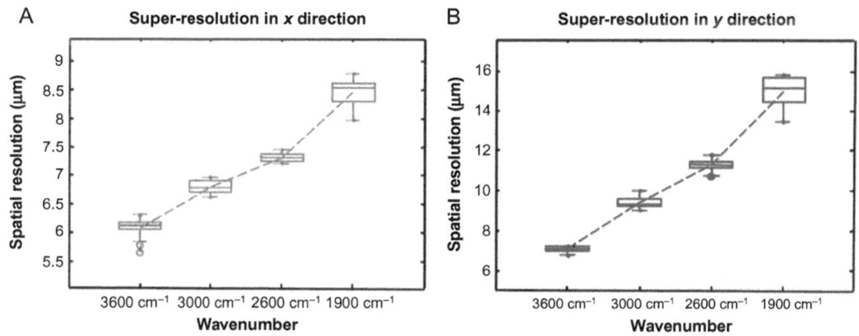

FIG. 11 The spatial resolution in *x* and *y* directions for different wavenumber after applying super-resolution.

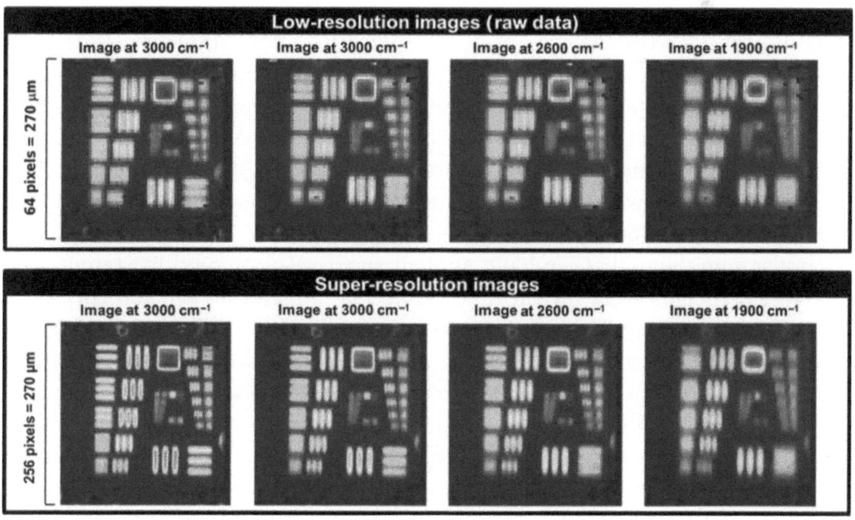

FIG. 12 Visualization of the spatial resolution improvement for different wavenumbers.

steepest. Fig. 11 shows the evaluation of the spatial resolution on the super-resolved images at different wavenumbers in *x* and *y* directions.

By comparison with previous results, one can observe a 30% improvement for the spatial resolution whatever the directions and wavenumber which is a rather spectacular result. It can be also noticed that the dispersion on the measurement of the spatial resolution is less important for super-resolved images than for raw LR images at a given wavenumber. This is mainly due to a higher image signal-to-noise ratio obtained with the super-resolution algorithm. Fig. 12 compares the chemical images of the target sample before and after the super-resolution concept considering different wavenumbers. It corresponds to the observation of an area of the target from element 1 group

6 (LW of 7.81 μm) to element 3 group 9 (LW of 0.78 μm). It demonstrates the visual improvements on this ideal sample with higher contrast and spatial resolution beyond a pixel density increase.

The aim of this part was to prove by using a well-defined sample that spatial resolution can be increased with super-resolution concept. The natural follow-up is now to explore the potential on real and more complex samples.

4.5 Super-Resolution in MIR Imaging for Real Sample Exploration

The pharmaceutical industry often uses MIR imaging for the characterization of complex solid samples at the micron scale. We propose here to study a plastic packaging which has contained tablets. Specifically, the objective is to characterize the distribution of cellulose solid particles, the excipient of the tablet, remaining on the polymer surface. Fig. 13 presents all the spectra acquired by the MIR instrument during one FPA acquisition. We can notice that some spectral contributions are negatives due to dead pixels on the detector. This instrumental failure was observed on first-generation FPA detectors. Despite major spectral contribution of the polymer on the whole spectral domain, it has been possible to find a specific contribution of the cellulose at 2560 cm^{-1} from which a chemical image has been generated. It is a rather blurred representation of the particles distribution.

In the same way that we have applied super-resolution on the target sample, 16 LR images were therefore generated by integration at 2560 cm^{-1} from 16 shifted FPA acquisitions. Fig. 14 compares a raw FPA image and the super-resolution one generated from the whole set. We can notice that the super-resolution has increased pixel density but also spatial resolution. Indeed, some cellulose aggregates on the LR map are separated into several subparticles on the super-resolved one. Moreover, the smaller particles are not observed on the raw FPA image.

This industrial example shows the interest that can bring the super-resolution concept in order to spatially characterize a complex sample. Pharmaceutical industry often uses spectroscopic imaging to estimate a particle-size distribution or detect minor compounds present on few pixels. Obviously, the direct use of a low resolution for such investigations would generate biased estimations. Thus super-resolution can be a good solution to address this problem.

5 OPTIMIZATION AND STUDY OF THE SUPER-RESOLUTION CONCEPT IN NEAR-INFRARED SPECTROSCOPY [19]

NIR spectroscopy is widely used for the characterization of raw materials, final or in-process products in many industrial or scientific fields [20]. This molecular spectroscopy is generally chosen for its speed, low cost, and

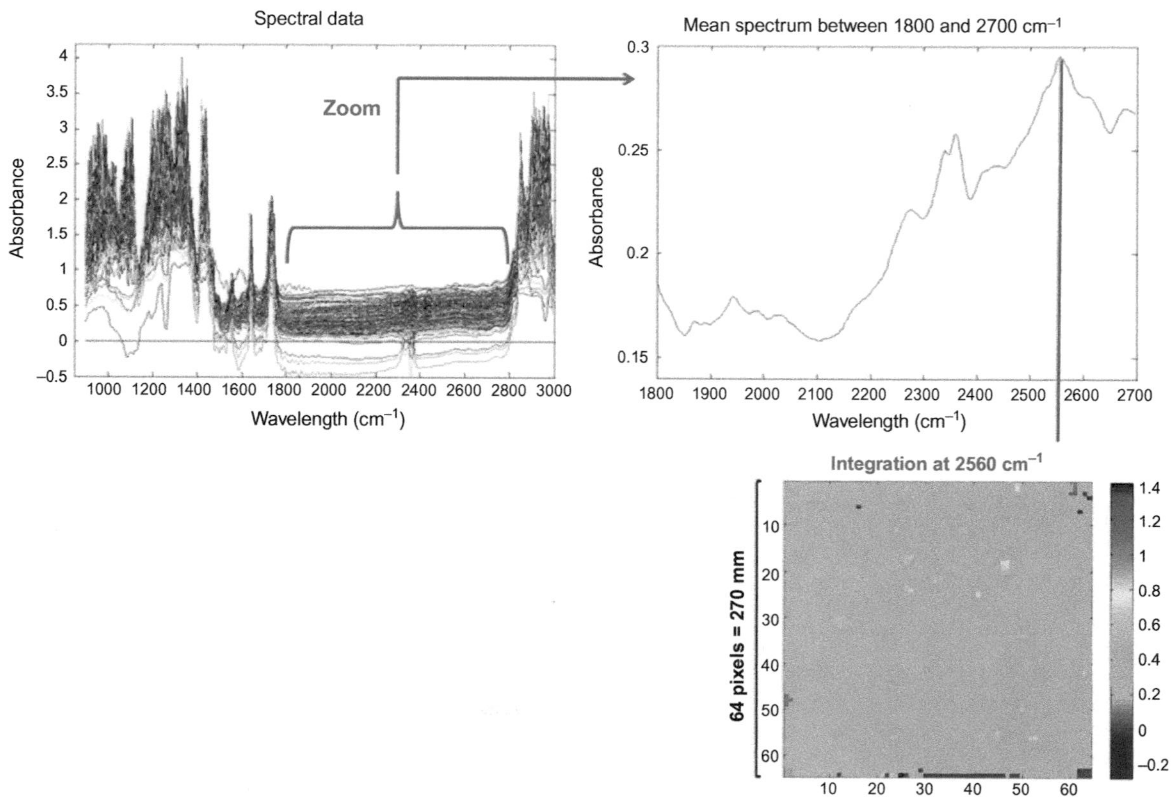

FIG. 13 Spectra measured by the MIR instrument with the cellulose/polymer sample. A low-resolution image of the cellulose is obtained by signal integration at 2560 cm^{-1}.

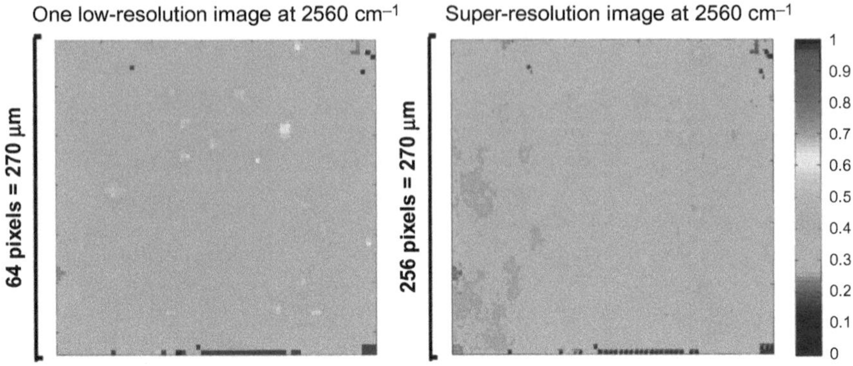

FIG. 14 Comparison of a low-resolution and the super-resolved one. A pharmaceutical industry case.

nondestructive characteristic. Furthermore, NIR spectroscopy does not need sample preparation, and even sometimes analyses can be done directly through the final packaging. This approach is a valuable tool in global analysis for quantification and classification purposes. Moreover, the development of chemical imaging has gifted this spectroscopy an additional dimension. Chemical imaging systems effectively complement chemical identification by acquiring spatially located spectra that enable visualization of chemical compound distributions. For example, it is obvious in the pharmaceutical industry that such spectroscopic techniques provide relevant information about the distributions of excipients and active pharmaceutical ingredient in complex mixtures [21]. Indeed, it is a good way to assess the product's behavior during manufacture and more importantly its physical attributes such as dissolution properties or stability. NIR spectroscopic imaging is a perfect methodology for enhancing drug quality and understanding process. However, the current trend is toward trying to increase spatial resolution to observe always more details of samples but also to keep the highest FOV which is rather incompatible. The aim of the next part is to show that super-resolution can be a good solution to reach these two goals. We will illustrate the potential of the method with the analysis of a pharmaceutical tablet by NIR imaging.

5.1 NIR Imaging Instrumentation

The aim of this part is to describe the instrumental setup used in this study. A Spectral Dimensions Sapphire NIR-CI 2450 spectrometer (Malvern Instruments, Olney, MD) equipped with an InSb FPA detector (320 × 256 pixels) is used. Spectral data cubes of the sample are acquired in reflectance mode in the 1150–2450 nm spectral range with a 10 nm spectral resolution. The total acquisition time is approximately 5 min per data cube. The FOV is set to

12.54 mm × 10 mm corresponding to a pixel size of 39.2 μm on the sample plane. In this condition, it is possible to observe the whole tablet surface. However, even with its 81,920 pixels the detector does not allow us to observe micron size details on the surface. After spectral acquisition, reflectance values (*R*) are converted into absorbance ones (*A*) using Eq. (10),

$$A = -\log(R) \tag{10}$$

where

$$R = (S - D)/(B - D) \tag{11}$$

R is obtained by processing the sample (*S*), dark (*D*), and background (*B*) for each wavelength of the data cube. The background acquisition corresponds to the spectral analysis of a high reflectance ceramic standard. The dark acquisition is necessary to take into account the signal one can measure when the FPA detector is not illuminated. In a last step, an LR image (raw image) is generated with classical signal integration from the spectral data cube.

5.2 Spatial Resolution Evaluation Before and After Super-Resolution

Considering previous work in the MIR spectral range, it is again possible to use the 1951 USAF resolution target sample as an ideal sample to measure the spatial resolution before and after super-resolution. Indeed, this target sample is particularly well suited for such NIR experiments because contrasted images are obtained due to the high reflectivity difference between metal and glass in the NIR spectra range. Moreover, the reflectivity remains approximately constant on the whole spectral range which enables the evaluation of the spatial resolution for various wavelengths. Fig. 15 presents the step-edge criteria used for the intrinsic spatial resolution evaluation of the system. The large FOV of the detector allows us to observe elements 1–9 with only one acquisition. At a wavelength equal to 1300 nm, spatial resolutions in *x* and *y* directions are almost the same, i.e., around 90 μm. We can see here that even with a rather short photon wavelength of 1.3 μm, the spatial resolution we can reach is absolutely not in the same order of magnitude. It is mainly due to the detector pixel size. In order to apply the super-resolution concept, it is necessary to use several FPA acquisitions slightly shifted from each other with a motion step among them lower than the pixel size of the raw images.

Because the pixel size of raw image is 39.2 μm × 39.2 μm, a subpixel shift multiple of 10 μm in the *x* and *y* directions is used to acquire 16 data cubes from the target sample. The 16 LR images are then generated with the conventional integration procedure for a specific wavelength of the compound of interest. A HR chemical image is then obtained with the super-resolution concept from all available LR images. Fig. 16 presents integrated LR images

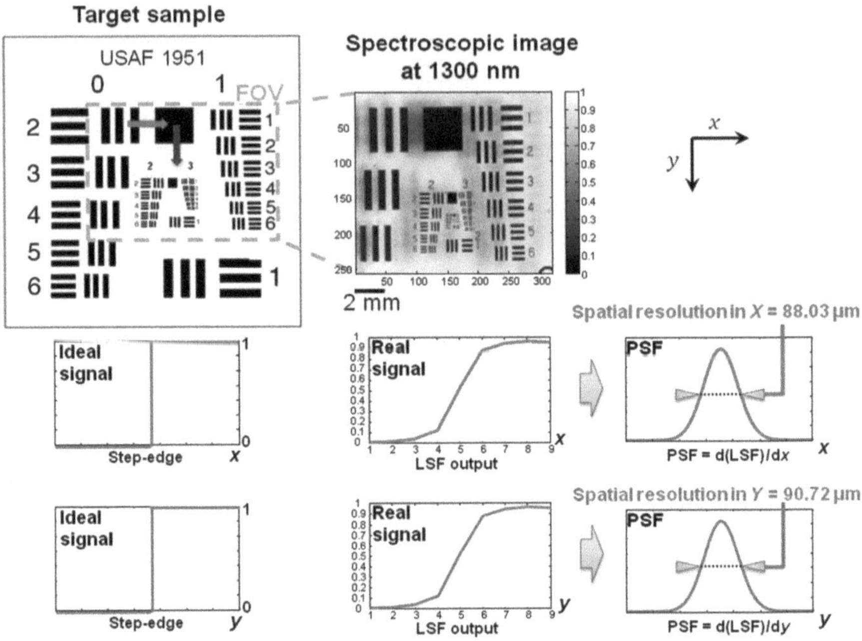

FIG. 15 The spatial resolution evaluation on an NIR spectroscopic image at 1300 nm.

at 1300 and 2100 nm with their corresponding HR images when super-resolution is applied. From a general point of view, we can observe first that fine structures of metal patterns seem to be better observed when super-resolution is applied. The pixel density is also increased by a factor of 16. In agreement with the diffraction limit formula (Eq. 7), the spatial resolution seems to be higher when a lower wavelength is used.

Thus step-edge method criterion is applied on the super-resolved images. Fig. 17 proposes a quantitative evaluation of spatial resolution at different wavelengths when super-resolution is used or not. Impressive results can be observed because the spatial resolution is almost divided by a factor of two when the super-resolution is applied. Nevertheless, the usual trend of a diffraction-limited optical system against wavelength is not demonstrated. Indeed, the linear relation between spatial resolution and wavelength is not observed as it should. Furthermore, the spatial resolution should be up to few times the wavelength depending on the usual numerical aperture range of the system. This result can be explained again by the FPA detector pixel size. Again we do not have a diffraction-limited system but a pixel size-limited one [2]. As a result, the resolution is really far from the wavelength in this NIR spectral range when a multichannel detector is used.

It should be pointed out that super-resolution performance depends on the number of available LR images. In our case, the lowest possible step for the

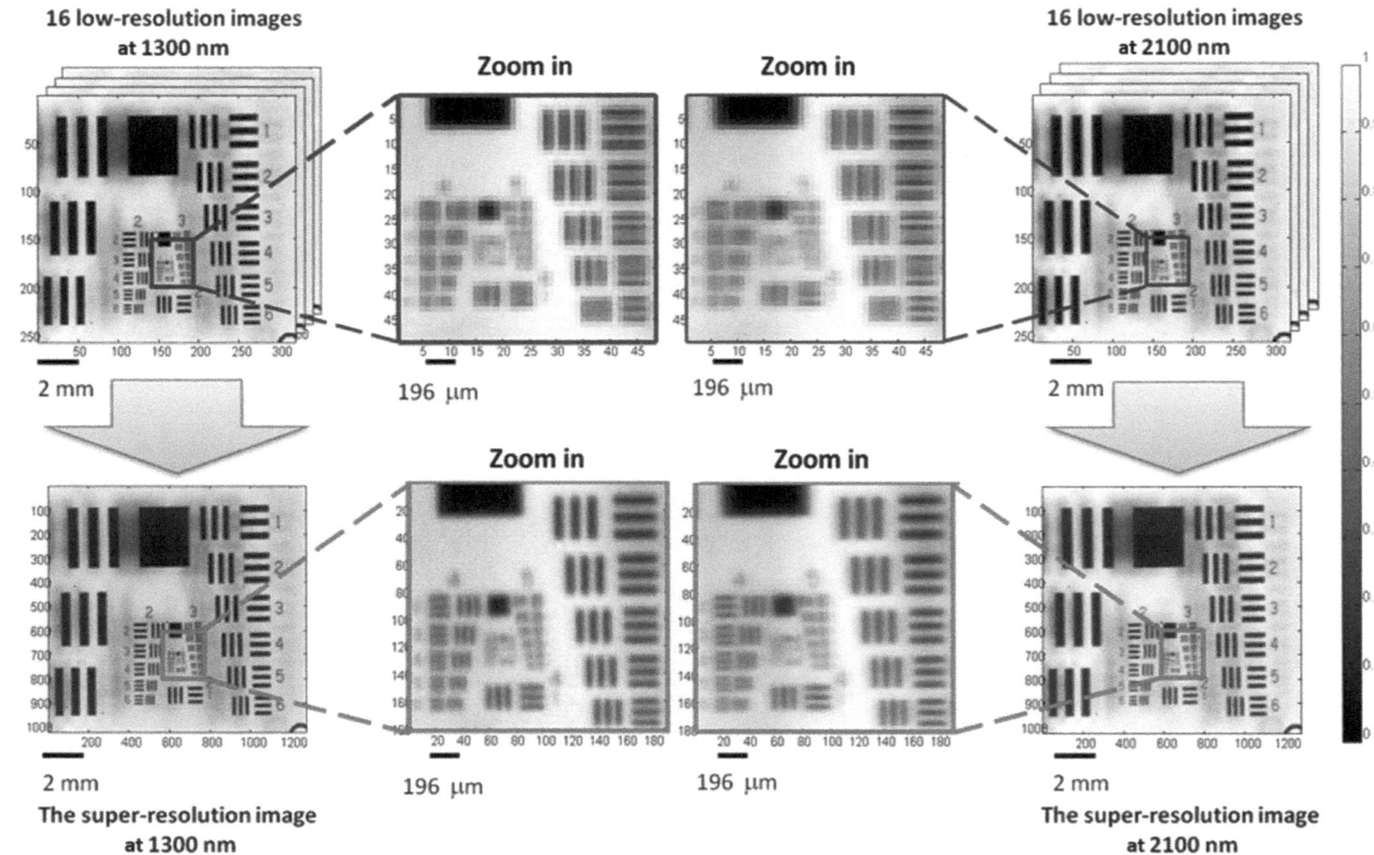

FIG. 16 The super-resolution concept applied on NIR target sample images.

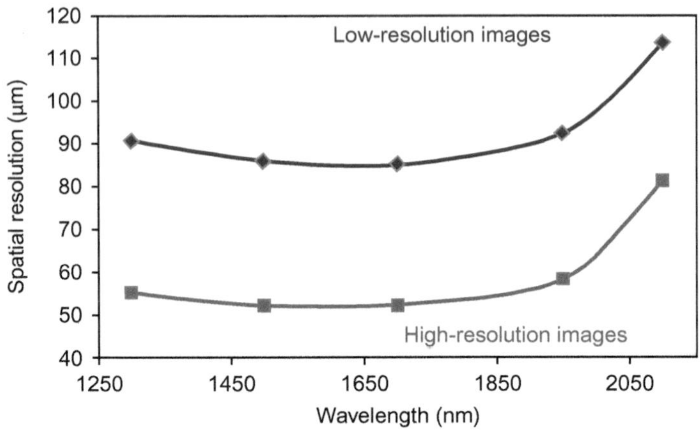

FIG. 17 Evaluation of the spatial resolution before (in *blue* (*dark gray* in the print version)) and after (in *red* (*gray* in the print version)) super-resolution.

chosen microscope stage was 10 μm. Therefore the number of possible subpixel positions (i.e., of LR images) was limited to 16. We are convinced that spatial resolution could be even lower than 50 μm using a more accurate stage. Thus the instrumental setup and experimental conditions are not negligible to obtain the best spatial resolution with super-resolution. In the next part, real pharmaceutical samples are explored with the super-resolution concept.

5.3 The Super-Resolution Concept Applied on Pharmaceutical Samples in NIR Imaging

In pharmaceutical industry, NIR imaging is widely used for characterization of raw materials and in-process or final products. This increasing interest in the use of NIR chemical imaging in pharmaceutical industry can be explained by instrumental developments made around the 2000s [22]. Prior to this period, the point mapping technique is the most used to generate NIR chemical images. Even if the point mapping technique can have advantages, the strongest argument in favor of this wide field imaging system is a real faster acquisition time, making it an instrument of choice for the PAT initiative (process analytical technology) and pharmaceutical Quality by Design [23]. As a consequence, two solid forms are investigated in this part. First, a powder mixture of cellulose and talc is used as representative of raw materials. Second, a tablet containing caffeine and lactose is selected as representative of a final product. The experimental conditions are the same as the target sample. Indeed, 16 data cubes shifted with a multiple of 10 μm between each other in x and y directions are used in order to generate 16 LR images. All 16 raw images are generated from the integration of a characteristic wavelength of a compound of interest. Figs. 18 and 19 present the results of

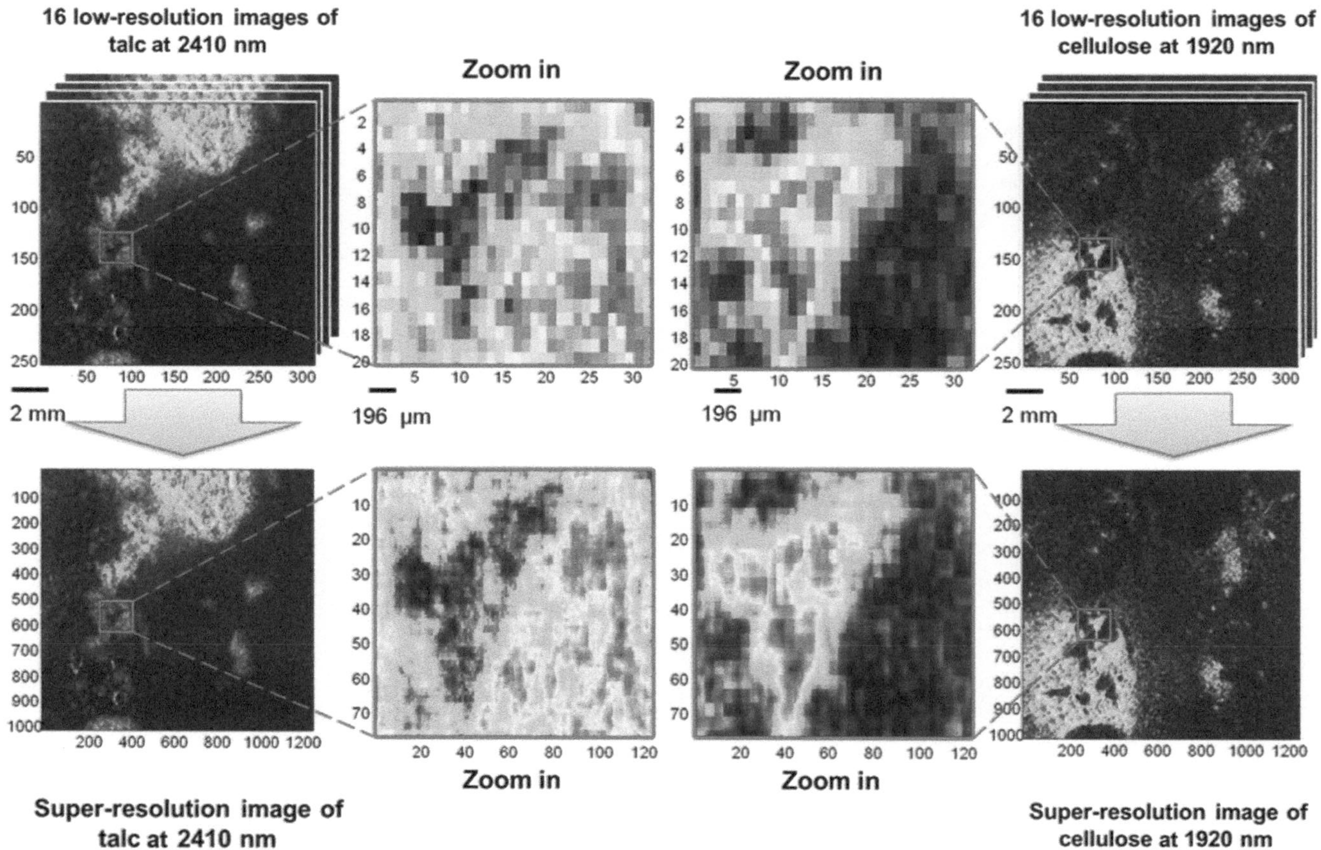

FIG. 18 Super-resolution for powder analysis. *Hot colors* (*black* in the print version) in images indicate the presence of cellulose.

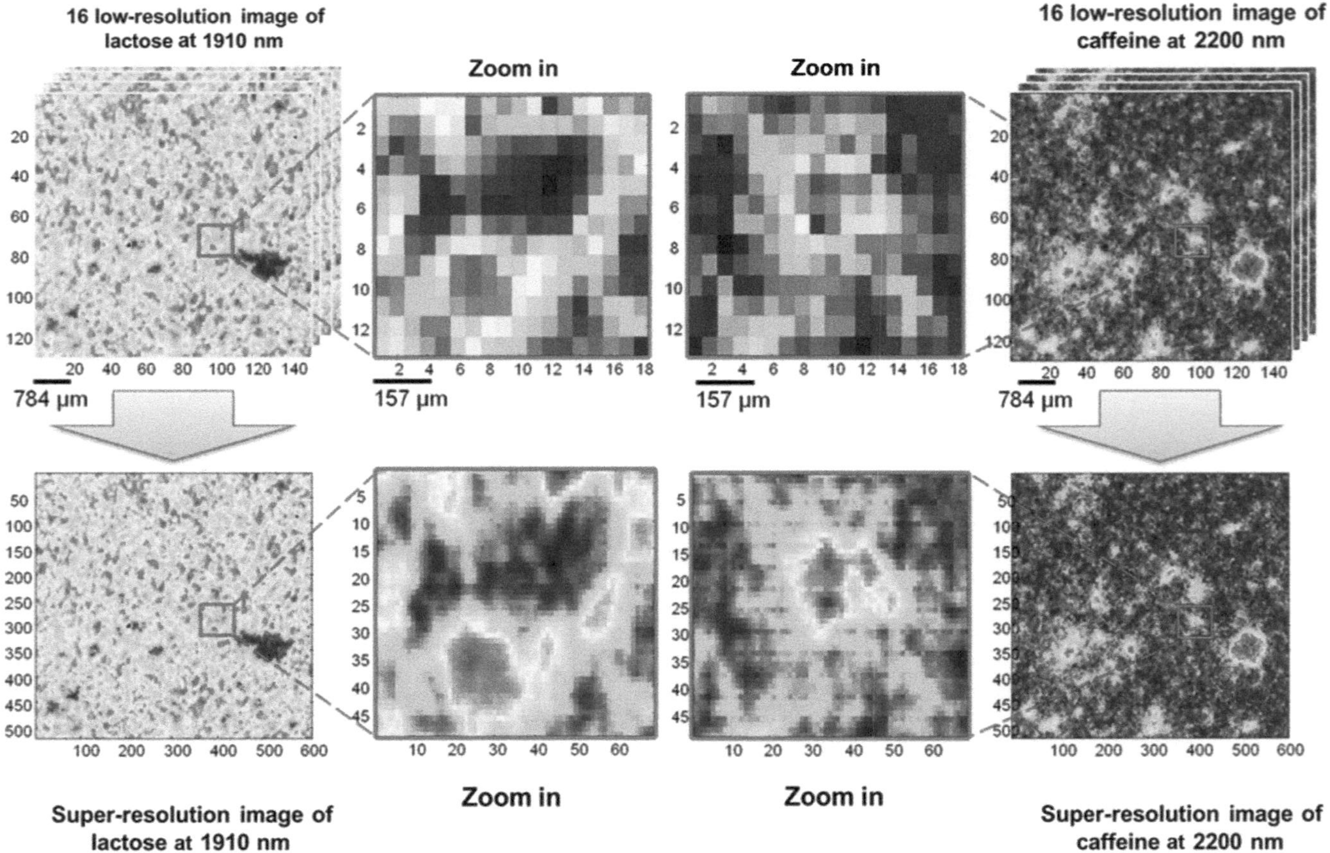

FIG. 19 Super-resolution for tablet analysis. *Hot colors* (*black* in the print version) in images indicate the presence of caffeine.

super-resolution applied to the powder mixture and the tablet, respectively. From a global point of view, it is first difficult to observe real differences between LR images and super-resolution ones mainly due to the high FOV. However, real differences can be observed when zoom in. It is obvious that using LR images can give biased vision of the sample fine structures. Even worse, wrong results will be obtained from algorithm using such LR images for the estimation of physical properties like particle-size distribution. Thus super-resolution could be a good way to keep simultaneously a high FOV with a better spatial resolution.

6 OPTIMIZATION AND STUDY OF THE SUPER-RESOLUTION CONCEPT FOR RAMAN CONFOCAL IMAGING [24]

Previous results showed that it was possible to increase by about 30% the spatial resolution in MIR and NIR spectroscopic imaging. In this part, we propose to explore the super-resolution concept in confocal Raman imaging which is even more challenging. Indeed, the intrinsic spatial resolution in such Raman imaging experiment is usually considered around 1 μm. The main aim of this work is to prove that super-resolution can propose solution for the analysis of environmental samples like dust particles. Indeed, many industries contribute to the air quality deterioration with dust emissions. These aerosol particles aggregate and react with tropospheric reagents leading to complex heterogeneous and multiphasic phenomena which occur at aerosol size scale [25]. The impact of atmospheric aerosols on visibility and climate depends on the chemical composition and size of the particles. Detailed knowledge of the aerosol composition is thus needed for source identification and apportionment. Moreover, their bioavailability and toxicity properties depend on the size, morphology, and chemical composition of individual aerosol particles [26,27].

The proposed chemometric strategy applied on aerosol samples will present a potential solution for obtaining molecular information on micron- and submicron-sized particles under in situ conditions. The validation of the approach will be divided into two parts. First, we will use as for the previous parts a target sample to measure the intrinsic spatial resolution of the Raman instrument and finally evaluate quantitatively the contribution of the super-resolution concept. In a second part, we will test the super-resolution approach coupled with the Multivariate Curve Resolution Alternating Least Squares (MCR-ALS) [28] for the exploration of atmospheric particles.

6.1 Raman Instrumentation

In this study, all Raman hyperspectral images are acquired with the LabRAM HR confocal scanning spectrometer manufactured by Horiba Jobin Yvon Scientific Company (Villeneuve d'Ascq, France). The spectrometer is coupled confocally with an Olympus BX 40 high-stability microscope equipped with

100× objective (NA = 0.9). This instrument is equipped with a 1200 grooves/mm holographic grating that enables a spectral resolution of 0.5 cm^{-1}. A liquid nitrogen-cooled CCD detector is used in the spectrometer. Raman backscattering is obtained with a 532 nm excitation wavelength (25 mW) supplied by a solid-state laser. The samples are placed under the microscope on an XYZ piezo motorized stage from PI (Physik Instrument). This stage offers movement up to 100 μm in the three space directions with a step that can achieve the nanometer (if necessary). All the instrumental setup is placed on an optical table with a vibration control system ST-UT2 from Newport Corporation (Irvine, CA, USA). This device can be activated when required for the experiment.

6.2 The Intrinsic Spatial Resolution Evaluation with a Target Sample

As usual, a target sample is used for two objectives. The first one is to evaluate the spatial resolution along the x and y directions (of the sample plane) with a quantitative criterion for different instrumental setups. The second objective is to estimate the new spatial resolution for the same instrumental setups after applying the super-resolution algorithm. The purpose of these steps is also to select experimental conditions for which the spatial resolution is the best before and after super-resolution process. Because the previous USAF target is not compatible with Raman experiment, it is first necessary to develop a new one. Thus electron lithography is used to draw submicron gold patterns on a silicon wafer surface [29]. A Scanning Electron Microscopy (SEM) image of the target is given in Fig. 20A. The target structure is inspired by the Microscopy Resolution Test Chart (NBS 101A).

It consists of five bars regularly spaced horizontally and vertically. The distance between two bars corresponds to the width of a bar. This five bars mask is replicated in a homothetic way in order to observe widths from 1 μm to 62.5 nm. Position markers are also added on the target to locate the different patterns when they become too small and almost invisible with a visible imaging system. Silicon is selected because it has very good Raman scattering properties, while gold patterns exhibit no signal. In order to develop this new target, a polymethyl methacrylate (PMMA) electron-sensitive resist solution is first spin-coated onto Si sample and baked to leave a hardened thin film on the surface (Fig. 20B). An electron beam scans the sample following the five bars mask and caused PMMA chain scissions. These fragmented polymer chains are then dissolved by a developer solution (Fig. 20C and D). A 50 nm thickness of gold is then deposited by evaporation over the entire sample surface (Fig. 20E). The final stage, called lift-off, is the immersion of the sample in acetone to remove any remaining traces of PMMA and finally reveal the gold patterns (Fig. 20F). In this work, the step-edge criterion is applied on an Au location marker from the silicon wafer to determine the

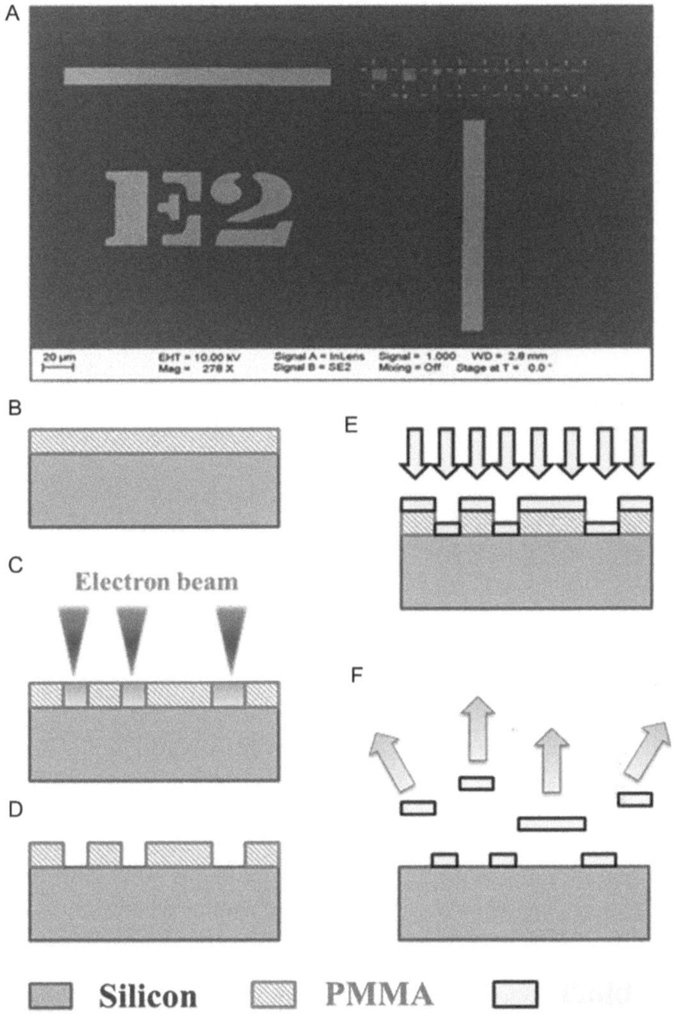

FIG. 20 Development of a resolution target with electron lithography. (A) Scanning Electron Microscopy (SEM) image of the target, (B) PMMA electron-sensitive resist solution spin-coated onto Si sample, (C) an electron beam scans the sample following the five bars mask, (D) fragmented polymer chains dissolved by a developer solution, (E) gold deposit, and (F) lift-off, i.e., remove off any traces of PMMA to reveal gold patterns.

spatial resolution before and after a super-resolution algorithm. The relatively large width of this Au object allows us to select two areas from which spatial resolution in the x and y directions can be determined (Fig. 21A and B). The typical silicon Raman band at 520.7 cm^{-1} is then integrated to generate maps. Due to the large analysis area, eight LSF profiles can be measured with one of these presented in Fig. 21C. Each corresponding PSF function is then used as

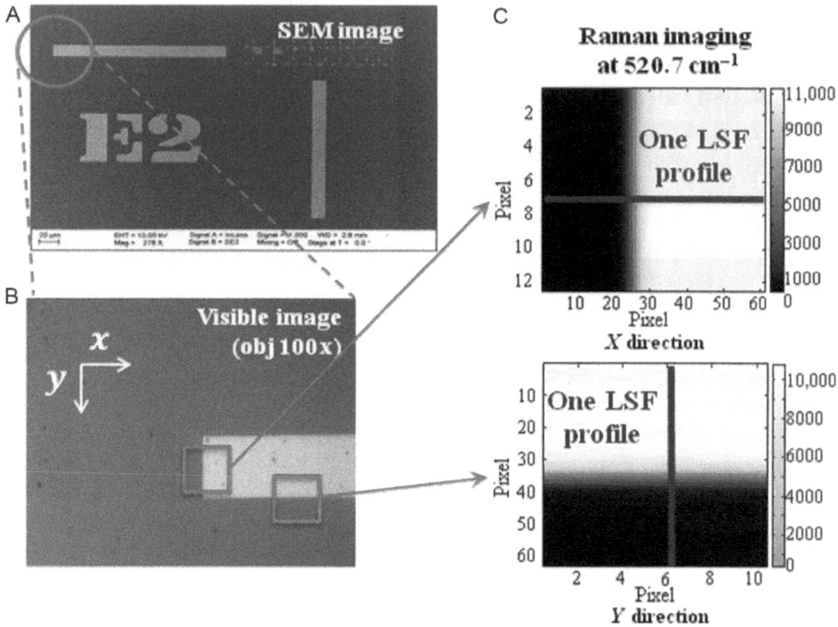

FIG. 21 Spatial resolution evaluation for Raman imaging. (A) and (B) regions of interest for spatial resolution evaluation, and (C) LSF profiles measurements in x and y directions.

a measure of the resolution by calculating the FWHM with an associated standard deviation. This procedure is also used on the super-resolved images considering different instrumental setups.

In confocal systems, pinholes used on the optical pathway have a direct impact on spatial resolution. Thus four experiments with different pinhole sizes were performed to measure the intrinsic spatial resolution of the Raman instrument. The fifth experiment has been setup in order to observe the influence of the antivibration control system on the optical table and by extension to the spatial resolution. The mapping size was 10×60 pixels and 60×10 pixels in x and y direction, respectively, with a step size of 100 nm. The spatial resolution calculations have been done on eight LSF derivative profiles for each mapping. Table 2 gives an estimation of intrinsic spatial resolution of the Raman instrument for different experimental conditions.

First, we notice that the spatial resolution is significantly different in the x and y directions when the pinhole size is between 300 and 1000 μm (Experiments #1 to #3). This is explained by the fact that the instrument is not in a real confocal mode in these conditions. Only a 200-μm pinhole gives almost the same spatial resolution in each direction. Experiments #4 and #5 show fairly equal spatial resolution, demonstrating that an active antivibration system does not change the intrinsic spatial resolution for conventional Raman measurements. Consequently, the intrinsic spatial resolution can be

TABLE 2 The Intrinsic Spatial Resolution of the Raman Instrument Measured for Different Experimental Setups

			Intrinsic spatial resolution			
Experiments	*Pinhole size (μm)*	*Antivibration system state*	*Intrinsic spatial resolution along x-direction (μm)*	*Standard deviation* σ_x *(μm)*	*Intrinsic spatial resolution along y-direction (μm)*	*Standard deviation* σ_y *(μm)*
#1	1000	ON	0.47	0.01	0.77	0.01
#2	500	ON	0.46	0.02	0.78	0.02
#3	300	ON	0.52	0.01	0.66	0.02
#4	200	ON	0.59	0.02	0.56	0.01
#5	200	OFF	0.57	0.01	0.62	0.01

considered around 0.59 μm in each direction. Moreover, the calculations of the standard deviation confirm the excellent repeatability of the step-edge criterion which does not exceed 20 nm. It is also interesting to note that the intrinsic spatial resolution estimated experimentally is much worse than the one we can obtain theoretically with the diffraction limit formula of a confocal system (Eq. 12).

$$\text{diffraction limit} \approx 0.46 \times \frac{\lambda}{\text{NA}} \tag{12}$$

Indeed for a 532-nm wavelength laser and a 100× objective (NA = 0.9) it would predict a diffraction limit of 272 nm. One more, we show that spatial resolution must be estimated experimentally in order to obtain an unbiased vision of the instrumental performance. In order to apply super-resolution, a set of LR images was acquired by inducing a subpixel shift (lower than the pixel size) in the x and y directions between them. In this way, 100 images with a pixel size of 1 μm shifted from 100 nm from each other in x and y directions were acquired. All data cubes were integrated at 520.7 cm^{-1} to obtain 100 LR images. The super-resolution algorithm was then used to generate an HR image. Table 3 presents the spatial resolution of the super-resolved image calculated with the step-edge criterion.

As we can see, the new spatial resolution is always better than the intrinsic spatial resolution for a given setup. However, there is still a significant difference between the x and y directions for a pinhole size from 300 to 1000 μm. Again only a 200-μm pinhole size provides an equivalent spatial resolution in each direction. In this condition, the spatial resolution is around 200 nm. It corresponds to an improvement of at least 65% compared with the intrinsic spatial resolution in the same experimental conditions. When considering our previous results of super-resolution in NIR and MIR spectroscopic imaging, we can state that these new results are by all means the most spectacular. We can also notice that the resolution values of experiment #5 are not indicated in Table 3. Indeed, the super-resolution algorithm failed to find an HR image in these conditions (Fig. 22A).

In fact, experiment #5 does not allow us to retrieve the pattern while experiment #4 does. This fact indicates that even using an accurate piezo stage is not enough when applying super-resolution. The antivibration system seems to be necessary to obtain the highest reproducibility in position. In general, when we present for the first time super-resolution results, it is just considered as an oversampling procedure. That is why in Fig. 22B we have compared the super-resolution image from experiment #1 with the integrated image directly obtained with the same pixel density. In this last case, the new mapping step over the 5-bars pattern is 100 nm with a 1000-μm pinhole size and activated antivibration system. It is obvious that the spatial resolution and the contrast are significantly improved when super-resolution algorithm is used. It is also observed in the associated LSF profiles. This first part focused

TABLE 3 The Spatial Resolution of the Instrument After Applying the Super-Resolution Concept on the "Target" Sample for Different Settings

			Spatial resolution after using super-resolution			
Experiments	*Pinhole size (μm)*	*Antivibration system state*	*New spatial resolution along x-direction (μm)*	*Standard deviation σ_x (μm)*	*New spatial resolution along y-direction (μm)*	*Standard deviation σ_y (μm)*
#1	1000	ON	0.18	0.01	0.26	0.02
#2	500	ON	0.19	0.01	0.33	0.01
#3	300	ON	0.20	0.02	0.25	0.01
#4	200	ON	0.19	0.01	0.22	0.01
#5	200	OFF	—	—	—	—

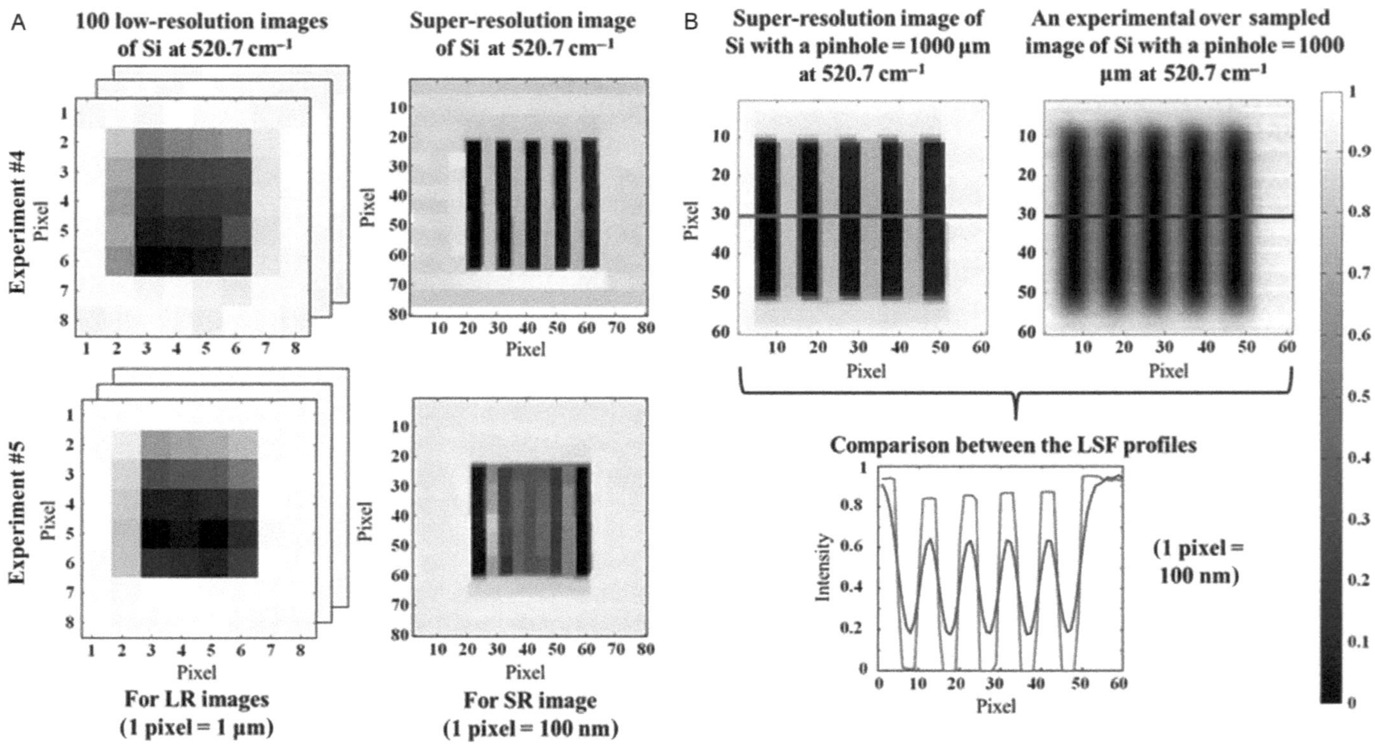

FIG. 22 Super-resolution concept applied on the target sample.

on the target sample was essential to prove the concept and determine optimized environmental and instrumental conditions in order to achieve the best spatial resolution when using the super-resolution concept. For the following aerosol study, the experimental conditions will be fixed to a 200-μm pinhole and an activated antivibration system.

6.3 Spectral and Spatial Characterizations of Aerosols

Atmospheric aerosols can be considered as heterogeneous entities resulting from atmospheric heterogeneous processes that occur at the particle-size scale [23]. The impact of atmospheric aerosols on visibility and climate change as well as their toxicity depends on the size, the chemical composition, and the chemical mixing state (i.e., the spatial distribution of species within single particle) of the aerosols [24,25,30]. So far, a detailed knowledge about the aerosol composition is lacking for health risk assessment and to evaluate the environmental impact of aerosols. The aim of this part is then to characterize aerosol particles in order to identify all the pure components by Multivariate Curve Resolution and alternating Least Squares (MCR-ALS) [31–33] in the collected sample. The information in each specific compound set of maps will be directly used with a super-resolution postprocessing step.

The sample of aerosol was collected in the courtyard of a lead-acid battery-recycling factory located in Toulouse, France. The particles were collected by impaction on silver sheets using a cascade impactor (Dekati Ltd., Kangasala, Finland) operating at a flow rate of 10 L/min. The impactor setup consists of three successive stages with aerodynamic cut-off diameters (AD) which correspond to the size segregation of 2.5 μm < AD < 1 μm (stage 1), 1 μm < AD < 0.1 μm (stage 2), and AD < 0.1 μm (stage 3). The collection time was 1 h to ensure the particle dispersion on the collecting substrate. In this study, only stage 2 was examined to demonstrate the feasibility of the analytical methodology considering submicron-sized particles. As a first step, we explore the sample of collected aerosols in order to select a region of interest (Fig. 23). We optimize experimental conditions of acquisition for this area

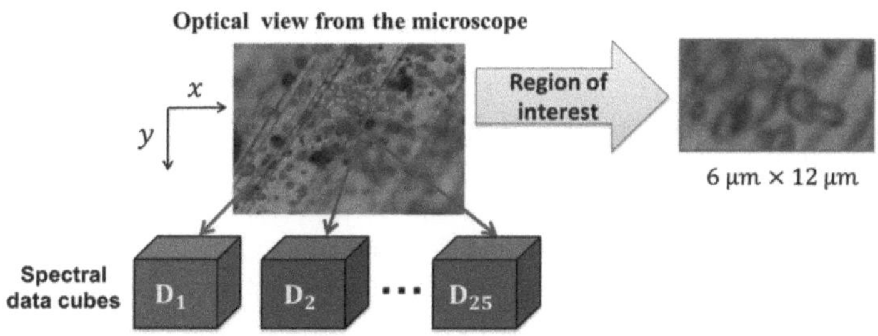

FIG. 23 The 25 spectral data acquired on the region of interest.

of 6 μm × 12 μm to obtain the best signal-to-noise ratio. Acquisition time is thereby fixed to 40 min per data cube. We successively acquire 25 data cubes shifted with a multiple of 200 nm in *x* and *y* directions.

MCR-ALS is performed on the 25 submatrices **D** structured as a column-wise augmented matrix, since the spectral direction is common to all data cubes (from 900 to 1190 cm^{-1}) (Fig. 24). Therefore the multiset data structure contains the 1800 spectra (25 × 6 × 12). The aim of MCR-ALS is to calculate a bilinear decomposition of the spectral matrix **D** into a pure concentration matrix **C** and a pure spectral matrix **S**. In a final step, the **C** matrix is refolded in order to generate concentration maps for each pure contribution. Corresponding spectral profiles are used for molecular identification. Looking under the hood of MCR-ALS, after detecting a rank of six for **D** matrix (i.e., six pure contributions) by Singular Value Decomposition (SVD) [34], initial estimates of pure spectra are obtained by SIMPLe-to-use Interactive Self-modeling Mixture Analysis (SIMPLISMA) applied to the full set of LR images. The nonnegativity constraint is used in the resolution analysis on the concentration profiles and spectra. The normalization constraint is also applied in $\mathbf{S}^T$. We refer to Chapter 2 for further details about the implementation of the algorithm and the application of constraints. Fig. 24 shows the results of the six extracted chemical compounds. For each chemical contribution, 25 LR images are generated from the refolded concentration matrix denoted **C**. Pixel size is 1 μm for each LR image. Same figure exhibits interesting pure spectral profiles.

First, contribution #6 seems to be due to uncorrected fluorescence effect on the boundary of the spectral domain. Contributions #1 to #5 are pure chemical compounds or mixtures. The identification of these five contributions has been made based on the literature [35–37]. The particles collected are mainly composed of sea salt-derived particles (i.e., Na_2SO_4 and $CaSO_4{\cdot}2H_2O$), or reacted particles (i.e., $NaNO_3$) in aggregation with submicron metal-rich particles ($PbO{\cdot}PbSO_4$). In addition, internal mixing of sea salt-derived particles is observed (i.e., Na_2SO_4 and $CaSO_4{\cdot}2H_2O$). The cocrystallization during the sea salt droplets evaporation in the atmosphere explains the formation of mixed solid particles. The distribution maps show a rough localization of the pure chemical compounds. They are blurred and present low spatial resolution as expected from the large 1 μm pixel size in the original maps.

To obtain a single distribution map per chemical contribution with higher spatial resolution, the super-resolution postprocessing step was applied separately to each set of 25 LR distribution maps. In other words, a super-resolution map for each contribution is obtained from the fusion of 25 maps (6 × 12 pixels) with a pixel size of 1 μm and shifted with a motion step of 200 nm between them in the *x* and *y* directions. These new super-resolution maps (60 × 30 pixels) have then a pixel size equal to 200 nm. As for the target sample analysis, we present in Fig. 25 images of the five components obtained from a single data block with an oversampling step of 200 nm on the same aerosol sample area.

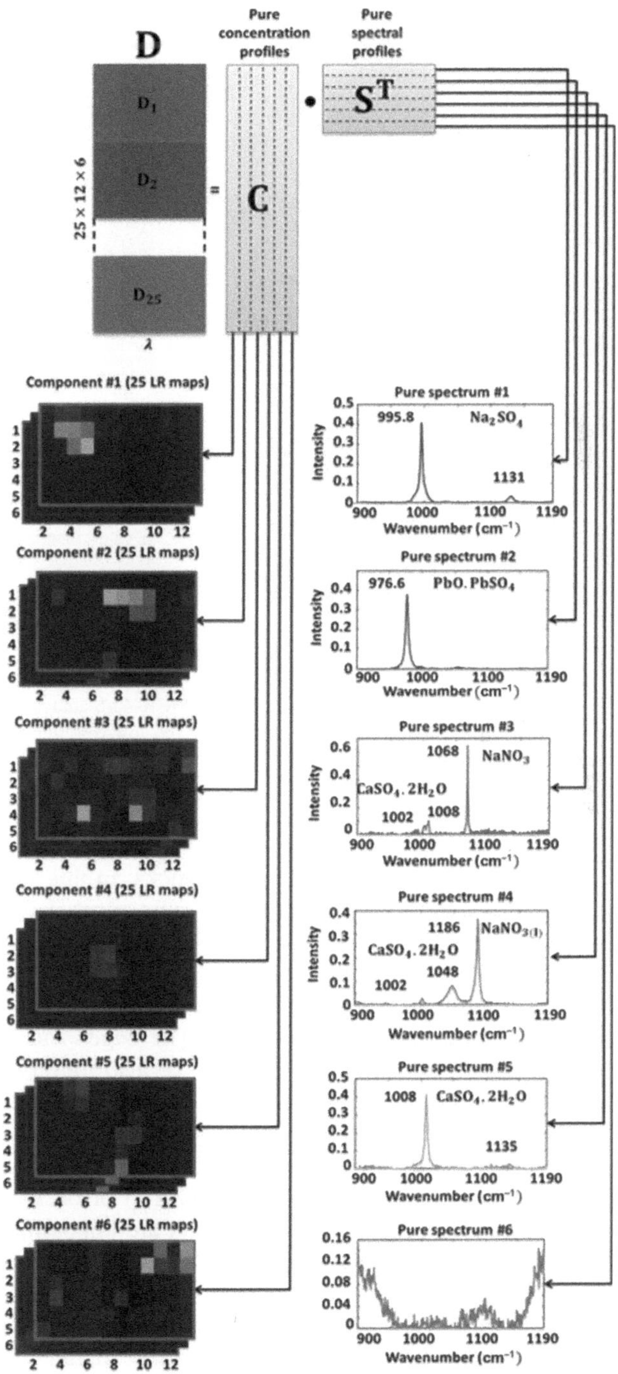

FIG. 24 MCR-ALS analysis of the multiset data structure.

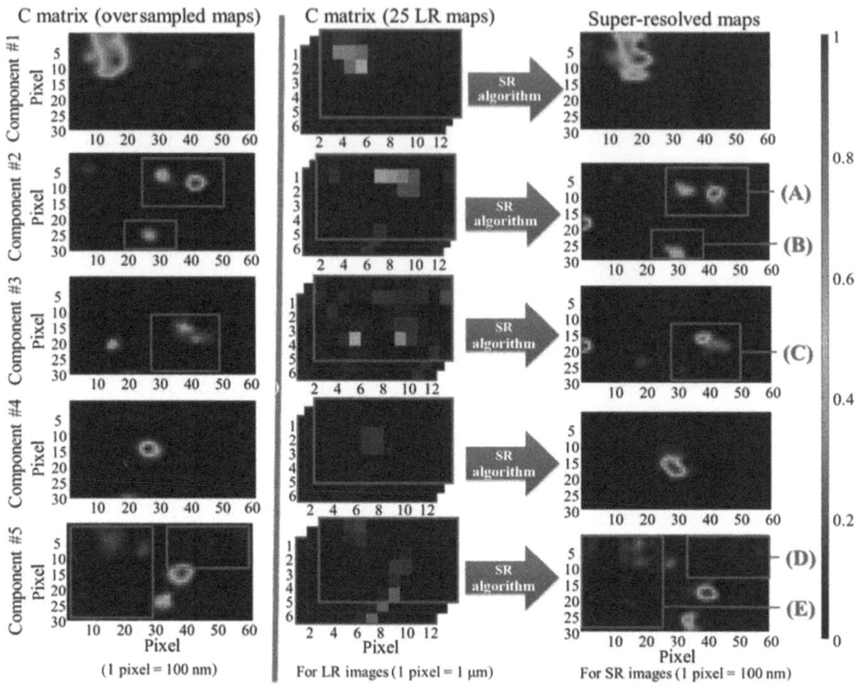

FIG. 25 Comparison of the distribution maps obtained from MCR-ALS analysis of oversampled images (*left plots*), MCR-ALS multiset structure analysis of the low-resolution images (*center plots*), and the super-resolution images (*right plots*).

The map of each chemical compound was extracted with the same MCR-ALS procedure previously used on the 25 data blocks. In this way, the oversampled and super-resolution images have the same pixel density for each pure contribution. Considering the results, one can notice directly the improvement in spatial resolution and contrast on the super-resolution results compared with oversampling procedure. Indeed, the component #1 is much better described on the super-resolution image. The edges and the shape of the aerosol aggregates are better defined. It is also observed for area (A) and (B) of the component #2. This is even more evident on the component #4. A single particle is detected on the oversampled map, while two stuck particles seem to be observed on the super-resolution image. Some new aggregates can even be observed on the component #3 (C) or on the component #5 (D) and (E) for the super-resolution images that are impossible to observe on the oversampled images. These results clearly evidence that MCR-ALS methodology coupled with the super-resolution concept leads to richer information on physicochemical description of single particles collected in industrial area than that obtained with the classical Raman imaging approach. In particular, the significant improvement of the species distribution maps reveals aggregation of submicronic particles with a single composition, whereas classical Raman imaging shows a unique particle with a micronic size composed of several species in internal mixing.

7 CONCLUSION

The current trend is toward increasing spatial resolution of spectroscopic imaging systems for different reasons. First, always more details about the samples are needed in order to obtain an unbiased representation of the sample fine structure. Second, there are more and more multimodal spectroscopic experiments with different intrinsic spatial resolutions. Simultaneous analysis of the data forces us to obtain same resolutions for the different spectroscopic modes which is not always possible due to theoretical or instrumental reasons. It is in these senses that super-resolution can be an interesting solution. However, it is obvious that super-resolution and a spectral unmixing technique such as MCR-ALS have to be coupled when complex samples are analyzed. In conclusion, we have demonstrated the capability of super-resolution in vibrational spectroscopy, but other spectroscopic imaging techniques would benefit from its use in near future.

REFERENCES

[1] Krishnan K, Powell JR, Hill SL. Infrared microimaging. In: Humeci HJ, editor. Practical guide to infrared microspectroscopy. Practical spectroscopy series, , vol. 19. New York: Marcel Dekker; 1995. p. 85–110.

[2] Salzer R, Siesler HW. Infrared and Raman spectroscopic imaging. Weinheim: Wiley-VCH; 2014.

[3] Paesler M, Moyer PJ. Near-field optics: theory, instrumentation, and applications. New York: John Wiley; 1996.

[4] Janesick JR. Scientific charge-coupled devices. Bellingham, WA: SPIE Press; 2001.

[5] Tsai RY, Huang TS. Multi-frame image restoration and registration. Adv Comput Vis Image Process 1984;1:317–39.

[6] Farsiu S, Robinson MD, Elad M, Milanfar P. Fast and robust multiframe super-resolution. IEEE Trans Image Process 2004;13:1327–44.

[7] Brown JL. Multi-channel sampling of low-pass signals. IEEE Trans Circuits Syst 1983;28:101–6.

[8] Ur H, Gross D. Improved resolution from subpixel shifted picture. CVGIP Graph Model Image Process 1992;54:181–6.

[9] Lee H. Maximum entropy and Bayesian methods. Kluwer Academic Publishers; 1996. p. 305.

[10] Farsiu S, Robinson MD, Elad M, Milanfar P. Advanced and challenges in super-resolution. Int J Imaging Syst Technol 2004;14:47–57.

[11] Nguyen N, Milanfar P, Golub G. A computationally efficient image superresolution algorithm. IEEE Trans Image Process 2001;10:573–83.

[12] Lasch P, Naumann D. Spatial resolution in infrared microspectroscopic imaging of tissues. Biochim Biophys Acta 2006;1758:814–29.

[13] Stelzer EHK. Contrast, resolution, pixilation, dynamic range and signal-to-noise ratio: fundamental limits to resolution in fluorescence light microscopy. J Microsc 1998;189:15–24.
[14] Levenson E, Lerch P, Martin M. Infrared imaging: synchrotrons vs. arrays, resolution vs. speed. Infrared Phys Technol 2006;49:45–52.
[15] Ippolito SB, Thorne SA, Eraslan MG, Goldberg BB, Ünlü MS. High spatial resolution subsurface thermal emission microscopy. Appl Phys Lett 2004;84(2231):4529–31.
[16] Offroy M, Roggo Y, Milanfar P, Duponchel L. Infrared chemical imaging: spatial resolution evaluation and super-resolution concept. Anal Chim Acta 2010;674:220–6.
[17] Sommer AJ, Katon JE. Diffraction-induced stray light in infrared microspectroscopy and its effects on spatial resolution. Appl Spectrosc 1991;45:1633–40.
[18] Teramae N, Nishikoda T. Infrared microscopy. Tokyo: IPC Publisher; 2003.
[19] Offroy M, Roggo Y, Duponchel L. Increasing the spatial resolution of near infrared chemical images (NIR-CI): the super-resolution paradigm applied to pharmaceutical products. Chemom Intell Lab Syst 2012;117:169–82.
[20] Burns DA, Ciurczak EW. Handbook of near-infrared analysis. CRC Press; 2001. p. 814.
[21] Ciurczak EW, Igne B. Pharmaceutical and medical applications of near-infrared spectroscopy. CRC Press; 2014. p. 172.
[22] Gendrin C, Roggo Y, Collet C. Pharmaceutical applications of vibrational chemical imaging and chemometrics: a review. J Pharm Biomed Anal 2008;48(3):533–53.
[23] Cox Gad S. Pharmaceutical manufacturing handbook: regulations and quality. Wiley-Blackwell; 2008. p. 856.
[24] Offroy M, Moreau M, Sobanska S, Milanfar P, Duponchel L. Pushing back the limits of Raman imaging by coupling super-resolution and chemometrics for aerosols characterization. Sci Rep 2015;5:12303.
[25] Sobanska S, Falgayrac G, Laureyns J, Brémard C. Chemistry at level of individual aerosol particle using multivariate curve resolution of confocal Raman image. Spectrochim Acta A 2006;64:1102–9.
[26] Sobanska S, Falgayrac G, Rimetz-Planchona J, Perdrixc E, Brémard C, Barbillat J. Resolving the internal structure of individual atmospheric aerosol particle by the combination of Atomic Microscopy, ESEM-EDX, Raman and ToF-SIMS imaging. Microchem J 2014;114:89–98.
[27] Batonneau Y, Brémard C, Gengembre L, Laureyns J, Le Maguer A, Le Maguer D, et al. Speciation of PM10 sources of airborne nonferrous metals within the 3-km zone of lead/zinc smelters. Environ Sci Technol 2004;38:5281–9.
[28] de Juan A, Tauler R. Multivariate curve resolution (MCR) from 2000: progress in concepts and applications. Crit Rev Anal Chem 2007;36(3–4):163–76.
[29] Gentili M, Giovannella C, Selci S. IB and X-ray lithographies (NATO Science Series E). Springer; 1994.
[30] IPCC (Intergovernmental Panel on Climate Change). Changes in atmospheric constituents and in radiative forcing. Climate change in 2007, New York: Cambridge University Press; 2007.
[31] de Juan A, Tauler R, Dyson R, Marcolli C, Rault M, Maeder M. Spectroscopic imaging and chemometrics: a powerful combination for global and local sample analysis. Trends Anal Chem 2004;23:70–9.
[32] de Juan A, Maeder M, Hancewicz T, Duponchel L, Tauler R. Chemometric tools for image analysis in infrared and Raman spectroscopic imaging. Weinheim, Germany: Wiley-VCH; 2009. p. 65–106.

[33] Jaumot J, Gargallo R, de Juan A, Tauler R. A graphical user-friendly interface for MCR-ALS: a new tool for multivariate curve resolution in MATLAB. Chemom Intell Lab Syst 2005;76:101–10.

[34] de Juan A, Maeder M, Hancewicz T, Duponchel L, Tauler R. Chemometric tools for image analysis. In: Salzer R, Siesler HW, editors. Infrared and Raman spectroscopic imaging. Weinheim, Germany: Wiley-VCH; 2009. p. 65–106 [chapter 2].

[35] Batonneau Y, Brémard C, Laureyns J, Merlin JC. Microscopic and imaging Raman scattering study of PbS and its photo-oxidation products. J Raman Spectrosc 2000;31:1113–9.

[36] Batonneau Y, Brémard C, Laureyns J, Merlin JC, Winding W. Polarization effects of confocal Raman microspectrometry of crystal powders using interactive self-modelling analysis. J Phys Chem B 2003;107:1502–13.

[37] Uzu G, Sobanska S, Sarret G, Sauvain JJ, Pradere P, Dumat C. Characterization of lead-recycling facility emissions at various workplaces: major insights for sanitary risks assessment. J Hazard Mater 2011;186:1018–27.

Chapter 16

Multivariate Curve Resolution for Magnetic Resonance Image Analysis: Applications in Prostate Cancer Biomarkers Development

J.M. Prats-Montalbán, E. Aguado Sarrió and A. Ferrer[1]
Multivariate Statistical Engineering Group, Universitat Politècnica de València, Valencia, Spain
[1]*Corresponding author: e-mail: aferrer@eio.upv.es*

Chapter Outline

1 INTRODUCTION

Traditionally, multivariate curve resolution (MCR) models have been applied for chemistry applications [1–4]. From early 2000s, the applications of these models on images have increased, mainly in chemistry and pharmaceutical environments [5,6] favored by the similitude between the image data structure (after proper unfolding) and the data arrays commonly analyzed, where each row containing the profile along the columns variables (e.g., wavelengths) of each observation (pixels vs samples).

Data Handling in Science and Technology, Vol. 30. http://dx.doi.org/10.1016/B978-0-444-63638-6.00016-4
Copyright © 2016 Elsevier B.V. All rights reserved.

The wide applicability of MCR on chemical images comes from the fact that the spectrum at each pixel location can be modeled as the weighted sum of its pure chemical compounds present in the whole data structure (i.e., the unfolded image), where this weight is the relative concentration of each chemical compound at that pixel location.

This concept can be easily extrapolated to other environments, as is the case of medical magnetic resonance (MR) images, where the registered signal, which depends on the type of sequence (or study), is also a weighted sum of the relative relevance of each "pure" physiological behavior (phenomenon) at that pixel location.

There are many studies where these tools can be applied. In this chapter, we present the application of MCR on dynamic contrast enhanced (DCE) and diffusion weighted (DW) MR images for the development of new images (imaging biomarkers) able to detect and locate tumors at early stage, in the case of prostate cancer (PCa). The main objective has been to create images with clinical interpretation that will help doctors in diagnosis and prognosis tasks.

PCa is the most common in men over 50 malignant tumors. While it is true that PCa mortality rates are falling dramatically, increased life expectancy and population above 50 years age is increasing the number of detected cases, so efforts are focusing on its early diagnosis and effective treatment.

Today, early diagnosis of PCa is still done after the joint assessment of digital rectal examination (DRE) and the prostate-specific antigen (PSA). However, DRE has low predictive value and great inter- and intraexaminer variation. Regarding PSA, it is accepted as normal values lower than 4 ng/mL. Nevertheless, it is estimated that 32% of men with positive biopsies for tumor have PSA levels below 4 ng/mL (false negatives) and 79% of men without PCa have PSA levels above 4.1 ng/mL (false positives). After a DRE detection or an elevated PSA value, the confirmation test of PCa is the transrectal ultrasound-guided prostate biopsy. Although false positive biopsies hardly occur, it is very insensitive and reaches up to 50% false negative rate, so often it has to be repeated. In addition, the widespread use of PSA serum has caused PCa overdetection with the subsequent overtreatment, affecting up to 90% of PCa patients at early stage. So far, despite multiple discriminant histological studies with biomarkers, it has not been possible to clearly differentiate those biologically harmless and not aggressive PCa from those that will progress and cause significant morbidity and mortality.

Radical prostatectomy and radiation therapy are the current curative treatment options for localized PCa, but have a nonnegligible urinary, sexual, and intestinal morbidity. As a result, we face the dilemma of treating PCas that would have never caused death, with the added consequence of worsening the quality of life of these patients.

Focal therapy is an alternative between active surveillance (stressful for patients) and radical therapy (with their described side effects). This way, by effectively acting on the tumor, the rest of the gland is preserved and the side effects are reduced. Its effectiveness requires accurate information on the location and spread of PCa within the gland (PCa intraglandular map) and the degree of aggressiveness. Conventional prostate biopsy provides limited information on the PCa intraglandular map. Furthermore, results from conventional MRI T2 usually not provide enough information for proper guiding of the radiation therapy [7]. It is therefore necessary to develop new procedures to accurately detect and locate the tumors [8]. And for this, it is mandatory to understand the nature and behavior of the tumoral processes.

The main indicators of a tumor process are vascular tissue and cell proliferation. Angiogenesis (creation of new vascular tissue) and neovascularization (development of the existing ones) are phenomena that occur when a group of growing cells presents abnormally high oxygen demands. In these circumstances, new vessels and capillaries are formed, in order to allowing tumor growth. This new vascular network is chaotic and does not present a normal structure. They are fragile and highly permeable to macromolecules, due to the presence of large gaps between vessels and endothelial cells. Furthermore, they have a very unstable blood flow. These newly formed vessels are biological markers of malignancy, since vascular proliferation is a descriptor used for histological tumor grade, with cell proliferation and necrosis.

In order to improve diagnosis, tracking, and tracing of these malignant tumor processes, new magnetic resonance imaging (MRI) techniques have been developed to acquire complementary information to that obtained from conventional MRI (not able to discriminate the spread of invasive part of the tumor): DCE-MRI and DW-MRI [9,10].

2 STATE-OF-THE-ART METHODS

2.1 Dynamic Contrast Enhanced-Magnetic Resonance Imaging

The analysis of tumor angiogenesis can be performed from DCE-MR images, by studying the perfusion process after the administration of an intravenous contrast. In these studies, an exogenous contrast media is administered intravenously and diffuses from the capillary network into the extravascular extracellular space (EES), establishing a dynamic relationship between the image signal intensity changes and the amount of contrast media that passes and diffuses into a certain tissue. From these dynamics, perfusion models are created, which provide quantitative and reproducible information relevant to characterize tumors or other tissues with pathological changes associated with some microvascular components.

In order to obtain these quantitative measurements, it is necessary to characterize the intensity vs time curves (enhancement curve) associated to each pixel of the image. Out of the different approaches proposed to achieve this evaluation, mathematical pharmacokinetic models have become the most popular in recent years due to their ability to provide physiologically meaningful parameters [11].

Pharmacokinetic modeling of DCE-MR images is based on the application of a mathematical analysis to tissue enhancement curves on a pixel-by-pixel basis.

Here, each individual pixel enhancement curve is fitted to the pharmacokinetic model chosen. In this work, the Tofts model [11], which is one of the most commonly used, is presented in Eq. (1):

$$C(x,y,t) = K^{\text{trans}} \int_0^t C_{\text{AIF}}(\tau) e^{-k_{ep}\tau} d\tau \tag{1}$$

where $C(x,y,t)$ is the tissue enhancement curve (i.e., a vector) at pixel with coordinates (x,y), $C_{\text{AIF}}(t)$ is the enhancement curve of the arterial input function (AIF, the artery that supplies most blood to the tissue of interest), k_{ep} is the washout constant, which measures the average time spent in EES or contrast transfer coefficient between EES and plasma; and K^{trans} is the transfer constant, which evaluates vascular permeability or transfer coefficient between plasma and EES.

The model needs as input the C_{AIF} and the tissue enhancement curves $C(x,y,t)$. This C_{AIF} is usually extracted by manually placing regions of interest (ROIs) in the artery (i.e., one of the iliac arteries for the prostate) and the tissue of interest (i.e., the prostate in this study). Then the pharmacokinetic parameters k_{ep} and K^{trans} can be obtained by curve fitting algorithms for each pixel in the image area under study. This fact is of prior relevance, since the accuracy in the computation of these parameters depends on the proper determination of the AIF: the higher the uncertainty in the determination of the AIF, the higher the uncertainty in the estimation of the pharmacokinetic parameters.

One characteristic of these models is the lack of a priori knowledge about the tissue vascular environment, which leads to a series of assumptions that condition the use of different pharmacokinetic modeling approaches [12]. Accordingly, and depending on the tissue dynamics patterns (Fig. 1), the pharmacokinetic model may provide biased measurements which may not properly reflect the true physiology of the tissue [7]. This variability affecting the calculations and interpretation of the results provided by these parametric models has limited their applicability and widespread use in clinical practice, as it is difficult to obtain a validation and standardization of the parameters that ensure reproducibility among different studies. It is therefore desirable to have a priori knowledge about the dynamics of tissue to easily interpret the information provided by the pharmacokinetic parameters. Thus it becomes necessary to have models that are able to directly extract such a priori knowledge from the data sequences and create robust models that are able to cope with external artificial variability that can affect the biomarker computation.

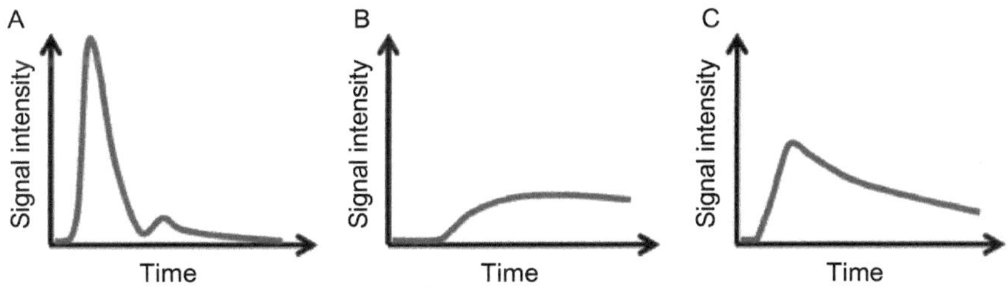

FIG. 1 Dynamic patterns after the injection of a contrast media in dynamic contrast-enhanced magnetic resonance images. (A) Abrupt initial peak and fast washout: AIF. (B) Slow progressive enhancement without washout. (C) Delayed fast initial enhancement and slow washout. Curves (B) and (C) are often analyzed using the same pharmacokinetic model, which may lead to biased results in the estimated parameters.

In this case, when a tumor is evolving in the prostate, three different dynamic behaviors can be expected:

1. *Type AIF*: abrupt initial peak and fast washout (blood flowing within the arteries), corresponding to the dynamics pattern in the artery (Fig. 1A).
2. *Type NT*: slow progressive enhancement without washout, corresponding to a nontumoral tissue (Fig. 1B).
3. *Type VT*: delayed fast initial enhancement and slow washout, corresponding to a highly vascular tissue, such as a tumor (Fig. 1C).

Furthermore, it is of primary importance to check if some of the underlying dynamics is related to the AIF, obtained from the tumor-feeding artery [13–16]. As commented before, the determination of this curve is often based on the visual inspection of a volume of images and the manual selection of the ROI including the artery. This approach introduces a user-dependent bias into the analysis, penalizing the reproducibility of the results. Therefore being able to extract this AIF in a fast and objective way, moreover being patient-focused, would result in major benefits related to radiologist time, reproducibility, and personalized medicine.

2.2 Diffusion Weighted-Magnetic Resonance Imaging

Another way of facing the oncogenesis (combination of cellularization and vascularization) process is by studying the tissue local diffusion process [17], which is a physical process that occurs due to the thermal agitation of the water molecules inside the human body. These translational displacements depend, among other factors, on the tissue structure according to the cellular organization. Tumors are characterized by an increased cellularity, with rests of necrosis and fibrosis, all reducing the mobility of water molecules. When the tissue is highly cellularized, the molecules have more restrictions to movement due to a decreased interstitial space and higher cell membrane interfaces. However, when the tissue is highly vascularized, molecules are in a nonrestricted high-velocity environment within the vessels, and the spatial movements are random with fewer restrictions in all spatial directions.

The diffusion process can be evaluated with DW-MRI. This noninvasive technique provides high-resolution images that are sensitive to the water molecules movement inside the tissues. Depending on the configuration of the MR equipment and based on the duration and the amplitude of the applied magnetic field gradient, image acquisition is associated to a parameter known as b-value [8]. The signal of the image decreases with the increase in the b-value acquired. This attenuation depends on the characteristics of the tissue, being stronger if the tissue is vascularized and much more moderate if it is highly cellular. The range of different signal attenuations between these two types of tissues at the same b-value is the basics to study the different behaviors in the diffusion process.

In order to model the signal decay s of the diffusion process signals, spectra can be fitted with different models. For those based on the isotropic decay of the signal as a function of the value b (as in the prostate), the most widely used is the monoexponential model, whose associated parameter, the apparent diffusion coefficient (ADC) integrates the global effect of the slow displacements (intracellular and interstitial) as well as fast movements due to extravascular diffusion:

$$\frac{s}{s_0} = e^{-b(\mathrm{ADC})} \tag{2}$$

where s_0 is the initial value of the signal s when the b-value equals zero. The ADC values express the average distance that the water molecules cover within a voxel at a certain time. It is related with the cell density, the permeability of the membranes, and the tortuosity of the intercellular interstitial space. It is called "apparent" because it reflects several different mechanisms, as it is a combination of the two phenomena expressed before: (i) the movement associated to the water molecules (Brownian movement) inside the cellular tissue, slow diffusion (typical in cellular tissue) and (ii) the intravascular movement in the microcapillaries vessels, fast diffusion (typical in vascularized tissue). Fast diffusion may produce an overestimation of the real diffusion values if not properly considered.

This way, the monoexponential model does not take into account the different mechanisms of the diffusion process. A more advanced biexponential mathematical model called intravoxel incoherent motion (IVIM) [18] is able to separate these two effects from the signal decay in the voxel with increasing values of b [12,19]. This allows quantifying and segregating the slow diffusion (real diffusion), due to extravascular spread (cellularization); from fast diffusion (pseudo-perfusion), due to vascularization [20]. This helps in a more accurate detection of tumors, since they are related to low-real diffusion values and higher pseudo-perfusion values. The expression of the IVIM model is shown below:

$$\frac{s}{s_0} = (1-f)e^{-bD} + fe^{-b(D+D^*)}. \tag{3}$$

The signal decays s are normalized with s_0 as in the monoexponential approach. Three different parameters must be estimated: the slow diffusion coefficient (D), the fast diffusion coefficient (D^*) and the vascular fraction (f), which relates to the proportion of vascular tissue in a voxel. This way, the normalized signal s/s_0 is modeled as a weighted average of the slow diffusion (water movement inside the cellular tissue, characterized by D) and the fast diffusion (water movement inside the vascular tissue, characterized by $D+D^*$). The slow diffusion behavior is weighted by $(1-f)$ and the fast diffusion term by f, since the fast diffusion is related to the vascular tissue. Despite the fact that the IVIM model is theoretically more appropriate according to physiological criteria, the monoexponential model is, nowadays, the most widely used in medical practice to model the diffusion process, because of the high computational demand of the IVIM biomarkers acquisition process, as well as their complex interpretability.

Note that the IVIM model is not a typical biexponential model because the two exponential decays are not independent as they are complementary weighted by the vascular fraction, f. Furthermore, the normalization of the spectra causes a distortion, modifying the shape of the original curve (Fig. 2A) and decreasing the signal-to-noise ratio (Fig. 2B). Thus, this standardization reduces the variability range masking the different behaviors present in the signal decays. All these concepts shown above and the difficulty in the interpretation of the results provided by these biomarkers (such as D, D^*, and f parameters) have limited their applicability in clinical practice.

Furthermore, these biomarkers are obtained, as in the case of DCE-MRI, from a generally pixel-by-pixel modeling, which do not take advantage of the relation between pixels with the same behavior, increasing the uncertainty in their estimation, and degrading the corresponding imaging biomarkers

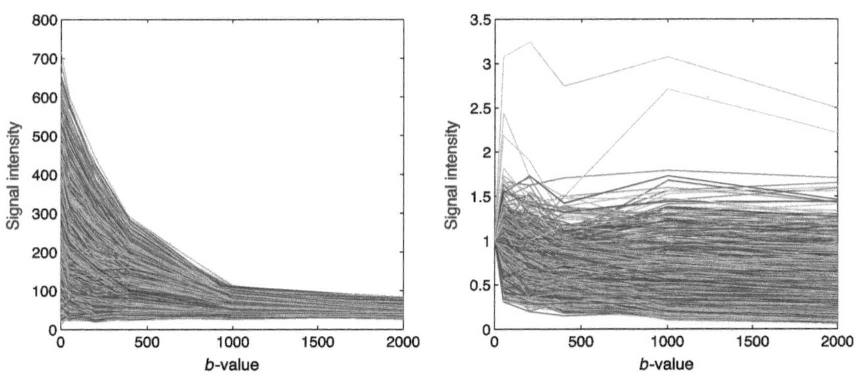

FIG. 2 Signal attenuation spectra in a DW-MRI case. (A) Before normalization. (B) After normalization.

(images built from each D, D^*, and f parameters at each pixel location) used for clinical purposes.

The combined analysis of morphological MR sequences with one or more of the functional sequences (perfusion, diffusion) is known as multiparametric MR.

The different characteristics and parameters extracted by these techniques can be spatially interpreted as imaging biomarkers [12,20]. An imaging biomarker can be defined as an extracted feature of medical imaging, which can be measured objectively (i.e., give quantitative information) and acts as an indicator of a normal biological process, a disease or a response to a therapeutic intervention [21].

However, analysis of imaging biomarkers poses different problems:

- These imaging biomarkers are obtained in a univariate way by fitting them to some mathematical models, not taking advantage of the internal correlation structures between pixels.
- The simultaneous evaluation of several biomarkers is difficult, even more when considering the different combinations that may be related to a tumor. This can be complicated and require a priori knowledge.
- The types of biomarkers to analyze depend on the studied organ (prostate, breast, brain, etc.).
- Some of the obtained parameters (e.g., the transfer coefficient of K^{trans}) have a complex physiological interpretation [22], since it mixes permeability and vascular flow. Therefore, it seems interesting to obtain biomarkers having an easy clinical interpretation.
- Likewise, there are different models of different levels of complexity both for perfusion and diffusion. Such models are justified in certain cases (e.g., IVIM in diffusion). In other cases (e.g., certain pharmacokinetic models [12]) the complexity has not been justified yet. Therefore, it becomes necessary to have models that are able to extract such a priori knowledge from data directly extracted from the sequences.
- In addition to this, the a priori knowledge about the correlation between biomarkers (combinations of values between them) when determining the existence of a tumor may not be enough, since there may be other combinations that also lead to the detection of a tumor; which results in false negatives, that is, in nondetected tumors. This possibility increases with the number of biomarkers.
- Also these biomarkers may suffer from bias in their estimation due to measurement errors introduced by the different devices (ghosts) and/or reproducibility errors associated with the values of some model input function from which are obtained (e.g., the AIF).

3 NEW BIOMARKERS DEVELOPMENT

One way to address the above issues and challenges is to analyze the imaging biomarkers in a multivariate way. Multivariate statistical projection models

[4,23,24] allow analyzing large data sets, obtaining simplified structures that help to understand the relationships between the studied variables and the underlying physiological phenomena.

The application of these techniques in images is called multivariate image analysis (MIA) [25,26]. MIA is the application of multivariate statistical projection models on data structures from images. In the case of medical images, the observations are formed by each of the pixels of the images, while columns contain the signal of each pixel at each time in the case of the pharmacokinetic perfusion models, or b-values in diffusion models.

The application of MIA in oncology [27–29] allows extracting the sources of variation from a relevant number of sequenced images from different individuals, providing new statistical models that help in describing and explaining the behavior differences between healthy and cancer-affected tissues, as well as the physiological phenomena undergoing, better than the mathematical imposed model. These models reduce the uncertainty in the estimation of the parameters obtained from the computational models for the treatment of medical images [30–32]. They provide new biomarkers, complementary to those commonly used and with clinical interpretation [27,29] in the case of using MCR models [33], which directly point out to the pixels related to each physiological behavior and areas of starting active tumor growth, facilitating the clinical diagnosis, as well as other ROIs in an objective and reproducible way (e.g., AIF [29,34]).

This might be at least partially due to the fact that "the wealth of physiological assumptions likely oversimplifies an exceptionally complex system [28]," varying from patient to patient. Thus, this procedure comes out as a proper way to deal with this natural variation and seek, in the case of the pharmacokinetic models, for the best candidate pixels for the estimation of the AIF, thus reducing the uncertainty in the estimation of the pharmacokinetic parameters.

The main and the most widespread MIA tool is principal component analysis (PCA) [24]. However, two problems arise when PCA is applied on MRI data: (i) no prior information can be included in the model and (ii) the orthogonality of the principal components (PCs) is a limitation to model the different diffusion behaviors that are not necessarily orthogonal. In order to overcome these drawbacks, it is possible to use more flexible models, as is the case of MCR models [4,35]. MCR does not impose these types of restrictions, but others related to some a priori knowledge, when available; hence being able to provide more clinically (or physiologically) interpretable results. Applications of MCR models on biomedical images can be found in Refs. [27,29,36].

For applying MIA, since we are focusing on the pixels behaviors, it is necessary to unfold each image ($n_1 \times n_2 \times J$) ($n_1 \times n_2$=image size, $J=t$ in DCE-MR, $J=b$ in DW-MR), thus forming a 2D $\mathbf{S}$ ($(n_1 \times n_2) \times J$) matrix that can be analyzed by MCR. Furthermore, since it is necessary to analyze the

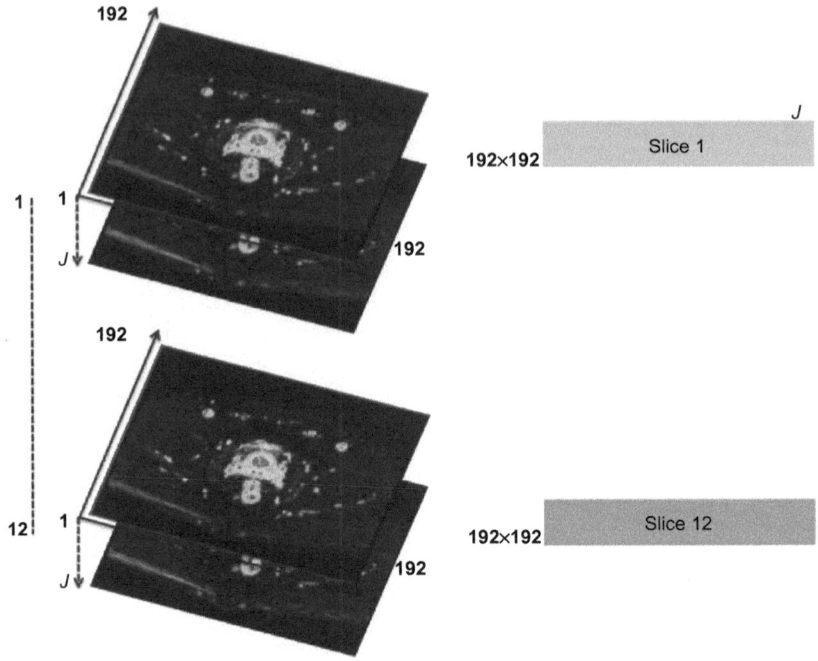

FIG. 3 Data structure for MCR analysis. Images are 192 × 192 pixels size, with a spatial resolution of 1,5625 × 1,5625 mm^2 per pixel.

whole organ volume, each slice (12 in this case, both for DCE-MR and DW-MR) needs to be stacked one below the other, forming the data structure shown in Fig. 3. This type of unfolding, keeping the J variables in columns, has to be applied because the underlying physiological phenomena are assumed to be the same throughout the whole prostate volume.

The following sections will show how these issues have been addressed. We will show how MCR can help in developing new biomarkers easy to interpret, both in perfusion and diffusion. In the case of perfusion, we will show how these new biomarkers are related to those obtained from the pharmacokinetic models. We will also provide a methodology for reducing the uncertainty in the estimation of the pharmacokinetic models (filtering). Finally, for the diffusion sequence, a method for validating the different models used will be shown.

3.1 Materials and Methods

The database consists of DCE-MRI and DW-MRI acquired from patients with proven prostate carcinoma. The different types of MRI sequences are held throughout the volume of the organ, at different spatial levels corresponding to different sections of the human body (the number of slices depends on the organ studied, 12 slices in the case of the prostate) for different times (perfusion) or for different values of the *b* parameter (diffusion), building the data structure shown in Fig. 3. All images were anonymized and transferred to a dedicated workstation for postprocessing. For each slice, images with a resolution of 192×192 pixels were acquired with the following characteristics:

- For the DCE-MR images, a high-temporal resolution sequence (3 Tesla magnet, 3D spoiled gradient echo sequence, repetition time 3 ms, echo time 1.7 ms, flip angle 40°, 3 s per image, total acquisition time 5′30″, 47 nonequally spaced temporal samples contrast agent Gd-DOTA 0.2 mL/kg injected at 4 mL/s with 40 mL of saline flush at the same rate). The output of this sequence was a series of 564 images (12 slices $\times$ 47 time steps) showing the pelvis area perfusion dynamics after the contrast media administration.
- For the DW-MR images, six different *b*-values (0, 50, 200, 400, 1000, and 2000 s/mm^2). The output for each patient was a series of 72 (12 slices $\times$ 6 *b*-values) images, showing the pelvis area and decay of the signal with the *b*-value.

All calculations were done using Matlab (The Mathworks Inc., Natick, MA, USA). Multivariate statistical analyses were performed using SIMCA (MKS Umetrics AB, Umea, Sweden). MCR-alternating least squares (ALS) algorithms were self coded or downloaded from http://mcrals.info/ [37].

3.2 Dynamic Contrast Enhanced-Magnetic Resonance Imaging

By applying the unfolding procedure shown in Fig. 3, an **S** matrix (($12 \times 192 \times 192$) rows $\times$ 47 columns) is obtained for the DCE-MR images related to each patient. This way, MCR can be applied. In this case, an MCR-ALS [4,35,38,39] algorithm has been used.

The idea behind MCR, traditionally applied in analytical chemistry, can be easily transferred to the concept of dynamic perfusion phenomena, stating that the pixel enhancement curve $C(x,y,t)$ is a linear combination of the different "pure" dynamic behaviors existing in it. The lack of orthogonality restriction in the pure dynamic behaviors improves the model in terms of physiological interpretability.

MCR-ALS is an iterative method that performs a bilinear decomposition of matrix **S** by means of an ALS optimization (see Chapter 2):

$$\mathbf{S} = \mathbf{C}\mathbf{D}^{\mathrm{T}} + \mathbf{E} \tag{4}$$

In this particular application, $\mathbf{D}^{\mathrm{T}}$ is a matrix containing in its rows each one of the dynamic behaviors modeled, **C** gathers in its rows the relative importance of each modeled dynamic behavior for each pixel of the image, and **E** is a residual matrix. Be aware that this notation is not the commonly used in MCR, where **D** usually relates to the data matrix and **S** to the pure spectra.

However, the iterative process can provide infinite solutions for the same data matrix. This problem can be solved by imposing other type of constraints commonly related to prior knowledge about the problem faced, so that it is possible to obtain easier-to-interpret solutions, which also tend to be unique when the constraints introduced under the hypothesized assumptions are sensible [40].

MCR relies in the correct determination of the number of “real” dynamic behaviors in a data set (the image in this case). When some a priori information is available, it can be used as an initial estimation. This a priori knowledge can be checked by using some tool able to show up the relevant sources of information in the time-sequenced images. One possible way to do this is by applying PCA on the data set, and taking a look at the number of latent variables (PCs) with the highest variances. Once the number of likely dynamic behaviors present in the image is determined, the purest dynamics in the raw data can be sought using e.g., simple-to-use interactive self-modeling mixture analysis (SIMPLISMA) [41] or more recent approaches proposed by Ref. [42] (see Chapter 3). ROIs of loadings extracted from PCA (not the best option in this case, because not all the purest dynamic behaviors could be a priori known to be located at some pixels nor negative dynamics behaviors could happen in reality), genetic algorithms [43–45], or orthogonal projection approach [46]. In order to ease the calculations, this step can be made on some slice known to retain these main behaviors.

The previous steps only help on the initial estimation of the dynamic behaviors. However, these dynamic behaviors do not need to be present in all pixels. In order to find out the likely number of dynamic behaviors present in a pixel, the fixed size image window-evolving factor analysis [47,48] can be applied, as in Ref. [29]. Nevertheless, since this step is focused on finding a good estimation of the AIF, the experience in this problem has led to impose the following constraints: nonnegativity on the pixel intensity values, because the intensity in a pixel has to be nonnegative, and nonnegativity on the dynamic profiles.

Regarding the use of additional constraints, as in this case that related to the form of the AIF and its parameterization, since different types of equations might be used [49,50], it is preferably not to impose any a priori form that might bias the results, and leave the model free to fit the best AIF in each case. The MCR model built this way provides the matrices $\mathbf{D}^{\mathrm{T}}$ and **C** previously commented, gathering the dynamic behaviors found, as well as the corresponding relative importance of each dynamic behavior at each pixel location, respectively. By folding the **C** matrix back into the original $n_1 \times n_2$ spatial dimension, new images (known as distribution maps) are obtained, which permit to locate those pixels more related to each of the corresponding dynamic behaviors provided by the model.

The whole sequential procedure is summarized in the following steps:

(a) Determination of the approximate number of dynamic behaviors by PCA, according to the number of relevant singular values. If a priori knowledge is available, it should be used.
(b) Determination of the pixels with purest dynamic behaviors.
(c) Application of the MCR-ALS with the additional constraints of nonnegativity both on the pixel intensities and the dynamic behaviors.

Applying these first three steps on the whole pelvic area is necessary in order to obtain an accurate AIF.

(d) For improving the precision in the prostate area, local MCR models were built for each of the prostates analyzed. The same ROIs defined at the pharmacokinetic modeling step were used here for segmentation purposes. The dynamic behaviors obtained in the previous MCR model, with the exception of the AIF (which is present in the arteries, but not in the prostate) can be implemented as initial guess. The same constraints should be imposed.

In order to obtain an accurate initial estimation of the dynamic behaviors and for simplicity reasons, it is recommended to first apply these four steps on some slice assumed to be representative of the prostate, afterward extending the results of step 4 to the whole prostate volume.

To illustrate the application of the MCR-based methodology on the DCE-MR images, one case is discussed in detail on a patient with a carcinoma in the translational gland (inner part of the prostate).

As commented before, the first step in any MCR analysis on DCE-MR images deals with assessing for the number of significant dynamic behaviors in the image. Fig. 3 presents the variances of the 2nd to 10th PCs from the fitted PCA model. It can be seen that from PC 5 in advance, all the PCs have similar amount of explained variance, and even PC 4 is doubtful. This means that not only the three a priori assumed physiological dynamic behaviors may be present in the MR images but also some more unknown behavior. Therefore both possibilities, using three (results not shown) and four behaviors, should be checked.

Assuming four potential dynamic behaviors, the purest ones to be introduced in the model are found. Fig. 5 shows the results corresponding to the case under study for the four dynamic behaviors indicated by PCA in Fig. 4. As can be seen, not only the types AIF, NT, and VT can be guessed but also an (assumed) artificial effect, linked to a contrast media arrival (type CMA) (Fig. 5).

The next step consists on applying the nonnegativity constraints on the pixels' intensities and on the dynamic behaviors in the MCR model, matrices **C** and **D**, respectively. Fig. 6 shows the obtained dynamic behaviors, very similar to those provided by the purity-based methods. The main difference

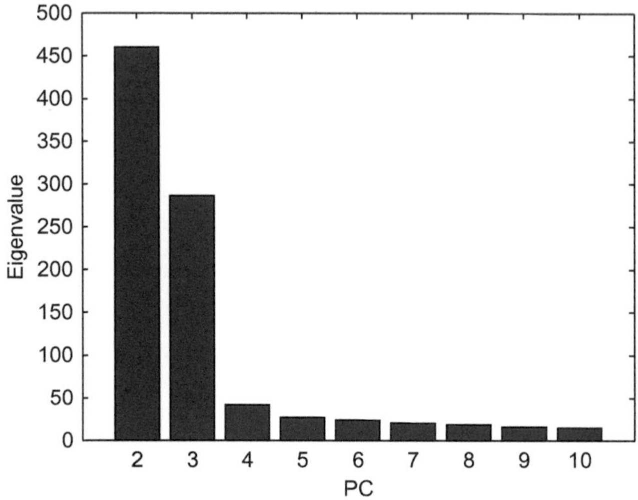

FIG. 4 Variance related to each PC, from the second PC to the 10th.

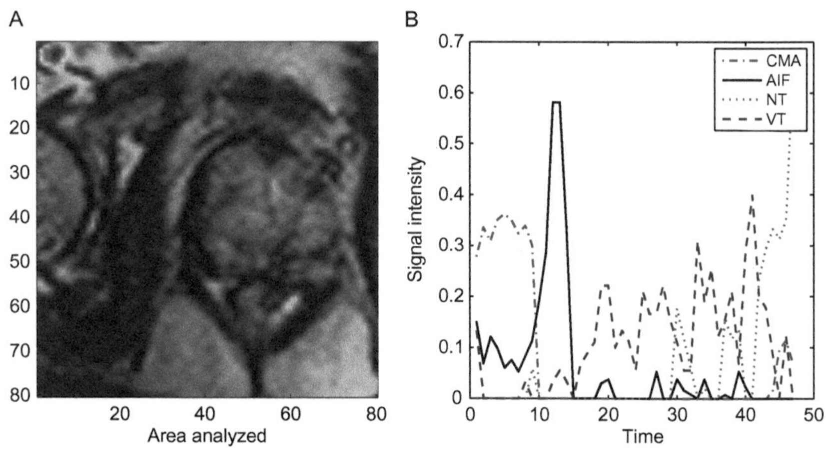

FIG. 5 (A) Image area selected for finding the initial estimates of the physiological dynamic behaviors. (B) The corresponding found initial guesses: Nonphysiological Contrast Media Arrival (type CMA, *dotted-dashed green* (*gray* in the print version)) effect, type AIF (*solid black*), type NT (*dotted blue* (*dark gray* in the print version)), and type VT (*dashed red* (*light gray* in the print version)) initial guess dynamic behaviors.

would be in the estimation of the AIF, which is quite important for the pharmacokinetic modeling, as already stated. It is also possible to see that VT behavior is delayed with respect the AIF (about 3–4 s) and presents much larger amplitude.

In order to visualize where these dynamic behaviors are more predominant, it is necessary to fold back each of the columns of the **C** matrix into their

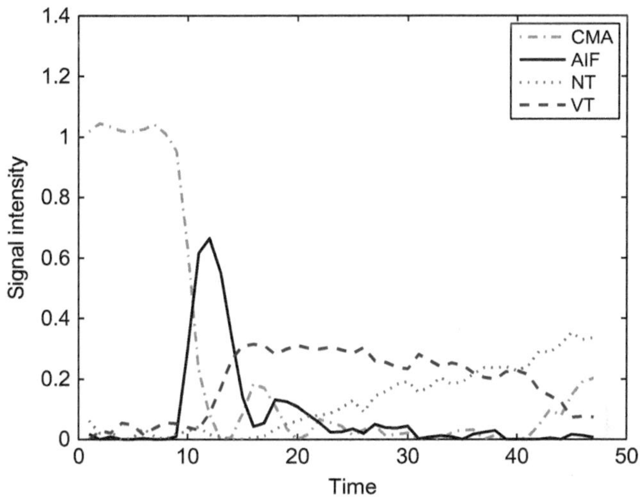

FIG. 6 Dynamic behaviors provided by the MCR model: type AIF (*black solid line*), type NT (*blue* (*dark gray* in the print version) *dotted line*), type VT (*red* (*light gray* in the print version) *dashed line*), and nonphysiological Contrast Media Arrival (type CMA) effect (*green* (*gray* in the print version) *dashed-dotted line*).

original spatial dimension, $n_1 \times n_2$, obtaining the dynamic behaviors distribution maps (Fig. 7). By inspecting both figures, we can appreciate that VT behavior (Fig. 7D) is slightly happening in some internal area of the prostate, and it is an indicator of a highly vascularized region. In addition, Fig. 7B provides the pixels that show the arterial perfusion process sought, corresponding to the iliac arteries. Furthermore, there is a progressive enhancement in the rest of the prostate, as well as in the rest of the tissue, which is in accordance to the healthy-like contrast uptake process (type NT) (Fig. 7C). Finally, it can be assessed that there is a clear additional nonphysiological type CMA effect (Fig. 7A) of no physiological interest, but also captured by the model. This nondesired effect is probably inherent to MR studies, which use an exogenous contrast media. The arrival of the contrast media in the tissue of interest changes its magnetic properties and the signal intensity, so that the images are slightly affected too, though the MR equipment usually autoscales them to minimize the effects. Also, this fourth dynamic is predominant at certain low-vascular peripheral regions (see Fig. 7A), which correspond to some specific elements of the MR receiving coil. Nevertheless, it would be convenient to subtract this CMA from the data structure before the analysis, in order to better determine the AIF, as well as to improve the pharmacokinetic biomarkers calculation. From Fig. 7, it is possible to appreciate that the CMA effect seems to affect, to some extent, the VT and NT types in the peripheral areas.

Up to this step, it must be stated that PCA is also able to extract out the AIF function in a very short time [34], but not the segregation between the

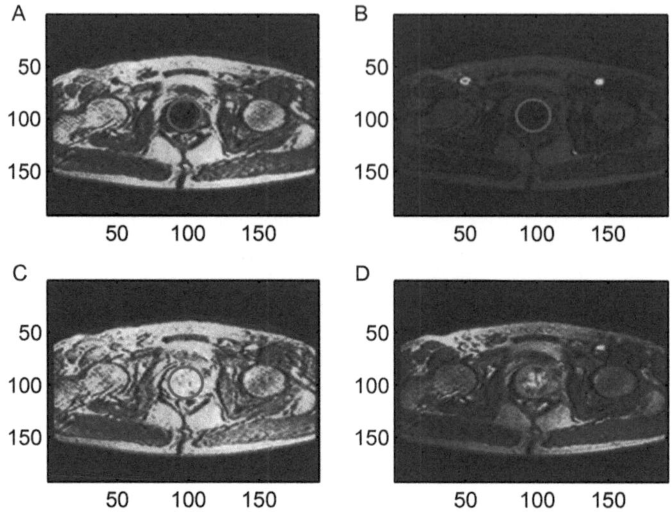

FIG. 7 Distribution maps of the dynamic behaviors shown in Fig. 5. (A) Nonphysiological Contrast Media Arrival (Type CMA) effect. (B) Type AIF. (C) Type NT. (D) Type VT. *Red* (*gray* in the print version) *circle* surrounds the prostate area.

behaviors in the prostate, nor the existence of the type CMA behavior. Furthermore, the fact of imposing orthogonality often forces the AIF dynamic to disaggregate on more than one PC (although this is not a limitation for the procedure related to the segmentation of the iliac arteries [34]).

The dynamic behaviors previously detected may still be estimated more precisely when focusing into the prostate area local model. In order to deal with this issue, MCR local models for the prostate area are fitted. This approach dismisses the AIF pure arterial behavior, since it is not present in the prostate ROI. Fig. 8A shows the final optimized dynamic behaviors and distribution maps for the prostate area MCR local model. Furthermore, MCR bands [40] can be also applied in order to check for possible ambiguities, (see Chapter 4) and study the feasibility of the final dynamic behaviors obtained. As shown in Fig. 8B, the bands are overlapping with the proposed solution. In other cases [29], the bands are very close to the found dynamics, which gives confidence (joint to the clinical validation) to the results.

As can be seen by comparing Figs. 6 and 8, all dynamic behaviors remain essentially the same. Only the nonphysiological CMA effect has slightly changed, losing a peak at time point 17. On the other hand, Fig. 9 shows the benefits of focusing on the prostate: now, we can see the VT dynamic distribution map (Fig. 9C) clearly showing up the pixels with higher intensities, i.e., more related to this type of behavior.

All these steps validate to some extent the methodology proposed. However, they serve as an initial guess for the real goal, consisting on applying

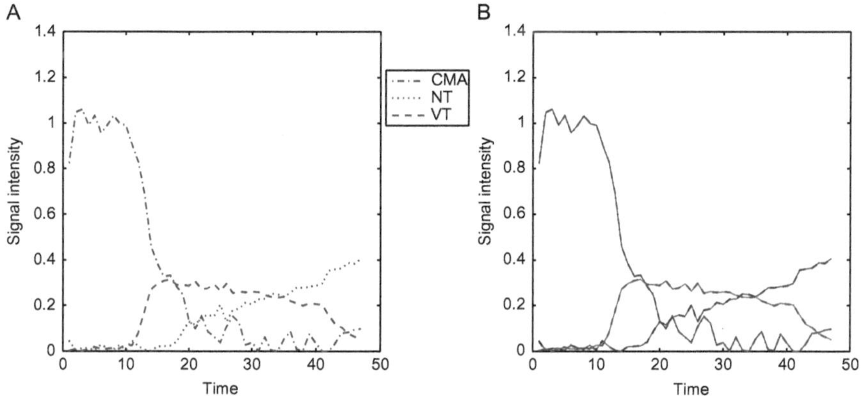

FIG. 8 (A) Dynamic behaviors provided by the prostate area MCR local model: type NT (*blue* (*dark gray* in the print version) *dotted line*), type VT (*red* (*light gray* in the print version) *dashed line*), and type CMA (*green* (*gray* in the print version) *dotted-dashed line*). Note that type A has not been included as there are no arteries in the prostate area. (B) Feasible bands.

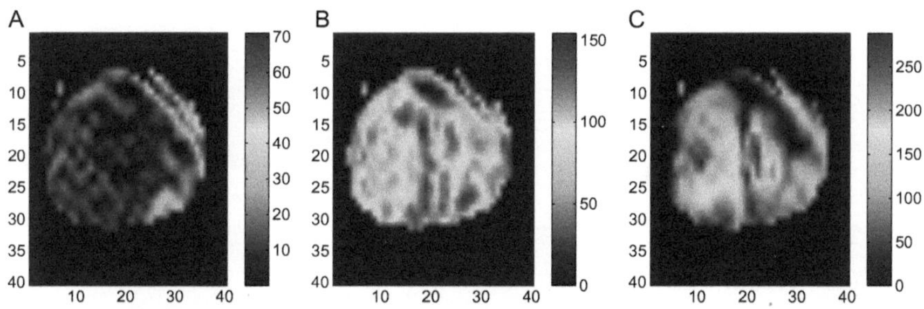

FIG. 9 Distribution maps of the dynamic behaviors from prostate MCR local model shown in Fig. 7. (A) Type CMA. (B) Type NT. (C) Type VT. Note that type AIF has not been included as there are no arteries in the prostate area.

the MCR models to the whole data structure, as indicated in Fig. 3. By stacking all the slices one below the other and using these found dynamic behaviors as initial guess on the masked selected pixels (corresponding to the prostate area), and applying MCR with the commented nonnegativity constraints both on the intensities and the dynamics behaviors, Fig. 10 presents the final global optimal solution (a) as well as the corresponding feasible bands (b). There is only a very small variability at the end of the VT dynamics behavior. Moreover, these dynamic behaviors are very similar to those used as initial guess, shown in Fig. 8A, indicating that behaviors do not change that much throughout the prostate volume.

The different distribution maps are shown for all 12 slices: Fig. 11, corresponding to the CMA artificial dynamic behavior; Fig. 12, corresponding to the NT dynamic behavior; and Fig. 13, corresponding to the VT dynamic

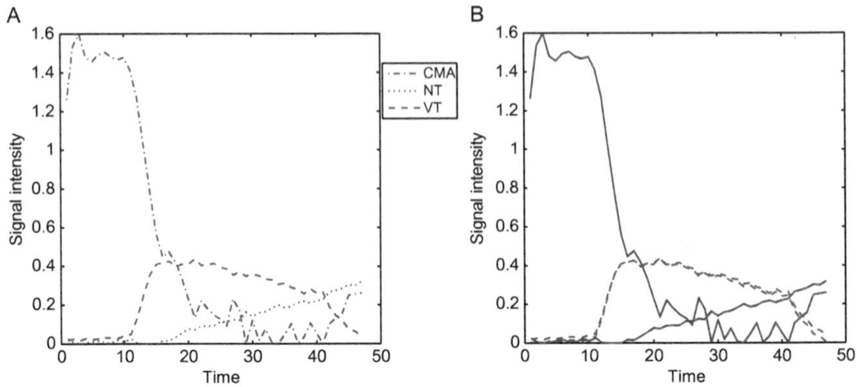

FIG. 10 (A) Dynamic behaviors provided by the prostate area MCR global model: type NT (*blue* (*dark gray* in the print version) *dotted line*), type VT (*red* (*light gray* in the print version) *dashed line*), and type CMA (*green* (*gray* in the print version) *dotted-dashed line*). Note that type A has not been included as there are no arteries in the prostate area. (B) Feasible bands.

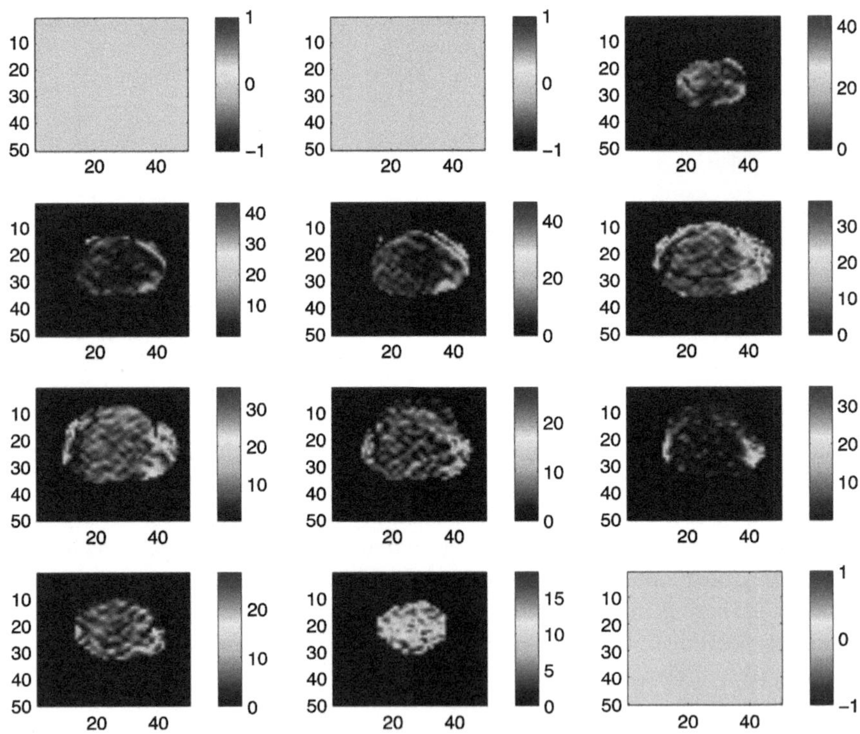

FIG. 11 CMA distribution maps for the 12 slices analyzed, from left to right and from top to bottom. Note that slices 1, 2, and 12 do not have prostate area masked. Masks are the union of the peripheral and inner prostate zones, so some slices present prostate pixels unselected.

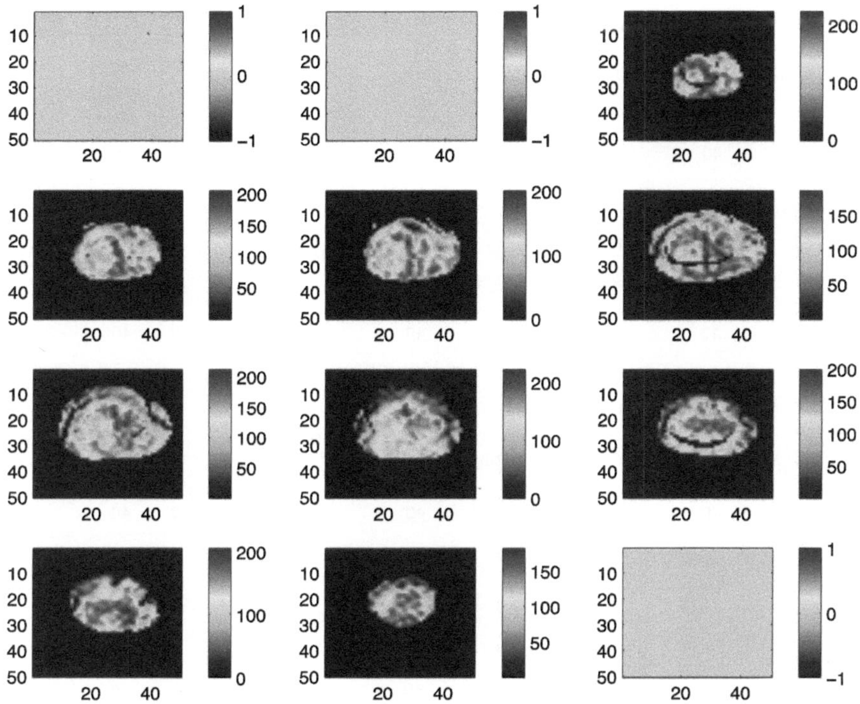

FIG. 12 NT distribution maps for the 12 slices analyzed, from left to right and from top to bottom. Note that slices 1, 2, and 12 do not have prostate area masked. Masks are the union of the peripheral and inner prostate zones, so some slices present prostate pixels unselected.

behavior. The model gathers a 94.6% of the variability in the data. This is extendable to all cases analyzed. Fig. 14 provides the residuals sum of squares (RSS) images that signal the pixels with some abnormal deviations with respect to the modeled dynamic behaviors.

Using this analysis scheme it is thus possible to obtain new and easy-to-interpret biomarkers MCR-based biomarkers, which provide a direct clinical interpretation of the underlying events of interest. Moreover, it is possible to detect (whenever existing) nondesired artificial effects, hence being able to extract them out from the pure physiological dynamic behaviors and improving the estimation, not only of the VT and NT dynamics but also of the AIF. Finally, it must be stated that using some kind of population dynamics behavior as initial guess provides essentially the same results.

3.3 Diffusion Weighted-Magnetic Resonance Imaging

The DW-MRI acquisition sequence is performed along the volume of the studied organ, at spatial planes corresponding to different slices of the human body. For each slice, images are taken with different b-values, obtaining a 3D

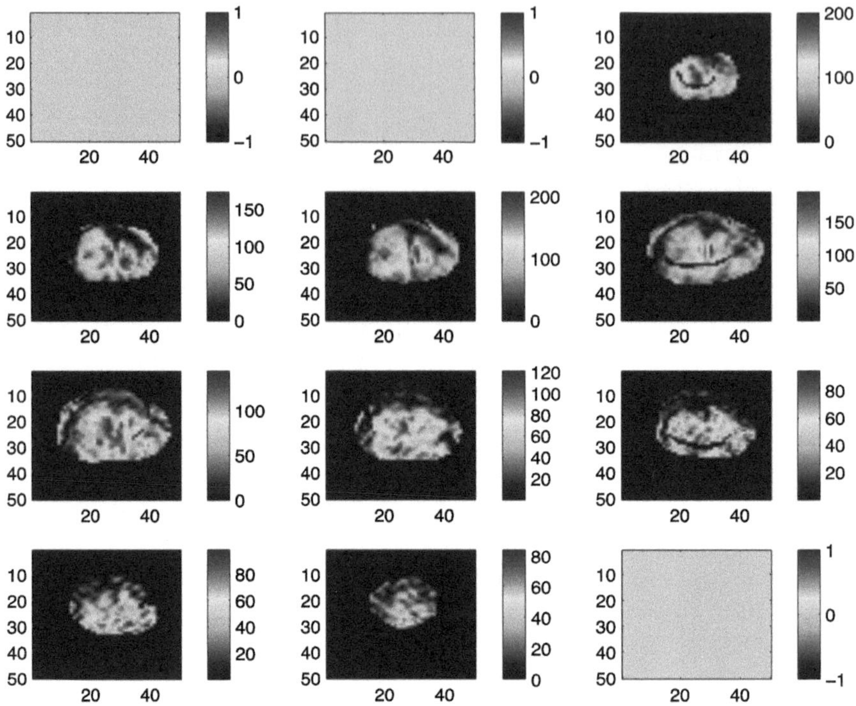

FIG. 13 VT distribution maps for the 12 slices analyzed, from left to right and from top to bottom. Note that slices 1, 2, and 12 do not have prostate area masked. Masks are the union of the peripheral and inner prostate zones, so some slices present prostate pixels unselected.

12 slices × (192 × 192) rows × 6 columns data structure as in Fig. 3. This way, a signal decay *s* is extracted from each pixel of the image, associated to the different *b*-values. The number of *b*-values varies among clinical studies, reaching up to 10 values for the clinical setting [51]. In our study, 6 *b*-values were used for prostate imaging based on previous experience. In all cases, local models were built by removing the pixels that do not pertain to the prostate zone with manual masks provided by the doctors. This way, the interpretation of the results is improved and the computational time reduced.

In order to develop the new biomarkers in DW-MRI, it is first required to choose which *n*-exponential model should be selected. As for DCE-MRI, using MCR as the multivariate statistical projection model in MIA can help in deciding the number of exponential decays (i.e., behaviors) to use, providing new nonparametric models that can explain the principal diffusion behaviors extracted, helping specialists to detect and characterize early tumors in the prostate, and providing new imaging biomarkers that may complement those commonly used for clinical diagnosis.

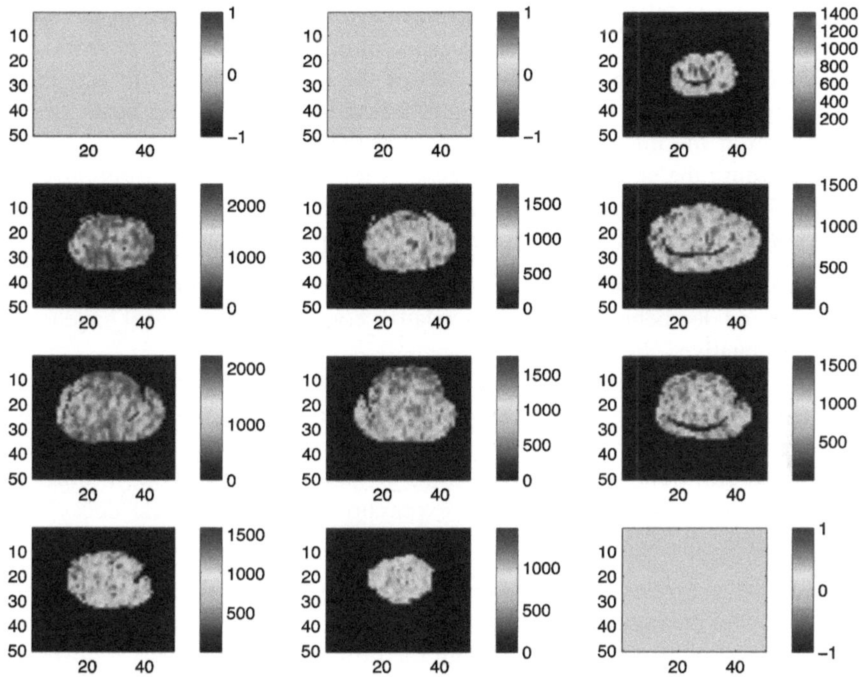

FIG. 14 RSS distribution maps for the 12 slices analyzed, from left to right and from top to bottom. Note that slices 1, 2, and 12 do not have prostate area masked. Masks are the union of the peripheral and inner prostate zones, so some slices present prostate pixels unselected.

As commented before, in the diffusion process, the studied phenomena are those related to slow diffusion, associated to cellularization, and fast diffusion, associated to vascularization. Assuming that the signal spectrum in a pixel j can be expressed as a weighted sum of different decreasing exponential functions modeling the different phenomena of the diffusion process, Aguado-Sarrió *et al.* [27] propose the following model:

$$s_j = \sum_{i=1}^{I} c_{ij}\left(\alpha_i \mathrm{e}^{-\beta_i b}\right); \quad \alpha_i, \beta_i, c_{ij} \geq 0 \tag{5}$$

where I stands for the number of exponential functions used. In this case, models using 1, 2, and 3 exponential functions are studied. The triexponential approach is proposed in order to model a possible additional behavior with physiological meaning that might remain in the residuals of the biexponential approach.

The main differences regarding the theoretical models outlined previously are the no-normalization of the spectra (using s instead of s/s_0) and the independence between the α_i coefficients, which are not forced to sum up 1.

As for DCE-MRI, the MCR-ALS has been used (see Eq. 4). In diffusion studies, matrix **S** contains the signal spectrum *s* for each pixel in rows, $\mathbf{D}^T$ is a matrix containing in its rows each of the diffusion behaviors modeled, **C** gathers in its rows the relative contribution of each modeled behavior for each pixel of the image, and **E** is a residual matrix.

Regarding the constraints to be used, in the case of the diffusion process, the following ones can be applied successively in order to validate the theoretical models most commonly applied in clinical practice.

(a) Nonnegativity constraints are applied both in **D** and **C** matrices, because the modeled behaviors and their relative contribution in a pixel have to be nonnegative.
(b) Unimodality constraints where only one maximum or minimum is admitted are imposed only with **D**, due to the fact that the modeled behaviors are monotonically decreasing.
(c) Shape constraints are applied on the **D** matrices inside ALS, in order to obtain a specific mathematical expression (i.e., exponential decay) for the modeled behaviors.

Including these constraints, the model can select the solutions with more physiological meaning, discarding the rest lacking any clinical interpretation.

The sequential procedure to obtain the different MCR models follows:

1. Fit MCR models with only nonnegativity constraints in **C** and **D** with one, two, and three components.
   Select the model with most appropriate number of components according to the similarity of the corresponding fitted behaviors with the exponential decay (i.e., expected physiological behavior). This is Model 0.
2. Fit a new MCR model including nonnegativity constraints in **C** and **D** and unimodality constraints (only one maximum or minimum is allowed for each behavior) in **S** using the number of components decided in step 1. This is Model 1.
3. Check the adequacy of Model 1 by fitting the behaviors obtained (rows of $\mathbf{D}^T$ matrix in step 2) with an exponential function and projecting the predicted exponential behavior onto the original data matrix **S** to obtain the new **C** matrix. This is Model 2.
4. Finally, fit a new MCR model including nonnegativity constraints in **C** and **D** and shape (i.e., exponential decay) constraints in **D** using the number of components decided in step 1. This is Model 3.

The MCR models built this way provide the matrices $\mathbf{D}^T$ and **C** previously commented, gathering at each pixel location the behaviors (types of exponential decays) found and their relative contribution, respectively. By relocating the pixels in each column of **C** matrix into their original spatial dimension (192×192 pixels), the distribution maps are obtained, locating those pixels more related to each of the corresponding behaviors provided by the model.

It must be stressed that these distribution maps from **C** matrix, not the prostate diffusion behaviors modeled in $\mathbf{D}^T$ matrix, are the ones performing as imaging biomarkers. This is a conceptual difference with the clinical theoretical models method, where the biomarkers (e.g., D, D^*, and f in the case of the IVIM model) are extracted from the behavior modeled at each pixel location.

This tool can help the doctors to locate areas of the prostate that may be potentially affected by a tumoral process. Furthermore, the distribution maps of the RSS were used as a validation method, as they show the pixels that are not well fitted by the model. In the following, we illustrate the application of the MCR-based methodology on the DW-MR images, on a patient with a tumor on the peripheral gland.

3.3.1 Model 0

Initially, the only constraint considered is nonnegativity in **C** and **D** matrices. Figs. 15 and 16 show the results with one and two components, respectively. In the one-component model, the fitted behavior follows an exponential decay pattern. As can be seen in the distribution map shown in Fig. 15A, the pixels related to this behavior are distributed in the whole image, being the RSS low (Fig. 15B). However, with two components (Fig. 16), the residuals are even lower (Fig. 16C), i.e., the model explains 99% of the variability, and the behaviors are more physiologically meaningful because of the form of the obtained curves (Fig. 16D): a quick fall related to fast diffusion (d2) and a slow fall associated to slow diffusion (d1). Nevertheless, there are still two "artifacts" (slight increases of signal intensity) present in both behaviors, at lower *b*-values in the pure diffusion (slow diffusion, d1) component and at higher *b*-values in the perfusion component (fast diffusion, d2). Regarding the distribution maps, they show that the component related to diffusion (Fig. 16A) is more representative in the whole zone of the prostate than that of the perfusion component (Fig. 16B), which is more localized.

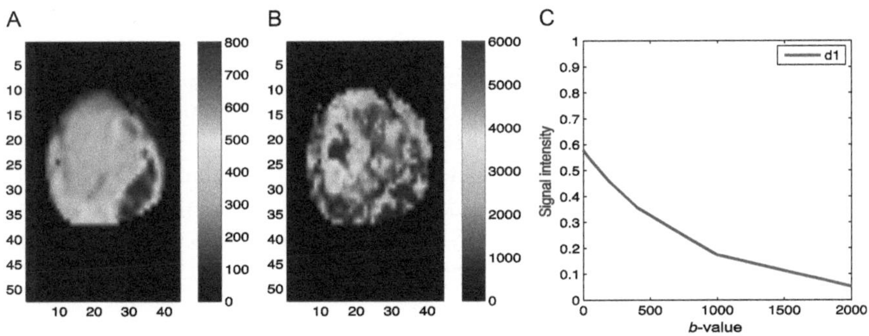

FIG. 15 MCR Model 0 with one component (95% of explained variability). (A) Score distribution map related to the modeled behavior d1. (B) RSS distribution map. (C) Modeled behavior d1 (*solid blue* (*dark gray* in the print version) *line*).

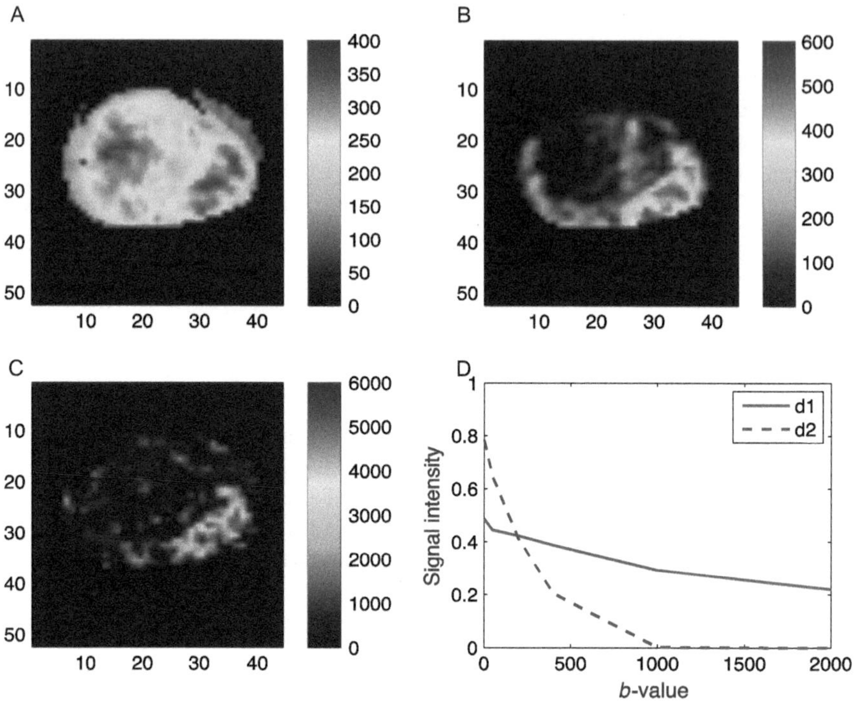

FIG. 16 MCR Model 0 with two components (99% of explained variability). (A) Score distribution map related to first modeled behavior d1 (slow diffusion, *solid blue* (*dark gray* in the print version) *line*). (B) Score distribution map related to second modeled behavior d2 (fast diffusion, *dashed green* (light gray in the print version) *line*). (C) RSS distribution map. (D) Modeled behaviors.

By comparing both models (with one and two components), it can be seen that the behavior fitted in the one-component model (Fig. 15C) is a combination of the behaviors modeled in the two-component model (Fig. 16D). This can also be observed in the score distribution maps (Figs. 15A and 16A and B). Therefore, the one-component MCR model is not able to explain all the physiological behaviors present in the diffusion studies due to its simplicity.

Fig. 17 shows the modeled behaviors from the MCR model with three components. The behaviors associated to components 2 and 3 present shapes that make no sense in diffusion studies because the signal spectra cannot increase as the *b*-value goes up (it is assumed that signal always attenuates when the *b*-value increases). Thus in the following, the two-components model is selected as the most appropriate to study the diffusion process.

3.3.2 Model 1

The next step consists of including new constraints in the model. The first of these is unimodality (in **D** matrix) in order to provide new behaviors more

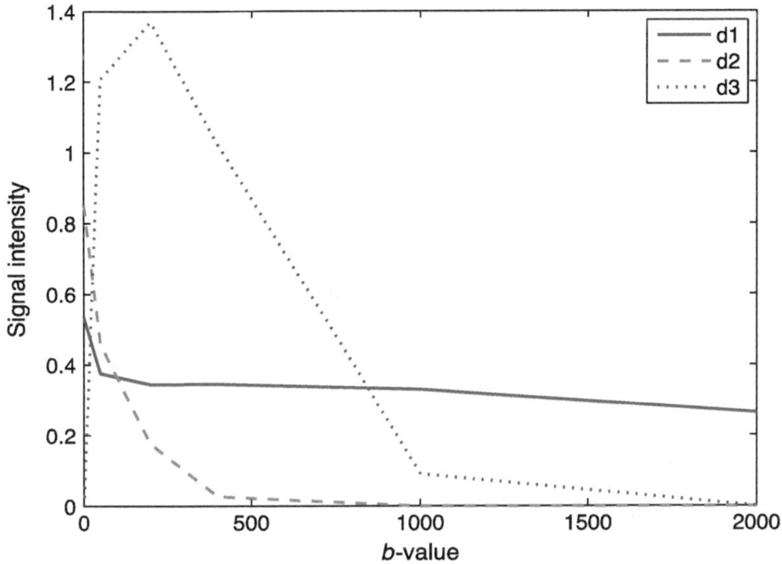

FIG. 17 Modeled behaviors from the MCR Model 0 with three components. The *red* (*light gray* in the print version) *dotted line* has no clinical meaning.

similar to the exponentially decay shape. Actually, monotonicity constraints should be introduced, but this is accomplished in the following steps (with the shape constraints introduced in the ALS algorithm when imposing the exponential decays), so it was preferred to slowly adding constraints, in order to carefully validate the models. The inclusion of this type of constraint allows removing the artifact present in the perfusion component at high *b*-values (a slightly increase of d2 from 1500 to 2000 in the *b*-value) that appears in Model 0 (Fig. 16D). Nevertheless, there is still one artifact (maybe introduced by the image acquisition system) at low *b*-values for the first behavior (slow diffusion component), which still accomplishes the unimodality constraint. After careful discussion with doctors, it was decided that this artifact could not happen from a physiological point of view. Furthermore, adding a new component did not isolate this behavior (results not shown).

3.3.3 Model 2

With the aim of removing the second artifact and checking the adequacy of the biexponential model in diffusion MR studies, a fitting of the behaviors is proposed using: (1) the results provided by Model 1 (the pure behaviors saved in the **D** matrix) and (2) the information of the clinical theoretical models that proposes two decreasing exponential behaviors. Therefore, using fitting algorithms, both behaviors are tuned to exponential expressions $\alpha \exp(-\beta b)$. Furthermore, the goodness of fit is high (95%) obtaining two different pure exponential decays related to diffusion (d1) and perfusion (d2), respectively.

The exponential decay shape is obtained in the following way: for each one of the behaviors obtained from the previous model, initial parameters α_i and β_i ($i=1,2$) are fitted using the exponential decays shown in equation 5. Afterward, using these exponential functions and the b-values, the "pure" d behaviors are predicted, and arranged in the **D** matrix. Once the exponential shape of the behaviors is obtained, the next step was to project the original data matrix on the pseudoinverse of the new fitted **D** to get matrix **C**, and from it the corresponding distribution maps.

Note that all these calculations are not performed within the ALS algorithm. It is just a fit of the $\mathbf{D_1}$ matrix obtained from previous Model 1 on two exponential decays, thus obtaining the new $\mathbf{D_2}$, afterward computing $\mathbf{C_2}$. This also means that this new $\mathbf{C_2}$ matrix might provide negative values, since no negativity constrain is applied.

3.3.4 Model 3

Finally, once the biexponential model is validated, the shape constraints for the **D** matrix were included in the iterative process of the MCR-ALS algorithm, assuming classical exponential expressions $\alpha_i \exp(-\beta_i b)$ for both behaviors, and nonnegativity for **S** and **D** matrices. Details can be found in Ref. [27]. For every case, the exponential parameters, α_i and β_i ($i=1,2$), obtained from Model 2 have been used as an initial approximation.

As can be seen in Fig. 18D, the fitted behaviors are two different strictly decreasing exponentials, one related to slow diffusion (slow attenuation, d1) and other related to fast diffusion (fast attenuation, d2). Furthermore, the goodness of fit is at least 99% in all the cases under study, which derives in low-RSS distribution maps (Fig. 18C).

From all these results, it can be concluded that the monoexponential model seems too simple when trying to obtain all the behaviors present in the images. The triexponential approach is not useful either, as it provides behaviors that are not physiologically interpretable. On the other hand, the biexponential model has shown up as a sensible approach for modeling the diffusion process. This model has been sequentially improved by comparing the new results provided by each further constrained model with the previous one. These conclusions are significantly related to the existing knowledge about the behavior of diffusion in tissues, where the biexponential approach through the IVIM theory is preferred in medical imaging research to explain the behavior of water molecules in tissues like tumors, where not only a cellular component but also a microvascular behavior is appreciated.

Fig. 19 summarizes the two modeled behaviors obtained from Models 1, 2, and 3 and Fig. 20 shows the distribution maps for these models. In all three models, the fast diffusion component presents an exponential decay, even in Model 1 where no shape is imposed. Regarding the slow diffusion component,

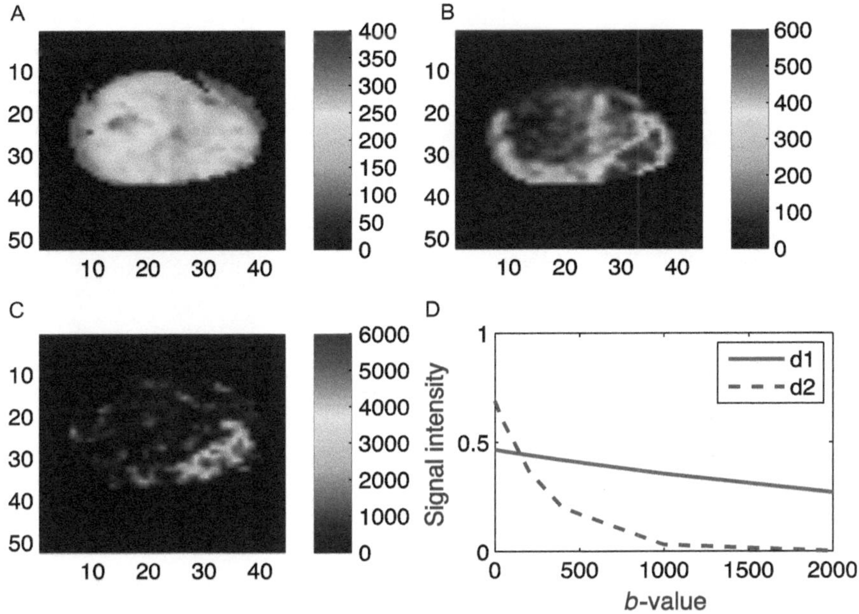

FIG. 18 MCR Model 3 (99% of explained variability). (A) Score distribution map related to d1 (slow diffusion, *solid blue* (*dark gray* in the print version) *line*). (B) Score distribution map related to d2 (fast diffusion, *dashed green* (*gray* in the print version) *line*). (C) RSS distribution map. (D) Modeled behaviors.

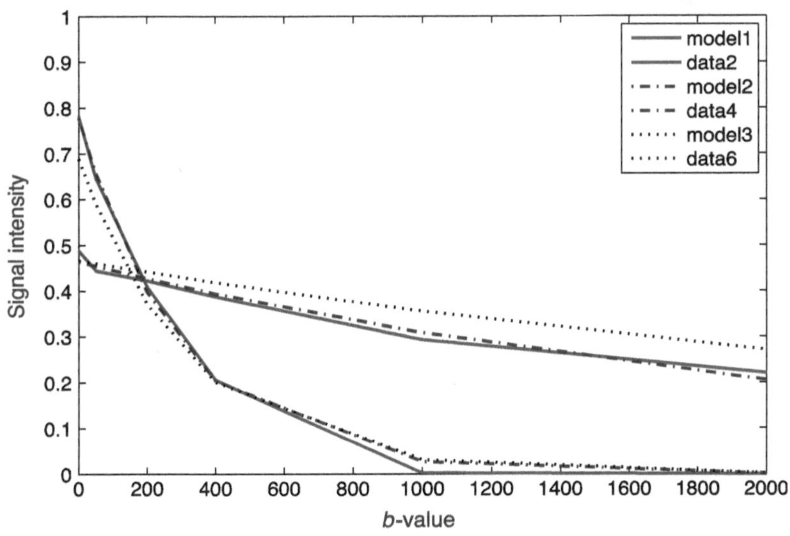

FIG. 19 Comparison between the behaviors provided by models 1 (*solid red* (*gray* in the print version) *line*), 2 (*dashed blue* (*dark gray* in the print version) *line*), and 3 (*dotted black line*).

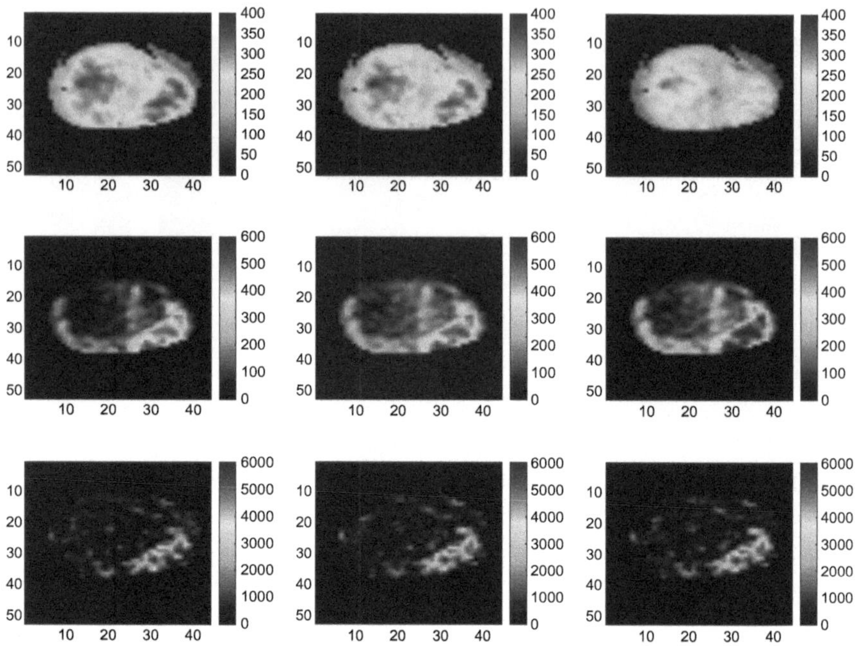

FIG. 20 Comparison between the distribution maps provided by the three different models. (*Top*) d1 related to slow diffusion, as the main behavior of normal prostate tissue. (*Middle*) d2 related to fast diffusion, related to highly vascularized regions, as tumors in the peripheral gland, like in this patient with a tumor in the peripheral gland. (*Bottom*) RSS distribution maps. (*Left* to *right*) Models 1, 2, and 3.

it also presents an exponential decay afterward removing the artifact present at low *b*-values (Models 2 and 3). This can be seen in the Fig. 21 where the higher residuals of Model 3 related to low *b*-values match this artifact in Model 1. Therefore, the pixels mainly related to this phenomenon are those having higher (negative) residuals in Model 3, as shown in Fig. 20, third column, third row. On the other hand, the distribution maps observed in Fig. 20 remain almost the same in all three models.

This way, the proposed procedure provides sensible imaging biomarkers based on the biexponential model, which complement the IVIM biexponential model, using the distribution maps as a quantification of each behavior gathered in **S** matrix at each pixel location. Still, both the MCR-based modeling and the theoretical biexponential modeling show two basic differences: first, the normalization of the spectra and second, the fact that the two factors weighting the exponentials do not have to be necessarily complementary. Furthermore, the fitting of the IVIM model is fitted pixel-by-pixel, unable to take profit of the global information that can be obtained using MCR models.

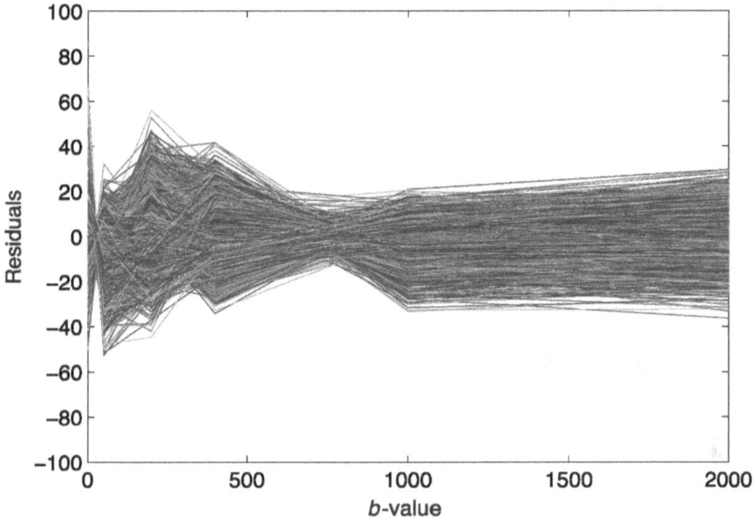

FIG. 21 Plot of the residuals obtained from Model 3 vs the *b*-values.

4 CONCLUSIONS

This chapter shows the capability of MCR models to extract behaviors with clinical meaning from the DCE-MR and DW-MR images by including a priori knowledge.

MCR lets to directly locate and grade the intensity of these behaviors in the images, providing new imaging biomarkers, complementary to those obtained from the theoretical models, to improve clinical diagnosis.

MCR also helps in segregating those artifacts that may introduce uncertainty in the estimation of any biomarker, and provides an evaluation tool for assessing the appropriateness of theoretical models, with a data driven model methodology that allows incorporating prior knowledge in a sequential fashion.

ACKNOWLEDGMENTS

The authors thank Drs. Roberto Sanz-Requena, Ángel Alberich-Bayarri, and Luis Martí-Bonmatí for the joint work performed, and Prof. Anna de Juan for providing help, comments, and code for developing our own MCR software. Research in this study was partially supported by the Spanish Ministry of Economy and Competitiveness and FEDER funds (DPI2014-55276-C5-1R).

REFERENCES

[1] de Juan A, Tauler R. Multivariate curve resolution (MCR) from 2000: progress in concepts and applications. Crit Rev Anal Chem 2006;36(3–4):163–76.

[2] Gargallo R, Tauler R, Cuesta-Sanchez F, Massart DL. Validation of alternating least-squares multivariate curve resolution for chromatographic resolution and quantitation. Trends Anal Chem 1996;15(7):279–86.

[3] Smilde AK, Tauler R, Saurina J, Bro R. Calibration methods for complex second-order data. Anal Chim Acta 1999;398(2–3):237–51.

[4] Tauler R. Multivariate curve resolution applied to second order data. Chemom Intell Lab Syst 1995;30:133–46.

[5] de Juan A, Maeder M, Hancewicz T, Tauler R. Local rank analysis for exploratory spectroscopic image analysis. Fixed size image window-evolving factor analysis. Chemom Intell Lab Syst 2005;77(1–2):64–74.

[6] Gallagher NB, Shaver JM, Martin EB, Morris J, Wise BM, Windig W. Curve resolution for multivariate images with applications to TOF-SIMS and Raman. Chemom Intell Lab Syst 2004;73:105–17.

[7] Lüdemann L, Prochnow D, Rohlfing T, Franiel T, Warmuth C, Taupitz M, et al. Simultaneous quantification of perfusion and permeability in the prostate using dynamic contrast-enhanced magnetic resonance imaging with an inversion-prepared dual-constrast sequence. Ann Biomed Eng 2009;2(37):749–62.

[8] Le Bihan D. Molecular diffusion nuclear magnetic resonance imaging. Magn Reson Q 1991;7:1–30.

[9] Giannarini G, Petralia G, Thoeny HC. Potential and limitations of diffusion-weighted magnetic resonance imaging in kidney, prostate, and bladder cancer including pelvic lymph node staging: a critical analysis of the literature. Eur Urol 2012;61:326–40.

[10] Heijmink SW, Fütterer JJ, Strum SS, Oyen WJ, Frauscher F, Witjes JA, et al. State-of-the-art uroradiologic imaging in the diagnosis of prostate cancer. Acta Oncol 2011;50(Suppl 1): 25–38.

[11] Tofts PS, Brix G, Buckley DL, Evelhoch JL, Henderson E, Knopp MV, et al. Estimating kinetic parameters from dynamic contrast-enhanced T1-weighted MRI of a diffusable tracer: standardized quantities and symbols. J Magn Reson Imaging 1999;10:223–32.

[12] Sourbron SP, Buckley DL. Tracer kinetic modelling in MRI: estimating perfusion and capillary permeability. Phys Med Biol 2012;57(2):R1–R33.

[13] McGrath DM, Bradley DP, Tessier JL, Lacey T, Taylor CJ, Parker GJ. Comparison of model-based arterial input functions for dynamic contrast-enhanced MRI in tumor bearing rats. Magn Reson Med 2009;61:1173–84.

[14] Meng R, Chang SD, Jones EC, Goldenberg SL, Kozlowski P. Comparison between population average and experimentally measured arterial input function in predicting biopsy results in prostate cancer. Acad Radiol 2010;17:520–5.

[15] Port RE, Knopp MV, Brix G. Dynamic contrast-enhanced MRI using Gd-DTPA: interindividual variability of the arterial input function and consequences for the assessment of kinetics in tumors. Magn Reson Med 2011;45:1030–8.

[16] Yang C, Karczmar GS, Medved M, Oto A, Zamora M, Stadler WM. Reproducibility assessment of a multiple reference tissue method for quantitative dynamic contrast-enhanced MRI analysis. Magn Reson Med 2009;61:851–9.

[17] Charles-Edwards EM, De Souza NM. Diffusion-weighted magnetic resonance imaging and its application to cancer. Cancer Imaging 2006;6:135–43.

[18] Le Bihan D, Breton E, Lallemand D, Grenier P, Cabanis E, Laval-Jeantet M. MR imaging of intravoxel incoherent motions: application to diffusion and perfusion in neurologic disorders. Radiology 1986;161:401–7.

[19] Le Bihan D, Breton E, Lallemand D, Aubin ML, Vignaud J, Laval-Jeantet M. Separation of diffusion and perfusion in intravoxel incoherent motion MR imaging. Radiology 1988;168(2):497–505.

[20] Leach MO, Brindle KM, Evelhoch JL, Griffiths JR, Horsman MR, Jackson A, et al. Assessment of antiangiogenic and antivascular therapeutics using MRI: recommendations for appropriate methodology for clinical trials. Br J Radiol 2003;76(Spec No 1):S87–91.

[21] Atkinson AJ, Colburn WA, DeGruttola VG, Demets DL. Biomarkers and surrogate endpoints: preferred definitions and conceptual framework. Clin Pharmacol Ther 2001;69:89–95.

[22] Donaldson SB, West CML, Davidson SE, Carrington BM, Hutchison G, Jones AP, et al. A comparison of tracer kinetic models for T1-weighted dynamic contrast-enhanced MRI: application in carcinoma of the cervix. Magn Reson Med 2010;63:691–700.

[23] Geladi P, Kowalski BR. Partial least-squares regression: a tutorial. Anal Chim Acta 1986;185:1–17.

[24] Jackson JE. A user's guide to principal components. New York: Wiley; 2003.

[25] Geladi P, Grahn H. Multivariate image analysis. Chichester, England: Wiley; 1996.

[26] Prats-Montalbán JM, de Juan A, Ferrer A. Multivariate image analysis: a review with applications. Chemom Intell Lab Syst 2011;107(1):1–23.

[27] Aguado-Sarrió E, Prats-Montalbán JM, Sanz-Requena R, Alberich-Bayarri A, Martí-Bonmatí L, Ferrer A. Prostate diffusion weighted-magnetic resonance image analysis using multivariate curve resolution methods. Chemom Intell Lab Syst 2015;140:43–8.

[28] Bruwer MJ, MacGregor JF, Noseworthy D. Dynamic contrast-enhanced MRI diagnosis in oncology via principal component analysis. J Chemom 2008;22:708–16.

[29] Prats-Montalbán JM, Sanz-Requena R, Martí-Bonmatí L, Ferrer A. Prostate functional magnetic resonance image analysis using multivariate curve resolution methods. J Chemom 2014;28(8):672–80.

[30] Buckley DL. Uncertainty in the analysis of tracer kinetics using dynamic contrast-enhanced T1-weighted MRI. Magn Reson Med 2002;47:601–6.

[31] Henderson E, Sykes J, Drost D, Weinmann HJ, Rutt BK, Lee TY. Simultaneous MRI measurement of blood flow, blood volume, and capillary permeability in mammary tumours using two different contrast agents. J Magn Reson Imaging 2000;12:991–1003.

[32] Oto A, Yang C, Kayhan A, Tretiakova M, Antic T, Schmid-Tannwald C, et al. Diffusion-weighted and dynamic contrast-enhanced MRI of prostate cancer: correlation of quantitative MR parameters with Gleason score and tumor angio-genesis. Am J Roentgenol 2011;197:1382–90.

[33] de Juan A, Maeder M, Hancewicz T, Duponchel L, Tauler R. Chemometric tools for image analysis. In: Salzer R, Siesler HW, editors. Infrared and Raman spectroscopic imaging. Weinheim, Germany: Wiley-VCH; 2009. p. 65–112.

[34] Sanz-Requena R, Prats-Montalbán JM, Martí-Bonmatí L, Ferrer A, Alberich-Bayarri A, García-Martí G, et al. Automatic individual arterial input functions calculated from PCA outperform manual and population-averaged approaches for the pharmacokinetic modeling of DCE-MR images. J Magn Reson Imaging 2015;42:477–87.

[35] Tauler R, Smilde AK, Kowalski BR. Selectivity, local rank, three-way data analysis and ambiguity in multivariate curve resolution. J Chemom 1995;9:31–58.

[36] Piqueras S, Duponchel L, Tauler R, de Juan A. Resolution and segmentation of hyperspectral biomedical images by multivariate curve resolution-alternating least squares. Anal Chim Acta 2011;705:182–92.

[37] Multivariate Curve Resolution Homepage, http://www.mcrals.info/.

[38] de Juan A, Tauler R. Chemometrics applied to unravel multicomponent processes and mixtures: revisiting latest trends in multivariate resolution. Anal Chim Acta 2003;500:195–210.

[39] Jaumot J, Gargallo R, de Juan A, Tauler R. A graphical user-friendly interface for MCR-ALS: a new tool for multivariate curve resolution in MATLAB. Chemom Intell Lab Syst 2005;76:101–10.

[40] Jaumot J, Tauler R. MCR-BANDS: a user friendly MATLAB program for the evaluation of rotation ambiguities in multivariate curve resolution. Chemom Intell Lab Syst 2010;103:96–107.

[41] Windig W, Guilment J. Interactive self-modeling mixture analysis. Anal Chem 1991;63:1425–32.

[42] Windig W. Two-way data analysis, detection of purest variables. In: Brown SD, Tauler R, Walczak B, editors. Comprehensive chemometrics, vol. 2. Amsterdam: Elsevier; 2009. p. 275–307.

[43] Chtioui Y, Bertrand D, Barba D. Feature selection by a genetic algorithm. Application to seed discrimination by artificial vision. J Sci Food Agric 1998;76:77–86.

[44] Leardi R. Nature-inspired methods in chemometrics: genetic algorithms and artificial neural networks. Data handling in science and technology, vol. 23. Amsterdam: Elsevier; 2003.

[45] Wang M, Zhou X, King R, Wong STC. Context based mixture model for cell phase identification in automated fluorescence microscopy. BMC Bioinformatics 2007;8(32):1–12.

[46] Cuesta Sánchez F, Toft J, van den Bogaert B, Massart DL. Orthogonal projection approach applied to peak purity assessment. Anal Chem 1996;68:79–85.

[47] de Juan A, Maeder M, Hancewicz T, Tauler R. Use of local rank-based spatial information for resolution of spectroscopic images. J Chemom 2008;22:291–8.

[48] Ked HR, Massart DL. Peak purity control in liquid chromatography with photodiode-array detection by a fixed size moving window evolving factor analysis. Anal Chim Acta 1991;246:379–90.

[49] Orton MR, d'Arcy JA, Walker-Samuel S, Hawkes DJ, Atkinson D, Collins DJ, et al. Computationally efficient vascular input function models for quantitative kinetic modelling using DCE-MRI. Phys Med Biol 2008;53:1225–39.

[50] Parker GJM, Roberts C, Macdonald A, Buonaccorsi GA, Cheung S, Buckley DL, et al. Experimentally-derived functional form for a population-averaged high-temporal-resolution arterial input function for dynamic contrast-enhanced MRI. Magn Reson Med 2006;56:993–1000.

[51] Lemke A, Stieltjes B, Schard LR, Faun FB. Toward an optimal distribution of b values for intravoxel incoherent motion imaging. Magn Reson Imaging 2011;29:766–76.

Chapter 17

Endmember Library Approaches to Resolve Spectral Mixing Problems in Remotely Sensed Data: Potential, Challenges, and Applications

B. Somers*[,1], L. Tits[†], D. Roberts[‡] and E. Wetherley[‡]
*Division of Forest, Nature, and Landscape, KU Leuven, Leuven, Belgium
[†]Flemish Institute for Technological Research (VITO), Remote Sensing Unit, Boeretang 200, BE-2400 Mol, Belgium
[‡]University of California, Santa Barbara, CA, United States
[1]Corresponding author: e-mail: ben.somers@ees.kuleuven.be

Chapter Outline

Data Handling in Science and Technology, Vol. 30. http://dx.doi.org/10.1016/B978-0-444-63638-6.00017-6
Copyright © 2016 Elsevier B.V. All rights reserved.

1 INTRODUCTION

Earth observation or remote sensing, i.e., the interpretation of air- and space-borne sensor data to study and monitor processes on the earth surface, has become an indispensable tool in environmental monitoring and assessment studies at different spatial and temporal scales [1]. The value of remote sensing technology has been demonstrated in climate change research [2], terrestrial ecosystem monitoring and management [3], precision agriculture and production monitoring [4], natural hazard risk assessment [5], forest inventories and forest health assessments [6], water quality assessment and fresh water management [7], geological mapping [8], and mapping of the urban environment [9].

Despite the continuous evolution toward platform-sensor systems with a higher accuracy and detail in spatial, temporal, spectral, and radiometric resolution, the full potential of remotely sensed data analysis for monitoring processes on the earth surface has not been fully realized. One reason is that no matter the spatial resolution of the image data, the spectral signatures collected in natural environments are invariably a mixture of the signatures of the various materials found within the spatial extent of the ground instantaneous field of view of the imaging instrument. This composite nature of observed spectra inevitably complicates image interpretation [10].

Over the past decades, a number of image analysis techniques to resolve spectral mixing problems in remotely sensed data have been proposed (e.g., [10]). Spectral mixture analysis (SMA) is one of the most common unmixing techniques utilized in many applications [11]. The technique assumes that the spectral signature collected by an imaging spectrometer over one pixel, vector $\boldsymbol{r}$, can be expressed as a (non)linear combination of endmembers (i.e., spectral properties of the pure ground components contained within the image pixel, such as vegetation or soil) weighted by their corresponding subpixel abundances. Under the linear mixture model assumption, $\boldsymbol{r}$ can be modeled using the following expression:

$$\boldsymbol{r} = \sum_{z=1}^{p} \boldsymbol{M}_z \cdot \boldsymbol{f}_z + \boldsymbol{\varepsilon} \quad (1)$$

where M is a $n \times p$ matrix of which each column vector $\boldsymbol{M_z}$ corresponds to the spectral signal of a specific endmember or ground cover class, z, and each row corresponds to a specific waveband. $\boldsymbol{f}$ is a column vector $[f_1, \ldots, f_p]^{\mathrm{T}}$ that denotes the subpixel cover fractions occupied by each of the p endmembers in the pixel. The portion of the spectrum that cannot be modeled is expressed as a residual term, $\boldsymbol{\varepsilon}$. Two physical constraints are generally imposed, i.e., the abundance nonnegativity constraint (ANC) $\boldsymbol{f} \geq 0$ and the abundance sum-to-one constraint (ASC) $\sum_{z=1}^{p} f_z = 1$.

The implementation of SMA involves two steps: (i) *endmember definition/identification*, i.e., identifying the unique signatures of the pure ground components or endmembers, $\boldsymbol{M}_z$; and (ii) *abundance estimation*, i.e., estimating the proportions of different endmembers for each pixel, f_z, by model inversion [12]. The spectral signatures of the endmembers may be derived from field or laboratory measurements, obtained using ground-based or portable spectroradiometers, or derived directly from the image data themselves or simulated using radiative transfer models (RTMs; [4]; see also Section 3.1). Commonly used inversion techniques for solving (1) are singular value decomposition [13], Gramm–Schmidt orthogonalization [14], maximum-likelihood [15], or least squares regression analysis [16]. The accuracy of SMA is often assessed based on the fit between the modeled and observed mixed spectral signals and quantified using an error metric such as the residual term $\boldsymbol{\varepsilon}$ or the root mean square error (RMSE) [17].

SMA is often prone to residual spectral error caused by ambient (e.g., inaccurate atmospheric correction), sensor specific (e.g., insufficient signal-to-noise ratio), and model structure input noise (e.g., multiple scattering effects causing nonlinear mixing behavior). Yet, the most expressed source of error is the lack of ability to account for the temporal and spatial spectral variability typical for natural materials [18,19]. Think of how leaf biochemistry within a single tree or tree species in a forest can change over short spatial and temporal scales, or how heterogeneous the soil conditions can be in a single agricultural parcel (Fig. 1). Also changing illumination conditions will cause similar materials to spectrally differ throughout or between images.

In addition to the spectral variability of a given material within and between images, also the number of components within a mixed pixel can vary throughout the image. Determining the exact value of p in (1) is key in SMA, as well as to determine which components are present in the mixed pixel [17]. The most widely used SMA approach consists of employing a single set of endmembers on the whole image, and using all available

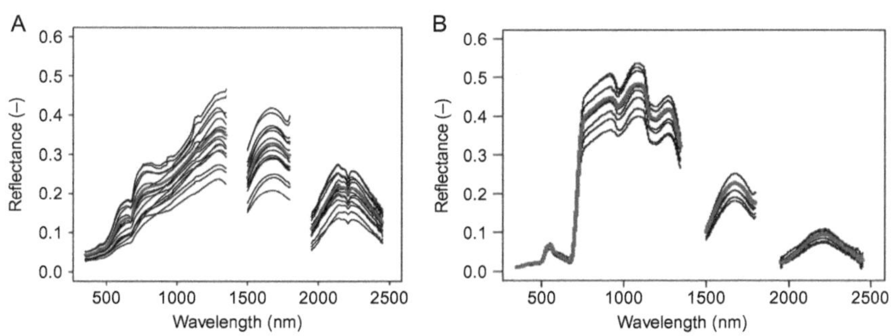

FIG. 1 Variability in (A) soil and (B) orange tree reflectance as observed in a 1 ha citrus orchard in Wellington, South Africa. If not properly accounted for in the endmember definition, this spatial spectral variability causes significant errors in SMA. In conventional SMA applications each endmember class (i.e., vegetation and soil in this case) is characterized by a fixed image-wide endmember spectrum (e.g., the *red* (*gray* in the print version) spectrum), ignoring as such the spatial and temporal spectral variability.

endmembers at the same time, thus using a value *p* that is larger than the actual number of presented components within one pixel. In the decomposition of a mixed pixel it is mathematically feasible to use as many endmembers as spectral bands available, though even with a small number of endmembers, noisy and highly correlated data exert a significant influence on the abundance estimates of the lowest contrast materials [20]. A value *p* lower than the actual number of components, however, will result in a poor representation of the mixed signature [21]. Fractions of the unmodeled endmembers will be partitioned or allocated to incorrect endmembers, consequently yielding inferior abundance estimates for these components [20]. When accurate unmixing of the image scenes is required, this spectral variation within and among endmembers should thus be accounted for.

In conventional SMA applications, however, each endmember class (e.g., vegetation or soil) is characterized by a fixed image-wide endmember spectrum (Fig. 1, red spectrum) thereby ignoring endmember variability and resulting in significant unmixing errors when applied in heterogeneous landscapes. Solutions to overcome endmember variability in SMA have been presented in recent literature (see [19,22] and references therein) and techniques can generally be grouped in two broad categories: (i) methods that consider the endmembers as statistical distributions, where a sample from these distributions can be viewed as a possible variation on the spectral signature of the material being represented; and (ii) endmember library-based SMA approaches taking advantage of (extensive) spectral libraries containing representative instances of all endmembers present within the scene (cf. the soil and orange tree endmember library in Fig. 1) (e.g., [17]). In addition to these two broad categories, multivariate curve resolution (MCR) methods are finding their way into remote sensing applications, albeit very slowly and still only limited to experimental settings [23,24]. As MCR is already discussed extensively in the other chapters, it will not be elaborated on in this chapter. In addition, considering endmembers as statistical distributions is also a theoretical concept which has only been introduced very recently, gaining considerable popularity in the signal processing community, yet the transfer to remote sensing applications remains limited. A more comprehensive description of the concepts of these methods can be found in Zare and Ho [22]. The focus of this chapter will be on library-based SMA approaches, which have now been applied for land use and land cover monitoring for nearly two decades.

This chapter first provides a brief overview of the theoretical concepts of some widely applied endmember library-based SMA approaches. Second, common challenges associated with the establishment and implementation of endmember libraries in SMA will be discussed. This chapter concludes by illustrating the relevance of library-based SMA approaches in two case studies, i.e., fire severity and postfire recovery in a Mediterranean-climate ecosystem (Section 4.1) and early-stress detection in fruit orchards (Section 4.2).

2 ENDMEMBER LIBRARY-BASED SMA APPROACHES

This section gives a brief overview of the theoretical principles of commonly used library-based SMA approaches.

2.1 Iterative Mixture Cycles

Today, multiple endmember spectral mixture analysis or MESMA is by far the most widely used library-based SMA technique. The approach, developed by Roberts *et al.* [17], introduced the use of iterative mixture analysis cycles in SMA. Endmembers are allowed to vary on a per-pixel basis to better cope with the effects of endmember variability in the abundance estimation step. The MESMA procedure can briefly be described as follows: (i) the conventional SMA algorithm is run in iterative fashion; (ii) for each run different endmember combinations, selected from a spectral library (of hundreds of spectra) accounting for (each) plausible endmember condition, are used to decompose each pixel; (iii) the model with the best fit, i.e., with the lowest RMSE in the reconstruction of the original pixel, is adopted. For a full detailed description of the MESMA approach the reader is redirected to Roberts *et al.* [17]. MESMA has been successfully tested in a range of natural (e.g., [17,25]) and urban (e.g., [26,27]) environments using both multi- (e.g., [25]) and hyperspectral optical (e.g., [28]), and thermal (e.g., [29]) imagery.

In parallel with MESMA, alternative iterative mixture analysis cycles approaches have been introduced. Examples are the Monte Carlo spectral unmixing model (AutoMCU; [30]) and Bayesian spectral mixture analysis (BSMA; [31]). Similar to MESMA, AutoMCU calculates the subpixel abundances for each pixel using a large number of endmember combinations, in this case randomly selected from a spectral database. Subsequently, a minimum, mean, and maximum cover fraction image is produced. This iterative mixture analysis technique provides a conceptual framework for understanding the endmember variability and quantifying the introduced uncertainties (in an automated way) into fractional estimates. BSMA works similarly to the other iterative mixture analysis cycles, i.e., each pixel is unmixed with randomly selected combinations of endmember signatures; however, endmember signatures are no longer treated as constants but represented by probability density functions. BSMA, as such, takes into account the probabilities of spectral signatures, instead of assuming equal probabilities for all endmembers. Both approaches have been successfully applied in different biomes on the planet using a variety of air- and space-borne sensors (e.g., [32,33]). Fig. 2 schematically illustrates the basic concepts of iterative mixture analysis cycles.

2.2 Sparse Unmixing

An alternative for the iterative mixture analysis cycles are the unmixing algorithms that exploit the *sparsity* of the mixtures [34]. Similar to MESMA and

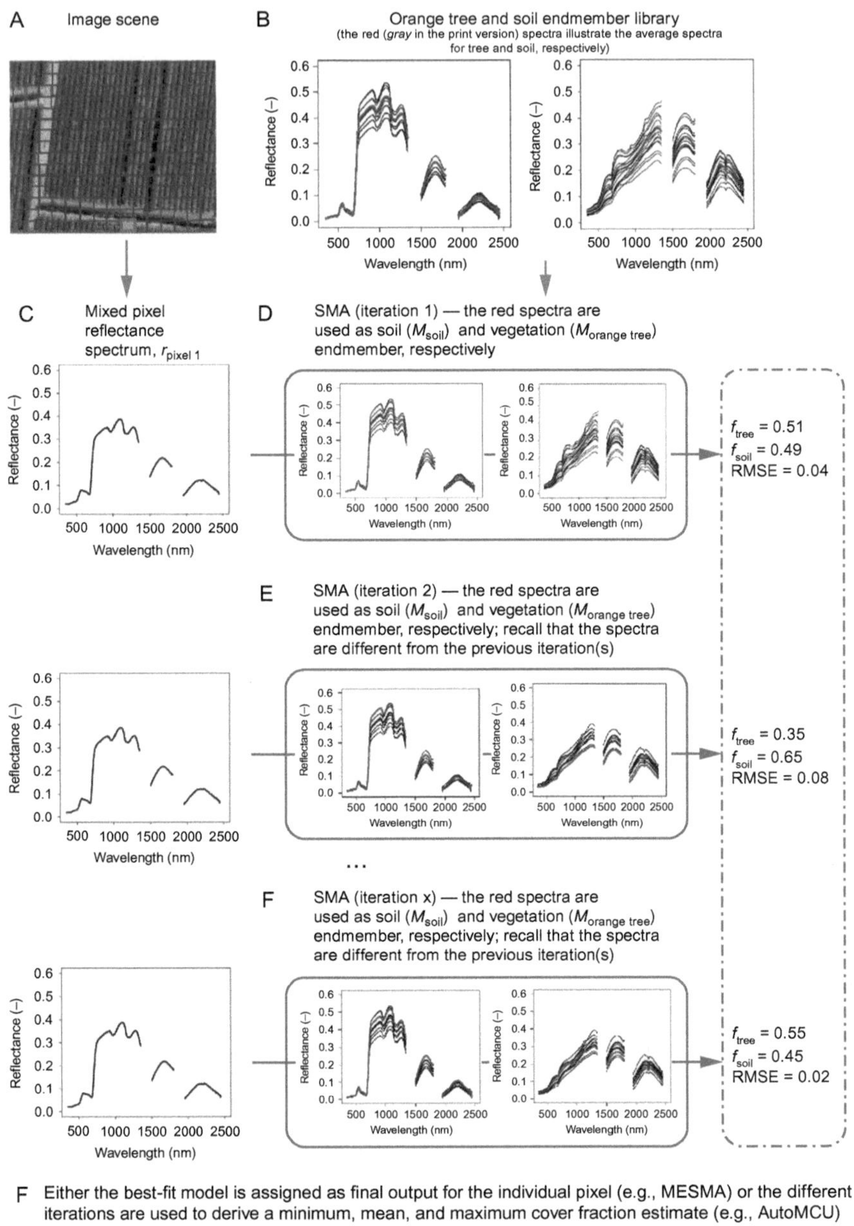

FIG. 2 Schematic illustration of the iterative mixture analysis process. Starting from the image scene (A) and the endmember library (containing a set of spectra accounting for (each) plausible endmember condition) (B) each individual image pixel (C) is unmixed in an iterative fashion (D–F). In each iteration a different set of endmember spectra (*red*, *gray* in the print version) is used to decompose the mixed pixel spectrum. Finally, either the best-fit model is used to assign the pixel's fraction estimate (e.g., MESMA) or the output of the different iterations is used to derive a minimum, mean, and maximum cover fraction estimate for the pixel (e.g., AutoMCU).

other iterative mixture analysis cycles, sparse unmixing algorithms make use of large spectral libraries. Yet, instead of using random combinations of endmembers in an iterative unmixing cycle, sparse unmixing assumes that the endmembers contributing to the observed spectrum of a pixel are present in the spectral library. Sparse unmixing, as such, uses all endmember spectra contained within the library in one single SMA cycle (see Fig. 3 for a schematic illustration of the process). Instead of redefining M in (1) for each iteration as is done for the iterative mixture cycles, M is defined as the entire spectral library. As the number of actual endmembers on the ground (i.e., contained within a single pixel) is much smaller than the number of spectra in the library, the corresponding vector of fractional abundances $\boldsymbol{f}$ contains only a few nonzero values (i.e., it is a sparse vector). The goal of sparse unmixing is to find a reduced set of endmembers present in the mixture which reconstructs the observed pixel spectrum with high accuracy, i.e., the lowest

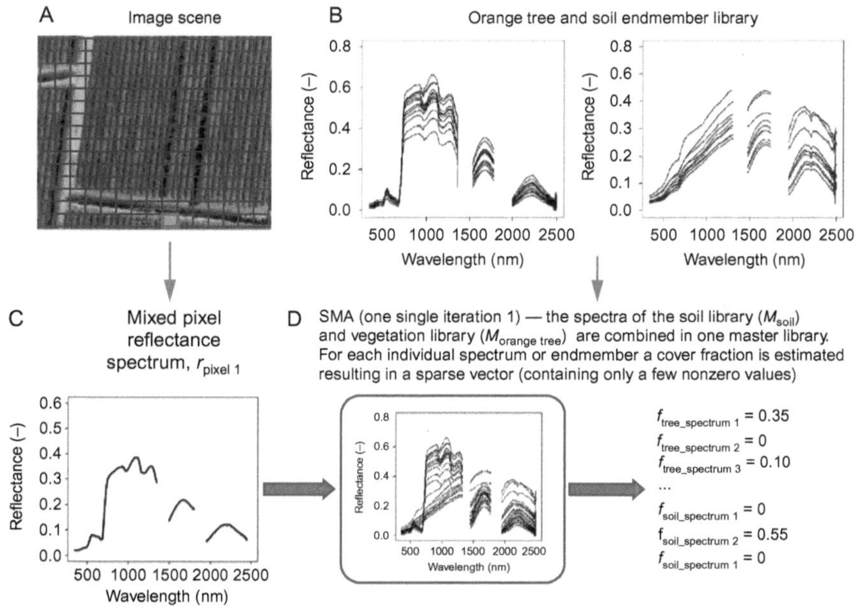

FIG. 3 Schematic illustration of the sparse unmixing process. Starting from the image scene (A) and the endmember libraries for soil and orange tree in this example (i.e., each library contains a set of spectra accounting for (each) plausible endmember condition) (B) each individual image pixel (C) is unmixed. Yet, as opposed to the iterative mixture analysis cycles (see Fig. 2), in sparse unmixing all image pixels are unmixed in one single iteration. This is done by defining a master endmember library composed of, in this two-class example, the endmember libraries of orange tree and soil (D). Each individual endmember spectrum contained within the master library is now considered an individual endmember term in the mixture model. For each endmember a fraction estimate is provided. The goal of sparse unmixing is to find the reduced set of endmembers present in the mixture which reconstructs the observed pixel spectrum with high accuracy, i.e., the lowest number of active signatures ($f_z > 0$) in M with the lowest modeling error.

number of active signatures ($f_z > 0$) in M with the lowest modeling error. For a full description of the mathematical background of spare unmixing the reader is redirected to Iordache *et al.* [34]. In summary, the sparse unmixing procedure can conceptually be described as follows: (i) the unmixing is formulated as a sparse regression problem; and (ii) the *alternating direction method of multipliers* (ADMMs; [35]) is used to solve the obtained optimization problem. ADMM represents a class of algorithms which decomposes a hard optimization problem into subproblems which are easier to solve, by introducing new variables in the objective function. In recent years, several algorithms exploiting this sparsity characteristic have been proposed to resolve spectral mixing problems in remotely sensed data (e.g., [36,37]).

2.3 Machine Learning-Based Estimation of Land Cover Fractions

Another group of algorithms (indirectly) relying on spectral libraries to resolve the mixing problem in remotely sensed data is the machine learning or regression-based approaches. Examples include support vector machines (SVM), random forests, and multiple layer perceptrons [38,39]. SVM is a commonly used supervised two-class classifier, where SVM defines a hyperplane that separates the two classes based on a training set. In the context of spectral unmixing, the two SVM classes are the endmember class of interest containing pixels from mixing endmembers at specific proportion values vs the unfavorable endmember classes of those coming from other proportion choices. Unlike the previous methods that rely solely on endmember spectra for land cover fraction estimation, machine learning approaches require a vast amount of training data describing all types of mixtures actually occurring within a scene. Since it is not straightforward to develop such a comprehensive training set including all types of mixtures, a spectral endmember library is used as the basis to construct synthetically mixed training data for regression-based unmixing [9,40,41]. The procedure can be summarized as follows: (i) in an iterative fashion two endmember spectra are randomly selected from a spectral endmember library accounting for (each) plausible endmember condition; (ii) according to a predefined mixing interval, mixed pixel spectra are modeled based on (1). The mixing interval refers to the number of intermediate mixtures within the given fraction range between 0% and 100%; (iii) all pure original and binary-mixed spectra are combined in a single spectral library. This library, together with the associated mixing fractions between 0% and 100%, constitutes the training data for regression-based unmixing. Given an unknown pixel for unmixing, it is evaluated through all of the SVM classifiers and the SVM that gives the largest classification margin will identify the corresponding candidate as the unmixed proportion solution. This approach was successfully applied to Hymap data collected over Berlin to quantify urban land cover using SVR [9], but requires further validation for other applications.

One advantage of SVM unmixing is that spectral variability is automatically taken care of when creating synthesized data for SVM training. Due to the discretization of proportion values, however, this approach produces a finite and possibly limited number of abundance proportion choices only, which might not be acceptable in some applications.

3 CHALLENGES OF ENDMEMBER LIBRARY-BASED SMA APPROACHES

3.1 How to Build Comprehensive Endmember Libraries for SMA?

Previous studies (see references in Section 2) clearly illustrated the effectiveness of endmember library-based SMA approaches in a variety of earth observation applications. Yet, the premise of these library-based unmixing techniques is the availability of a spectral library containing representative instances of all endmembers present within the scene. In other words, the library should incorporate the spectral variability of the endmembers. It is clear that the accuracy of the abundance estimation heavily depends on the quality of the library, or in other words how well the library accounts for the spectral endmember variability observed within the image [4].

Building these comprehensive endmember libraries is a crucial yet not straightforward aspect of the unmixing process. Spectral endmember libraries can be *derived from field (or laboratory) measurements* obtained using ground-based or portable spectroradiometers (e.g., [17,30]). Spectra from field and laboratory measurements can be controlled well and measured accurately, yet they may not match those in the image due to differences in sensors, atmospheric effects, and illumination conditions [42]. Field-based libraries are also often not well suited to cover or sample sufficient spatial variability in the endmembers. Moreover, in certain landscapes (e.g., difficult to access or remote areas like tropical forests, wetlands, or roof tops) it is often (too) costly or labor intensive to collect field data and establish extended spectral libraries.

As an alternative, endmembers can also be *extracted from image data* themselves and this both in a supervised as well as in an unsupervised way. In the supervised approach endmember reference polygons are located in the image and used to construct training and validation libraries [43]. Reference polygons can be identified in the field, or using other image sources, such as Google Earth, and typically are required to consist of at least 70% of one material. Limitations of this approach include endmember classes that are often not evenly distributed across the landscape, the assumption that reference polygons are temporally invariant and limitations in portability outside of the training image. Finally, it is often difficult to identify enough reference polygons for each class to ensure adequate random sampling between training and validation data (similar to standard training of a classifier).

In addition, a number of advanced analysis techniques to automatically locate and extract endmembers from image data themselves have been described. Techniques in this category include the pixel purity index (PPI), N-FINDR, iterative error analysis (IEA), vertex component analysis (VCA), and many others. A thorough discussion of these techniques are outside the scope of this chapter. Theseira *et al.* [44], Plaza *et al.* [12], and Veganzones and Grana [45] have systematically discussed and compared different image endmember extraction techniques. Image endmembers have the advantage of being collected at the same scale and under the same environmental conditions as the remotely sensed data and can, thus, be more easily associated with spectral components in the image scene. Yet, the classic endmember extraction techniques (as referred to earlier) fail to address endmember variability as they traditionally extract image-wide endmember spectra [46]. This means that only one single standard endmember spectrum is defined for each of the endmember classes or scene components (e.g., vegetation or soil) and subsequently assigned to all image pixels during the abundance estimation step of SMA. In an attempt to address this issue, Bateson and Curtiss [47] and Bateson *et al.* [48] introduced the concept of *endmember bundles*, a way to extract endmember libraries from image data to as such cope with endmember variability in SMA. From the principal component analysis (PCA) eigenvectors of the image data, bundles of vectors are identified to represent the endmembers. A linear programming routine is then used to determine a minimum, mean, and maximum fraction for each endmember. A similar endmember bundles extraction approach was developed by Somers *et al.* [46] and Zare *et al.* [49]. Here the authors ran the traditional endmember extraction techniques (as referred to earlier; e.g., PPI, NFINDR, VCA) in (randomly) selected subsets of the original spectral image, and subsequently grouped the extracted samples of pure materials in a bundle using a clustering technique. The output is a spectral library of pure materials, extracted automatically from the input scene. The spectral library was subsequently used to feed a library-based SMA algorithm.

A remaining downside of image endmember extraction techniques is that they require the availability of a sufficient number of pixels comprised purely of each dominant endmember class or ground component to be able to properly cover the endmember variability present in the scene. This is particularly problematic in case the spatial resolution of the image exceeds the ground extent of endmembers or ground components such that mixed pixels prevail. Although this problem is less pronounced by using spectral libraries built from field measurements, as mentioned earlier establishing such a library can be very time consuming and intensive. Especially if the variation in both time and space is considered [19]. Take for example an agricultural setting in which the aim is to monitor crop status throughout the growing season (see also Case study 2, Section 4.2) [50]. The reflectance of the canopies will not only vary due to differences in, for example, crop management, soil texture, and planting distance but also vary due to phenological changes that occur throughout a growing season. In this setting both the image extracting techniques and field measurements may not be adequate for the inclusion of sufficient variation within the spectral library. A possible solution is the use of *datasets of simulated spectra* consisting of the full range and combination of expected field conditions, also sometimes referred to as look-up tables (LUTs) [4]. A LUT is commonly established using an RTM, which is available for a range of ground components such as vegetation [51], soils [52], and snow [53]. Important is that the RTM inputs cover the full range and combination of expected field conditions. The number of spectral signatures in an LUT can as such become very large [4].

3.2 How to Deal with Extensive Spectral Libraries?

The amount of detail included in the spectral library will determine to a large extent the accuracy of the unmixing methodology. Roberts *et al.* [17], for example, demonstrated a decrease in accuracy due to shortcomings in the spectral library, including a lack of species diversity and scale problems, in which field spectra measured for the library were not representative at the scale of the satellite imagery. However, a broad variation in endmembers included in the library will potentially hamper the performance of the unmixing methodologies as well. Several pitfalls have been identified over the years, which need to be taken into account when applying or developing spectral unmixing techniques incorporating endmember variability.

A first drawback is the computational burden associated with these unmixing techniques. Simple SMA techniques have become very efficient over time in order to be able to process large amounts of remote sensing data in a timely manner, especially the geometric-based unmixing techniques [54]. However, the unmixing techniques described in Section 2 all need additional processing steps in order to take the endmember variability into account. These additional steps will result in a longer processing time and can as such hamper the operational applicability of these methods. Especially the iterative mixture cycles are known for their slow processing times, given that each possible endmember combination must be evaluated for each pixel, and this for different complexity levels (i.e., 2-, 3-,4-endmember models) [55]. *Library or dictionary pruning* techniques are therefore often enforced to optimize endmember library definition. Library pruning identifies from the extended library those endmembers that actually are present in the image under consideration. A first group of library optimization methods only use the information available in the library itself. Common examples are count-based endmember selection (COB), endmember average RMSE (EAR), mean average spectral angle (MASA), and iterative endmember selection (IES) [43]. These approaches optimize a library in order to explain as much of the library as possible, while keeping the library size as small as possible. A second group of library optimization methods, including e.g., the Music-CSR algorithm [36], uses the information from the image to discard those signatures from the library that do not contribute to explain the variability in the image. Although the computation burden is less pronounced in sparse unmixing, these latter group of library pruning techniques based on the sparsity concept are still often applied [56].

The second drawback is commonly referred to as the *ill-posedness* effect, a very common problem when applying model inversion. It refers to the phenomenon where more than one input combination can lead to the same modeled output, or in the case of SMA, where more than one endmember combination can result in the same mixed signature [57]. This poses problems as it has the consequence that (i) one of the models with the lowest RMSE is not necessarily the best model (i.e., the endmember combination might not be representative for the mixed pixel), and (ii) there might not be one ideal solution, but multiple adequate solutions [4]. This is conceptually illustrated in Fig. 4, where the lowest possible modeling error is shown in function of the number of spectral signatures included in the vegetation endmember library to account for its spectral variability in the field. The 50–50% soil-vegetation mixture is unmixed with a fixed soil endmember (not the correct one to account for ambient noise, and to avoid the modeling error becoming zero).

To circumvent this problem, Petrou and Foschi [58] introduced the fuzzy linear unmixing approach for two-endmember mixtures. For a given set of pure endmember spectra, all possible combinations of endmembers were considered for multiple mixtures. As such, the convex hull of the two pure class clusters is filled with mixture clusters of known proportions. In the same space, the mixed pixels are projected and the number of model samples from

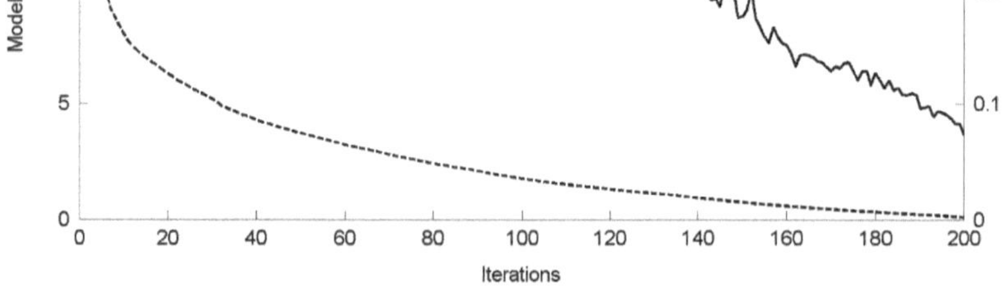

FIG. 4 Modeling error and estimated subpixel abundance error in function of the number of signatures included in the endmember libraries. Although the modeling error will decrease with an increasing number of endmember signatures, the abundance error will not necessarily decrease.

each of the possible mixture combinations that are within a predefined distance are counted. These samples are then used to compute the probability and confidences for each mixture combination.

3.3 Preprocessing of Endmember Libraries as a Way to Further Increase Unmixing Accuracy?

Recall that the use of endmember libraries (as opposed to the use of a single spectral representation per endmember class or ground component) and the associated endmember library-based SMA techniques (cf. Section 2) are a way to reduce the errors in SMA related to endmember variability.

On top of that it has been demonstrated that targeted preprocessing of the endmember libraries prior to SMA can further reduce the impact of endmember variability and thus increase unmixing accuracy. As such it has been shown in many applications that a careful selection of wavelengths, robust against spectral variability (i.e., minimizing intra- and maximizing interclass variability), can significantly improve subpixel quantification of fractional material cover (e.g., [6,30,59]). Often band selection techniques are implemented to identify (directly) from the image data, a subset of spectral features that explain most of the spectral variability (e.g., MNF, PCA). Yet, these solutions do not necessarily minimize endmember variability and therefore do not systematically improve unmixing accuracy [60]. By contrast, endmember libraries offer the opportunity to directly identify or calculate the spectral subset that minimizes intra- and maximizes interclass variability between and among the endmembers/target classes of interest and several studies have shown the added value of implementing spectral band selection prior to library-based SMA [6]. Alternative spectral transformations to reduce endmember variability that can be applied on the libraries prior to unmixing are spectral weighting, endmember normalization, derivative and wavelet analysis, or a combination of these. A full overview of approaches to address endmember variability in SMA can be found in Somers *et al.* [19] and Zare and Ho [22].

4 APPLICATIONS OF ENDMEMBER LIBRARY APPROACHES TO RESOLVE SPECTRAL MIXING PROBLEMS IN REMOTELY SENSED DATA

4.1 Case Study 1: MESMA for Studying Fire Severity and PostFire Recovery in a Mediterranean-Climate Ecosystem

Wildfire represents one of the most severe forms of disturbance altering natural landscapes. The severity of fire depends on a combination of fuel properties including live fuel moisture and fuel loads, weather including air temperature, relative humidity and winds, and terrain, where fires tend to burn more rapidly as they move up slope [61]. SMA has played an important role in all stages of wild fire assessment, including assessment of prefire fuel moisture and fuel condition [62], active fire properties [63], fire severity [64], and postfire recovery [65]. A particular challenge in assessing fire severity and postfire recovery is endmember variability, in which spectral separability between many postfire products, including dead crowns (nonphotosynthetic vegetation), exposed soil, and ash, can be low and these materials occur in multiple forms [66].

In the following case study, we illustrate the use of MESMA to quantify prefire fuel conditions, fire severity, and postfire recovery in the Santa Ynez Mountains, north of the cities of Santa Barbara and Goleta, California. Imagery were acquired by the airborne visible infrared imaging spectrometer (AVIRIS), prior to three major fires, the July 1–28 2008 Gap Fire (3862 ha), November 13–17 2008 Tea Fire (785 ha, 210 structures), and May 5–18 2009 Jesusita Fire (3534 ha). All three fires burned following a relative hot, dry year and two (Tea and Jesusita) burned under extreme meteorological conditions, including high downslope winds.

AVIRIS imagery were acquired prior to the fires in 2004, between fires (March 10, March 31 2009) and during multiple campaigns following the fires in June 2009, August 2009, April 2010, July 2011, April 2013, June 2013, and November 2013. A key objective of the acquisitions was to quantify prefire fuel conditions, fire severity, and postfire recovery. Prefire imagery reveal a mixed landscape consisting of mixed urban, natural, and anthropogenic vegetation along a coastal plain (Fig. 5) abutted against a steep south facing range consisting of a mixture of schlerophyllous shrubs, senesced grassland, and oak woodlands in more mesic river channels. Dominant surface fractions based on MESMA include green vegetation (GV; green in Fig. 5) in the front range with variable levels of nonphotosynthetic vegetation (NPV; red in Fig. 5) modeling drought-deciduous shrubs and grasslands, and large patches of senesced grass along the coastal plain modeled with high NPV fractions. This model, consisting of NPV, GV, soil, and ash (Fig. 6), leaves much of the urbanized landscape unmodeled because impervious materials were not included in the model.

Following the Jesusita fire, spectral libraries were developed from reference polygons designed to capture spectral variability in GV, NPV, soil, and ash (Fig. 6). Optimal endmember sets were identified using EAR, MASA, and COB (EMC; see Section 3.2) with June 2009 and August 2009 imagery, using June endmembers to unmodel spring images and August endmembers to unmodel summer and fall imagery. The greatest diversity was observed in the GV endmembers, requiring 11 endmembers (six from June, five from August), followed by nine Ash spectra (three from June, six from August), seven NPV (four from June, three from August), and four soils, with a pair from each image date.

Fire severity was high throughout much of the burn scar (Fig. 7). The difference Normalized Burn Ratio (dNBR) is a common ratio-based method for assessing fire severity [67]. The dNBR is an adaptation of the normalized

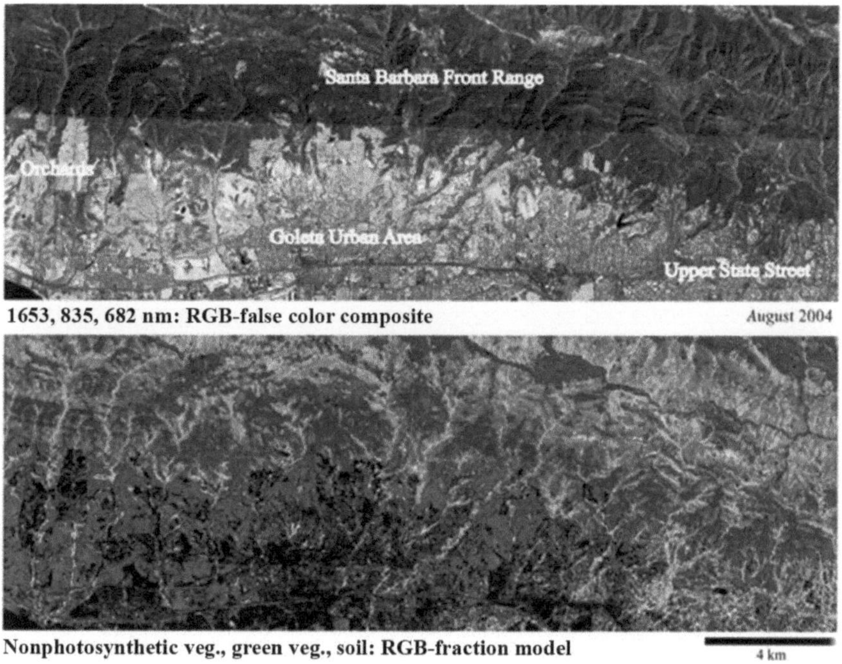

FIG. 5 AVIRIS false color composite showing 1653, 835, and 682 nm in RGB and MESMA fractions for NPV, GV, and soil (RGB). This image was acquired August 2004 from an ER-2 at an elevation of ~20 km generating 18 m spatial resolution.

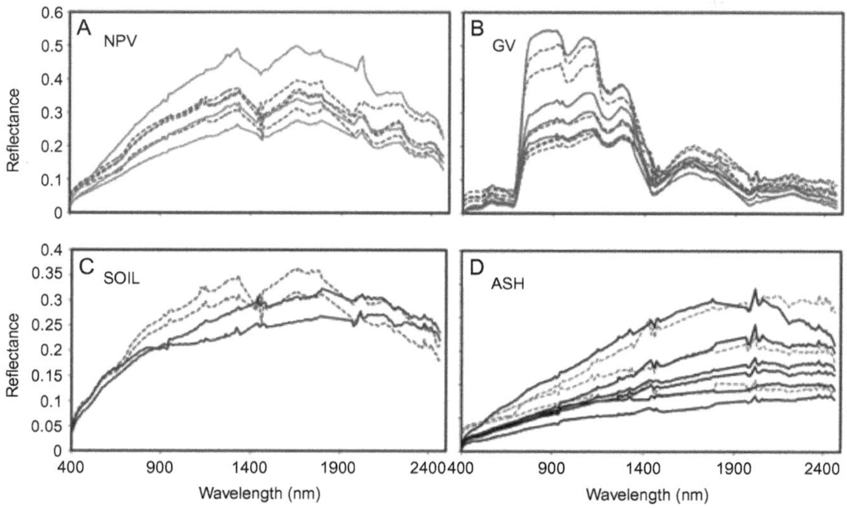

FIG. 6 Endmember spectra selected using EMC for use in MESMA. *Dashed lined spectra* were extracted from the June 2009 image, whereas *solid lines* show spectra extracted from August 2009. NPV (A) is discriminated from SOIL (C) and ASH (D) by prominent lignocellulose bands in the SWIR. SOIL and ASH are spectrally similar, but ASH tends to have lower reflectance and a more systematic increase in reflectance toward longer wavelengths than soils. GV (B) spectra differ primarily in the NIR, with subtle differences in the visible and SWIR.

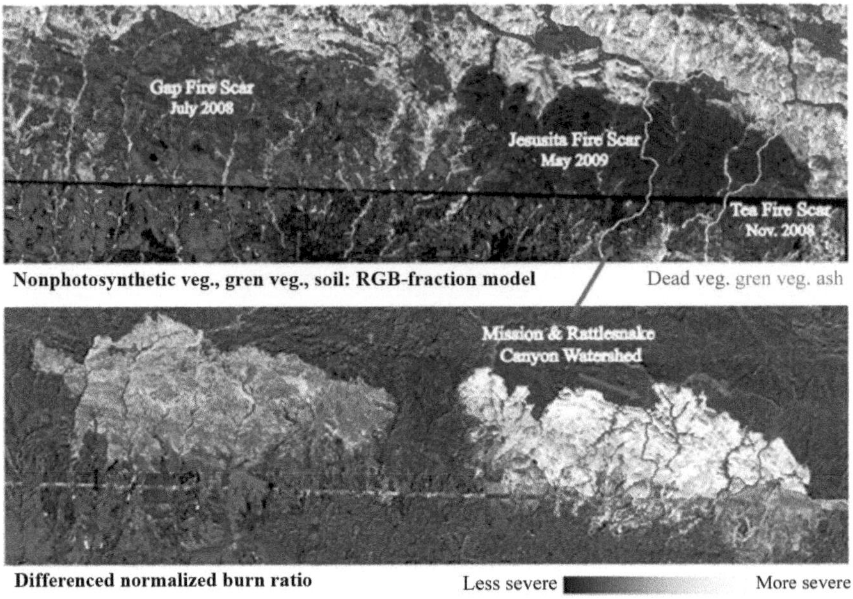

FIG. 7 August 2009 mixing model (*top*) and difference Normalized Burn Ratio (*bottom*). Recent fire scars appear in *blue* (*dark gray* in the print version), modeled primarily as ASH. Senesced grasslands outside of the fire scars were modeled with a high NPV fraction, appearing *red* (*gray* in the print version). Within the fire scars, unburned riparian areas were modeled as having a high GV fraction. Different levels of severity are mapped in riparian areas, with lower severity occurring in Rattlesnake Canyon (eastern portion of the fire scar), where canopies are mostly intact and modeled as GV, and higher severity fires in Rattlesnake Canyon (*left*), in which canopies lower in the water shed are modeled as dominated by NPV.

burn ratio, which is based on the difference between NIR and SWIR2 reflectance (NIR-SWIR2), divided by the sum of reflectance at these two wavelengths. Fires, because they remove vegetation that has high NIR reflectance and low SWIR, and replace them with dead vegetation, ash, or soils with lower NIR and higher SWIR, decrease NBR. The dNBR is calculated as the difference between prefire NBR and postfire NBR, and thus is a positive number. It is typically multiplied by 1000. High dNBR throughout much of the Jesusita Fire scar is modeled as large fractions of ash (Fig. 7). Low dNBR in the Jesusita Fire scar represents unburned riparian zones. Lower ash and dNBR in the Gap Fire is a product of lower fire severity and partial recovery between July 2008 and August 2009. dNBR in the Tea Fire scar is a product of partial recovery, as shown by a high NPV fraction in the postfire imagery, a product of annual vegetation growing in the recent fire scar. NPV modeled in the Tea Fire scar illustrates a weakness in the dNBR, which cannot distinguish between a change in reflectance due to fire, from senesced annual vegetation in a fire scar.

A major focus on postfire recovery is in how fire severity impacts water sheds that feed into the coastal zone south of the city of Santa Barbara. Fires have the potential of enhancing sediment input in to coastal waters through erosion, changing the transport of nutrients downstream and impacting canopy overlying streams, enhancing algae growth. Fractions of NPV, GV, soil, and ash are shown within the boundaries of two fire-impacted water sheds, Mission Canyon and Rattlesnake Canyon (Fig. 8).

Prior to the fires, the watersheds were dominated by GV, with close to 25% NPV (typical of mid-summer chaparral; [62]) with minor amounts of exposed soil. Ash is nearly absent, which is expected given the unburned nature of the watershed. Immediately before the Jesusita Fire, but after the Tea Fire, GV fractions peak, with considerably lower fractions of NPV, typical of late-winter chaparral, in which plants have begun to flush new leaves, and senesced material is lowest in canopies. Minor increases in ash and soil fraction in the watershed likely is a product of small sections of the watershed that were burned by the Tea Fire. Following the Jesusita Fire, the watersheds are dominated by the ash fraction. The GV fraction has decreased substantially, concentrated almost exclusively in the more mesic riparian zones. An increase in soil fraction is also observed, but only minor increases in NPV, likely a product of leaf mortality in scorched canopies. The latter observation

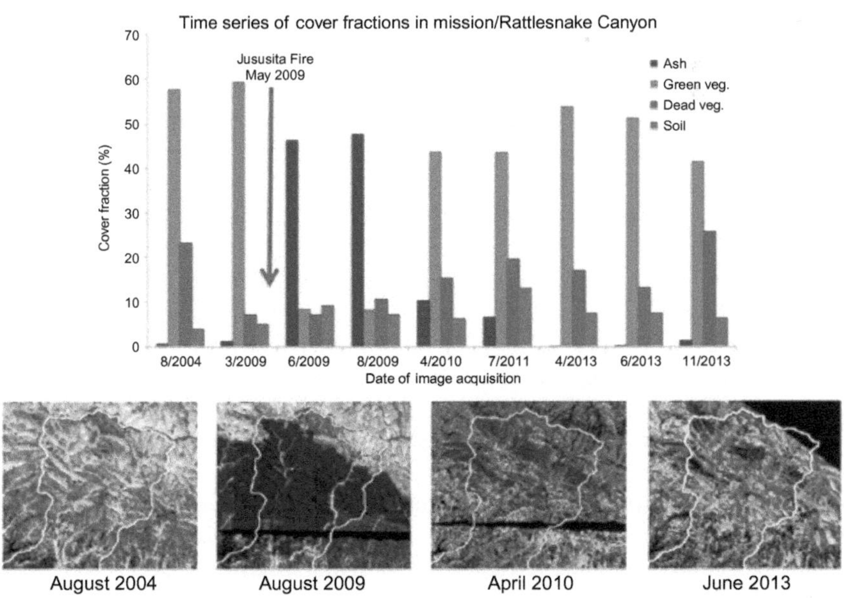

FIG. 8 Showing temporal changes in surface composition between 2004 and 2013. Fractions of NPV, GV, soil, and ash averaged over the Mission Canyon and Rattlesnake Canyon Water sheds are shown for seven AVIRIS images acquired between 2004 and 2013. NPV–GV–Ash images of the water sheds are shown below for four dates.

is supported by an increase in NPV between June and August, likely reflecting continued leaf mortality in scorched crowns.

Subsequent fraction images show a rapid loss of the ash fraction, decreasing to 10% by April 2010, and less than 5% by 2011. Substantial increases in NPV fraction in 2010 and 2011 are a product of senesced annual vegetation within the fire scars. By 2013, GV fraction remained below prefire levels and NPV fraction remained elevated, likely a product of the extended drought that has persisted from 2012 to 2015. The impact of phenology is also very apparent in 2013, in which substantial changes in GV and NPV are observed, reflecting leaf flushing in spring and leaf senescence in summer and fall. Interannual changes in precipitation and phenology remain challenges in postfire research, in which changes over multiple years may reflect postfire recovery, or changes in annual precipitation or difference due to the timing of image acquisition. In this example, postfire recovery appears delayed, likely resulting from the extended drought.

4.2 Case Study 2: MESMA for Studying Crop Vigor State at a Subpixel Level

4.2.1 The Mixture Problem in a Precision Farming Context

In a precision farming context, within-field variability in crop vigor is taken into account to optimize management practices. This requires accurate and consistent information on soil and plant conditions across the farm, and at temporal and spatial scales that match rapidly evolving capabilities to vary cultural procedures, irrigations, and agrochemical inputs [68].

Worldwide, many studies have been conducted to determine the variability in crop yield using remotely sensed measurements. Recent studies have focused on the determination of the spatial variability of soil nutrients, water stress, location-based crop diseases, and other parameters associated with in-field variability [69]. Hyperspectral remote sensing has been widely used in the prediction of chlorophyll content and related stresses such as nitrogen deficiency [70,71], in the estimation of leaf biomass [70], and in the detection of the plant's water status [72]. The biophysical and biochemical variables of interest can be retrieved from hyperspectral remote sensing data through a wide range of algorithms. Typical examples include the derivation of vegetation indices (VI), inversion of RTMs, or machine learning techniques [73].

Despite all of this potential, different processes and technological limitations still restrict the accuracy by which the information on the vigor status of the vegetation can be retrieved from hyperspectral imagery [68]. The main issue remains the orbital height, together with the trade-off between the spatial and spectral resolution of sensors, which results in a coarse spatial resolution of space-borne hyperspectral sensors. Current (i.e., Hyperion) and upcoming (e.g., the Hyperspectral Infrared Imager (HyspIRI) and the

Environmental Mapping and Analysis Program (EnMAP)) hyperspectral satellites have a spatial resolution of 30 or 60 m, which exceeds in many cases the size of the objects of interest. Consequently, the reflectance signal of a pixel is the integrated result of spectral contributions of different subpixel elements building up a pixel footprint. This is especially true in perennial cropping systems such as orchards due to the discontinuous open canopies.

Due to the subpixel spectral contribution of the background soils, weeds, and shadows, the extraction of the biophysical parameters from the measured hyperspectral signature is hampered. This is illustrated in Fig. 9. In a virtual orchard developed by Stuckens *et al.* [74], in which trees with reduced water content were included. The RGB images of the hyperspectral scenes, created with ray-tracing simulations, are shown for three different pixel sizes, together with the VI maps of the same hyperspectral scene. The red colors depict the stressed trees present in the orchard. In the 10 and 50 cm resolution images, the stressed trees can be clearly delineated. In the 200 cm resolution image,

FIG. 9 Influence of spatial resolution on the information extraction from hyperspectral imagery. The panels in the *top row* show RGB images of a virtual orchard developed by Stuckens *et al.* [74], where trees with reduced water content were included. The images were rendered in a ray-tracing simulation with pixel sizes increasing from 10 to 200 cm. The panels in the *bottom row* show the derived maximum difference water index [72] VI maps, with the *red* (*gray* in the print version) colors depicting the stressed regions. While in the first two VI maps the stressed trees can be clearly delineated, in the 200 cm image the *color* (*different gray shades* in the print version) of the VI is mainly determined by the contribution of the soil rather than the (health) status of the tree.

however, mixed pixels occur especially in the middle of the image, hampering an accurate interpretation of the VI. The color of the pixels in this image is mainly determined by the degree of mixing rather than the state of the trees present in the pixels. This will lead to a misinterpretation of the VI values in the image, which on its turn will result in a poor decision making by the farmer.

4.2.2 Signal Unmixing as a Solution

To derive accurate information on the spatial and temporal variability of crop condition from mixed hyperspectral data, it is thus required that the mixed pixel signatures are unmixed into crop-related and noncrop-related components. Traditional SMA approaches are designed to provide information on the areal extend of the subpixel components (i.e., their subpixel cover fraction f), but not on the subpixel spectral signature of the components. Endmember extraction techniques (see Section 3.1), on the other hand, are designed to extract the pure signatures of the different components, but only at an image level. However, in a precision farming context, spectral unmixing needs to be taken to a new level as the pure crop signature is required at the pixel level (i.e., pixel-specific crop signatures need to be identified). This requires a new unmixing concept, which is referred to as *Signal Unmixing* (*SU*; [50,75,76]). In Fig. 10, a schematic overview is given of the concepts of SU in a precision farming context.

The crop signature extracted from the mixed pixel can be used to derive information on the vigor of the crop, through the conventional methods, like VI or radiative transfer modeling, as mentioned earlier. The better the crop signature is extracted, the more reliable the information on the crop vigor can be determined. As opposed to the traditional SMA approaches, SU methods therefore aim to minimize the error of the extracted signatures rather than the estimated cover fractions. The endmember variability methodologies mentioned in this chapter have been developed for the estimation of the subpixel cover fractions, where the spectral variability is considered a nuisance. The aim of these techniques is thus to minimize the influence of the spectral variability, in order to obtain better estimates of the cover fractions. For SU, however, the variability in the spectral signature of the crop is the main interest, as the pure spectral signature of the crop is sought per pixel. Library-based SMA approaches, however, open opportunities for the extraction of the pure crop signature. Somers *et al.* (2009) introduced the first SU method, based on MESMA. The premise is that for perennial cropping system, the mixture problem can be simplified to the extent that the different subpixel components are known exactly (e.g., tree and soil in Fig. 10), rendering among-endmember variability insignificant. This simplification of the mixture problem allows the definition of very detailed tree and soil libraries, as to describe the complete spectral variation which is to be expected in the mixed pixel. The key is then to find the set of endmember signatures which represent the pure components'

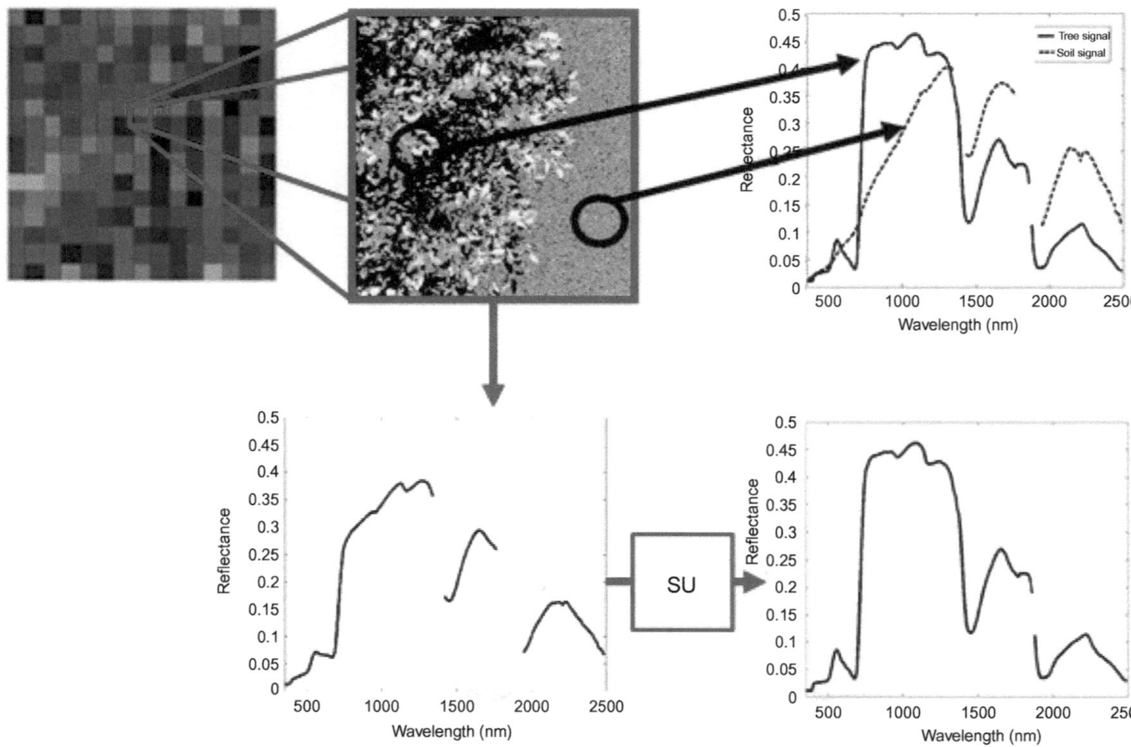

FIG. 10 Overview of the composition of a mixed pixel in a hyperspectral image of an orchard, and the concept of signal unmixing (SU), where the pure tree signature is extracted from the mixed pixel signature.

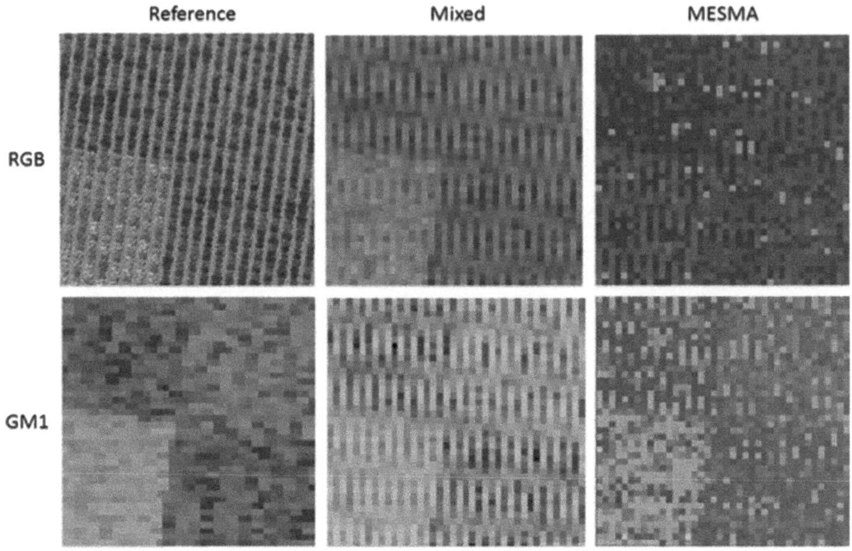

FIG. 11 Concept of SU for the improved estimation of plant vigor. The *top row* depicts the hyperspectral images, shown in RGB, while the *bottom row* depicts the GM1, a common chlorophyll index calculated by dividing the reflectance in bands 750 and 550 nm. The *first column* shows a high-resolution image where the chlorophyll-stressed region is clearly visible, while in the 2 m resolution mixed image, all the highly mixed pixels are also denoted as stressed. The *last column* shows the SU results based on MESMA, with each pixel only consisting of vegetation.

signatures within the pixel as close as possible. In Fig. 10, the improved monitoring of the tree vigor status is shown for a 2 m resolution image for a simulated citrus orchard (Fig. 11B). In the orchard a chlorophyll-stressed block was present (bottom left), clearly visible in the high-resolution reference image (Fig. 11A) and corresponding chlorophyll map (Fig. 11D). Due to the background contamination, all highly mixed pixels are also denoted as chlorophyll-stressed in the 2 m resolution image (Fig. 11E), but this effect can be strongly mitigated through the application of an SU method (Fig. 11C), with the stressed region clearly visible in the chlorophyll index map of the MESMA result (Fig. 11F).

4.2.3 Challenges of Signal Unmixing

In order to incorporate the full range of tree vigor states which may be encountered in the field, extremely large libraries are needed for SU (e.g., 30,000 tree signals used in Tits et al. (2012)). The resulting high collinearity between signals, as well as the more pronounced ill-posedness effects and the very large processing times, makes it extremely difficult to find this optimal solution. It is therefore key to use a library which offers a great variability close to the optimal solution, but to restrict the range within which signatures are included in the library as much as possible. Library pruning methods (see Section 3.2) are used to reduce the number of redundant signatures from the library, as such making the SU method more efficient and accurate.

In addition to library pruning, the incorporation of ancillary data into the SU unmixing chain has proven its worth in MESMA-based SU. As ancillary data provide more information regarding the components within the pixel, the number of possible solutions can be strongly reduced, thus mitigating ill-posedness effects. In addition, the libraries can be optimized using this ancillary data, strongly reducing computation times. In Tits *et al.* [50], for example, MESMA was combined with a soil moisture model, using in situ measured soil moisture and soil texture to provide pixel-specific soil signatures. This rendered the tree signature to be the only unknown spectrum in the tree/soil mixtures, which in turn strongly reduced ill-posedness effects, especially in the shortwave infrared spectral region (1500–2500 nm) where canopy moisture content and soil moisture content have a very similar effect on the spectral characteristics. In addition, computation time was reduced by a factor of 300, i.e., the size of the soil library.

5 CONCLUSIONS

This chapter introduced the concept of spectral libraries to address the mixing problem in remotely sensed data. The theoretical concepts of the most common approaches were briefly introduced and based on two illustrative cases the effectiveness of endmember library-based spectral unmixing was demonstrated. Note, however, that in the literature this concept of library-based unmixing has also successfully been used in many other applications (cf. citations throughout the text). Despite the potential of the approach to better account for the spatial and temporal spectral variability observed in natural environments, the availability and optimal use of extended libraries (needed for effective unmixing results) is an element of continuous attention for the remote sensing community, with still some room for improvement and some un(der)explored ground. Finally, it was pointed out that proper preprocessing of the endmember library, through library pruning and/or spectral feature selection, is considered a key element to fully address endmember variability issues and thus provide optimal unmixing results in an operational setting.

ACKNOWLEDGMENTS

The research presented in this paper is funded by the Belgian Science Policy Office in the framework of the STEREOII (projects Vegemix (SR/67/146) and ReMEDy (SR/67/164)) and STEREOIII (projects VIPER2 (SR/XX/171) and UrbanEARS (SR/00/307)) programs.

Analysis of AVIRIS imagery, used to map prefire fuels, fire severity, and postfire effects, was supported in part by funding from the National Aeronautics and Space Administration (Award number NNX12AP08G) and the U.S. National Science Foundation's Long Term Ecological Research Program (Award number OCE1232779).

REFERENCES

[1] Adams JB, Gillespie AR. Remote sensing of landscapes with spectral images: a physical modeling approach. Cambridge, UK: Cambridge University Press; 2006. 362 pp.

[2] Asner GP, Heidebrecht KB. Spectral unmixing of vegetation, soil, and dry carbon cover in arid regions: comparing multispectral and hyperspectral observations. Int J Remote Sens 2002;23:3939–58.

[3] Roder A, Udelhoven T, Hill J, del Barrio G, Tsiourlis G. Trend analysis of Landsat-TM and -ETM+ imagery to monitor grazing impact in a rangeland ecosystem in Northern Greece. Remote Sens Environ 2008;112:2863–75.

[4] Tits L, Somers B, Coppin P. The potential and limitations of a clustering approach for the improved efficiency of multiple endmember spectral mixture analysis in plant production system monitoring. IEEE Trans Geosci Remote Sens 2012;50:2273–86.

[5] Eckmann TC, Still CJ, Roberts DA, Michaelsen JC. Variations in subpixel fire properties with season and land cover in Southern Africa. Earth Interact 2010;14(6):1–29.

[6] Somers B, Asner GP. Tree species mapping in tropical forests using multi-temporal imaging spectroscopy: wavelength adaptive spectral mixture analysis. Int J Appl Earth Obs Geoinf 2014;31:57–66.

[7] Mertes LAK. Remote sensing of riverine landscapes. Freshw Biol 2002;47:799–816.

[8] Bedini E. Mapping lithology of the Sarfartoq carbonatite complex, southern West Greenland, using HyMap imaging spectrometer data. Remote Sens Environ 2009;113:1208–19.

[9] Okujeni A, van der Linden S, Tits L, Somers B, Hostert P. Support vector regression and synthetically mixed training data for quantifying urban land-cover. Remote Sens Environ 2013;137:184–97.

[10] Keshava N, Mustard JF. Spectral unmixing. IEEE Signal Process 2002;19:44–57.

[11] Adams JB, Smith MO, Johnson PE. Spectral mixture modeling—a new analysis of rock and soil types at the Viking Lander-1 site. J Geophys Res Solid Earth Planets 1986;91:8098–112.

[12] Plaza A, Martinez P, Perez R, Plaza J. A quantitative and comparative analysis of endmember extraction algorithms from hyperspectral data. IEEE Trans Geosci Remote Sens 2004;42:650–63.

[13] Boardman JW. Inversion of imaging spectrometry data using singular value decomposition, In: Proceedings of IGARSS, Vancouver, CA; 1989. p. 2069–72.

[14] Adams JB, Sabol D, Kapos V, Filho RA, Roberts DA, Smith MO, et al. Classification of multispectral images based on fractions of endmembers: application to land-cover change in the Brazilian Amazon. Remote Sens Environ 1995;52:137–54.

[15] Settle J. On the effect of variable endmember spectra in the linear mixture model. IEEE Trans Geosci Remote Sens 2006;44:389–96.

[16] Shimabukuro YE, Smith J. The least-squares mixing models to generate fraction images derived from remote sensing multispectral data. IEEE Trans Geosci Remote Sens 1991;29:16–20.

[17] Roberts DA, Gardner M, Church R, Ustin S, Scheer G, Green RO. Mapping chaparral in the Santa Monica Mountains using multiple endmember spectral mixture models. Remote Sens Environ 1998;65:267–79.

[18] Roberts DA, Smith MO, Sabol DE, Adams JB, Ustin S. Mapping the spectral variability in photosynthetic and non-photosynthetic vegetation, soils and shade using AVIRIS, In: Summaries 3rd annual JPL airborne geoscience workshop: vol. 1, AVIRIS. In: Pasadena, CA. June 1 and 2, 1992; 1992. p. 38–40.

[19] Somers B, Asner GP, Tits L, Coppin P. Endmember variability in spectral mixture analysis: a review. Remote Sens Environ 2011;115:1603–16.

[20] Garcia-Haro FJ, Sommer S, Kemper T. A new tool for variable multiple endmember spectral mixture analysis (VMESMA). Int J Remote Sens 2005;26:2135–62.

[21] Sabol DE, Adams JB, Smith MO. Quantitative sub-pixel spectral detection of targets in multispectral images. J Geophys Res 1992;97:2659–72.

[22] Zare A, Ho KC. Endmember variability in hyperspectral analysis: addressing spectral variability during spectral unmixing. IEEE Signal Process Mag 2014;31(1):95–104.

[23] Tits L, Somers B, Saeys W, Coppin P. Site specific plant state monitoring through hyperspectral alternative least squares unmixing. IEEE J Sel Top Appl Earth Observ Remote Sens 2014;7:3606–18.

[24] Zhang X, Tauler R. Application of multivariate curve resolution alternating least squares (MCR-ALS) to remote sensing hyperspectral imaging. Anal Chim Acta 2013;762:25–38.

[25] Sonnentag O, Chen JM, Roberts DA, Talbot J, Halligan KQ, Govind A. Mapping tree and shrub leaf area indices in an ombrotrophic peatland through multiple endmember spectral unmixing. Remote Sens Environ 2007;109:342–60.

[26] Franke J, Roberts DA, Halligan K, Menz G. Hierarchical multiple endmember spectral mixture analysis (MESMA) of hyperspectral imagery for urban environments. Remote Sens Environ 2009;113:1712–23.

[27] Powell RL, Roberts DA. Characterizing variability of the urban physical environment for a suite of cities in Rondonia, Brazil. Earth Interact 2008;12(13):1–32.

[28] Dennison PE, Roberts DA. Endmember selection for multiple endmember spectral mixture analysis using endmember average RMSE. Remote Sens Environ 2003;87:123–35.

[29] Eckmann TC, Roberts DA, Still CJ. Estimating subpixel fire sizes and temperatures from ASTER using multiple endmember spectral mixture analysis. Int J Remote Sens 2009;30:5851–64.

[30] Asner GP, Lobell DB. A biogeophysical approach for automated SWIR unmixing of soils and vegetation. Remote Sens Environ 2000;74:99–112.

[31] Song CH. Spectral mixture analysis for subpixel vegetation fractions in the urban environment: how to incorporate endmember variability? Remote Sens Environ 2005;95: 248–63.

[32] Byambakhuu I, Sugita M, Matsushima D. Spectral unmixing model to assess land cover fractions in Mongolian steppe regions. Remote Sens Environ 2010;114:2361–72.

[33] Dobigeon N, Moussaoui S, Coulon M, Tourneret JY, Hero AO. Joint Bayesian endmember extraction and linear unmixing for hyperspectral imagery. IEEE Trans Signal Process 2009;57:4355–68.

[34] Iordache D, Biocas-Dias J, Plaza A. Sparse unmixing of hyperspectral data. IEEE Trans Geosci Remote Sens 2011;49:2014–39.

[35] Eckstein J, Bertsekas D. On the Douglas-Rachford splitting method and the proximal point algorithm for maximal monotone operators. Math Program 1992;5:293–318.

[36] Iordache D-M, Bioucas-Dias J, Plaza A, Somers B. Music-csr: hyperspectral unmixing via multiple signal classification and collaborative sparse regression. IEEE Trans Geosci Remote Sens 2014;52:4364–82.

[37] Iordache D-M, Bioucas-Dias J, Plaza A. Collaborative sparse regression for hyperspectral unmixing. IEEE Trans Geosci Remote Sens 2014;52:341–54.

[38] Baraldi A, Binaghi E, Blonda P, Brivio PA, Rampini A. Comparison of the multilayer perceptron with neuro-fuzzy techniques in the estimation of cover class mixture in remotely sensed data. IEEE Trans Geosci Remote Sens 2001;39:994–1005.

[39] Brown M, Gunn SR, Lewis HG. Support vector machines for optimal classification and spectral unmixing. Ecol Model 1999;120:167–79.

[40] Bovolo F, Bruzzone L, Carlin L. A novel technique for subpixel image classification based on support vector machine. IEEE Trans Image Process 2010;19(11):2983–99.

[41] Mianji F, Zhang Y. SVM-based unmixing-to-classification conversion for hyperspectral abundance quantification. IEEE Trans Geosci Remote Sens 2011;49(11):4318–27.

[42] Tompkins S, Mustard JF, Pieters CM, Forsyth DW. Optimization of endmembers for spectral mixture analysis. Remote Sens Environ 1997;59:472–89.

[43] Roth KL, Dennison PE, Roberts DA. Comparing endmember selection techniques for accurate mapping of plant species and land cover using imaging spectrometer data. Remote Sens Environ 2012;127:139–52.

[44] Theseira MA, Thomas G, Taylor JC, Gemmell F, Varjo J. Sensitivity of mixture modeling to endmember selection. Int J Remote Sens 2003;24:1559–75.

[45] Veganzones MA, Grana M. Endmember extraction methods: a short review. In: Knowledge-based intelligent information and engineering systems, 12th international conference, KES 2008; 2008. p. 400–7.

[46] Somers B, Zortea M, Plaza A, Asner GP. Automated extraction of image-based endmember bundles for improved spectral unmixing. IEEE J Sel Top Appl Earth Observ Remote Sens 2012;5:396–408.

[47] Bateson CA, Curtiss B. A method for manual endmember selection and spectral unmxiing. Remote Sens Environ 1996;55:229–43.

[48] Bateson CA, Asner GP, Wessman CA. Endmember bundles: a new approach to incorporating endmember variability into spectral mixture analysis. IEEE Trans Geosci Remote Sens 2000;38:1083–94.

[49] Zare A, Gader P, Casella G. Sampling piecewise convex unmixing and endmember extraction. IEEE Trans Geosci Remote Sens 2013;51:1655–65.

[50] Tits L, Somers B, Stuckens J, Farifteh J, Coppin P. Integration of in situ measured soil status and remotely sensed hyperspectral data to improve plant production system monitoring: concept, perspectives and limitations. Remote Sens Environ 2013;128: 197–211.

[51] Verhoef W. Light scattering by leaf layers with application to canopy reflectance modeling: the SAIL model. Remote Sens Environ 1984;16(2):125–41.

[52] Jacquemoud S, Baret F, Hanocq JF. Modeling spectral and bidirectional soil reflectance. Remote Sens Environ 1992;41(2):123–32.

[53] Wiscombe WJ, Warren SG. A model for the spectral albedo of snow. I: pure snow. J Atmos Sci 1980;37(12):2712–33.

[54] Heylen R, Burazerović D, Scheunders P. Fully constrained least squares spectral unmixing by simplex projection. IEEE Trans Geosci Remote Sens 2011;49(11):4112–22.

[55] Okin GS, Roberts DA, Murray Be, Okin WJ. Practical limits on hyperspectral vegetation discrimination in arid and semiarid environments. Remote Sens Environ 2001;77(2):212–25.

[56] Iordache D-M, Tits L, Plaza A, Bioucas-Dias J, Somers B. A dynamic unmixing framework for site specific monitoring of plant production systems. IEEE J Sel Top Appl Earth Observ Remote Sens 2014;7:2016–34.

[57] Price JC. How unique are spectral signatures? Remote Sens Environ 1994;49(3):181–6.

[58] Petrou M, Foschi PG. Confidence in linear spectral unmixing of single pixels. IEEE Trans Geosci Remote Sens 1999;37(1):624–6.

[59] Somers B, Delalieux S, Verstraeten WW, van Aardt JAN, Albrigo G, Coppin P. An automated waveband selection technique for optimized hyperspectral mixture analysis. Int J Remote Sens 2010;31:5549–68.

[60] Miao X, Gong P, Swope S, Pu R, Carruthers R, Anderson GL, et al. Estimation of yellow starthistle abindance through CASI-2 hyperspectral imagery using linear spectral mixture models. Remote Sens Environ 2006;101:329–41.

[61] Pyne SJ, Andrews PL, Laven RD. Introduction to wildland fire. 2nd ed. New York: John Wiley & Sons, Inc.; 1996. 769 pp.
[62] Roberts DA, Dennison PE, Peterson S, Sweeney S, Rechel J. Evaluation of AVIRIS and MODIS measures of live fuel moisture and fuel condition in a shrubland ecosystem in Southern California. J Geophys Res Biogeosci 2006;111:G04S02. http://dx.doi.org/10.1029/2005JG000113. 16 pp.
[63] Dennison PE, Charoensiri K, Roberts DA, Peterson SH, Green RO. Wildfire temperature and land cover modeling using hyperspectral data. Remote Sens Environ 2006;100:212–22.
[64] Veraverbeke S, Hook SJ. Evaluating spectral indices and spectral mixture analysis for assessing fire severity, combustion completeness and carbon emissions. Int J Wildland Fire 2013;22:707–20.
[65] Riano D, Chuvieco E, Ustin SL, Zomer R, Dennison PE, Roberts DA, et al. Assessment of vegetation regeneration after fire through multitemporal analysis of AVIRIS images in the Santa Monica Mountains. Remote Sens Environ 2002;79:60–71.
[66] Quintano C, Fernando-Manso A, Roberts DA. Multiple endmember spectral mixture analysis (MESMA) to map burn severity levels from Landsat images in Mediterranean countries. Remote Sens Environ 2013;136:76–88.
[67] Key CH, Benson NC. Landscape assessment (LA). FIREMON: fire effects monitoring and inventory system: Gen. Tech. Rep. RMRS-GTR-164-CD. Fort Collins, CO: US Department of Agriculture, Forest Service, Rocky Mountain Research Station; 2006.
[68] Pinter Jr PJ, Hatfield JL, Schepers JS, Barnes EM, Moran MS, Daughtry CS, et al. Remote sensing for crop management. Photogramm Eng Remote Sens 2003;69(6):647–64.
[69] Panda SS, Hoogenboom G, Paz JO. Remote sensing and geospatial technological applications for site-specific management of fruit and nut crops: a review. Remote Sens 2010;2(8): 1973–97.
[70] Haboudane D, Miller JR, Pattey E, Zarco-Tejada PJ, Strachan IB. Hyperspectral vegetation indices and novel algorithms for predicting green LAI of crop canopies: modeling and validation in the context of precision agriculture. Remote Sens Environ 2004;90(3):337–52.
[71] Zarco-Tejada P, Miller J, Morales A, Berjon A, Aguera J. Hyperspectral indices and model simulation for chlorophyll estimations in open canopy tree crops. Remote Sens Environ 2004;90:463–76.
[72] Dzikiti S, Verreynne JS, Stuckens J, Strever A, Verstraeten WW, Swennen R, et al. Determining the water status of Satsuma mandarin trees [Citrus Unshiu Marcovitch] using spectral indices and by combining hyperspectral and physiological data. Agric For Meteorol 2010;150:369–79.
[73] Verrelst J, Rivera JP, Alonso L, Moreno J. ARTMO: an automated radiative transfer models operator toolbox for automated retrieval of biophysical parameters through model inversion. In: Proceedings of EARSeL 7th SIG-imaging spectroscopy workshop, Edinburgh, UK; 2011. p. 11–3.
[74] Stuckens J, Somers B, Delalieux S, Verstraeten WW, Coppin P. The impact of common assumptions on canopy radiative transfer simulations: a case study in citrus orchards. J Qual Spectrosc Radiat Transfer 2009;110:1–21.
[75] Tits L. Hyperspectral signal unmixing techniques for site specific plant state monitoring. Ph.D. thesis, KU Leuven, Group Science, Technology & Engineering, Arenberg Doctoral School; 2013.
[76] Somers B, Delalieux S, Verstraeten WW, Coppin P. A conceptual framework for the simultaneous extraction of sub-pixel spatial extent and spectral characteristics of crops. Photogramm Eng Remote Sens 2009;75:57–68.

Chapter 18

Spectral–Spatial Unmixing Approaches in Hyperspectral VNIR/SWIR Imaging

N. Gorretta*[,1] and C. Gomez†
*IRSTEA UMR ITAP, Montpellier Cedex 5, France
†IRD, UMR LISAH (INRA-IRD-SupAgro), Montpellier, France
[1]Corresponding author: e-mail: nathalie.gorretta@irstea.fr

Chapter Outline

1 INTRODUCTION

Visible and near-infrared (VNIR, 400–1000 nm) and short-wave infrared (SWIR, 1000–2500 nm) hyperspectral remote sensing technology is characterized by instrumental parameters, including the spectral resolution, revisit time, signal-to-noise ratio (SNR), and spatial resolution. Typically, the spatial resolution of the remote sensing sensors is coarser than the scale of the spatial heterogeneity of the ground surface, so a mixture of disparate components is contained in each pixel. The spectral unmixing technique refers to any process that separates the pixel spectra from a hyperspectral image into a collection of component spectra or spectral signatures, called endmembers and a set of fractional abundances, one set per pixel (e.g., Chapter 1). The endmembers are assumed to represent the pure materials present in the image and the set of abundances, or simply abundances, the percentage of each endmember that

Data Handling in Science and Technology, Vol. 30. http://dx.doi.org/10.1016/B978-0-444-63638-6.00018-8
Copyright © 2016 Elsevier B.V. All rights reserved.

is present in a pixel [1]. As noted by Dobigeon *et al.* [2] among others, spectral unmixing is an important issue for the processing of hyperspectral imaging, not only in the remote sensing community but also in the signal and image processing community, and has been a very active research area in recent years. Solving this problem can indeed provide answers to various important issues such as classification [3], material quantification [4], and subpixel detection [5].

A combination of the reflectance spectra of individual components can be an areal mixture or an intimate mixture. In case of an areal mixture, the spectrum of a mixed pixel is a linear combination of the endmember spectra, weighted by the areal coverage of each endmember in the pixel [6]. A linear mixing model may therefore be adapted to the areal mixture. In case of an intimate mixture, such as sand grains of different compositions in a beach deposit, the incident light is multiply scattered at the surface, so it interacts with more than one endmember, and the resulting spectrum is a nonlinear mixture of the components [7]. Nonlinear spectral mixing effects can be a crucial component in many real-world scenarios, such as planetary remote sensing, intimate mineral mixtures, vegetation canopies, and urban scenes. A linear mixing model may not be adaptable to an intimate mixture, so nonlinear models have to be developed. Nevertheless, the use of nonlinear models usually requires more knowledge of the material than available, such as scattering coefficients, optical constants of materials, and grain size (e.g., Chapter 6 of this book N. Dobigeon; Ref. [8]). In spite of the more realistic representation of the mixture by nonlinear models, the most used spectral unmixing model is the linear mixing model.

For the sake of simplicity, we use the following notations throughout this chapter. Capital bold characters will be used for matrices, e.g., **Z**; lower case bold characters for vectors, nonbold characters will be used for scalars, e.g., matrix elements z_{ij} or indices i. Under the linear mixture assumption, the observed spectrum $\mathbf{x}_i$ ($p \times 1$ dimensional vector) for a given pixel i of a hyperspectral image with p wavelengths is defined as:

$$\mathbf{x}_i = \mathbf{M}\boldsymbol{\alpha}_i + \mathbf{n}_i \tag{1}$$

where $\mathbf{M} = [\mathbf{m}_1, \ldots, \mathbf{m}_N]$ is a $p \times N$ matrix containing spectra of N endmembers (where $\mathbf{m}_i$ is a $p \times 1$ dimensional vector and denotes the ith endmember signature), $\boldsymbol{\alpha}_i$ is a $N \times 1$ dimensional abundance vector, and $\mathbf{n}_i$ is a $p \times 1$ dimensional noise vector (Fig. 1). A large variety of techniques based on linear models have been proposed to obtain endmembers and their abundances in hyperspectral imagery (e.g., Chapter 6 of this book, N. Dobigeon).

Spectral unmixing techniques comprise three steps. The first is optional and is a dimensionality reduction step, which can allow the estimation of the number N of endmembers. The second step consists of the extraction and identification of the endmember spectra in the studied area (which could

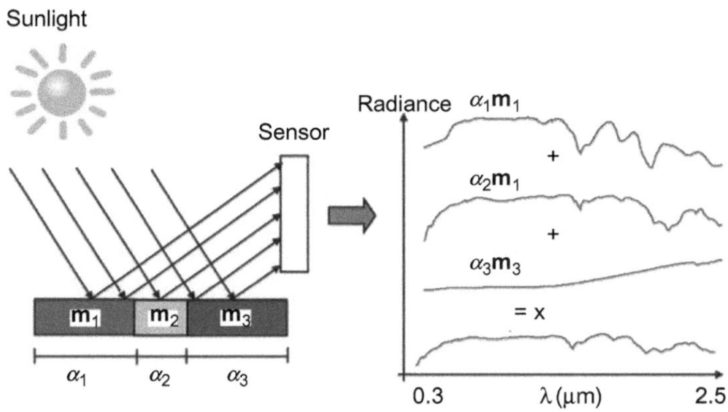

FIG. 1 Schematic view of a linear mixing process without noise: signal recorded by a hyperspectral sensor over a scene composed of three materials (m_1, m_2, and m_3). *Inspired by Bioucas-Dias JM, Plaza A, Dobigeon N, Parente M, Du Q, Gader P, Chanussot J. Hyperspectral unmixing overview: Geometrical, statistical, and sparse regression-based approaches. IEEE J Sel Topics Appl Earth Observ Remote Sens 2012;5(2):354–79.*

be the whole image or a part of the image), and the third step is an inversion step for the abundance estimation of the endmembers in each pixel. It should be noted that the extraction and inversion steps can be consecutive or simultaneous.

The success of the spectral unmixing is dependent on the choice of the method for the mixing model (e.g., the choice between linear and nonlinear mixtures) and the selection of the number and type of endmembers. In particular, this selection may prove to be difficult because of (i) the spectral within-class (also called intraclass) variations, which refers to relative differences in spectral signatures within the same land cover class, and (ii) the spectral between-class (also called interclass) variations, which refers to relative differences between two or more different land cover classes [9].

Until recently, spectral mixture analyses were always based on the implicit assumption that the spatial distribution of endmembers is random, so that each pixel is independent. However, this approach violates one of the basic tenets of geography, the direct relationship between distance and likeness, as well as one of ecological theory that elements of an ecosystem close to one another are more likely to be influenced by the same generating process and will therefore be similar [10]. In other words, a spatial autocorrelation exists between pixels, reflecting the tendency of neighboring pixels to present similar characteristics. Furthermore, as emphasized by Plaza *et al.* [11], the combination of spatial and spectral information leads to a new interpretation of the endmember concept, related to the capacity of a particular signature to describe various mixed pixels coherently in both spectral and spatial terms.

In vegetation studies, the number of independent vegetation variables that influence the spectra is small [12,13], so the reflectance of different vegetation species is close [14]. As a consequence, vegetation species identification by spectral mixture analysis involves an endmember extraction with low spectral between-class variations, which is expected to be difficult. Taking into account spatial information, with a local selection of vegetation species endmembers (e.g., within a neighborhood) rather than one that is global (e.g., across the whole image), may decrease the impact of the between-class variability for endmember extraction and thereby for the selection of an endmember combination.

In soil studies, the endmember associated with a "soil" class over a study area covered by remote sensing imagery has to be representative of all soils at the image scale, even if the study area is covered by different pedological contexts, soil rugosity, humidity, etc. Therefore, (1) the identification of one or several spectral endmembers representative of this "soil" class could be difficult at the image scale, and (2) the identification and quantification of this class over each mixed pixel may therefore be inaccurate. At the same time, semivariograms of soil properties show strong spatial dependences [15,16]. As mentioned previously, taking into account spatial information, with a local selection of soil endmembers (e.g., within a neighborhood) rather than a global selection (e.g., across the whole image), may reduce the impact of the within-class variability for endmember identification and the selection of an endmember combination [17].

As a last example, in urban studies, if spectral information alone is used to identify and quantify materials in mixed pixels, the roofs of private houses and larger buildings will be assigned to the same endmember, which means the same type of structure. Taking into account spatial information such as the sizes of roofs may allow increasing the number of endmembers and improving the precision of the endmember identification [18].

Considering these aspects and the strong increase in spatial resolution due to recent developments in sensory technology, researchers have acknowledged the benefits of including spatial information in the classification process, and so in the unmixing process. Hence, instead of only considering isolated pixels, referred to as "pixel-based" or "pixel-by-pixel" classification, some developments have been initiated to introduce spatial context into multiband image classification using the extraction and classification of homogeneous objects classifier [19,20]. Since then, many studies have proposed new algorithms taking into account the spectral and spatial dimensions of the data, especially in the classification and segmentation domains [18,21–23]. In these fields, numerous works have been published, especially in the remote sensing community (more than 250 articles have been published on this topic since 2005). More recently, some authors (e.g., Refs. [11,24,25]) have investigated the use of the spatial dimension of hyperspectral images to improve the unmixing process.

This chapter aims to provide a review of methods that have been developed for spectral–spatial unmixing. To the best of our knowledge, few spectral–spatial unmixing methods have considered the nonlinear approach, so only linear methods are considered in this chapter. Section 2 introduces the background of linear unmixing approaches by describing the different steps of an unmixing process, i.e., dimension reduction, endmembers extraction, and the ultimate abundance estimation steps. In Section 3, the available spectral–spatial unmixing approaches developed in the literature are described. The description of these approaches is ordered according to the different stages in which the spatial information is accounted for. Finally, Section 4 concludes this chapter with a draw of the highlights of the spectral–spatial unmixing approach.

2 BACKGROUND ON LINEAR SPECTRAL UNMIXING

Unmixing approaches described in the remainder of this chapter are focused on exploiting the linear model. As noted by Bioucas-Dias and Plaza [26], despite its simplicity, the linear mixing model is an acceptable approximation of the light scattering mechanisms in many real scenarios. The usual linear unmixing approaches that are mentioned in the following sections are roughly described in this paragraph. For more details, readers are directed to read some dedicated chapters of this book (e.g., Chapter 6 of this book, N. Dobigeon) and the references cited throughout this section.

As seen in Section 1, a standard unmixing process includes three main steps (Fig. 2). The first one is an optional dimensionality reduction step, which allows the identification of an appropriate subspace with a lower dimensionality, strongly connected with the estimation of the number N of endmembers present in the image. The second step consists of extracting the N endmember spectra and identifying them. The last step addresses, by using the endmember spectra given by the previous step, the estimation of their abundance by solving a constrained optimization problem.

2.1 Dimension Reduction Step

A reduction of the original spectral dimension is performed to obtain a minimal representation of the signal in a lower-dimensional space that sufficiently retains the requisite information for the successful unmixing in the lower dimension [27]. The identification of this subspace enables an accurate representation of the data in a low-dimensional space, yielding gains in computational time and complexity, in data storage, and in SNR [26].

Numerous methods have been proposed in the literature to reduce the dimension space of VNIR/SWIR hyperspectral data. Among these, linear projection techniques such as principal component analysis (PCA) [28], maximum noise fraction (MNF) [29], noise adjusted principal components

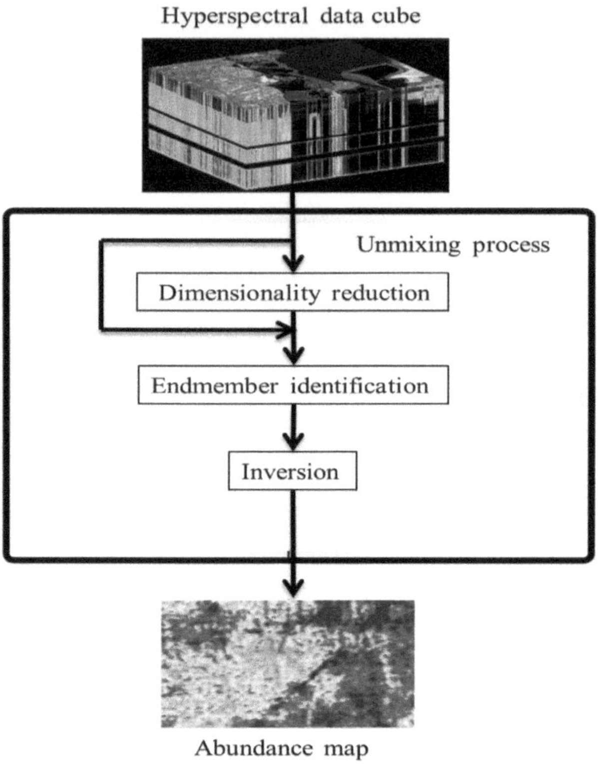

FIG. 2 Conceptual diagram of end-to-end spectral unmixing.

(NAPC) [30], projection pursuit (PP) [31], and independent component analysis (ICA) [32] have been intensively used. These methods derive their transformations from statistical information (second-order or all moments) of the data and can be further differentiated by the optimization criterion that is utilized [27]. Indeed, PCA seeks the projection that allows the best data representation in the least squares sense (variance maximization), MNF and NAPC optimize the SNR, PP generalizes the PCA allowing the use of a more sophisticated criterion than the decorrelation [33] and usually looks for directions maximizing the kurtosis of the projected values (non-Gaussian values), and ICA searches for projection such that the probability distributions of the data along these projections are fourth-order statistically independent. Other approaches, such as the minimum description length [34], Akaike information criterion (AIC) [35], and hyperspectral signal identification by minimum error (HySime) [36], have also been used to estimate the signal subspace.

Readers interested in dimension reduction methods may refer to some reviews dedicated to this topic [37–39].

Moreover the dimensional reduction step may be strongly connected to the estimation of the number of endmembers N. Indeed each spectrum can be linearly transformed to the reduced space by using the first N eigenvectors of the scene covariance. In this reduced space, and under the constraint of sum-to-one and nonnegative abundance of endmembers, the data may be shrink wrapped by a simplex, and endmembers may be estimated from the N+1 vertices of the final simplex enclosing the data [40] (Fig. 3). Several approaches were developed for estimating the number of endmembers N using this concept of dimensional space, including the most successful approaches which are the virtual dimensionality [41] concept and the HySime [36].

2.2 Endmembers Spectra Extraction and Their Identification

The endmember spectra may be derived from spectral libraries built from field or laboratory spectral measurements [42]. They can also be simulated using radiative transfer models [43], and they can be derived directly from the hyperspectral data themselves [44] (Chapter 17). A number of advanced analysis techniques to locate and extract endmembers from hyperspectral data have been developed.

The endmembers extraction from hyperspectral data is mainly performed by a geometry-based approach and less frequently by a statistics-based approach [1,45]. The linear mixing model assumes that the hyperspectral data can be expressed in the form of a linear combination of endmembers weighted by their corresponding abundances (Eq. 1). The fundamental assumption of

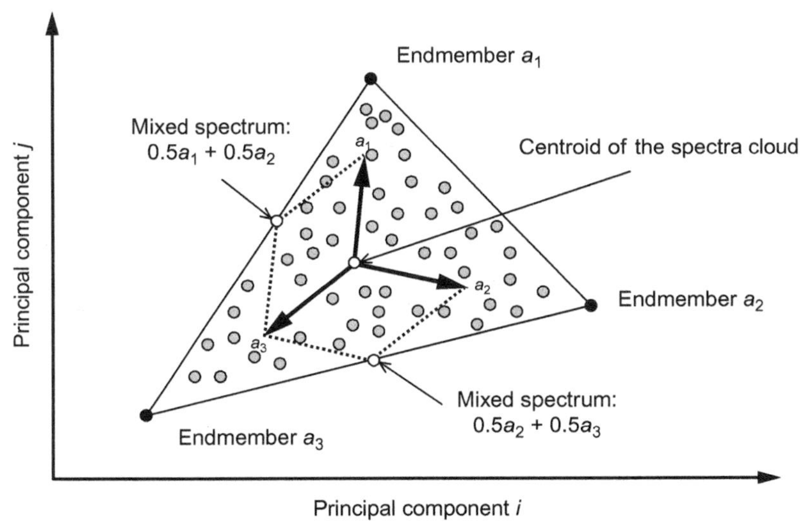

FIG. 3 Scatterplot of N-dimensional spectral data illustrating the physical interpretation of a mixture model based on endmembers, with $N=2$.

the geometry-based approach is that the endmembers are pure substances in a scene whose spectra reside at the extremities of the volume occupied by the data. Consequently, all mixed pixels that occupy the interior of the volume are linear mixtures of endmembers [1]. For example, for a scene comprising N materials, all pure components will be located at the vertexes of a simplex with N vertexes, itself located in an $N-1$ dimensional space (e.g., Ref. [1]; Fig. 3). The pure pixel-based algorithm is the most used class of the geometry-based approach. It assumes the presence of at least one pure pixel per endmember in the image, meaning that there is at least one spectral vector on each vertex of the data simplex [1]. Such endmember extraction algorithms based on a convex geometry as the pixel purity index (PPI) [46], N-FINDR [47], vertex component analysis (VCA) [48], orthogonal subspace projection (OSP) [49], and simplex growing algorithm (SGA) [50] search for the set of pixels that define the simplex with the maximum volume inscribed within the data set. By assuming that each pixel in a hyperspectral image can be identified by a point in an $N-1$ dimensional space, the PPI algorithm generates a large number of random $N-1$ dimensional vectors named "skewers" through the $N-1$ dimensional space. All pixels in the image are then projected on each of these skewers, and those falling at the extreme of the direction of each skewer are counted. After many repeated projections onto different skewers, those pixels that give counts above a cutoff threshold are considered potential endmembers. Among these, manual selection is then carried out using an N-dimensional (N-D) interactive visualization tool [51]. The N-FINDR method is an iterative simplex volume expansion approach in which several combinations of a set of endmembers are successively tested to obtain the largest simplex volume. The SGA algorithm improves upon the N-FINDR method by including a process of growing a simplex one vertex at a time until it reaches the desired number of vertices [50]. The OSP and VCA algorithms use successive orthogonal projections to find simplex vertices. Conversely of these earlier approaches, the minimum volume transform algorithm [63] offers to recover the lowest simplex that contains the entire dataset.

When the spectral mixtures are highly mixed, the geometry-based methods yield poor results, and in these cases, statistical methods are a good alternative, despite their higher computational complexity [1]. Statistical methods are thus based on statistical knowledge about the sought endmembers and their abundances. For example, in a Bayesian framework, by the knowledge of each endmember's Gaussian distribution, an estimation of endmembers and abundances is conducted by generating the posterior distribution of abundances and endmember parameters using the Gaussian model. Dependent component analysis (DECA) [53] and piecewise convex endmember detection [54] are some examples of such an approach. An overview of these methods was proposed in Bioucas-Dias *et al.* [1].

After the endmember spectra extraction, the identification of the extracted endmember spectra may be conducted by comparing them to reference spectra (laboratory or field) using spectral distance calculation between each endmember spectrum and reference spectra, e.g., the spectral angle mapper (SAM) developed by Kruse *et al.* [55], the spectral correlation mapper developed by Carvalho and Menezes [56] and used in Gomez *et al.* [57], or visual inspection as used in Andreou and Karathanassi [58].

2.3 Abundance Estimation

An inversion step for the abundance estimation of the endmembers is performed immediately following the endmember extraction step. The abundance of the endmembers can be estimated using a variety of approaches, including least squares [59]. Based on the extracted endmember spectra, the least squares approaches solve an optimization problem that minimizes the errors between the mixed pixel spectra and a linear combination of endmember spectra (Eq. 1), following:

$$\mathbf{min}\left(\sum_{i=1}^{I}\|\mathbf{x_i} - \mathbf{M}\boldsymbol{\alpha}_\mathbf{i}\|\right) \tag{2}$$

Most of the time, this optimization problem is constrained by a sum-to-one abundance of endmembers (sum-to-one constrained least squares, SCLS), a nonnegative abundance of endmembers (nonnegativity constrained least squares, NCLS), or both (fully constrained least squares, FCLS) in each mixed pixel [4].

The three steps of the unmixing process (Sections 2.1–2.3) have been applied in different material quantification studies, such as lithological quantification [57], postfire soil burn severity quantification [60], flower quantification [61], and vegetation quantification [62].

2.4 Alternative Linear Unmixing Methods

Alternative linear unmixing approaches are available in which the spectral unmixing problem is processed in a semisupervised way by assuming that the observed signatures can be expressed in the form of linear combinations of a number of pure spectral signatures known in advance (e.g., spectra collected on the ground by a field spectrometer or available in an open spectral library [1]). Due to the increasing availability and dimensionality of spectral libraries, this problem calls for efficient sparse regularization, resulting in the development of sparse unmixing approaches. By using this large library, the objective of the sparse regression unmixing is to find the optimal subsets of spectral signatures (i.e., endmembers) to model each mixed pixel in the hyperspectral scene. This is done by resolving an optimization problem to address the sparsity of the fractional abundance vectors. Several optimization algorithms have been proposed such as the orthogonal matching pursuit [63] and the basis pursuit [64]. More recently, to mitigate the computational complexity of these algorithms, others based on augmented Lagrangian multipliers such as SUnSAL (sparse unmixing by variable splitting and augmented Lagrangian) and C-SUnSAL (constrained SUnSAL [65,66]) were proposed.

However, sparse unmixing still remains a challenging task due to the usually high correlation of spectra within the spectral library.

In others linear unmixing approaches the endmember extraction and abundance estimation steps are performed simultaneously. The blind source separation (BSS) technique [67,68] allows solving a spectral unmixing problem when neither the endmember spectra nor the abundance of the endmembers in the mixture are known. The existing BSS techniques differ by their algorithms (batch type, neural network, etc.) and can be addressed in blind or semiblind contexts using second-order or higher-order statistics. BSS methods can also be divided into four classes that differ in their hypotheses: (i) the ICA, which supposes that the source signals S_i are statistically independent from each other, and at most one source can have a Gaussian distribution (Refs. [69,70], e.g., Chapter 7 of this book, D. Rutledge); (ii) the nonnegative matrix factorization, which supposes that the source S_i and the mixture matrix are nonnegative [71]; (iii) the sparse component analysis (SCA), which supposes that the source signals S_i have a certain parsimony in the analysis domain [72]; and (iv) the Bayesian methods (e.g., Chapter 8 of this book, D. Brie). In remote sensing, BSS is mainly used for the unsupervised classification of rock, vegetation, and urban areas [73,74]. It is less frequently used for the identification and quantification of natural materials present in mixed pixels [57,75,76].

Recently introduced in remote sensing image analysis, multivariate curve resolution [77] (Chapter 2) shares the same overall objective as ICA (i.e., decomposition of data into a bilinear model). However, while the ICA imposes statistical constraints (signal independence), MCR maximizes the explained variance in the data by imposing on the component profiles to follow statistically or chemically meaningful constraints [78].

2.5 Difficulties in Spectral Unmixing

Selecting endmembers involves identifying both the number and types of endmembers and their corresponding spectral signatures [79] (Chapter 17 of this book, Somers *et al.*). Endmember extraction, which identifies spectrally singular and distinctive spectral signatures from hyperspectral images, plays an extremely important role in spectral mixture analysis because improperly selected endmembers often lead to inappropriate interpretation of hyperspectral images [80]. Indeed, too many or too few estimated endmembers leads to abundance maps that are physically inaccurate by producing overfitting and large residual errors in the spectral unmixing results, respectively [81]. Moreover, the selection of poor endmember pixels could also give worse results. For example, classical spectral methods based on convex geometry are based on the assumption that the vertices of the simplex can be represented by corresponding pixels in the scene. Some of them (e.g., SMACC, N-FINDR, VCA, SGA, and Max-D) select one pixel to represent one endmember, leading to the inclusion of outlier pixels as endmembers [24,82].

Moreover, in various fields, and more specifically when hyperspectral images of natural scenes need to be processed, the main issue remains the variability of endmembers [79] due to changing environmental conditions (illumination, atmospheric correction, etc.) and/or physical differences in broadly defined ground substances (e.g., crops, soil; [45]). Indeed, the lack of ability to account for, among others, spatial variability between and among endmembers has been acknowledged as a major shortcoming of the conventional spectral mixture approaches [79].

3 USE OF SPATIAL INFORMATION IN UNMIXING PROCESS

As we have noted in the previous sections, most of the techniques proposed in the literature for spectral unmixing rely on the use of spectral information alone. As one of the most essential properties of hyperspectral images is the availability of multivariate information (i.e., spectra dimension) coupled with a 2-D representation, some authors have investigated the use of spatial information in the process of hyperspectral image unmixing. In this section, we successively review spectral–spatial unmixing methods according to the two main stages in which spatial information is accounted for (Fig. 1; Section 2), i.e., endmember extraction (Section 3.1) and map abundance estimation (Section 3.2). As noted in the previous sections, the review will be confined to the case of linear unmixing.

3.1 Use of Spatial Information for Endmember Extraction

To circumvent some issues of endmember extraction presented in Section 2.2 (e.g., high variability of endmembers, changing environmental condition, etc.), some authors have proposed the inclusion of spatial constraints to retrieve more accurate endmembers.

Obviously, all the spectral–spatial identification endmember techniques developed and described in the literature are based on spatial dependence principles but are more or less refined. The essential difference between those approaches is then in the type of spatial dependence concepts used and, in the case where more than one spatial dependence concept is involved, how to combine them. This spatial dependence concept is also related to the endmember variability problem, which has gradually been introduced in the spectral–spatial approach for endmember identification. Furthermore, it is worthwhile noting that the tools used to set up this spectral–spatial combination in the proposed endmember identification processes are also uneven. This is particularly cogent in regard to the image analysis methods generally used to take into account the spatial dimension of the data from basic tools (e.g., histogram segmentation) to more sophisticated tools as morphological operators or image segmentation methods (e.g. region growing segmentation, watershed, etc.).

The following subsections describe some published spectral–spatial endmember identification approaches. In the first section, some unsupervised methods are presented that integrate homogeneity of class (or object or cluster) and class dissimilarity concepts to select pure endmembers (Section 3.1.1). Then, some works are presented that use spectral–spatial constraints in an identification endmember scheme to either discard spurious pixels or search for atypical endmember signatures over the image (low contrast or rare endmember; Section 3.1.2). Finally, the last subsection is dedicated to the review of some works dealing with the variable endmember problem thanks to spatial constraint integration (Section 3.1.3).

3.1.1 Toward Pure Pixel Identification

Plaza *et al.* [11] conducted pioneering work in this research field by introducing the automated morphological endmember extraction (AMEE) in 2002. To obtain a method that integrates both spatial and spectral responses in a simultaneous manner and to allow a local-to-global approach in the search for endmembers (i.e., to find endmembers in local areas and then to group them into global endmembers), they used morphological tools and associated spatial kernels (or structuring elements, in the morphology jargon). In the first instance, this approach is based on the implementation of new morphological operators (dilatation and erosion operators) in an N-D space (extended morphological operator or EMO). This necessitates the definition of a vector-ordering relation that allows for the determination of the maximum and minimum elements in any family of N-D vectors [83]. To achieve this prerequisite and obtain a low computational complexity, a local eccentricity measure for a particular N-D point of the kernel is defined by calculating the distance between every spectral point within the kernel and the centroid to the kernel data cloud (spectral angle). Using this distance, the dilatation operator will identify the most distant pixel to the centroid (maximization of distance), while the erosion operator will identify the closest element to the centroid (minimization of distance; Fig. 4).

To respect the physical meaning of linear unmixing, a spectral purity index, i.e., the morphological eccentricity index (MEI), is defined by subtracting the outputs provided by the dilatation and erosion operations (evaluation strategy). This eccentricity index is calculated for each pixel within the kernel (window). Plaza *et al.* [11] proposed to use an iterative process where the MEI

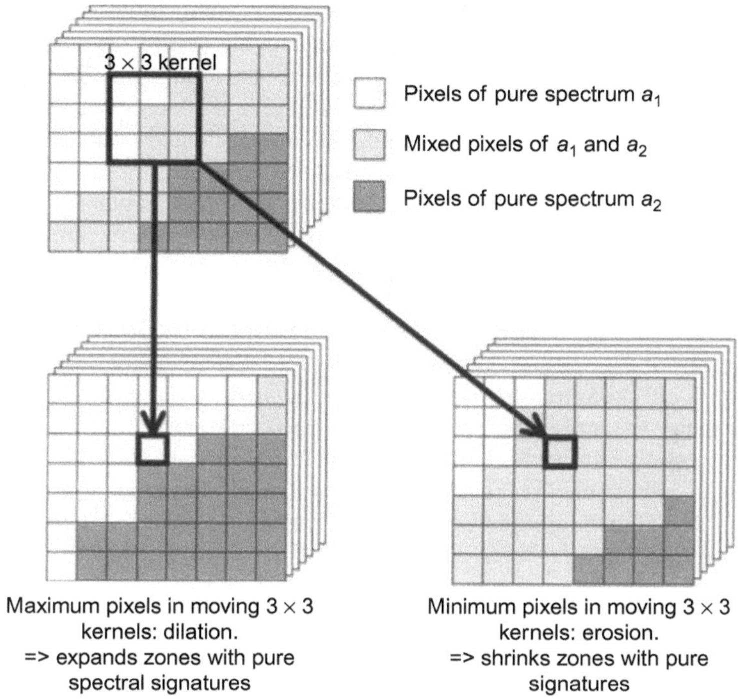

FIG. 4 Local and global effect of applying spatial–spectral dilation and erosion operations to a simple hyperspectral image.

values of each pixel are iteratively calculated using different kernel sizes. The associated values of the selected pixels at subsequent iterations are updated by means of newly obtained values as a larger spatial context is considered. Finally, a grayscale image (MEI map) is obtained and segmented using an Otsu algorithm [84]. The final endmember selection is then obtained by refining the previous segmented map using a region growing approach [85] to aggregate nearby and sufficiently similar pixels. The mean spectra obtained for the resulting regions are then considered to be endmembers, and the abundances are calculated using some classical tools (see Section 2.3).

The authors noted that this method may easily be reconfigured to a wide range of applications due to the inherent modularity of the morphological tools used (kernel size and shape). It presents some limitations, particularly a significant increase in processing time as the maximum size of the spatial neighborhood becomes large [25]. Finally, they addressed some interesting challenges for improving this method in evaluating different distance measures in the extension of morphological operations.

To better model natural scenes with step discontinuities and multiscale features and regions, Martín and Plaza [86,87] proposed an adaptive approach where a spatial preprocessing stage prior to endmember extraction is conducted (region-based spatial preprocessing or RBSPP strategy). They assume that pure signatures are less likely to be found in transition areas and more likely to be present in well-defined spatial regions. Then, their preprocessing scheme searches first for spectrally pure and spatially homogeneous regions and then selects a set of representative regions in spectral–spatial terms. The first step of this approach is a dimensionality reduction transformation (PCA or MNF, see Section 2.1) allowing the projection of the initial data in a new subspace with a dimension equal to the virtual dimensionality (VD; see Section 2.1) [41]. A clustering process, i.e., ISODATA [88], is then applied to these reduced data to obtain several clusters that are spectrally distinct. Pixels belonging to the same cluster are spectrally similar. A cluster selection strategy is then applied by using an OSP method on the mean spectra of the previously obtained clusters to find a set of spatially representative regions with associated spectra that are both spectrally pure and orthogonal. A classical spectral linear unmixing approach (see Section 2.2) is then applied on the pixels belonging to the selected clusters to obtain final endmembers.

Shortly after the previous publication, Martín and Plaza [24] proposed an improvement of the RBSPP process in which both the spectral information and the spatial arrangement of the pixel vectors are taken into account jointly. The new proposed method, called the joint spectral–spatial preprocessing (SSPP) module, overcomes some drawbacks of the RBSPP method, by using a pixel-based approach instead of the region-based approach used in RBSPP. Indeed, the use of a region-based approach in RBSPP sometimes leads to discarding some pure pixels belonging to a nonpure region (discarded region in the process described earlier) or retaining a mixed pixel belonging to a pure region. By discarding those pixels that are not sufficiently homogeneous (in the spatial sense) or pure (in the spectral sense), the RBSPP method is also less sensitive to noise. The main idea of this work is to coupled spatial and spectral information extracted from the initial image using a spatial smoothing and a clustering step, respectively, to identify a subset of spatially homogeneous and spectrally pure pixels from a scene. The SSPP module is divided into five main steps. In the first step, the original hyperspectral image is spatially filtered using a multiscale Gaussian filtering controlled by an input parameter. In the second step, a spatial homogeneity index is derived for each pixel of the image (root mean squared error or RMSE between the original image and the filtered one). In the third step, a spectral purity index is calculated for each pixel by using the projection values on the first principal components obtained by applying a PCA on the original image. In the fourth step, which is performed simultaneously to the two first steps, a clustering approach using an ISODATA algorithm is conducted on the reduced original image (PCA or MNF), resulting in the definition of several clusters. Finally, in the fifth and last step, endmembers are identified when they are selected as both homogeneous pixels and spectrally pure pixels. This identification was performed using the spatial homogeneity and spectral purity indexes calculated in steps 2 and 3, respectively, and the image cluster (spectral classification map) determined in step 4.

The authors noted that the use of the SSPP method as the preprocessing of the endmember extraction step allows the significant reduction of the processing time of the endmember extraction methods themselves. Indeed, the SSPP module, by discarding pixels that are not sufficiently homogeneous (in the spatial sense) or pure (in the spectral sense), reduces the number of pixels that the endmember extraction methods have to evaluate. Still, as noted by the authors, the SSPP approach may be penalized by its disability to select anomalous endmembers, due to the main assumption that endmembers are more likely to be located in spatially homogeneous areas rather than in transition areas. Finally, they addressed some interesting future work for improving this method, in adding a definition of a second filter specially designed to look for anomalies.

The spatial algorithm proposed by Zortea and Plaza [89] can be combined with an existing unmixing spectral algorithm (N-FINDR, VC, see Section 2.2), as it is used as a preprocessing module. The method estimates, for each input pixel vector, a scalar factor that relates to the spectral similarity between the pixel and its spatial neighbors. This scalar is then used to weight the importance of the spectral information associated with the pixel in term of its spatial context. Once the factor has been calculated, every pixel is displaced to the center centroid of the data depending on this factor. An important difference between this approach and the RBSPP proposed by Martín and Plaza [87] is the quality of the extracted regions. Indeed, in the approach proposed by Martín and Plaza [87], the spatial information is used as a guide to exploit the spectral information more effectively by directing the endmember search process to the subset of images that are both spatially homogeneous and spectrally pure, as opposed to the method proposed by Zortea and Plaza [89], which mainly accounts for the spatially homogeneous nature of the regions instead of their spectral purity [90]. Finally, Zortea and Plaza [89] addressed some interesting ways to improve this method, for example, by incorporating a two-filter strategy looking at both highly homogeneous regions and single pixels that are distinct from their surroundings (anomalous pixels).

The spatial preprocessing method proposed by Thompson *et al.* [91] differs from the other spectral–spatial methods for endmember identification proposed in the literature, insofar as it uses the concept of a superpixel, which is a homogeneous image region comprised of several contiguous pixels having similar spectral values. This concept allows using the pixel connectivity to describe similarity relationships based on pairwise differences between neighborhood pixels. This representation exploits the idea that the physical features of a natural scene are spatially contiguous. Ideally, each superpixel corresponds to a single scene feature, and its component pixels are independent reflectance measurements of a single surface material. A superpixel's mean spectrum is a good estimator of this reflectance, and the set of all such superpixel spectra acts as a compact, noise-reduced description of the scene features [92]. These superpixels can be generated using a segmentation algorithm able to produce an oversegmentation of the image as, for example, watershed methods [93] or random field models. Due to its efficient computation and this ability to manage several spectrum divergence functions (e.g., Euclidian distance, SAM distance), a graph-based approach [94] was selected and used by the authors. This graph formulation allows describing an image with a connected graph, where vertexes represent the pixels (and associated spectrum) and the edges the neighborhood relationships between pixels. In their development, the authors propose to assign a weight to each edge between two pixels by using the Euclidian distance or the spectral angle distance (SAD) [95]. An iterative process then agglomerates pixels into large subgraphs (or clusters) with respect to a minimal internal weight in either subgraph and a maximum connecting weight between connected subgraphs (Fig. 5). By doing so, two neighborhood subgraphs could be merged whenever there is no evidence of a boundary between the two corresponding regions (e.g., sharp boundaries or arbitrary morphologies in real data representing natural scenes). To this end, a refinement step is carried out to exclude bad pixels (e.g., instrument error), and final homogenous superpixels are obtained. These superpixels can then be used as a set of candidate endmembers by a classical endmember extraction technique (e.g., N-FINDR).

The authors noted that this method seems especially relevant for the processing of poorly known or heterogeneous surfaces with high spatial fragmentation by reducing scene-specific noise, promoting the discovery of spatially contiguous spectral classes and allowing the discovery of highly mixed pixels. As also noted by the authors, this spatial preprocessing method may be penalized by masking spectral features when the refinement criterion is not satisfied. Finally, Thompson *et al.* [91] addressed some interesting ways of improving this method, for example, by evaluating learned or application-specific distance metrics to produce superior superpixels.

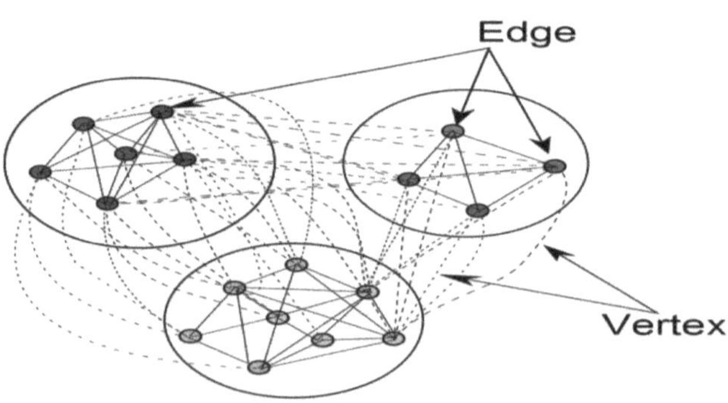

FIG. 5 Image graph formulation.

3.1.2 Toward Atypical Endmember Identification

As suggested by some authors in previously cited papers (e.g., Refs. [24,89]), their spectral–spatial endmember extraction algorithms may be penalized by the detection of anomalous pixels. Indeed, these algorithms are designed to be effective for extreme pixel detection, which may include outlier pixels.

Rogge *et al.* [25] introduced the spatial–spectral endmember extraction (SSEE) algorithm to detect low contrast endmembers, i.e., with minimal unique spectral information. Because the spectral contrast is variable in an image and so depends on the spatial neighborhood, the proposed approach works on image subsets to take advantage of the spatial characteristics of each endmember. This scheme allows a low contrast endmember in the whole image to have a higher spectral contrast in a specific subset image, thus facilitating its extraction. The SSEE method is divided into four steps. In the first step, a set of eigenvectors is determined, describing most of the variance of the data sets by using singular value decomposition (SVD) decomposition to subsets of an image (square windows of equal size and with no overlapping). For each subset, eigenvectors that account for 99% of the total spectral variance are retained for each subset and compiled into one vector file. These eigenvectors will likely be related to the local high contrast endmembers. A minimum of two eigenvectors are retained for each subset and stacked into one vector. In the second step, all the hyperspectral data are projected to this stacked vector. The pixels that lie on either end of the eigenvectors comprise the initial endmember set. In the third step, the image is scanned by using a sliding window centered on the pixel of interest (candidate endmembers), and each candidate endmember is compared to the other pixels of the subsets using a SAM distance. Pixels associated with a minimum distance are considered as candidate endmembers and are averaged to become the updated candidate endmembers. Multiple iterations of this process are realized to reduce the effect of noise. The end product is a set of endmembers that are defined both spectrally and spatially and take into account local spectral variance. Finally, the last step consists of a reordering of the endmember set based on spectral similarity and grouping into endmember classes based on their similar spatial coordinates. By constraining the averaging process (step 3 of the previous process) to include only spatially associated pixels, it should be possible to reduce the spectral contamination of spatially unrelated but spectrally similar endmembers [25].

Rogge *et al.* [25] noted that this method may be able to extract unique endmembers with subtle spectral variability that are not selected by other spectral unmixing approaches tested (iterative error analysis or IEA, PPI). As also noted by Rogge *et al.* [25], the SSEE algorithm is penalized by the size of the subsets and the SVD threshold value. Indeed, the more the size of the subset decreases, the greater the algorithm complexity is. The real benefit of SSEE comes when spectrally similar endmembers are spatially independent [25]. A major limitation of SSEE, as for AMEE, is that many parameters need to be finely tuned [96].

Zhang *et al.* [82] proposed including a spatial constraint in an algorithm built on a convex geometry and orthogonal projection common to classical linear spectral mixture analysis methods, i.e., SMA (see Section 2.3), to reduce their susceptibility to outliers and produce realistic endmembers. The proposed scheme, called the successive projection algorithm (SPA), attempts to define the vertexes of the simplex using not only one pixel but an average of multiple pixels that are spectrally distinct and spatially adjacent. Candidate pixels are selected based on their orthogonal projections onto the subspace spanned by the existing endmember (the first and second endmembers are the darker and brighter pixels). The noise is reduced by averaging spectrally similar target pixels within a local spatial window around the candidate [123]. Zhang *et al.* [82] noted that although this method was initially designed for geological applications, it may be extended to other applications such as plant community mapping. Finally, they addressed some interesting challenges for improving the SPA method, e.g., using the virtual dimensionality concept within SPA to better estimate the number of endmembers.

Recently, Erturk *et al.* [97] proposed using the SSPP algorithm [24] prior to spectral unmixing (FCLS) to guide the endmember extraction process to spatially homogeneous regions. However, to retain the anomalous/scarce pixels, which may include important endmembers (e.g., rare minerals in geological scenes, stressed crops in vegetation scenes, etc.), the endmembers extracted after the SSPP preprocessing step are used to reconstruct an image, and the pixels with a high reconstruction error (SAD between original pixels and reconstructed pixels) are considered to be anomalous endmember candidates. A pruning step is then conducted to select the final anomalous endmembers. As noted by the authors, the proposed approach allows the detection of subtle anomalous endmembers and has a low algorithmic complexity compared to other published methods coupling an SSPP preprocessing step and hyperspectral anomaly detection algorithm such as the Kernel-Reed-Xiaoli (Kernel-RX) algorithm [98].

3.1.3 Management of Variability of Endmembers

As explained in Section 1, the success of spectral unmixing arises from the selection of endmembers, which may be difficult because of (i) the spectral within-class variations, which refers to relative differences in the spectral signature within the same land cover class, and (ii) the spectral between-class variations, which refers to relative differences between two or more different land cover classes [9].

Mei *et al.* [80] proposed a spatial purity (SP)-based endmember extraction algorithm to alleviate the spectral variability of ground objects that occurs in real remote sensing images due, for example, to changes in illumination.

Based on an intensity-level and a feature-level measurement, an SP index is first determined for a spatial neighborhood (connected pixels in a kernel window) and is assigned to the pixel in the center of the spatial neighborhood. While the intensity-level measurement determines the purity of the spatial neighborhood in the original space of the hyperspectral image, the feature-level measurement is calculated in a transformed space (PCA or SVD), allowing the identification of specific feature component(s) in this feature domain for this spatial neighborhood. Therefore, a spectral neighborhood purity map (SNPM) is constructed, allowing the labeling of each pixel as either pure or mixed. A segmentation of this SNPM image leads to the extraction of endmember candidates. A refinement of this selection can be achieved by merging endmember candidates coming from two adjacent spatial neighborhoods. This last step can address the case of two endmember candidates from two adjacent regions that may belong to the same ground object (dealing with spectral variability). A major limitation of this algorithm is that many parameters need to be finely tuned to identify endmembers [96].

Deng and Wu [17] proposed a spatially adaptive spectral mixture analysis (SASMA) for mapping subpixel urban impervious surface distributions in an attempt to overcome endmember variability. Their idea is always to select endmembers locally (e.g., within a neighborhood) rather than globally (i.e., across the whole image) to take into account the principles of the spatial dependence of natural scenes [52]. The SASMA method involves a two-step process. First, an initial endmember extraction is conducted using a classification tree technique accounting for two spectral indexes, the biophysical composition index [99] and the normalized difference vegetation index [100], to discriminate typical land covers (bare soils and impervious surfaces) and soil and mixed land, respectively. A refinement step is then carried out on this initial endmember set to discard pixels labeled "soil" but with an insufficiently similar neighborhood. In the second step, using a local window, a distance-based weight was assigned to each neighboring endmember candidate, and an endmember signature was then synthesized as a weighted signature mean of neighboring endmember pixels. The weight of pixel within the local window not being identified as a candidate endmember is assigned as zero. The synthesized endmembers are then used as inputs to a linear mixture model for the abundance estimation.

3.2 Spatial Information in Abundance Estimation

The main objective of the unmixing process is to provide abundance maps of each component (endmember) over the study area. As explained in Section 2.3, the abundance of each component can be estimated using a variety of approaches, including least squares methods. Based on the extracted endmember spectra, the least squares approaches solve an optimization problem that minimizes the errors between the mixed pixel spectra and a linear combination of endmember spectra. One very natural approach to add a spatial constraint to an unmixing process is to work on the homogeneity of the abundance map, particularly because the abundances of neighboring pixels should differ as little as possible from each other [101].

3.2.1 Optimization of Objective Function by Spatial Constraint

The unmixing process can be interpreted as an optimization task and the unmixing objective function can be extended by a spatial regularization term. Many works investigated this solution using various methodologies under supervised, semisupervised, or nonsupervised schemes.

One of the methods investigated by Castrodad *et al.* [102] follows a supervised way by first using a sparse coding and dictionary learning to represent each endmember as a set of basis vectors of a dictionary or dictionary atoms (for C known classes in the image, each with an associated label $j \in [1, 2, \ldots, C]$, a class j is represented by set $\mathbf{r}_j$ basis vectors of the dictionary). These dictionaries are learned using a projected gradient iteration process and lead to atoms that are nonactual pure materials to account for material variability caused by factors such as noise or nonhomogeneous materials. Each class (a material present over the scene) is then represented by a set of vectors of the dictionary ($\boldsymbol{\psi}^j$ for the jth class) and a full dictionary is composed of the concatenation of all class-dependent-dictionaries ($\boldsymbol{\psi}$). An objective function to minimize could then be built by accounting for the available class dictionaries (Eq. 3). To take into account spatial information, a coherence constraint that depends on both the pixel's shape and the coefficient vector similarities is introduced (third term of Eq. 3). A weighting function is thus built using a Gaussian kernel to compare each pixel with a weighted linear combination of its neighbors. This weighting function is close to one if the pixels are very similar and near zero if they are orthogonal. The authors noted the importance of the dictionary-based class representation, independent of the spatial information consideration. Finally, Castrodad *et al.* [102] addressed some future works, e.g., considering adaptive sampling techniques where subsampling is avoided in areas of limited spatial coverage or high spectral uncertainty.

$$\sum_{i=1}^{k} \|\mathbf{x}_i - \psi\boldsymbol{\alpha}_i\|_2^2 + \gamma_S \sum_{i=1}^{k} S(\boldsymbol{\alpha}_i) + \gamma_G \sum_{i=1}^{k} \vartheta(\boldsymbol{\alpha}_i, w_i) \tag{3}$$

$$\text{s.t.} S(\boldsymbol{\alpha}_i) \le t$$

With $\|.\|_2^2$ the L_2 norm, γ_S a nonnegativity term that balances the complexity and the fitting quality, $S(\boldsymbol{x}) = \sum_i x(i)$ and $t \epsilon \mathfrak{R}^+$ is a threshold that controls the sparsity of the coefficients, γ_G is a term that controls the influence of the exploitation of spatial structure of the image (if $\gamma_G = 0$, each pixel is treated independently for each other), ϑ is a function that controls spatial and spectral coherence in the modeling and coding process. This function ϑ is defined following:

$$\vartheta(\alpha_i, w_i) = \left\| B(\alpha_i - \sum_{l \in \eta} w_{il}\alpha_l) \right\|_2^2 \tag{4}$$

where η denotes the neighborhood of the ith pixel, $\mathbf{w}_{il}$ defines a weighting Gaussian function, and $\mathbf{B}$ is a $C \times r$ dimensional matrix, with r the set of basis vector of the $\boldsymbol{\psi}$ dictionary atoms.

Iordache *et al.* [66] addressed the linear unmixing problem in a semisupervised fashion, where a large number of endmember candidates can potentially model the mixed pixel spectra and can be solved by linear sparse regression (see Section 2.4). The fractional abundances are then obtained by minimizing the number of nonzero components of N (number of endmembers), subject to the minimization of the classical objective function (Eq. 1; Section 2.3), often containing a quadratic data term and a sparsity-inducing regularizer, usually, the L1 norm [103]. However, due to the high correlation of spectral libraries, sparse regression techniques have mixed results. To address such difficulties, Iordache *et al.* [104] propose a new algorithm called "sparse unmixing via variable splitting augmented Lagrangian and total variation" (SUnSAL-TV), in within which a regularizing term is included in the classical sparse unmixing formulation using a total variation regularizer to take into account spatial homogeneity (i.e., two neighboring pixels are more likely to have similar fractional abundances for the same endmembers). The authors noted that this method only uses first-order neighborhood systems in the definition of the total variation regularizer, although second-order neighborhood systems may be well suited for piecewise affine surfaces. Finally, Iordache *et al.* [66] addressed some interesting future work for improving on the SUnSAL-TV method, focusing, e.g., on a full evaluation of isotropic versus nonisotropic total variation regularization defined in the spatial dimension. However, the total variation regularizer used is known to preserve edges really well but also to approximate linear transitions by step [101]. This can be detrimental for the unmixing of natural scenes. To address this problem, Bauer *et al.* [101] proposed to investigate a Hessian regularizer (i.e., second derivatives) of the pixel absorbance with an L_2 norm.

If the number of endmembers, their spectra, and their abundances are not known at all, the spectrum unmixing falls into the class of a BSS problem (unsupervised case, see Section 2.4). By considering linear hyperspectral unmixing as BSS problem, Jia and Qian [105] introduced the spectral and spatial complexity blind source separation algorithm (SSCBSS). The complexity concept originates in predictability, i.e., in the ability of previous observations to predict a new observation. In the SSCBSS algorithm, the authors extended a complexity-based BSS algorithm (i.e., complexity pursuit algorithm [106]) to the 2-D spatial domain by using spatial predictability to describe the local spatial interactions (i.e., absorbance for a pixel is related to the abundance of its neighboring pixels). Moreover, by assuming that the abundance statistics of each endmember vary smoothly from pixel to pixel, a spatial correlation term is added to an objective function to achieve the spectral smoothness. Even through this SSCBSS method is based on a more reasonable assumption than other techniques (such as BSS techniques), in that the statistics of fractional abundance vary smoothly due to the low spatial resolution of the hyperspectral imagery [105], this method may provide unmixing results with negative abundance values [107].

In addition to these publications, some proceedings were published on the spatial information in abundance estimation, developing interest in the remote sensing community on this topic and increasing algorithm development. Zare [108] proposed adding spatial information during a classical spectral unmixing process by adding a smoothing spatial term to the objective function to minimize (see Section 2.3). The smoothing spatial term then evolves into two terms. The first term, independent of the spatial context, concerns the classical squared error between the input mixed pixel spectra and the estimated linear combination of endmember spectra. The second term includes a spatial term that promotes setting the abundance of the current pixel with a value calculated by weighting the abundance values of similar neighborhood pixels with their relative distance. A parameter is used to trade off the error term and the spatial smoothing term.

$$\sum_{i=1}^{K} \|x_i - M\alpha_i\|_2^2 + \gamma G_i^t \alpha_i \tag{5}$$

$$G_{iz} = \sum_{\substack{j \in \eta_i \\ i \neq j}} \frac{1}{d_{ij}+1} \left(1-\alpha_{zj}\right)^2 \frac{1}{\|x_j - M\alpha_j\|_2^2 + 1} \tag{6}$$

where K is the number of pixels in the image, η_i is the neighborhood around pixel i, d_{ij} is the distance in pixel space between spatial location of pixel i and pixel j, α_{zj} is the proportion of the zth endmember in the jth pixel, $(.)^t$ is the transposed of (.), G_i is the vector of G values (Eq. 6) computed centered on pixel i, and γ is a parameter used to trade off between the squared error term and the spatial smoothing term.

In order to solve linear unmixing equation (Eq. 1) under nonnegativity and additivity constraints (Section 2.3), Song *et al.* [109] used genetic algorithm [110]. The objective function is thus built using three distinct terms. The first is equal to the mean square error of mixed spectral values, the second is a global Moran's I term (Eq. 7) used to characterize the spatial consistency of abundance for each endmember, and the last one is a local Moran's I term (Eq. 8) used to represent the spatial autocorrelation of abundances between a pixel and its neighborhood [45].

$$\Pi_{k,i} = \frac{\sum_{m=1}^{M}\sum_{n=1}^{N} w_{mn}(x_{m,i} - \bar{x})(x_{n,i} - \bar{x})}{\sum_{r=1}^{M+N} (x_{r,i} - \bar{x})^2 \sum_{m=1}^{M}\sum_{n=1}^{N} w_{mn}} \tag{7}$$

$$\Pi_{m,i} = \frac{M(x_{m,i} - \bar{\mathbf{x}})^2 \sum_{n=1}^{N} w_{m,n}(x_{m,i} - \bar{x})(x_{n,i} - \bar{x})}{\sum_{m=1}^{M} (x_{m,i} - \bar{x})^2} \tag{8}$$

where $w_{m,n}$ is the weight represented by $1/d_{m,n}$, $d_{m,n}$ is the distance between the mth and nth pixel, $\mathbf{x}_{m,i}$ is the fraction of the ith endmember at the m-th pixel, $\mathbf{x}_{n,i}$ is the fraction of the ith endmember at the nth pixel, N is the number of the neighbor pixels surrounding the m-th pixel, and M is number of pixels falling into the current zth land cover class (authors considered Z land cover classes in the image, each class corresponding to a subset of the endmembers).

Ammanouil *et al.* [111] proposed a method very similar to the one proposed by Iordache *et al.* [104]. However, to circumvent some limits of the SUnSAL-TV algorithm and specifically the limited local spatial similarity due to the TV regularizer (only four neighborhood pixels are taken into account in this regularization process), they used a graph Laplacian regularizer to address pixel similarity over the whole image. First, the hyperspectral image is mapped to a graph (see Section 3.1) where the edges represent the neighborhood relationship between pixels (two nodes connected are then likely to have similar abundances). A graph Laplacian regularization scheme is then applied on this graph, which promotes smoothness in the estimated abundance maps and collaborative estimation between homogeneous areas of the image.

3.2.2 Spatial Constraints in Abundance Unmixing

Another way for the unmixing process addresses simultaneously endmember identification and corresponding abundances estimation for each pixel [112,113] or uses specific tools that allow properly describing neighborhood dependence between image pixels, i.e., using Markov random fields (MRFs).

To develop effective methods for mapping urban surface cover types, Roessner *et al.* [112] early investigated the use of unmixing methods with spatial constraints. To address wide and complex urban surfaces representing a large number of spectrally similar endmembers (main types of urban land use such as nonindustrial development, industry and business, transportation, wetlands, forest, and water) but also many mixed pixels, they developed a method that combines the advantages of classification with linear unmixing.

Indeed, on the one hand, mixing effects cause problems for classification approaches that require spectral templates for all spectrally distinct surfaces and their mixtures occurring in the image data. On the other hand, due to their theoretical foundations, the spectral unmixing approaches require less a priori information [112]. They thus propose an iterative unmixing procedure using a growing approach. Using some knowledge about the scene materials, the endmembers pixels are defined, identified over the image and used as seed pixels in the growing process [114]. The resulting seedlings serve as spectral and spatial starting points for the selection of endmember combinations for nonidentified and mixed pixels occurring in the neighborhood of the spectrally pure pixels (seeds). Successively, for all nonidentified pixels, all possible combinations of endmembers present in a predefined neighborhood are tested. The most likely endmember combination is then selected to define the abundance for the corresponding pixel.

Zhang *et al.* [113] introduced a spatial constraint in the MCR process to "better" resolve hyperspectral images, i.e., to resolve the bilinear model (see Section 2.2). Indeed, MCR solutions are generally not unique, and they have a certain amount of ambiguity. The incorporation of any preliminary information, either chemical or mathematical, can positively influence the resolution of the image and decrease the mathematical rotation ambiguities. The resolution of hyperspectral images by MCR-ALS evolves an optimization process by ALS, where most applied constraints are nonnegative in abundances and endmembers (or spectral profiles). However, local rank and selectivity constraints must also be applied to the pixel in the abundance matrix to efficiently reduce rotational ambiguities. In addition to global rank constraints (e.g., the number of endmembers or compounds over the images), Zhang *et al.* [113] proposed to apply local rank constraints in the MCR-ALS process to improve the remote sensing hyperspectral imaging resolution. Fixed size image window-evolving factor analysis (FSIW-EFA) is then investigated to explore the local pixel complexity in images [115]. FSIW-EFA performs multiple local SVD by moving small windows over the whole image. The singular values obtained through these local analyses are then considered to extract significant contributions to the signal related to the presence of chemical constituents (large singular values) and to describe the experimental noise (small singular values). A local rank map could then be constructed and allows the identification of absent compounds in pixels with low compound overlap. The introduction of this information to the resolution process (constraint) helps to increase the performance of the resolution and decrease the ambiguity linked to the final solutions [116]. Recently, Hugelier *et al.* [117] proposed to include spatial constraint into the MCR-ALS process. By refolding, individually, the columns of the concentration matrix into the corresponding component distribution map during the ALS optimization step, they could applied any kind of spatial data processing as a constraint. In order to capture large-scale patterns in distribution map and to discard final spatial structures, they proposed an implementation of their generic scheme using a shape smoothness constraint [118].

Eches *et al.* [119] proposed using a Bayesian framework where MRFs are used to exploit the spatial correlation of pixels. The hyperspectral image is first partitioned in homogeneous regions in which the abundance vectors share the same first- and second-order statistical moments (mean and covariance). By doing so, the image is modeled using hidden labels whose spatial dependencies follow a Poots–Markov field. These labels associated with each pixel of the image form a set of random variables, and so a random field. An MRF is then defined where the conditional distribution of a pixel given the other pixels depends only on its neighbors (a four pixel–pixel structure or first-order neighborhood is considered in this work). To satisfy the positivity and sum-to-one constraints, the abundances are reparametrized using logistic coefficients exhibiting the same first- and second-order moments. The joint posterior distribution of the unknown parameters (i.e., noise variance, label vector, and logistic coefficient vectors) and hyperparameters (i.e., mean and variance of logistic coefficients associated with each defined region or class) is then derived, and a Markov chain Monte Carlo method was then used to generate samples. These latter were then used to compute the Bayesian estimators of the image labels and class parameters.

In a recent paper, Eches *et al.* [120] used a similar approach, i.e., a joint unmixing and segmentation algorithm, but they introduced a novel approach to address the spatial correlation between adjacent pixels of an hyperspectral image, allowing the computational cost of MRFs to be reduced significantly. A new relation between spatially close pixels (or sites) related to their similarity is built using a filter defined in the context of mathematical morphology, i.e., the self-complementary area filter [121]. The sites of the MRF (regions) are then built using this filter, and the resulting similar regions are then characterized by close abundances sharing the same means and covariance.

4 CONCLUSIONS

Spectral unmixing refers to the process that separates the mixed pixel spectra from a hyperspectral image into a collection of component spectra and a set of fractional abundances of these components. Spectral unmixing is an important issue for the processing of hyperspectral imaging, but the spatial dependence among pixels is neglected in many spectral unmixing methods. Over the past several years, new unmixing techniques have been developed, taking into account the combination of both spatial and spectral information, which leads to new endmember identification and abundance estimation. This chapter was dedicated to the presentation of new developments that incorporate spatial information, inherent to remotely sensed imagery, to improve the performance of spectral unmixing.

In spite of the more realistic representation of the spectral mixture by nonlinear models, the most used spectral unmixing model to estimate component abundances in hyperspectral imagery is the linear mixing model. To the best of our knowledge, the predominance of the use of the linear model compared to the nonlinear models is still pregnant in spectral–spatial unmixing technique development. As a consequence, this review was limited to the presentation of linear spectral unmixing.

The spatial information was used both in the endmember extraction step [11] (Section 3.1) and the abundance estimation step (Section 3.2). We observed that the spectral dimension reduction step (Section 2.1) is weakly used as a preliminary step in the spectral–spatial unmixing technique. Moreover, from the literature, the spatial information seems to be less complex to implement in the endmember extraction step than in the abundance estimation step. However, in both cases, all spectral–spatial unmixing techniques are sensitive to several parameters, such as the size of the windows (e.g., Ref. [80]), which implies the sensitivity of performance results.

The choice of the endmember extraction method may depend on the context and application of the user project, i.e., the study aims, the addressed object diversity, and the knowledge of the study area. In spectral unmixing, the choice of the method is driven by the presence or absence of pure pixels. However, in spectral–spatial unmixing approaches, the choice of method is also driven by the between-class and intraclass variations and the presence of either spurious pixels or atypical signatures over the image (low contrast or rare endmembers). In our opinion, the spectral–spatial methods for abundance estimation are less driven by the context than mathematical concepts to approach the optimal solution (e.g., semisupervised approach by sparse regression, MCR-ALS method with reduction of rotational ambiguities).

Finally, the spatial information consideration in the unmixing process, both in endmember extraction and in abundance estimation, allows the improvement of the resulting performance [122]. This is particularly true when a high spectral autocorrelation exists in the hyperspectral image [122]. Nevertheless, some of the developed spectral–spatial unmixing approaches are highly time consuming and involve strong mathematical concepts, which can reduce the diffusion and use of these approaches in larger scientist communities for real-world applications. A diffusion of software packages (R or Matlab languages) addressing the spectral–spatial unmixing algorithms would increase this use, following the diffusion of spectral unmixing algorithms.

We hope this chapter review will open up both the use of spectral–spatial unmixing applications in a variety of contexts (geology, agriculture, urban studies, etc.) and the development of spectral–spatial unmixing technique in regard to the needs of these contexts.

REFERENCES

[1] Bioucas-Dias JM, Plaza A, Dobigeon N, Parente M, Du Q, Gader P, et al. Hyperspectral unmixing overview: geometrical, statistical, and sparse regression-based approaches. IEEE J Sel Top Appl Earth Obs Remote Sens 2012;5:354–79.

[2] Dobigeon N, Moussaoui S, Coulon M, Tourneret J-Y, Hero A. MCMC algorithms for supervised and unsupervised linear unmixing of hyperspectral images. EAS Publ Ser 2013;59:381–401.

[3] Chang C-I, Ji B. Weighted abundance-constrained linear spectral mixture analysis. IEEE Trans Geosci Remote Sens 2006;44:378–88.

[4] Heinz DC, Chang C-I. Fully constrained least squares linear spectral mixture analysis method for material quantification in hyperspectral imagery. IEEE Trans Geosci Remote Sens 2001;39:529–45.

[5] Manolakis D, Siracusa C, Shaw G. Hyperspectral subpixel target detection using the linear mixing model. IEEE Trans Geosci Remote Sens 2001;39:1392–409.

[6] Singer RB, Mccord TB. Mars: large scale mixing of bright and dark surface materials and implications for analysis of spectral reflectance. In: Lunar and planetary science conference proceedings; 1979. p. 1835–48.

[7] Hapke B. Bidirectional reflectance spectroscopy: 1. Theory. J Geophys Res Solid Earth (1978–2012) 1981;86:3039–54.

[8] Heylen R, Parente M, Gader P. A review of nonlinear hyperspectral unmixing methods. IEEE J Sel Top Appl Earth Obs Remote Sens 2014;7:1844–68.

[9] Zhang J, Rivard B, Sánchez-Azofeifa A, Castro-Esau K. Intra-and inter-class spectral variability of tropical tree species at La Selva, Costa Rica: implications for species identification using HYDICE imagery. Remote Sens Environ 2006;105:129–41.

[10] Tobler WR. A computer movie simulating urban growth in the Detroit region. Economic geography 1970;46:234–40.

[11] Plaza A, Martinez P, Pérez R, Plaza J. Spatial/spectral endmember extraction by multidimensional morphological operations. IEEE Trans Geosci Remote Sens 2002;40:2025–41.

[12] Price J. Variability of high-resolution crop reflectance spectra. Int J Remote Sens 1992;13:2593–610.

[13] Price JC. How unique are spectral signatures? Remote Sens Environ 1994;49:181–6.

[14] Portigal F, Holasek R, Mooradian G, Owensby P, Dicksion M, Fene M. Vegetation classification using red edge first derivative and green peak statistical moment indices with the Advanced Airborne Hyperspectral Imaging System(AAHIS). In: 3rd international airborne remote sensing conference and exhibition—development, integration, applications & operations; 1997. Copenhagen, Denmark.

[15] Bhatti A, Mulla D, Frazier B. Estimation of soil properties and wheat yields on complex eroded hills using geostatistics and thematic mapper images. Remote Sens Environ 1991;37:181–91.

[16] Sudduth K, Kitchen N, Sadler E, Drummond S, Myers D. VNIR spectroscopy estimates of within-field variability in soil properties. Springer Netherlands: Proximal Soil Sensing; 2010. p. 153–63.

[17] Deng C, Wu C. A spatially adaptive spectral mixture analysis for mapping subpixel urban impervious surface distribution. Remote Sens Environ 2013;133:62–70.

[18] Chanussot J, Benediktsson JA, Fauvel M. Classification of remote sensing images from urban areas using a fuzzy possibilistic model. IEEE Geosci Remote Sens Lett 2006;3:40–4.

[19] Kettig RL, Landgrebe D. Classification of multispectral image data by extraction and classification of homogeneous objects. IEEE Trans Geosci Electron 1976;14:19–26.

[20] Landgrebe DA. The development of a spectral-spatial classifier for earth observational data. Pattern Recogn 1980;12:165–75.
[21] Fauvel M, Tarabalka Y, Benediktsson JA, Chanussot J, Tilton JC. Advances in spectral-spatial classification of hyperspectral images. Proc IEEE 2013;101:652–75.
[22] Gorretta N, Rabatel G, Fiorio C, Lelong C, Roger J-M. An iterative hyperspectral image segmentation method using a cross analysis of spectral and spatial information. Chemom Intell Lab Syst 2012;117:213–23.
[23] Tarabalka Y, Rana A. Graph-cut-based model for spectral-spatial classification of hyperspectral images. In: International geoscience and remote sensing symposium (IGARSS); 2014. p. 3418–21.
[24] Martín G, Plaza A. Spatial-spectral preprocessing prior to endmember identification and unmixing of remotely sensed hyperspectral data. IEEE J Sel Top Appl Earth Obs Remote Sens 2012;5:380–95.
[25] Rogge DM, Rivard B, Zhang J, Feng J. Iterative spectral unmixing for optimizing per-pixel endmember sets. IEEE Trans Geosci Remote Sens 2006;44:3725–36.
[26] Bioucas-Dias JM, Plaza A, Dobigeon N, Parente M, Du Q, Gader P, Chanussot J. Hyperspectral unmixing overview: Geometrical, statistical, and sparse regression-based approaches. *IEEE J Sel Topics Appl Earth Observ iRemote Sens* 2012;5(2):354–79.
[27] Keshava N. A survey of spectral unmixing algorithms. Lincoln Lab J 2003;14:55–78.
[28] Strang G. Linear algebra and its applications academic. New York: Cengage Learning; 1976.
[29] Green AA, Berman M, Switzer P, Craig MD. A transformation for ordering multispectral data in terms of image quality with implications for noise removal. IEEE Trans Geosci Remote Sens 1988;26:65–74.
[30] Lee JB, Berman M. Enhancement of high spectral resolution remote-sensing data by a noise-adjusted principal components transform. IEEE Trans Geosci Remote Sens 1990;28:295–304.
[31] Friedman JH, Tukey JW. A projection pursuit algorithm for exploratory data analysis. IEEE Trans Comput 1974;c-23(9):881–90.
[32] Hyvarinen A. Independent component analysis: a new concept. Signal Process 1994;297–314.
[33] Lee JA, Lendasse A, Verleysen M. Nonlinear projection with curvilinear distances: isomap versus curvilinear distance analysis. Neurocomputing 2004;57:49–76.
[34] Schwarz G. Estimating the dimension of a model. Ann Stat 1978;6:461–4 [Find this article online].
[35] Akaike H. A new look at the statistical model identification. IEEE Trans Autom Control 1974;19:716–23.
[36] Bioucas-Dias JM, Nascimento JM. Hyperspectral subspace identification. IEEE Trans Geosci Remote Sens 2008;46:2435–45.
[37] Carreira-Perpinán MA. A review of dimension reduction techniques: Tech. Rep. CS-96-09, 9. Department of Computer Science, University of Sheffield; 19971–69.
[38] Fodor IK. A survey of dimension reduction techniques: Technical report UCRL-ID-148494, Lawrence Livermore National Laboratory; 2002.
[39] Van Der Maaten LJ, Postma EO, Van Den Herik HJ. Dimensionality reduction: a comparative review. J Mach Learn Res 2009;10:66–71.
[40] Plaza AJ, Chang C-I. High performance computing in remote sensing. Boca Raton, FL: Taylor & Francis; 2007.

[41] Chang CI, Du Q. Estimation of number of spectrally distinct signal sources in hyperspectral imagery. IEEE Trans Geosci Remote Sens 2004;42:608–19.
[42] Asner GP, Lobell DB. A biogeophysical approach for automated SWIR unmixing of soils and vegetation. Remote Sens Environ 2000;74:99–112.
[43] Eckmann TC, Roberts DA, Still CJ. Using multiple endmember spectral mixture analysis to retrieve subpixel fire properties from MODIS. Remote Sens Environ 2008;112:3773–83.
[44] Bateson C, Asner GP, Wessman C. Endmember bundles: a new approach to incorporating endmember variability into spectral mixture analysis. IEEE Trans Geosci Remote Sens 2000;38:1083–94.
[45] Shi C, Wang L. Incorporating spatial information in spectral unmixing: a review. Remote Sens Environ 2014;149:70–87.
[46] Boardman JW. Automating spectral unmixing of AVIRIS data using convex geometry concepts. In: Summaries 4th annual JPL airborne geoscience workshop. JPL Publication 93-26; 1993. p. 11–4.
[47] Winter ME. N-FINDR: an algorithm for fast autonomous spectral end-member determination in hyperspectral data. In: SPIE's international symposium on optical science, engineering, and instrumentation. Bellingham: International Society for Optics and Photonics; 1999. p. 266–75.
[48] Nascimento JM, Dias JMB. Vertex component analysis: a fast algorithm to unmix hyperspectral data. IEEE Trans Geosci Remote Sens 2005;43:898–910.
[49] Harsanyi JC, Chang C-I. Hyperspectral image classification and dimensionality reduction: an orthogonal subspace projection approach. IEEE Trans Geosci Remote Sens 1994;32:779–85.
[50] Chang C-I, Wu C-C, Liu WM, Ouyang YC. A new growing method for simplex-based endmember extraction algorithm. IEEE Trans Geosci Remote Sens 2006;44:2804–19.
[51] Plaza A, Martínez P, Pérez R, Plaza J. A quantitative and comparative analysis of endmember extraction algorithms from hyperspectral data. IEEE Trans Geosci Remote Sens 2004;42:650–63.
[52] Deleted in review.
[53] Nascimento JMP, Bioucas-Dias JM. Hyperspectral unmixing algorithm via dependent component analysis, In: Geoscience and remote sensing symposium, 2007 (IGARSS 2007), IEEE international; 2007. p. 4033–6. 23–28 July.
[54] Zare A, Gader P. PCE: piecewise convex endmember detection. IEEE Trans Geosci Remote Sens 2010;48:2620–32.
[55] Kruse F, Lefkoff A, Boardman J, Heidebrecht K, Shapiro A, Barloon P, et al. The spectral image processing system (SIPS)—interactive visualization and analysis of imaging spectrometer data. Remote Sens Environ 1993;44:145–63.
[56] Carvalho Jr O, Menezes P. Spectral correlation mapper (SCM): an improving spectral angle mapper. In: Ninth JPL airborne earth science workshop. Pasadena: JPL Publication 00-18; 2000.
[57] Gomez C, Le Borgne H, Allemand P, Delacourt C, Ledru P. N-FindR method versus independent component analysis for lithological identification in hyperspectral imagery. Int J Remote Sens 2007;28:5315–38.
[58] Andreou C, Karathanassi V. Estimation of the number of endmembers using robust outlier detection method. IEEE J Sel Top Appl Earth Obs Remote Sens 2014;7:247–56.
[59] Shimabukuro YE, Smith J. The least-squares mixing models to generate fraction images derived from remote sensing multispectral data. IEEE Trans Geosci Remote Sens 1991;29:16–20.

[60] Robichaud PR, Lewis SA, Laes DY, Hudak AT, Kokaly RF, Zamudio JA. Postfire soil burn severity mapping with hyperspectral image unmixing. Remote Sens Environ 2007;108:467–80.
[61] Landmann T, Piiroinen R, Makori DM, Abdel-Rahman EM, Makau S, Pellikka P, et al. Application of hyperspectral remote sensing for flower mapping in African savannas. Remote Sens Environ 2015;166:50–60.
[62] Qu L, Han W, Lin H, Zhu Y, Zhang L. Estimating vegetation fraction using hyperspectral pixel unmixing method: a case study of a karst area in China. IEEE J Sel Top Appl Earth Obs Remote Sens 2014;7:4559–65.
[63] Pati YC, Rezaiifar R, Krishnaprasad PS. Orthogonal matching pursuit: recursive function approximation with applications to wavelet decomposition. In: Conference record of the twenty-seventh Asilomar conference on signals, systems and computers. vol. 1; 1993. p. 40–4. 1–3 November.
[64] Chen SS, Donoho DL, Saunders MA. Atomic decomposition by basis pursuit. SIAM J Sci Comput 1998;20:33–61.
[65] Bioucas-Dias JM, Figueiredo MAT. Alternating direction algorithms for constrained sparse regression: application to hyperspectral unmixing. In: 2010 2nd workshop on hyperspectral image and signal processing: evolution in remote sensing (WHISPERS); 2010. p. 1–4. 14–16 June.
[66] Iordache MD, Bioucas-Dias JM, Plaza A. Sparse unmixing of hyperspectral data. IEEE Trans Geosci Remote Sens 2011;49:2014–39.
[67] Hérault J, Ans B. Circuits neuronaux à synapses modifiables: décodage de messages composites par apprentissage non supervisé. C R Acad Sci 1984;299:525–8.
[68] Hérault J, Jutten C, Ans B. Détection de grandeurs primitives dans un message composite par une architecture de calcul neuromimétique en apprentissage non supervisé. In: 10° Colloque sur le traitement du signal et des images, France. GRETSI, Groupe d'Etudes du Traitement du Signal et des Images; 1985.
[69] Hyvärinen A, Oja E. A fast fixed-point algorithm for independent component analysis. Neural Comput 1997;9:1483–92.
[70] Hyvärinen A, Oja E. Independent component analysis: algorithms and applications. Neural Netw 2000;13:411–30.
[71] Lee DD, Seung HS. Algorithms for non-negative matrix factorization. Adv Neural Inf Proces Syst 2001;556–62.
[72] Georgiev P, Pardalos P, Theis FJ, Cichocki A, Bakardjian H. Sparse component analysis. In: Pardalos P, Boginski V, Vazacopoulos A, editors. A new tool for data mining, Data Mining in Biomedicine. Berlin: Springer; 2007. p. 91–116.
[73] Lennon M, Mercier G, Mouchot M, Hubert-Moy L. Spectral unmixing of hyperspectral images with the independent component analysis and wavelet packets. In: Geoscience and remote sensing symposium, 2001 (IGARSS'01) IEEE 2001 international, IEEE; 2001. p. 2896–8.
[74] Nascimento JM, Dias JM. Independent component analysis applied to unmixing hyperspectral data. In: Remote sensing. International Society for Optics and Photonics; 2004. p. 306–15.
[75] Bayliss JD, Gualtieri JA, Cromp RF. Analyzing hyperspectral data with independent component analysis, In: 26th AIPR workshop: exploiting new image sources and sensors, international society for optics and photonics; 1998. p. 133–43.
[76] Moussaoui S, Hauksdottir H, Schmidt F, Jutten C, Chanussot J, Brie D, et al. On the decomposition of Mars hyperspectral data by ICA and Bayesian positive source separation. Neurocomputing 2008;71:2194–208.

[77] Zhang X, Tauler R. Application of multivariate curve resolution alternating least squares (MCR-ALS) to remote sensing hyperspectral imaging. Anal Chim Acta 2013;762:25–38.

[78] Ruckebusch C, Blanchet L. Multivariate curve resolution: a review of advanced and tailored applications and challenges. Anal Chim Acta 2013;765:28–36.

[79] Somers B, Asner GP, Tits L, Coppin P. Endmember variability in spectral mixture analysis: a review. Remote Sens Environ 2011;115:1603–16.

[80] Mei S, He M, Wang Z, Feng D. Spatial purity based endmember extraction for spectral mixture analysis. IEEE Trans Geosci Remote Sens 2010;48:3434–45.

[81] Canham K, Schlamm A, Basener B, Messinger D. High spatial resolution hyperspectral spatially adaptive endmember selection and spectral unmixing. In: Proceedings of SPIE—the international society for optical engineering; 2011.

[82] Zhang J, Rivard B, Rogge D. The successive projection algorithm (SPA), an algorithm with a spatial constraint for the automatic search of endmembers in hyperspectral data. Sensors 2008;8:1321–42.

[83] Lambert P, Chanussot J. Extending mathematical morphology to color image processing. In: CGIP'00-1st international conference on color in graphics and image; 2000. p. 158–63.

[84] Otsu N. A threshold selection method from gray-level histograms. Automatica 1975;11:23–7.

[85] Zhu SC, Yuille A. Region competition: unifying snakes, region growing, and Bayes/MDL for multiband image segmentation. IEEE Trans Pattern Anal Mach Intell 1996;18:884–900.

[86] Martín G, Plaza A. Spatial-spectral preprocessing for volume-based endmember extraction algorithms using unsupervised clustering. In: 2nd workshop on hyperspectral image and signal processing: evolution in remote sensing (WHISPERS), IEEE; 2010. p. 1–4.

[87] Martín G, Plaza A. Region-based spatial preprocessing for endmember extraction and spectral unmixing. IEEE Geosci Remote Sens Lett 2011;8:745–9.

[88] Richards AJ. Remote Sensing Digital Image Analysis. An Introduction. Verlag: Springer; 1983.

[89] Zortea M, Plaza A. Spatial preprocessing for endmember extraction. IEEE Trans Geosci Remote Sens 2009;47:2679–93.

[90] Martín G. Design and implementation of new methods for spatial preprocessing prior to spectral unmixing of remotely sensed hyperspectral data. Universidad de Extremadura; 2013.

[91] Thompson DR, Mandrake L, Gilmore MS, Castaño R. Superpixel endmember detection. IEEE Trans Geosci Remote Sens 2010;48:4023–33.

[92] Gilmore MS, Thompson DR, Anderson LJ, Karamzadeh N, Mandrake L, Castaño R. Superpixel segmentation for analysis of hyperspectral data sets, with application to Compact Reconnaissance Imaging Spectrometer for Mars data, Moon Mineralogy Mapper data, and Ariadnes Chaos, Mars. J Geophys Res Planets 2011;1991–2012:116.

[93] Beucher S. The watershed transformation applied to image segmentation. Scanning Microsc Suppl 1992;299.

[94] Felzenszwalb PF, Huttenlocher DP. Efficient graph-based image segmentation. Int J Comput Vis 2004;59:167–81.

[95] Chang C-I. Hyperspectral imaging: techniques for spectral detection and classification. New York, NY: Springer Science & Business Media; 2003.

[96] Li H, Zhang L. A hybrid automatic endmember extraction algorithm based on a local window. IEEE Trans Geosci Remote Sens 2011;49:4223–38.

[97] Erturk A, Cesmeci D, Gullu MK, Gercek D, Erturk S. Endmember extraction guided by anomalies and homogeneous regions for hyperspectral images. IEEE J Sel Top Appl Earth Obs Remote Sens 2014;7:3630–9.

[98] Erturk A, Cesmeci D, Gerçek D, Gullu MK, Erturk S. Integrating anomaly detection to spatial preprocessing for endmember extraction of hyperspectral images. In: Geoscience and remote sensing symposium (IGARSS), 2013 IEEE international, IEEE; 2013. p. 1087–90.
[99] Deng C, Wu C. BCI: a biophysical composition index for remote sensing of urban environments. Remote Sens Environ 2012;127:247–59.
[100] Rouse JW, Haas RH, Schell JA, Deering DW. Monitoring vegetation in the great plains with ERTs. In: proceedings of the Third ERTS Symposium; NASA SP-351, Washington, DC, USA: NASA; 1973. p. 309–17.
[101] Bauer S, Neumann F, Puente León F. Spatial regularization for the unmixing of hyperspectral images. In: Proceedings of SPIE—the international society for optical engineering; 2015.
[102] Castrodad A, Zhengming X, Greer JB, Bosch E, Carin L, Sapiro G. Learning discriminative sparse representations for modeling, source separation, and mapping of hyperspectral imagery. IEEE Trans Geosci Remote Sens 2011;49:4263–81.
[103] Iordache M-D, Bioucas-Dias JM, Plaza A. Collaborative sparse regression for hyperspectral unmixing. IEEE Trans Geosci Remote Sens 2014;52:341–54.
[104] Iordache M-D, Bioucas-Dias JM, Plaza A. Total variation spatial regularization for sparse hyperspectral unmixing. IEEE Trans Geosci Remote Sens 2012;50:4484–502.
[105] Jia S, Qian Y. Spectral and spatial complexity-based hyperspectral unmixing. IEEE Trans Geosci Remote Sens 2007;45:3867–79.
[106] Jia S, Qian Y. Improved stone's complexity pursuit for hyperspectral imagery unmixing. In: 18th international conference on pattern recognition, ICPR 2006, IEEE; 2006. p. 817–20.
[107] Jia S, Qian Y. Constrained nonnegative matrix factorization for hyperspectral unmixing. IEEE Trans Geosci Remote Sens 2009;47:161–73.
[108] Zare A. Spatial-spectral unmixing using fuzzy local information. In: International geoscience and remote sensing symposium (IGARSS); 2011. p. 1139–42.
[109] Song X, Jiang X, Rui X. Spectral unmixing using linear unmixing under spatial autocorrelation constraints. In: International geoscience and remote sensing symposium (IGARSS); 2010. p. 975–8.
[110] Golberg DE. Genetic algorithms in search, optimization, and machine learning. Boston, MA, USA: Addison-Wesley Longman Publishing Co., Inc.; 1989.
[111] Ammanouil R, Ferrari A, Richard C. A graph Laplacian regularization for hyperspectral data unmixing. In: IEEE international conference on acoustics, speech and signal processing (ICASSP); 2015. p. 1637–41. 19–24 April.
[112] Roessner S, Segl K, Heiden U, Kaufmann H. Automated differentiation of urban surfaces based on airborne hyperspectral imagery. IEEE Trans Geosci Remote Sens 2001;39:1525–32.
[113] Zhang X, Juan AD, Tauler R. Local rank-based spatial information for improvement of remote sensing hyperspectral imaging resolution. Talanta 2016;146:1–9.
[114] Zucker SW. Region growing: childhood and adolescence. Comput Graphics Image Process 1976;5:382–99.
[115] De Juan A, Maeder M, Hancewicz T, Tauler R. Local rank analysis for exploratory spectroscopic image analysis. Fixed size image window-evolving factor analysis. Chemom Intell Lab Syst 2005;77:64–74.
[116] De Juan A, Maeder M, Hancewicz T, Tauler R. Use of local rank-based spatial information for resolution of spectroscopic images. J Chemometr 2008;22:291–8.

[117] Hugelier S, Devos O, Ruckebusch C. On the implementation of spatial constraints in multivariate curve resolution alternating least squares for hyperspectral image analysis. J Chemometr 2015;29:557–61.

[118] Hugelier S, Devos O, Ruckebusch C. Constraining shape smoothness in multivariate curve resolution–alternating least squares. J Chemometr 2015;29:448–56.

[119] Eches O, Dobigeon N, Tourneret JY. Enhancing hyperspectral image unmixing with spatial correlations. IEEE Trans Geosci Remote Sens 2011;49:4239–47.

[120] Eches O, Benediktsson JA, Dobigeon N, Tourneret JY. Adaptive Markov random fields for joint unmixing and segmentation of hyperspectral images. IEEE Trans Image Process 2013;22:5–16.

[121] Soille P. Beyond self-duality in morphological image analysis. Image Vis Comput 2005;23:249–57.

[122] Plaza A, Martin G, Plaza J, Zortea M, Sanchez S. Recent developments in spectral unmixing and endmember extraction. Berlin, Germany: Springer-Verlag; 2011.

[123] Thompson DR, Mandrake L, Gilmore MS, Castaño R. Superpixel endmember detection. Geoscience and Remote Sensing, IEEE Transactions on 2010;48(11):4023–33.

Chapter 19

Sparse-Based Modeling of Hyperspectral Data

R. Calvini*, A. Ulrici* and J.M. Amigo[†,1]
**University of Modena and Reggio Emilia, Reggio Emilia, Italy*
[†]*Faculty of Sciences, University of Copenhagen, Frederiksberg C, Denmark*
[1]*Corresponding author: e-mail: jmar@food.ku.dk*

Chapter Outline

1 INTRODUCTION

1.1 Hyperspectral Imaging in Context

Hyperspectral imaging (HSI) has rapidly emerged as a nondestructive method for quality analysis and control in different fields, including food science [1–3], pharmaceutical research [4–7], and cultural heritage [8,9] among others.

HSI integrates conventional spectroscopic systems (e.g., near infrared or Raman) with imaging techniques in order to combine spectral and spatial information from a sample. The resulting hyperspectral images (also known as *hypercubes* or *datacubes*) are three-dimensional data structures, containing two spatial dimensions (X–Y) and one spectral dimension (λ). Consequently, each pixel located in the X–Y spatial domain contains a spectrum which reflects its chemical characteristics [10]. Due to its attributes, HSI takes the advantages from both spectroscopic and imaging methods. Like single-point

Data Handling in Science and Technology, Vol. 30. http://dx.doi.org/10.1016/B978-0-444-63638-6.00019-X
Copyright © 2016 Elsevier B.V. All rights reserved.

spectroscopy, HSI is a nondestructive technique and, in many applications, it does not require any sample preparation; it also offers the possibility to perform the chemical mapping of the sample under investigation, i.e., to use the spatial information in order to locate specific chemical components within the image scene, based on their spectral features.

HSI images contain a wide amount of data, since each sample can be composed by thousands of spectra. For example, a HSI system with a spatial resolution of 230×230 pixels will produce a hyperspectral image containing 52,900 spectra. Depending of the type of spectral device, each collected spectrum can contain more than 100 wavelengths, which means more than 529×10^4 data points for a single image. On the one hand, this huge amount of data represents an advantage since it allows a detailed representation of the analyzed samples. On the other hand, this aspect has also been addressed as the *curse* of dimensionality since such high-dimensional data involve issues related to data handling, storage, and analysis [11]. In this sense, multivariate image analysis is an essential tool for the analysis of hyperspectral images and the use of chemometric methods allows facing different problems: data compression, image exploration, and classification or quantification [12–14].

1.2 Feature Extraction Methods

Principal component analysis (PCA) [15] has been reported many times in the context of hyperspectral image analysis as one of the main tools for exploration and feature extraction. It gives a qualitative evaluation of the main sources of variability and finds similarities and dissimilarities between the objects presented in a surface. The main issue with PCA, though, is that the interpretation of the obtained spectral loadings is sometimes cumbersome and not straightforward (as it is in resolution methods like multivariate curve resolution). It is however possible to simplify the chemical interpretation of the results by using variable reduction methods that select the most relevant spectral regions according to a specific target. Many mathematical strategies have been proposed in order to perform variable selection, such as genetic algorithms [16], interval partial least squares (iPLS) [17,18], wavelet transform [19,20], successive projection algorithm [21], and uninformative variable elimination [22]. All these methods work in different manners and are suitable for different problems. Though very effective, they require long computational times or the optimization of numerous parameters. Nevertheless, the key point of all the above-mentioned methods is that they need a specific target in order to select the spectral areas. This is a major problem in PCA, since it is an unsupervised methodology that uses no external information about the category of the objects in the image.

Sparse methods have been proposed for facing problems related to calibration and classification of high-dimensional data, mainly in the field of bioinformatics where the variables generally are thousands of genes and it is expected that there are few correlations between variables and many zeros. They have also found a wide use in statistic learning [23], analysis of biological data [24], metabolomics [25], and genomics [26]. Some applications are also available in the literature where sparsity is used for the analysis of spectroscopic data for food analysis and control, for example, regarding food-borne bacterial species [27], white wines discrimination [28], or virgin olive oil adulteration [29]. Furthermore, sparse methods were also successfully applied to the analysis of HSI data in order to identify relevant wavelengths in a well-defined classification problem [30].

Generally, sparse methods perform variable selection by forcing the model coefficients related to noise or to uninformative variables to be equal to zero; the level of sparsity to induce on the model is a user-defined parameter that has to be optimized. Therefore, the resulting sparse model will have only few active (nonzero) variables with a gain in model performances and interpretability.

In this chapter, we show the benefits and drawbacks of sparse PCA-based models, when they are used to highlight spectral zones of interest in hyperspectral images. After an overview of the theory behind sparsity and sparse methods, two practical applications concerning the choice of a proper sparsity level will be presented. In particular, a sparse version of principal component analysis (sPCA) will be considered. The choice of the sparsity level to be induced in the model is not straightforward and, thus, we will provide some guidance to the reader in order to face this issue.

2 THEORY: SPARSE METHODS

The term *sparse* is referred to a matrix in which most of the elements are equal to zero. Similarly, in the context of *sparse methods* this term refers to the estimated parameter vector of a model, e.g., loading vector, which is forced to contain many zeros. Therefore, the key aspect of sparse methods is the possibility to perform variable selection by indirectly forcing to be equal to zero the model coefficients related to variables containing only noise or being not relevant for the problem at hand. As a matter of fact, classical methods (e.g., PCA or PLS) usually set the contribution of uninformative variables only to small absolute values, but not exactly to zero. In this manner, many variables with small contributions can have a considerable global influence on the model and thus lower the predictive ability and the interpretability of the model [31].

In general, sparse methods are extensions of the corresponding classical methods and sparsity is achieved by adding a penalty term to the objective function of the considered method, where the objective function is the function that has to be maximized or minimized in the computation of a model. Therefore, the level of sparsity to be induced in the model is a user-defined parameter that needs to be properly optimized, in the same manner as the number of model components.

Different penalization approaches have been proposed depending on the data requirements and analytical problems. The Ridge regression adds a penalization on the sum of squared values of a vector (or L_2 norm), resulting in a shrinkage of the coefficients even if generally it does not set them exactly to zero, thus it does not actually lead to a sparse solution [32a]. The least absolute shrinkage and selection operator (Lasso) applies a constraint to the sum of absolute values of a vector (or L_1 norm) forcing several coefficients of the vector to be equal to zero [32b]. The elastic net can be considered as a compromise between Ridge regression and the Lasso since it selects variables like the Lasso and shrinks coefficients like Ridge regression; in particular, it turned out to be effective when the number of variables is much higher than the number of samples [33].

It is worth mentioning that several sparse variants of other classical methods have been reported for regression and classification aims. For regression problems, Chun and Keles proposed a sparse version of PLS based on the elastic net approach [34], while Lê Cao *et al.* proposed a sPLS method using the Lasso penalty [35]. Furthermore, Lê Cao *et al.* introduced an extension of their sPLS for a sparse variant of partial least squares discriminant analysis [36], and Clemmensen *et al.* proposed a sparse version of linear discriminant analysis [37]. More recently, Burden and Winkler presented a sparse version of support vector machine for classification purposes [38].

2.1 The Lasso

The Lasso approach was initially proposed for linear regression models. It consists in adding an L_1 norm penalty to the least squares criterion, where the L_1 norm refers to the sum of the absolute values of a vector [32]. The (usual) linear regression problem can be formulated as

$$\boldsymbol{y} = \boldsymbol{X}\boldsymbol{b} + \boldsymbol{e} \tag{1}$$

where $\boldsymbol{X}$, with size $\{n,m\}$, is the descriptor matrix, $\boldsymbol{y}$ $\{n,1\}$ is the response vector, $\boldsymbol{b}$ $\{m,1\}$ is the regression vector, and $\boldsymbol{e}$ $\{n,1\}$ is the error term. The ordinary least squares (OLS) criterion minimizes the residual sum of squares between $\boldsymbol{y}$ and $\boldsymbol{Xb}$.

The Lasso approach constrains the regression vector estimate as follows:

$$\arg\min_{\boldsymbol{b}}\left(\|\boldsymbol{y} - \boldsymbol{X}\boldsymbol{b}\|_2^2\right) \tag{2}$$

subject to:

$$\|\boldsymbol{b}\|_1^1 \leq c \tag{3}$$

where $\|\cdot\|_2^2$ refers to the sum of squared elements of a vector, or L_2 norm, and $\|\cdot\|_1^1$ refers to the sum of absolute values of the elements of a vector, or L_1 norm. The Lasso penalty forces the sum of the absolute values of the regression vector to be lower than the tuning scalar c, and it gives sparse solutions since, for an appropriate value of c, some coefficients are shrunk and others are set to zero.

The scalar c controls the level of coefficients shrinkage toward zero, and it should be set in the range between zero and the L_1 norm of the regression vector obtained with OLS ($0 \leq c \leq \|\boldsymbol{b}_{\text{OLS}}\|_1^1$). When c is equal to the L_1 norm of $\boldsymbol{b}$, the Lasso approach converges to the same result of OLS, while setting smaller values of c leads to sparse solutions. Therefore, the Lasso method allows to perform both coefficients shrinkage and variable selection, where the smaller is c, the higher the sparsity induced on the model.

2.2 Sparse Principal Component Analysis

sPCA estimates a PCA-based model in which sparsity is induced on the model parameters. Several sparse variants of PCA were proposed with sparsity induced only on the loading vectors [39–42] or on both scores and loadings [43,44].

One of the algorithms that can be used to calculate the sPCA model is the alternating shrunken least squares (ASLS), which consists in calculating a PCA model in which sparsity is induced on the loadings [45]. For a fixed number of components A, the ASLS algorithm estimates a solution to the following penalized optimization problem:

$$\arg\min_{\boldsymbol{T},\boldsymbol{P}}\left(\|\boldsymbol{X}-\boldsymbol{T}\boldsymbol{P}^T\|_F^2\right) \tag{4}$$

subject to:

$$\|\boldsymbol{p}_i\| \leq c \text{ and } \|\boldsymbol{p}_i\|_2^2 = 1\ (1 \leq i \leq A) \tag{5}$$

where $\|\cdot\|_F^2$ is the squared Frobenius norm of the matrix (sum of squared elements), $\boldsymbol{T}$ is the scores matrix, $\boldsymbol{p}_i$ are the column vectors of the normalized loadings matrix $\boldsymbol{P}$ for the ith component, and c is the L_1 norm constraint on each loading vector.

Eq. (4) represents the objective function of the standard PCA model consisting on the minimization of the residuals. Therefore, the ASLS algorithm adds an L_1 norm constraint ($\|\boldsymbol{p}_i\| \leq c$) on each normalized ($\|\boldsymbol{p}_i\|_2^2 = 1$) loading vector in order to achieve sparsity.

The scalar c, from here onward referred to as sparsity constraint, is a positive tuning parameter which controls the sparsity level of the model; thus it influences the number variables forced to be equal to zero. In particular, the sparsity constraint may range between 1 and the square root of the number of variables: the lower the value of c, the higher the sparsity induced on the loadings [46]. A sparsity constraint equal to 1 corresponds to a high sparsity level, where only one variable is selected for each component. On the other hand, if c is equal to the square root of the number of variables, the loadings are the same as those obtained with PCA. However, in practice, convergence to the unconstrained PCA model can be achieved also for values of c smaller than the square root of the number of variables. Indeed, for a specific data set and for a given number of principal components the maximum value of the sparsity constraint is equal to the maximum L_1 norm of the loading vectors of the PCA model.

Another aspect to be considered is that the ASLS algorithm calculates simultaneously all the components by iterating between scores and loadings until convergence, in such a way that the L_1 norm constraint is fulfilled for each component. Conversely to PCA, sPCA loading vectors are not orthogonal and the sparse principal components (sPCs) are not nested, which means that sPCA may give different loading vectors according to the number of components used to build the model, even for the same value of the sparsity constraint.

Therefore, the choice of the optimal combination of the tuning parameters, i.e., number of sPCs and sparsity constraint, is a critical point in the construction of a sPCA model.

In the following examples, sPCA models were computed using a freely available routine[a] written in Matlab language [45].

3 SPARSE-BASED IMAGE EXPLORATION: APPLICATIONS

Like PCA, sPCA can be used for image exploration, with the advantage of enhancing the interpretability of the model by forcing to zero the loading coefficients of noisy or uninformative variables. However, the sparsity level to induce on the model is a user-defined parameter that needs to be carefully tuned in order to avoid the deletion of features bearing useful information for the problem at hand. To this aim, it is therefore necessary to define a parameter which relates the sparsity level to the desired performance of the sPCA model.

Frequency distribution curves calculated from the score vectors of different sPCA models can be used in order to solve this issue. As a matter of fact, successful approaches based on frequency distribution curves of quantities derived from PCA models have been already reported in the literature for the characterization of whole hyperspectral images [47,48] and for the discrimination of objects within the image scene [49].

In this chapter, we describe the use of frequency distribution curves of the sPCA scores in order to optimize the c value and the number of sPCs. In particular, the applications described in the following sections are related to two different issues that generally may occur when dealing with HSI data: the discrimination between two groups of homogeneous samples and the identification of outlier regions within the image scene (i.e., of outlier pixels). These two

[a]sPCA routine is downloadable from http://models.life.ku.dk/sparsity.

issues require different ways of using the frequency distribution curves in order to properly set the sPCA parameters. In the case of the discrimination between groups of objects, it is necessary to enhance the separation between the frequency distribution curves of the different groups. Conversely, for the identification of outlier regions, the optimal situation is the one leading to the detection of the highest possible amount of outlier pixels.

3.1 Detection of Differences Between Groups of Homogeneous Samples: Arabica and Robusta Green Coffee

3.1.1 Description of the Data Set

A total of 26 different batches of green coffee, including 12 batches of *Coffea arabica* (usually referred to as Arabica coffee) and 14 batches of *Coffea canephora* (usually referred to as Robusta coffee), were considered in this study. From each batch, an amount equal to about 500 g of green coffee beans was sampled and stored in a sealed package until the day of analysis. From each sample, three aliquots of 70 g of randomly selected beans were kept, and two images were acquired, shuffling the beans between the two acquisitions; this procedure was repeated twice in different days. All the samples were acquired in random order, and between the different acquisition sessions, the packages were sealed again and stored at room temperature.

The hyperspectral images were acquired using a desktop near infrared (NIR) Spectral Scanner (DV Optic) embedding a Specim N17E reflectance imaging spectrometer, coupled to a Xenics XEVA 2608 camera (320×256 pixels) and covering the 955–1700 nm spectral range with a spectral resolution of 5 nm.

Before further analysis, the hyperspectral images were preprocessed using an internal calibration in order to reduce the variability among images over time, as described in Ref. [50]. In addition, the pixels related to the background were removed from each image using a thresholding procedure based on the most discriminant wavelength between sample and background (this wavelength was selected using the Fisher ratio).

Then, 2 batches (1 of Arabica coffee and 1 of Robusta coffee) were randomly selected and assigned to the test set, while the remainder 24 batches and the corresponding hyperspectral images were used in order to calculate a training set of 288 average spectra (one average spectrum for each training set image). The training set of average spectra was then used in order to calculate different sPCA models considering different combinations between sparsity and number of principal components. The influence of the different combinations of the model parameters on the frequency distribution curves of the scores was evaluated using a test image obtained by merging together one image of Arabica coffee and one image of Robusta coffee belonging to the test batches. The sPCA models were calculated on the spectra preprocessed using standard normal variate (SNV) followed by first derivative and mean center.

3.1.2 Approach on Frequency Distribution Curves

The loadings, calculated from each sparse model applied to the training set, were used in order to obtain the corresponding scores of the test image. Since the test image was constructed by merging together one image of Arabica coffee and one image of Robusta coffee, it was possible to calculate two separate frequency distribution curves, i.e., one for the scores of Arabica pixels and one for the scores of Robusta pixels. In particular, the frequency distribution curves of Arabica and Robusta pixels were calculated for each sPC, and the intersection point of the two curves was considered as a threshold value used to calculate the percentage of Robusta pixels overlapping those of Arabica, and vice versa. In fact, in this case the best sparse model was identified as the one leading to the highest separation between the two frequency distribution curves. As an example, Fig. 1 reports the frequency distribution curves of the sPC 1 score values (sPCA model with 1 sPC and $c=10$) calculated for the pixels belonging to Arabica and Robusta coffee beans. The pixels of Arabica coffee beans whose score values were higher than the threshold value (black dash-dotted line) were considered as "Arabica overlapped" pixels, i.e., pixels overlapping the distribution of Robusta coffee; analogously, the pixels of Robusta coffee beans whose score values were lower than the threshold value were considered as "Robusta overlapped" pixels, i.e., pixels overlapping the distribution of Arabica coffee.

3.1.3 Results

Different sPCA models were built considering all the combinations from 1 to 4 sPCs and values of the sparsity constraint ranging from 1 to 10, spaced by 0.25. For each tested model, the frequency distribution curves of the scores

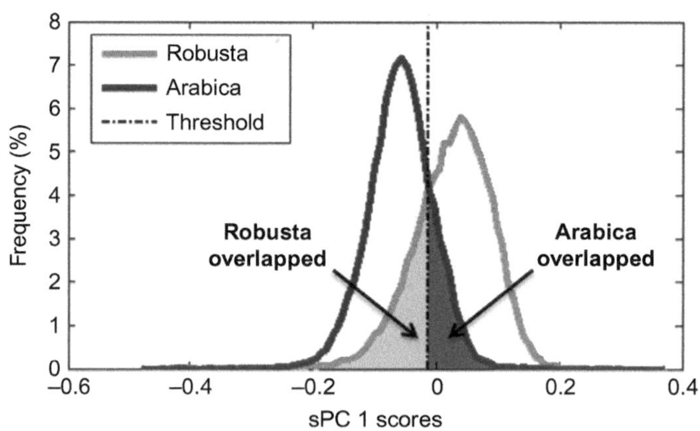

FIG. 1 Frequency distribution curves of Arabica and Robusta pixels on sPC 1; the *black dash-dotted line* represents the threshold value, calculated as the intersection point of the two frequency distribution curves. The *filled areas* represent the Arabica and Robusta overlapped pixels.

of Arabica and Robusta pixels belonging to the test image were calculated componentwise, and then the percentage of overlapped pixels for both classes was calculated for each component. Each graph reported in Fig. 2 shows how, for a given sPC, the percentage of overlapped pixels varies as a function of the sparsity constraint; each row of graphs corresponds to a sPCA model calculated using a given number of sPCs (from 1 to 4). For all the models, sPC 2 is always the sparse component leading to the better separation between Arabica and Robusta pixels. Interestingly, considering more principal components in the model allows decreasing the percentage of overlapped pixels on sPC 2, but the additional components bear information which is not pertinent to the separation between Arabica and Robusta pixels. Therefore, the best compromise between a parsimonious model and a low percentage of overlapped pixels (i.e., higher separation between the two groups) was found using a sPCA model calculated with 2 sPCs and a value of c equal to 4 (indicated by the black arrow in Fig. 2). In this situation the percentage of overlapped pixels of Arabica coffee is equal to 5.2%, while the percentage of overlapped pixels of Robusta coffee is equal to 8.2%.

The sPC 2 score image obtained from this sPCA model is reported in Fig. 3. The separation between Arabica and Robusta pixels is evident; in particular, the majority of Robusta coffee pixels have positive sPC 2 score values, while all the pixels of Arabica coffee have negative sPC 2 score values.

Furthermore, it is also important to evaluate the sparse loading vector of sPC 2 in order to assess the spectral variables which are mainly involved in the separation between Arabica and Robusta coffee. In Fig. 4, the PC 2 loading vector calculated with PCA is reported together with the sPC 2 loading vector of the selected sPCA model (2 sPCs, $c=4$). The variables retained by the sPCA model are related to the C–H aromatic (1143 nm) and aliphatic (1195–1225 nm) second overtones and to the O–H first overtone of aliphatic (1410 nm) alcohol [51]. The number of selected variables on sPC 2 is equal to 22 out of the 150 original variables, while the overall number of nonzero variables of the sPCA model is equal to 38.

3.1.4 Stability of Variable Selection

The proper selection of the sPCA model allowed to better separate the pixel clusters of Arabica and Robusta coffee beans; furthermore, it also allowed to identify more clearly the relevant spectral regions, thus increasing the interpretability of the model. As far as this latter aspect is concerned, a further check was made in order to evaluate the stability of the variable selection performed by sPCA, i.e., to assess whether sPCA tends to converge toward the selection of the same spectral regions when varying the number of sPCs and the sparsity constraint value.

To this aim, different sPCA models were tested, corresponding to all the combinations of the number of sPCs retained in the model (ranging from

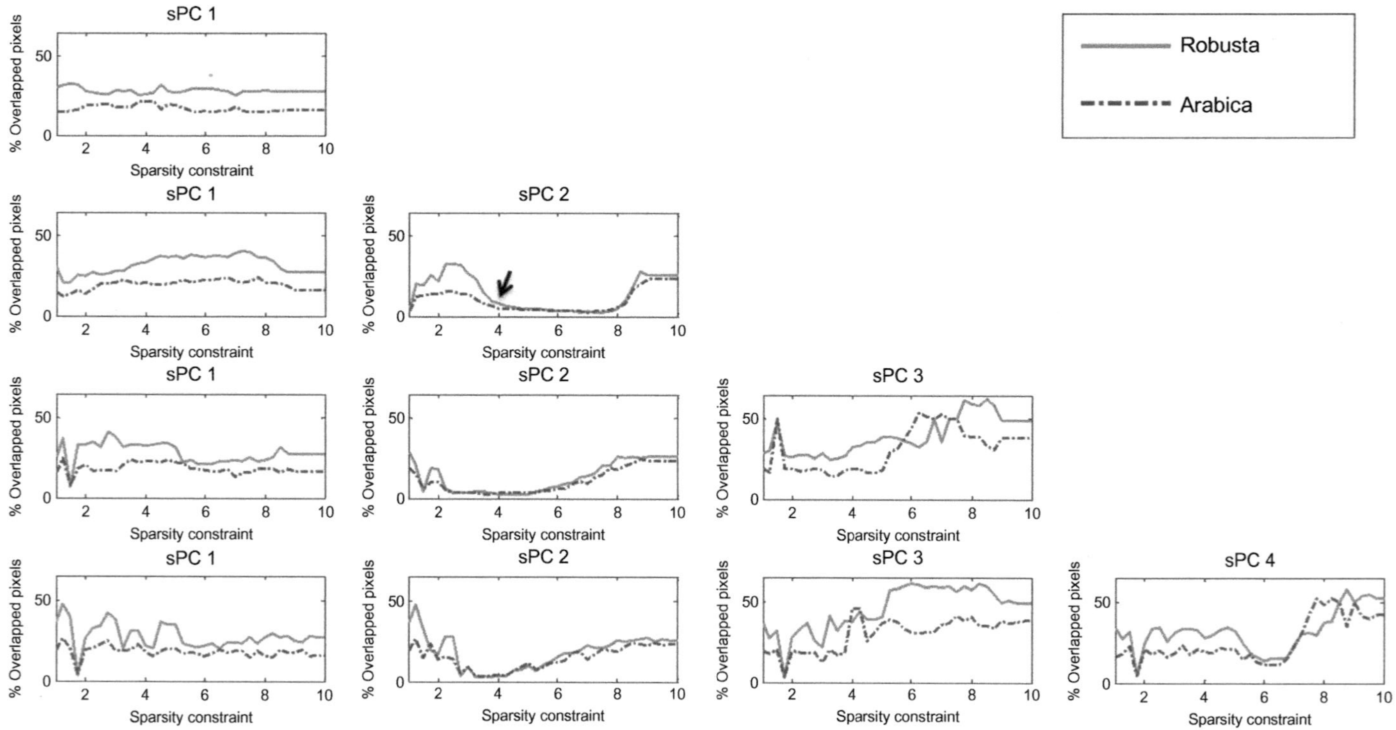

FIG. 2 Percentage of Arabica and Robusta overlapped pixels on each component according to different sparsity constraint values and different number of principal components.

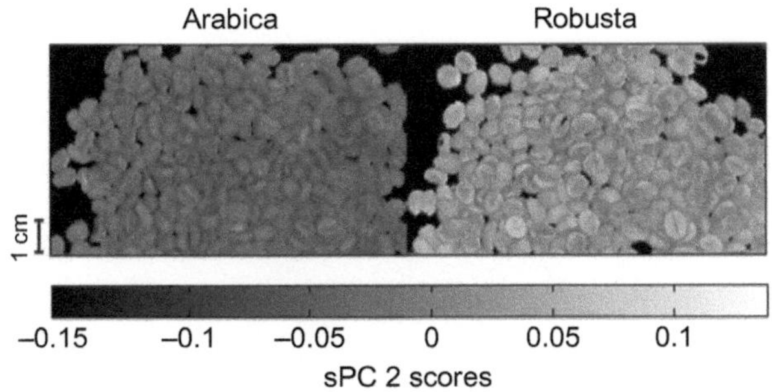

FIG. 3 sPC 2 score image of the sPCA model calculated with 2 sPCs and a value of *c* equal to 4.

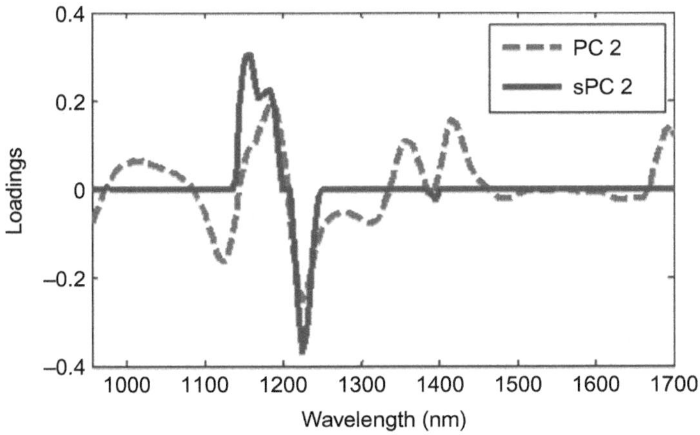

FIG. 4 PC 2 loading vectors of PCA and sPCA models.

1 to 6) and of sparsity constraint values (ranging from 1 to 10, with steps of 0.25). Then, for each number of sPCs retained in the sPCA model, the frequency of selection of the spectral variables was calculated by varying the values of *c* (Fig. 5A). In the same manner, the frequency of selection of the spectral variables was also calculated for each fixed value of *c* and varying the number of sPCs (Fig. 5B). In the latter case the frequency of selection ranges from 0, in correspondence to those variables that have never been selected, to 6, in correspondence to those variables that have always been selected for all the six sPCA models calculated including a number of sPCs ranging from 1 to 6.

Considering both Fig. 5A and B, it is possible to observe that the different sPCA models calculated with different combinations of the model parameters converge to the selection of the same relevant spectral regions. Furthermore, the most frequently selected variables include the relevant regions identified with the best sPCA model reported in Section 3.1.3.

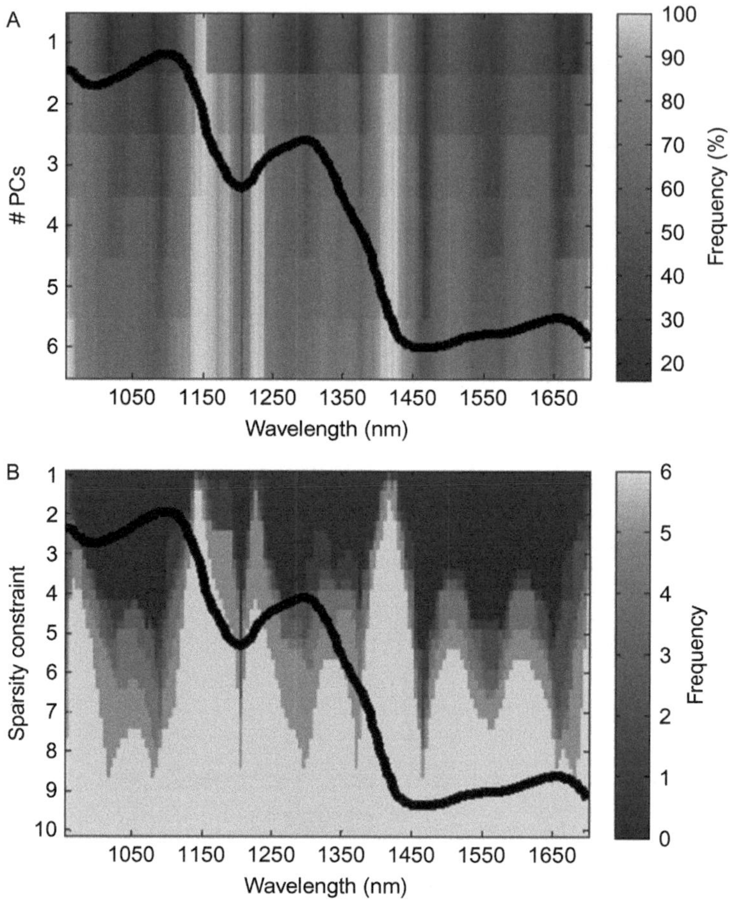

FIG. 5 Frequency of variable selection of the sPCA models calculated considering different combinations of the model parameters; in (A) the frequency of selection was calculated for each fixed number of sPCs and changing the value of *c* and in (B) the frequency of selection was calculated for each fixed value of *c* and changing of number of sPCs. The *black curve* superimposed to the images represents the average NIR spectrum of the training set.

3.2 Detection of Outliers: Plastics Pieces

3.2.1 Description of the Data Set

Plastic waste objects made of polystyrene (PS), polyethylene (PE), and polyethylene terephthalate (PET) were chopped in order to obtain plastic pieces of the different polymers. Afterward, 20 hyperspectral images were acquired as follows:

- 16 images of different groups of PS pieces
- 1 image of PS containing one piece of PE

– 1 image of PS containing two pieces of PE
– 1 image of PS containing one piece of PET
– 1 image of PS containing two pieces of PET.

The hyperspectral images were acquired using a desktop NIR Spectral Scanner (DV Optic) with a Specim N17E reflectance imaging spectrometer, coupled to a Xenics XEVA 2608 camera (320×256 pixels) and working in the 955–1700 nm spectral range with a spectral resolution of 5 nm. Before further analysis, the pixels related to the background were removed using the same procedure described in Section 3.1.1.

From each image, the average spectrum was calculated obtaining a data set composed of 20 spectra and 150 variables (Fig. 6). Considering the average spectra data set reported in Fig. 6, it is not possible to observe appreciable differences in the spectral shape between the average spectra calculated from images containing only PS and from those containing also other polymers (PE or PET).

PCA and sPCA were then applied to the average spectra data set preprocessed with SNV and mean center.

3.2.2 Approach on Frequency Distribution Curves

Different sPCA models were tested using a fixed number of 4 sPCs and evaluating different values of the sparsity constraint, *c*. The sPCA models were calculated on the average spectra data set and then they were applied to three test images: one image with only pieces of PS, one image with PS and two pieces of PE, and one image with PS and two pieces of PET.

Along the sPCs able to differentiate PS from the other polymers, the pixels related to PE or PET should behave as outliers with respect to those related

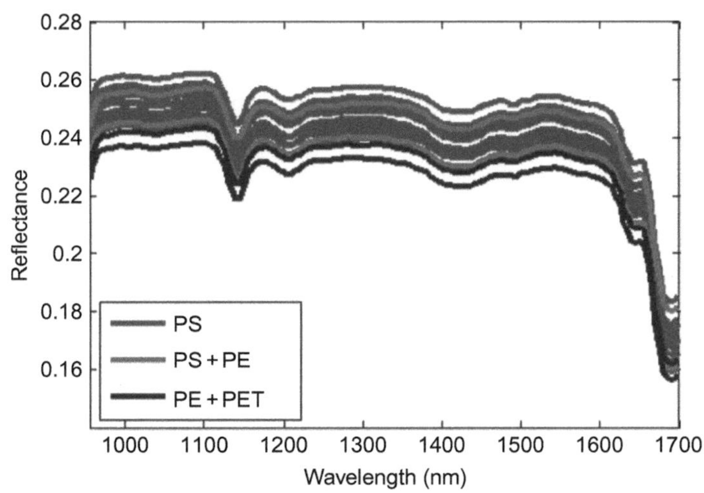

FIG. 6 Average spectra of plastics data set.

to PS. In other words, their score values should fall at extreme values with respect to the distribution of the scores of PS pixels. Therefore, in order to identify these outlier pixels, boundaries were estimated using the frequency distribution curve of the sPC score values of PS pixels and used as thresholds. In particular, the procedure consisted in the following steps:

1. The frequency distribution curve of the scores of the image with only PS was calculated for each sPC.
2. The frequency distribution curve of the considered PS image was used to calculate two threshold values for each component, namely $TR_{0.5}$ and $TR_{99.5}$, corresponding to the 0.5th and to the 99.5th percentile, respectively.
3. For the two test images containing PE and PET, the percentage of outlier pixels was calculated componentwise as the percentage of pixels whose score values were lower than $TR_{0.5}$ or greater than $TR_{99.5}$.

As an example, the frequency distribution curves of the scores of sPC 3 calculated using a sparsity constraint value equal to 11 are reported for the three images in Fig. 7. The black dash-dotted lines correspond to the threshold values, defined as the 0.5th and 99.5th percentile values of the frequency distribution curve of the sPC 3 scores obtained for the PS image. Therefore, the pixels whose score values fall at the left of the first line ($TR_{0.5}$) or at the right of the second line ($TR_{99.5}$) have been considered as outlier pixels.

3.2.3 *Results*

First, standard PCA was applied to the preprocessed average spectra data set and the scores and loadings of PC 3 and PC 4 are reported in Fig. 8. From the score plot (Fig. 8A) it is possible to observe that the samples containing only PS are separated from those containing also the other polymers. In particular,

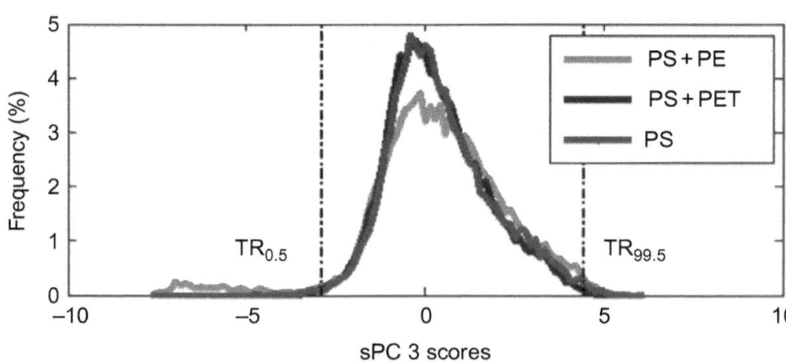

FIG. 7 Frequency distribution curves of sPC 3 scores of the three considered images and threshold values ($TR_{0.5}$ and $TR_{99.5}$) calculated from the frequency distribution curve of PS image.

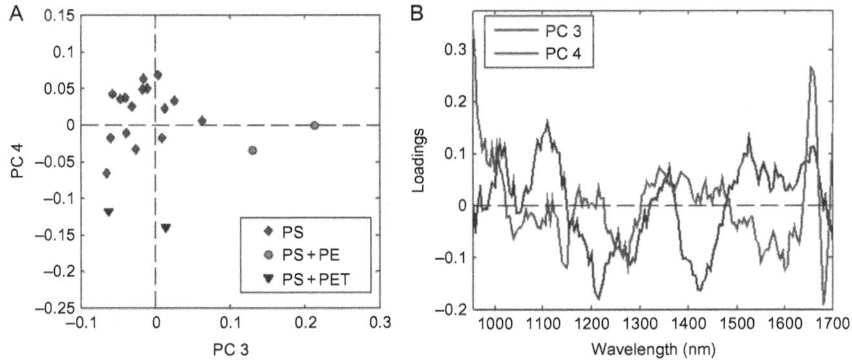

FIG. 8 PC 3 and PC 4 score plot (A) and corresponding loading vectors (B) of PCA model.

PC 3 allows differentiating PS from PS+PE, while PC 4 allows differentiating PS from PS+PET. However, the interpretation of the corresponding PC 3 and PC 4 loading vectors (Fig. 8B) is not straightforward; indeed PC 3 and PC 4 loadings are very noisy and it is difficult to assess the most relevant variables.

Therefore, sPCA was used in order to increase the interpretability of the model. Different sPCA models were calculated considering 4 sPCs and values of the sparsity constraint c ranging from 1 to 11 (with steps of 0.25). Fig. 9 shows how the percentage of outlier pixels of each test image varies depending on the sparsity constraint value, separately for each sPC. As observed in standard PCA, in general sPC 3 and sPC 4 are the components with the higher number of outlier pixels, which corresponds to a higher ability to separate the different polymers. Moreover, bearing in mind that when the sparsity constraint is set to its maximum value ($c=11$ in this case) the results of sPCA are equal to those of standard PCA, it is also possible to notice that sparsity allows to better highlight the differences between the polymers. In other words, higher percentages of outlier pixels can be observed when c is much lower than 11; in particular, the best compromise between high sparsity and high percentage of outlier pixels on sPC 3 and sPC 4 can be reached with a value of c equal to 2.75. With this value of the sparsity constraint, also sPC 2 has an influence on the separation of PE from the other polymers.

In order to verify that the outlier pixels are mainly ascribable to the presence of pieces of PE and PET, Fig. 10 reports the reconstruction of the score images of the three test images, where the pixels detected as outliers with the best sPCA model (4 sPCs and $c=2.75$) are represented in white color. Indeed, on sPC 3, and partially on sPC 2, the outlier pixels correspond to the two pieces of PE, while on sPC 4 the outlier pixels correspond to the two pieces of PET. In particular, the percentage of outlier pixels within the PS+PE

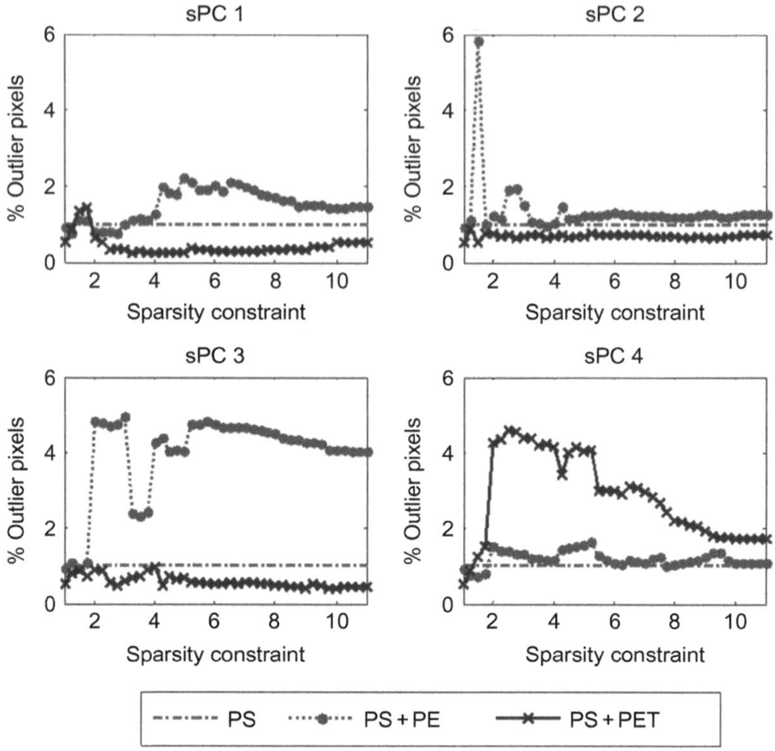

FIG. 9 Percentage of outlier pixels for PS, PS + PE, and PS + PET images for different values of the sparsity constraint.

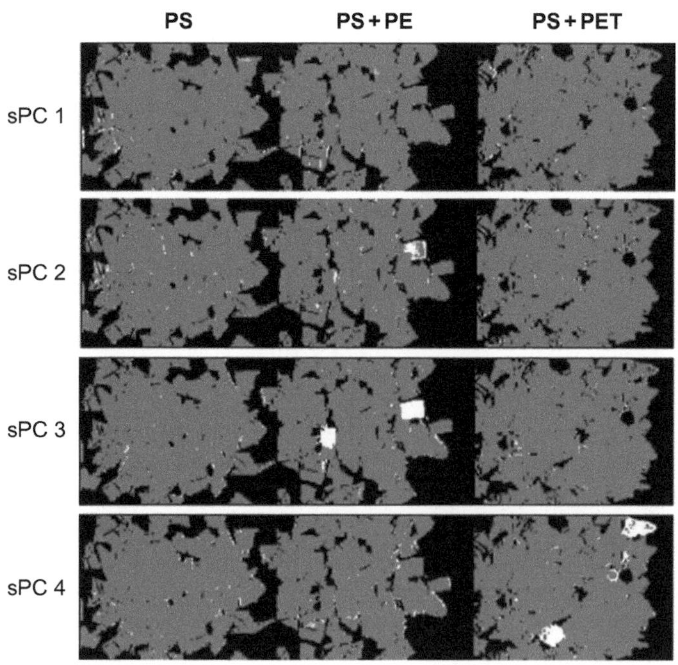

FIG. 10 Reconstruction of the score images of the three test images: the pixels related to the background are represented in *black color*, the pixels related to PS are represented in *gray color*, and the pixels detected as outliers are represented in *white color*.

image calculated on sPC 2 and sPC 3 is equal to 2.0% and 5.0%, respectively, while the percentage of outlier pixels within the PS + PET image calculated on sPC 4 is equal to 4.6%.

Considering the variables selected with this sPCA model, the overall number of nonzero variables on the loading vectors is equal to 30 out of the 150 original spectral variables. In particular, the sparse loadings of sPC 2, sPC 3, and sPC 4 were compared to the pure spectra of PS, PE, and PET (Fig. 11). Interestingly, the most relevant region selected on sPC 2 is related to the C–H aromatic second overtone (1143 nm) which corresponds to an absorption band common to PS and PET but not present in PE spectra. On the other hand, the regions selected on sPC 3 are around 1200 nm, corresponding to absorption bands of PE (C–H aliphatic second overtone). Moreover, the spectral regions at 1650 and 1680 nm have high relevance on sPC 4. These spectral regions mainly differentiate the spectra of PS and PET and are related to the C–H aromatic stretching first overtone [51,52].

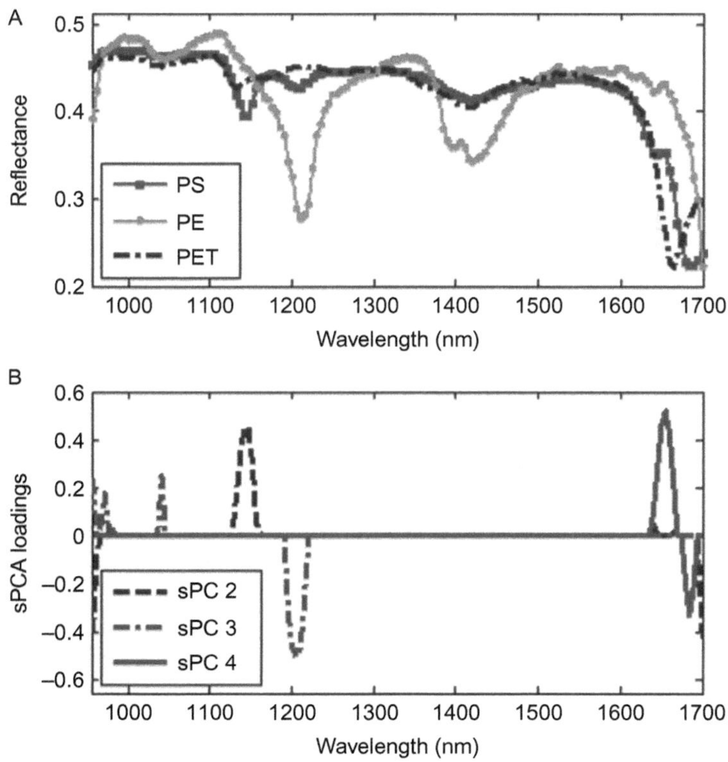

FIG. 11 In (A) PS, PE, and PET pure spectra and in (B) sparse loading vectors of sPC 2, sPC 3, and sPC 4.

Finally, considering the score plot of sPC 2, sPC 3, and sPC 4 of the average spectra data set (Fig. 12) it is possible to observe a separation between the samples of only PS and those containing also other polymers, even if this is less clear respect to what observed in standard PCA (Fig. 8A).

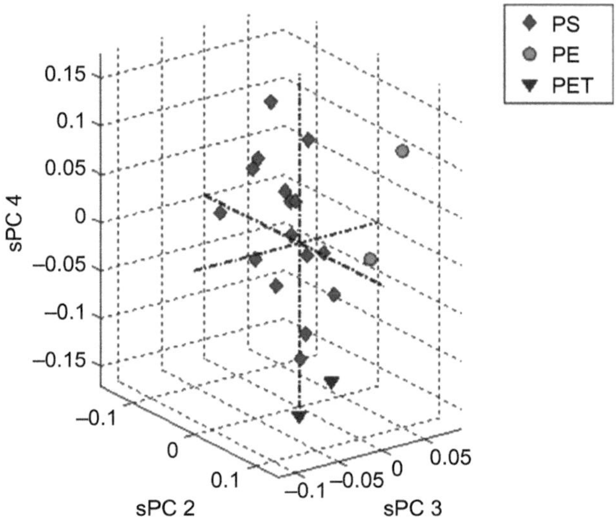

FIG. 12 sPC 2, sPC 3, and sPC 4 score plot of the sPCA model calculated with 4 sPCs and a value of c equal to 2.75.

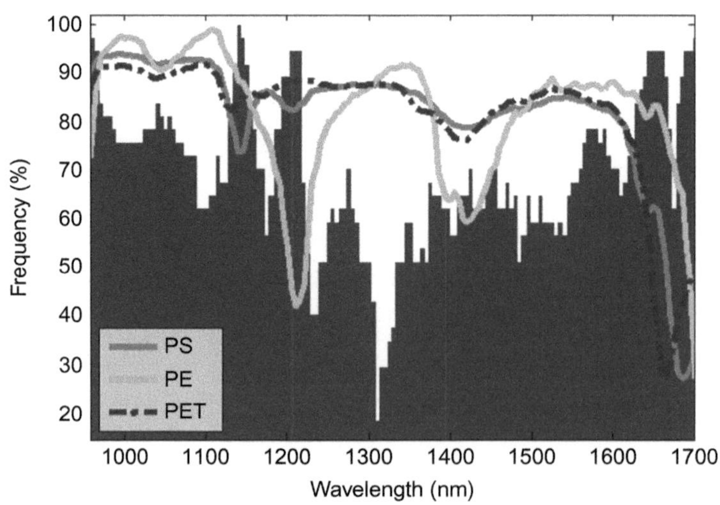

FIG. 13 Frequency of variable selection of sPCA models calculated considering 4 sPCs and values of c ranging from 1 to 11.

3.2.4 Stability of Variable Selection

Also in this case the stability of variable selection performed with sPCA was investigated. In particular, different sPCA models were calculated considering 4 sPCs and changing the value of the sparsity constraint from 1 to 11 (with steps of 0.25). The overall frequency of selection of the variables is reported in Fig. 13, together with the spectra of the three polymers. The most frequently selected variables correspond to regions where the difference between spectral signatures of the considered polymers is greater. In addition, the most frequently selected regions converge to the variables selected with the best sPCA model obtained in Section 3.2.3 (Fig. 11B).

4 CONCLUSIONS

Dealing with high-dimensional data is a challenging issue, and the use of classical chemometric tools can lead to multivariate models influenced by a huge amount of variables, thus resulting of difficult interpretation. HSI data are an example of high-dimensional data, since each image is composed by tens of thousands of pixel spectra. As a first very simple approach, PCA is generally used for unsupervised data exploration of the images before applying more complex regression or classification methods. However, sometimes the interpretation of the outcomes of a PCA model can be not straightforward, in particular as for the identification of relevant variables. For these reasons, a sparse-based PCA model can combine the advantages of classical PCA, giving also more interpretable results even from a chemical point of view.

Since sPCA is an unsupervised method, the main problem consists in the proper optimization of the level of sparsity to induce on the model. In this chapter, we proposed to use the frequency distribution curves of the score vectors of each component as a way to solve this issue. In this manner, it was possible to observe the influence of different sparsity levels on the effectiveness of the resulting models to enhance data interpretation.

Two different case studies were considered: the separation among Arabica and Robusta green coffee and the identification of PE and PET plastic pieces among pieces of PS.

The first case study concerned the discrimination between homogeneous samples of Arabica and Robusta coffee, where the identification of the sparsity level leading to the best separation between the two groups allowed also to highlight the relevant spectral variables.

The second case study concerned the identification of pixels related to the presence of pieces of different plastic types among pieces of PS. In this case, the optimal sparse model allowed enhancing the identification of outlier pixels of PE and PET. Furthermore, the comparison between the sparse loading vectors and the pure spectra of the considered polymers showed that the selected wavelengths were those matching the characteristic spectral bands of PE, PET, and PS, thus confirming the chemical relevance of the variables selected by sPCA.

In addition, the stability of variable selection was assessed by estimating the convergence of the variables selected with different sparse models: for both the case studies, the most frequently selected variables converged to those selected with the optimal sPCA model.

REFERENCES

[1] Elmasry G, Kamruzzaman M, Sun DW, Allen P. Principles and applications of hyperspectral imaging in quality evaluation of agro-food products: a review. Crit Rev Food Sci Nutr 2012;52:999–1023.

[2] Gowen AA, O'Donnell C, Cullen PJ, Downey G, Frias JM. Hyperspectral imaging—an emerging process analytical tool for food quality and safety control. Trends Food Sci Technol 2007;18:590–8.

[3] Lorente D, Aleixos N, Gómez-Sanchis J, Cubero S, García-Navarrete OL, Blasco J. Recent advances and applications of hyperspectral imaging for fruit and vegetable quality assessment. Food Bioprocess Technol 2012;5:1121–42.

[4] Amigo JM, Cruz J, Bautista M, Maspoch S, Coello J, Blanco M. Study of pharmaceutical samples by NIR chemical-image and multivariate analysis. Trends Anal Chem 2008;27:696–713.

[5] Amigo JM. Practical issues of hyperspectral imaging analysis of solid dosage forms. Anal Bioanal Chem 2010;398:93–109.

[6] Gowen AA, O'Donnell CP, Cullen PJ, Bell SEJ. Recent applications of chemical imaging to pharmaceutical process monitoring and quality control. Eur J Pharm Biopharm 2008;69:10–22.

[7] Roggo Y, Edmond A, Chalus P, Ulmschneider M. Infrared hyperspectral imaging for qualitative analysis of pharmaceutical solid forms. Anal Chim Acta 2005;535:79–87.

[8] Kubik M. Hyperspectral imaging: a new technique for the non-invasive study of artworks. In: Creagh D, Bradley D, editors. Physical techniques in the study of art, archaeology and cultural heritage, , vol. 2; Amsterdam, The Netherlands: Elsevier; 2007. p. 199–259.

[9] Sciutto G, Oliveri P, Prati S, Quaranta M, Lanteri S, Mazzeo R. Analysis of paint cross-sections: a combined multivariate approach for the interpretation of μATR-FTIR hyperspectral data arrays. Anal Bioanal Chem 2013;405:625–33.

[10] Amigo JM, Babamoradi H, Elcoroaristizabal S. Hyperspectral image analysis: a tutorial. Anal Chim Acta 2015;896:34–51.

[11] Burger J, Gowen A. Data handling in hyperspectral image analysis. Chemometr Intell Lab 2011;108:13–22.

[12] Amigo JM, Martì I, Gowen A. Hyperspectral imaging and chemometrics: a perfect combination for the analysis of food structure, composition and quality. In: Marini F, editor. Data handling in science and technology, vol. 28; Amsterdam, The Netherlands: Elsevier; 2013. p. 343–70.

[13] de Juan A, Tauler R, Dyson R, Marcolli C, Rault M, Maeder M. Spectroscopic imaging and chemometrics: a powerful combination for global and local sample analysis. Trends Anal Chem 2004;23:70–9.

[14] Prats-Montalbán JM, De Juan A, Ferrer A. Multivariate image analysis: a review with applications. Chemometr Intell Lab 2011;107:1–23.

[15] Geladi P, Grahn H. Multivariate image analysis. Chichester, England: John Wiley & Sons Ltd.; 1996.

[16] Leardi R. Application of genetic algorithm—PLS for feature selection in spectral data sets. J Chemometr 2000;14:643–55.

[17] Leardi R, Norgaard L. Sequential application of backward interval partial least squares and genetic algorithms for the selection of relevant spectral regions. J Chemometr 2004;18:486–97.

[18] Norgaard L, Saudland A, Wagner J, Nielsen JP, Munck L, Engelsen SB. Interval partial least-squares regression (iPLS): a comparative chemometric study with an example from near-infrared spectroscopy. Appl Spectrosc 2000;54:413–9.

[19] Cocchi M, Seeber R, Ulrici A. WPTER: wavelet packet transform for efficient pattern recognition of signals. Chemometr Intell Lab Syst 2001;57:97–119.

[20] Jouan-Rimbaud D, Walczak B, Poppi RJ, de Noord OE, Massart DL. Application of wavelet transform to extract the relevant component from spectral data for multivariate calibration. Anal Chem 1997;69:4317–23.

[21] Araújo MCU, Saldanha TCB, Galvão RKH, Yoneyama T, Chame HC, Visani V. The successive projections algorithm for variable selection in spectroscopic multicomponent analysis. Chemometr Intell Lab 2001;57:65–73.

[22] Centner V, Massart DL, de Noord OE, de Jong S, Vandeginste BM, Sterna C. Elimination of uninformative variables for multivariate calibration. Anal Chem 1996;68:3851–8.

[23] Friedman JH. Fast sparse regression and classification. Int J Forecast 2012;28:722–38.

[24] Lê Cao KA, Martin PG, Robert-Granié C, Besse P. Sparse canonical methods for biological data integration: application to a cross-platform study. BMC Bioinformatics 2009;10:34.

[25] Allen GI, Maletić-Savatić M. Sparse non-negative generalized PCA with applications to metabolomics. Bioinformatics 2011;27:3029–35.

[26] Colombani C, Croiseau P, Fritz S, Guillaume F, Legarra A, Ducrocq V, et al. A comparison of partial least squares (PLS) and sparse PLS regressions in genomic selection in French dairy cattle. J Dairy Sci 2012;95(4):2120–31.

[27] Karaman I, Qannari EM, Martens H, Hedemann MS, Knudsen KEB, Kohler A. Comparison of Sparse and Jack-knife partial least squares regression methods for variable selection. Chemometr Intell Lab Syst 2013;122:65–77.

[28] Wang R, Zeng W, Ming J. Discrimination of the white wine based on sparse principal component analysis and support vector machine. In: Wong WE, Zhu T, editors. Computer Engineering and Networking: Proceedings of the 2013 International Conference on Computer Engineering and Network (CENet2013). Switzerland: Springer International Publishing; 2014. p. 695–702. http://dx.doi.org/10.1007/978-3-319-01766-2_80.

[29] Kunz MR, Ottaway J, Kalivas JH, Georgiou CA, Mousdis GA. Updating a synchronous fluorescence spectroscopic virgin olive oil adulteration calibration to a new geographical region. J Agric Food Chem 2011;59(4):1051–7.

[30] Calvini R, Ulrici A, Amigo JM. Practical comparison of sparse methods for classification of Arabica and Robusta coffee species using near infrared hyperspectral imaging. Chemometr Intell Lab 2015;146:503–11.

[31] Filzmoser P, Gschwandtner M, Todrov V. Review of sparse methods in regression and classification with application to chemometrics. J Chemometr 2012;26:42–51.

[32a] Hoerl AE, Kennard RW. Ridge regression: Biased estimation for nonorthogonal problems. Technometrics 1970;12:55–67.

[32b] Tibshirami R. Regression shrinkage and selection via the lasso. J R Stat Soc 1996;58:267–88.

[33] Zou H, Hastie T. Regularization and variable selection via the elastic net. J R Stat Soc B 2005;67:301–20.

[34] Chun H, Keles S. Sparse partial least squares regression for simultaneous dimension reduction and variable selection. J R Stat Soc B Stat Methodol 2010;72:3–25.

[35] Lê Cao KA, Rossouw D, Robert-Granié C, Besse P. A sparse PLS for variable selection when integrating omics data. Stat Appl Genet Mol Biol 2008;7. Art. 35. http://dx.doi.org/10.2202/1544-6115.1390.

[36] Lê Cao KA, Boitard S, Besse P. Sparse PLS discriminant analysis: biologically relevant feature selection and graphical displays for multiclass problems. BMC Bioinformatics 2011;12:253–69.

[37] Clemmensen L, Hastie T, Witten D, Ersbøll B. Sparse discriminant analysis. Technometrics 2011;53:406–13.

[38] Burden FR, Winkler DA. Relevance vector machines: sparse classification methods for QSAR. J Chem Inf Model 2015;55:1529–34.

[39] Jolliffe IT, Trendafilov NT, Uddin M. A modified principal component technique based on the LASSO. J Comput Graph Stat 2003;12:531–47.

[40] Shen H, Huang JZ. Sparse principal component analysis via regularized low rank matrix approximation. J Multivar Anal 2008;99:1015–34.

[41] Witten DM, Tibshirami R, Hastie T. A penalized matrix decomposition, with applications to sparse principal components and canonical correlation analysis. Biostatistics 2009;10:515–34.

[42] Zou H, Hastie T, Tibshirani R. Sparse principal component analysis. J Comput Graph Stat 2006;15:265–86.

[43] Bro R, Papalexakis EE, Acar E, Sidiropoulos ND. Coclustering—a useful tool for chemometrics. J Chemometr 2012;26:256–63.

[44] Lee M, Shen H, Huang JZ, Marron JS. Biclustering via sparse singular value decomposition. Biometrics 2010;66:1087–95.

[45] Rasmussen MA, Bro R. A tutorial to the Lasso approach to sparse modelling. Chemometr Intell Lab 2012;119:21–31.

[46] Gürdeniz G, Hansen L, Rasmussen MA, Acar E, Olsen A, Christensen J, et al. Patterns of time since last meal revealed by sparse PCA in an observational LC-MS based metabolomics study. Metabolomics 2012;9:1073–81.

[47] Ferrari C, Foca G, Ulrici A. Handling large datasets of hyperspectral images: reducing data size without loss of useful information. Anal Chim Acta 2013;802:29–39.

[48] Ferrari C, Foca G, Calvini R, Ulrici A. Fast exploration and classification of large hyperspectral image dataset for early bruise detection on apples. Chemometr Intell Lab 2015;146:108–19.

[49] Kucheryavskiy S. A new approach for discrimination of objects on hyperspectral images. Chemometr Intell Lab 2013;120:126–35.

[50] Ulrici A, Serranti S, Ferrari C, Cesare D, Foca G, Bonifazi G. Efficient chemometric strategies for PET–PLA discrimination in recycling plants using hyperspectral imaging. Chemometr Intell Lab 2013;122:31–9.

[51] Shenk JS, Workman JJ, Westerhaus MO. Application of NIR spectroscopy to agricultural products. In: Burns DA, Ciurczak EW, editors. Handbook of near infrared analysis. 3rd ed. Boca Raton: CRC Press; 2008. p. 347–86.

[52] Tatzer P, Wolf M, Panner T. Industrial application for inline material sorting using hyperspectral imaging in the NIR range. Real-Time Imaging 2005;11:99–107.

Index

Note: Page numbers followed by "*f*" indicate figures, "*t*" indicate tables, and "*s*" indicate schemes.

D

H

I

J

K

L

M

N

O

P

T

FSC
www.fsc.org
MIX
Papier | Fördert gute Waldnutzung
FSC® C083411

Kolibri 360 GmbH
Ferdinand-Jühlke-Straße 7
99095 Erfurt, Deutschland
produktsicherheit@kolibri360.de